MEDICAL ENTOMOLOGY

Medical Entomology

A Textbook on Public Health and Veterinary Problems Caused by Arthropods

Revised Edition

Edited by

Bruce F. Eldridge and John D. Edman

Department of Entomology,
Center for Vectorborne Diseases,
University of California, Davis, USA

SPRINGER SCIENCE+BUSINESS MEDIA, B.V.

A C.I.P. Catalogue record for this book is available from the Library of Congress.

ISBN 978-1-4020-1413-0 ISBN 978-94-007-1009-2 (eBook)
DOI 10.1007/978-94-007-1009-2

Cover photo produced with the kind permission of Robert Gwadz.

Printed on acid-free paper

First published 2000
Revised edition 2004

Table of Contents

Contributors

Dr. Thomas R. Burkot
Division of Vector-borne Infectious Diseases
Centers for Disease Control and Prevention
Fort Collins, Colorado

Dr. Jonathan F. Day
Florida Medical Entomology Laboratory
University of Florida
Vero Beach, Florida

Dr. John D. Edman
Department of Entomology and
Center for Vectorborne Diseases
University of California
Davis, California

Dr. Bruce F. Eldridge
Department of Entomology and
Center for Vectorborne Diseases
University of California
Davis, California

Dr. Dana A. Focks
Center for Medical, Agricultural, and
Veterinary Entomology
US Department of Agriculture
Gainesville, Florida

Dr. Lane D. Foil
Department of Entomology
Louisiana State University
Baton Rouge, Louisiana

Dr. Kenneth L. Gage
Division of Vector-borne Infectious Diseases
Center for Disease Control and Prevention
Fort Collins, Colorado

Dr. J. Richard Gorham
Department of Biometrics and
Preventive Medicine
Uniformed Services University of the
Health Sciences
Bethesda, Maryland

Dr. Patricia M. Graves
Division of Vector-borne Infectious Diseases
Center for Disease Control and Prevention
Fort Collins, Colorado

Dr. Sidney E. Kunz
US Department of Agriculture (retired)
US Livestock Insects Laboratory
Kerrville, Texas

Dr. Philip G. Lawyer
Department of Biometrics and
Preventive Medicine
Uniformed Services University of the
Health Sciences
Bethesda, Maryland

Dr. James B. Lok
Department of Pathobiology
School of Veterinary Medicine
University of Pennsylvania
Philadelphia, Pennsylvania

Dr. Peter V. Perkins
US Army (retired)
Gainesville, Florida

Dr. Joseph Piesman
Division of Vector-borne Infectious Diseases
Center for Disease Control and Prevention
Fort Collins, Colorado

Dr. William S. Romoser
Department of Biological Sciences
Ohio University
Athens, Ohio

Dr. Glen A. Scoles
School of Medicine, Epidemiology and Public Health
Yale University
New Haven, Connecticut

Dr. Thomas W. Scott
Department of Entomology
University of California
Davis, California

Dr. James F. Sutcliffe
Biology Department
Trent University
Peterborough, Ontario

Dr. Walter J. Tabachnick
Florida Medical Entomology Laboratory
University of Florida
Vero Beach, Florida

Dr. Edward D. Walker
Department of Entomology
Michigan State University
East Lansing, Michigan

Dr. Donald E. Weidhaas
US Department of Agriculture (retired)
Gainesville, Florida

Dr. Stephen K. Wikel
Department of Entomology
Oklahoma State University
Stillwater, Oklahoma

Preface

The subject of medical entomology continues to be of great importance. Arthropodborne diseases such as malaria, yellow fever, dengue and filariasis continue to cause considerable human suffering and death. Problems in animal production, wildlife and pets of humans caused by arthropods still exact a large economic toll. In the past 2 decades, the invasion of exotic pests and pathogens has presented new problems in several countries, including the USA. For example, the year 1999 saw the invasion of the eastern USA by *Aedes japonicus*, an Asian mosquito, and West Nile virus, a mosquito-transmitted African arbovirus related to St. Louis encephalitis virus.

At the same time old and new health problems with arthropods occur, the traditional approaches to arthropod control have become more limited. Arthropod resistance to chemical pesticides by arthropods, a diminishing interest on the part of pesticide producers to develop new products for public health and veterinary uses, and the regulatory restrictions on pesticide use worldwide forces medical entomologists to seek entirely new methods to control arthropod vectors.

The field of medical entomology has expanded in recent years to include a number of new approaches and disciplines. When the widely-used textbook of medical entomology by James and Harwood was released in 1979, molecular biology was in its infancy. There were many examples of applications of genetics to medical entomological problems, but little understanding of the expression of genes for factors such as insecticide susceptibility, vector competence, host preference and similar important phenomena. Now, a variety of new methods are available to study genes, and to genetically alter important characteristics in vectors as a potential means of controlling human and animal diseases.

Many of the traditional tasks of medical entomologists continue to be important. Arthropod systematics is important because the need for accurate identification of arthropods is vital to an understanding of natural disease cycles. Systematics has been made even more challenging because of the current appreciation of the number of groups of sibling species among vectors of important disease pathogens. New molecular tools are assisting in separating these forms.

The ecology of arthropod-mediated problems is still the fundamental component of medical entomology. New tools available to researchers also abound here, perhaps none more important than the desktop computer.

With the enormous increase in the literature associated with medical entomology – there are several new journals that have appeared during the past 20 years – it may be unrealistic to cover the entire field of medical entomology in a single textbook. However, we felt that a textbook of medical entomology aimed at advanced undergraduate and graduate students was lacking, especially one that recognized modern approaches to the field. Further, we sought to emphasize the ecology of arthropod-mediated diseases rather than the arthropods themselves. To produce such a book, information traditionally available, such as keys to the identification of medically-important arthropods, had to be left out. Fortunately, there are other sources of this information, most notably the excellent treatment by Lane and Crosskey entitled "Medical insects and arachnids." For advanced students interested in the application of modern biological approaches to vector biology there is the book edited by Beaty and Marquardt entitled "The biology of disease vectors." Neither of these books stresses the epidemiology of arthropodborne diseases and other public health and veterinary problems with arthropods. We believe this textbook will fill that gap, while serving as an overall introduction to the subject.

Another book that should be mentioned is Service's "Medical entomology for students." This book provides a concise treatment of public health entomology for students, including physicians, nurses and public health officials, who seek a brief introduction to medical entomology.

We hope our text will provide adequate coverage of the subject of medical entomology on a worldwide basis. It is inevitable that a book edited and written mainly by Americans will have an American bias, especially in terms of the examples offered, but we have tried to minimize that.

Finally, we hope that those interested in public health and veterinary problems will find an adequate treatment of both those subjects under the overarching title of medical entomology.

This book could never have been produced without the substantial assistance of many people. Nancy Dullum was an enormous help in typing and retyping manuscripts, in handling much of the administrative responsibility for contacting authors and in coordinating corrections. James Freeman edited all the chapters for grammer and syntax. Kenneth Lorenzen prepared many of the computer generated figures and did the final editing and checking of manuscripts. Carol S. Barnett prepared several original drawings.

We are especially appreciative of the many people who reviewed individual chapters. Their comments were extremely helpful, and often uncovered important references in the literature. Those reviewing one or more chapters were: M.F. Bowen, Patricia A. Conrad, Anthony J. Cornel, J. Gordon Edwards, Durland Fish, John E. George, Nancy C. Hinkle, Harry K. Kaya, Marc J. Klowden, Robert S. Lane, Chester W. Moore, Mir S. Mulla, Frederick A. Murphy, Roger S. Nasci, William K. Reisen, Ronald M. Rosenberg, Michael W. Service, Walter J. Tabachnick, Edward D. Walker, Robert K. Washino and Robert A. Wirtz.

We are indebted to many other individuals and organizations who furnished illustrations. They are mentioned in the figure captions.

Finally, we wish to acknowledge the support and encouragement of our late friend and colleague, E. Paul Catts, who was originally one of the authors and the illustrator for this text. This book is dedicated to him.

Bruce F. Eldridge and John D. Edman
Davis, California, December 1999

A NOTE CONCERNING THE SECOND PRINTING

We are indebted to many individuals who reviewed the first printing of this book and pointed out numerous errors and inconsistencies. We hope we have removed most of them. We have also corrected several figures and their captions and added several new ones.

BFE
JDE
Davis, California, October 2003

Chapter 1

Introduction to Medical Entomology

BRUCE F. ELDRIDGE AND JOHN D. EDMAN
University of California, Davis

At the end of the 20th Century about 6 billion people lived in the world. Over 40% inhabited areas, mostly in the tropics, where they were at risk to malaria, a mosquitoborne disease. About 120 million new clinical cases of malaria occur each year, with over 1 million deaths among young children in Africa alone. About 300 million people probably are infected with the malarial parasite at any given time (World Health Organization 1990). In each year of the past 2 decades of the 20th Century, there was an epidemic of dengue, another mosquitoborne disease, in some part of the world, with each epidemic resulting in thousands of cases, many of them fatal in children (Fig. 1.1). Arthropods continually attack cattle, a primary source of food for many people of the world. These attacks can involve transmission of disease pathogens, which in turn can result in death, lower body weight, reduced milk production and slower development (Sellers 1981). Even when pathogen transmission does not take place, there can be direct injury and reduced economic return because of severe annoyance and blood loss (Steelman 1976). These examples emphasize that arthropod-mediated diseases are not phenomena associated only with the past. There have been spectacular advances in our under-

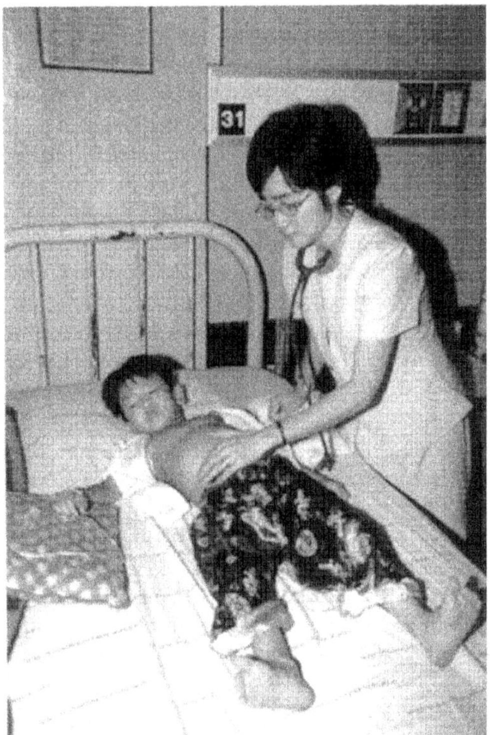

Figure 1.1. Young patient with dengue, Thailand (courtesy of World Health Organization).

B.F. Eldridge and J.D. Edman (eds.), Medical Entomology, Revised Edition, 1–12.

standing of these diseases, and significant advances in technology aimed at disease detection, control and prevention. In spite of this progress, debilitating and often fatal diseases associated with insects and other terrestrial arthropods continue to plague people, their domestic animals and pets, and wildlife. Rapid transportation of people, domestic animals and commodities, and new or reemerging pathogens and their vectors put nearly everyone at risk for contracting arthropodborne diseases.

The field of study devoted to understanding, preventing and controlling these diseases is **medical entomology**.

Medical entomology investigates the relationship of insects and other arthropods to the health of humans, domestic animals and wildlife. This branch of science has 2 primary subdivisions: **public health entomology**, the study of arthropods and human health; and **veterinary entomology**, the study of arthropods and their effects on pets, livestock and wildlife. The distinction between public health entomology and veterinary entomology is not well marked, because in many instances the same species of arthropods cause diseases in both human hosts and other vertebrate animals. Furthermore, the approaches used to solve research questions in the 2 subdisciplines often are similar, if not identical. Most arthropodborne infectious diseases are **zoonoses**, i.e., diseases such as plague that primarily affect animals other than human beings, but that also can cause human ailments. Further, many zoonoses are transmitted by arthropods. This close relationship among arthropodborne infectious diseases, various species of vertebrate animals, and human beings makes futile any attempt to divorce the related subjects of public health and veterinary entomology.

Not all authorities accept this definition of medical entomology. McClelland (1987) restricts the term "medical entomology" to human health considerations. Harwood and James (1979) recognized "medical entomology" and "veterinary entomology" as separate disciplines. We believe that such distinctions serve no useful purpose, and obscure the many similarities and interrelationships involved with diseases of human beings and other vertebrate animals.

Some entomologists consider infestations of household pests such as ants, silverfish, house flies and cockroaches to fall within the scope of medical entomology. Certainly, where insects such as roaches and house flies contaminate human or animal food with infectious disease organisms, this linkage is true. However, studies on the biology and control of household pests overlap the disciplines of industrial pest control and urban entomology and, as a practical matter, receive greater emphasis in those well-established disciplines.

THE ROLE OF ARTHROPODS IN HUMAN AND ANIMAL HEALTH

Medical entomology addresses the important role played by arthropods associated with diseases in people and certain other vertebrate animals. A **disease** is any condition that represents a departure from a normal or healthy condition in an organism. Under this definition, a wide variety of conditions qualify as diseases, including traumatic conditions such as broken bones, inflammatory skin conditions caused by toxic substances, or infections caused by pathogens. Such infections are called infectious diseases, and most frequently come to mind when thinking about diseases. Arthropods can cause diseases directly (2-component relationships), or can serve as vectors or hosts of pathogenic microorganisms (3-component relationships). Arthropods serving as vectors of disease pathogens represent by far the most significant aspect of the relationship between arthropods and vertebrate animals. In many such cases the arthropod serves as a host, at times the primary host, for the pathogen as well. However, asso-

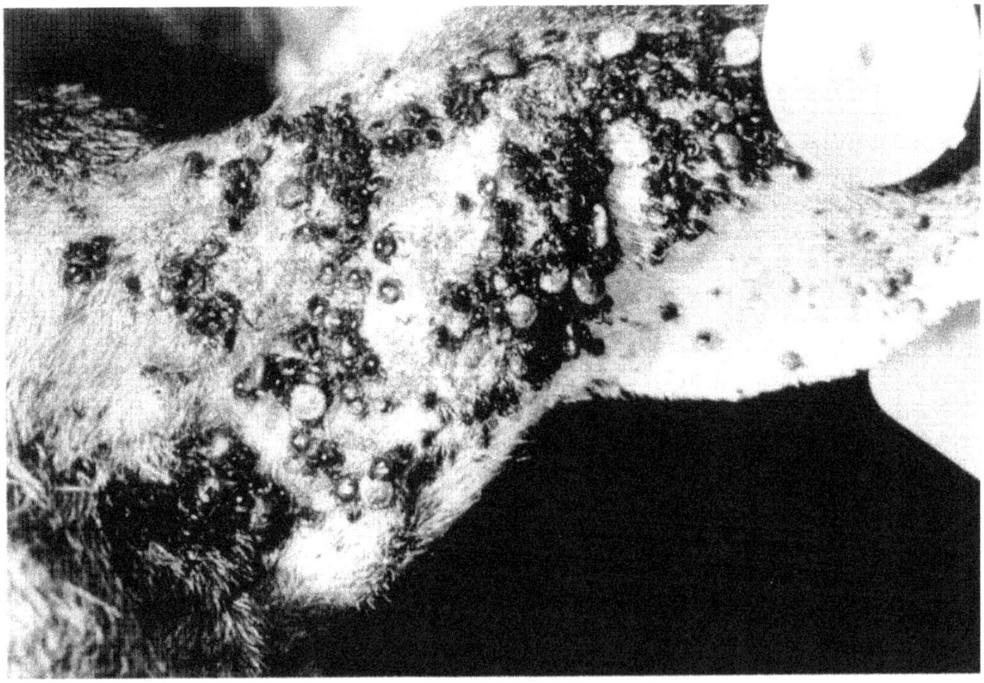

Figure 1.2. Ear of white-tailed deer infested with adult and immature lone star ticks, *Amblyomma americanum* (courtesy of Chris Santos).

ciating with arthropods can cause allergic reactions, and secondary infections can occur due to severe scratching of the affected areas of the skin, particularly among children.

Arthropods as Direct Causes of Disease

The ways in which arthropods directly affect the health of people and other animals vary. Medical entomologists have used many different schemes to classify these relationships. Some classifications recognize the type of injury produced, others the role that arthropods play as parasites of vertebrate animals. **Parasites** are organisms that live at the expense of other species of organisms (called **hosts**). **Pathogens** are organisms whose presence in another organism has the potential to cause disease (i.e., a pathogenic condition). The definitions of pathogen and parasite overlap, although by convention organisms classified as animals (e.g., protozoans, helminths, arthropods) are referred to as parasites, whereas viruses, rickettsiae, fungi and bacteria are termed pathogens. The following paragraphs cover arthropods as parasites as well as the types of injury they may cause by biting, stinging, burrowing, and otherwise coming into contact with animal hosts. These activities result in a wide range of pathogenic conditions from actual or imagined skin irritation to death from destruction of vital organs.

Ectoparasitoses

These are diseases caused by various kinds of contact between arthropods and the external body surfaces of hosts. Arthropod bloodfeeding, burrowing, crawling, or scraping at and just beneath the skin surface cause adverse reactions. There are many species of ectoparasitic arthropods. Ticks, fleas and sucking lice are examples of bloodsucking ectoparasites (Fig. 1.2). Scabies mites are an example of arthropods that burrow just beneath the skin of vertebrate animals, including humans. Reactions can include dermatoses, allergic reactions, and loss of efficiency and productivity. In cattle, the activities of ectoparasites cause weight loss and lowered milk production (Steelman 1976).

Endoparasitoses

Some arthropods invade tissues or body cavities of vertebrate hosts. One example is the "chigoe" or "chigger" flea, the female of which becomes imbedded as a result of the swelling of the host's tissues surrounding the feeding site of the flea. Vertebrate animals, including humans, are attacked in tropical and subtropical areas of the world, including parts of the USA (Jellison 1959). Some insects invade living tissue of vertebrates. When the invasion is by fly maggots, the condition is called **myiasis**. Many species of fly are involved, and most vertebrate animals are susceptible. Although humans can suffer from myiasis, it is much more common in livestock and wildlife.

Envenomization

This is the introduction of **venom** or other **toxins** by arthropods through stinging or biting. Toxins are substances produced by a variety of organisms. These substances often are proteins that cause poisonous reactions in other animals. Venoms are also toxins, but usually are secreted by arthropods and other animals such as snakes for defensive purposes or to kill prey. Wasps, bees and spiders are arthropods that produce venom.

Allergic reactions

Most animals have physiological mechanisms that defend against the introduction of foreign, or nonself, substances. A foreign substance that results in the production of antibodies by the animal is called an **antigen**. Most antigens are proteins. Antigens that produce unusually strong defensive reactions in animals are called **allergens** and the conditions in animals are **allergies**. These hypersensitive reactions often are associated with adverse symptoms such as itching, redness, swelling and rash. Repeated exposure to the same antigen usually causes allergy. Venoms of arthropods also can act as allergens, and in rare instances, can result in a particularly serious condition known as **anaphylaxis**. Untreated, an anaphylactic reaction can be fatal. In some hypersensitive individuals, fragments of arthropod integument may be allergenic. A small mite in house dust (the house dust mite) frequently causes allergy. Cockroaches can cause serious allergies in humans. Mosquito bites can result in allergic reactions to some people because they inject salivary fluids containing antigens.

Annoyance

Arthropods can cause considerable discomfort and annoyance to people and other animals merely by their presence and normal activities, even when they produce no serious physical harm. The larger the number of arthropods the greater the possibility of severe annoyance. Examples of annoyance caused by arthropods are flies that surround and land on cattle, predaceous wasps that land on exposed food at picnics, and insects such as chironomid midges, mosquitoes and black flies that splatter on automobile windshields. Annoying attacks by bloodsucking arthropods may not result in transmission of disease pathogens, but as discussed earlier, they can cause allergic reactions.

Delusory parasitosis

One of the most unusual types of disease associated with arthropods is not caused by arthropods at all, at least not directly. Certain people may develop a psychopathic condition that is manifested by a strong sense of being infested by arthropods. This condition may be accompanied by actual symptoms such as skin rash, redness, abrasions and secondary infections. However, the arthropods may not be the actual cause of the symptoms; rather they may be the result of self-inflicted scratching, or the unwise application of fluids. Once it is determined that arthropods are not directly involved, responsibility for treatment passes from the medical entomologist to the psychiatrist or other health care professional.

Details concerning the role of arthropods as direct causes of human and animal diseases are discussed in Chapter 4.

Arthropods as Vectors and Hosts of Pathogenic Microorganisms

For little over a century, scientists have known that arthropods could serve as vectors and intermediate hosts of pathogens of humans and other vertebrate animals. Since the late 19th Century scientists discovered that mosquitoes and other bloodfeeding flies, ticks and bloodfeeding mites, fleas, lice and kissing bugs were vectors of organisms causing infectious diseases. Malaria, trypanosomiasis, leishmaniasis, filariasis, dengue and yellow fever are mosquitoborne diseases, and are among the most serious parasitic ailments of humans. Many arthropodborne infectious diseases affect livestock and wildlife. Some of these diseases are zoonoses, with humans sometimes serving only as dead-end hosts of the pathogens.

Arthropods as vectors

The most significant role played by arthropods in human and animal health is as vectors of pathogenic microorganisms. In epidemiology, a vector is any agent by which a pathogen is transmitted from one host to another. In medical entomology, the agent of transmission, or vector, is an insect or an arachnid (ticks and mites). The most common type of transmission occurs when an infected bloodsucking insect or other arthropod probes or feeds on a human or other vertebrate host. However, there are other types of transmission mechanisms involving arthropods, discussed in detail in Chapter 5.

Arthropods as hosts

Some instances of disease transmission involve no arthropod vector, but arthropods can serve as intermediate hosts of a disease pathogen. An example is the double-pored dog tapeworm, *Dipylidium caninum*. Here, larval fleas that live in the bedding of dogs and cats ingest eggs of the tapeworm, and the dogs and cats become infected when they ingest the biting adult fleas during grooming. Small aquatic arthropods in the Class Crustacea serve as intermediate hosts of other parasitic flatworms. Parasitic flatworms and roundworms commonly are referred to as **helminths**.

Origins of Arthropod-Vertebrate-Pathogen Relationships and History of Early Work

Arthropods have been on the earth much longer than humans or any mammals or birds. The long evolutionary history of insects stretches back at least into the Devonian period, about 450 million years ago. Diptera, the insect order to which flies and mosquitoes belong, are believed to have first appeared no later than the Triassic period, about 200 million years ago. Mosquitoes probably arose later in this same period, but the first fossil specimens recognizable as such are associated with the Eocene epoch of the Quaternary Period (40–60 million years ago). Some of these specimens look remarkably similar to present-day *Culex* species. The first *Aedes* fossils are found in sediments from the middle Oligocene Epoch, about 30 million years ago (Edwards 1932). Most entomologists believe that the bloodsucking habit in mosquitoes had already developed by the time birds and mammals arose late in the Mesozoic Era (70–230 million years ago). They support this assertion by noting species of mosquitoes that do not feed on warmblooded animals, and lack certain sensory receptors found in species that do. It seems likely that biting and crawling arthropods have adversely affected people for as long as they have been on the earth.

Figure 1.3. Sir Patrick Manson in 1908 (courtesy of Wellcome Institute Library, London).

Early writings made numerous references to what today we consider medically-important arthropods. Greenberg (1971) provided a detailed review of early references to **synanthropic** (closely associated with humans) flies. Busvine (1976) is another useful source of early writings about associations between arthropods and people. For the most part, insects were regarded as annoying pests because, like mosquitoes and other bloodsucking arthropods, they bite, or because, like flies and roaches, they contaminate food. Researchers did not understand the role of microorganisms and other parasites in diseases until after the middle of the 19th Century; it took even longer to discover the role of certain arthropods as vectors of disease pathogens.

Several scientific developments presaged the discovery that arthropods transmitted human and animal disease pathogens. One was the invention of the microscope by Antony Van Leeuwenhoek (1632–1723). Another was a series of discoveries by scientists in the middle of the 19th Century that linked microorganisms to various human and animal diseases. A classic paper by Robert Koch in 1876 explained the relationship between *Bacillus anthracis* and the disease anthrax. Before this time, the only organisms known to cause human infectious diseases were helminths, although the association between many of these parasites and certain human diseases often were overlooked (Hoeppli 1959).

Sir Patrick Manson (Fig. 1.3) is generally regarded to have begun the modern period of medical entomology. Service (1978) provides an excellent description of this research and the other research on arthropodborne diseases that followed. In 1877, Manson, working in China, demonstrated that mosquitoes became infected with immature stages of filarial worms (microfilariae) in the process of taking a bloodmeal. Thus, he was the first person to prove an association between insects and parasites causing human and animal diseases.

Figure 1.4. Conquerers of yellow fever (courtesy of Wyeth-Ayerst Laboratories, Philadelphia).

A number of landmark discoveries concerning the role of arthropods as vectors of microbial pathogens followed Manson's research.

In 1891, 2 employees of the United States Department of Agriculture, Theobald Smith and F.L. Kilbourne, released the results of their epoch-making discovery. Working in Texas, they demonstrated the transmission of Texas cattle fever (piroplasmosis) by the cattle tick, *Boophilus annulatus*. This discovery represented the first demonstration of the transmission of a pathogenic microorganism from an infected arthropod to a vertebrate animal, and also the first demonstration of vertical transmission from an adult female arthropod to its progeny via infected eggs (transovarial transmission) of a microorganism pathogenic to vertebrate animals.

The discovery that yellow fever virus, which causes an often-fatal disease of people, is transmitted by a mosquito represents a fascinating story in the history of medicine (Fig. 1.4). Carlos Finlay, a Cuban physician, hypothesized that the yellow fever mosquito, now known as *Aedes aegypti*, was the vector of the disease. Scientists of the Yellow Fever Commission, under the leadership of Major Walter Reed, working in Havana, Cuba in 1900 proved this hypothesis. James Carroll, A. Agramonte, and Jesse Lazear participated in a series of experiments that demonstrated conclusively that *Ae. aegypti* mosquitoes could transmit the virus from infected patients in hospitals to uninfected volunteers, and that the disease could not be transmitted through direct contact with contaminated food or bedding, nor with infected patients. Dr. Lazear lost his life by permitting infected mosquitoes to feed upon him.

In 1895, David Bruce discovered that trypanosome parasites (later named *Trypanosoma brucei* in his honor) present in the blood of cattle caused nagana, a fatal disease of cattle and horses. In 1896, he demonstrated that the parasites were transmitted by flies in the genus *Glossina* (tsetse). A few years later (1903), Bruce and David Nabarro reported that tsetse transmitted the related parasite *T. gambiense* to humans, resulting in African sleeping sickness.

The story of the discovery of the cause and mode of transmission of malaria extends over a number of years. In 1894, Charles Laveran became the first to observe malarial parasites in the blood of infected patients. This was followed by the discovery, in 1897 by Sir Ronald Ross, of developing malarial parasites in mosquitoes that he characterized as "dapple-winged." We now know he was referring to mosquitoes in the genus *Anopheles*. A year later, Ross discovered that *Culex* mosquitoes transmitted parasites causing bird malaria, and the Italian scientist Battista Grassi demonstrated the complete developmental cycle of malarial parasites in *Anopheles* mosquitoes. Shortly thereafter, L.W. Sambon and G.C. Low, following a protocol developed in collaboration with Sir Patrick Manson, demonstrated that the human malarial parasite is transmitted by anopheline mosquitoes. Manson had called attention to the fact that not all species of mosquitoes were able to serve as vectors. That only anopheline mosquitoes transmitted human malarial parasites was confirmed by Malcom Watson in Malaya and by Sir Rickard Christophers in India (Philip and Rozeboom 1973). These discoveries laid the basis for the concept of vector competence developed much later (see Chapter 6). Unfortunately, this entire chapter in the history of medical entomology was marred in later years by conflicting claims of originality between Ross and Grassi, culminating in controversy and bitterness (Garnham 1971).

Figure 1.5. William C. Reeves, medical entomologist (courtesy of University of California).

H. Graham made the significant discovery in Beruit in 1902 that mosquitoes were the vectors of the then unknown organism causing dengue (Graham 1903).

During the period 1895-1910, fleas were incriminated as vectors of plague based on the research of a number of investigators, including W.G. Liston (1905) and D.B. Verjbitski (1908, cited in Hirst 1953). The work of the Commission for the Investigation of Plague in India provided confirmation of this connection by 1910 (Hirst 1953).

Carlos Chagas, a Brazilian parasitologist, discovered that a disease prevalent in South America was caused by a trypanosome parasite he named *Trypanosoma cruzi* in honor of his colleague Oswaldo Cruz. In 1908, he reported that the triatomid bug *Panstrongylus megistus* transmitted these parasites, causing the disease now called Chagas disease.

In 1926, D.B. Blacklock reported the results of his research in Sierra Leone that demonstrated that black flies (Simuliidae) vectored the parasite causing onchocerciasis, or river blindness (Blacklock 1926).

In the 1930s, research began on the epidemiology of a group of viruses that cause a variety of disease symptoms in human and domestic animal hosts. These viruses are grouped under the general term **encephalitides**. Not until the years just preceding World War II were various species of mosquitoes incriminated as vectors of these viruses. Then, William McD. Hammon and William C. Reeves conducted their pioneering research in the Yakima Valley of Washington that led to the understanding of the basic ecology of western equine encephalomyelitis and St. Louis encephalitis (Fig. 1.5). They demonstrated that the mosquito *Culex tarsalis*, which feeds mainly on birds, was the primary vector of the viruses causing both diseases.

Breakthrough research of this type has continued to the present. The decade of the 1980s saw the isolation and identification of the causal agents of Lyme borreliosis and ehrlichiosis, and the incrimination of several species of ticks as vectors. The recent awareness of these diseases has alerted experts to the possibility that global warming, deforestation, and other environmental changes may continue, and perhaps increase, the threat of emerging infectious diseases during the coming decades (Lederberg et al. 1992, Morse 1993).

CURRENT STATUS OF MEDICAL ENTOMOLOGY

In the USA, medical entomology is taught primarily in schools of public health or veterinary medicine, and in colleges of agriculture and natural sciences. In addition to colleges and universities where medical entomology is offered as a subject, employers of medical entomologists include government departments of agriculture, military, public health and foreign services. Many state or provincial health departments and local agencies such as mosquito, vector, or malaria control programs employ medical entomologists. Job opportunities also exist with the World Health Organization and other international agencies and foundations.

A number of professional organizations welcome membership by medical entomologists. There are at least 4 professional journals devoted exclusively to some aspect of medical entomology: *Medical and Veterinary Entomology* (published by the Royal Entomological Society), the *Journal of Medical Entomology* (published by the Entomological Society of America), the *Journal of Vector Ecology* (published by the Society for Vector Ecology) and the *Journal of the American Mosquito Control Association*. Individual articles on medical entomology subjects also appear in many other scientific journals.

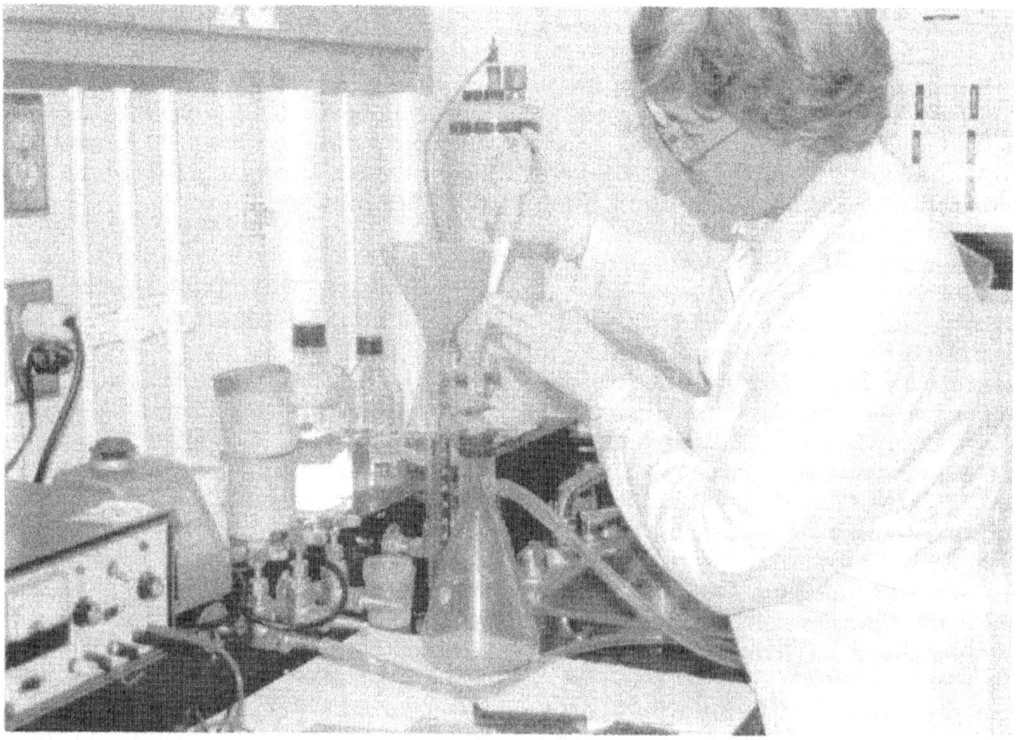

Figure 1.6. A scientist using molecular biology methods to conduct research in medical entomology (photograph by Bruce F. Eldridge).

MEDICAL ENTOMOLOGY AND OTHER DISCIPLINES

Medical entomology often is regarded as a branch of epidemiology or applied ecology. Over the past 10–15 years, emphasis in the field has broadened considerably to include sub-organismal approaches, especially those using tools of molecular biology. Presently, the scope of activities within the field are very broad and include ecology, systematics, genetics, behavior, evolution, physiology, parasitology, toxicology, biological control and molecular biology (Fig. 1.6). Because of this gradual evolution in the scope and practice of medical entomology, there have been some changes in the disciplinary designations of people who address problems in medical entomology. For example, scientists studying the gene structure of anopheline vectors of malaria might not call themselves medical entomologists in spite of the fact that their activities would fall within the scope of medical entomology through the understanding and management of the arthropods that are responsible.

In part to differentiate ecologically-based activities in medical entomology, the career designation "vector ecologist" has come into vogue in recent years. Regardless of the job titles applied to persons engaged in various activities in medical entomology, the discipline is active and vigorous, and remains focused on reducing or eliminating arthropodborne diseases through the management of the arthropods that are responsible.

FUTURE TRENDS IN MEDICAL ENTOMOLOGY

A number of trends occurring on a world-wide basis suggest that the field of medical entomology will continue to evolve so that a description of the field written 15–20 years from now probably will differ considerably from one written today. Up to the present time, solutions to problems in medical entomology often have been sought in a relatively vertical, single-objective framework. If mosquitoes transmit disease-causing pathogens, then the obvious solution is to eliminate the mosquitoes. Henceforth, any such solution will have to consider all consequences of intervention, including adverse consequences for the total environment.

In the USA, a number of significant pieces of federal legislation now influence decisions on the control of arthropods of medical importance. The Federal Insecticide, Fungicide and Rodenticide Act placed restrictions on the use of chemical pesticides. The Federal Clean Water Act requires detailed reviews of measures such as ditching, draining and filling of wetlands to eliminate mosquito breeding. The Federal Endangered Species Act limits the application of pesticides of all kinds within the range of "threatened" or "endangered" species.

Preventing arthropodborne diseases in the world has become a difficult challenge. On the one hand, scientific and technical advances have produced exciting new tools with which to study and manage arthropodborne diseases. On the other hand, economic constraints and public policies for environmental management have rendered obsolete the simple approaches to controlling arthropod populations as formerly practiced. Modern approaches will be discussed in detail in the following chapters against a backdrop of contemporary environmental science.

REFERENCES

Blacklock, D.B. 1926. The development of *Onchocerca volvulus* in *Simulium damnosum*. Ann. Trop. Med. Parasitol. **29:**1–47.

Busvine, J.R. 1976. Insects, hygiene and history. University of London, London. 262 pp.

Edwards, F.W. 1932. Diptera, Family Culicidae. Fasc. 194. Pp. 1–258 *in* P. Wytsman (ed.), Genera insectorum. Desmet Verteneuil, Brussels.

Garnham, P.C.C. 1971. Progress in parasitology. Athlone Press, London. 224 pp.

Graham, H. 1903. The dengue. A study of its pathology and mode of transmission. J. Trop. Med. (London) **6:**209–214.

Greenberg, B. 1971. Flies and disease. Vol. 1, Ecology, classification, and biotic associations. Princeton University Press, Princeton, New Jersey. 856 pp.

Harwood, R.F. and James, M.T. 1979. Entomology in human and animal health, 7th ed. Macmillan, New York. 548 pp.

Hirst, L.F. 1953. The conquest of plague; a study of the evolution of epidemiology. Clarendon Press, Oxford. 478 pp.

Hoeppli, R. 1959. Parasites and parasitic infections in early medicine and science. University of Malaya Press, Singapore. 526 pp.

Jellison, W.L. 1959. Fleas and disease. Annu. Rev. Entomol. **4:**389–414.

Lederberg, J., Shope, R.E. and Oaks, S.C., Jr. 1992. Emerging infections. National Academy Press, Washington, DC. 294 pp.

Liston, W.G. 1905. Plague, rats and fleas. J. Bombay Nat. Hist. Soc. **16:**253–273.

McClelland, G.A.H. 1987. Medical entomology, an ecological perspective, 11th ed. University of California, Davis. 433 pp.

Morse, S.S. 1993. Emerging viruses. Oxford University Press, New York. 317 pp.

Philip, C.B. and Rozeboom, L.E. 1973. Medico-veterinary entomology: a generation of progress. Pp. 333-359 *in* R.F. Smith, T.E. Mittler and C.N. Smith (eds.), History of entomology. Annual Reviews, Inc., Palo Alto, California.

Sellers, R.F. 1981. Bluetongue and related diseases. Pp. 567–584 *in* E.P.J. Gibbs (ed.), Virus diseases of food animals: a world geography of epidemiology and control. Academic Press, New York.

Service, M.W. 1978. A brief history of medical entomology. J. Med. Entomol. **14:**603–626.

Steelman, C.D. 1976. Effects of external and internal arthropod parasites on domestic livestock production. Annu. Rev. Entomol. **21:**155–178.

World Health Organization. 1990. Control of leishmaniasis. WHO Tech. Rep. Ser. No. 793. World Health Organization, Geneva. 158 pp.

Chapter 2

Introduction to Arthropods: Structure, Function and Development

WILLIAM S. ROMOSER
Ohio University, Athens

THE PHYLUM ARTHROPODA

The phylum Arthropoda is the largest assemblage in the Animal Kingdom, the number of arthropod species outstripping all others many times over. Of particular interest in medical entomology are the more than 17,000 species of bloodsucking (hematophagous) insects and the 25,000 or so species of ticks (Beaty and Marquardt 1996). The enormous success of this group is reflected in the seemingly endless variety of niches occupied, their 600 million-year evolutionary time span, and their high biomass in various ecosystems.

Arthropods (Fig. 2.1) have an **exoskeleton** of hardened plates separated by regions of flexibility that allow movement. Arthropods are thought to have evolved from annelid-like creatures made up of repeating body segments (**somites** or **Y**) capped anteriorly by a preoral segment (the **acron** or **prostomium**) and posteriorly by a segment containing the anus (**telson** or **periproct**). Whether the key event in the evolution of arthropods, i.e., the development of paired, jointed appendages (**arthropodization**), occurred once (monophyly) or more than once (polyphyly) has caused debate. Interpretation of evidence has shifted periodically from one point-of-view to the other. Currently, the monophyletic origin holds sway (Romoser and Stoffolano 1998). In any case, it is clear that there are at least 4 major groups of invertebrates with jointed appendages. During the evolution of the major arthropod groups, the primitive metameres and their accompanying appendages fused in various combinations, producing distinct body regions (**tagmata**). The appendages took on specialized functions appropriate to a given region (tagma); e.g., several pairs of appendages have given rise to the extant insect mouthparts.

Other arthropod characteristics include the presence of chitin, a nitrogenous polysaccharide that is a polymer of n-acetyl-glucosamine, in the exoskeleton; an open circulatory system; and commonly, specialized excretory structures, Malpighian tubules, which open into the alimentary canal.

Arthropod groups that contain members of medical importance include the crustaceans, chelicerates, myriapods and insects. An additional major group of arthropods, the trilobites, has long been extinct, and is known only from fossils.

The Class **Crustacea** (Fig. 2.2a), lobsters, crabs, copepods, etc., are mostly aquatic gill-

B.F. Eldridge and J.D. Edman (eds.), Medical Entomology, Revised Edition, 13–51.

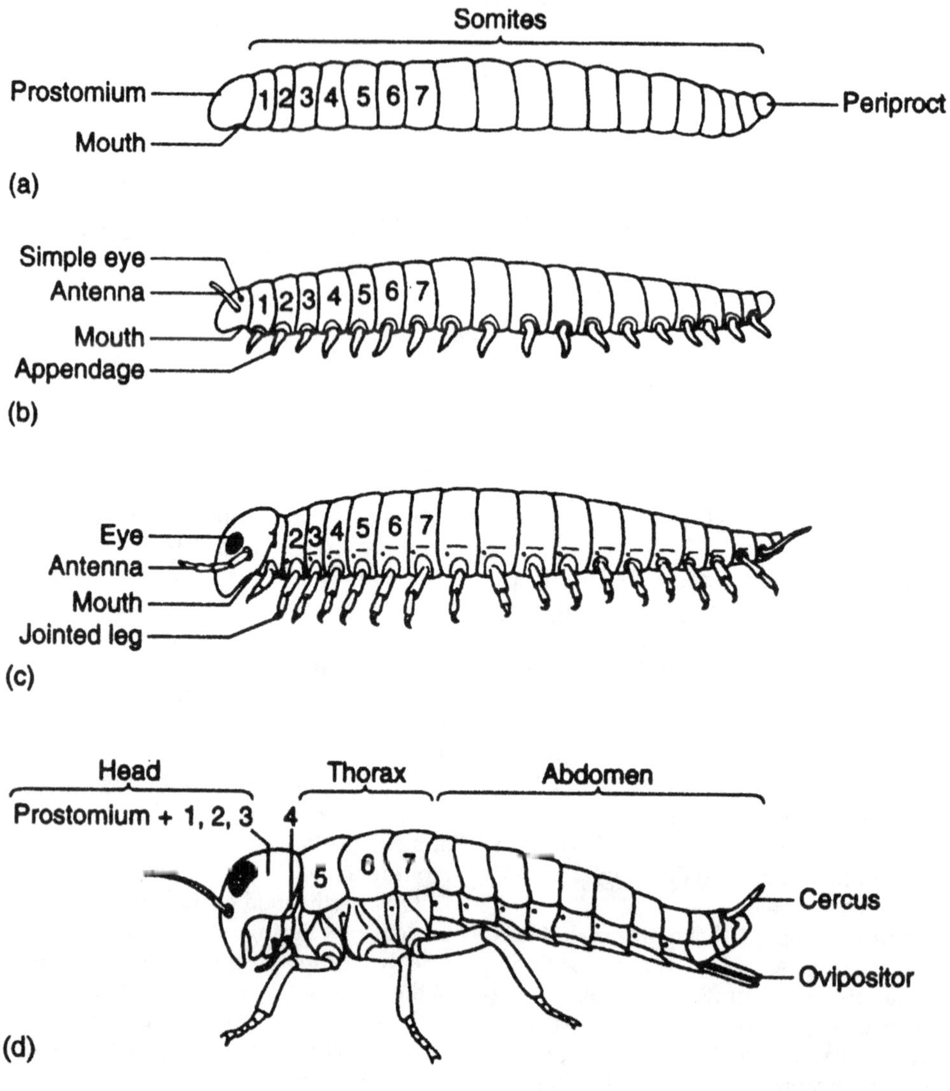

Figure 2.1. Hypothetical evolution of insect form from annelid-like ancestor (modified from Romoser and Stoffolano 1998, after Snodgrass 1935).

breathers, commonly with 2 well-defined body regions: a cephalothorax and an abdomen. They possess 2 pairs of cephalic appendages, antennae, and several pairs of branching appendages on the remaining cephalothoracic segments. A few have abdominal appendages as well. Calcium salts that impart considerable rigidity often impregnate the exoskeleton.

The Class **Arachnida** (spiders, mites and ticks, scorpions and relatives; Fig. 2.2b-d) also are characterized by 2 body regions: a cephalothorax and an abdomen. They lack antennae and typically bear 6 pairs of appendages. The first pair, the **chelicerae**, are jawlike or may bear fangs. The second pair of appendages, the **pedipalps**, are variable in form, but typically not leglike. For example, the pedipalps in scorpions are clawlike (chelate). The remaining 4 pairs of appendages are legs.

The **myriapods** (Fig. 2.2e,f) have 2 body regions: a **head** and **trunk**. Their elongate, wormlike bodies typically bear 9 or more pairs of appendages. Two classes, **Diplopoda** (millipedes) and **Chilopoda** (centipedes), are of minor medical significance. Millipedes have cylindrical or slightly flattened bodies, with more than 30 pairs of legs, typically 2 legs per body segment, and short antennae. Found in moist habitats, they are mostly scavengers, but a few are predatory. Several millipede species secrete a vesicating (blistering) substance from dermal glands. Centipedes have fewer legs than millipedes, usually more than 15, but with one pair of legs on each body segment. They are predatory and the appendages on the first body segment bear poison fangs with which they capture and subdue prey.

The Class **Insecta** (Fig. 2.2g) contains the vast majority of medically important arthropods. Members of this class are characterized by 3 body regions (head, thorax and abdomen), 3 pairs of thoracic appendages, compound eyes and commonly, simple eyes (ocelli).

THE CLASS INSECTA

The Integument

The general body covering of an insect, the integument (Figs. 2.3, 2.11, 2.12), serves many roles, not the least of which is to form a supportive and protective shell, the exoskeleton. The exoskeleton is made up of hardened, or **sclerotized**, plates (**sclerites**) separated by flexible ("membranous") regions, **conjunctivae** (or **arthrodial membranes)**, or by grooves (**sutures**). These grooves may or may not indicate infolding of the integument. This structure of sclerites and conjunctivae or sutures allows flexibility as well as the protection inherent in overlapping, hardened plates.

Although the integument is multilayered and complex, overall it is formed into a continuous sheet in which the relative hardness varies spatially. The **epidermis**, a single layer of cells that synthesizes and deposits the cuticle, determines this variation.

Muscles attach directly to the inside of the integument and certain muscles are opposed by cuticular elasticity when they contract. Any protuberances of the exoskeleton directed externally or internally are continuous with the integument, and modified forms of cuticle line parts of the alimentary canal, tracheae and ducts of the reproductive system. Various single-celled or multicellular **dermal glands** may be found interspersed among the epidermal cells of the integument. As a group, these carry out a variety of functions, such as secreting the outermost layer of cuticle and secreting defensive substances and various pheromones.

The insect integument bears an assortment of external processes, such as setae, spines and minute fixed hairs, as well as infoldings. **Setae** typically are hairlike, but may be branched, modified into flattened scales, or take on any of several other forms. They usually articulate in a socket, one cell forming the projected portion (hair or otherwise) and another cell forming the

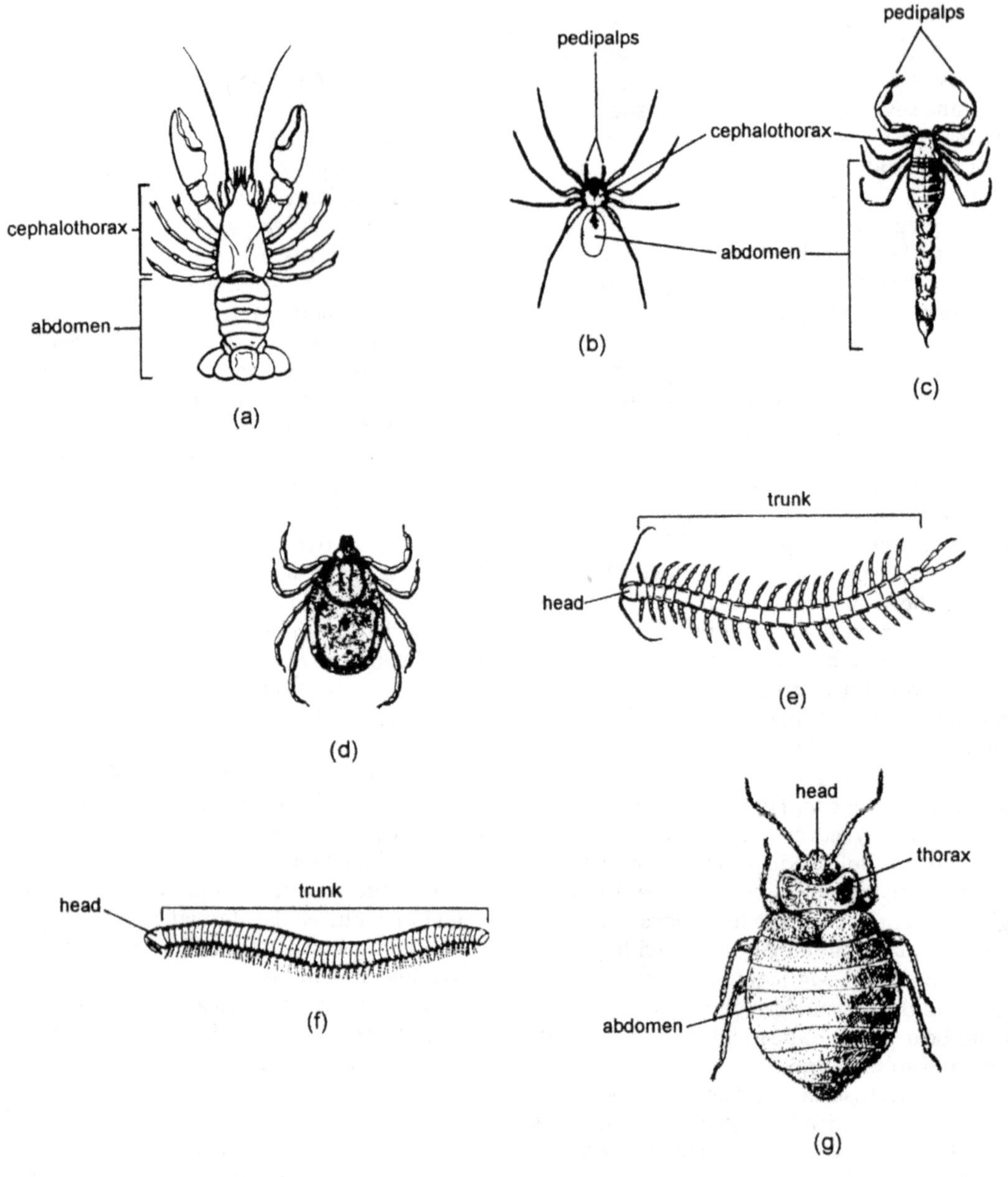

Figure 2.2. Representative arthropods: (a) Crayfish; (b) spider; (c) scorpion; (d) hard tick; (e) centipede; (f) millipede; (g) bed bug (from US Public Health Service 1969).

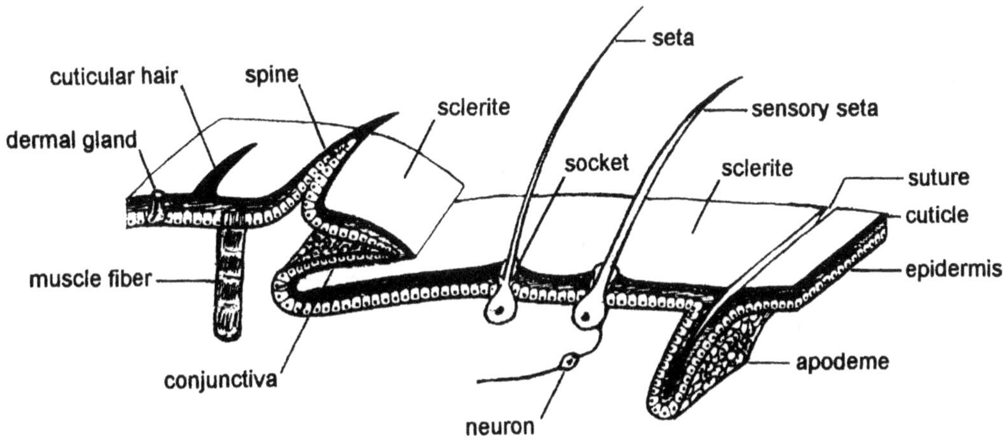

Figure 2.3. Diagram of typical insect integument (modified from Metcalf, Flint and Metcalf 1962).

socket. Setae often are pigmented and impart characteristic colors and patterns on the body and appendages. Coloration of this kind can be seen, for example, with mosquitoes. Setae also may be connected to a neuron and act as sensory structures. **Spines** are multicellular processes, the cuticle of which is continuous with that of the integument. Regions of cuticle commonly are covered with various tiny cuticular hairs or **microtrichia**. Infoldings of the body wall may be spinelike **apophyses** or ridgelike **apodemes**. These infoldings provide structural support for the exoskeleton as well as extensive surface areas for the attachment of muscle fibers.

Generalized Insect Form

An effective approach to learning insect morphology is to use a generalized form as a basis for comparison with actual insects. Such a generalized insect is shown in Fig. 2.4. The closest real insect to this generalized form might be a grasshopper or a cockroach. More detailed information on insect structure may be found in Chapman (1982), Richards and Davies (1977), Romoser and Stoffolano (1998), and Snodgrass (1935, 1959).

As mentioned in the introduction, the insect body is divided into 3 regions: the head, thorax and abdomen. The head and thorax are box-like structures, various primitive metameres having coalesced to form structural/functional units. In contrast, the insect abdomen is composed of very flexible segments capable of expanding. The rigidity of the head and thorax is correlated with the strong muscle contractions associated with mouthpart movement and locomotion, respectively.

In addition to the mouthparts, the head bears a pair of **compound eyes**, simple eyes (**dorsal ocelli**) and a pair of **antennae**, all of which are sensory organs. The head thus is specialized for ingestion of food and sensation of the environment. The thorax evolved from 3 primitive segments that are evident as the **prothorax, me-**

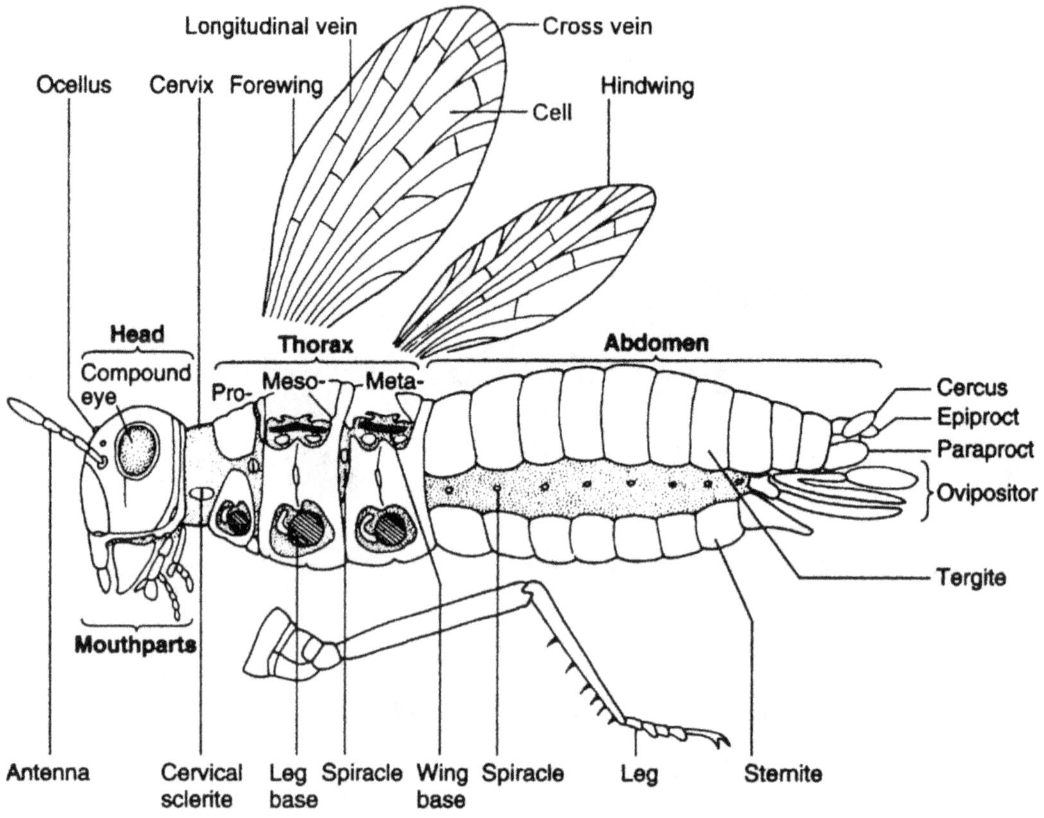

Figure 2.4. Generalized insect (from Romoser and Stoffolano 1998).

sothorax and **metathorax**. Typically, each thoracic segment bears a pair of legs. Wings, when present, are borne on the meso- and/or metathorax. The abdomen in the most primitive insects is composed of 11 segments. The last segment is divided into a dorsal **epiproct** and lateroventral **paraprocts** and bears a pair of appendages, the **cerci** (singular, **cercus**). Segments 8 and/or 9 bear the external **genitalia**.

Head and mouthparts

The head capsule

The head can be viewed as consisting of 2 parts: the rigid head capsule and the **mouthparts**, which articulate with the **head capsule** (Figs. 2.4, 2.5). The mouthparts are moveable while the head capsule provides rigid points of attachment for mouthpart musculature. The head capsule is joined with the prothorax by means of the membranous **cervix.** The cervix contains small plates, the **cervical sclerites**, which provide articular points for both the posterior head capsule and the anterior edge of the prothorax. Skeletal muscles, as well as the alimentary canal, the aorta and nerves, enter the head capsule from the thorax via the opening

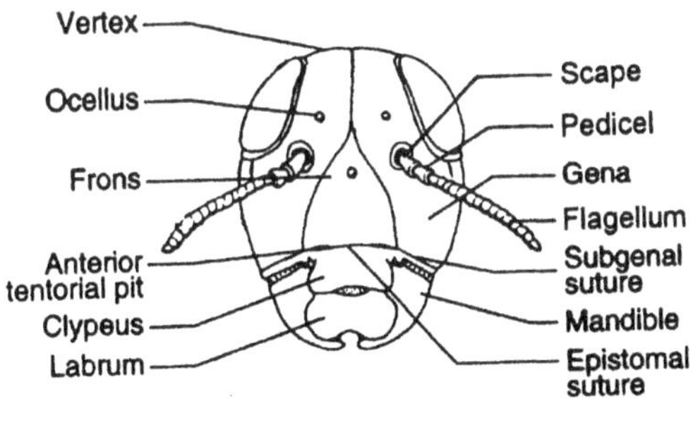

Figure 2.5. Generalized insect head (from Snodgrass 1935).

at the back of the head capsule called the **foramen magnum**. Sutures on the head capsule, some of which probably reflect primitive intersegmental lines, are useful in comparative morphological studies of the various groups of insects. Roughly delineated regions may be recognized on the head capsule. The region between the compound eyes is referred to as the **frons**, the lateral regions below the compound eyes, the **genae** (singular, **gena**), and between the compound eyes on the dorsum of the head, the **vertex**.

During the development of the head capsule, an internal strutwork forms from 2 anterior invaginations and 2 posterior invaginations. These invaginations meet and fuse, forming the **tentorium**. This structure provides internal support for the head capsule and additional surface area for muscle attachment. In the fully formed head, external pits remain from the invaginations that formed the tentorium, 2 **anterior tentorial pits** and 2 **posterior tentorial pits**. The **epistomal suture** runs horizontally between the anterior tentorial pits and the lobe-like **clypeus,** which is located immediately beneath the epistomal suture.

The compound eyes, located on each side of the head capsule, are made up of eye units called **ommatidia**. The externally apparent part of each ommatidium is the **corneal lens**. Insects are thought to perceive the external world as a mosaic, each ommatidium sensing a portion of the environment. Thus, the resolving power of a compound eye may be a function of the number of ommatidia: the greater the number, the greater the resolving power. The 3 or fewer simple eyes, or dorsal ocelli, are found on the frons.

The pair of antennae characteristic of the insect head may have evolved from the appendages of primitive metameres. Whatever their origin, antennae are sensory structures and are composed of the basal **scape**, the **pedicel** and the distal **flagellum**, which is relatively long and composed of several small **antennomeres**. The pedicel typically houses a highly specialized sensory structure, **Johnston's organ**. The basal scape articulates with the head capsule in the antennal socket.

Mouthparts

In the more generalized, primitive insects, the mouthparts are of the chewing, or mandibulate, type (Fig. 2.6). They are composed of an anterior **labrum** ("upper lip") that articulates with the clypeus. Two pairs of mouthpart appendages articulate with the head capsule behind the labrum: the more anterior, tooth-like **mandibles** and the more posterior, tearing **maxillae**. The posteriormost mouthpart is the **labium** that forms the "lower lip." The mouthparts, and possibly the labrum, may have evolved from appendages associated with primitive metameres. If the labium is visualized as being bisected, the structure of each half parallels the structure of a maxilla. The labium thus is considered to have arisen by the medial fusion of 2 maxilla-like primitive appendages. Likewise, the labrum may be viewed as representing medially fused, primitive basal leg segments. The tonguelike **hypopharynx** is located centrally and is surrounded by the mouthparts. The **true mouth** is located dorsally at the base of the hypopharynx while the opening of the common salivary duct is located at the ventral/basal extreme of the hypopharynx. The cavi-

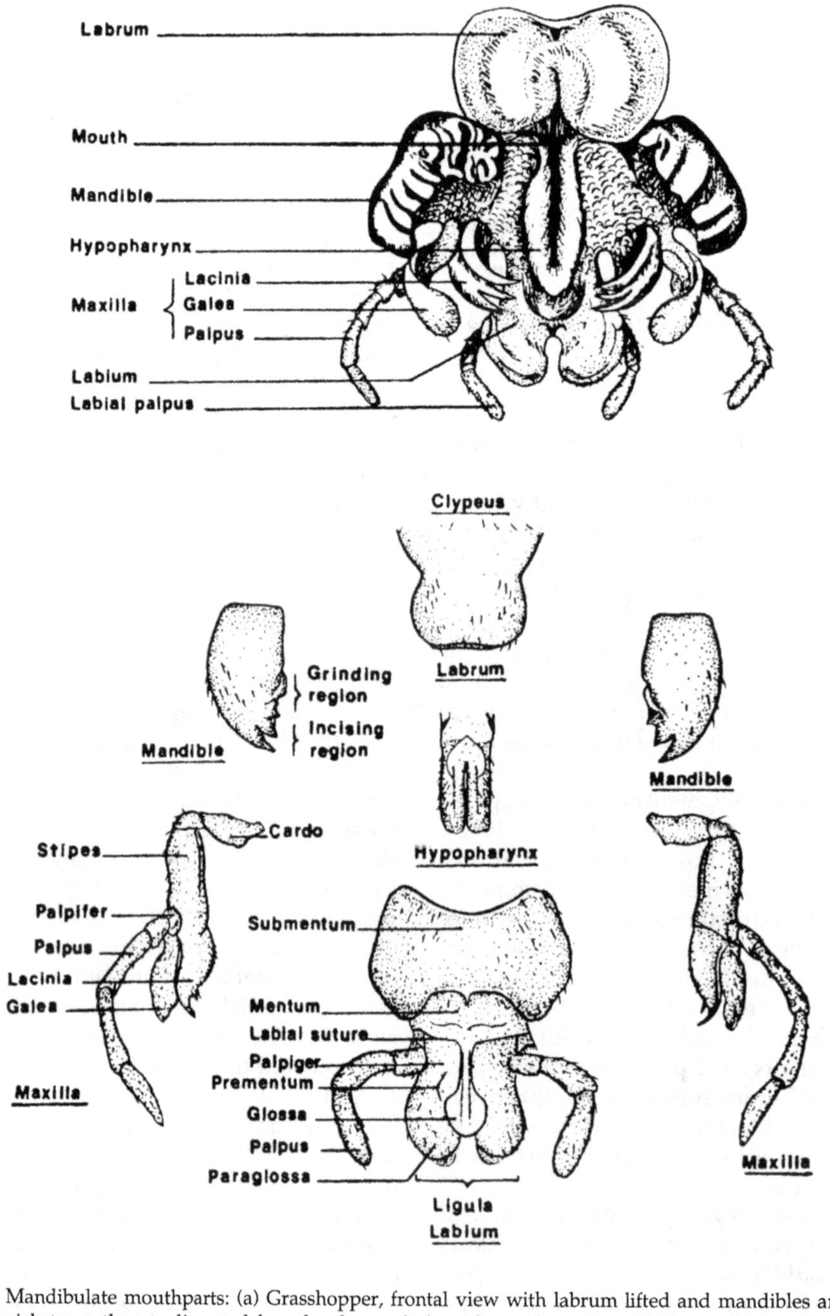

Figure 2.6. Mandibulate mouthparts: (a) Grasshopper, frontal view with labrum lifted and mandibles and maxillae spread; (b) cricket mouthparts dissected from head capsule (a, redrawn from Metcalf, Flint and Metcalf 1962; b, from Harwood and James 1979).

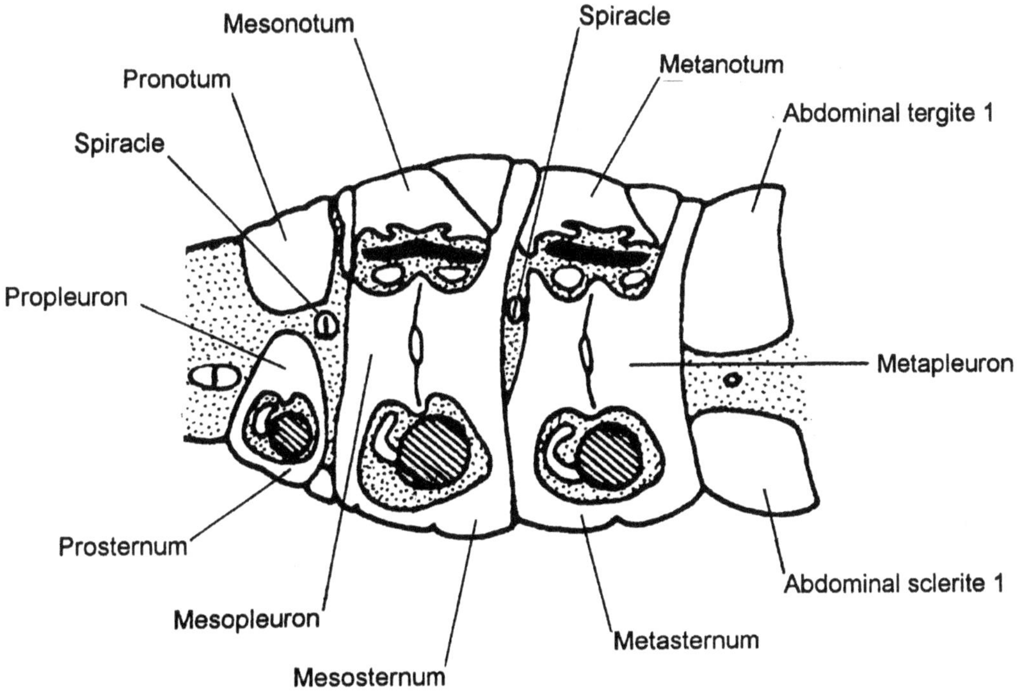

Figure 2.7. Generalized insect thorax (modified from Romoser and Stoffolano 1998).

ties formed by the hypopharynx and the surrounding mouthparts are the dorsal **preoral cavity**, or **cibarium**, and the ventral **salivarium**.

Thorax

Of the 3 thoracic segments, the **prothorax** is the smallest, the meso- and metathoracic segments being fused into a structure that bears the wings and associated musculature (Fig. 2.7). Together, the **mesothorax** and **metathorax** are referred to as the **pterothorax** ("wing-bearing thorax"). Each thoracic segment consists of a dorsal **notum**, bilateral **pleura** (singular, **pleuron**) and the ventral **sternum**. Each may be further subdivided by sutures into smaller sclerites. The legs articulate in membranous regions on the pleura and each leg is composed of a number of segments. External openings to the tracheal (ventilatory) system, **spiracles**, typically are found in the pleura between the pro- and meso- and the meso- and metathoracic segments.

As with the head capsule, infoldings and invaginations in the thoracic segments provide additional structural rigidity and increased surface area for muscle attachment.

Wings

Insect wings (Fig. 2.4) evolved on a thorax originally designed for walking and running

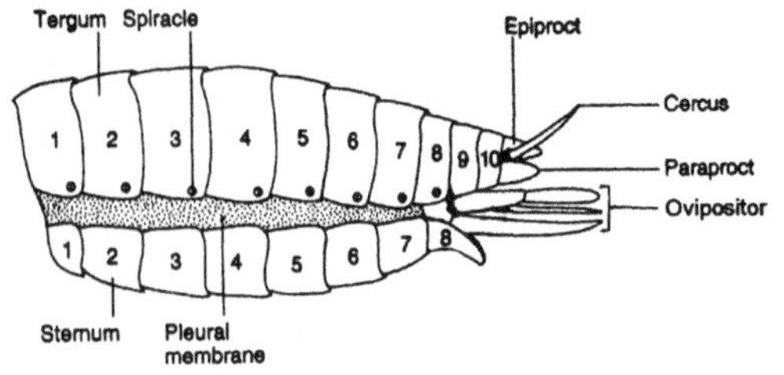

Figure 2.8. Generalized insect abdomen (from DuPorte 1961).

and are not derived from primitive walking appendages, but rather as evaginations of the integument. Wings articulate between the thoracic nota and pleura on the meso- and metathorax. Since wings arise as integumental evaginations, they are composed of 2 layers of integument. During development, tracheae grow into the space between these integumental layers and the wing veins form along these tracheae. Thus, spaces along each wing vein remain continuous with the hemocoel. The wing veins typically are more rigid than the adjacent integument and thus provide additional support. **Cross-veins** between the **longitudinal veins** divide wings into **cells**. Patterns of wing veins and cells vary between different groups of insects and are, therefore, valuable in insect identification.

Abdomen

Abdominal segments (Fig. 2.8) are much simpler than thoracic segments, each being composed of a dorsal tergum and a ventral sternum with the 2 plates being separated by the very flexible, laterally located, pleural membranes. Successive abdominal segments are separated by intersegmental membranes. The combination of pleural and intersegmental membranes imparts flexibility to the abdomen, both dorsoventrally and longitudinally. This arrangement facilitates expansion of the stomach with food, or ovaries with yolk-filled eggs. Spiracles typically are present in the pleural regions on the first 8 segments.

Along the thorax, the abdominal segments can be grouped into **pregenital** (1-7), **genital** (8 and 9) and **postgenital** segments (10 and 11). The external ovipositor in females and aedeagus in males have been derived from appendages associated with the primitive genital segments. Together, the genital and postgenital segments often referred to as **terminalia**.

Variation in Insect Structure

A fascinating array of structural modifications accompanies the seemingly endless spectrum of terrestrial niches in which insects may be found. Their impressive evolutionary success is probably in large part due to the structural plasticity associated with their segmented bodies, segmented appendages, and the integument to which muscles can attach at virtually any point. Structure also typically varies between the sexes of a given species and it usually is possible to identify the sex of a given insect by structural differences, i.e., **sexual dimorphisms**.

Head structures

Head structures of particular interest are the eyes, antennae and mouthparts. Compound eyes may vary in general form, number of ommatidia, and position along the midline of the vertex. The simple eyes, or dorsal ocelli, may be absent or 2 or 3 in number and may vary in position relative to one another and to the compound eyes. Antennae can be classified roughly as being one of the several types shown in Figure 2.9.

Mouthparts may be well-developed or reduced. Among those that are well-developed are 2 general types: **mandibulate** and **haustellate**. Two well-defined "toothlike" mandibles

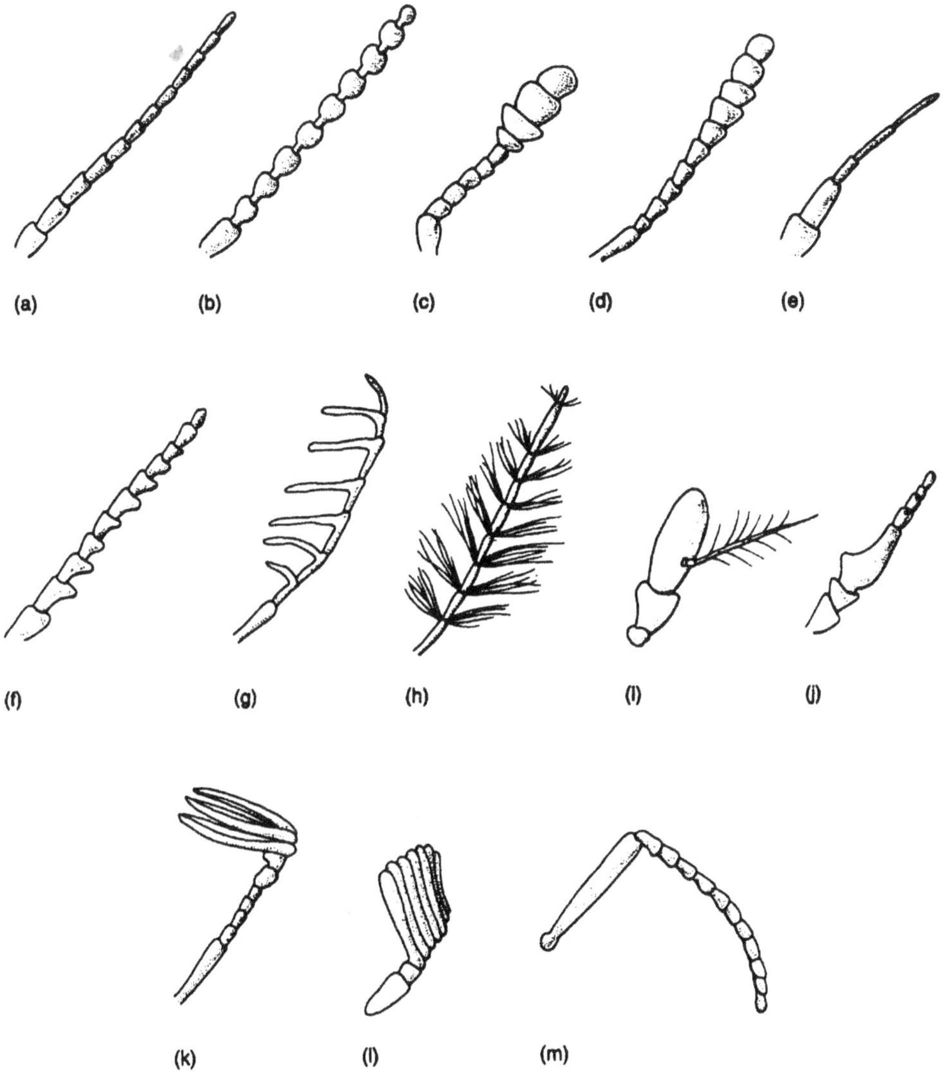

Figure 2.9. Variation in insect antennae: (a) Filiform; (b) moniliform; (c) capitate; (d) clavate; (e) setaceous; (f) serrate; (g) pectinate; (h) plumose; (i) aristate; (j) stylate; (k) lamellate; (l) flabellate; (m) geniculate (from Romoser and Stoffolano 1998).

usually are apparent in insects with mandibulate mouthparts. Typically, mouthparts that fall into this category are used to chew. However, there are many modifications that depart from strictly chewing. Haustellate mouthparts, on the other hand, typically have sucking functions, and the mandibles, although homologous with those in insects with the mandibulate type, are highly modified or have been reduced or lost altogether. In many insects with haustellate

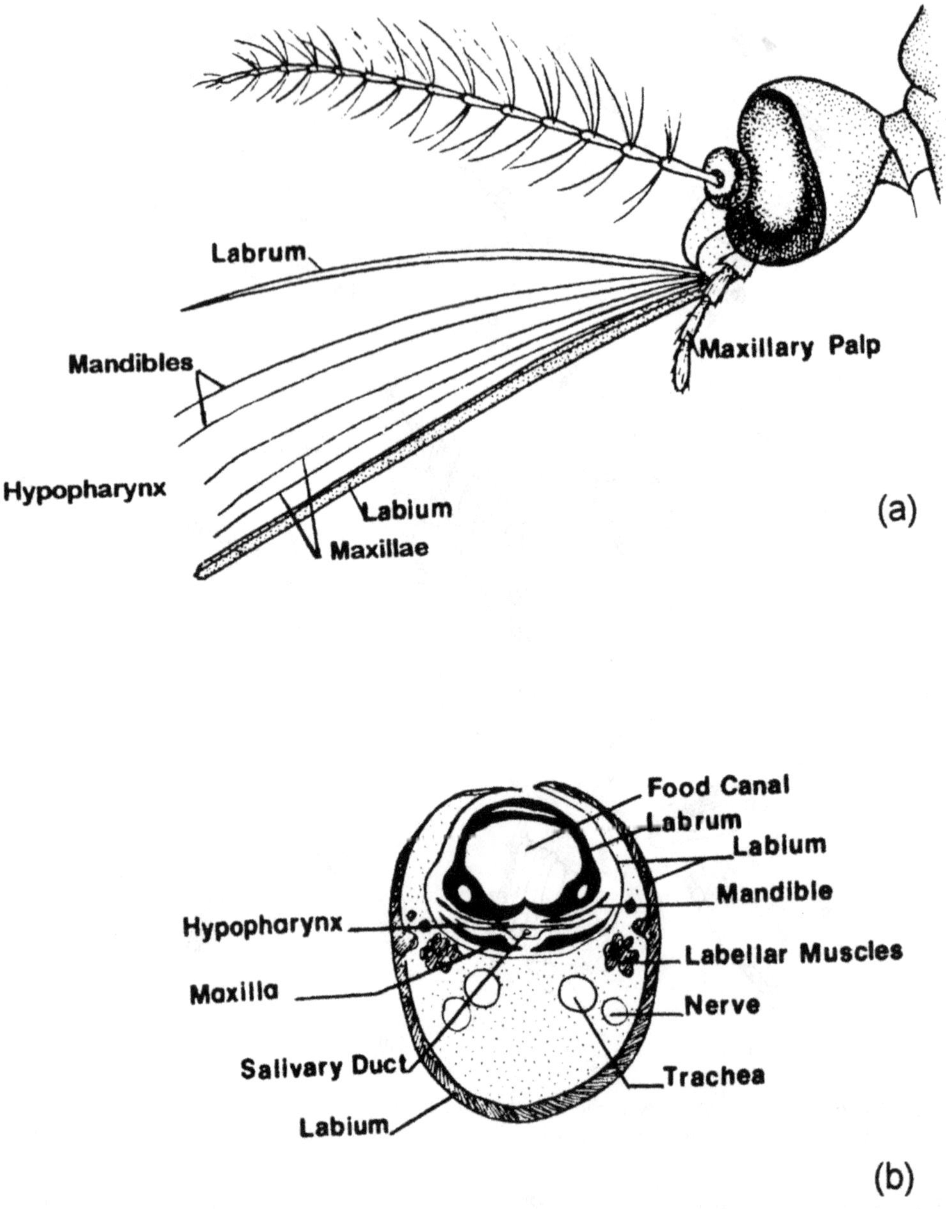

Figure 2.10. Piercing-sucking mouthparts of a mosquito: (a) Stylets removed from labial sheath (labium); (b) cross-section showing stylet positions within labial sheath (from Tipton 1974).

mouthparts, the various components are designed to penetrate plant or animal tissue and suck sap or blood (Fig. 2.10). These are referred to as **piercing-sucking** mouthparts and the mandibles, maxillae and hypopharynx have been formed into needle- or sword-like **stylets** that facilitate piercing. In these insects, other mouthparts have been modified to serve as sheaths containing the stylets, which usually are held together in a bundle or **fasicle**. Within the fasicle of stylets are food and salivary channels. Not every insect capable of piercing and sucking has the full complement of mouthparts, and in these forms, like the biting muscoid flies (e.g., the stable fly, *Stomoxys calcitrans*), modifications of the remaining mouthparts enable piercing. Some insects, like butterflies, skippers and moths (order Lepidoptera) and non-biting true flies (order Diptera), have **non-stylate**, haustellate mouthparts that enable them to suck fluids available without penetration of tissues, e.g., flower nectar. In insects with sucking mouthparts, the cibarium commonly has become modified as a pump. The pharynx, a portion of the alimentary canal, also is modified as a sucking pump in many insects.

Thoracic structures

As with the head, virtually every thoracic structure varies among the members of the Insecta.

In the generalized insect form, legs have a basic walking (cursorial) function. However, in many of the more specialized insects, legs display modifications, the functions of which usually are obvious from their appearance. There are **raptorial** forelegs used in prey capture; **fossorial** legs used in digging; **natatorial** legs used in swimming; and **saltatorial** legs used in jumping. Variations in the tarsi and tarsal claws are common. For example, among the true flies there may be a spine-like or bulbous **arolium** between the claws, or there may be pad-like **pulvilli** associated with the claws.

Flight is one of the premier adaptations of insects, and wings display an enormous array of specializations. The most obvious variation in wings is their presence or absence. Most insects have one or 2 pairs of wings, but some lack wings. The absence of wings is primitive in insects whose ancestors lacked wings, but winglessness also may be a secondary condition in insects that have winged ancestors. The later condition prevails in ectoparasitic insects such as lice, bed bugs and fleas. In the true flies, the mesothoracic wings have been reduced to small knob-like organs called **halteres**, which are flight-stabilizing organs. Wings also vary relative to their relationship to one another. In some insects with 2 pairs of wings, the wings operate together as a unit by virtue of a **coupling mechanism**. For example, in the order Hymenoptera (ants, bees, wasps and relatives), the fore- and hindwings are united by a series of hooks. Further wing variations include size, patterns of venation, texture, coloration, presence or absence of scales and hairs, and resting position.

Abdominal structures

In most adult insects the pregenital abdominal segments are fairly uniform, but the genital and postgenital segments vary greatly, particularly relative to the external genitalia, cerci, and epi- and paraprocts. The terminalia are useful particularly in species identification because they often are involved in reproductive isolation. Another important abdominal variation is in the number of segments. More modified insect forms have reduced the number of abdominal segments and the higher winged forms usually have 10 or fewer, and lack cerci. Other variations include size and the occurrence of pregenital appendages, particularly in larvae.

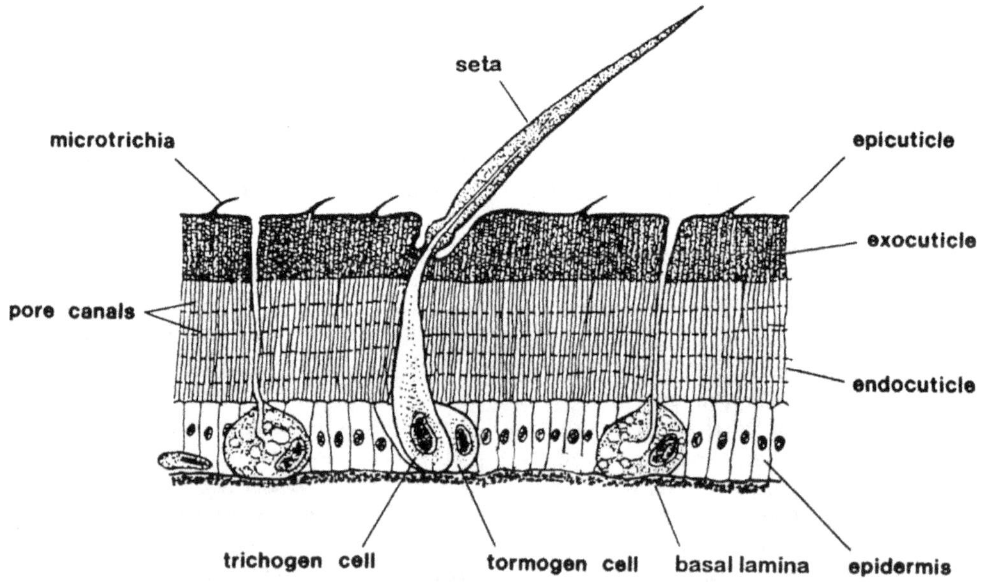

Figure 2.11. Diagrammatic section of insect cuticle (modified from Davies 1988).

General body form

Overall body form may vary dramatically in insects adapted for life in very specific habitats. For example, among ectoparasitic insects such as fleas, the body may be bilaterally flattened, an adaptation that favors movement through host hair. Or the body may be dorsoventrally flattened, as in bed bugs, a characteristic that enables them to hide in very small cracks and crevices.

External integumentary processes

Spines, setae and other external integumentary structures associated with a given species typically are arranged in characteristic, constant patterns. The study of such patterns for identification purposes is called **chaetotaxy**.

Internal Structure and Physiology

Knowledge of insect physiology is essential to understanding medically important insects. It provides the foundation necessary to understand, for example, how blood is digested and assimilated, interactions between ingested microorganisms and insects, why certain chemicals act as toxins to insects, how novel approaches to insect control might be developed, and how particular kinds of insects deal with specific environments.

As with all living things, the cell is the fundamental unit that sustains life, and a multicellular organism is the totality of expression of its variously specialized cells. Cells need a fa-

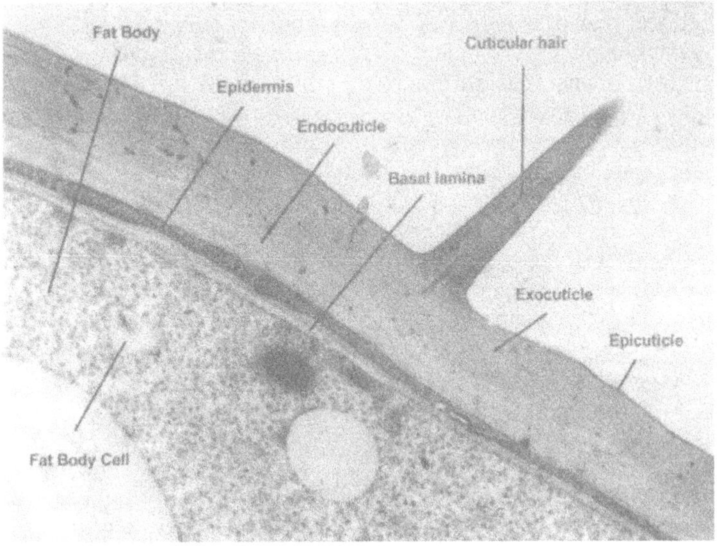

Figure 2.12. Electron photomicrograph (24,500X) of mosquito cuticle (courtesy of Kriangkrai Lerdthusnee).

vorable environment, access to reduced carbon and oxygen, a means by which waste material such as the nitrogenous end-products of protein metabolism and CO_2 are removed, and protection from forces that would interfere with optimal functioning. In multicellular organisms, the various organ systems ultimately are responsible for meeting these needs.

Useful treatments of insect physiology may be found in Beaty and Marquardt (1996), Chapman (1982), Kerkut and Gilbert (1985), and Romoser and Stoffolano (1998).

The integumentary system

A preceding section stressed the skeletal function of the integument. However, by virtue of its location between the external and internal environments, the integument plays roles far beyond that of external structure and it is, in fact, a dynamic system.

The integument is composed of 3 basic layers (Figs. 2.11, 2.12): the epidermis, a single layer of cells and 2 noncellular layers, the **cuticle** and the **basal lamina**. The cuticle is on the outside and is secreted by the epidermal cells. The basal lamina is formed on the hemocoel side of the epidermis, is 0.5 μm or less in thickness, and in electron photomicrographs appears as a continuous, amorphous granular layer.

Under the light microscope 3 different layers of insect cuticle can be seen: the innermost endocuticle, the exocuticle, and the very thin (0.03–4.0 μm thick) outer layer, the epicuticle. The endo- and exocuticles may or may not be pigmented, and the epicuticle is unpigmented. Cuticle in the more flexible integumental region is called arthrodial membrane (conjunctiva, Fig. 2.3).

Smaller organisms, such as insects, have a relatively large surface area per unit of volume compared to larger organisms and hence smaller organisms have a greater tendency to lose water. Although thin, the epicuticle is responsible for preventing water loss by evaporation from the body surface.

The cuticle, like the skin in vertebrates, is the first line of defense against the invasion of parasites. Not surprisingly, microbes that attack insects (e.g., certain fungi) synthesize and secrete the enzyme chitinase, which enables them to penetrate the chitinous insect cuticle. Likewise, the cuticle and epidermis determine the kinds of molecules that can enter from the environment making integumental permeability characteristics of special interest to toxicologists. Contact pesticides are those that can penetrate the insect cuticle.

As the seat of body coloration, the integument is important in providing camouflage, signals of recognition to the opposite sex, warnings to would-be predators, and in the dynamics of body heat and temperature. Coloration may be due to pigments synthesized by the epidermal cells and deposited in the cuticle, or it may be due to the light diffracting or scattering effects of cuticular layers or striations. Alternatively, it may be a combination of pigments and physical effects. Pigment granules in epidermal cells, or even in fat body cells, may show through thin cuticle and thereby provide coloration.

Depending upon the nature of the outermost cuticular layer, a region of cuticle may be wettable (hydrophilic) or non-wettable (hydrophobic). Here again, the relatively large surface area to volume ratio of insects comes into play, but with mass more important than volume. This characteristic makes insects especially liable to becoming stuck to water surfaces when exposed cuticular regions are hydrophilic; it also enables some insects to move on the water surface and facilitates such things as opening spiracular flaps (e.g., in mosquito larvae) when exposed cuticle is hydrophobic.

Finally, external sensory structures are by virtue of their location intimately involved with the integument. Thus, the corneal lenses of each ommatidium of a compound eye are made of modified cuticle. Various external processes actually may be mechanical or chemical sensilla, having not only hair- and socket-forming cells, but also involvement with sensory neurons.

Coordination and effector systems: nerves, muscles and glands

The nervous system monitors both the internal and external environments and brings about responses appropriate to the conditions prevailing in these environments. The endocrine system and muscles provide the means by which the nervous system brings about these responses and thus are called effectors.

The nervous system

The single nerve cell, or **neuron**, is the basic functional unit of the nervous system. A neuron (Fig. 2.13) is made up of a cell body and one or more elongate **axons**. Typically associated with the cell body are numerous branches called **dendrites**. Other branches, called **terminal arborizations**, also occur at the end of the axon. The terminal arborizations of one neuron are located close to the dendrites of an adjacent neuron (the synaptic cleft) and they usually communicate by means of molecules called **neurotransmitters**, such as acetylcholine. Neurotransmitters are secreted from the ends of the terminal arborizations and stimulate an electrical wave, or action potential, in the next neuron via its dendrites. Neurons may be associated with specialized sensory structures, with muscles, with glands, or with each other.

Neurons are organized into nerves, several fibers running along a path together, and **ganglia**, clumps of neurons. **Glial cells** invest nerves and ganglia, providing functional support. **Afferent nerves** carry input to the central nervous system from the sensory organs that monitor the external as well as the internal environment, while **efferent nerves** stimulate muscle contraction or glandular secretion. Neurons between afferent and efferent pathways are called interneurons or association neurons.

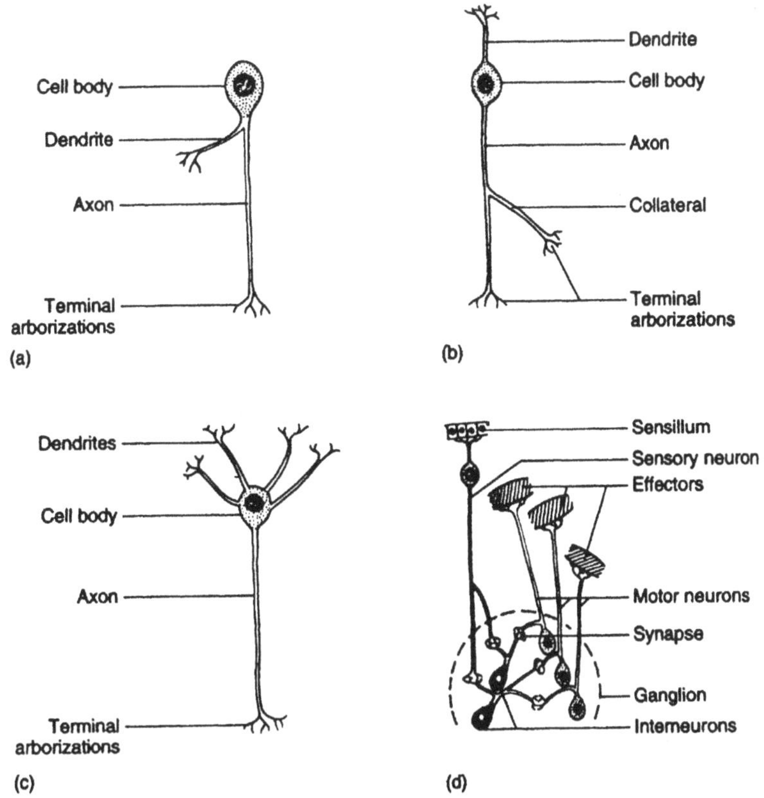

Figure 2.13. Neuron types: (a) Unipolar; (b) bipolar; (c) multipolar; (d) relationship among sensory, motor, and interneurons (from DuPorte 1961).

In ganglia, the neuron cell bodies are peripheral and their axons intermingle centrally, forming the neuropile. Each ganglion is ensheathed in connective tissue layers that help support nervous function.

Fused, paired ganglia and the nerves that run between them make up the **central nervous system** (Fig. 2.14). In the head, the brain, which is dorsal to the alimentary canal, has evolved from at least 3 primitive ganglia and receives input from the sensory organs (antennae, compound eyes and ocelli) on the head as well as from the rest of the central nervous system. A pair of bilateral nerves provides continuity between the brain and the subesophageal ganglion, located ventral to the esophagus. The subesophageal ganglion receives sensory input from, and provides motor output to, the mouthparts and is connected via the **paired ventral nerve cord** to a chain of ganglia that runs along the ventral midline of the body. In more primitive insects like cockroaches, the ventral chain ganglia are found in each thoracic segment and several of the abdominal segments. Each ganglion provides coordination for its respective segment, but there also is interaction among the

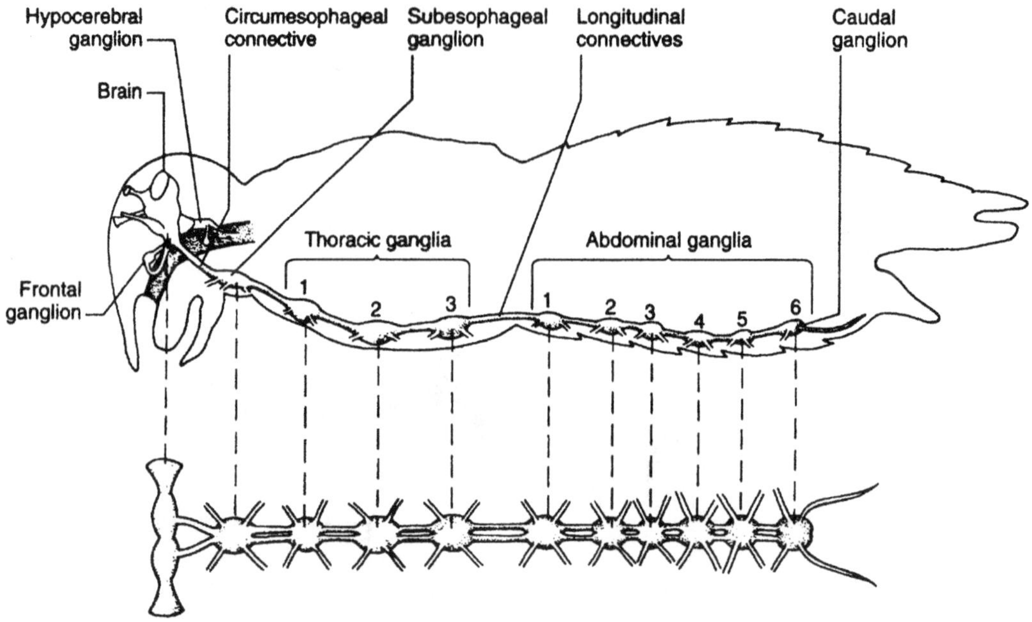

Figure 2.14. Generalized insect nervous system (from Romoser and Stoffolano 1998).

ganglia as well as with the brain. The last ganglion in the ventral chain, often called the caudal ganglion, coordinates the reproductive organs and the hindgut.

A complex of small ganglia and nerves regulates endocrine function and coordinates alimentary functions. This complex, usually referred to as the **stomatogastric nervous system,** ultimately connects to the brain via the recurrent nerve, which runs beneath the brain, and the frontal ganglion, which lies immediately dorsal to the alimentary canal and just anterior to the brain.

The endocrine system

The endocrine system (Fig. 2.15) is composed of several specialized cells and tissues that may be found throughout the insect body, frequently in close association with the nervous system. **Endocrine glands**, unlike **exocrine glands**, usually are ductless, their secretions being released directly into the hemolymph or into storage organs (**neurohemal organs**) for later release. Endocrine secretions are called **hormones** (Greek, "to excite"). When released, hormones circulate throughout the insect body cavity, but they have specific target cells and tissues that respond by virtue of receptor molecules on their plasma membranes. The endocrine system is like the nervous system in that it brings about adjustment to external or internal changes, but typically the response is slower than that brought about by nervous transmission.

Several endocrine cells and tissues have been identified, but the best known are the **thoracic**

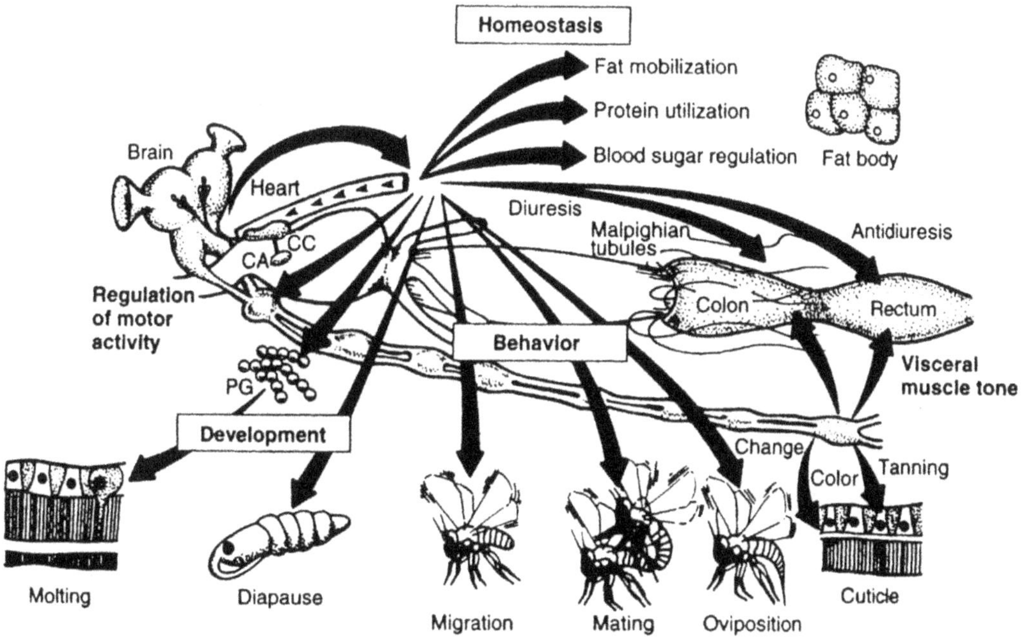

Figure 2.15. Endocrine regulated processes in insects (from Cook and Holman 1985, with permission from Pergamon Press Ltd., Headington Hill, Oxford, UK).

glands, the **corpora allata** and the **corpora cardiaca**. In addition to these, highly specialized neurosecretory cells derived from neurons can be found in the brain, the ventral chain ganglia, and other locations. These secrete neurohormones and link the nervous and endocrine systems. The thoracic glands (ecdysial glands) are found in association with tracheae in the thoracic region of immature insects, except in insects in the Subclass Apterygota. These glands secrete the hormone **ecdysone**, which initiates molting. The corpora allata are associated with the stomatogastric nervous system and secrete juvenile hormone as well as act as storage organs for some brain hormones. The corpora cardiaca also are closely associated with the stomatogastric nervous system and act as neurohemal organs by storing and releasing certain brain hormones.

Many functions are regulated, at least in part, by the endocrine system, e.g., molting, growth and development, polymorphism, egg development and yolk deposition, behavior relative to seasonal changes (e.g., migratory behavior and winter dormancy), reproductive behavior, hardening and darkening of the cuticle (sclerotization and melanization, respectively) and osmoregulation.

Muscles

Muscles (Fig. 2.16) are made up of elements that convert chemically stored energy to effect contraction. Muscles are either anchored to the integument (**skeletal muscle**, Fig. 2.16a) or form

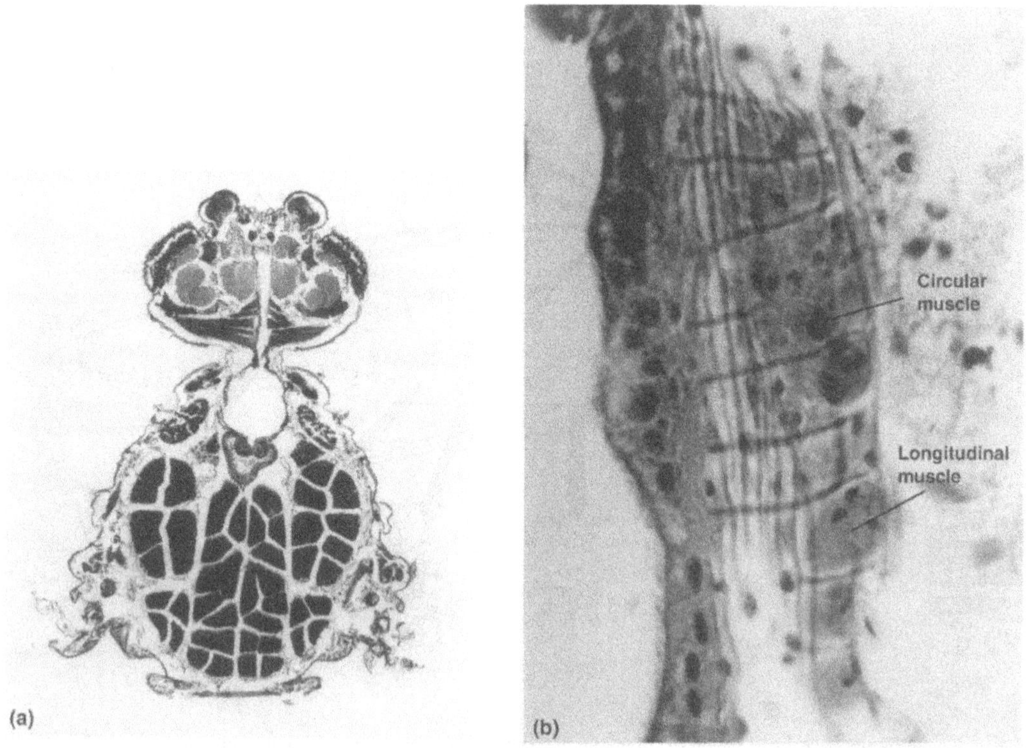

Figure 2.16. Photomicrographs of mosquito muscles: (a) Transverse section showing skeletal muscles of the head and thorax; (b) section of midgut showing visceral muscles (photographs by W.S. Romoser).

into regular or irregular networks around various viscera (**visceral muscle**, Fig. 2.16b). Thus, contraction produces such actions as skeletal movement, e.g., appendicular movement and locomotion, and peristalsis, regular waves of contraction that propel, for example, food along the alimentary canal. Although sometimes difficult to discern under the microscope, all insect muscles are striated and typically colorless or weakly colored.

The 2 basic sets of wing muscles deserve special mention. **Direct flight muscles** attach to the wing bases and can contribute to the longitudinal twisting of the wings during flight as well as flexion of the wings back over the abdomen in many insects. Wing twisting also is a function of the mechanical relationship between the wings and the pterothorax. Direct flight muscles provide the main force for flight in some insects, e.g., dragonflies and damselflies. However, in most higher insects, the **indirect flight muscles** provide the main force for wing flapping and hence for flight. There are 2 groups of indirect flight muscles: the dorso-ventrals (tergosternals) that run between the terga and sterna of wing-bearing thoracic segments, and the longitudinals that run from anterior to posterior in the wing-bearing segments.

The mechanical design of the thorax is such that contraction of the dorsoventral flight muscles causes the terga to depress and the wings to elevate, while contraction of the longitudinal flight muscles causes an arching of the thorax that results in depression of the wings. Thus, the 2 sets of indirect flight muscles act as antagonists in effecting flight. In many insects, both nerve impulses and stretching cause these flight muscles to contract. Again, due to the design of the thorax (in this case the elasticity of the thoracic cuticle), contraction of one set of muscles induces not only wing movement, but also a stretching of the antagonistic set of muscles at the end of the wing movement. A single nervous impulse prompts several oscillating wingbeats, enabling the insects to achieve high wingbeat frequencies and rapid flight.

Maintenance systems

The systems discussed in this section get energy and molecular building blocks in the form of food, as well as oxygen, to cells, and remove metabolic wastes (including nitrogenous wastes and carbon dioxide) from cells.

Alimentary system

The alimentary system (Figs. 2.17, 2.18, 2.23) acquires and processes food and eliminates undigested wastes. The alimentary canal, a tube beginning with the anterior mouth and ending with the posterior anus, is composed of a continuous one-cell-thick epithelial layer. On the body cavity (hemocoel) side of this epithelial cell layer is the noncellular basal lamina.

Three fundamental regions are recognizable in the insect alimentary canal (Fig. 2.17): the **foregut**, **midgut** and **hindgut**. The fore- and hindguts arise as ectodermal invaginations during development and are lined with modified cuticle (intima). The intima is continuous with the cuticle of the integument. Typically, the foregut intima is impermeable, while the hindgut intima has regions that are permeable to inorganic ions and small molecules. The midgut arises internally (probably endodermally), hooking up with the anterior and posterior invaginations, completing the tube. Each gut region has a complement of regularly or irregularly arranged circular and longitudinal muscles that effect peristalsis. Extrinsic muscles that attach to a gut region and an area of cuticle provide suspension, and in the anterior part of the foregut, pumping action. A complement of tracheae and tracheoles also is associated with the alimentary canal.

The fore-, mid- and hindguts typically are subdivided into specialized areas. The true mouth can be found at the base of the hypopharynx and is surrounded by the mouthparts. In insects with piercing-sucking mouthparts, the cavity immediately anterior to the true mouth, the **cibarium**, may be modified into a pumping structure, e.g., in mosquitoes (Fig. 2.18).

The **pharynx** (Figs. 2.17, 2.18,) is the first structure immediately posterior to the true mouth. Probably every insect pharynx is capable of a degree of pumping action, but in many insects the pharynx is highly specialized as a pump, e.g., in mosquitoes (Fig. 2.18). The associated cuticle in insects with cibarial and/or pharyngeal pumps often is highly elastic and acts antagonistically with the extrinsic muscles that dilate the pump lumen.

Lying posterior to the pharynx, the **esophagus** (Figs. 2.17, 2.18) is a conducting tube that commonly opens into an expanded region, the **crop**, which stores food. In some insects, the crop is in the form of a blind sac, or **diverticulum**, which extends well into the abdomen, e.g., in tsetse. Three such diverticula are found in mosquitoes (Fig. 2.18): 2 dorsal sacs that inflate between the 2 sets of wing muscles and a larger ventral sac that extends into the abdomen. The **proventriculus** (Fig. 2.17) lies posterior to the opening of the crop and in more generalized insects, like cockroaches, the intima contained in this structure has hardened tooth-like projections that carry out a secondary chewing function. The **stomodeal valve** (Fig. 2.17) lies between the proventriculus and the midgut and regulates the passage of food from the foregut

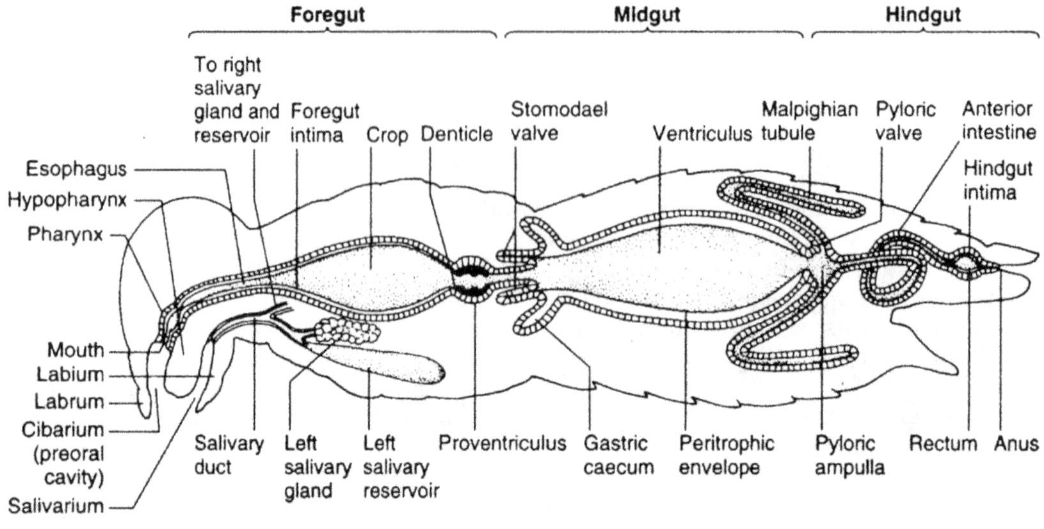

Figure 2.17. Generalized insect alimentary canal (from Romoser and Stoffolano 1998).

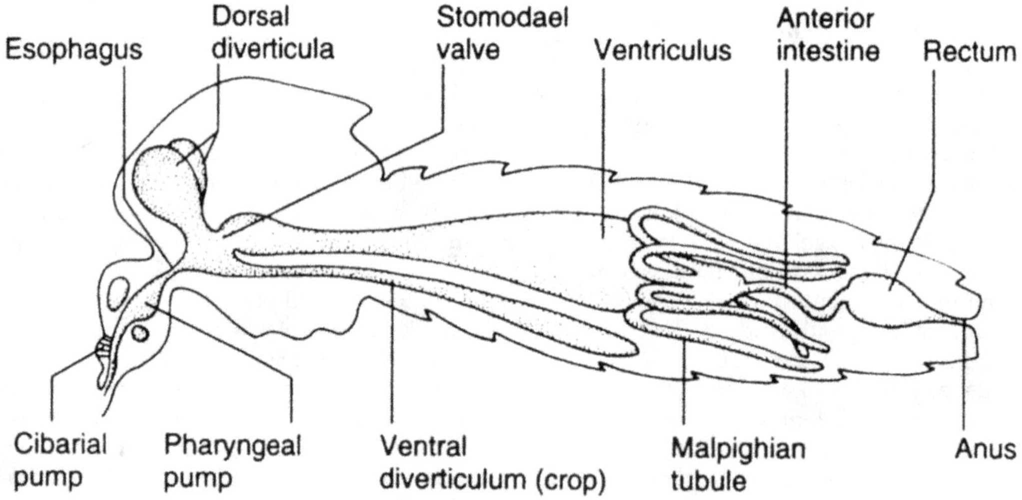

Figure 2.18. Mosquito alimentary canal (from Romoser and Stoffolano 1998).

to the midgut. This complex structure forms as an intussusception between the developing foregut and midgut and the foregut-midgut junction is located within it. In the order Diptera, the proventriculus with teeth as just described is not found and the term "proventriculus" is applied to the stomodeal valve (Fig. 2.18).

The enzymatic breakdown of large molecules in food into absorbable-size molecular units (digestion), and the absorption of these units, occurs in the midgut. Three major structures make up the midgut in most insects: the **cardia**, the **gastric caecae** and the **ventriculus**. The cardia is the outside layer of the stomodeal valve and its anterior extreme delineates the beginning of the midgut. The gastric caecae (Fig. 2.17), when present, are outpocketings of the midgut epithelium. The ventriculus (Fig. 2.17) is the largest part of the midgut and in some insects, e.g., true bugs, structural and functional subdivisions can be recognized. The gastric caecae secrete digestive enzymes and act as absorptive structures for water, inorganic ions and certain other molecules. Three basic cell types are present in the midgut epithelium: columnar digestive/absorptive cells, regenerative cells and endocrine cells. The columnar cells have well-developed microvilli on the luminal side. These finger-like projections greatly increase the absorptive surface area. On the body-cavity side, the columnar cells have an extensively folded plasma membrane, forming a series of interconnecting channels and collectively referred to as the basal labyrinth. Regenerative cells usually are located at the bases of the columnar cells, but also may be present in clumps or nests. They replace the columnar cells as they break down and are involved in reconstructing the midgut during metamorphosis in many insects. In most insects, the food bolus is surrounded by a noncellular chito-proteinaceous sheet referred to as the **peritrophic membrane** (Fig. 2.17) or peritrophic envelope. Typically, it is formed by specialized cells in the cardial epithelium, e.g., in mosquito larvae, or by delamination from most or all of the midgut cells, e.g., adult mosquitoes in response to a bloodmeal. The function of the peritrophic membrane is not well understood and many hypotheses have been advanced, including that of protection of the midgut epithelial cells from abrasion by food particles, protection of the insect from parasitic invasion, an ultrafilter and so on. For additional information on the peritrophic matrix, see Jacobs-Lorena and Oo (1996).

The hindgut forms a conducting tube between the midgut and the anus. Depending on the species of insect, a varying number of slender, elongate blind tubes, the **Malpighian tubules** (Figs. 2.17, 2.18), open into the lumen immediately posterior to the pyloric valve that separates the mid- and hindguts. The region into which these tubules open is called the pyloric ampulla. In its simplest form, the hindgut posterior to the pyloric valve consists of an elongate anterior intestine and the posterior, well-muscled rectum. In some insects, the anterior intestine is further differentiated into subregions. Several pad-like structures, the rectal papillae, may protrude into the rectal lumen. In addition to its role in conducting undigested wastes from the midgut, the hindgut, along with the Malpighian tubules, makes up the excretory system. The Malpighian tubule cells, which are microvillate on their luminal sides, pump water, various inorganic ions, amino acids and nitrogenous wastes (such as uric acid) into their lumens. These molecules then are washed down into the hindgut lumen and are propelled peristaltically to the rectal lumen. Within the rectal lumen and with the help of the rectal papillae, the molecules still needed by the insect are reabsorbed back into the body cavity and metabolic wastes egested along with undigested wastes from the midgut. Needed water in undigested wastes from the midgut also is absorbed in the insect rectum.

Depending on the insect, various specialized cells or tissues may be present in the alimentary canal that support the lives of symbiotic microorganisms that may provide specific digestive enzymes or other needed molecules (e.g., certain vitamins).

See Romoser (1996) for more information on the insect alimentary canal and its involvement with ingested parasites.

Salivary glands

The **salivary glands** (Fig. 2.17) usually function in association with the alimentary process. These glands are derived from the labial segment of the head and usually lie in the anterior part of the thorax, lateroventral to the foregut. They vary in structure but usually fall into one of 2 categories: acinar (grape-like) and tubular. These bilateral glandular masses communicate with a common salivary duct via lateral salivary ducts, and the common salivary duct in the more primitive insects (e.g., cockroaches) opens into the salivarium ventral to the hypopharynx. More advanced insects with piercing-sucking mouthparts, like mosquitoes, usually have a salivary channel within one of the mouthparts within the fasicle of stylets. In addition to the ducts, a salivary reservoir may be present to store salivary secretions. Insect saliva has been shown to have many functions, depending on the species. These functions include moistening of the mouthparts, serving as a medium for digestive enzymes, anticoagulants or agglutinins, acting as a food solvent, secretion of antimicrobial substances, and so on.

Circulatory system

Insects have an open circulatory system (Fig. 2.19), i.e., the blood (**hemolymph**) bathes the tissues directly within the body cavity. The insect body cavity, or **hemocoel**, is bounded by the epidermal cells, the tissues of ectodermal origin that form from invaginations or evaginations of ectodermal tissues (e.g., the tracheae and the fore- and hindguts), and by the midgut. This arrangement differs from vertebrates whose body cavity is lined with tissues of mesodermal origin, i.e., a true coelom. Although the insect has an open circulatory system, there are pumping and conducting elements. The most obvious is the **dorsal vessel** that runs along the dorsal midline from the posterior end of the abdomen to the head. This vessel is closed posteriorly and opens at the posterior base of the brain just above the esophagus. The tissue that makes up the dorsal vessel is contractile and peristaltic waves typically move from posterior to anterior. Two regions of the dorsal vessel are discernible: the posterior heart and the anterior aorta. The heart is characterized by the presence of pairs of openings, usually bilateral, called ostia. Hemolymph may enter or exit the heart via the ostia, but in the generalized picture, it enters as a peristaltic wave passes. The nature of the ostia is such that they close when hemolymph already within the heart passes by them. Thus, the general tendency is for hemolymph to enter the heart in the abdomen and be pumped anteriorly. Periodically spaced, fibromuscular alary muscles connect bilaterally with the heart and attach to the integument at the opposing ends. Contraction of the alary muscles opens the heart lumen. The aorta is contractile, but has no ostia or alary muscles. It opens at the base of the brain, and hemolymph exits the dorsal vessel at this point. The pumping of the heart has been studied physiologically and its rhythmic pulsations generally are thought to be under neurohormonal control. Accessory pulsatile organs provide the force necessary to drive hemolymph into the extremities of the antennae, wings and legs, which are so designed that hemolymph enters in certain points and exits at others.

Insect blood usually is colorless and is called hemolymph because it carries out most of the functions of both blood and lymph, which are separate fluids in vertebrates. While hemolymph contains both cellular and fluid portions, as in vertebrates, it differs from vertebrate blood in that except over very short distances, it does not transport oxygen to the tissues. Such transport is carried out by the tra-

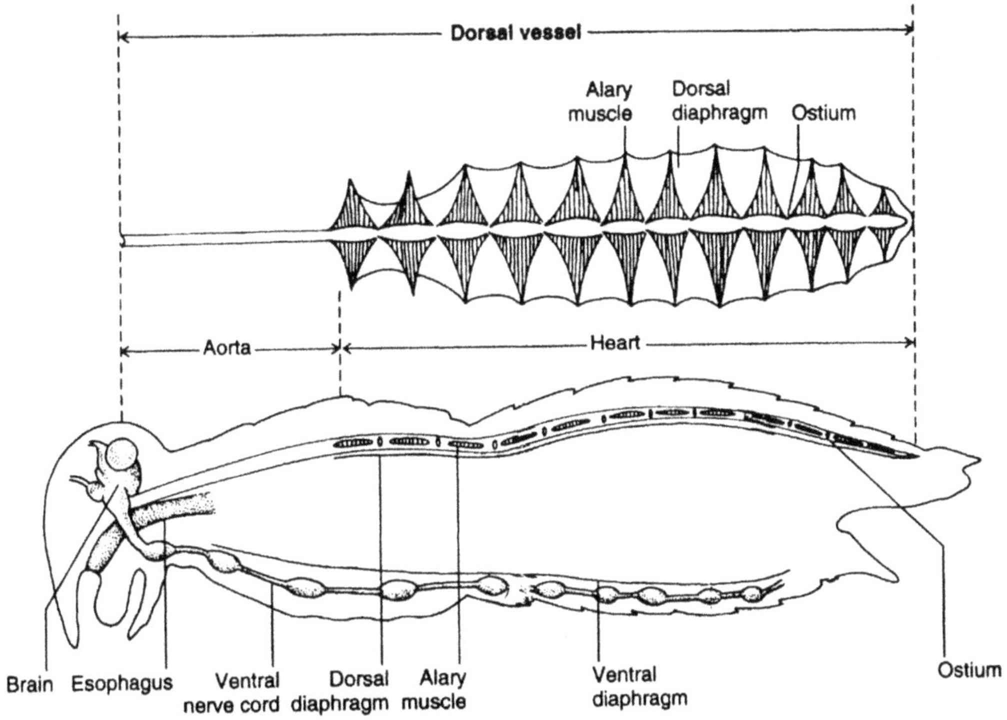

Figure 2.19. Generalized insect circulatory system (from Romoser and Stoffolano 1998).

cheal system. The characteristics such as solute concentration and pH are closely regulated, as the hemolymph provides the growth and maintenance medium for all of the cells of an insect. Hemolymph carries out many important functions, not the least of which is that of transporting molecules from digested food materials that have been absorbed into the midgut cells and released into the hemocoel. Thus, it serves as a reservoir for raw building materials. The hemolymph also transports metabolic wastes such as CO_2, nitrogenous wastes generated by protein metabolism, and so on, from the tissues where they are produced to the Malpighian tubules or other excretory organs. The hemolymph also commonly acts a hydraulic medium. Contractions of muscles associated with the body wall can direct the incompressible hemolymph to accomplish specific tasks. For example, the proboscis of the house fly is extended by hydraulic (hemostatic) pressure being applied in the head region and then is retracted by contraction of muscles in the proboscis and head. Hydraulic pressure also comes into play with the splitting and shedding of old cuticle during molting. Hemolymph, being present throughout the hemocoel, lubricates the various tissues and organs as they move about and as the insect itself moves.

Among the functions of the cellular fraction of the hemolymph, the **hemocytes**, are hemostasis (blood clotting) at wound sites, wound repair, detoxification of toxic molecules and attack of microbial and parasitic invaders through phagocytosis or sequestration by the formation of nodules or by encapsulation. Nodule formation and encapsulation are associated with bacterial invasion in numbers too high to be neutralized by phagocytosis alone or with large invaders such as helminths. Inducible antibacterial and antifungal proteins, e.g., cecropins and defensins, also have been discovered in the hemolymph.

The **fat body**, **nephrocytes** and **oenocytes** appropriately are considered in association with the hemolymph. The fat body (Figs. 2.12, 2.23a) is a diffuse organ made up of regularly placed packets of individual cells called **adipocytes**. The name "fat body" is inaccurate since it synthesizes, stores and releases many important molecules, including lipids, proteins and carbohydrates, and functions in a fashion somewhat analogous to the vertebrate liver. Fat body cells are involved in intermediary metabolism, synthesis of egg yolk proteins, storage of nutrients for growth and development, and so on.

Nephrocytes are stationary cells that take in colloidal-size materials by the process of endocytosis. These cells may occur singly or in groups, and those associated with the dorsal vessel are called **pericardial cells**.

Oenocytes are found in nearly all insects and often are arranged in segmental packets. Although they may synthesize and secrete material deposited in the epicuticle, their function remains unclear.

Ventilatory system

Oxygen reaches the tissues of insects via the tracheal system (Fig. 2.20) of branching tubes. Their openings to the outside world are called spiracles. As they divide and subdivide, penetrating deeper and deeper into the tissues, tracheae become progressively smaller in diameter. Ultimately, the smallest tubes end in single cells called **tracheoblasts**. Tracheoblasts have several thin projections on the order of 1.0 µm or less in diameter that are called **tracheoles**. All insect tissues are close to a tracheole and it is from these tubes that oxygen and carbon dioxide exchange occurs. Tracheoles also may arise directly from a trachea. Tracheae arise as ectodermal invaginations and as such are lined with a modified cuticle, the tracheal intima. The intima is thrown into supportive helical folds called taenidia.

In many flying insects, some tracheae may be enlarged into air sacs. These structures have a variety of functions including increasing the tidal flow in actively ventilating insects, providing room for expansion associated with midgut filling and egg development, and reducing specific gravity, which aids flight.

Spiracles provide access to oxygen, but given the enormous surface area provided by the finely divided tracheae and tracheoles and given the relatively large surface area to volume ratio, there is great potential for loss of water (transpiration) through the spiracles. An evolutionary solution to this problem is the presence of various spiracular closure mechanisms that remain closed, or partially closed, except when replenishment of oxygen is needed. Spiracles are under neural control and respond to changes in carbon dioxide levels in the hemolymph.

The organization of the tracheal system is complex. Typically, the spiracles open into lateral longitudinal trunks. In addition, there may be paired dorsal and ventral longitudinal trunks. The various paired trunks are connected by transverse tracheal commissures. Spiracles vary from a single pair, such as those at the posterior end of the body in mosquito larvae, to 2 thoracic and several abdominal pairs. In many aquatic insects and those with endoparasitic larvae, there are no functional spiracles, and oxygen enters and carbon dioxide exits directly through the integument.

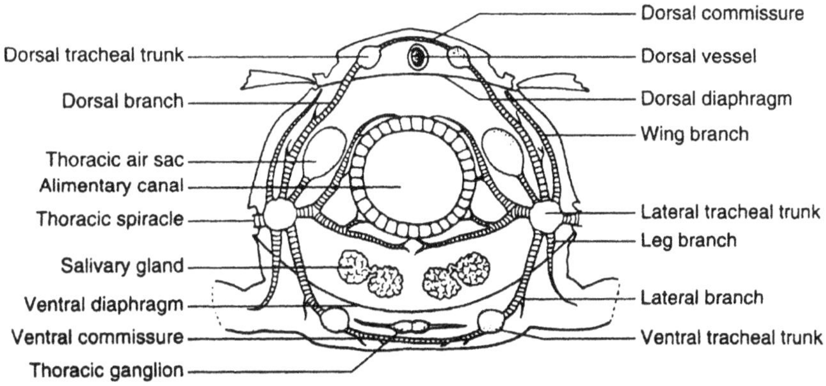

Figure 2.20. Diagrammatic cross-section of insect thorax showing major tracheal branches (from Essig 1942).

Diffusion of gases along concentration gradients is sufficient in smaller insects and in the deeper tracheae of larger, active insects. However, larger active insects may ventilate actively, gas movement being aided by muscular contraction and overall body movement.

Excretory system

The hemolymph bathes all the tissues in the body cavity and thus provides the "culture medium" for these tissues. Consequently, the characteristics of the hemolymph such as osmotic concentration, pH and so on must be maintained at constant levels. This is the role of the excretory system (Figs. 2.17, 2.18). Maintenance of a constant internal environment that is at least partly under endocrine control, is accomplished by the removal of metabolic wastes and excesses and the regulation of ions and water. Metabolic wastes include the end-products of amino acid/protein metabolism, e.g., uric acid. Hemolymph is filtered in the Malpighian tubules and needed materials are reabsorbed in the rectum. The number of Malpighian tubules varies among different insect species, ranging from 2 to more than 200. These tubules usually have muscles and are capable of some movement within the hemocoel. The Malpighian tubule epithelium is one-cell thick, the cells being comparatively large and possessing microvilli on the luminal side. As with the other epithelia in the hemocoel, a basal lamina invests the tubules on the hemocoel side. The epithelium is richly supplied with mitochondria, presumably to provide the energy needed for the active transport of ions.

The kind and extent of changes that need to be effected by the excretory system depend, in part, on the kind of food ingested. For example, insects that feed on vertebrate blood must conserve sodium by reabsorption in the rectum. The kind and extent of changes also depend upon the nature of the external environment of an insect. Different situations pose different excretory problems. For example, terrestrial insects continually are challenged by the tendency to lose water through the spiracles and across the integument. Therefore, mechanisms for conserving water are essential. Such mechanisms include the ability to regulate opening and closing of spiracles, rectal reabsorption of water, production of metabolic water with production of very dry feces, uric acid, which requires relatively little water to egest, and so on. Aquatic insects in saline environments, such as some mosquito larvae, are susceptible to loss of water and buildup of excess inorganic ions. They counteract this danger by drinking large amounts of water, and removing ions by active transport across membranes of the Malpighian tubules and rectum. Conversely, aquatic insects in fresh water risk flooding and loss of inorganic ions. They drink little water and actively absorb chloride, sodium and potassium ions from the surrounding water (Bradley 1987).

Reproduction, Growth and Development

Insects are fundamentally bisexual, egg-depositing organisms. Here, we consider the systems and processes that are involved in producing and bringing together the sperm and egg, and in the subsequent growth and development of the egg into sexually mature adults.

The reproductive system

Both the male and female reproductive systems are located in the posterior part of the abdomen and the external genitalia, considered to have been derived from primitive segmental appendages, are associated with the 8th and 9th abdominal segments.

Male reproductive system

The male reproductive system produces, stores and delivers prodigious numbers of microgametes called **spermatozoa**. The male reproductive system, shown in Fig. 2.21 in generalized form, consists of bilateral, paired **testicles**. A testicle is made up of one or more testicular follicles invested in a **follicular epithelium** that absorbs nutrients from the hemolymph and provides them for the developing gametes. The meiotic cell divisions that produce mature spermatozoa occur in a distal to proximal sequence in each follicle. The testicles are connected to the rest of the male system by means of conducting tubes, each follicle opening via a **vas efferens** into a common **vas deferens**. The vas deferens from each side join the medial **ejaculatory duct** that opens at the base of the **aedeagus,** or **penis**. The vas deferens may be thrown into convoluted folds (**epididymis**) or have an expanded region (**seminal vesicle**) in which spermatozoa are stored. Muscles associated with the various ducts propel spermatozoa along the lengths of the ducts as well as out of the insect during copulation. Other structures usually associated with the male system are the accessory glands that secrete seminal fluid and probably substances that provide signals for the female system, e.g., stimulation of contractions of the female reproductive ducts that might aid in propulsion of spermatozoa. Some insects produce sperm in groups and house it in a proteinaceous secretion of the accessory glands called a **spermatophore**. The spermatophore may be deposited directly into the female during copulation or may be deposited on a substrate to be picked up by the female.

Female reproductive system

The female reproductive system is shown in Fig. 2.22 in generalized form. Like the testes, the **ovaries** are bilaterally located. Each ovary may consist of a varying number of units called **ovarioles**. The ovarioles are collectively invested in an **ovarian sheath**. The number of ovarioles present varies with the species, the extreme of one being found in tsetse and the opposite extreme of 1,000 to 2,000 in termite queens. Each ovariole consists of a distal **germarium** in which the cells proliferate mitotically, producing the cells that will develop into **oocytes** (egg cells), **nurse cells** and follicular epithelial cells. Proximal to the germarium in each ovariole are a number of successively more developed ovarian follicles, each invested in a follicular epi-

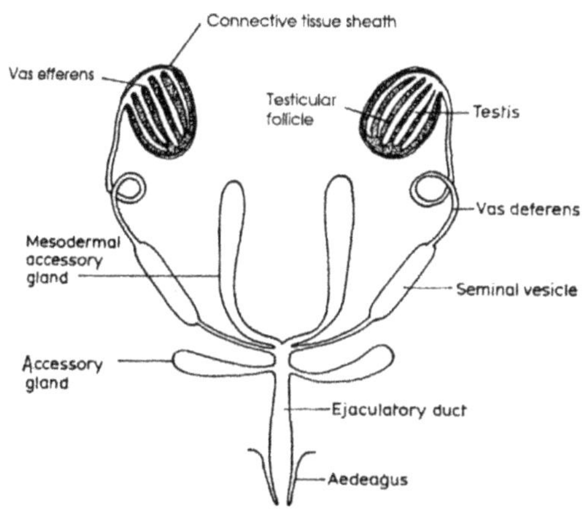

Figure 2.21. Diagram of male reproductive system (from Davies 1988).

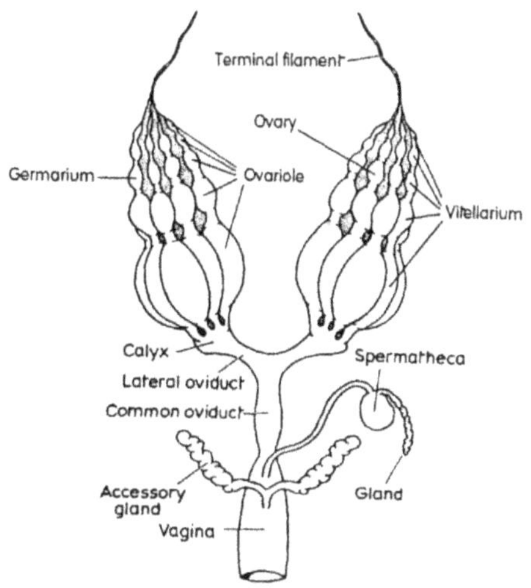

Figure 2.22. Diagram of female reproductive system (from Davies 1988).

thelium. A duct located at the base of each ovariole opens into a common region, the **calyx**, which in turn communicates with the **lateral oviduct**. The 2 lateral oviducts open into the **common oviduct** and this system of tubes with its investing muscles conducts vitellogenic eggs (i.e., those with yolk) into the central portion of the reproductive system, the **vagina**, which opens to the outside of the insect. The vagina also receives the male penis during copulation. The vagina typically bears one or more evaginated structures, the **spermathecae**, in which sperm are stored until fertilization. Thus, insect females typically have the ability to control when, and in some cases if, fertilization occurs. One or more accessory glands also may be associated with the ducts in the female system. Accessory glands carry out such functions as secretion of gelatinous materials that protect clutches of deposited eggs and secretions used to glue eggs to the substrate.

The female reproductive system is designed to:

(a) receive and store spermatozoa,
(b) produce **eggs**,
(c) provide eggs with the nutriment (**yolk**) necessary to support embryological development,
(d) provide eggs with a protective outer shell or chorion that resists transpiration, but at the same time facilitates gaseous exchange,
(e) bring eggs and spermatozoa together, effecting fertilization, and
(f) deposit the eggs in an environment that will support the newly hatched insects.

As with the other physiological/behavioral processes, the processes directly involved with reproduction are closely regulated. The expansion caused by the process of **vitellogenesis** (yolk synthesis and deposition in the eggs) is accommodated by the tremendous flexibility and expandability of the insect abdomen.

Egg development

The processes that lead to the development of a mature egg ready to be fertilized can be grouped under the heading of "**oogenesis**" and includes the processes of egg differentiation, vitellogenesis (Fig. 2.23a-c), **ovulation** (exit of an egg from an ovariole), and finally, around the time of fertilization, completion of the reduction divisions of meiosis, producing the haploid egg nucleus. Yolk proteins (**vitellogins**) are synthesized in the fat body and transferred via the hemolymph to the ovaries, where they are taken up by endocytosis and deposited as yolk.

Insect eggs usually are elongate, ovoid structures surrounded by a cuticle-like chorion secreted by the follicular epithelia. The chorion protects the egg from water loss while allowing gaseous exchange. Within the confines of the chorion, the yolk dominates the volume. Small cytoplasmic regions are present containing the egg nucleus (nuclear cytoplasm) and around the periphery of the yolk (periplasm).

Copulation, insemination and fertilization

The male and female gametes are brought into proximity during copulation as the male deposits semen in the female (insemination). Such internal fertilization protects against water loss. Gametes, being very small, are especially susceptible to drying. Fertilization occurs by 2 sequential events: first the release of the spermatozoa from the spermatheca and second the entry of the egg by a sperm passing through the **micropyle**, a pore in the egg. Insect spermatozoa are filamentous to facilitate passage through the very narrow micropyle. This is another mechanism by which the drying tendency is avoided, i.e., by having filamentous sperm, the micropyle can be very narrow, thus minimizing water loss.

Embryonic and postembryonic development

Since most insects develop eggs with chorions and deposit them externally (**oviposition**), embryogenesis usually occurs in the egg when it is separate from its parent. The energy for embryogenesis comes from yolk. When embryogenesis is complete and cues are received signaling that environmental conditions are favorable, an insect egg undergoes eclosion (hatching) and the first immature stage emerges. The immature stage, or larva, then proceeds to eat and grow, growth being facilitated by **molting**, a periodic shedding of partially digested old cuticle and deposition of new. Molting involves the following processes: separation of the old cuticle from the epidermis (**apolysis**), partial digestion and resorption of old cuticle, secretion of new cuticle, and the shedding of remaining old cuticle (**ecdysis** or **eclosion**). A newly ecdysed, or **teneral**, individual is soft and pale, but soon the processes of hardening (**sclerotization**) and darkening (**melanization**) occur. Newly emerged insects gulp air or water, expanding the body prior to hardening.

Following hatching of the egg, insects grow through a series of immature stages, each separated by a molt, and ultimately reach the adult stage and become sexually mature. Each stage, including the adult, is referred to as an **instar** and often can be recognized on the basis of size and various characteristic anatomical traits. For example, mosquitoes have 4 larval instars each distinct from the others. The instar is the insect itself and the time spent as a given instar is called the stadium for that instar. The adult instar is called an **imago** and when referring to an adult trait the term **imaginal** is used. The strict definition of an instar has been debated but it may be viewed as an insect either between successive ecdyses or between successive apolyses. If one subscribes to the later definition, the insect is referred to as **pharate** after apolysis but before ecdysis. Thus, in the case of a biting gnat larva, for example, after apolysis in the 3rd larval instar and before ecdysis, the form is referred to as a pharate 4th instar.

Metamorphosis

Insects vary in the degree of change they undergo between eclosion and reaching the adult stage. The change may be minimal, con-

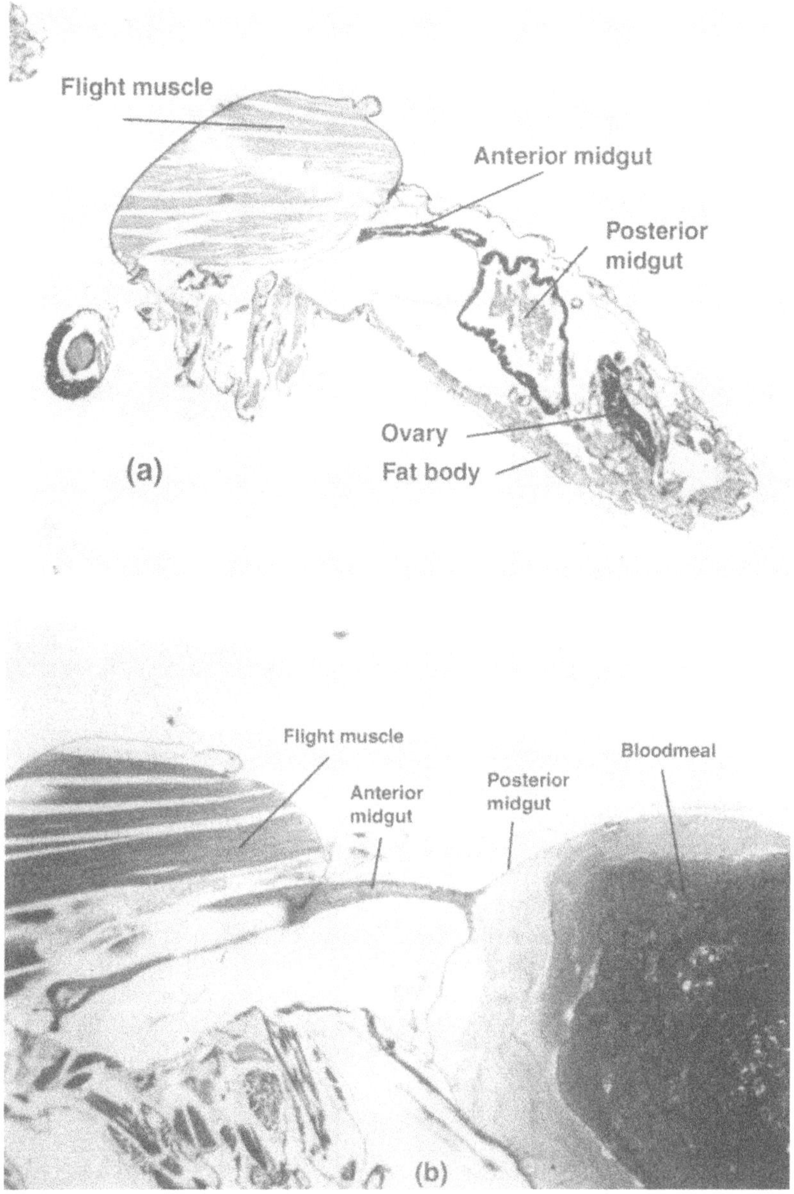

Figure 2.23. Photomicrographs of saggital sections of a mosquito: (a) Prior to bloodmeal—note that midgut is undistended and ovaries small; (b) immediately following bloodmeal—note that posterior midgut is distended with blood (photographs by W.S. Romoser).

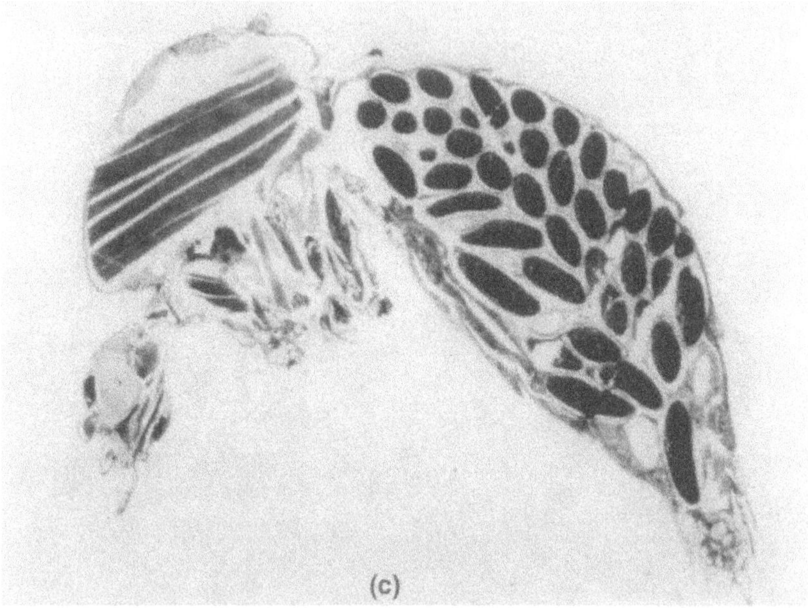

Figure 2.23 (continued). Photomicrographs of saggital sections of a mosquito: (c) Following digestion of blood—note abdomen is now filled with vitellogenic eggs and the posterior midgut is compressed by the fully expanded ovaries (photograph by W.S. Romoser).

sisting only of a gradual size increase from molt to molt, with the adult looking essentially like a large larva except that it is sexually mature. Or the change may be more distinct, with **wingpads** appearing in late larval instars and with the greatest change occurring between the last larval instar and the adult. Or the change may be dramatic, the wingless larvae appearing completely different from the adult and the adult being "assembled" during a single stage, the **pupa**, inserted between the larval and adult stages. Insects that undergo the first mentioned minimal change, are referred to as undergoing **no metamorphosis,** or **ametabolous** development. Insects that undergo some change have **simple metamorphosis** (Fig. 2.24). There are **paurometabolous** and **hemimetabolous** variations of simple metamorphosis. Finally, those that exhibit dramatic major change undergo **complete metamorphosis** or **holometabolous** development (Fig. 2.25). In insects that undergo complete metamorphosis, the larval instars appear similar to one another and the only evidence of the future adult stage is the presence of **imaginal buds** or **disks** that form internally in association with the integument. These disks will form into the antennae, mouthparts, legs and wings that are apparent in the pupal stage and reach full development by the adult stage. By contrast, in insects that undergo simple metamorphosis, the antennae, legs and mouthparts are already present in the larval instars and the wings appear as wing pads only in the later larval instars. Based on the differences in wing development, insects that undergo simple metamorphosis are referred to as **exopterygotes** ("external wings") and those with complete metamorphosis as **endopterygotes** ("internal

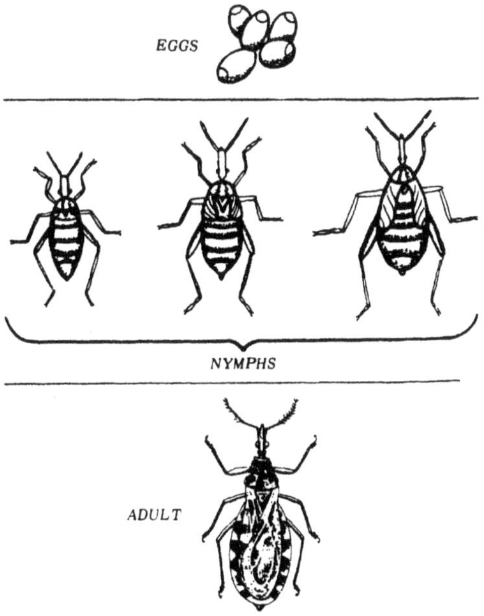

Figure 2.24. Simple metamorphosis illustrated by a triatomid bug in the family Reduviidae (from US Public Health Service 1969).

wings"). In the past, insects that undergo simple metamorphosis and have aquatic larvae were said to undergo hemimetabolous development. Conversely, insects with simple metamorphosis having both terrestrial-aerial larvae and adults were said to be paurometabolous. These terms now are rarely used, and Hemimetabolous larvae are sometimes called **naiads** and paurometabolous larvae, **nymphs**. More commonly, both types of larvae are called nymphs.

Control of growth, development and metamorphosis

Growth, development and metamorphosis are under endocrine control. Molting is initiated when glands in the prothorax secrete the molting hormone (also called ecdysone or prothoracic gland hormone). Another hormone, juvenile hormone (JH), which is synthesized and secreted by the corpora allata, determines the nature of a given molt. Juvenile hormone sustains the larval stage. When it is present, a given molt will be from one larval instar to the next. The titer of JH gradually diminishes over the course of larval development. At the time of the larval-adult molt in insects with simple metamorphosis, or the larval-pupal adult in insects with complete metamorphosis, only ecdysone is secreted, and metamorphosis commences. The timing of secretion of ecdysone and hence the timing of molting is determined by the secretion of brain hormone (**prothoracicotropic hormone,** or PTTH) that is synthesized in neurosecretory cells in the brain and is stored in the corpora allata. The target tissues of PTTH are the glands and this hormone induces activity of the prothoracic glands and the secretion of ecdysone. At least one hormone, eclosion hormone (a neuropeptide) and probably 2, is involved in the process of ecdysis in insects. This hormone is synthesized in the brain and influences functions such as ecdysial behavior.

The appropriate timing of synthesis and release of particular amounts of hormones accomplishes physiological regulation. Chemicals that have hormonal effects are called **insect growth regulators** (IGRs). Examples of IGRs include the juvenile hormone analogues methoprene and kinoprene. Synthetic IGRs are used to disrupt these processes and thereby control insects.

Ecological synchronization of life cycles

Evolution has synchronized insect life cycles with the environmental changes that occur in a given region. During winter in temperate zones, a cold-hardy insect may enter a resting state or diapause (as eggs, larvae, pupae or adults). An insect might become torpid and resistant to drying in a region with a regularly periodic dry season. An insect may begin to display migratory behavior in response to seasonal changes. Diapause and migration thus are mechanisms that have evolved to enable insects to deal with

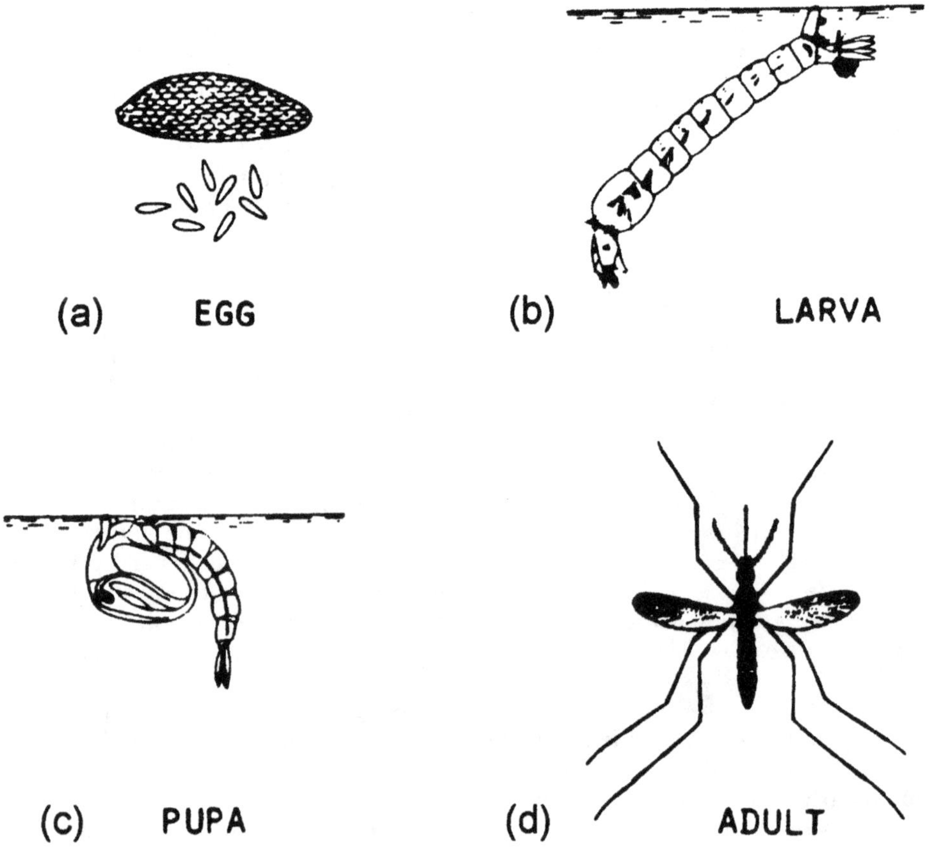

Figure 2.25. Complete metamorphosis illustrated with a mosquito (from US Public Health Service 1969).

deleterious seasonal environmental conditions. The study of the relationships between the insect life cycle and seasons is called **phenology**. As with most other processes, diapause and migratory behavior are under endocrine and neurological control.

THE CLASS ARACHNIDA

Acari Structure and Function

The Acari (from the Greek "akares," small) are ticks and mites included in the class Arachnida. The Acari possess the typical arachnid traits: they lack antennae, and as adults, bear 6 pairs of appendages (chelicerae, palps and 4 pairs of jointed legs). Several members of this group are of medical importance. Information presented in this section is based on Sauer and Hair (1986) and Evans (1992).

Exterior Structures

The integument

The integument of mites and ticks is similar to that found in insects, comprising a single layer of cells (epidermis) that secretes, on the outside, a noncellular cuticle and which is separated from the hemolymph by a basal lamina. Numerous specialized cells, including dermal glands that open to the outside through minute pores, are interspersed among the epidermal cells. As with other arthropods, the degree of cuticular sclerotization varies temporally and spatially, and among different species.

The mite and tick cuticle often is pigmented, and in some species with pale bodies the color of ingested food may show through, e.g., the red to brown to black appearance of a bloodmeal. Likewise, in pale forms pigment within the epidermal cells may impart color.

As with other arthropods, mites and ticks grow in association with periodic episodes of molting. However, molting has not been well studied in this group.

Other structures

Unlike other arachnids, except spiders, mites and ticks differ in lacking evident body segmentation, although there may be sutures. The acarine body (Fig. 2.26) is divisible in various ways. Fundamentally, it can be divided into the anterior **gnathosoma** and posterior **idiosoma**. The idiosoma is divided further into the leg-bearing region (**podosoma**) and the remaining portion (**opisthosoma**).

The gnathosoma often is called the **capitulum**, especially in ticks. The **epistome** and **subcapitulum** (**basis capituli**) form the upper and lower walls of the gnathosoma, respectively, and connect to the esophagus. A median **hypostome** extends anteriorly from the subcapitulum and in ticks, for example, serves as an anchoring organ, a process facilitated by posteriorly directed teeth. Closely associated with the gnathosoma are the paired, and sometimes retractable, chelicerae (variously modified piercing, cutting, tearing appendages) and the paired **palps** (sensory appendages).

The idiosoma bears 3 (in larvae) or 4 (in nymphs and adults) pairs of legs as well as the external openings (**stigmata** or spiracles) of the respiratory system and various sensory and secretory structures, e.g., some bear simple eyes. Each leg is composed of a basal coxa, followed by the trochanter, femur, patella, tibia, tarsus, caruncle and claw. The stigmata may or may not be present and vary in number and position, features used to divide the Acari into subgroups.

Internal Structure and Physiology

Coordination and effector systems

The central nervous system is composed of a ganglionic mass (**brain**) that surrounds the esophagus and gives rise to nerves that extend to various areas within the body (Fig. 2.27). This mass is divided by the esophagus into 2 regions: the **pre-** or the **supraesophageal** and **post-** or **subesophageal**. The preesophageal region innervates the photoreceptors (in some cases, eyes) and the gnathosoma, while the postesophageal region innervates the remaining parts of the body. Mites and ticks possess a variety of sensory structures on the body surface similar to sensory structures in insects.

Endocrine regulation of physiological processes in mites and ticks has not been studied as intensively as it has been in insects. However, there are clearly identifiable neuroendocrine complexes and evidence of neurosecretion. It seems likely that processes such as molting,

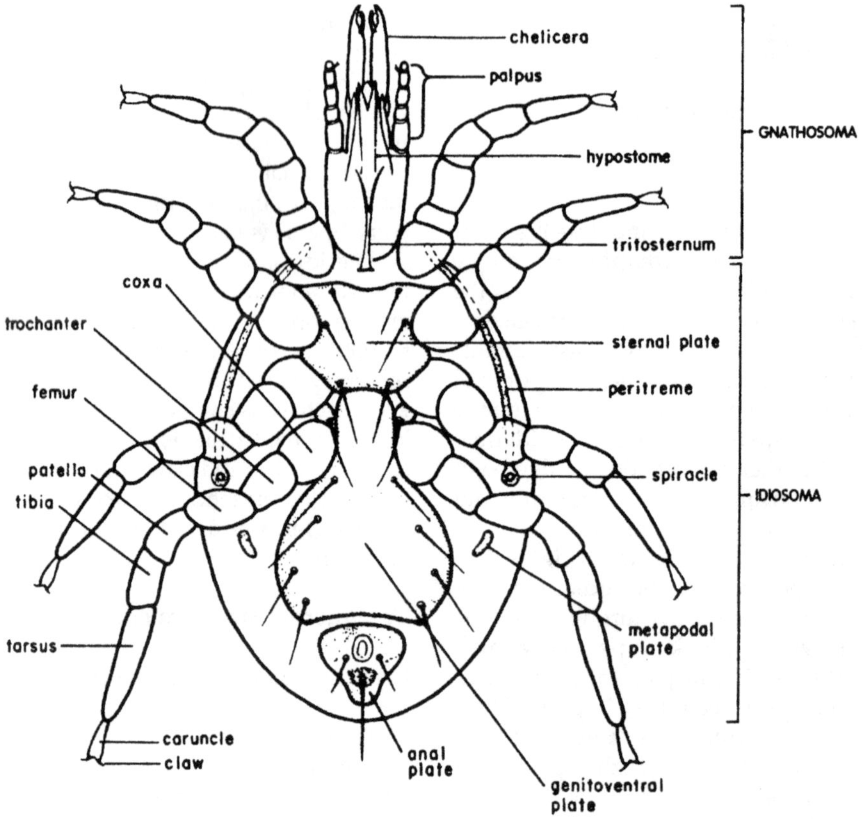

Figure 2.26. External anatomy of a mite (from US Public Health Service 1969).

diapause and reproduction are regulated in a fashion similar to that in insects.

All acarine muscles are striated and there is a complex skeletal musculature involved with moving of the various appendages. The chelicerae in most mites and ticks are retracted by the action of skeletal muscles and protracted by hydrostatic pressure translated through the hemolymph by contraction of muscles in the idiosoma. Longitudinal, circular and irregular networks of visceral muscle effect peristalsis in various tubular organs such as the alimentary canal.

Maintenance systems

The embryogenesis and functional subdivision of the acarine alimentary canal closely resembles that of insects. There are 3 basic gut regions. The fore- and hindguts are of ectodermal origin and are lined with a chitinous intima; the midgut originates endodermally. In the fully developed alimentary canal (Fig. 2.27) the mouth opens into a muscular pharynx, followed by the esophagus, which passes through the ganglionic mass. Interestingly, as with insects, the foregut forms an intussusception with

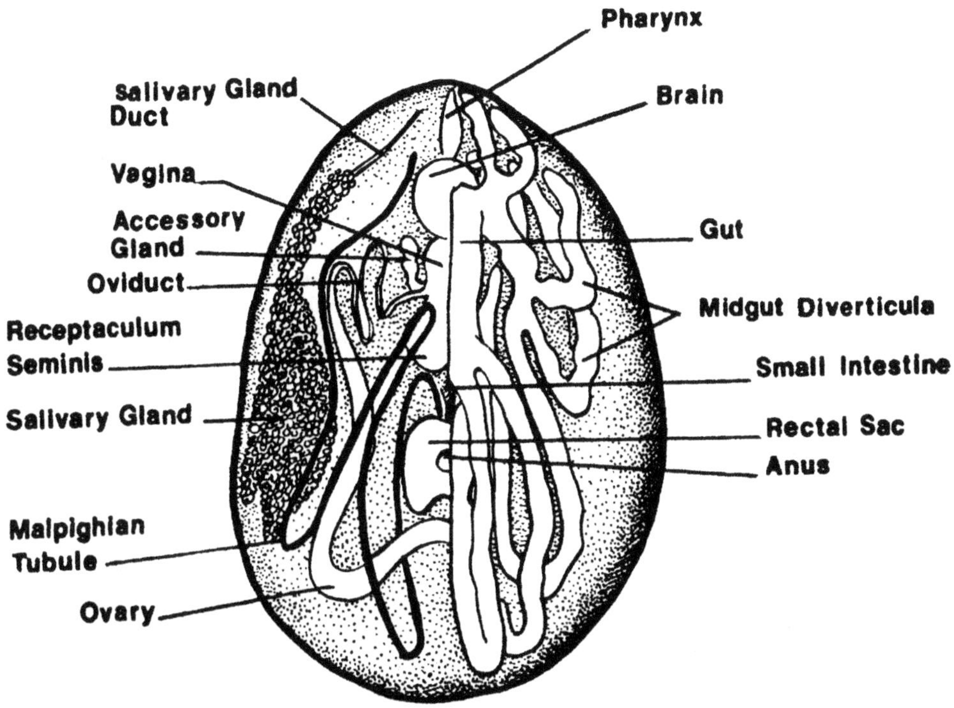

Figure 2.27. Internal anatomy of a tick (from Balashov 1972).

the midgut, and muscles lie between the folded foregut walls, providing a sphincter effect. This probably helps regulate movement of material from the foregut into the midgut as well as prevent regurgitation of midgut contents into the foregut.

The midgut (Fig. 2.27) often is diverticulate, i.e., it has several finger-like projections, and often is capable of great expansion, especially in vertebrate bloodfeeding forms. The midgut is the region where most digestion and absorption occurs and this organ is composed of a single layer of epithelial cells separated from the hemolymph by a basal lamina. In addition to the ventriculus (stomach) *per se*, there often is a **postventricular region** within the midgut that is composed of a **colon** and a **postcolon**. A pyloric sphincter apparently controls movement of materials from the ventriculus into this region. Commonly, a pair of Malpighian tubules opens into the postventricular region. The hindgut is short, consisting of the **anal atrium,** which opens through the slit-like anus. Bilaterally located glandular masses are found in the posterior of the acarine body and communicate by means of lateral ducts either with the buccal/subcapitular cavity or near the coxal bases of the front pair of legs. Those that open into the capitulum often are called salivary glands, e.g., in ixodid ticks (subclass Ixodida). While these glands have been studied extensively in ixodid ticks, little is known about their function in other groups.

Mites and ticks have an open circulatory system in which the more or less clear hemolymph circulates in the hemocoel. The hemolymph comprises a cellular and fluid portion and plays the same physiological roles as in other arthropods. Fat body occurs near the central ganglion and in association with the coxae and tracheae.

Contraction of body muscles and movements of viscera effect circulation in the smaller Acari, but in ticks and certain mites this process is aided by a dorsal, longitudinal heart. Two regions are recognizable in the heart: an anterior portion with longitudinally oriented muscle fibers and a posterior region with radially arranged muscle fibers and openings (ostia) into the hemocoel.

Ventilation is achieved by a system of **tracheae** that open to the outside via openings called stigmata.

Malpighian tubules that open into the hindgut in several groups, including ixodid ticks, eliminate nitrogenous wastes. In Acari that lack Malpighian tubules, a portion of the midgut carries out the necessary function. Water and ion balance involve several organs, including **coxal glands**, salivary glands, **genital papillae** and **claparede organs**. In ixodid ticks, much of the excess water in bloodmeals is excreted back into the host via the salivary glands.

Reproduction and Development

The germinal tissues (**testes** and ovaries; Fig. 2.27) are of mesodermal origin and communicate with the outside by means of ducts that originate from mesodermal tissue (**oviducts** in the female, **vasa deferentia** in the male) and ectodermal tissue (vagina; ejaculatory duct). Although the ovaries and testes are primitively paired structures, they are fused into a single median mass in certain groups. In most cases, the external genital opening (**gonopore**) is located ventrally. In males, there may be a seminal vesicle inserted between the vasa deferentia and the ejaculatory duct, a portion of which may be elaborated into an intromittent organ (penis or aedeagus). Accessory glands may open into the seminal vesicle or ejaculatory duct and contribute to the production of seminal fluid. In females, a seminal receptacle may be associated with the vagina and accessory glands may be present. In some Acari there is a sperm receptacle or spermatheca.

Sperm, usually within a spermatophore, are introduced into the female by direct deposition or are deposited on the substrate and then taken up by the female. Parthenogenesis is common among the Acari.

Acarines go through, at most, 6 stages during their life cycle: prelarva, larva, protonymph, deuteronymph, tritonymph and adult. As with other arthropods, each stage is separated by a molt and considerable variation exists in their life cycles. As the nymphal stages resemble one another, 4 distinct phases in the life cycle are recognized: **prelarva**, **larva**, **nymph** and **adult**. The prelarva and larva have 6 pairs of legs while the nymphs and adults have 8 pairs. Mites and ticks do not continue to molt once they have reached the adult stage. Mite life cycles vary greatly. Chapter 3 provides more information on the life cycle.

REFERENCES

Balashov, Y.S. 1972. Bloodsucking ticks (Ixodoidea) – Vectors of diseases of man and animals (translation). Misc. Publ. Entomol. Soc. Am. **8**:161-376.

Beaty, B.J. and Marquardt, W.C. (eds). 1996. The biology of disease vectors. University Press of Colorado, Niwot, Colorado. 632 pp.

Bradley, T.J. 1987. Physiology of osmoregulation in mosquitoes. Annu. Rev. Entomol. **32:**439-462.

Cook, B.J. and Holman, G.M. 1985. Peptides and kinins. Pp. 531-593 *in* Kerkut, G.A. and Gilbert, L.I. Comprehensive insect physiology, biochemistry and pharmacology, Vol. 11.

Chapman, R.F. 1982. The insects — structure and function, 3rd ed. Harvard University Press, Cambridge, Massachusetts. 919 pp.

Davies, R.G. 1988. Outline of entomology, 7th ed. Chapman and Hall Publishing, New York and London. 408 pp.

DuPorte, E.M. 1961. Manual of insect morphology. R.E. Krieger Publ. Co., Huntington, New York. 224 pp.

Essig, E.O. 1942. College entomology. Macmillan, New York. 900 pp.

Evans, G.O. 1992. Principles of acarology. CAB International, Wallingford, Oxon, UK. 563 pp.

Harwood, R.F. and James, M.T. 1979. Entomology in human and animal health, 7th ed. Macmillan, New York. 548 pp.

Jacobs-Lorena, M. and Oo, M.M. 1996. The peritrophic matrix of insects. Pp. 318-332 *in* B.J. Beaty and W.C. Marquart (eds.). The biology of disease vectors. University Press of Colorado, Niwot, Colorado.

Kerkut, G.A. and Gilbert, L.I. (eds). 1985. Comprehensive insect physiology, biochemistry and pharmacology. 13 vol. Pergamon Press, New York.

Metcalf, R.E., Flint, E.E. and Metcalf, R.R. 1962. Harmful and useful insects, 4th ed. Macmillan, New York. 780 pp.

Richards, O.W. and Davies, R.G. 1977. Imm's general textbook of entomology, 10th ed. Vol. 1, Structure, physiology and development. Chapman and Hall, London and Wiley, New York (Halstead Press). 418 pp.

Romoser, W.S. 1996. The vector alimentary system. Pp. 298-317 *in* B.J. Beaty and W.C. Marquart (eds.), The biology of disease vectors. University Press of Colorado, Niwot, Colorado.

Romoser, W.S. and Stoffolano, J.G., Jr. (eds). 1998. The science of entomology, 4th ed. WCB/McGraw-Hill, Dubuque, Iowa. 605 pp.

Sauer, J.R. and Hair, J.A. (eds). 1986. Morphology, physiology, and behavioral biology of ticks. John Wiley & Sons (Halstead Press), New York. 510 pp.

Snodgrass, R.E. 1935. Principles of insect morphology. McGraw-Hill Book Co., New York. 667 pp.

Snodgrass, R.E. 1959. The anatomical life the mosquito. Publication 4388, The Smithsonian Institution, Washington, DC 87 pp.

Tipton, V.J. (ed.). 1974. Medical entomology, mediated instructional kit. Entomological Society of America and Brigham Young University. College Park, Maryland and Provo, Utah.

US Public Health Service. 1969. Pictorial keys to arthropods, reptiles, birds and mammals. Publication No. 1955, U.S. Department of Health, Education and Welfare, Communicable Disease Center, Atlanta, Georgia.

Chapter 3

Introduction to Arthropods: Systematics, Behavior and Ecology

WILLIAM S. ROMOSER
Ohio University, Athens

ARTHROPOD SYSTEMATICS

General Principles

The variety of living forms bewilders us with its diversity. However, more detailed inspection reveals strong threads of unity running throughout. **Systematics** identifies these threads and uses them to devise useful groupings, or classifications, of organisms. These groupings enable biologists to identify specific kinds of organisms, i.e., to recognize whether or not a given organism has been previously described and where it fits relative to other organisms. Thus, 3 fundamental tasks of biosystematics are to **establish identifications, provide descriptions** and **erect classifications.** Modern biology recognizes organisms as the products of millions of years of evolution and therefore classifications are viewed logically as reflecting, to a greater or lesser extent, **phylogenetic relationships**. As with all biological sciences, systematics constantly changes. New species are described and classifications are erected, modified and corrected. Each change moves our understanding of phylogeny closer to the truth. For additional information on the role of systematics in entomology, see Danks (1988).

The fundamental unit of living things is the **individual** organism. Individual organisms typically exist in interacting groups called **populations**. Populations of individual organisms capable of mating and producing viable offspring are recognized collectively as **species**. By making detailed comparisons based on many kinds of organismal characteristics (morphological to molecular), systematists organize species into successively larger groupings based on similarities. Such groupings, including species, are called **categories** that are arranged in a **hierarchy** of increasingly inclusive groups (Fig. 3.1). In any classification scheme, the actual organisms assigned to a given category constitute a **taxon** (pl., **taxa**). Once erected, classifications facilitate identification and in so doing provide a "place" to file biological information about a given taxon. Thus, classifications form the basis for storage and retrieval of biological information. They also provide a basis for communication among biologists.

Another critical task of systematics is to develop and maintain the procedures for naming and grouping organisms, i.e., the task of dealing with **nomenclature**. The International Code

B.F. Eldridge and J.D. Edman (eds.), Medical Entomology, Revised Edition, 53–97.

Kingdom	Animalia
Phylum	Arthropoda
Class	Insecta
Order	Diptera
Suborder	Nematocera
Family	Culicidae
Subfamily	Culicinae
Tribe	Culicini
Genus	*Aedes*
Subgenus	*(Stegomyia)*
Epithet (species)	*aegypti*

Figure 3.1. Naming conventions for biological organisms.

of Zoological Nomenclature published by the International Commission on Zoological Nomenclature, founded in 1895, spells out these rules. The most recent edition, the 3rd, was published in 1985; the 4th edition is well underway. Because this publication is extensive and complex, a few nomenclatural details deserve mention here. Each species of organism is named by a **binomen** that consists of the **genus** and **specific epithet**; e.g., the formal name of the Asian tiger mosquito is *Aedes albopictus*. A comprehensive checklist of the scientific names and the taxonomic position of the medically most important arthropods may be found in Pittaway (1991). The categories above the generic level are uninomials and there are accepted conventions regarding the suffixes of some of these names; e.g., family names end in "-idae" while most order names of insects end in "-ptera" (Fig. 3.1). Arthropods also may be assigned common names. In some cases these names are vernacular, while others are formally recognized. For example, all members of the order Diptera are referred to as true flies and all members of the order Hemiptera, true bugs. Species of arthropods of medical (or economic) significance sometimes are assigned formal common names, e.g., the yellow fever mosquito, *Aedes aegypti*. The Entomological Society of America publishes an extensive listing of accepted common names. For consideration of the biosystematics of blood-feeding insects, see Service (1988).

Cryptic Species and Species Complexes

Two groups of closely similar organisms are recognized as distinct species if they are reproductively isolated from one another, regardless of whether they live in the same geographic locale (**sympatric**) or in distinctly separate geographic locales (**allopatric**). The mechanisms of this reproductive isolation may act before mating occurs (e.g., due to behavioral differences that result in a failure to mate, or structural differences that make copulation impossible), or after mating, due to various cellular and genetic incompatibility mechanisms. Despite this concept of the biological species, tests regarding the ability to interbreed seldom are feasible and species designations historically have rested firmly on morphological bases (**morphospecies**).

Because subtle biological differences can prevent interbreeding, it is not surprising that similar species can be difficult or impossible to distinguish on purely morphological grounds. Such species are referred to as **sibling** or **cryptic** species. Among the arthropods, there are many sibling species, some of which involve several different species (so-called **species complexes**). Often, this may cause confusion in the use of species names in the literature. When authors refer to a complex of species, especially when the name of the complex has been well established previously as a species name, they may use the designation **sensu lato** (i.e., "in the broad sense," abbreviated s.l.). When referring to a species member of a complex bearing the same name, they may use the designation **sensu stricto** (i.e., "in the strict sense," abbreviated s.str.). Thus, *Anopheles maculipennis* s.l. would refer to the *An. maculipennis* group of sibling

Table 3.1. Major divisions of the important orders of the Class Insecta.

Subclass	Section	Order	Common names
Apterygota		Collembola	Springtails
		Thysanura	Silverfish and relatives
Pterygota	Paleopterous Exopterygotes	Odonata	Dragonflies and damselflies
		Ephemeroptera	Mayflies
	Neopterous Exopterygotes	Plecoptera	Stoneflies
		Phasmida	Walking sticks
		Orthoptera	Grasshoppers and relatives
		Dermaptera	Earwigs
		Isoptera	Termites
		Blattaria	Cockroaches
		Mantodea	Praying mantids
		Psocoptera	Barklice and booklice
		Mallophaga	Chewing lice
		Anoplura	Sucking lice
		Thysanoptera	Thrips
		Hemiptera	True bugs
		Homoptera	Cicadas, leafhoppers and relatives
	Neopterous Endopterygotes	Coleoptera	Beetles
		Neuroptera	Dobsonflies, lacewings
		Mecoptera	Scorpionflies
		Siphonaptera	Fleas
		Diptera	True flies
		Tricoptera	Caddisflies
		Lepidoptera	Butterflies and moths
		Hymenoptera	Bees, wasps and relatives

species; *An. maculipennis* s.str. would refer to the species belonging to the group.

Sibling species and species complexes are of particular interest in medical entomology because species within a complex may vary in traits that influence their ability to transmit pathogens. These traits include behavioral differences (e.g., host preference, tendency to feed indoors versus outdoors, choice of oviposition sites and so on) and physiological differences that directly affect their ability to transmit a given pathogen (**vector competence**). For example, scientists in early 1900s in Europe recognized that the distribution of malaria and of the anopheline vector *Anopheles maculipennis*, did not coincide. This phenomenon was referred to as "anophelism without malaria" and remained a puzzle until all life stages (particularly the eggs) had been examined closely and new technology that provided information based on other than morphological traits had been developed. Eventually, it became evident that the 7 species previously recognized as *An. maculipennis* differed in behavioral and physiological traits important for transmitting malaria. Thus, in order to understand the dynamics of a given vectorborne disease system, we must know exactly which species are vectors.

The identification of species complexes sometimes has used subtle morphological characters in other than adult stages, as was the case in early studies of the *An. maculipennis* complex. Differences in egg colors and patterns provided the first signs of differences within the group. But separation of the members of species complexes became possible when techniques were developed that facilitated comparisons based on non-morphological characters. Such techniques involve observation of chromosomal characteristics (**cytospecies**), biochemical characteristics (recognition of isoenzyme differences), identification of cuticular hydrocarbons by means of gas chromatography, and most recently, studies of nucleic acids using methods of molecular biology. The latter techniques are especially valuable since they discriminate at the genome level, can be applied to any life stage and require only small samples of hemolymph, leaving material available for other kinds of analyses, e.g., the identification of pathogens.

Several species complexes of medically important arthropods are known and no doubt many more await discovery. The most intensively studied and best known are those in mosquitoes (family Culicidae) and black flies (family Simuliidae). Members of these complexes vary considerably in vector competence relative to pathogens transmitted. Among the mosquitoes are the already mentioned *An. maculipennis* complex (at least 7 species) and the *Anopheles gambiae* complex (6 species). Two subspecies have been identified in *Aedes aegypti* (*Ae. aegypti aegypti* and *Ae. aegypti formosus*). Some systematists consider *Culex pipiens* and *Culex quinquefasciatus* to be subspecies, while others treat them as distinct species. Several complexes have been identified among the black flies, the best known being the *Simulium damnosum* complex, which includes 40+ species. Species complexes also have been discovered among the tsetse (family Muscidae, subfamily Glossinae) and among the Acari (e.g., the *Ornithodoros moubata* group, the *Rhipicephalus sanguineus* group and the *Hyalomma marginatum* group). Additional information on sibling species and species complexes can be found in Black and Munstermann (1996), Crosskey (1990), Lane and Crosskey (1993) and Service (1988).

The Orders of Insects

The current major subdivisions of the class Insecta are based on key events in their evolutionary history and on their major life cycle patterns (Romoser and McPheron 1998). The first insects were wingless, terrestrial, comparatively small scavenging forms, perhaps similar to the extant wingless Thysanura (silverfish and firebrats) and Archaeognatha (jumping bristletails). Early in their evolution they developed wings, which arose on the 2 posterior thoracic segments as bilateral evaginations superimposed on segments originally designed for walking. In the most primitive winged insects, the wings at rest were held laterally or together over the dorsum. This position is referred to as the **paleopterous** ("ancient wing") condition. The advent of wings on animals of relatively small size enabled insects to fill a multitude of niches. Once flight occurred, those possessing this ability comprised a monophyletic group. Flight favored escape from predators, dispersal and genetic mixing by increasing distributional ranges. A subsequent major event in insect wing evolution was the ability to fold the wings back over the abdomen when not in use, the so-called **flexion mechanism**. This ability facilitated running and hiding while still providing the advantages of wings. Insects with the ability to flex the wings are referred to as possessing the **neopterous** ("new wing") condition. Another major event in insect evolution was the development of complete metamorphosis. This ability is considered to have arisen once and therefore all insects that display it form a monophyletic group.

Insects are divided into 2 subclasses on the basis of wings, the prefix or suffix "pter," meaning wing, being used in the names. Insects considered to be primitively wingless, i.e., derived from ancestors that never had wings, are placed in the subclass Apterygota, while those insects that have wings, or wingless insects with winged ancestors, are placed in the subclass Pterygota. Essentially no metamorphosis occurs during the life cycle of insects in the Apterygota, i.e., they are ametabolous. Pterygote insects can be subdivided into groups on the basis of the presence or absence of wing flexion and kind of metamorphosis they undergo, i.e., simple metamorphosis (exopterygote) or complete metamorphosis (endopterygote). Thus, within the subclass Pterygota, we have **paleopterous exopterygotes, neopterous exopterygotes** and **neopterous endopterygotes**. More than 90% of insects can flex their wings and undergo complete metamorphosis, i.e., they are neopterous endopterygotes. The orders of insects may be conveniently grouped based on these designations (Table 3.1).

BLOODFEEDING INSECTS

The Bloodfeeding Habit

Taking vertebrate blood as nutriment (**haematophagy**) is spread widely within the class Insecta and clearly evolved independently several times (Table 3.2). Through convergent evolution, many aspects of the structure and function of bloodfeeding insects are similar. Bloodfeeding insects are found among both the exopterygotes and endopterygotes. In the exopterygote bloodfeeders (bed bugs, kissing bugs and sucking lice), both sexes of all stages take blood. In endopterygotes, with few exceptions, bloodfeeding is confined to the adult stage, and sometimes solely the female. Lehane (1991) discussed in detail many aspects of bloodfeeding in insects.

Different groups (Table 3.2) utilize nutrients from the bloodmeal in various ways that may involve one or more of the following: synthesis of egg yolk material (e.g., mosquitoes; see Fig. 2.23), growth and development (e.g., sucking lice) and source of energy for mobility and maintenance of life (all bloodfeeders). Only adult female mosquitoes take bloodmeals as well as flower nectar; males feed solely on nectar. Mosquito larvae are aquatic filter-feeding forms. In this case, the primary utilization of blood is in the development of eggs. On the other hand, blood provides all of the nutritional needs for insects like kissing bugs in which both sexes and nymphs as well as adults bloodfeed.

Because vertebrate blood lacks certain nutrients, particularly B vitamins, insects that rely solely on blood typically harbor symbiotic microorganisms that provide crucial missing nu-

Table 3.2. Feeding habits and symbionts in bloodfeeding insects.

Taxon	Larval food	Adult food		Use of blood*	Symbionts
		Male	Female		
Anoplura	Blood	Blood	Blood	1,2,3	Ventral wall of midgut, ovaries
Cimicidae	Blood	Blood	Blood	1,2,3	Fat body
Triatominae	Blood	Blood	Blood	1,2,3	Some in blood storage area of midgut lumen
Siphonaptera	Miscellaneous organic debris	Blood	Blood	1,2,3	Some in lumen of midgut
Psychodidae (Phlebotominae)	Miscellaneous organic debris	Nectar	Blood and nectar	3	None known
Culicidae	Microplankton	Nectar	Blood and nectar	3	Some within germinal cells
Simuliidae	Microplankton and decaying organic matter	Nectar	Blood and nectar	3	None known
Ceratopogonidae	Microbial growth, some scavengers and predators	Nectar	Blood and nectar	3	None known
Tabanidae	Known and possible predators	Nectar and pollen	Blood and nectar	2,3	None known
Muscidae (Stomoxyinae)	Decaying organic matter	Blood and nectar	Blood and nectar	2,3	None known

*Code: 1 = growth and development, 2 = maintenance, 3 = egg development.

trients. These symbionts may be free-living or maintained in specialized cells (**mycetocytes**) or tissues (**mycetomes**).

The bloodfeeding habit may have stemmed from long-term associations between a vertebrate and a pre-bloodfeeding insect such as might have occured between scavenging insects that were lured to a vertebrate's nest or burrow by virtue of a sheltered, organic matter-rich, humid environment or by predatory insects attracted to a vertebrate nest due to other insects congregated there (Lehane 1991). Some insects may have been preadapted for bloodfeeding by already having piercing-sucking mouthparts used to drain hemolymph from their prey. These insects already may have had in place physiological characteristics in their alimentary system that made them capable of digesting and assimilating vertebrate blood. This pre-adaption could have happened in the kissing bugs, whose relatives in the same family are hemolymph-sucking insect predators. Scavenging insects

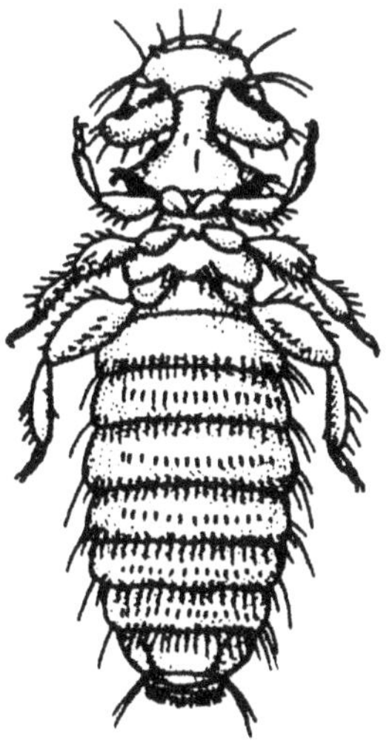
(a)

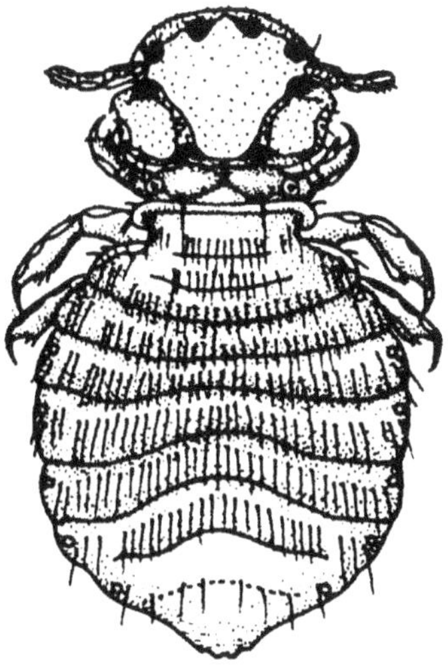
(b)

Figure 3.2. Chewing lice: (a) Amblycera; (b) Ischnocera (from Communicable Disease Center 1969).

may have been attracted to sloughed skin or scabs and then developed the ability to induce bleeding or actually suck blood with their existing mouthpart structures. For more information on the evolution of bloodfeeding, see Waage (1979).

Bloodfeeding Groups

Details of the morphology, systematics and biology of bloodfeeding arthropods may be found in Beaty and Marquardt (1996), Borror et al. (1989), Durden and Romoser (1998), Kettle (1994), Lane and Crosskey (1993), Lehane (1991) and Smith (1973).

Lice

Two orders of insects belong in this group: the chewing lice (order **Mallophaga**) and the sucking lice (order **Anoplura**). Members of these closely related orders are vertebrate ectoparasites. Some systematists lump them into the single order **Phthiraptera**. Both orders show strong morphological affinities with the order Psocoptera (the booklice and barklice). Members of this order are small, free-living scavengers that feed on organic matter under bark, in vegetation and in vertebrate nests. Scavenging in the nests of birds and mammals probably led to the ectoparasitic way of life. For more information on lice, consult Marquardt (1996), Marshall (1981), Durden and Musser (1994), Kim and Ludwig (1978) and Kim et al. (1986).

Both chewing and sucking lice display traits associated with ectoparasitism, such as their small size, winglessness, reduced antennae, dorsoventrally flattened bodies, commonly with grasping tarsi, and reduction or loss of many structures, including the compound eyes, ocelli and cerci.

Mallophaga

Mallophaga (Fig. 3.2) means literally "wool eaters." These insects have chewing mouthparts. Superficially they resemble the sucking lice, but have a triangular head that is broader than the thorax, whereas the sucking lice have just the opposite condition, i.e., a narrow head and broad thorax. Chewing lice have well-developed mandibles and short, 3- to 5-segmented antennae.

Three suborders of chewing lice are recognized: **Amblycera**, **Ischnocera** and **Rhyncophthirina**. Amblycera and Ischnocera predominate. They can be separated by the fact that Amblycera have clubbed (capitate) antennae concealed in grooves along the head and maxillary palps, while the Ischnocera have free, unclubbed antennae and lack maxillary palps. Rhyncophthirina is an extremely small group, but of systematic interest because the chewing mouthparts are borne on the end of an elongate projection of the head capsule (**rostrum**), a condition which may represent an intermediate condition between the chewing and sucking lice.

The majority of chewing lice hosts are birds, but nearly 15% of the known species parasitize mammals (e.g., cattle, horses, sheep, dogs, cats). They do not attack humans, but may occur accidentally, e.g., a horse louse crawling onto a horseback rider. However, one species, the common dog louse, *Trichodectes canis*, serves as an intermediate host for the double-pored dog tapeworm *Dipylidium caninum*. This parasite can infect a human if an infected louse is accidentally ingested.

Chewing lice feed on a wide variety of organic matter, including feathers, skin, sebum from hair follicles and blood. Typically, they use their mandibles to pick away material from the host. In some species, the mandibles are sharp enough to induce bleeding. Eggs are attached to the feathers or hairs of the host; after they hatch, chewing lice have 3 nymphal stages before becoming adults. Being permanent residents on a host, these lice rely largely on direct contact between their hosts for dispersal. However, some may attach to a biting fly as it feeds on their host and "hitch a ride" to the next host the fly feeds on. This phenomenon is called **phoresy**. There are cases where more than one species of chewing louse may infest the same host species and the lice may prefer living on specific parts of the host's body. Chewing lice of domestic animals rarely constitute a major problem unless in excessive numbers. Heavy infestations in poultry may cause restlessness and even a drop in egg production.

Anoplura

In contrast to the chewing lice, the Anoplura have piercing-sucking mouthparts that are retracted into the head when not in use. They have short 3- to 5-segmented antennae and one-segmented tarsi typically adapted for grasping hairs (Fig. 3.3). Like the chewing lice, they have 3 nymphal instars. Unlike the chewing lice, they are exclusively bloodsucking parasites of mammals. Sucking lice are highly host specific, reflecting their coevolution with mammalian hosts. Several major pests of domestic animals are in this order as well as 2 species that attack humans. Among the domestic animal pests are the hog louse, *Haematopinus suis*, the horse sucking louse, *Haematopinus asini*, and the long-nosed ox louse, *Linognathus vituli*. The 2 species of sucking lice that attack man are *Pediculus humanus* (family **Pediculidae;** Fig. 3.3a) and *Pthirus pubis* (family **Pthiridae;** Fig. 3.3b).

Pediculus humanus is divisible on morphological grounds into 2 subspecies: *P. humanus humanus*, the body louse, and *P. humanus capitis*, the head louse. The body louse favors body regions that are in frequent contact with cloth-

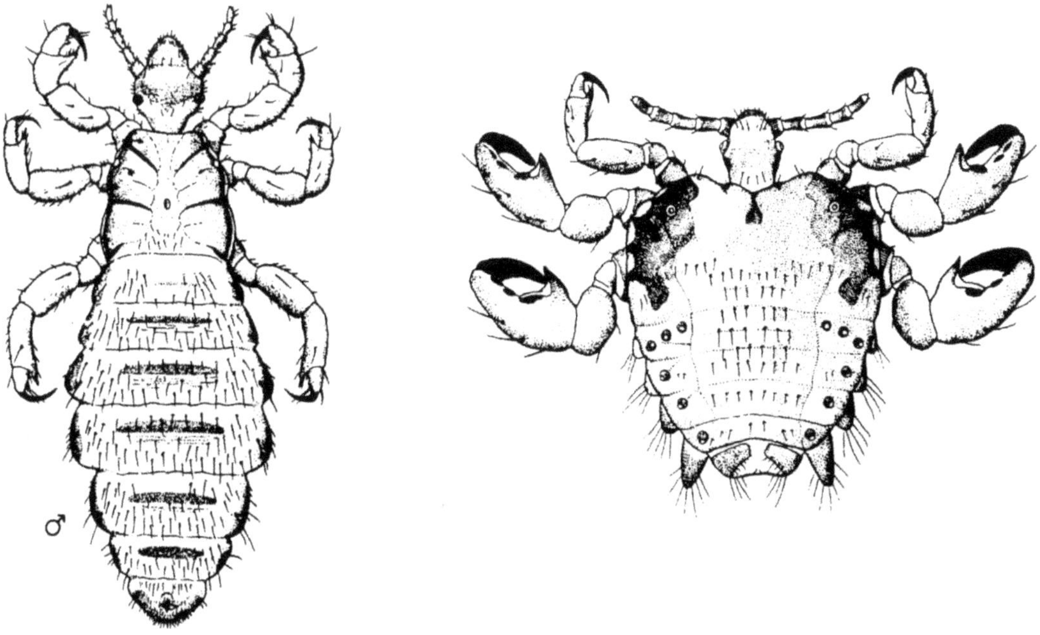

(a) (b)

Figure 3.3. Sucking lice: (a) The body louse, *Pediculus humanus humanus;* (b) the pubic louse or crab louse, *Pthirus pubis* (from Lane and Crosskey 1993).

ing, e.g., the neck, armpits and crotch. Body louse eggs are attached to clothing and this, in addition to direct contact, is the way the lice spread. In contrast, head lice favor the head region where they attach their eggs to hairs. Head lice are spread by contact with eggs attached to stray hairs on objects like combs and hats. Infestations cause intense itching and large ones can cause anemia. Body lice serve as vectors of several human pathogens, including *Rickettsia prowazekii,* the cause of epidemic typhus.

Pthirus pubis, the pubic louse or crab louse, infests areas of the body that have coarse hair, particularly the pubic region and the armpits. Eggs, or nits, are attached to the host's hairs. Nymphs and adults are spread mainly by direct contact such as sexual intercourse (hence the vernacular name "papillons d'amour," meaning "butterflies of love"). These lice cause intense itching.

Kissing bugs and bed bugs

These bloodfeeding insects are members of families in the order **Hemiptera**, the true bugs. This is the largest order of exopterygote insects. True bugs (Fig. 3.4) are so-named on the basis of their forewings, which are hardened basally, and membranous apically, called **hemelytra** ("half-wings"). The wings flex back over the abdomen at rest, with the membranous hindwings held beneath the forewings. Some are wingless or have shortened wings. True bugs have piercing-sucking mouthparts com-

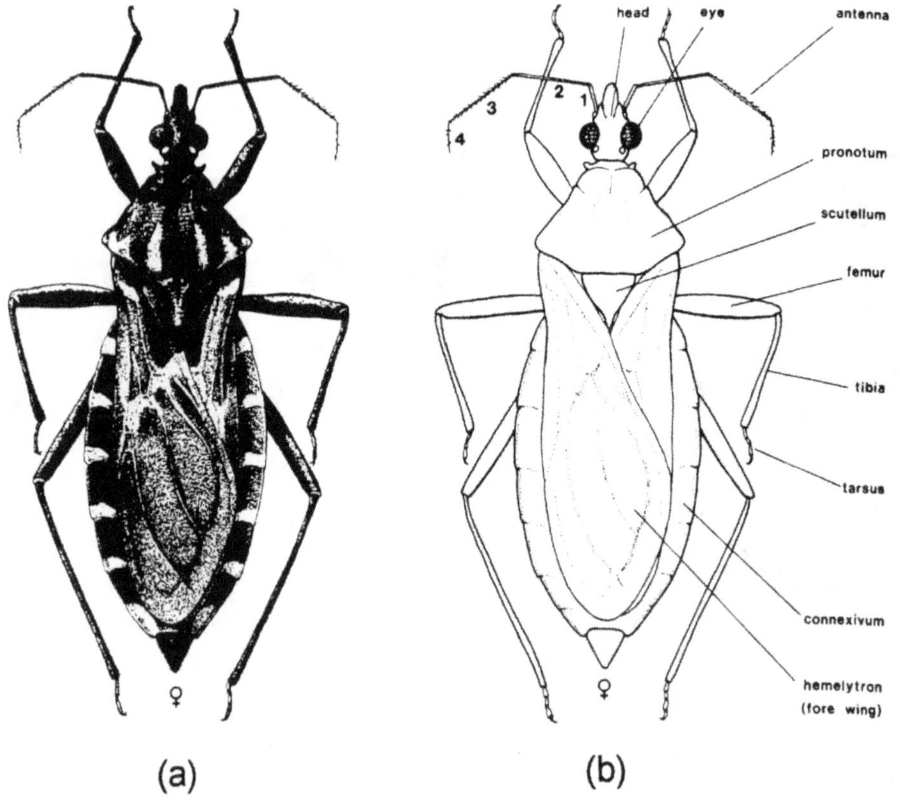

Figure 3.4. A triatomid bug, *Panstrongylus megistus*: (a) General appearance of adult female; (b) morphological features (from Lane and Crosskey 1993).

posed of a fasicle of stylets encased in a labial sheath and degenerate palpi. The pronounced **pronotum** and small, triangular **scutellum** lie between the forewings at rest. Most have well-developed compound eyes and 2 or no ocelli, and their relatively long antennae are composed of 4 or 5 segments. None have cerci. Many true bugs possess glands that synthesize and release repugnatorial substances. For example, a large infestation of bed bugs can be detected by a characteristic odor. For more information on the true bugs, see Dolling (1991).

Hemiptera are found in a wide variety of habitats and display different food preferences, ranging from green plants to other insects to vertebrate blood. Many beneficial insect predators are in this order. The bloodfeeding habit can be found in the families **Reduviidae** (conenose or kissing bugs) and **Cimicidae** (bed, bird and bat bugs).

Kissing bugs

Most members of the family Reduviidae are predators, feeding on other insects. However, members of the subfamily **Triatominae** (Fig. 3.4) are ectoparasites of vertebrates, with both sexes of nymphs and adults taking bloodmeals. The winged adults can fly. As with the bed bugs, they live and deposit their eggs in crevices around human abodes, some species favoring thatched roofs. And again as with the bed bugs, they visit hosts at night only to feed. They are not particularly host-specific and most will take a bloodmeal from a human as readily as from a rodent. Many species in this group defecate on the host during bloodfeeding, a habit that favors the transmission of *Trypanosoma cruzi*, the etiological agent of Chagas disease, a major health problem in Central and South America. Common vector species include *Rhodnius prolixus*, *Triatoma dimidiata* and *Panstrongylus megistus*. Lent and Wygodzinsky (1979) and Schofield (1988) provided extensive information on triatomid bugs.

Bed bugs

Members of the family Cimicidae (Fig. 3.5) are ectoparasites of birds and various mammals including bats and rodents. They are wingless, oval, dorsoventrally flattened insects with prominent 4-segmented antennae. There are 5 nymphal instars before the adult stage. Unfed individuals are yellowish or brownish, but become a dark brownish red after a bloodmeal. Bed bugs differ from lice in that they do not remain on the host after feeding. To take their nocturnal bloodmeals they approach the host, stick out the proboscis and penetrate the skin. When not feeding they hide in cracks in furniture and molding. They also deposit their eggs in these locations. Both sexes of nymphs and adults take bloodmeals and blood is the sole food of these insects.

Bed bugs largely depend on passive transport in furniture and luggage. Bed bugs and a few close relatives display a bizarre method of copulation, apparently unique among the members of the class Insecta. The male has an elongate, daggerlike penis that is inserted directly through the integument of the female into the hemocoel. The region of penetration on the ventral surface of the female abdomen is identifiable by the presence of a characteristic notch.

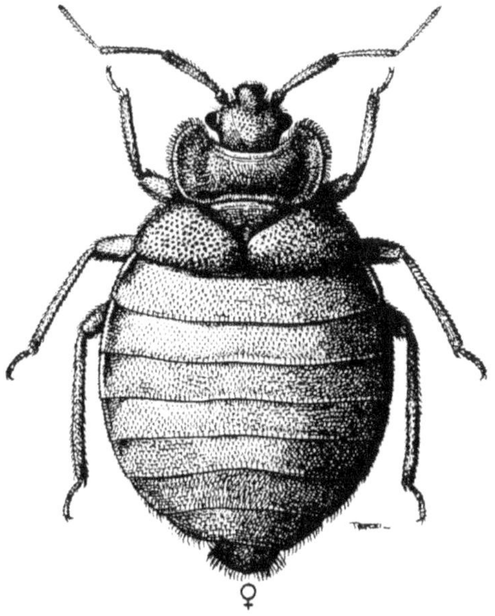

Figure 3.5. Adult female bed bug, *Cimex lectularius* (from Lane and Crosskey 1993).

Among the 90 or so species in the family Cimicidae, 2 attack humans: *Cimex lectularius* and *Cimex hemipterus*. Some other species that are ectoparasites of birds (e.g., *Oeciacus vicarius*) will bite humans who accidentally come in contact with infested nests. *Cimex lectularius* has a wide distribution, while *C. hemipterus* is more or less tropical. Although large infestations can cause anemia in children, and possible allergic problems, bed bugs are not important vectors of disease agents.

Fleas

Like lice, adult fleas (order **Siphonaptera**) display distinctive adaptations associated with an ectoparasitic life (Fig. 3.6a). These similarities include very small size; hard, bilaterally compressed bodies with many posteriorly-directed spines; lack of compound eyes; and the ability to jump. They have piercing-sucking

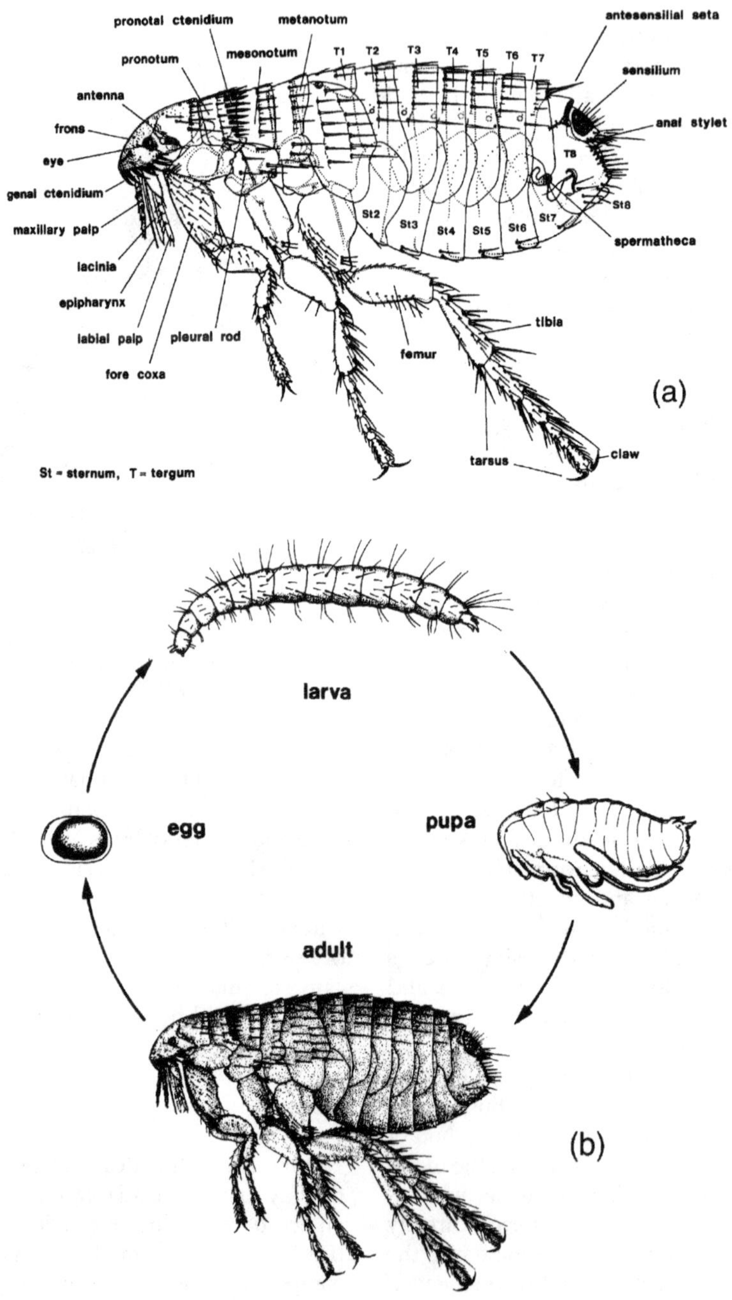

Figure 3.6. (a) Morphological features of a generalized flea; (b) typical flea life cycle stages (from Lane and Crosskey 1993).

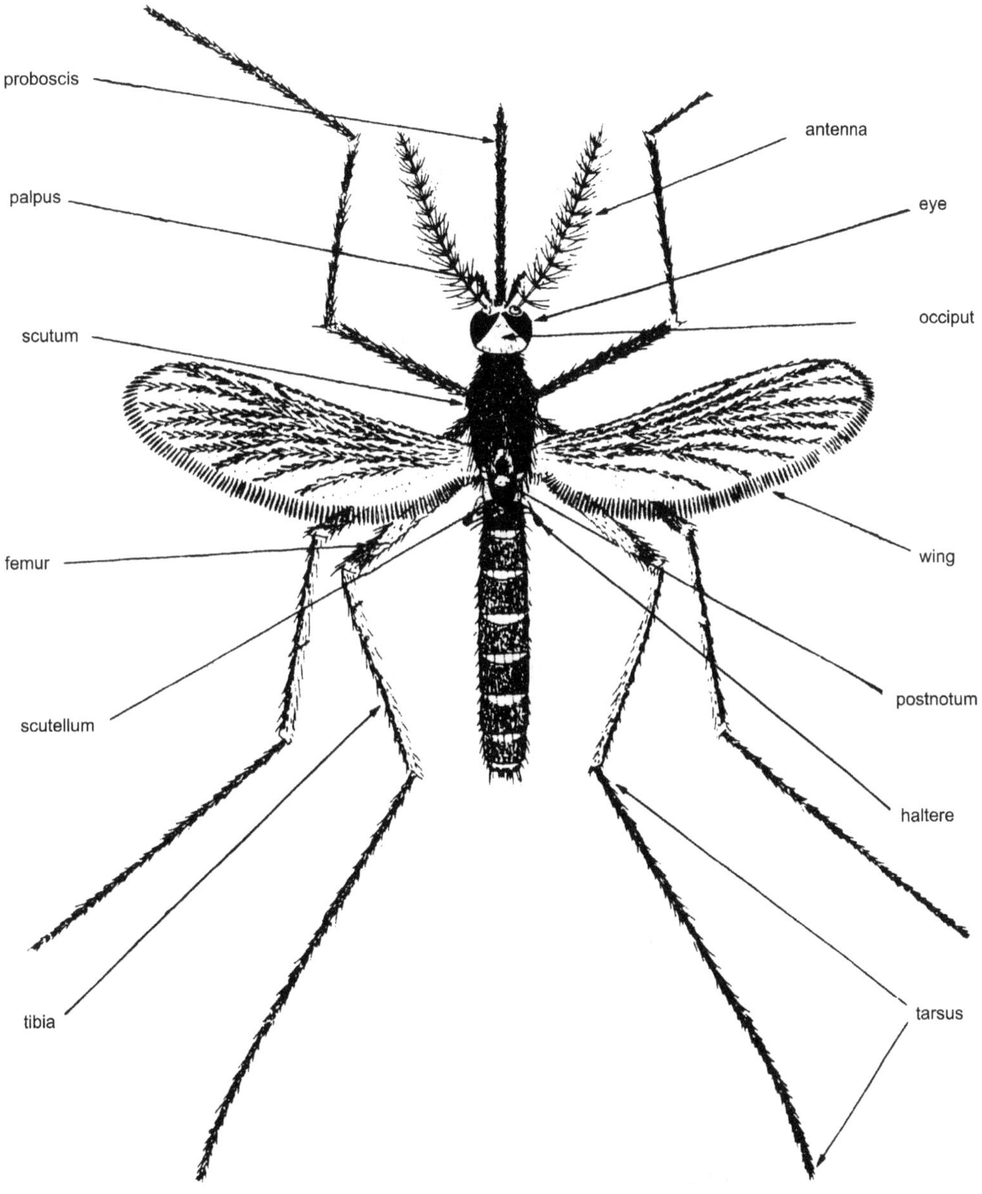

Figure 3.7. Morphological features of an adult mosquito (from Restifo 1982).

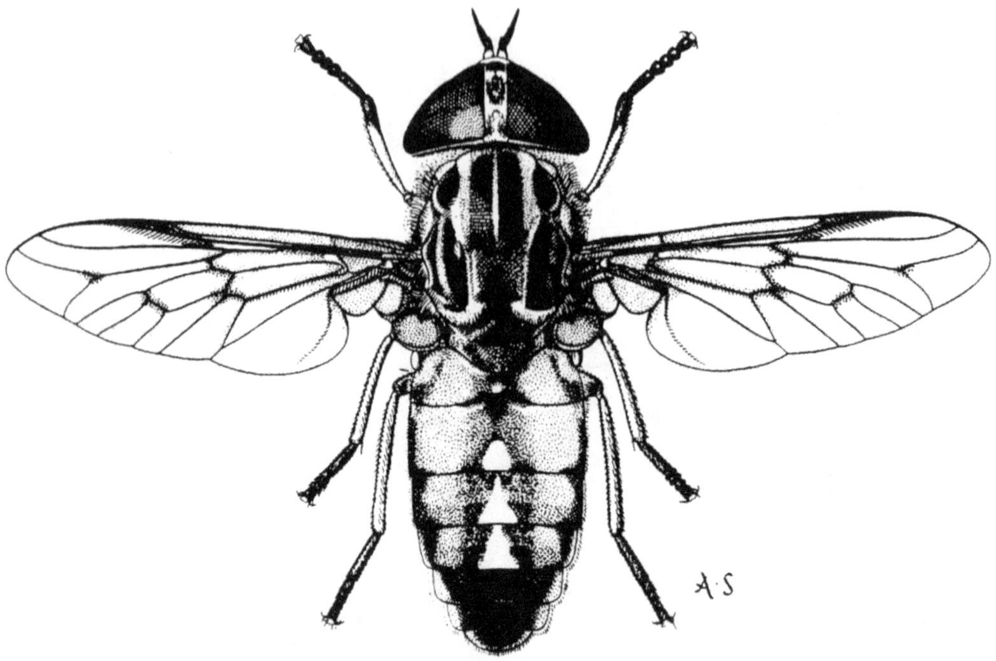

Figure 3.8. Horse fly, *Tabanus fraternus* (from Lane and Crosskey 1993).

mouthparts with which both sexes of adults obtain bloodmeals. Fleas undergo complete metamorphosis and their larvae differ dramatically from the adults. Whereas the adults live on their vertebrate hosts, with few exceptions the larvae are free-living scavengers in the host nest or burrow. The larvae (Fig. 3.6b) are wormlike (vermiform) with well-developed heads. They pupate in silken cocoons in which they can remain for extended periods. Adults emerge from the cocoons in response to stimuli given off by a potential vertebrate host (CO_2, vibrations, shadows). They attack warmblooded animals, e.g., birds and mammals.

Biting flies

The order **Diptera**, the true flies (including mosquitoes; Fig. 3.7), is the 4th largest in the class Insecta and its members occupy many different niches and habitats. Several families of flies have acquired the vertebrate bloodfeeding habit and consequently have become involved with the transmission of microorganisms that cause disease. Aside from the transmission of disease-causing microbes, bloodfeeding flies occurring in large populations can annoy humans and domestic animals by their feeding activities alone. Further, larvae in several families invade animal tissues. On the positive side, many flies are beneficial because of their predatory (e.g., robber flies, family Asilidae) and parasitoid (e.g., the tachinid flies, family Tachinidae) habits.

The true flies are characterized by a single pair of mesothoracic wings with the hindwings reduced to gyroscopic structures called **halteres**. Several species are apterous. The true flies range in size from minute to large insects, with legless and sometimes headless larvae. The adult

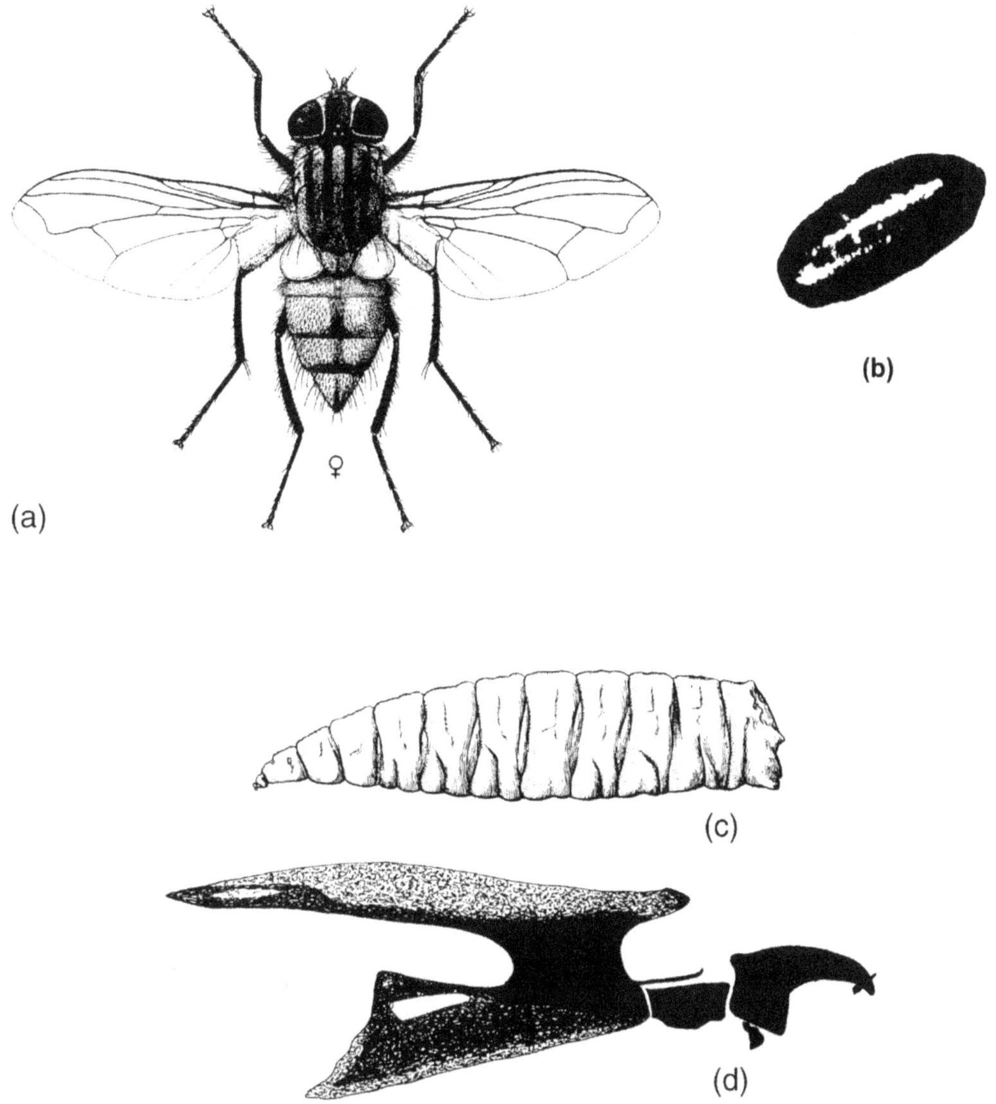

Figure 3.9. (a) Adult female house fly, *Musca domestica;* (b) puparium; (c) maggot (3rd-stage larva) of the blow fly *Calliphora vicina;* (d) enlarged view of the cephalopharyngeal skeleton and mouthhooks (a, c and d from Lane and Crosskey 1993, b from USDA).

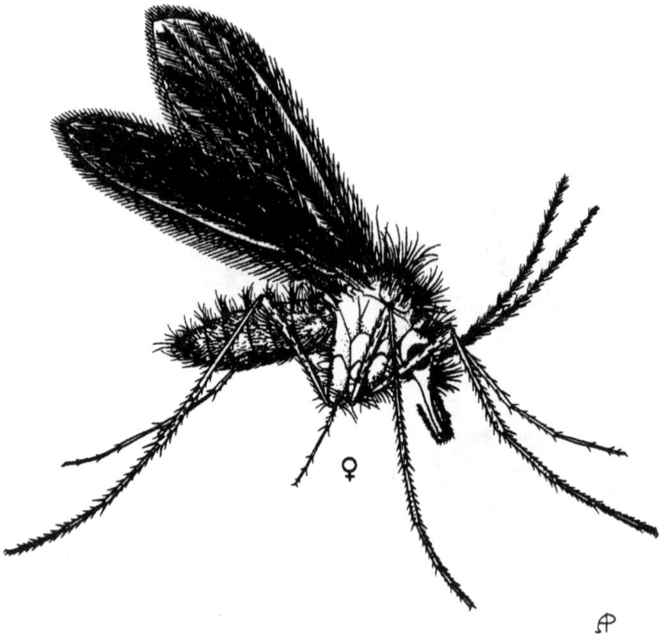

Figure 3.10. An adult female phlebotomine sand fly, *Phlebotomus papatasi* (from Lane and Crosskey 1993).

mouthparts vary, but are fundamentally sucking structures. The larvae, on the other hand, have various modifications of mandibulate mouthparts. Adult flies typically have well-developed compound eyes and most have 3 dorsal ocelli.

Three suborders are recognized: Nematocera, Brachycera and Cyclorrhapha. The **Nematocera** ("thread-horned;" mosquitoes, black flies and relatives; Figs. 3.7, 3.11) are called the long-horned flies in reference to the relatively long antennae, typically filiform, found in the adult stage. The members of this group are mostly mosquito- or gnat-like in appearance. The larvae have well-developed heads and the pupae are obtect (without free appendages).

The **Brachycera** ("short-horned;" horse flies, robber flies and relatives; Fig. 3.8) are called the short-horned flies in reference to the relatively short (shorter than the thorax), typically stylate antennae found in the adults. In contrast to the nematocerous flies, the larvae have weakly-developed heads and the pupae are exarate (with free appendages).

The **Cyclorrhapha** ("circular seam;" tsetse, house flies, stable flies and relatives; Figs. 3.9a, 3.14) also have short, typically aristate antennae, but their name is based on the appearance of the opening through which they escape from the pupal case or puparium (Fig. 3.9b). This structure is formed from the somewhat contracted, hardened and darkened, next-to-last larval exuviae. Pupation occurs within the puparium and when ready to emerge, the pharate adult directs pressure against one end of the puparium by means of an airbag-like structure, the **ptilinum,** that is inflated from the front of its head by hydrostatic pressure. This pressure causes a small portion of the puparium to break free due to a preformed line of weakness. The

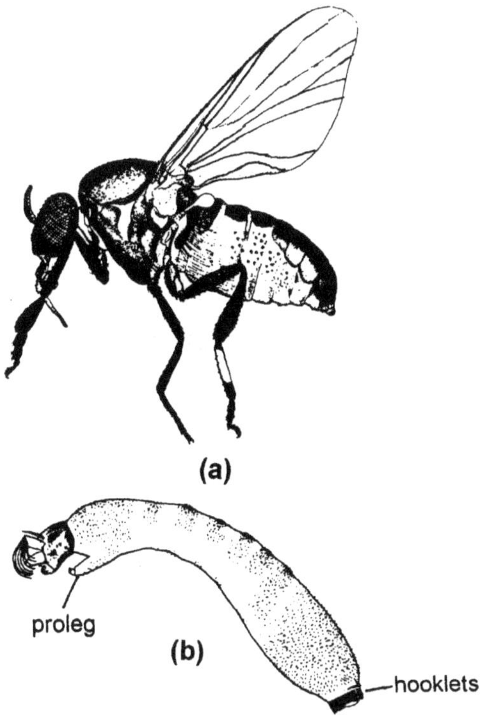

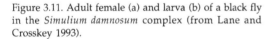

Figure 3.11. Adult female (a) and larva (b) of a black fly in the *Simulium damnosum* complex (from Lane and Crosskey 1993).

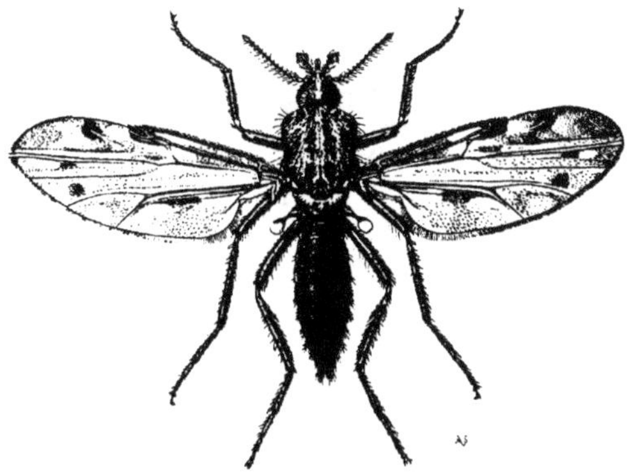

Figure 3.12. An adult female biting midge, genus *Culicoides* (from Lane and Crosskey 1993).

adult also uses the ptilinum to push its way up through the environmental material in which pupation has occurred. Larvae (Fig. 3.9c) in this suborder are vermiform ("maggots"), lacking heads altogether. The mouthparts articulate with a sclerotized, melanized **cephalopharyngeal skeleton** (Fig. 3.9d).

Flies of medical importance are found in each suborder and the following paragraphs provide a brief synopsis of the major medically significant families.

Medically important nematocerous Diptera are found in several families. The **Psychodidae** (moth and sand flies) are minute, hairy flies, one subfamily of which are vertebrate bloodfeeders, the **Phlebotominae** ("vein cutters" or sand flies; Fig. 3.10). These flies transmit several species of pathogenic protozoans in the genus *Leishmania* that cause potentially serious human diseases, collectively referred to as leishmaniasis, as well as viruses that cause sand fly fever in humans. Members of the other major subfamily, the **Psychodinae** (moth flies), breed in habitats containing decaying organic matter such as sink drains, sewers and water treatment plants.

Members of the black fly family **Simuliidae** (Fig. 3.11) are voracious bloodfeeders and transmit the causative agent of onchocerciasis, the filarial worm *Onchocera volvulus*. They can seriously annoy wild and domestic animals and humans by their bloodfeeding habit alone. Black

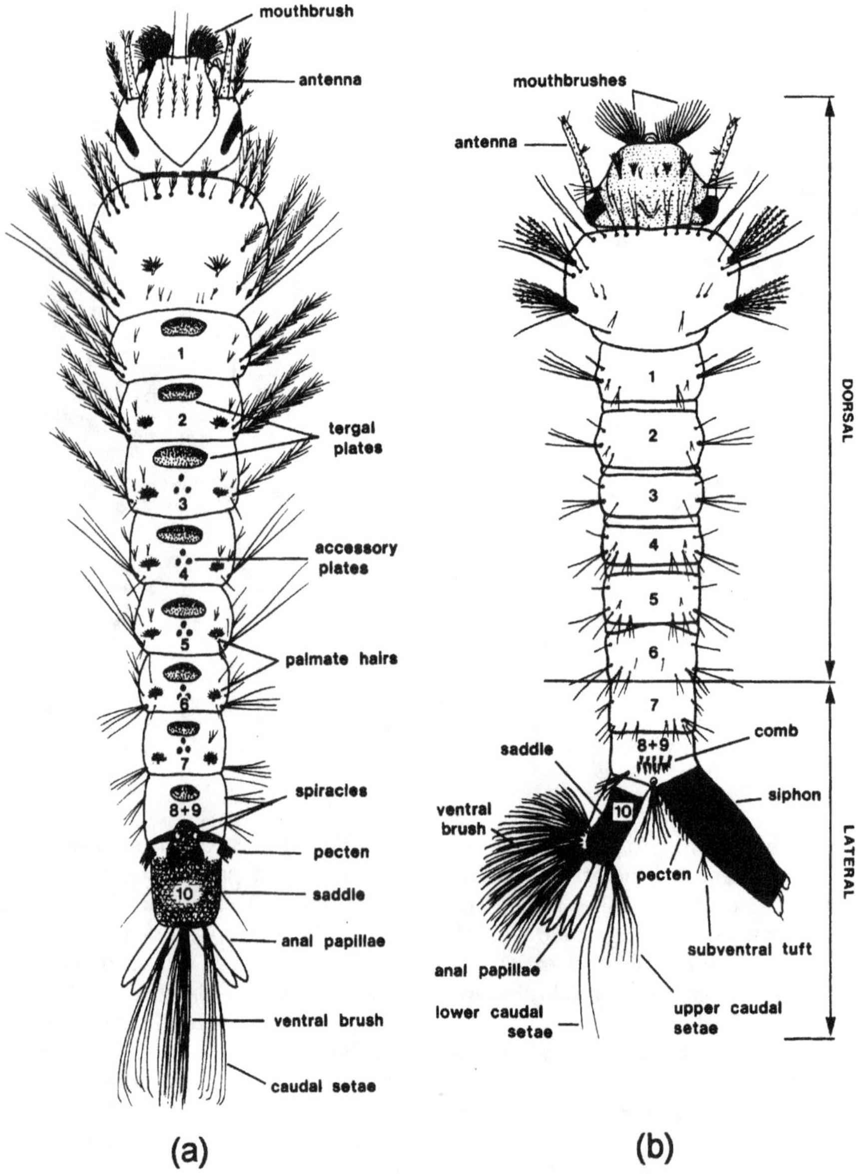

Figure 3.13. Morphological features of anopheline (a) and culicine (b) mosquito larvae (from Lane and Crosskey 1993).

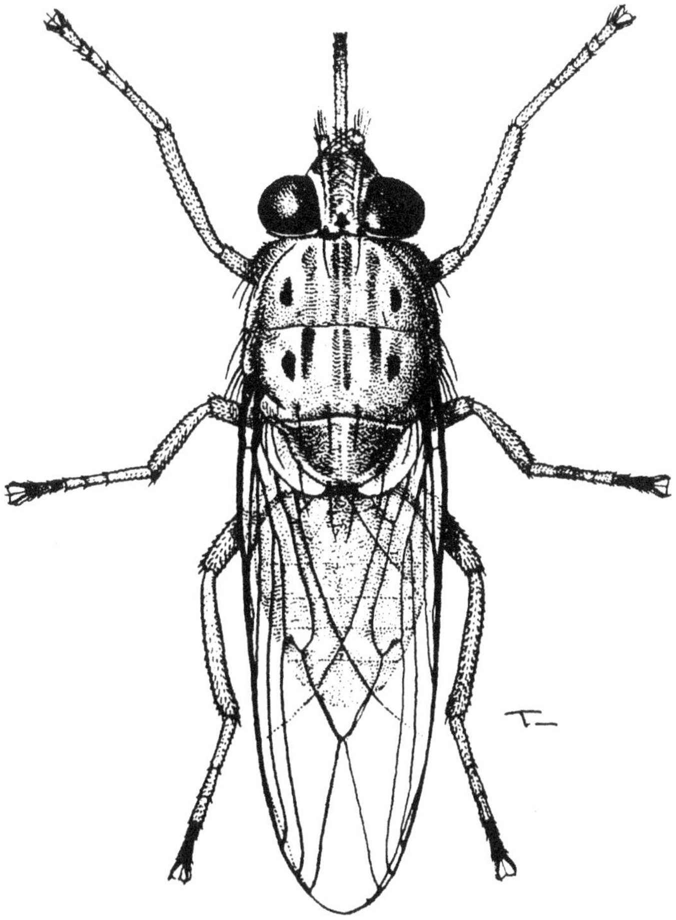

Figure 3.14. A tsetse, *Glossina longipennis*, in resting attitude (from Lane and Crosskey 1993).

fly adults (Fig. 3.11a) have a robust appearance. The larvae (Fig. 3.11b) are found in well-oxygenated, rapidly moving streams and rivers where they filter organic particles into the mouth. They attach themselves to submerged objects such as logs and rocks by means of a circle of hooklets that they are able to release, and by using a combination of the hooklets and the single median **proleg** on the thorax, they are able to move in an inchworm-like fashion.

The biting midges, or no-see-ums (family **Ceratopogonidae;** Fig. 3.12), are minute (less than 3 mm long) and the bloodsucking forms have profoundly annoying bites that seem disproportionate to their size. Some biting midges prey on other insects. The worm-like larvae live in aquatic or other moist habitats. Wings of adult midges have pigmented spots that characterize the various species. Some of the blood-feeding species act as vectors of disease agents, e.g., bluetongue virus.

Mosquitoes (family **Culicidae;** Fig. 3.7) are arguably the most medically important arthropod group. These flies are vectors of the etio-

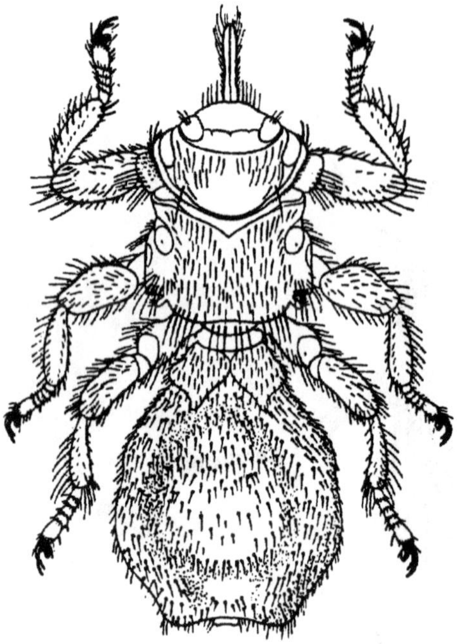

Figure 3.15. The sheep ked, a hippoboscid fly (from Richards and Davies 1977).

logic agents of malaria, filariasis and a host of arboviruses (yellow fever, dengue, various encephalitides and so on). Adult mosquitoes have variously pigmented scales on the body as well as on the wings and legs. The color and arrangements of these scales help to identify mosquitoes. Mosquito larvae (Fig. 3.13) are aquatic and mostly filter-feeding detritivores. However, a few, e.g., members of the genus *Toxorhynchites*, are beneficial predators, feeding on other mosquito larvae. In the bloodfeeding species, the females take the bloodmeals as well as feed on flower nectar, while the males are solely nectar feeders. For extensive information on the biology and ecology of mosquitoes, see Clements (1992) and Service (1993).

Bloodfeeding brachycerous flies are found in the families **Tabanidae** (horse flies, deer flies and clegs) and **Rhagionidae** (snipe flies). Tabanid flies (Fig. 3.8) are comparatively large flies, typically dark-bodied with pale patches, although some are lighter in color and others even metallic green or blue. While many species take human blood, these flies transmit few disease pathogens biologically. A notable exception is the filarial worm *Loa loa*, which flies in the genus *Chrysops* transmit. Tabanids also serve as vectors for *Trypanosoma evansi*, the causative agent of surra, a disease of livestock and other domestic animals in the tropics and subtropics. Tabanids probably are involved in the mechanical transmission of diseases such as tularemia. Adult females take meals of vertebrate blood, while the males are restricted to flower nectar. The larvae typically are aquatic.

Most snipe flies are predatory, feeding on other insects, but a few species bite humans and they may be involved in the mechanical transmission of disease organisms.

Families of the suborder Cyclorrhapha that contain vertebrate bloodfeeders are Muscidae, Hippoboscidae, Nycteribiidae and Streblidae. Among the **Muscidae**, stable flies (*Stomoxys* spp.), horn flies (*Haematobia* spp.), and several species in the genus *Glossina* (tsetse) deserve mention. Stable flies and horn flies feed on various wild and domestic mammals and, although not their primary source of blood, will attack humans. These flies can severely annoy, but except for the possibility of mechanical transmission, they apparently do not transmit disease-causing microbes.

In contrast to the other muscid flies, tsetse (subfamily **Glossinae**; Fig. 3.14), which are found in a large belt across the middle of the African Continent, constitute much more than nuisances. They are vectors of the protozoans that cause African sleeping sickness in humans (*Trypanosoma brucei gambiense* and *T. brucei rhodesiense*) and nagana in cattle (*T. brucei brucei*). Both sexes of tsetse take bloodmeals and even the larvae rely solely on the nutrients in blood. The larvae do so because they are retained within the reproductive chamber of the

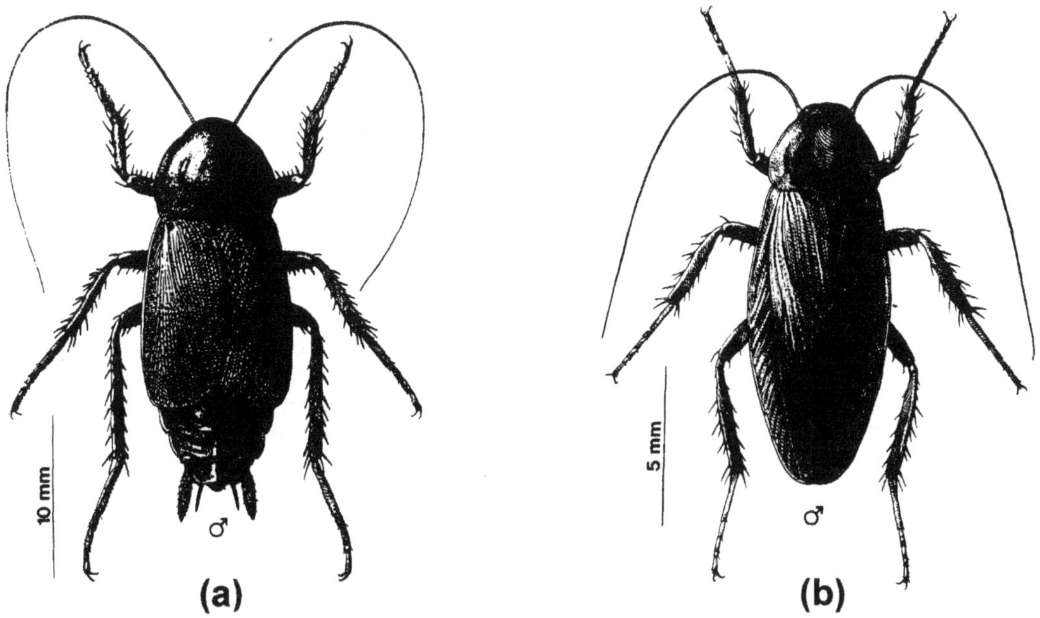

Figure 3.16. Adult male cockroaches: (a) The oriental, or common, cockroach, *Blatta orientalis*; (b) the German cockroach, *Blattella germanica* (from Lane and Crosskey 1993).

maternal parent until shortly before they are ready to pupate.

The families **Hippoboscidae, Nycteribiidae** and **Streblidae** sometimes are grouped as **Diptera Pupipara** since they, like the tsetse, retain larvae in their bodies until they are ready to pupate. Many of these bloodfeeding flies are wingless and feed on various mammals and birds. For example, the wingless "sheep ked" (Fig. 3.15) attacks sheep and goats.

Other Insects of Medical Importance

In addition to disease transmission through bloodfeeding activities, many insects may transmit disease pathogens indirectly through their various activities or they may harm or annoy humans or other vertebrate animals. Two orders of insects, Diptera and Blattaria, contain members associated with habitats characterized by large numbers of microorganisms, especially bacteria.

The order Diptera already has been introduced. Members of several families within this order are important detritivores that break down decaying organic matter. In this capacity, they play important roles in the functioning of ecosystems. However, these same habits bring them into contact with myriads of bacteria (e.g., those in fecal material and garbage) and with humans and other vertebrates. Most of these flies are in the suborder Cyclorrhapha in the families Muscidae (e.g., house flies, *Musca domestica* and *Fannia canicularis*; the latrine fly, *Fannia scalaris*), **Calliphoridae** (blow flies) and **Saracophagidae** (flesh flies). These flies can spread microorganisms through direct contamination of food items and consequent involve-

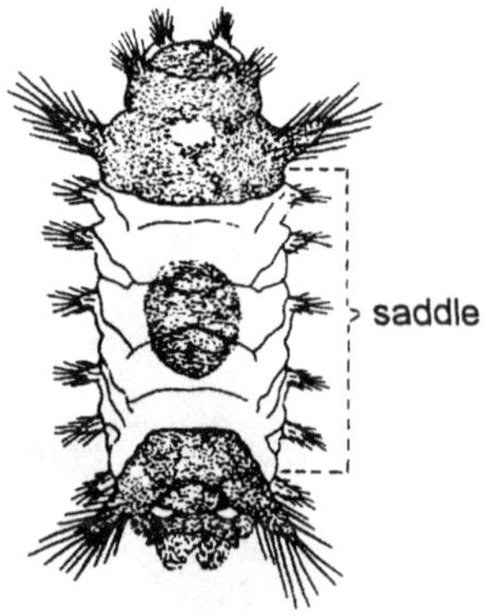

Figure 3.17. Urticating saddleback caterpillar, *Sibine stimulae* (from Communicable Disease Center 1969).

ment in the spread of foodborne illnesses like dysentery, typhoid and so on. Flies in several families, especially the family **Chloropidae** (frit flies and relatives) are attracted to body secretions (e.g., tears and mucous) as well as to open sores, and they tend to congregate around the mouth, nose, genitalia and anus. These habits set them up as the probable transmitters of conjunctivitis ("pinkeye"), yaws and bovine mastitis.

Members of the order **Blattaria** (the cockroaches; Fig. 3.16), are characterized by a dorsoventrally flattened body that tends to be dark, although some in the tropics are brightly colored. They have chewing mouthparts, well-developed compound eyes, 2 or no dorsal ocelli, long, threadlike (filiform) antennae and 2 sets of wings. The forewings are parchment-like (tegmina) while the hindwings are more membranous and broader than the forewings, beneath which they are folded at rest. The pronotum is large and shield-like, nearly covering the dorsum of the head from above. Small cerci are present in both sexes and the males have, in addition, a pair of styli on the 9th abdominal segment.

Cockroaches are omnivorous, feeding on decaying plant and animal matter. They tend to be nocturnal and are quick, agile runners, seldom taking flight. They undergo simple metamorphosis and many are oviparous or ovoviviparous. In oviparous species like *Periplaneta americana*, the American cockroach, the eggs are contained in an "egg purse" or **ootheca**. In the case of *P. americana*, the ootheca is cemented to the substrate, while others carry it in a specialized part of the reproductive system.

Several species of cockroaches have become **domiciliary** (adapted to living in human habitations), surviving on garbage, fecal material and food. Because of this diet, they, along with the filth flies, are common suspects in outbreaks of foodborne illness. Among the common domiciliary species in the USA are *P. americana*; *Blatta orientalis*, the Oriental cockroach (Fig. 3.16a); *Blattella germanica*, the German cockroach (Fig. 3.16b); and *Supella longipalpa*, the brown-banded cockroach.

Members of several orders have medical significance other than bloodfeeding and associating with filth. For example, insects that are predominately plant-feeders or predators on other insects, especially those with piercing-sucking mouthparts, will on occasion bite humans or other vertebrates, causing discomfort. Bites of insect predators usually are defensive in nature. For example, many of the predatory aquatic true bugs, such as backswimmers and giant water bugs, can inflict painful bites if handled carelessly. In the case of plant feeders, "bites" probably are associated with reflected light cues and perspiration that induce attempts to "feed." Many families of true bugs (order Hemiptera), the leafhoppers and relatives (order Homoptera), the thrips (order Thysanoptera) and the butterflies and moths (order Lepidoptera) bother humans and other animals in this way.

The **Homoptera** resemble the Hemiptera, but lack the hemelytra characteristic of the true bugs. They all feed on plants with piercing-sucking mouthparts that are capable of piercing or abrading human skin. The thrips (order

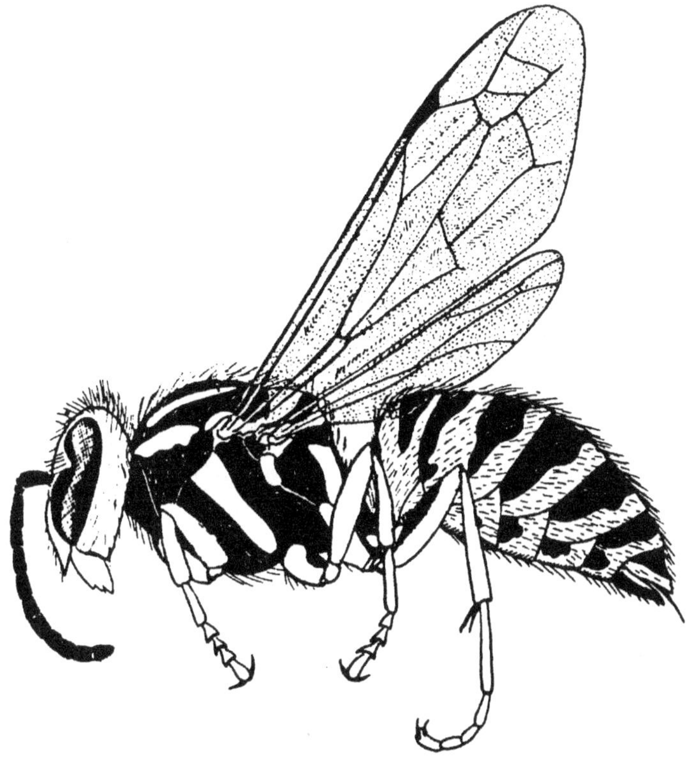

Figure 3.18. Yellow jacket, *Vespula squamosa* (from Communicable Disease Center 1969).

Thysanoptera) are plant-feeders or prey on other insects. An outstanding feature of these insects is their pair of thin elongate wings outlined with well-developed hairs; hence the name Thysanoptera, which means "tassel-wing." When their "rasping-sucking" mouthparts are applied to human skin, the itch can be maddening. Butterflies and moths (order **Lepidoptera**) are characterized by the presence of 2 pairs of characteristically-shaped wings covered with scales (modified setae), which impart an impressive variety of patterns and colors. A few species in Africa and Malaysia are attracted to the eyes of various animals, including humans, where they ingest lachyrmal secretions (tears). Some of these even have been observed to take bloodmeals (Bänzinger 1971).

When their body secretions act as toxins, irritants and/or allergens, insects also cause problems for humans. Two orders, Lepidoptera and Hymenoptera, contain members having structures that can deliver toxic secretions. Among the Lepidoptera, the larvae (**caterpillars**) of several families possess stinging or **urticating hairs**. These hairs may be associated with dermal glands that secrete toxic materials that cause severe itching or painful burning and blistering. Examples include the slug caterpillars (family **Limacodidae**; Fig. 3.17) and puss caterpillars (family **Megalopygidae**).

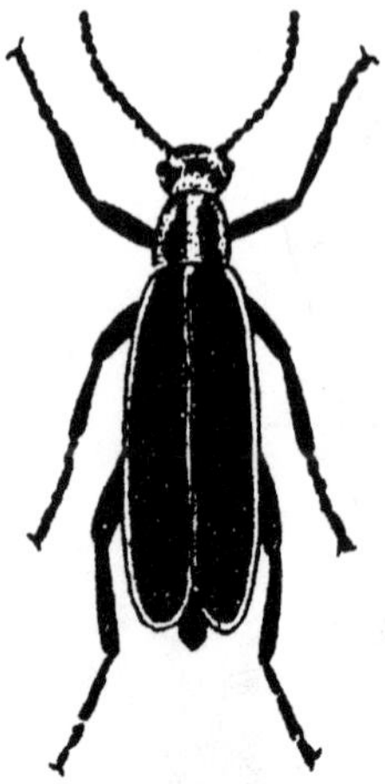

Figure 3.19. A blister beetle, family Meloidae (from Communicable Disease Center 1969).

Members of the order **Hymenoptera** (ants, bees, wasps and relatives; Fig. 3.18) are characterized by having 2 pairs of membranous wings with the hindwings smaller and joined to the forewings by means of a line of small hooks. Further, most Hymenoptera have a narrow petiole separating the thorax and abdomen, the so-called "wasp waist." Many groups of Hymenoptera can sting via modified ovipositors that have evolved into efficient, complex toxin delivery organs. Many insects in this order are social and congregate in large nests, increasing the potential for multiple stings.

Insects in a number of orders produce defensive secretions that irritate, or sometimes blister. For example, some members of the order **Coleoptera** (blister beetles in the family **Meloidae**; Fig. 3.19) demonstrate the phenomenon of **reflex bleeding**. In response to disturbance, these insects release droplets of hemolymph from the body that may contain active vesicating molecules. For example, blister beetles produce a substance called cantharidin. This substance severely blisters the skin and if ingested acts as a genitourinary irritant. The infamous Spanish fly, *Lytta vesicatoria*, is a blister beetle. Coleoptera have mandibulate mouthparts and the forewings are hardened cases, **elytra**, which protect the membranous hindwings.

Theoretically, any insect secretion can be allergenic. Allergy to insects is something of an occupational hazard for entomologists who are prone to allergies. The toxins of stinging insects can be severe allergens, as can the airborne particles that result from the breakdown of dead, dried insects. Mayflies (order **Ephemeroptera**) are one example of this. These insects have aquatic larvae. The adults can be recognized by 2 pairs of membranous wings with reticulate venation and the hindwings much smaller than the forewings, well-developed compound eyes, short bristlelike antennae and 2 or 3 long caudal filaments. When they emerge to mate they go through a unique winged, subadult stage (**subimago**). This stage molts once and the adults enjoy a brief mating period and then die. Mayflies often occur in exceptionally large populations. For allergy-prone persons, they can cause severe conjunctivitis, rhinitis and even asthma, as has been observed in regions around the Great Lakes. In the case of stinging insects, the allergic response may take on the severe form of anaphylactic shock that can lead to death if not treated immediately.

Finally, as with many species of true flies, non-bloodfeeding insects may invade organs and tissues of humans or other vertebrates directly. Insects may be accidentally ingested and cause considerable gastrointestinal irritation. Invasion by fly larvae is called **myiasis** and can lead to severe injury and even death. The author once observed maggots in a stool from a young woman who had eaten a sandwich from a vending machine. An invasion also may represent a facultative or obligate response as part of the fly's normal life cycle. Flies in several families (e.g., Calliphoridae, Sarcophagidae), attracted to secretions associ-

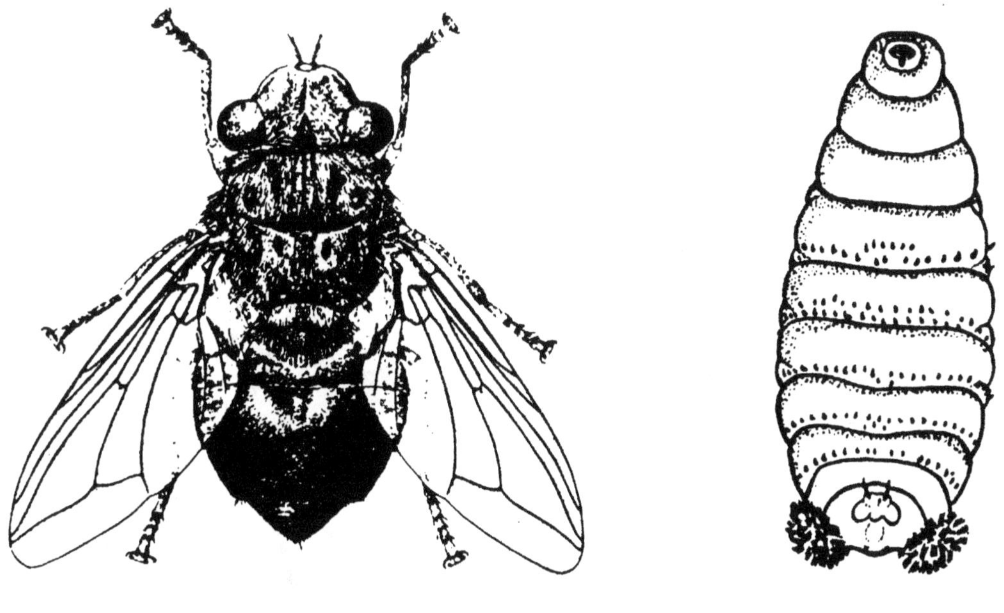

(a) (b)

Figure 3.20. The human bot fly, *Dermatobia hominis*, family Cuterebridae: (a) Adult female (drawing by A. Cushman, USDA, from Harwood 1979); (b) larva (from Communicable Disease Center 1969).

ated with open wounds or nasal discharge, may deposit eggs that hatch into larvae that invade. Several families contain members in which the tissue invasion is an obligate part of the life cycle. This is the case in the screwworm fly, *Cochliomyia hominivorax* (family **Calliphoridae**) and members of the families **Gasterophilidae** (horse bot flies) and **Oestridae** (warble flies and bot flies; Fig. 3.20). In the tropics, the so-called human bot fly, *Dermatobia hominis* (family Oestridae) is said to deposit its eggs on bloodfeeding insects, usually mosquitoes, and when these mosquitoes feed on livestock or humans, the eggs hatch and penetrate the skin. An interesting description of an experience with bot flies can be found in Forsyth and Miyata (1984) in a chapter entitled "Jerry's maggot."

Invasions of the body by beetles have been recorded and for the most part result from accidental ingestion. However, beetles have crawled into the noses and ears of sleeping persons, and possibly into the anus. Invasion by beetles is sometimes referred to as **canthariasis**.

Arachnids

The preceding chapter described the defining characteristics of the chelicerates. Among the arthropods, the class Arachnida contains several medically important groups. The class Arachnida is divided into subclasses that contain approximately 65,000 species. However, only those in the subclasses Araneae (spiders), Scorpiones (scorpions) and Acari (mites and ticks) contain members of major interest in medical entomology. Although members of the subclass **Solifugae** (windscorpions) are not venomous, the larger members are capable of in-

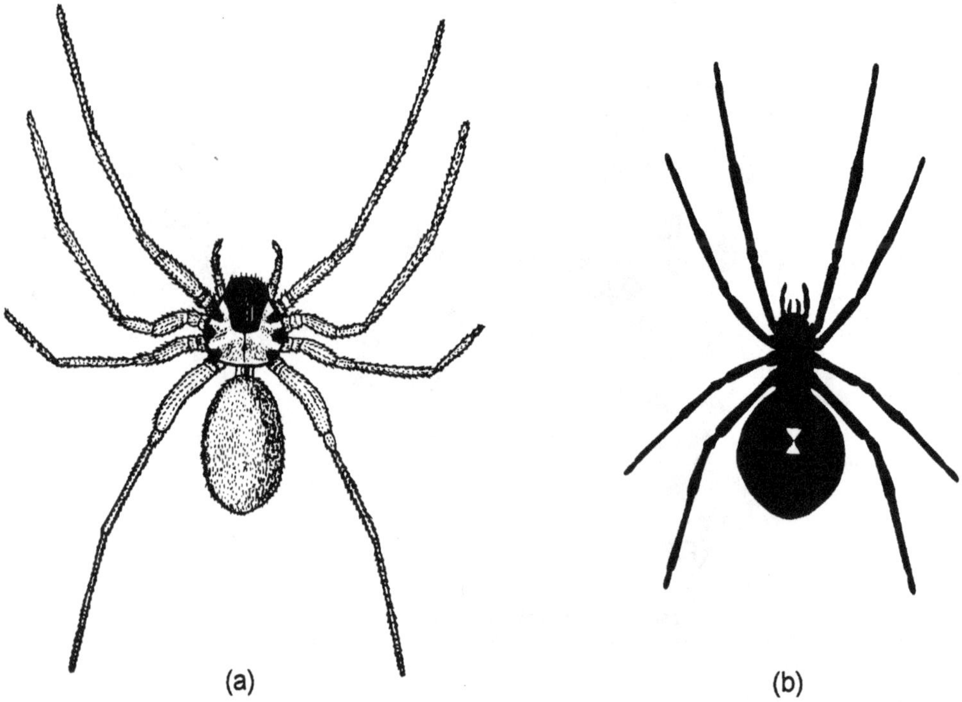

Figure 3.21. Spiders: (a) Brown recluse, *Loxosceles reclusa*, dorsal view; (b) black widow, *Latrodectus mactans*, ventral view (from Communicable Disease Center 1969).

flicting painful bites, and the large vinegaroon (subclass **Uropygi**, the whipscorpions) *Mastigoproctus giganteus* found in Florida and the Southwestern USA, is purported to spray a skin-irritating defensive substance.

Subclass Araneae

Spiders (Fig. 3.21) are predatory arthropods that feed mostly on insects. There are approximately 30,000 species found worldwide, with an estimated 2,500 species living in North America. In spiders, a stalk-like pedicel joins the 2 tagmata (cephalothorax and abdomen). Among the 6 pairs of appendages associated with the cephalothorax are the anterior chelicerae, which are composed of 2 parts: a distal fang that articulates with a larger basal segment. The fang contains an opening through which secretions of the venom glands can be released. The chelicerae in most spiders are prehensile and when brought to bear on a prey item, come together in a horizontal, forceps-like fashion. Some more primitive spiders apply the fangs vertically. In addition to capturing and immo-

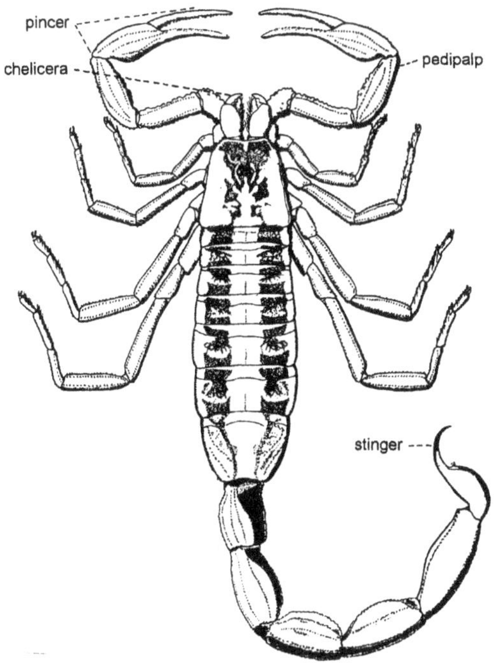

Figure 3.22. A scorpion, *Centruroides vittatus* (from Communicable Disease Center 1969).

bilizing prey, the chelicerae are used as organs of defense.

The pedipalps are leg-like sensory structures, which in male spiders insert semen into the female during copulation.

The remaining 4 pairs of appendages on the cephalothorax are walking appendages. A transverse furrow on the ventral side of the abdomen contains openings for the genitalia and for the book lungs, specialized ventilatory structures. Located on the posterior ventral tip of the abdomen are 2–6 (typically 6) spinnerets, organs associated with silk-producing cells from which silken threads emanate. This silk is used to form draglines, webs, wrap prey and to "balloon." The later is a form of aerial locomotion used by small spiderlings in which they become airborne and drift on wind currents by means of a long silk streamer.

The vast number of spiders play an important role in the functioning of nature and most are unable to penetrate human skin with their poison fangs. However, a few, probably fewer than 40 species, are capable of harming humans. Especially notable among these are members of the widely-distributed families Loxoscelidae (recluse spiders; e.g., *Loxosceles reclusa*, the brown recluse; Fig. 3.21a) and Theridiidae (the comb-footed spiders; e.g., *Latrodectus mactans*, the black widow; Fig. 3.21b). Two other families with dangerous species are the Dipluridae (funnel-web spiders; e.g., *Atrax robustus*, the Sydney funnel-web spider) found in Australia and Tasmania, and Ctenidae (wandering spiders; e.g., *Phoneutria nigriventer*, the Brazilian "banana spider") found in South America.

Subclass Scorpiones

Scorpions (Fig. 3.22) are found in arid tropical and subtropical regions. The number of described species is approximately 1,400. Scorpions differ somewhat from spiders in structure. The cephalothorax (prosoma) and abdomen are joined broadly between the 6th and 7th abdominal segments, and the abdomen, divided into the broad mesosoma and narrow metasoma, terminates with the stinging apparatus (telson). The chelicerae are small and jaw-like, while the pedipalps are large and bear well-developed claws. As with spiders and other arachnids, the remaining 4 pairs of appendages that articulate with the cephalothorax are walking legs. Comb-like organs, pectines (singular, pecten), probably tactile in function, are found on the ventral side of the mesosoma. Keegan (1980) provided detailed information on medically important scorpions.

Scorpions are nocturnal predators that feed on insects and spiders by catching them with the pedipalps, immobilizing them with stings from the telson and chewing them with the chelicerae. The stinger also is used as an organ of defense whether the victim is a potential preda-

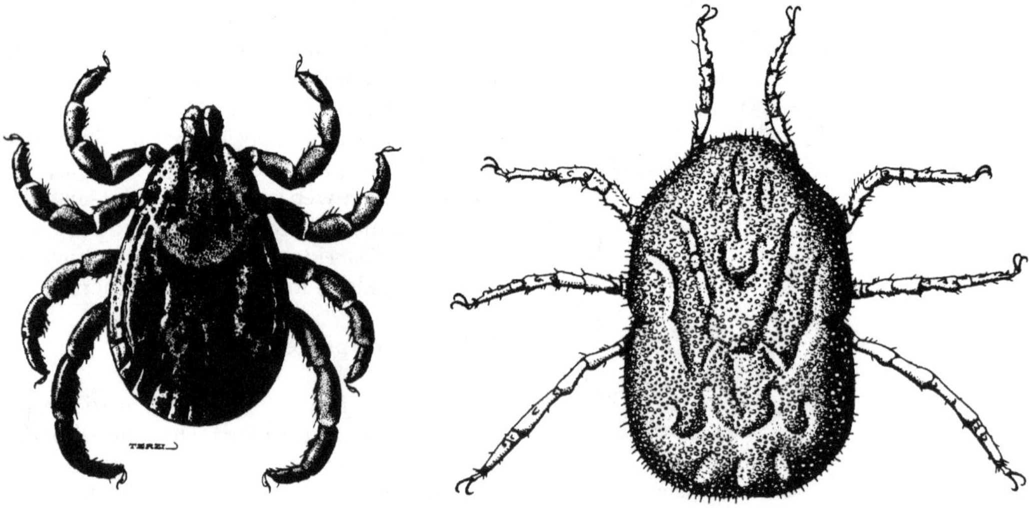

(a) (b)

Figure 3.23.Ticks: (a) Adult hard tick, *Dermacentor andersoni*, family Ixodidae; (b) adult soft tick, *Ornithodoros moubata*, family Argasidae (from Lane and Crosskey 1993).

tor or an unfortunate human being. Most scorpions can deliver a painful sting and the venom of some species can be deadly. Most of the dangerous scorpions are in the family Buthidae; important genera include *Buthus, Androctonus, Parabuthus* and *Centruroides*. Among the known species, the number capable of inflicting lethal stings is probably fewer than 25.

Subclass Acari

The preceding chapter outlined the external structure and defining characteristics of this group. The taxonomy of the Acari has been in continual flux and there is no widely accepted classification scheme for the higher taxa. Here we follow the approach outlined in Evans (1992). The Acari represent a monophyletic group divisible into 2 superorders, **Anactinotrichida** and **Actinotrichida**, on the basis of several morphological traits. The Anactinotrichida includes the orders Notostigmata, Holothyrida, Ixodida (= Metastigmata) and Mesostigmata (= Gamasida). The Actinotrichida includes the orders Prostigmata (= Actineidida), Astigmata (= Acaridida), and Oribatida. In other schemes, the Notostigmata are recognized as **Opilioacariformes**, and the Holothyrida, Ixodida and Mesostigmata are grouped as

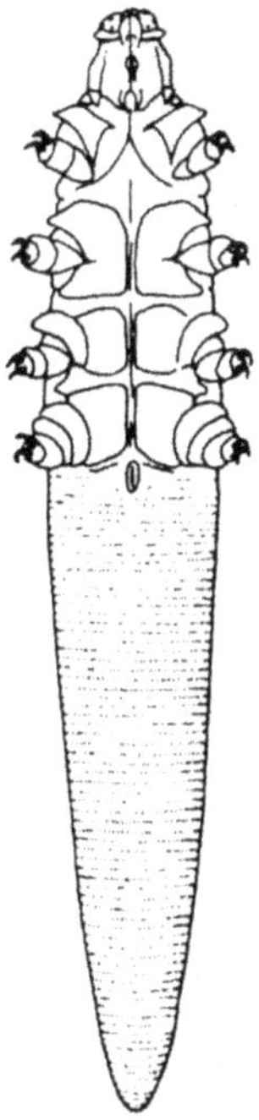

Figure 3.24. Follicle mite, *Demodex* spp. (from Lane and Crosskey 1993, adapted from Krantz 1978).

Parasitiformes. The Prostigmata are recognized as **Trombidiformes** and the Astigmata and Oribatida are grouped as **Sarcoptiformes**. The orders Ixodida, Mesostigmata, Prostigmata, Astigmata and Oribatida include medically significant members.

The 800 or so members of the **Ixodida** are called "ticks" (Fig. 3.23). The largest members of the Acari, they are bloodfeeding ectoparasites of a wide variety of vertebrates and transmit many pathogenic microorganisms, including viruses, rickettsiae, bacteria and protozoa. The Ixodida is divided into 3 superfamilies: **Ixodoidea** (the "hard" ticks; Fig. 3.23a), **Argasoidea** (the "soft" ticks; Fig. 3.23b) and **Nuttallielloidea**. The last is represented by one species and occurs only in South Africa and Tanzania, while the Ixodoidea and Argasoidea are distributed worldwide and contain many genera and species.

The hard ticks are well-sclerotized and characterized by a dorsal scutum and an anteriorly directed gnathosoma. The soft ticks are less sclerotized, lack a dorsal scutum, and the gnathosoma is more or less ventrally located and usually not visible from above. The hard ticks attach to their hosts and take some time to engorge, while the soft ticks are relatively rapid feeders. Important genera of Ixodoidea include *Dermacentor, Amblyomma, Haemaphysalis, Boophilus, Rhipicephalus* and *Hyalomma*. Important genera of Argasoidea include *Argas, Otobius* and *Ornithodoros*.

Only 3 basic instars are recognized in the tick life cycle: larva, nymph and adult. However, among the soft ticks there are additional nymphal instars that resemble one another. Ticks are obligate vertebrate bloodfeeders, one or more bloodmeals preceding each molt, with hard ticks (superfamily Ixodoidea) feeding once during each stage and soft ticks typically taking several bloodmeals per instar. Ticks may change individual hosts or host species during their life cycle. Most species of soft ticks take bloodmeals from the same host several times or have several hosts of the same or different species. Hard ticks are limited to a maximum of 3 host species, i.e., a different host for each instar. Thus, there are one-host, 2-host and 3-host hard tick life cycles. Between times on the host, ticks are free-living but do not feed or drink. As with insects, the tick life cycle is closely attuned to environmental changes. For example, in the northern Temperate Zone the entire life cycle may take 2–3 years to complete. Some ticks can live for several years without a bloodmeal.

The small mesostigmatid mites (order **Mesostigmata**) range in size from 200–2,500 µm. Many are pale in color with little sclerotization, but typically there are brownish sclerites ("shields") on the dorsum of the body. They can be differentiated from ticks by a number of

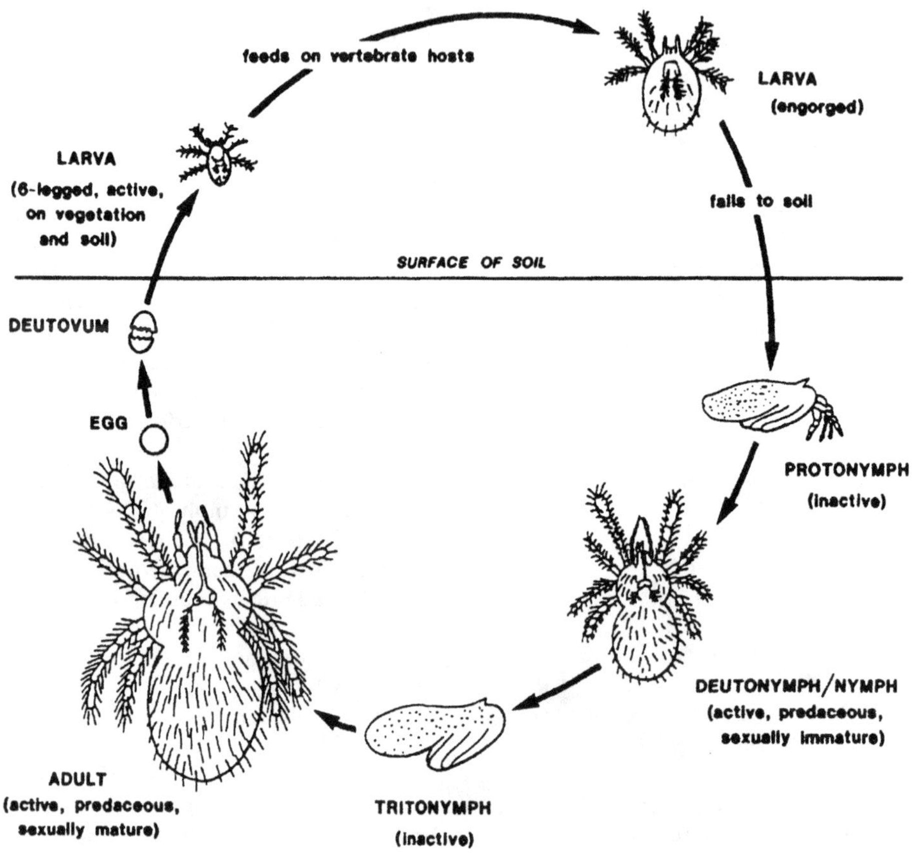

Figure 3.25. Life cycle of a trombiculid mite (from Lane and Crosskey 1993).

traits, including the presence of a single pair of stigmata (spiracles) located on the venter at some point posterior to the bases (coxae) of the 2nd pair of legs. A length of trachea (**peritreme**) usually is evident, running in an anterior direction from each stigma. Three families contain members that attack humans: the **Dermanyssidae** (bloodsucking parasites of birds and mammals; *e.g., Dermanyssus gallinae*, the red poultry mite); **Macronyssidae** (bloodsucking mites of birds, mammals and reptiles; e.g., *Ornithonyssus bacoti*, the tropical rat mite); and **Laelapidae** (bloodsucking parasites of rodents or free-living forms; e.g., *Haemogamasus* and *Laelaps*).

The order **Prostigmata** includes a wide variety of forms with a rather large size range (100 µm–16 mm). Stigmata, often with peritremes, are found between the bases of the chelicerae or at least well anterior on their weakly sclerotized bodies. The habits of these mites vary considerably because the group includes aquatic (Hydrachnidia = freshwater mites) as well as terrestrial, free-living forms with plant-feeders, predators and parasites. Among the large number of families, only 3 contain members of medi-

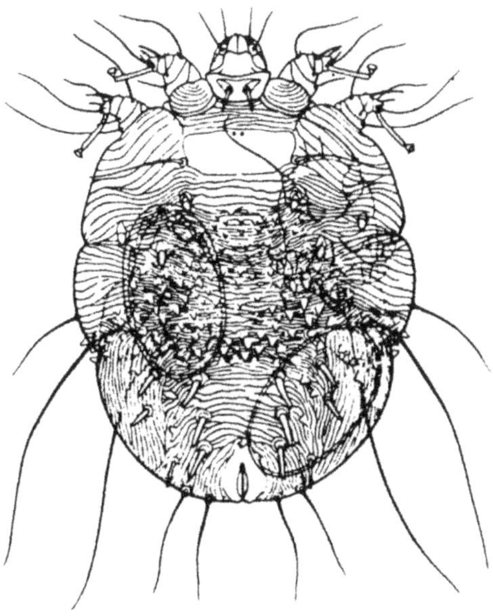

Figure 3.26. Scabies mite, *Sarcoptes scabiei*, family Sarcoptidae (from Lane and Crosskey 1993).

cal importance: Demodicidae, Trombiculidae and Pyemotidae.

Mites in the family **Demodicidae**, genus *Demodex* (*e.g.*, *D. folliculorum*, the human follicle mite; Fig. 3.24) are minute, with elongate bodies and attenuated legs. These mites live in hair follicles and sebaceous glands of many species of wild and domestic mammals as well as humans. Although usually harmless, they sometimes cause skin problems.

The **Trombiculidae** (Fig. 3.25) are tiny parasites of mammals, particularly rodents and insectivores, and birds. They range from white to red in color and sometimes are called chiggers, redbugs, harvest mites or scrub itch mites. Trombiculids usually are parasitic in the larval stage and free-living predators as nymphs and adults. When a larva feeds, a process that takes several days, it attaches by its chelicerae and the injected saliva partially digests host tissues. When feeding on humans, they can cause a maddening itch and dermatitis. In Japan, southeast Asia and Australia, the larvae serve as vectors of the pathogen that causes scrub typhus (*Orientia tsutsugamushi*).

Species in the family Pyemotidae can cause dermatitis when they get on the skin of humans who come into contact with materials (e.g., stored grain) infested with these mites. They normally attack insects and while very irritating do not establish themselves on mammalian skin.

Astigmatid mites (order **Astigmata)**, as the name describes, lack stigmata. Four families contain members of medical significance: Sarcoptidae, Pyroglyphidae, Acaridae and Glycyphagidae. The **Sarcoptidae** (e.g., *Sarcoptes scabiei*, the scabies mite; Fig. 3.26; *Notoedres cati*, feline mange mite) burrow into the skin of birds and mammals and remain on the host throughout their life cycles. The **Pyroglyphidae** are scavengers that infest stored food products and when on humans will feed on skin detritus. Certain species, such as *Dermatophagoides farinae* (the American house dust mite), are found in house dust as well as in beds and rugs. These mites form large populations and are so tiny that they are easily airborne and inhaled. Their bodies as well as fecal material contain potent allergens that can cause rhinitis and asthma, i.e., these mites produce the active principle associated with house dust allergy.

The **Acaridae** and **Glycyphagidae** infest various stored products (grain, dried fruits and cheeses). They can cause dermatitis in persons who handle these products, including "grocer's itch" caused by *Glycyphagus domesticus*. Mites in these families occasionally enter the respiratory or urinary tracts or are ingested accidentally, and can cause considerable irritation.

Oribatid mites (order Oribata) are free-living in soil and litter where they feed on fungi and other organic matter. Some species serve as intermediate hosts of tapeworms.

Myriapods

Of the 4 groups of myriapods, **Pauropoda**, **Symphyla**, **Diplopoda** (millipedes) and **Chilopoda** (centipedes), only the last 2 have members of medical significance. The **millipedes** feed on living or decaying plant material. A number of species have specialized dermal glands that produce noxious defensive substances, which in some cases can be sprayed for a distance. These secretions can cause burning, blistering, or staining of human skin and considerable irritation and possibly blindness if rubbed into the eyes. **Centipedes**, on the other hand, are predators in which the first pair of legs is modified so that they act as poison jaws to inject toxic venom. Their bites can be very painful, but usually are not fatal.

ARTHROPOD BEHAVIOR

In the previous section, the various systems were considered in isolation from one another. However, in nature, systems respond to changes (stimuli) in the internal and external environments by acting together in a regulated, integrated fashion. This total, dynamic responsiveness to change is termed behavior. Further, all organisms and the environment act together in the grand concert of life. Therefore, in order to understand behavior, we must view it in its ecological context. Of interest, then, is the adaptiveness of particular behavior patterns, i.e., the timing of appearance and appropriateness of particular behaviors in association with changes in the internal environment (nutritional state, readiness to mate) and the external environment (availability of a host, onset of potentially deleterious conditions). Behavior also may be viewed from the standpoint of evolution (the phylogenetic origins of specific patterns) as well as from the standpoint of mechanisms of neural, endocrine and genetic control. The approach below, based on Romoser and Stoffolano (1998), first will consider the kinds of behavior patterns followed by a broad treatment of the biological functions of behavior. For an excellent treatment of behavior of vector arthropods, see Klowden (1996).

Kinds of Behavior

Arthropods display various levels of **innate behavior** as well as different kinds of **learned behavior**, but not the ability to reason. Although flexible, innate behavior is characterized by a genetically determined, fixed response or a series of fixed responses to changes. Learned behavior, on the other hand, is brought about by experience with the external world, including other organisms, both conspecifics and other species. While the specific "lessons" learned by a particular arthropod are not genetic, the ability to learn is.

Innate behavior

Innate behavior patterns range from relatively simple reflexes to orientation behaviors, to the more complex **fixed action patterns** (**FAPs**, or stereotypic, preprogrammed motor patterns). **Reflexes** may involve only a part of the body, such as the proboscis extension reflex displayed by a house fly when it tastes something stimulating with its tarsal receptors, or they may involve the entire body, such as the righting reflex displayed with an insect is placed upside-down on a surface.

Orientation behaviors involve movements of the entire body that change the attitude or location of the body. These behaviors involve response to stimuli in the external environment as well as internal stimuli that provide information regarding the position (proprioreceptors) and state of the body. Some examples include the simple act of standing, the movement of the whole body in relation to the position of a potentially dangerous stimulus (e.g., excessive heat) or toward the position of a potentially beneficial stimulus (e.g., moisture or a host). Undirected locomotion in response to stimuli are called **kineses**, while directed responses, i.e., where the long axis of the body is positioned relative to a given stimulus, are called **taxes** (e.g., phototaxes in response to light, or geotaxes relative to gravity).

Fixed action patterns are characteristic sequences of behavior that usually require a state of internal readiness induced by particular hormones (primers) that in turn may represent responses to cues that impinge on sensory organs and the nervous system from without (e.g., light, moisture or temperature changes) and/or from within (e.g., an empty stomach, a full crop or a spermatheca full of spermatozoa). Readiness may be either a response to irregularly occurring environmental circumstances or a regular, periodic state controlled from within the arthropod. FAPs appear when an appropriate stimulus (**releaser**) is present. Readiness often is expressed by so-called **appetitive behavior**, such as an increase in locomotor activities, which increases the probability of encountering a releasing stimulus. When a releaser is encountered the FAP plays out, the so-called **consummatory act,** and reduces the state of readiness. A given behavior pattern may be divisible into a series of FAPs, each with a specific releasing stimulus, and a final one that completes a consummatory act and reduces or ends the state of readiness. An example: a bloodfeeding arthropod that displays appetitive behavior by seeking a vertebrate host. From a distance, particular stimuli characteristic of a potential host (such as an outline against a background) release locomotion toward the potential host, closer-in releasing stimuli may result in further approach by the "hungry" arthropod, and so on until the arthropod lands on the host. At this time further releasing stimuli induce behavior that may culminate in bloodfeeding, which in turn diminishes the state of readiness ("blood-hunger"). At any point during the sequence of releasing stimuli, if a given stimulus is "incorrect," the approach may be aborted.

Learned behavior

Learned behavior is behavioral change that results from experience with the environment. Thus, it involves the acquisition and storage of information that, after the experience, affects the arthropod's behavior. Examples of learned behavior are habituation, classical conditioning and trial and error learning. **Habituation** occurs when an organism's response to a given potentially noxious stimulus gradually diminishes as the stimulus turns out to be harmless or unavoidable. **Classical conditioning** involves 2 stimuli where one stimulus is known to induce a particular response and the other is known not to induce such a response. When the 2 stimuli are presented together repetitively to an organism, the organism associates the two such that later, the stimulus that originally failed to elicit a response, will do so in the absence of the other. **Trial and error learning** can be illustrated by putting a given animal in a maze and providing a strong positive stimulus (e.g., food) for a correct turn and/or a strong negative (e.g., electric shock) stimulus for an incorrect turn. Gradually, as trial and error learning occurs, the number of errors made in a given traverse of the maze diminishes.

Behavioral Periodicity

Many behaviors occur in regular cycles coordinated with regular environmental changes. These may be short-term, commonly in relation to the daily light-dark cycle, i.e., **circadian rhythms**. For example, bloodfeeding insects such as mosquitoes commonly exhibit daily biting patterns, e.g., at dawn and dusk (**crepuscular periods**), in conjunction with the activities of their vertebrate hosts. Alternatively, the behaviors may display long-term periodicities. Depending on geographic region and particular habitat, there are regular, highly predictable changes that have existed for very long periods. For instance, in the temperate regions of our planet the seasons march with highly predictable, regular cycles of cold and warm. Associated with these cold and warm cycles are regular, predictable changes in daylength. Associated with these environmental periodicities are periodicities in the patterns of behavior of arthropods that put them in synchrony with such things as the onset of potentially dangerous changes in temperature, availability of food (e.g., host vertebrates), mates and oviposition sites. Thus, many insects use the decreasing day-lengths of fall and winter as cues to enter a cold-resistant state or to migrate.

Communication

Communication is the process whereby some aspect of one organism (the **sender**) affects the behavior and/or physiology of another organism (the **receiver**). The sender and receiver may be members of the same or different species. Communicatory signals can be any aspect of an organism, ranging from direct contact to color, to form, to a chemical secretion or to a sound. The key is the elicitation of a response from the receiver. Arthropods have **sense modalities** capable of receiving stimuli in many forms including **tactile** (via hair sensilla), **visual** (via compound and simple eyes), **acoustical** (via various tympanic organs) and **chemical** (via chemosensilla). While all of these sense modalities play important roles in the lives of arthropods, the chemical sense stands out. Chemicals can effect communication over relatively long distances. They may act as signals for sexual readiness, defensive signals with repellent or even damaging effects, suitability as a host for a bloodfeeding arthropod, trails to food and calls to alarm. Chemicals involved in intraspecific communication are called **pheromones**. For example, when a honey bee colony is disturbed, workers that sense the disturbance release a chemical, an alarm pheromone, which alerts the other workers. Chemicals used in interspecific communication are called **allomones** and **kairomones**. Allomones benefit the sender; kairomones benefit the receiver. Thus, when an arthropod releases a repugnatorial substance from an exocrine gland and repels a would-be predator, the substance acts as an allomone. When a mosquito is attracted by the lactic acid in human perspiration, the lactic acid is acting as a kairomone.

The Biological Functions of Behavior

The behaviors briefly considered in this section combine the behaviors described above. Here we are concerned with the functional roles played by behavior during an arthropod's life cycle. These roles include behavior associated with feeding, escape and defense, environmental fluctuations, reproduction and group behavior.

Feeding Behavior

The behavior patterns associated with feeding are the means by which arthropods, all of which are at least secondary consumers, acquire both the simple and complex molecules that provide the building blocks and energy necessary to sustain life. The feeding activities of arthropods usually cause trouble for humans. Thus, the bloodfeeding activities of mosquitoes may infect humans with pathogens that cause diseases such as malaria, dengue and yellow fever.

Depending upon the source of nutriment, arthropods can be recognized as being saprophagous, phytophagous, mycetophagous, carnivorous or omnivorious. **Saprophagous** arthropods feed on dead, decaying organic matter such as rotting flesh (carrion), leaf litter and dung. Many apparently saprophagous arthropods actually feed on microorganisms found in decaying organic matter and perhaps should be called "microphagous." As such, these arthropods are important in the degradation side of the various mineral cycles. Examples of saprophagous arthropods that are of direct interest in medical entomology include the cockroaches and many of the so-called filth flies discussed earlier. **Phytophagous** arthropods feed on living plant materials; **mycetophagous** arthropods feed on fungi; and carnivorous arthropods feed on other animals (arthropods and otherwise). **Omnivores,** as their name denotes, display 2 or more of the other feeding preferences. Most insects, and many members of other arthropod groups, are phytophagous.

Of particular interest in medical entomology are groups of specialized carnivorous arthropods that can be grouped loosely as parasites. In this group we include the externally feeding ectoparasitic forms (ticks, bloodsucking bugs, lice, fleas and the bloodfeeding true flies such as mosquitoes and black flies) as well as the endoparasitic forms (fly maggots that invade wounds and other openings in vertebrates). Potential vertebrate hosts emit a wide spectrum of stimuli used by bloodfeeding arthropods for host location. These include visual, olfactory, tactile, thermal and hygro-stimuli. Other carnivorous arthropods, i.e., the predators and parasitoids, can be important in controlling arthropods of medical and veterinary importance.

Escape and Defensive Behavior

When confronted by a potential predator or other threat, an arthropod may respond in a variety of ways, including immediate escape, defense, or both.

Taking flight is probably the most common escape behavior. The ability to jump plays a major role in the case of fleas that can be there one moment and disappear instantly, the result of a fascinating mechanism. Some arthropods, like cockroaches, react by rapidly running off in response to a jet of air across their cerci.

Defensive approaches used by arthropods include an impressive array of tricks. Some benefit from camouflage, having bodies that resemble thorns, leaves or lichen in color and/or shape. Others use a combination of camouflage at rest, but when disturbed flash eyespots on their hindwings, apparently giving the impression of the eyes of an owl. Some synthesize noxious chemicals that can be sprayed or otherwise released. For example, as mentioned earlier, beetles in the family Meloidae respond to handling by reflex bleeding in which their leg joints release irritating hemolymph. Many insects are brightly colored, advertising that they taste bad (**aposematic coloration**). They may actually taste bad or they mimic other species that do. Other insects, such as the saddleback moths, have sharp spines with poison glands. Many wasps and bees store toxins in poison glands and have modified ovipositors capable of piercing vertebrate skin, thereby injecting doses of toxins. This stinging behavior can cause serious problems for vertebrates. When handled, some arthropods emit substances that induce a predator to let go.

Behavior Associated with Environmental Fluctuations

The environment of arthropods is subject to irregular as well as regular fluctuations, some of which may threaten survival. For example, an arthropod exposed to a less than optimal temperature or moisture level may move along a gradient until conditions are optimal.

Arthropods commonly respond to regular, deleterious environmental changes, such as in temperature or moisture, by entering a resistant physiological state of dormancy or by migrating. The cues that induce these behaviors may be potentially deleterious changes, e.g., decreasing temperatures themselves, or may be changes in other "predictable" environmental cues such as daylength. Migration is one form of dispersal, other forms of which include movements associated with searching for a mate or food, or movements along gradients of environmental factors such as temperature and moisture. Another form of dispersal is passive movement by being carried on air currents.

Dormancy in response to adverse environmental conditions ranges from quiescence to diapause. **Quiescence** is an immediate slowdown or arrest of metabolism and development (embryonic or larval) that reverses upon termination of the adverse conditions. **Diapause** is a more programmed, genetically determined response by which an arthropod enters a dormant, resistant (e.g., cold-hardy) state in response to environmental cues that indicate the onset of adverse conditions, but are not necessarily the adverse conditions themselves. Further, diapause doesn't necessarily terminate when the adverse conditions improve, but often only after a predetermined period of exposure to the adverse conditions and/or particular cues associated with the improvement of conditions. Diapause may be obligatory or facultative. When it is obligatory, arthropods enter this state each generation; when facultative, several generations can pass without entering diapause. Diapause may vary between species and among strains of the same species. Arthropods that display obligatory diapause undergo one generation per season (are **univoltine**). Those that display facultative diapause may undergo several generations per year and range from 2 (**bivoltine**) to several generations per year (**multivoltine**).

Reproductive Behavior

Mate location, courtship and mating, oviposition, and in some cases brood care, are the basic elements of reproductive behavior. There is considerable complexity and variation in reproductive behavior.

Mate location initially involves stimuli that are perceived over relatively long distances (such as those associated with sight, smell and sound) and that may operate alone or in combination with one another. Thus, mate location simply may be a matter of a male insect visually sensing and approaching any moving object within an appropriate size range. Alternatively, more specific visual stimuli may be required.

Following the initial encounter, other stimuli are necessary to release further approach behavior. These may involve particular stimuli, such as sex pheromones, or they may be a particular behavior displayed by the sex being pursued.

Environmental features also may be involved in mate location. For example, among the true flies, males may swarm about a given object or remain by a marker until a female approaches the same marker. The sexes also may get together at sites of other biological activities such as emergence or feeding sites.

Olfactory cues involve potent volatile substances called sex pheromones produced by specialized glands in males, females or both. These substances are extraordinarily active physiologically and can be detected in minute quantities. For example, a concentration of 1,000 molecules of the sex pheromone **bombykol**, produced by the female silk moth, *Bombyx mori*, stimulates a male. Most sex pheromones appear to be species-specific, but there are examples of nonspecificity. Acoustic signals used in mate location often have specialized sound-producing structures. In some insects, the sounds produced by beating wings may be involved. For instance, the sounds produced by female mosquito wingbeats elicit copulatory behavior in males that will react to the same frequency when produced by a tuning fork.

After mate location, often elaborate courtship behavior occurs and releases behavior that eventually leads to copulation. These behavior patterns may involve the same stimuli used in mate location, e.g., the wingbeat sound of female mosquitoes also releases copulatory behavior in the male. Sex pheromones may act similarly, e.g., female American cockroaches, *Periplaneta americana*, produce a sex pheromone that both attracts males and acts as an aphrodisiac. Courtship and mating patterns vary from simple to complex, depending upon the insects involved.

While insects may compete for mates in various ways before copulation, there are mechanisms of competition after mating. For instance, a copulating male may deposit mating plugs in the female's genital tract to block entry of sperm from another male. If the block is not present, multiple episodes with different males may occur (Thornhill and Alcock 1983).

As part of the elaborate mechanisms that regulate insect reproduction, male ejaculate may contain chemicals that signal mating has occurred, making the female unreceptive to the amorous advances of other males and releasing oviposition behavior.

The success of insect offspring depends to a large extent upon the ability of the female parent to place the eggs (oviposit) in a suitable habitat, particularly if the diet of a given species is highly specialized. As with mate location, various behavior-releasing stimuli are involved. For example, female mosquitoes respond to stimuli associated with the presence of water or places that have been repeatedly flooded and therefore emanate characteristic patterns of stimuli (Bentley and Day 1989).

While perhaps most insects simply deposit their eggs and depart, many carry out some brood care. This behavior reaches its zenith in the social insects. However, much brood care falls short of fully developed social behavior. For example, female earwigs (order Dermaptera) guard their eggs until hatching.

Behavior of Arthropods in Groups

In addition to solitary behavior, many insects organize into various groups. These range from simple aggregations to fully developed and complex societies with many intergradations. Simple aggregations may result from common attraction to a particular habitat based on particular stimuli such as temperature, light, food and specific chemicals. On the other hand, insects of the same species may be mutually attracted to one another. Examples of aggregations include those of cockroaches that form feeding aggregations or ladybird beetles that hibernate together in sheltered situations.

Truly social (eusocial) insects exist in only 2 orders: Isoptera (termites) and Hymenoptera (ants, bees, wasps and relatives). The Hymenoptera particularly interest medical entomologists because many sting humans and sometimes occur in very large populations.

Group behavior no doubt provides benefits such as the protection derived from collective displays as well as more efficient detection and utilization of food.

ARTHROPOD ECOLOGY

Ecology is the study of organisms in an environmental context. Knowledge about the ecology of arthropods helps us manage both harmful and beneficial species as well as assess the effects of human activities that could damage the environment. Under the rubric of ecology, we can appropriately examine arthropod population biology and the interactions of arthropods with both the abiotic and biotic components of their environment, keeping in mind that at any given moment all of these components act in concert.

Every arthropod is part of a complex **ecosystem** that may be viewed as a basic functional unit of the environment. An ecosystem consists of a **biotic component**, i.e., a community of living organisms, and an **abiotic component**, i.e., the nonliving aspects such as temperature, moisture, light and other physical/chemical parameters. An ecosystem can range anywhere from a rot cavity in a tree filled with water (the habitat of certain mosquito species) to the entire biosphere, the part of our planet where life exists. The biotic portion consists of **producers** (mostly green plants that carry out photosynthesis, thereby producing carbohydrates and other compounds), **consumers** (mostly animals that obtain energy in the form of molecular building blocks by eating producers, **herbivores;** by eating other consumers, **carnivores**; or by eating the remains of dead organisms, **detritivores**) and **decomposers** (mostly bacteria and fungi that recycle the complex organic molecules of usually dead producers and consumers back to simple inorganic nutrients usable by producers).

Every organism lives in a place, its **habitat**. This locale can vary from areas as small as the underside of a leaf to a pond or a house or an entire forest, depending on the species and life stage. Although the degrees of tolerance to particular environmental variables (e.g., temperature) may differ, the habitat of a given species typically remains similar throughout its distribution. Environmental variables are not distributed uniformly within a given ecosystem and therefore, if an arthropod species lives in a very limited portion of an ecosystem, we must consider its **microhabitat**.

All of the needed resources, both in nature and amount, comprise the **ecological niche** of a given species. Thus, a niche may be described as the total of the quality and quantity of food, shelter, breeding sites, temperature and humidity needed for a given species to survive and reproduce. As a group, arthropods fill just about every imaginable ecological niche.

Both biotic and abiotic components of the environment influence arthropod populations. One component determines the distributional limits of a given species. For example, a given maximum or minimum temperature necessary for reproduction and development or survival may determine the north-south distributional limits of a species.

Arthropod Populations

Organisms usually occur in relatively discrete groups or **populations**. Knowledge of population characteristics (genetic composition, sex ratio, age composition, dispersion or arrangement in space, biomass), population performance over time (i.e., dynamics, as determined by birth, death and migration rates) and of the environmental factors that directly influence composition and performance are important to medical entomology. Essential information, if available, may predict population performance and allow for the planning and execution of timely control measures. Genetic variation within a population, and factors that impact this variation, are particularly important in developing and understanding the roles of arthropods in disease transmission. Studies of this genetic variation and associated environmental factors fall under the rubric of population genetics. Excellent treatments of population biology and population genetics as they pertain to vectorborne diseases may be found in Black and Moore (1996) and Tabachnick and Black (1996).

Populations can be studied in the laboratory under controlled conditions or in the field under natural conditions. Studies in the laboratory help, but they often are too simplified and may not accurately represent what would happen under field conditions. Although learning about population performance under field conditions is preferable, most studies rely on sampling since it usually is unrealistic to study an entire natural population.

Mathematical modeling also studies populations. Modern computers and statistical methods facilitate this process. However, our overall knowledge of population performance generally relies on laboratory and field studies in conjunction with modeling.

In simple terms, population performance over time may be the result of interaction between 2 fundamental and opposing forces: an inherent capacity for most organisms to undergo exponential population growth (**biotic potential**) and factors in the environment that counteract this capacity (**environmental resistance**).

Exponential growth in the absence of environmental resistance is described by the equation:

$$dN/dt = N(b - d)$$

where **N** = population, **t** = time and **b** and **d** = average birth and death rates, respectively, per individual per unit; **dN/dt** = the instantaneous rate of increase.

The factors **b** and **d**, the average birth (natality) and death (mortality) rates respectively, can be reduced to:

$$r = (b - d)$$

where **r** is the **intrinsic rate of increase** (also known as the reproductive capacity) of a population.

This allows the equation:

$$dN/dt = rN$$

If **r** = 0, the population is stable; if **r** > 0, the population is growing; and if **r** < 0, the population is decreasing. If **r** remains constant, the population expands at an ever increasing rate (i.e., exponential growth). However, due to environmental resistance, **r** never remains constant, being subject to additive and subtractive density-dependent and density-independent factors described below.

Thus, a more realistic equation describing population growth is the so-called logistic model in which the changes in the instantaneous rate of increase are described by the equation:

$$dN/dt = r(K - N/K)N$$

where **K** = the **carrying capacity** of the environment.

As **N** aproaches **K**, the term **(K - N/K)** approaches zero and hence the instantaneous rate of increase approaches zero. Thus, the population reaches a constant level. At this point, the population is said to have reached the carrying capacity of the environment. While helpful in thinking about populations, this model of population growth is a simplification. It is limited by 4 assumptions: that reproductive potential of all individuals is equal; that the age distribution is constant with the same proportion of individuals reproducing at all times; that reproduction is not influenced by environmental factors; and that the carrying capacity does not change (Price 1997). Any one of these assumptions is rarely true under natural conditions. Further movement of individuals into and out of a population, immigration and emigration respectively, can exert a major influence.

As populations interact with the various environmental factors, they fluctuate in size. Some do so periodically in a regular or irregular fashion, while others fluctuate regularly about an average level. Such regular fluctuations may reflect **additive** or **subtractive** processes, i.e., processes that result in population increases or decreases, respectively. These processes may be **density-dependent** or **density-independent**, i.e., their action may or may not be related to factors that differ according to the density of a population, such as the influence of natural enemies and competition for food and shelter. Regular fluctuations of populations about a fairly constant average may indicate population regulation by a subtractive density-dependent process that acts with increasing power as population density increases, thus checking population growth. Seasonal changes typically are associated with arthropods that reproduce only at certain times of year such as those in the northern extremes of temperate climates.

Practical approaches to evaluating populations include "multifactor" studies that generate **life tables**. They describe population densities at different times during the life cycle and provide age-specific information regarding "key" environmental factors that account for mortality. When preliminary studies have indicated a few essential influences, the **key-factor** method may be applied. Population density is measured at one point in each generation, a procedure that requires less sampling by concentrating on periods in the life cycle when the "key" influences are operating.

Arthropods and Abiotic Factors

Temperature

Arthropods are fundamentally **poikilothermic**, body temperature tending to be the same as the ambient temperature. However, body temperature is not always the same as ambient. Absorption of radiant energy, heat produced metabolically and rapid muscle contractions can increase body temperature. Every arthropod species has a temperature range in which it is able to survive. Typically, this interval is within the extremes of zero and 50°C. Optimal temperatures for most species lie between 22°–38°C. These ranges vary from species to species, within a species and among individuals, according to physiological state. Tolerance of extreme temperature may vary seasonally, as with most arthropods found in temperate zones. Mechanisms that make these arthropods cold-hardy or result in migratory behavior respond to environmental cues such as daylength in the fall. Not surprisingly, tropical species tolerate less cold than temperate species. Arthropods avoid extreme temperatures and if exposed to a temperature gradient will move to the most optimal zone. Deleterious effects of high temperatures include protein denaturation and increased water loss through transpiration, while low temperatures may form ice crystals that destroy cellular structure.

Even if temperature does not kill an insect, it may impair duration of life, reproduction, development rate, mobility, horizontal and vertical distribution and rate of dispersal. In bloodsucking insects such as tsetse that depend on nutrient stores between bloodmeals, metabolic rates increase with temperature and high temperatures may shorten the survival period between meals. The human body louse does not deposit eggs below 25°C (Wigglesworth 1984).

Moisture

Water is critical to the survival and effective functioning of all living things, including arthropods. As with temperature, each species has an optimal moisture range below and above which it is negatively impacted. Below tolerable levels, individuals' activities are seriously impaired and with sufficient water loss, they can die. Excessive water loss is dangerous to arthropods because of their small size and their tendency to lose water through transpiration. Several factors influence arthropod water content, including temperature and air movement. Environmental moisture factors such as relative humidity and rainfall vary both temporally and spatially. Relative humidity varies with such factors as the time of day (typically higher at night), topography, season and vegetation. Too much environmental moisture, as well as too little, can be deleterious. For example, high moisture conditions favor fungi, bacteria and viruses that infect arthropods.

In addition to survival, factors that are negatively impacted by suboptimal moisture conditions include feeding, growth and development, reproduction, oviposition rates and duration of life. Therefore, moisture, like temperature, plays an important role in arthropod distribution.

Fascinating adaptations of arthropods enable them to cope with environmental moisture problems and a tendency to lose water. As the preceding chapter mentioned, the epicuticular wax layer, spiracular closure mechanisms, the production of relatively dry urine and feces and the chorion surrounding an egg illustrate such adaptations. Further, there are behavioral adaptations, the most obvious of which is drinking water. Other behaviors include orienting to optimal conditions in moisture gradients and entering a resting, desiccation-resistant stage during periods of low availability of environmental moisture. Arthropods living in regions where there have been regular periods of drought long enough for natural selection to have occurred possess this ability. The chironomid larva, *Polypedilum vanderplanki*, survives nearly complete dehydration for several years by entering a low metabolic state called cryptobiosis (Hinton 1960). Eggs of floodwater mosquitoes in the genera *Aedes, Psorophora* and others can withstand extensive dry periods.

Light

Light parameters such as photoperiod (daylength), illuminance and wavelength are relatively constant. Arthropods use them frequently as environmental cues. As discussed above, photoperiod changes in temperate regions signal the changing seasons, thereby providing consistent warnings of deleterious environmental changes.

Different wavelengths of reflected light provide potentially useful cues in mate or host location (plant or vertebrate). Cycles of bloodfeeding behavior sometimes are associated with the consistent periods of transition (**crepuscular periods**) between the light of day and dark of night, which occur at sunrise and sunset. Likewise, cloud cover may influence arthropod activity.

Other abiotic factors

Currents in air and water, gases dissolved in water, air composition, electricity, ionizing radiation and soil composition may influence arthropod activities. Air movement influences moisture by increasing the rate of transpiration, which could hurt or help under conditions of high temperature or humidity. Winds influence insect flight and dispersal. Under conditions of high wind, arthropods hunch down close to the substrate thereby diminishing the likelihood of being blown away. If the substrate is the surface of water, their exposure to predators may increase.

Water currents can influence the distribution of aquatic arthropods in a way similar to wind in terrestrial environments. Aquatic insects have adapted to water currents. For example, the ability of black fly larvae to affix their bodies to stationary submerged objects in fast-moving streams was discussed earlier. On the other hand, some aquatic insects such as mosquito larvae cannot survive in swiftly moving water. Water movement strongly influences the amount of dissolved oxygen and thereby influences the distribution of aquatic insects. Black fly larvae require a high concentration of dissolved oxygen and are found in rapidly moving streams while some mosquito larvae thrive at considerably lower oxygen concentrations.

Wind, water splashing and photosynthesis by aquatic plants also influence dissolved oxygen.

As with other living forms, arthropods suffer the negative effects of environmental pollutants, including agricultural and industrial chemicals, heavy metals and acid rain. These pollutants can affect arthropods directly by acting as toxins or indirectly by acting on host plants or animals, or predators and parasites.

Other abiotic factors that may affect arthropods include the ionization of air, atmospheric potential and high-energy electromagnetic radiation.

Arthropods and Biotic Factors

Biotic components include both intraspecific and interspecific interactions. **Intraspecific interactions** can be beneficial, involving cooperation, or deleterious, involving competition. Beneficial interactions reach their highest level in the social insects (orders Isoptera and Hymenoptera). Population density can influence the amount of competition. High population density may favor successful mate location and survival of potential predation, while low density may result in few fertilizations and lower fecundity. On the other hand, high population density may lead to competition for food and shelter, especially when resources are limited. Intraspecific interactions may, in turn, regulate population.

Competition, symbiosis, predator-prey interactions, herbivore-plant interactions and indirect interactions comprise types of interspecific interactions. When 2 different species overlap niches, they may compete for resources such as food and shelter. If one competing species has an advantage over another, it may lead to competitive displacement whereby one species completely displaces another (Price 1997). In fact, niche overlap logically would be expected to be selected against, thus providing a mechanism for segregation of species into a community segregated into different niches.

Two kinds of organisms may associate intimately. These associations are termed symbiosis. There are several different types of symbiosis, depending upon whether the participating species benefit, are harmed or left unchanged. Chapter 1 fully described these associations. The symbiotic relationships between microbes and insects whereby the microbes gain a place to live and disperse, and the insects gain some nutritive factor, are well known.

Parasitism is a type of symbiosis important in medical entomology. Among the many organisms that parasitize insects are insects themselves, microorganisms, mites and nematodes. Where microbes and nematodes parasitize both insects and vertebrates, the insects may act as vectors, carrying parasites to their vertebrate or plant hosts. More than a thousand microorganisms in insects (e.g., bacteria, viruses, protozoa and fungi), many of which cause disease, have been described. Microbes enter insects when the insects feed, through wounds in the integument or by active penetration of the integument (e.g., certain entomogenous fungi). Once a microbial infection is established in an insect, the infecting microbes may be passed transovarially (via the egg) to offspring, as with certain arboviruses (e.g., LaCrosse encephalitis virus) in mosquitoes, and rickettsiae (e.g., the causative agent of Rocky Mountain spotted fever) in ticks. Chapter 5 discusses the various kinds of arthropod transmission mechanisms.

Microbes that are pathogenic to arthropods are of particular interest due to their potential application in the biological control of their arthropod hosts. For example, the bacterium *Bacillus thuringiensis* has been used very successfully against a variety of insect pests and vectors. A particular variety, *B. thuringiensis israelensis*, is useful against mosquito and black fly larvae. Microbes can even influence the mating compatibility among different populations of the same species. Rickettsiae in the tribe Wolbachieae infect several mosquito species, lodging in the gonads. Specifically, *Wolbachia pipientis* infects the mosquito *Culex pipiens*, causing sexual incompatibility between geographic populations (Yen and Barr 1973). The biology of *Wolbachia* recently has been reviewed by Werren (1997).

Probably many of the relationships between arthropod species and between arthropods and microorganisms are commensal. A good example of commensalism is found in the phenomenon of phoresy, where an individual of one species attaches to an individual of another species, thereby gaining a mode of transportation. For example, chewing lice (order Mallophaga) may use their mouthparts to grasp louse flies (order Diptera, family Hippoboscidae), some of which are bird parasites, thus gaining a ride to a new host. The human bot fly, *Dermatobia hominis* (Fig. 3.20) may oviposit on other flies, such as mosquitoes and black flies, and in this way uses other insects to reach a potential vertebrate host (including humans). Predators of arthropods include arthropods themselves (various insects, mites, scorpions and pseudoscorpions), vertebrates (including birds, fish, reptiles, amphibians and mammals) and even a few plants (e.g., the Venus fly trap and pitcher plant). Arthropods also are subject to indirect effects from the activities of other organisms, notably those of humans. The manifold effects of "development" often are deleterious to harmless arthropods while sometimes enhancing the survival and spread of harmful ones. See, for example, the multi-authored text "Demography and vector-borne diseases" (Service 1989).

MOLECULAR BIOLOGY AND MEDICAL ENTOMOLOGY

Using a microbe (e.g., the bacterium *Escherichia coli*) as a molecular factory, it now seems possible to isolate specific genes and produce endless quantities of eucaryotic cell-derived genes (clones) and gene expression products (proteins). Reagents such as cloned genes and expressed proteins can then be used to locate given genes on a chromosome (*in situ* location) and when, where and to what extent given genes are expressed in a given organism. It also is possible to detect minute amounts of a given gene (theoretically from a single copy) using the ingenious technique of gene amplification by the **polymerase chain reaction (PCR)**. Further, current methods can introduce genes from one eucaryotic cell into another from a different species. As with virtually every other area of biological science, the exciting and powerful tools of molecular entomology are helping to solve old problems and define new ones in entomology. The ability to examine the very depths of life from the recipe written in the language of deoxyribonucleic acid (DNA) to the grand expression of this recipe in the multitude of living forms is revolutionary. It changes how we understand life, and challenges long-held notions of exactly what life is and what it means. This brief section provides just a few examples of developments in molecular entomology. The texts by Watson et al. (1987) and Lewin (1997) should be consulted for rigorous treatments of molecular biology, while Horodyski (1998), Hoy (1994) and Beaty and Marquardt (1996) discuss applications of molecular biology to insects. Specific treatments of molecular techniques may be found in Sambrook et al. (1989), Ausubel et al. (1987) and Crampton (1994), this last reference dealing specifically with applications to insect vectors of disease.

Examples of the application of molecular biology within the context of medical entomology include use in arthropod identification, analysis of mechanisms of insecticide resistance, identification of arthropodborne pathogens and genetic modification of arthropods for purposes of control.

A technique called RAPD (rapid amplification of polymorphic DNAs) technology can identify and use genus-specific or species-specific markers, and even differentiate between individuals of the same species. This PCR-based method has found application, for example, in the separation of cryptic (hidden) species in mosquitoes of the genus *Aedes* (Kambhampati et al. 1992). This technique is so sensitive that only small amounts of DNA are required; so small, in fact, that individual insects need not be killed to obtain samples.

Molecular techniques have been used to genetically dissect mechanisms of pesticide resistance, enhancing our understanding of this crucial process that threatens to make ineffectual one of the major methods of arthropod control. At the same time, understanding these mechanisms may enable us to incorporate the genes associated with pesticide resistance into beneficial insects, thereby facilitating the use of pesticides against dangerous vectors without harming insect predators of these vectors. Understanding pesticide resistance genes also may contribute to the development of new pesticides by genetically engineering cultured cells to express cloned resistance genes for pesticide testing.

The use of molecular "probes" to identify the presence of pathogen genomes in vector arthropods is finding wide use in medical entomology. For example, such probes have been used to identify the presence of specific arboviruses in vectors such as mosquitoes and ticks (Monroy and Webb 1994, Persing et al. 1990, Vodkin et al. 1994).

Genes can be introduced effectively into eucaryotic cells, so there are enormous possibilities for modifying cells to our advantage, e.g., in the genetic manipulation of arthropod populations by producing so-called transgenic animals. Genes could be introduced into vector arthropod populations to alter the fitness of individuals as a means of control, or genes that impact negatively on the competence of a given vector to transmit a given pathogen could be introduced.

Manipulation of insect symbionts also holds promise. For example, the bacterium *Rhodococcus rhodnii* is a gut symbiont of the bloodsucking true bug, *Rhodnius prolixus*, one of the major vectors of the protozoan *Trypanosoma cruzi* that causes Chagas disease. A method has been developed whereby this symbiont has been made to express a gene product that may impact negatively on the development or transmission of *T. cruzi* and hence reduce transmission of this parasite by the bug.

REFERENCES

Ausubel, F.M., Brent, R., Kingston, R.E., Moore, D.D., Seidman, J.G., Smith, J.A. and Struhl, K. 1987. Current protocols in molecular biology. John Wiley and Sons, New York. various pp.

Bänzinger, H. 1971. Skin-piercing bloodsucking moths. I. Ecological and ethological studies on *Calpe eustrigota* (Lepidoptera: Noctuidae). Acta Trop. **32**:125–144.

Beaty, B.J. and Marquardt, W.C. (eds). 1996. The biology of disease vectors. University Press of Colorado, Niwot. 632 pp.

Bentley, M.D. and Day, J.F. 1989. Chemical ecology and behavioral aspects of mosquito oviposition. Annu. Rev. Entomol. **34**:401–422.

Black, W.C., IV and Moore, C.G. 1996. Population biology as a tool for studying vector-borne diseases. Pp. 393–416 *in* B.J. Beaty and W.C. Marquardt (eds.), The biology of disease vectors. University Press of Colorado, Niwot.

Black, W.C., IV and Munstermann, L.E. 1996. Molecular taxonomy and systematics of arthropod vectors. Pp. 438–470 *in* B.J. Beaty and W.C. Marquardt (eds.), The biology of disease vectors. University Press of Colorado, Niwot.

Borror, D.J., Triplehorn, C.A. and Johnson, N.F. 1989. An introduction to the study of insects. 6th ed. Saunders College Publishing, Philadelphia. 875 pp.

Clements, A.N. 1992. The biology of mosquitoes. Vol. 1, Development, nutrition and reproduction. 2 vol. Chapman and Hall, London. 509 pp.

Crampton, J.M. 1994. Molecular studies of insect vectors of malaria. Adv. Parasitol. **34**:1–31.

Crosskey, R.W. 1990. The natural history of blackflies. Wiley, New York and Chichester, UK. 711 pp.

Danks, H.V. 1988. Systematics in support of entomology. Annu. Rev. Entomol. **33**:271–296.

Dolling, W.R. 1991. The Hemiptera. Oxford University Press, Oxford. 274 pp.

Durden, L.A. and Musser, G.G. 1994. The mammalian hosts of the sucking lice (Anoplura) of the world: a host-parasite list. Bull. Soc. Vect. Ecol. **19**:130–168.

Durden, L.A. and Romoser, W.S. 1998. Survey of the Class Insecta. I. Apterygota and Exopterygota and survey of Class Insecta II. Endopterygota. Pp. 340-416 *in* W.S. Romoser and J.G. Stoffolano (eds.), The Science of Entomology, 4th ed. WCB/McGraw-Hill, Dubuque, Iowa.

Evans, G.O. 1992. Principles of acarology. CAB International, Wallingford, Oxon, UK. 563 pp.

Forsyth, A.F. and Miyata, K.M. 1984. Tropical nature. Scribner, New York. 248 pp.

Harwood, R.F. and James, M.T. 1979. Entomology in human and animal health, 7th ed. Macmillan, New York. 548 pp.

Hinton, H.E. 1960. Cryptobiosis in the larva of *Polypedilum vanderplanki* Hint. (Chironomidae). J. Insect Physiol. **5**:286–300.

Horodyski, F.M. 1998. Molecular entomology. Pp. 506-520 *in* W.S. Romoser and J.G. Stoffolano (eds.), The science of entomology, 4th ed. WCB/McGraw-Hill, Dubuque, Iowa.

Hoy, M.A. 1994. Insect molecular genetics. Academic Press, San Diego, California. 546 pp.

Kambhampati, S., Black, W.C., IV and Rai, K.S. 1992. Ramdom amplified polymorphic DNA of mosquito species and populations. J. Med. Entomol. **29**:939–945.

Keegan, H.L. 1980. Scorpions of medical importance. University Press of Mississippi, Jackson. 140 pp.

Kettle, D.S. 1994. Medical and veterinary entomology, 2nd ed. Cambridge University Press, Cambridge, UK. 725 pp.

Kim, K.C. and Ludwig, H.W. 1978. The family classification of the Anoplura. Syst. Entomol. **3**:249–284.

Kim, K.C., Pratt, H.C. and Stojanovich, C.J. 1986. The sucking lice of North America. An illustrated manual for identification. Pennsylvannia State University Press, University Park. 241 pp.

Klowden, M.J. 1996. Vector behavior. Pp. 393–416 *in* B.J. Beaty and W.C. Marquardt (eds.), The biology of disease vectors. University of Colorado Press, Niwot.

Krantz, G.W. 1978. A manual of acarology, 2nd ed. Oregon State University, Corvallis. 509 pp.

Lane, R.P. and Crosskey, R.W. (eds). 1993. Medical insects and arachnids. Chapman and Hall, London. 723 pp.

Lehane, M.J. 1991. Biology of blood-sucking insects. Harper Collins Academic, London. 288 pp.

Lent, H. and Wygodzinsky, P. 1979. Revision of the Triatominae (Hemiptera, Reduviidae), and their significance as vectors of Chagas' disease. Bull. Am. Mus. Nat. Hist. **163**:123–520.

Lewin, B. 1997. Genes VI. Oxford University Press, Oxford. 1260 pp.

Marquardt, W.C. 1996. Hemimetabolic vectors and some cyclorraphan flies and the agents they transmit. Pp. 128–145 *in* B.J. Beaty and W.C. Marquardt (eds.), The biology of disease vectors. University Press of Colorado, Niwot.

Marshall, A.G. 1981. The ecology of ectoparasitic insects. Academic Press, London. 459 pp.

Monroy, A.M. and Webb, B.A. 1994. Detection of eastern equine encephalitis virus by polymerase chain reaction. Proc. N.J. Mosq. Contr. Assoc. **8**:103–111.

Persing, D.H., Telford, S.R., III, Rys, P.N., Dodge, D.E., White, T.J., Malawista, S.E. and Spielman, A. 1990. Detection of *Borrelia burgdorferi* DNA in museum specimens of *Ixodes dammini* ticks. Science **249**:1420–1423.

Pittaway, A.R. 1991. Arthropods of medical and veterinary importance: a checklist of preferred names and allied terms. CAB International, Wallingford, Oxon, UK. 178 pp.

Price, P.W. 1997. Insect ecology, 3rd ed. John Wiley and Sons, New York. 874 pp.

Restifo, R.A. 1982. Illustrated key to the mosquitoes of Ohio: adapted from C.J. Stojanovich's keys to the common mosquitoes of eastern United States. Biological Notes #17 (Ohio Biological Survey). Ohio State University and Ohio Dept. of Health. 17 pp.

Richards, O.W. and Davies, K.G. 1977. Imms general textbook of entomology, 10th ed. Vol. 2, Classification and biology. Chapman and Hall, London; John Wiley and Sons, New York. Distributed in USA by Halstead Press.

Romoser, W.S. and McPheron, B.A. 1998. Insect classification and evolution. Pp. 318-339 *in* W.S. Romoser and J.G. Stoffolano (eds.), The science of entomology, 4th ed. WCB/McGraw-Hill, Dubuque, Iowa.

Romoser, W.S. and Stoffolano, J.G., Jr. (eds). 1998. The science of entomology, 4th ed. WCB/McGraw-Hill, Dubuque, Iowa. 605 pp.

Sambrook, J., Fritsch.E.G. and Maniatis, T. 1989. Molecular cloning. Cold Spring Harbor Laboratory Press, Cold Spring Harbor, New York. 545 pp.

Schofield, C.J. 1988. Biosystematics of the Triatominae. Pp. 285–312 *in* M.W. Service (ed.), Biosystematics of haematophagous insects. Clarendon Press, Oxford.

Service, M.W. (ed.) 1988. Biosystematics of haematophagous insects. Clarendon Press, Oxford. 363 pp.

Service, M.W. (ed.) 1989. Demography and vector-borne diseases. CRC Press, Boca Raton, Florida. 402 pp.

Service, M.W. 1993. Mosquito ecology. Field sampling methods, 2nd ed. Chapman and Hall, London. 988 pp.

Smith, K.G.V. 1973. Insects and other arthropods of medical importance. British Museum (Natural History), London. 561 pp.

Tabachnick, W.J. and Black, W.C., IV. 1996. Population genetics in vector biology. Pp. 417–437 *in* B.J. Beaty and W.C. Marquardt (eds.), The biology of disease vectors. University Press of Colorado, Niwot.

Thornhill, R. and Alcock, J. 1983. The evolution of insect mating systems. Harvard University Press, Cambridge, Massachusetts. 547 pp.

US Public Health Service. 1969. Pictorial keys to arthropods, reptiles, birds and mammals. Publication No. 1955, U.S. Department of Health, Education and Welfare, Communicable Disease Center, Atlanta, Georgia.

Vodkin, R.J., Strait, T., Mitchell, C.J., McLaughlin, G.L. and Novak, R.J. 1994. PCR-based detection of arboviral DNA from mosquitoes homogenized in detergent. Biotechnol. **17**:111–116.

Waage, J.K. 1979. The evolution of insect/vertebrate associations. Biol. J. Linnean Soc. **12**:187–224.

Watson, J.D., Hopkins, N.H., Roberts, J.W., Steitz, J.A. and Weiner, A.M. 1987. Molecular biology of the gene. Benjamin/Cummings, Menlo Park, California. 1163 pp.

Werren, J.H. 1997. *Wohlbachia* run amok. Proc. Nat. Acad. Sci. **94**:11154–11155.

Wigglesworth, V.B. 1984. The principles of insect physiology, 8th ed. Chapman and Hall, London. 191 pp.

Yen, J.H. and Barr, A.R. 1973. The etiological agent of cytoplasmic incompatability in *Culex pipiens*. J. Invertr. Pathol. **22**:242-250.

Chapter 4

Direct Injury: Phobias, Psychoses, Annoyance, Allergies, Toxins, Venoms and Myiasis

JONATHAN F. DAY[1], JOHN D. EDMAN[2], SIDNEY E. KUNZ[3] AND STEPHEN K. WIKEL[4]
[1]Florida Medical Entomology Laboratory, Vero Beach, [2]University of California, Davis, [3]USDA Livestock Insects Laboratory, Kerrville, Texas and [4]Oklahoma State University, Stillwater.

INTRODUCTION

Arthropods cause a wide range of health problems in humans (Table 4.1). Arthropods injure hosts by (1) feeding on their fluids or tissues, (2) inflicting toxic stings or bites, (3) stimulating allergies and hypersensitivity reactions, (4) inducing phobias and psychoses and (5) accidental infestation and invasion. Arthropod injury is perceived falsely in entomophobia and delusory parasitosis, but these conditions still represent important health problems. Arthropod-caused injuries are usually the result of direct contact with arthropods but as in the case of airborne allergens, may result from indirect contact. Also, skin or tissue damage caused by arthropods can open the door to serious secondary bacterial infections.

Arthropods are the dominant animals on earth, comprising about 80% of all macroscopic species. Direct contact between arthropods and vertebrates, including humans, therefore is frequent, especially in warm weather. Injurious contact with arthropods has evolved from several different circumstances. Some arthropods have adapted to live near or inside vertebrate "nests," whereas others have evolved into true facultative or obligate parasites living on or in vertebrate hosts. In this case, contact with vertebrates is essential for their development and survival as species. All blood and tissue feeding arthropods fall into this category. These parasitic associations may be as short as the 3 minutes needed for kissing bugs to extract a bloodmeal from their sleeping host to permanent associations during all life stages, as in the human head louse. Some arthropods are parasites during only one life stage, as with larval bot flies tunneling into the skin of their hosts. Others are intermittent, changing hosts within each active stage, as with the larva, nymph and adult of many ticks.

Larval stages of some species of blow flies and flesh flies, and all species of warble flies, have evolved to invade vertebrate tissue, a condition called **myiasis**. Tissue loss caused by their feeding can cause severe injury, secondary infection and death to livestock and wildlife. Prior to its eradication from North America and much of Central America, the primary screwworm fly caused large losses to cattlemen in affected areas. Although less commonly involved than Diptera, other groups of arthropods also can cause injury by invasion of host tissue. For example, when species of Coleoptera are involved, the situation is called canthariasis.

B.F. Eldridge and J.D. Edman (eds.), Medical Entomology, Revised Edition, 99–149.

Table 4.1. Summary of direct effects of arthropods on humans and domestic animals.

Condition	Harmful effects	Arthropods responsible
Delusory psychoses	Irrational or destructive behavior	None: usually imagined to be skin-invading ectoparasites
Entomophobia/Arachnophobia	Accidents triggered by an unusual fear of arthropods	Many; usually spiders or Hymenoptera
Physical nuisance	Stress and mental anguish; reduced productivity	Any extremely abundant arthropod that interferes with normal activities
Contact irritation	Asthma and other respiratory allergies	Any abundant or domestic arthropod that produces airborne allergens
Feeding annoyance and blood loss	Pain, anemia, itching and other allergic skin reactions; stress, reduced fitness and death	Temporary and permanent ectoparasites
Envenomization	Prolonged pain, inflamation, cyto/neurotoxic symptoms; Arthus and anaphylactic allergic symptoms, including death	Arthropods with toxic stings, bites, setae, hemolymph, defensive secretions; usually Hymenoptera, spiders or scorpions
Myiasis	Prolonged pain, tissue and fluid loss; secondary infection; reduced fitness, death	Larval Diptera (Cyclorrhapha)

Besides feeding damage, arthropod parasites also can cause toxic reactions when large numbers of arthropods are involved or, as in unusual cases like tick paralysis, when feeding-associated secretions target critical physiological processes. Parasitic arthropods can induce immunologic reactions when hosts develop sensitivity to saliva or other allergens produced by the arthropod.

A second major form of injurious contact between arthropods and vertebrates results from the stings and bites of nonparasitic arthropods, especially social Hymenoptera and spiders. Venom (biological toxin) injected into vertebrates during these often accidental encounters with arthropods can cause acute poisoning or, if the host has a history of previous exposure to the venom, allergic reactions that can be fatal. Arthropods generally sting or bite vertebrates as defensive reactions to threats against the arthropods or their colonies. Defensive secretions also may be associated with fluid-filled hairs or other specialized delivery mechanisms such as reflex bleeding. In some cases, arthropod venom is not a defensive secretion but normally is used to immobilize prey.

In general, the medical importance of arthropods involved in all of these injuries is directly related to the frequency of contact with humans (or highly valued domestic animals) and the severity of the injury inflicted. The type of injury arthropods inflict depends on the nature of the contact. Damage takes many forms and results in a wide range of physical and mental symptoms. It is sometimes difficult to separate toxic symptoms associated with arthropod feeding or stinging from those induced by host immunologic responses to allergens in arthropod venom or saliva. Envenomization from arthropod stings and bites can cause severe complications, especially in individuals sensitized to specific arthropod allergens. Allergic symptoms normally result from repeated exposure to the same arthropod allergen (i.e., sensitization) and, in the extreme (i.e., hypersensitivity), life-threatening anaphylactic shock can result from a single arthropod sting or bite.

Except for a few highly venomous spiders and scorpions, severe toxic reactions, in the absence of allergic reactions, generally occur only when large numbers of stings are involved. Some non-stinging arthropods such as millipedes and blister beetles also have defensive secretions that can be toxic to vertebrates. Simply coming into contact with arthropods, their body parts or airborne particulates can cause allergic skin reactions or severe asthmatic reactions in sensitive individuals. Cockroaches and house dust mites are common household sources of such reactions.

Occasionally, biting flies are present in numbers sufficient to cause significant blood loss that can be fatal even in large vertebrates. Swarms of biting flies can block respiratory passages and large vertebrates can suffocate. More often, blood loss and overall aggravation result in lowered productivity among meat, milk or egg producing animals. Blood loss and annoyance in humans, domestic animals and wildlife also can result from contact with lice, cimicid bugs, triatomid bugs, ticks, chigger mites, itch or mange mites and fleas. Humans may be attacked accidentally by biting arthropods, especially mites and ticks that normally are associated with wild animals.

Biting arthropods cause annoyance that impacts tourism, recreation, land development and industrial or agricultural production. They affect homeowners by restricting outdoor activity and depressing home and land values. Production of essential products from agriculture, forestry and mining often are adversely limited by biting arthropods. Severe annoyance around homes and recreational areas has led to the development of entire industries centered on nuisance/vector control and personal protection. Some products can significantly protect vertebrates from biting pests while others are useless.

ECONOMIC IMPACT OF DIRECT EFFECTS OF ARTHROPODS

Effects on Humans

Although losses are surely substantial, it is difficult to estimate the economic impact of arthropods on humans caused by stinging, biting, annoyance and other conditions resulting directly from arthropod activities. It is especially difficult to make accurate loss estimates for human activities such as tourism, recreation, land resource development and agricultural productivity. However, it is useful to consider that according to the National Pediculosis Association (Newton, Massachusetts), over $150 million was spent in the USA in 1997 just for over-the-counter pediculicides and combs to eliminate head lice. Costs associated with allergies and toxic reactions from arthropods also are enormous. These costs include not only medical treatment and prophylaxis, but also public health preventive measures such as control of Africanized honey bees.

Undoubtedly, biting arthropod populations have an adverse effect on tourism and recreation, but there have been few studies to quantify this. Ironically, in some instances large populations of biting Diptera are considered to be a positive factor. At the Everglades National Park in southern Florida, *Aedes taeniorhynchus* mosquitoes serve as a major ecological focus of the park environment. Here, large-scale mosquito control is avoided, despite empty campgrounds during much of the year, and graphi-

Figure 4.1. Public exhibit on food webs at Everglades National Park (photograph by Jonathan F. Day).

cal displays along nature trails portray saltmarsh mosquitoes as key components of the food chain, an indication of their perceived importance to the park's ecology (Fig. 4.1).

Linley and Davies (1971) attempted to quantify the impact of ceratopogonid midges on tourism and recreation in Florida, the Bahamas and the Caribbean. Three species, *Culicoides furens, Culicoides barbosai* and *Leptoconops becquaerti,* were estimated to have an economic impact. Even today, biting midges in many coastal areas are a serious nuisance for which there is rarely an effective remedy. In southern Florida, mosquito control ditches constructed in saltmarshes during the late 1950s have silted in and now provide ideal *Culicoides* oviposition sites. In some situations, this has resulted in financial loss on the resale of homes and property because of persistently high biting midge populations.

Black fly infestations also can result in financial losses. In 1994 and 1995, emergences of black flies at a private South Carolina golf club caused a loss estimated at >$27,000 (Gray et al. 1996). A suppression program reduced adult black fly populations by 92% in 1995 and 88% in 1996, which resulted in acceptable black fly biting levels. The cost of suppression was approximately $12,500 per year.

It is difficult to estimate the number of visitors who fail to return to a resort, a state or national park, or even a particular region due to a bad experience with biting arthropods. Nevertheless, anecdotal accounts abound of such experiences and are accompanied by expressions of determination never to return.

Biting fly populations have influenced the course of land development, particularly in coastal habitats. New home construction along the barrier islands in southeastern Florida lagged behind development on the mainland. In Indian River County, Florida, earnest new home construction on the barrier islands was not attempted until the early 1960s, due mainly to unbearable mosquito and ceratopogonid midge populations. With the advent of modern mosquito control practices such as permanent control methods that included the construction of dikes to form saltmarsh impoundments, the barrier islands became habitable and construction flourished. Presently, the absence of effective ceratopogonid midge control and an increasing annual midge population are again making outdoor life on these islands more difficult.

Coastal habitats also can produce large numbers of other biting fly species. Greenhead flies (Tabanidae) present a major problem along the East Coast of the USA from Maine to South Carolina. Visually attractive box traps are sometimes placed in saltmarshes to reduce the biting annoyance of greenheads in nearby neighborhoods and beaches. Stable flies (Muscidae) are found in coastal habitats from Texas to New England. Adult seaweed flies (Coelopidae) can sometimes reach levels sufficient to drive people from beaches, especially along the Gulf Coast of Florida. There, a stable fly (*Stomoxys calcitrans*) control program relies on an aerial spray campaign to reduce the numbers of these biting flies during the summer.

Effects on Domestic Animals

The effects of biting arthropods on domestic animal production are more easily quantifiable. Steelman (1976) reviewed the financial effect of arthropod activities on livestock. Biting and other annoyance by arthropods can cause reduced milk, egg and meat production. Black flies, mosquitoes, stable flies (*S. calcitrans*), horn flies (*Haematobia irritans*), face flies (*Musca autumnalis*), house flies (*Musca domestica*), biting midges, lice and ticks all have been shown to reduce milk production in dairy cattle. Organophosphate-resistant ticks (*Boophilus microplus* and *Amblyomma cajennense*) caused a reduction in milk production by dairy cattle in Cuba (Cordoves et al. 1986). After dipping cattle in flumethrin to control ticks, milk yield rose by 18%, mortality rates due mainly to blood parasites in cows were halved, the abortion rate was reduced by 35%, the pregnancy rate rose by 28% and the annual cost for drug treatment was reduced by 32%.

Biting arthropods sometimes are present in numbers sufficient to reduce meat yield in beef cattle. Yield loss results from 2 factors: (1) increased grooming movements to escape biting arthropods that result in reduced caloric intake and (2) actual blood loss. Attack by bloodfeeding flies in the families Culicidae, Ceratopogonidae, Simuliidae and Tabanidae have been found to reduce the productivity of beef and dairy cattle in Russia. Attacks by these flies resulted in an average daily reduction in milk yield of 16.4% per cow and an average seasonal weight loss of 29.5% per animal in young cattle. During daytime observations, individual animals were estimated to be attacked by more than 10,000 biting flies (Muradov et al. 1980). In view of attack rates such as these, it is easy to understand why cows fail to produce milk and beef cattle fail to gain weight.

Stress caused by biting arthropods can be compounded by environmental conditions that result in even greater weight and productivity losses in domestic animals. Campbell (1988) reported that the adverse effect of stable flies on beef cattle weight gain and food assimilation was further aggravated by heat. Large stable fly populations cause cattle to bunch (Fig. 4.2), thus increasing heat stress, especially during summer months. Complex, multi-factored circumstances such as fly abundance, environmental temperature and the elevated body heat of densely grouped cattle make it difficult to incriminate the individual factors responsible for weight and productivity loss in animals exposed to high numbers of biting arthropods.

The ultimate loss to cattle ranchers is the death of stock animals due to blood loss or suffocation resulting from biting arthropods. Hollander and Wright (1980) estimated that during periods of high tabanid abundance in Oklahoma, blood loss in cattle due to fly bites averaged 200 ml/animal/day. Loss of blood due to attack by the saltmarsh mosquito, *Aedes sollicitans*, was linked to the deaths of 21 cattle during 4 nights in Brazzoria County, Texas (Abbitt and Abbitt 1981). In 1998, after a hurricane caused severe flooding in coastal Louisiana, massive broods of mosquitoes reportedly caused several deaths in beef cattle. Lone individuals such as penned bulls were most susceptible, presumably because they experienced greater biting pressure than animals bunched together. In many cases, suffocation, as well as blood loss, contributes to the death of animals exposed to biting flies. In addition to mosquitoes, other arthropods linked to domestic animal mortality include horse flies, ticks, lice and black flies. The deaths of 23 cows, one mule and one heifer in a small area of the Vosges Department of France were blamed on *Simulium ornatum* black flies. One dead cow was estimated to have received ~25,000 bites and the dead heifer received at least 60,000 bites (Noirtin and Boiteux 1979).

Losses in livestock due to myiasis have been substantial. There have been $800 million in losses annually to cattlemen from the southern USA from Florida to California and another $300 million in losses to Mexican ranchers (FAO 1998).

Effects on Wildlife

Biting arthropods can affect wildlife populations adversely. An important difference between wild and domestic animals is that for the most part, if the biting pressure becomes too great, wild animals can move whereas domestic animals generally are confined. As might be expected, biting arthropods have a greater impact on young animals than on healthy adults. Nestling birds are particularly at high risk to biting arthropods. High ectoparasite populations cause egg neglect and nest abandonment by adult birds and are directly responsible for the death of nestlings. In Florida, mourning dove chick mortality was observed on several occasions due to large numbers of bloodfeeding *Culex nigripalpus* mosquitoes.

Attacks by biting flies and other arthropod parasites force birds to devote a greater portion of their time budgets to avoidance and grooming. Cotgreave and Clayton (1994) studied 62 different bird species and found that the average bird spent 9.2% of the day in maintenance activities, 92.6% of which involved grooming. Bird species known to harbor parasitic lice spent more time grooming than did birds without lice. Swallows have been observed to abandon colonial nesting sites under bridges when high densities of swallow bugs (Cimicidae) in their nests affected the fitness and survival of their young.

In addition to grooming, wild animals have the option of leaving habitats, home and breeding ranges or nests when biting pressure becomes too high. The distribution, microhabitat choice and behavior of mountain caribou (*Rangifer tarandus*) in the Burwash area of southwestern Yukon, Canada, are influenced by *Aedes* mosquitoes and oestrid flies, especially *Cephenemyia trompe* and *Hypoderma tarandi*. Caribou reduce mosquito harassment by vertical migration up mountain slopes to elevations and temperatures where mosquito numbers and activity are reduced. Caribou also can reduce biting pressure by choosing microhabitats, such

Figure 4.2. Cattle forming a defensive circle to avoid attacks by biting flies (photograph by John D. Edman).

as patches of snow and exposed windy ridges where mosquito numbers are diminished. Recent studies in northern Norway indicate that caribou may stand on snow patches mainly for thermoregulation (Anderson and Nilssen 1998). Oestrids are more difficult for caribou to avoid and affect caribou energy budgets by reducing feeding and resting times and increasing the time spent standing and moving (Downes et al. 1986).

ENTOMOPHOBIA AND ARACHNOPHOBIA

Entomophobia and arachnophobia refer to an unusual and unreasonable fear of insects or spiders and scorpions. Public surveys show that after snakes, spiders are the most feared of all animals. Why some people fear animals out of proportion to the harm they might inflict is unclear, but such fears may have cultural origins. Biblical stories about serpents and plagues probably contribute to fears in the western world. There also may be a genetic component (Sagan 1977). Motion pictures and television reinforce many of these fears by creating vivid images of large deadly creatures that resemble the most repulsive looking arthropods (Fig. 4.3). Most rational people learn to deal with nature's less-loved creatures in a realistic way. However, others tend to panic, in some cases with tragic results. There are several news reports annually of serious or fatal accidents resulting from people being frightened or stung by bees or wasps inside their automobiles. Educating young people about arthropods and encouraging hands-on experience is the best antidote for such unwarranted fears (Berenbaum 1998).

DELUSORY PARASITOSIS

General medical practitioners, dermatologists, allergists, psychiatrists, medical entomologists and pest control specialists often deal with patients with **delusory** (or **delusional**) parasitosis (DP). Delusory parasitosis is defined as the condition in which patients complain of having the sensation that "bugs" are crawling, biting or burrowing into their skin (Mumford 1982). Arthropods normally are not involved, despite evidence of skin irritation and the fact that certain arthropods can cause symptoms that closely mimic many of those experienced by patients suffering from DP. The challenge for those who encounter patients presenting these symptoms is to quickly separate symptoms that are related to arthropods or some other medical condition from those that are only imagined and require treatment by mental health professionals. In mental health terms, DP is classified as an obsessive-compulsive disorder, a form of psychosis. More information on this disorder can be found in the web site and publication on Delusional Parasitosis produced by the Center for Biosystematics, University of California, Davis.

Many medical conditions, food and skin allergies, and prescription, over-the-counter and illicit drugs may mimic the symptoms described by DP sufferers. For example, insulin, estrogen, arthritis and hypertension medications, cocaine, soaps, latex, cosmetics and dyes all can cause itching, skin irritation, rashes and other sensations similar to those presented by DP sufferers. Medical conditions with symptoms that may mimic DP include diabetes, photosensitivity, thyroidism, hepatitis, and nutritional deficiencies or excesses. Such organic causes obviously demand proper diagnosis and appropriate treatment for a successful resolution. As part of this diagnostic process, it is important to be able to eliminate arthropods as a possible cause. This is a task for which most physicians are ill equipped.

Scabies mites are permanent ectoparasites that can be identified microscopically from skin scrapings taken in a physician's office. Most other skin irritations caused by temporary exposure to arthropods living in or around the home are best diagnosed by medical entomologists or qualified pest control professionals. Thorough inspection of the home, yard and pets of patients should reveal if arthropods such as bed bugs, mites, lice and fleas normally associated with pets, rodents or birds could be responsible for the dermatological symptoms. Generally, if arthropods are the cause, there are telltale signs (e.g., the bite pattern) or circumstances (e.g., only bitten at certain times of the day or in certain areas of the home or work place) that are at variance with the more stereotypic symptoms described by patients suffering from DP.

Familiar patterns of symptoms should alert knowledgeable professionals to the likelihood of DP. Patients usually claim that tiny bugs are biting or burrowing into their skin or body. These attacks normally occur when patients are at home and often while they are asleep. Sometimes attacks are linked to specific articles of clothing, certain activities or, in the case of cleptoparasitosis, to the home itself (Grace and Wood 1987).

The sensation of arthropod contact often results in chronic itching and mental anguish that can lead to self-abuse, including scratching the skin until raw. Self-treatment with bizarre or potentially harmful home remedies such as turpentine, gasoline or agricultural insecticides often follows, especially when proper diagnosis is delayed. Patients may insist adamantly and convincingly that their problems are real. People suffering from DP often appear com-

Figure 4.3. A poster advertising a movie in which insects attack humans. Movies and other popular media can invoke fear of insects and other arthropods well beyond their actual threat (courtesy of The Cabinet Productions).

pletely normal in other respects, a fact that often complicates proper and timely diagnosis.

Neuroleptic drugs, such as Haloperidol and Pimozide, generally relieve DP symptoms, but accompanying antidepressant drugs may be required in some patients (Frances and Munro 1989, Paholpak 1990). There is evidence that DP has a genetic basis (Gieler and Knoll 1990) and that it may be drug-induced (Marschall et al. 1991).

When physicians find no organic cause for DP, they sometimes accept the patient's diagnosis and recommend intervention by a pest control service or a medical entomologist. These individuals are not trained to provide the necessary psychiatric counseling and treatment of DP. Individuals with DP often are shuttled between entomologists and dermatologists without benefit of the mental health treatment that is needed.

ANNOYANCE CAUSED BY ARTHROPODS

Arthropods can cause a number of conditions in human and other vertebrate hosts ranging from annoyance to fatal allergic or toxic reactions. The most important groups of arthropods responsible for annoyance of humans or other animals caused by the bloodfeeding activities of ectoparasitic arthropods or from their crawling or burrowing activities are discussed below. Itching, rashes, welts and other manifestations of allergic reactions often heighten this annoyance. Allergic and toxic reactions to arthropods are discussed in other sections.

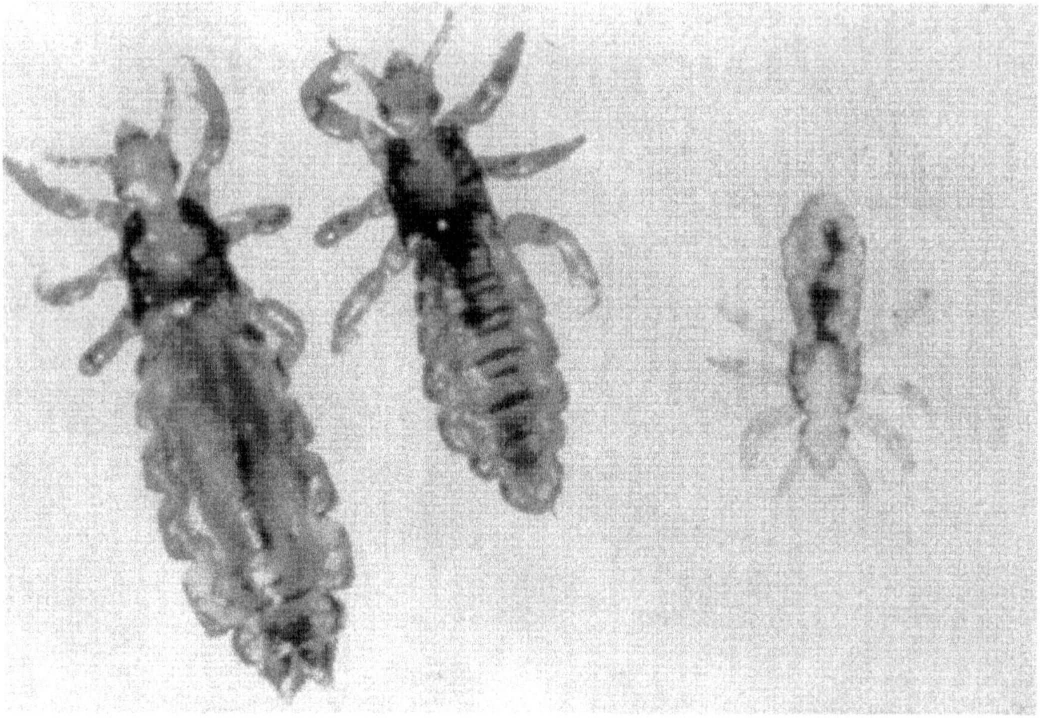

Figure 4.4. The human body louse: From left, adult female, adult male, nymph (courtesy of John D. Edman).

Lice

Lice are usually classified in 2 orders: Anoplura, the sucking lice, and Mallophaga, the chewing lice.

Chewing lice

Chewing lice are associated permanently with their vertebrate hosts but transfer from one host to another may occur when hosts come into close contact. Chewing lice feed on organic fragments of skin and feathers and on epidermal secretions. The chicken body louse (*Menacanthus stramineus*) can gnaw through the skin and feed on blood. Louse infestation in domestic animals may result in significant economic loss. Poultry infested with chewing lice have reduced egg production and weight gain. Sheep, cattle, swine and horses have associated species of chewing lice that, in cases of severe infestation, can cause reductions in body weight, pelt quality and milk production.

Sucking lice

All species of sucking lice feed on blood. The family Haematopinidae is a cosmopolitan group that includes species of lice found on cattle (*Haematopinus eurysternus*), pigs (*Haematopinus suis*) and horses (*Haematopinus asini*). Another cosmopolitan family, Linognathidae, includes species of lice that parasitize dogs (*Linognathus setosus*), sheep (*Linognathus ovillus*

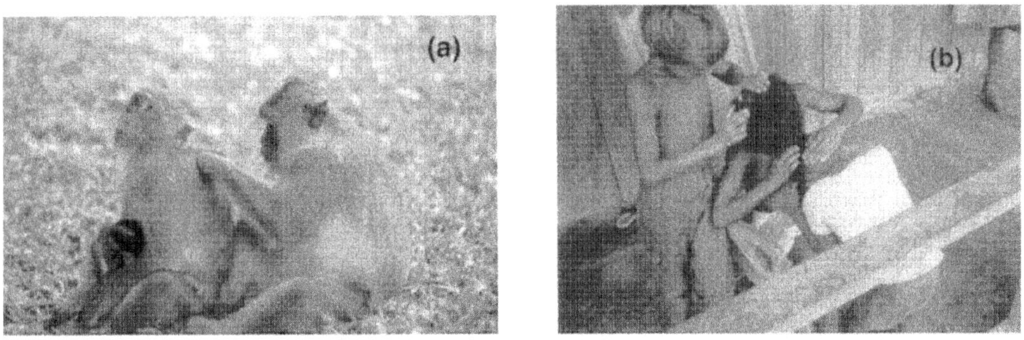

Figure 4.5. Monkeys (a) and humans (b) removing lice from one another (University of Massachusetts collection).

and *Linognathus pedalis*), cattle (*Linognathus vituli*) and goats (*Linognathus stenopsis*).

The family Pediculidae contains 2 subspecies in the genus *Pediculus* that are found exclusively on primates. *Pediculus humanus humanus* is the human body louse or "cootie", and *Pediculus humanus capitus* (head louse) parasitizes both humans and New World monkeys (Fig. 4.4). Morphologically, these subspecies are indistinguishable; however, they clearly are different in terms of behavior, microhabitat preference and vector status. The family Pthiridae includes *Pthirus gorillae*, found on gorillas, and *Pthirus pubis*, the pubic louse of humans.

Head lice, as their name implies, are found primarily on head hair (Fig. 4.5), but occasionally are found on other hairy regions of the body. Body lice feed on body regions where clothing fits tightly and infest the clothing during non-feeding periods. Pubic lice infest the pubic area, armpit and other coarse hair such as eyebrows. All 3 species of lice found on humans are transmitted by direct contact; head lice by sharing hats and combs, body lice by sharing clothing, and pubic lice through shared towels and close body contact. The life cycle of the pubic louse takes about one month to complete, whereas the life cycle of head and body lice is more rapid, with a generation time of about 3 weeks. Body lice attach their eggs (nits) to clothing and head lice attach their eggs near the base of individual hairs.

Pediculosis is the condition resulting from infestation by human sucking lice. The control of lice has been a problem faced by humans for centuries. Egyptians solved the problem by shaving their heads and bodies. Lice combs have been used for at least 3,500 years and remain effective today (Mumcuoglu 1996). Insecticidal treatments for the control of lice also have been used for centuries. Viper broth, herbal remedies, mercury, carbolic acid, cresol powder, naphthalene, sulfur and kerosene are all substances that once were used for the control of lice. Today, the active ingredients in products marketed internationally for the control of lice include organochlorines (lindane), organophosphates (malathion), carbamates (carbaryl) and pyrethroids (permethrin). Even though a large number of insecticides are now available for the treatment of lice, the number of cases has increased worldwide since 1965. In the USA, annual infestation rates number 6–12 million, mostly involving children. The increase in case numbers may be due in part to the large selection of over-the-counter pediculicides that often are ineffective. In laboratory tests, the greatest mortality achieved with pyrethroid products was 80% for adults and nymphs and 30% for eggs (Mumcuoglu 1996). Exposure to sublethal doses of insecticides can select quickly for resistance in populations of lice. Resistance to DDT, lindane, malathion and the pyrethroids have been reported worldwide.

The future control of pediculosis undoubtedly will be tied to better education, the judicious use of combs and development of pediculicides based on new, safe chemicals such as methoprene and avermectins. However, even with the introduction of new control products, lice will continue to be a major problem, especially head lice in children. Public education, personal hygiene and the frequent monitoring of school children for the presence of head lice will remain important components for the control of these pests.

Bed Bugs

These small, flattened hemipteran ectoparasites belong to the family Cimicidae. They are wingless nest parasites that feed on the blood of birds and mammals (Fig. 4.6).

There are 7 species of bed bugs associated with humans throughout the world, but 2 species predominate (Usinger 1966). *Cimex lectularius*, the human bed bug, is a cosmopolitan species, and *Cimex hemipterus*, the tropical bed bug, is found mainly in southern Asia and Africa. Bed bugs are gregarious and are associated most commonly with conditions of poor hygiene and overcrowding. They hide during the day, usually close to human sleeping quarters, and come out to feed at night. They are seldom seen, but emit a characteristic aroma that can permeate a room when the bugs are present in large numbers. Eggs are laid in crevices near daytime resting sites. After hatching, the bugs have 3 nymphal instars, each taking up to 3 bloodmeals. Cimicids have the ability to fast for extended periods when no hosts are available. Their bites usually do not cause immediate pain, an advantage for a species that feeds on sleeping humans. However, by morning, bites generally form reddened wheals that can cause extreme itching and irritation.

A 3-step process is necessary to control bed bugs: (1) launder all clothes and bedding that may contain eggs, nymphs or adults, (2) eliminate all potential resting places and (3) treat houses and furniture with a residual insecticide. Some bed bug populations now show resistance to organophosphate insecticides. During World War II, American troops in Italy continually were harassed by bed bugs until it was discovered that the bugs were living in the hollow supports for their folding cots.

Kissing Bugs

The neotropical subfamily Triatominae (kissing bugs, Fig. 4.7) of the family Reduviidae (assassin bugs) contains more than 5,000 described species of nest parasites that bloodfeed on vertebrates, including humans (Usinger 1944, Schofield 1994). The term "kissing bug" refers to the preference of these species to feed around the lips and eyes where the skin is thin. Like bed bugs, kissing bugs congregate in protected areas in and around bedrooms and feed at night on sleeping humans. Bites are not immediately painful, possibly because an anesthetic is injected along with the saliva. However, some people become hypersensitive to bites and develop severe swelling and itching at the site of the bite (Ribeiro 1995).

Source reduction is the most effective control strategy against reduviid bugs. The elimination of daytime hiding places (e.g., plastering walls, replacing dirt floors with concrete and replacing thatched roofs with sheet metal) all help to reduce the level of infestation. The use of residual insecticides sprayed on interior walls or incorporated into wall paints, and silicone dusts blown into crawl spaces are effective short-term control measures against reduviid bugs.

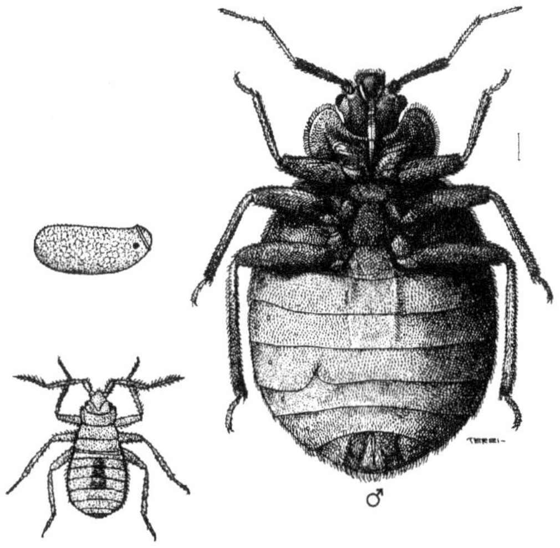

Figure 4.6. Life cycle stages of the common bed bug (from Lane and Crosskey 1993).

Fleas

Fleas are nest parasites belonging to the order Siphonaptera. Most occur on mammals, usually rodents. Their bilaterally compressed body and posteriorly directed heavy spines are adaptations that allow them to move quickly through dense hair and feathers. Adults of both sexes are exclusively bloodsucking ectoparasites that usually remain on hosts only long enough to obtain a bloodmeal. Many flea species live in animal nests and burrows during most of their adult lives.

Some flea species are host-specific (monoxenous). For example, the flea *Hystrichopsylla schefferi* is found only on beavers in the Pacific northwestern USA. Most flea species are polyxenous, attempting to feed on virtually any vertebrate host that comes within jumping range. Some species have been collected from as many as 20 different host species. Flea species that are important pests of humans and domestic animals are found in the families Pulicidae and Tungidae. Members of the family Pulicidae (approximately 180 species) are found worldwide. Important species include: *Xenopsylla cheopis* (the Oriental rat flea), *Pulex irritans* (the human flea), *Ctenocephalides canis* and *Ctenocephalides felis* (dog and cat fleas) and *Echidnophaga gallinacea* (the sticktight flea). The family Tungidae contains 10 species that are mostly tropical. One, *Tunga penetrans*, with local names such as jigger, chigoe and sandflea, (Fig. 4.8a) appears to burrow under the skin of its host. In fact, they do not burrow, but rather induce host tissue changes that cause the skin to swell and envelop them. They usually attach to human feet in areas where the skin is thin (Fig. 4.8b).

Because dogs and cats share our homes, so do their fleas. Dog and cat fleas do not hesitate to attack humans when their normal host is absent. Usually, the main flea on both cats and dogs in the New World is *C. felis*, while in the Old World the dog flea commonly is found on both cats as well as dogs. Adult fleas can remain quiescent in their cocoons for long periods and emerge synchronously in response to vibrations and other cues produced by the entrance of a potential host. This is an obvious adaptation that prevents adult fleas from emerging when there is no host in the "nest" and explains why homes may literally fill with hungry adult fleas after the occupants return from an extended absence.

Mites and Ticks

Ticks, mites, scorpions and spiders belong to the Class Arachnida and possess neither antennae nor mandibles. Ticks and mites are small arthropods belonging to the subclass Acari, which lack the cephalothorax of scorpions and spiders. Their life cycle consists of 4 stages: the egg, a hexapod larva, an octopod nymph (but with multiple nymphal instars in many mites)

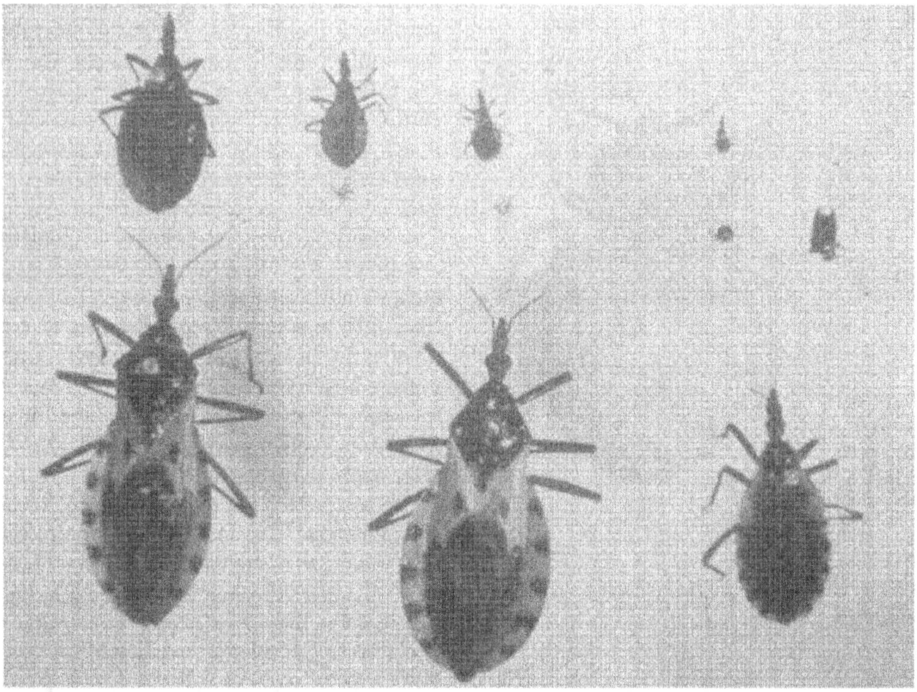

Figure 4.7. Various life cycle stages of the kissing bug, *Rhodnius prolixus* (from Tipton 1974).

and an octopod adult. Some mite and tick species have a wide-ranging effect on the health of humans and domestic animals.

Mites in the family Sarcoptidae are permanent ectoparasites of mammals and birds. These mites spend most of their life cycle buried under the host's skin, resulting in scabies or mange (a condition characterized by scabby eruptions and loss of hair) in mammals (Fig. 4.9a) and "scaly leg" in birds (Fig. 4.9b). *Sarcoptes scabiei* (the itch mite; Fig. 4.9c) is responsible for scabies in humans and sarcoptic mange in a variety of other mammals (Arlian 1989b). A second species, *Notoedres cati*, causes mange in cats and occasionally infests humans (Fig. 4.9d). Mites that cause mange generally are host-specific.

Sarcoptes scabiei has infested humans since antiquity (Mellanby 1972, Burgess 1994). Epidemics of human scabies occur approximately in 30-year cycles that are separated by 15 years (Orkin and Maibach 1978). Human infestations with *S. scabiei* are common, especially among homeless individuals and elderly invalids who receive poor care. Skin folds on the hands and elbows are favored points of entry for scabies mites in humans. As few as 20 mites can produce an intense immune response and associated itching. "Crusted," or "Norwegian" scabies results in individuals who have infestation rates of >1,000 mites. This condition produces scaly, crusted skin and infested individuals are extremely contagious. Mites are transferred from host to host by close and prolonged per-

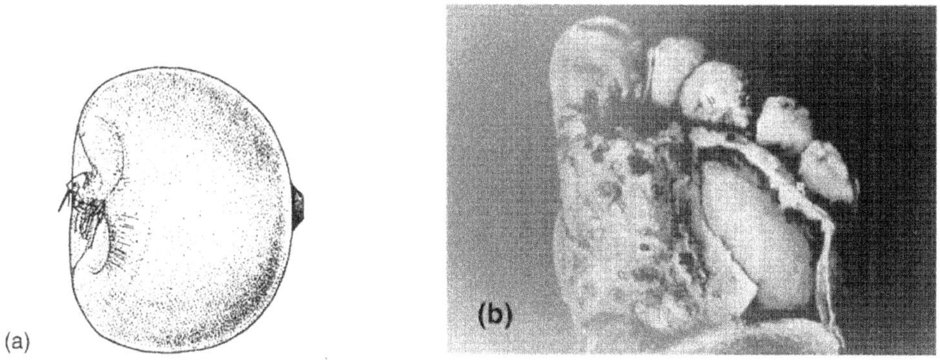

Figure 4.8. (a) The chigoe flea, *Tunga penetrans* (from Lane and Crosskey 1993); (b) damage to foot due to chiggoe flea infestation (courtesy of John D. Edman).

sonal contact. Infestations are treated with prescription skin lotions that contain an acaricide such as one of the pyrethroids.

Epizootics of scabies in animal species other than humans have been reported frequently (Arlian 1989b). There are a number of *S. scabiei* varieties that are considered to be host-specific (Arlian 1996). During a scabies infestation, gravid females burrow into the host's dermis and oviposit. The act of burrowing is painful and uncomfortable, but it is the host's allergic reaction to antigens produced by these mites that causes the most discomfort.

Heavy infestations of mites can result in lower productivity in domestic animals and changes in behavior and reduced survivorship in wild animals. For example, *S. scabiei*-infested red foxes in Norway were less active and traveled shorter distances than did healthy foxes (Overskaug 1994). Psoroptic mange in domestic and wild mammals also can have a serious impact (Thorne et al. 1982).

Most humans have a tiny hair follicle mite (*Demodex* spp.) that infests the sebaceous glands on their face and sometimes other regions (Fig. 4.10). These normally are benign but may cause skin acne in some individuals. *Demodex* mites can cause severe mange in certain other mammals, especially canids.

Chigger mites in the family Trombiculidae are cosmopolitan in distribution, especially in disturbed habitats. The 6-legged larval stages are parasites of mammals and birds (Fig. 4.11). Other stages are free-living. Larvae are extremely small, measuring only 0.25 mm long, and thus are difficult to see on host animals. Of the approximately 1,200 known species, about 20 attack humans. In the process of feeding, they may cause severe dermatitis.

Chigger larvae feed continuously on the host's epidermis by forming a feeding tube (stylostome) that extends into the skin from the point of attachment. Larvae feed by sucking liquefied host tissue that has been predigested by salivary enzymes injected by the chigger. Some humans are highly sensitive to chigger salivary allergens and severe reactions may result from chigger feeding. Mites resting in vegetation climb onto passing hosts and usually crawl up the host and begin to burrow in areas where clothing fits tightly (e.g., at the tops of socks, the backs of knees and the bands of undergarments). Many home-remedies, such as treating entry wounds with nail polish, are common folklore. However, by the time the allergic reaction begins, mites usually have dropped off. For humans that venture into chigger habitats, their best defense is to wear protective clothing

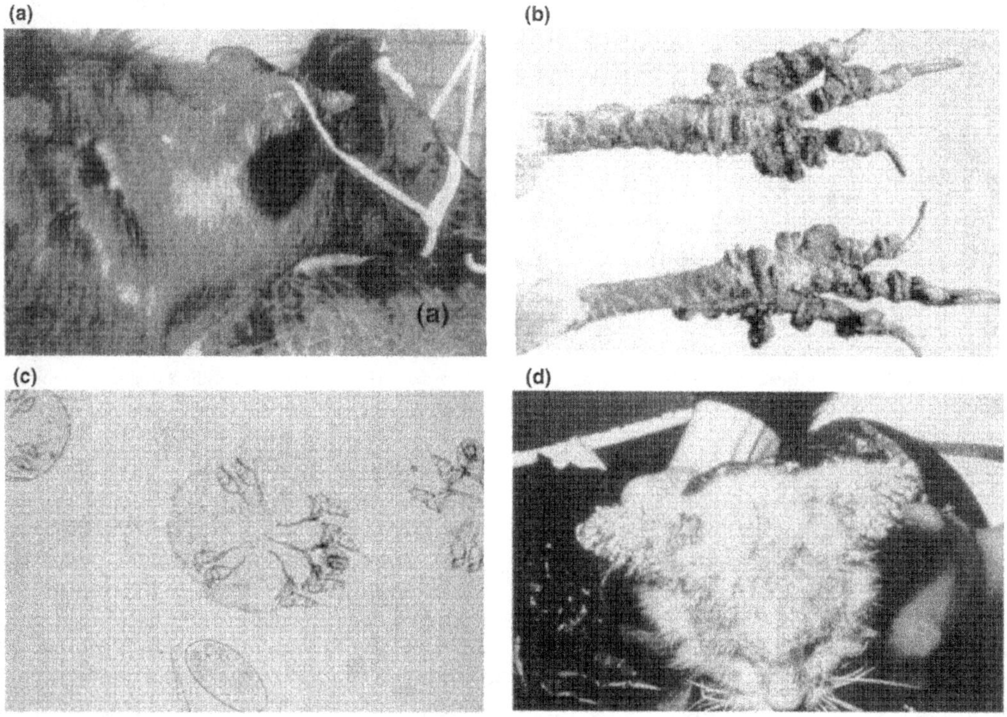

Figure 4.9. Mange and scabies: (a) Sarcoptic mange in a cow; (b) scaly leg in a chicken; (c) the scabies mite, *Sarcoptes scabiei*; (d) infestation in cat with *Notoedres cati* (courtesy of John D. Edman).

(boots and long pants), tuck pants into socks and spray boots and pants with a repellent containing at least 15% N,N-diethyl-meta-toluamide (DEET) as the active ingredient.

Mite infestations usually can be diagnosed by their location on the body and the appearance of the skin reactions. Inspecting the skin for burrows first identifies suspected scabies infestations; confirmation is made by identifying mites microscopically in skin scrapings. Once the infestation is confirmed, the entire body from the neck down is treated with insecticidal lotions or ointments such as 0.5% malathion, 25% benzyl benzoate (usually restricted to adults), 10% N-ethyl-o-crotontoluide (Crotamiton), sulfur ointment or tetraethylthiuram monosulphide (Tetmosol).

Handling wild birds (e.g., young birds that have fallen from their nest) and the nesting material of birds or rodents may result in accidental mite infestation. Chicken mites (*Dermanyssus gallinae*) infest wild birds and domestic poultry and readily bite humans when removed from their normal hosts. Similarly, the tropical rat mite (*Ornithonyssus bacoti*) occurs wherever rats are abundant (e.g., seaports, zoos, granaries and storehouses) and will feed on humans when rodents are unavailable. Humans frequenting fields and meadows also may accidentally encounter mites for which they are not the normal host. European harvest mites

(*Neotrombicula autumnalis*) infest small mammals and birds, but will bite humans who enter infested fields, especially during late summer and early autumn. Starved predatory grain or hay itch mites (*Pyemotes tritici*) attack people who come into contact with raw agricultural products.

Ticks (Fig. 4.12) belong to the suborder Ixodida that contains about 800 species in 3 families: Ixodidae (hard ticks), Argasidae (soft ticks) and Nuttalliellidae, a family containing the single species *Nuttalliella namaqua*. Only about 80 species of ticks feed on domestic animals, but these are responsible for significant economic loss both in developed and undeveloped countries. European breeds of cattle are particularly susceptible to blood loss and damage due to feeding ticks.

Ticks are particularly annoying because they remain attached for long periods of time, take large bloodmeals and are difficult to remove. Tick bites, especially those of nymphs and larvae, often are not noticed until after the tick has detached when there may be a severe toxic or allergic reaction at the bite site. Granulomatous lesions can form as a result of tick bites, especially at sites where clothing binds tightly to the skin. Tick bites sometimes are accompanied by intense itching and may take months to heal.

When a bloodfeeding tick is found, it usually is difficult to remove because the mouthparts, including a barbed hypostome that anchors it, are cemented firmly in the host's skin. There are numerous folklore remedies for tick removal ranging from hot matches to gasoline. However, the safest and most effective method is to grasp the tick as close to the skin surface as possible with angled forceps and exert a steady pulling motion until the tick is extracted. The residual wound should be treated with a topical antibiotic and covered.

Tick paralysis in humans, dogs, cattle and other animals including birds, sometimes results from tick engorgement. This can be a rapidly fatal, but easily reversible condition. The condition in humans is most common in children and has been associated with engorgement near the spinal column by at least 43 species of ixodid ticks and several *Argas* species worldwide. It is caused by the injection of a neurotoxin in the saliva during engorgement. The neurotoxin disrupts nerve synapses and blocks neuromuscular junctions. Ticks causing paralysis remain attached for long periods of time, sometimes more than a week. During that period, the host slowly develops fatigue, extremity numbness, fever and muscular pain. Paralysis is ascending, migrating rapidly from the legs up and can result in respiratory failure if the attached tick is not found and removed. Once the tick is removed, recovery usually is rapid. In humans, tick paralysis is associated most commonly with bites by *Dermacentor andersoni, Dermacentor variabilis* and *Amblyomma americanum* in the USA; *Ixodes holocyclus* in Australia; and *Ixodes rubicundus* in South Africa.

Lachrymal and Unusual Bloodfeeding Arthropods

Eye-moths in the families Pyralidae, Geometridae and Noctuidae feed on lachrymal secretions (moisture associated with the eyes, nose and mouth) of mammals, including humans. Slight pain is associated with feeding by these moths and eyes can become infected and inflamed from bacterial contamination. Some male moths bloodfeed. Bloodfeeding moths can be categorized by 2 basic feeding strategies: those that pierce the skin and those that scrape the skin to obtain blood. The noctuid moth *Calyptra eustrigata* from Southeast Asia has a strong proboscis that can pierce the skin of a number of mammalian species (Bänziger and Buttiker 1969).

Lachrymal and sweat-feeding flies in the family Muscidae are well-known nuisance pests. Those in the genus *Hydrotaea* feed on lachrymal secretions and lymph or even blood around fresh wounds. Some *Musca* species (in-

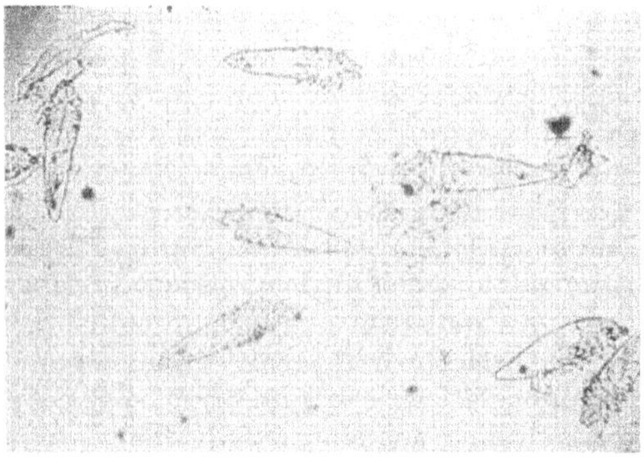

Figure 4.10. *Demodex*, the follicle mite (courtesy of John D. Edman).

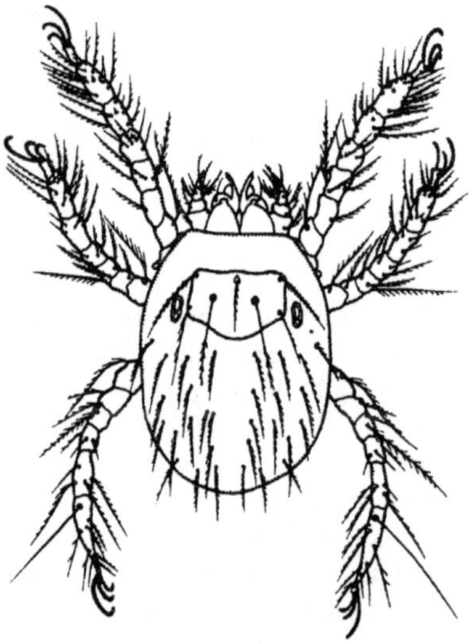

Figure 4.11. A chigger mite, *Leptotrombidium akamushi* (from Lane and Crosskey 1993).

cluding *Musca sorbens, Musca vetustissima* and *Musca autumnalis*) cause great annoyance by feeding on open sores, cuts and body secretions. Similarly, *Hippelates* eye gnats (family Chloropidae) are small (2 mm), shiny black Diptera that commonly feed on the lachrymal secretions of a variety of mammals, including humans.

ALLERGIC REACTIONS TO ARTHROPODS

Interactions of arthropods and their antigens with the vertebrate immune system are complex and varied. The term "allergic reaction," although convenient, reflects only one aspect of the immunologic relationship among vertebrates, arthropods and arthropodborne infectious agents. Allergic reactions include immediate and delayed responses and resulting sequelae to various arthropod venoms, saliva and inhaled or ingested antigens. Arthropod bites may stimulate a sequence of cutaneous reactions that change with exposure history.

Host-arthropod immunologic interactions also include development of host acquired resistance to infestation and arthropod circumvention of host defenses. Bloodfeeding insects, ticks and scabies mites stimulate development of immunity to infestation. Due to the longstanding and intimate association of many ectoparasitic arthropods with their hosts, it is not surprising that arthropods have developed countermeasures to host immune defenses. Immunosuppression of the host facilitates ectoparasite infestation and contributes to the successful transmission of vectorborne pathogens. Immunosuppressive proteins in arthropod saliva have been identified, purified and characterized for a few arthropods (Wikel and Bergman 1997). This emerging area of investigation is providing new insights into the role of arthropod modulation of host defenses in the facilitation of pathogen transmission (see Chapter 5).

Discovery that animals infested with biting arthropods can acquire resistance to repeated infestations has stimulated efforts to develop anti-arthropod vaccines. The first commercial anti-tick vaccine is now on the market. Novel technologies are being used to develop vaccines that not only limit the ability of arthropods to infest hosts, but also specifically target factors

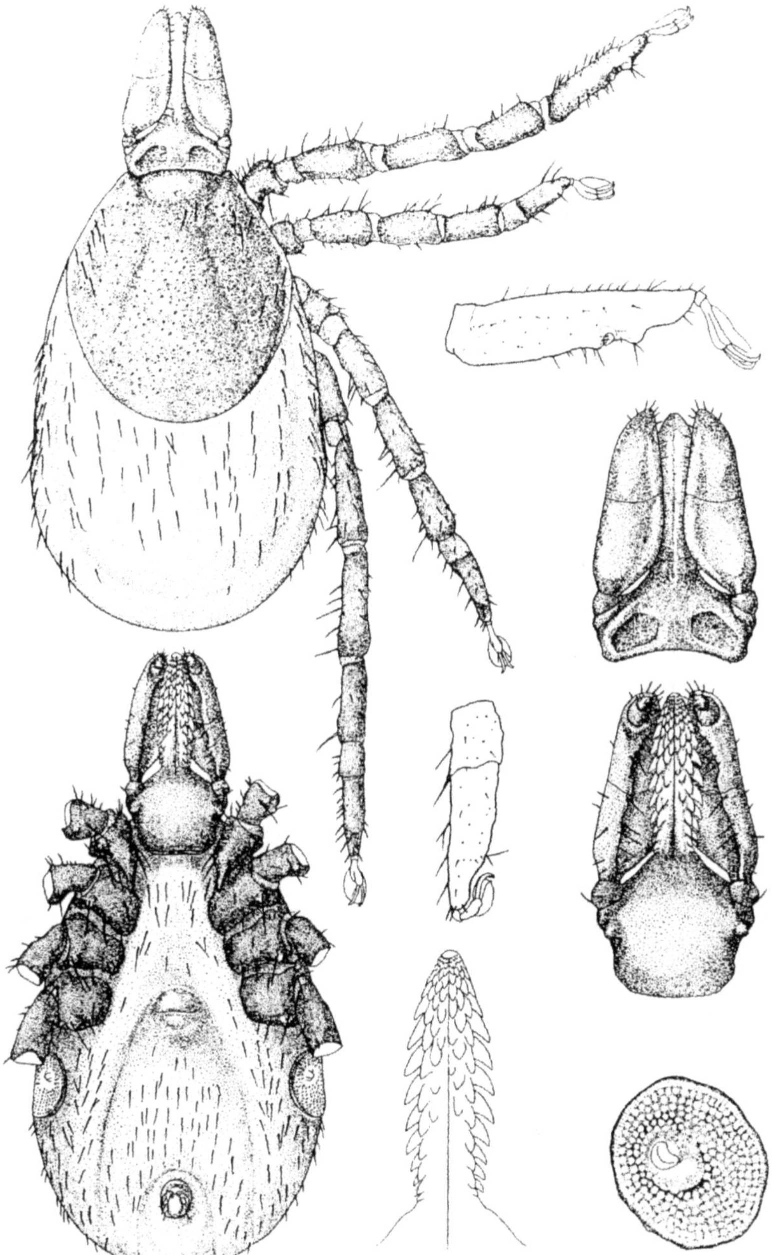

Figure 4.12. Female adult *Ixodes persulcatus*. Dorsal and ventral views plus details of hypostome and other structures (from US Army 1957).

essential for pathogen transmission (Wikel and Bergman 1997). Advances in immunology, molecular biology, biotechnology and vaccine research are enhancing understanding of the immunology of host-vector-pathogen interactions, and fostering development of new strategies to control arthropods and the diseases they transmit.

Allergic and Toxic Reactions to Insect Venoms

Stings, bites or other contact with arthropods can result in dermatitis, severe neurologic or cytologic symptoms, localized or systemic allergic reactions and secondary bacterial infections. Insects in the order Hymenoptera (bees, wasps and ants) most commonly are associated with venomous reactions. A local inflammatory response to the insect's venom is the most common cause of discomfort. A less common result is induction of a systemic allergic reaction that causes some individuals to become sensitized, usually following repeated encounters with a particular venom. Two basic manifestations of immunity to arthropods are immediate type hypersensitivity (ITH) and delayed type hypersensitivity (DTH). In DTH, activated macrophages respond to the antigen challenge by phagocytizing particulate antigens, secreting mediators that promote local inflammation and secreting cytokines and growth factors that facilitate immune clearance of the antigen. The term "delayed" refers to the fact that at least 24 hr elapse between the antigen challenge and the immune response.

In ITH, there often is a rapid release of mediators, such as histamine, causing dilation of local blood vessels and increased vascular permeability. Immediate (within an hour) swelling (wheal) and engorgement of vessels with blood (flare) is caused by escape of blood plasma into the surrounding tissue. Immediate type hypersensitivity in humans is the cause of allergies and asthma. Immediate systemic hypersensitivity is called anaphylaxis, a condition that can prove fatal if not treated rapidly. Human and animal deaths have resulted from stings by fire ants, honey bees and other stinging bees, hornets and wasps.

Female Hymenoptera are the only insects that possess a true stinger that has evolved from a modified ovipositor. Stinging insects include yellow jackets and New World hornets (*Vespula*), Old World hornets (*Vespa*), wasps (*Polistes*), honey bees (*Apis*), bumble bees (*Bombus*) and fire ants (*Solenopsis*). Honey bees, yellow jackets, bald-faced hornets, paper wasps and imported fire ants are the most common causes of allergic reactions to insect stings in North America. Honey bees leave their stingers in the skin, whereas other Hymenoptera can sting repeatedly. Fire ants use their mandibles to anchor themselves to their victim and sting multiple times while pivoting around the attachment site.

Honey bee venom contains approximately 50 mg of solids per droplet, which is thought to be the entire contents of the venom sac. Other Hymenoptera deliver smaller doses of venom protein per injection. Constituents of honey bee venom include histamine and other vasoactive substances, melittin (a polypeptide with detergent activity), apamin (a neurotoxic polypeptide), and the enzymes phospholipase A2, hyaluronidase and acid phosphatase (Habermann 1972). There is little, if any, cross-reactivity of honey bee antigens with those of the vespids. Vespid venom includes a major protein allergen (antigen 5) as well as phospholipase and hyaluronidase. Considerable venom cross-reactivity exists among vespids.

Reactions to stings and bites fall along a continuum ranging from virtually no reaction to localized reactions to systemic involvement. Hypersensitivity to arthropod venom ranges from mild, localized cutaneous reactions to severe systemic anaphylaxis and death. Excellent discussions of the etiology, pathogenesis, clinical presentation, diagnosis, and treatment of allergy to stinging insects can be found in reviews by Valentine (1995).

Approximately 50 deaths occur annually in the USA as a result of allergy to stinging insects. This estimate of fatalities undoubtedly is low because many sudden deaths occurring in the outdoors might be attributed inaccurately to other causes. Most deaths from arthropod-induced anaphylaxis result from one or 2 stings (Graft 1996). The absence of a history of sensitivity to insect stings, though rare, does not rule out the possibility of a fatal anaphylactic reaction.

In addition to anaphylaxis, death can result from toxic reactions to multiple insect stings. Death due to toxicity of insect venom occurs most frequently in young children and the elderly. McKenna (1994) estimated that the lethal dose of honey bee venom for 50% of the population (LD_{50}) is equivalent to 19 stings/kg body weight. This translates into 500 stings for the average child, 1,100 stings for the average adult female and 1,400 stings for the average adult male.

The African honey bee (*Apis mellifera adansonii*) from central and eastern Africa was released accidentally into Brazil in 1957. Initially, it was brought in for crossbreeding experiments to improve honey production of European honey bees (*Apis mellifera mellifera*) already there. However, the African queens escaped, mated with drones from local colonies and great concern arose due to the frequent aggressive attacks of hybrid 'Africanized' workers on humans and domestic animals. They spread southward in South America and northward into Central America (~400 km/yr), finally reaching the USA border in 1990. Now found in Texas, Arizona, New Mexico and California, they are expected to spread throughout the mid-temperate regions of the USA (Fig. 4.13). Sometimes erroneously referred to as 'killer' bees because of their aggressiveness, their stings are no more serious than other honey bees. Interestingly, the volume of the venom sac of the Africanized honey bee is slightly less than that of the European honey bee. However, they are more defensive of their hive and are able to mount large and sustained attacks that often result in multiple stings that can lead to serious toxic reactions. Within 6 years of their arrival in Mexico, at least 200 people had died as a result of stings by Africanized workers. Similar increases in deaths in people can be expected in the USA.

The red imported fire ant (*Solenopsis invicta*, Fig. 4.14) was introduced into the USA near Mobile, Alabama in the 1930s. It spread quickly throughout the southeastern states infesting over 9.3 million km^2 by 1985. Fire ants have their most adverse impact on birds that nest on or close to the ground and colonial nesting water birds. There is some evidence that there also may be a beneficial aspect to the introduction of fire ants. They are voracious predators on ticks, horn flies, mosquito eggs and many agricultural pests.

Allergic reactions to Hymenoptera venom can range from localized cutaneous responses at the site of the sting to systemic reactions of varying severity. Type I (Immunoglobin E-, or IgE-mediated) hypersensitivity, involving allergen-induced release of mediators from mast cells and basophils, is the basis for immediate localized and systemic reactions that occur within 1 hr of being stung (Valentine 1995). More prolonged cutaneous manifestations, persisting hours to days after the sting, also are initiated by IgE, and they represent a phenomenon known as the "allergic late phase response" (Lemanske and Kaliner 1993). While approximately 4% of the population of the USA is hypersensitive to Hymenoptera stings, 13% react to fire ant allergens and the rate is 58% in areas where the ants are endemic.

Differing opinions exist regarding diagnosis and treatment of insect sting allergy because the basis for effective immunotherapy is not fully understood. A positive diagnosis of allergy to insect venom is established by a skin test (reactivity and identification of venom-specific IgE) and/or radioallergosorbent test. Selection of specific venoms and the assessment of potential cross-reactions are important for diagnosis. Venoms are best for determining allergy to honey bees and vespids, whereas fire ant hypersensitivity can be tested with whole body extract. Acute sting reactions are treated according to the symptoms. Treatment for local reactions includes ice packs and antihistamines, whereas more extensive local reactions may require a course of corticosteroids. Systemic anaphylaxis is a life-threatening, acute situation involving hypotension, laryngeal edema, severe bronchospasm, increased vascular permeability, a vasodepressor effect and smooth muscle contraction. The treatment of systemic anaphylaxis involves the administration of epinephrine (a vasoconstrictor) and oxygen (Valentine 1995).

Immunotherapy (i.e., desensitization) usually can be used effectively to avoid anaphylaxis from Hymenoptera stings. Multiple mechanisms likely are involved in immunotherapy-induced protection against anaphylaxis. Whole body extracts of fire ants are used for immunotherapy, but are not recommended for honey bee, wasp, hornet and yellow jacket immunotherapy (Valentine 1995, Freeman 1997). Immunization with venom results in a lower incidence of anaphylaxis than does immunization with whole body extracts. A strong positive correlation exists between effective desensitization and the presence of venom-specific IgG "blocking" antibodies, particularly for immunoglobulins of the IgG4 subclass. Honey bee venom can be collected for desensitization by electroshocking live bees.

The most toxic arthropod bites are associated with spiders and scorpions. In terms of human death and sickness resulting from bites, the most dangerous spiders are found in 4 genera: *Atrax* (Dipluridae), *Harpactirella* (Barychelidae), *Loxosceles* (Sicariidae) and *Latrodectus* (Theridiidae). *Atrax robustus* is the Sydney funnel-web spider of Australia. The genus *Harpactirella* of South Africa contains 11 venomous species. *Loxosceles reclusa* is the brown recluse spider (Fig. 4.15a) and *Latrodectus mactans* is the black widow spider (Fig. 4.16). Both are found in North America and both are dangerous to humans in terms of the severity of their bites. There is a northern and southern subspecies of black widow that can be separated by the shape of the red hourglass on their underside. Two other species, the red widow and brown widow, are found only in Florida, and a 4th species occurs only in the western USA. Only female widow spiders, which are 3 times larger than males, inflict venomous bites. Their venom is 15 times more potent (dry weight) than the venom of the prairie rattlesnake. Females inject very small amounts of venom with each bite, so widow bites are rarely fatal. However, they can result in severe discomfort, with symptoms including abdominal pain and hardness, hypertension, muscular pain, agitation and irritability. When all symptoms are present or some symptoms persist (>12 hr) anti-venom treatments are recommended, especially in children.

Physical contact with arthropod body parts, hemolymph, poisonous spines and defensive secretions can cause contact dermatitis and local irritation. Conjunctivitis can be caused by beetles in the family Staphylinidae that swarm in large numbers and accidentally enter eyes. Once trapped in the eye, they release defensive secretions that cause intense burning and sometimes temporary blindness known as "Nairobi eye." Larvae of some beetles (Dermestidae), butterflies (Morphoidae and Nymphalidae) and moths (Arctiidae, Bombycidae, Eucliidae, Lasiocampidae, Limacodidae, Lithosiidae, Lymantriidae, Megalopygidae, Noctuidae, Notodontidae, Saturniidae, Sphingidae and Thaumetopoeidae) have urticating, or "nettling," hairs capable of penetrating the skin and causing a histamine release and dermatitis. Urticating hairs sometimes are connected to poison glands. Simply brushing against these

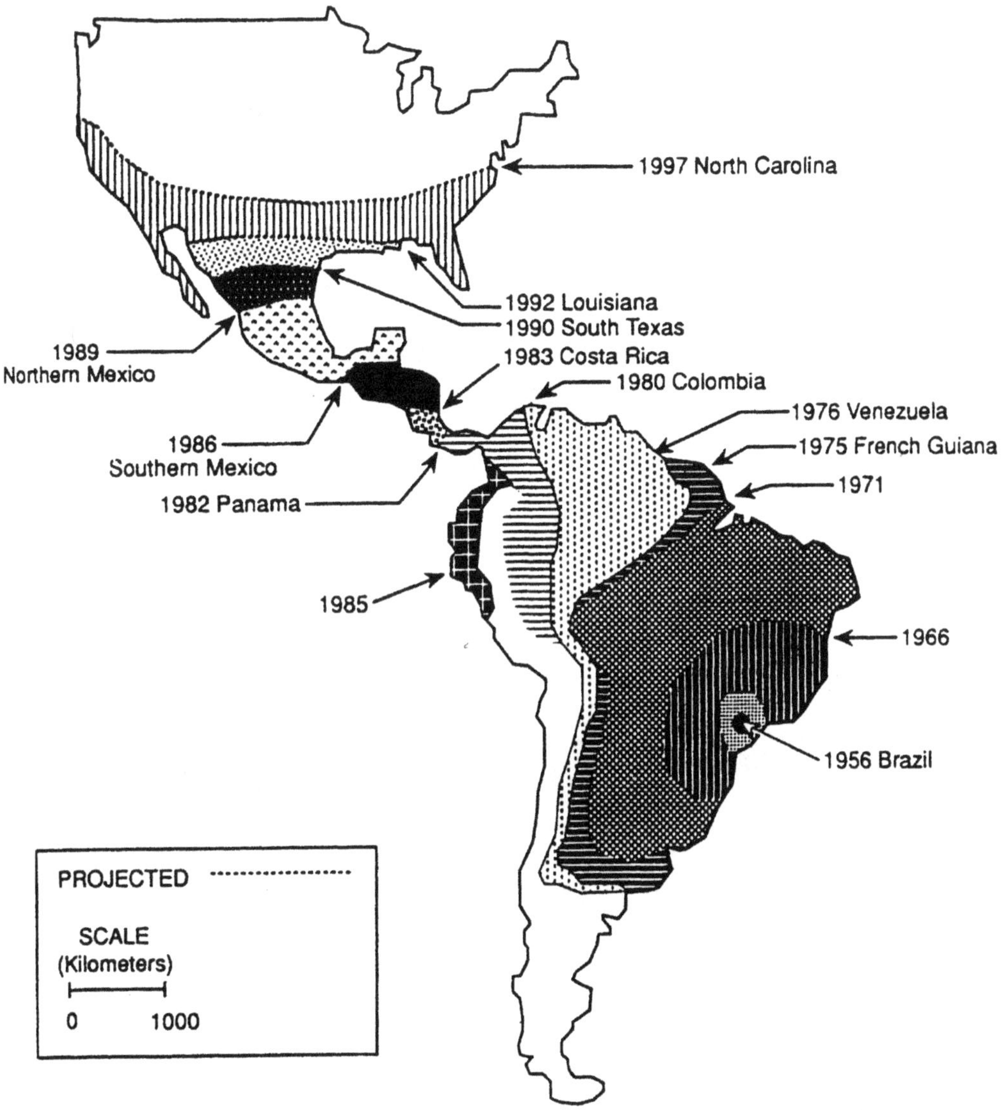

Figure 4.13. Distribution map of Africanized honey bees (from Winston 1992).

spines causes the release of toxin into the skin, and this often results in an intense burning sensation that may persist for more than an hour. Urticating hairs also are attached to cocoons. Hairs retain their urticating properties long after being shed by the insect. If urticating hairs

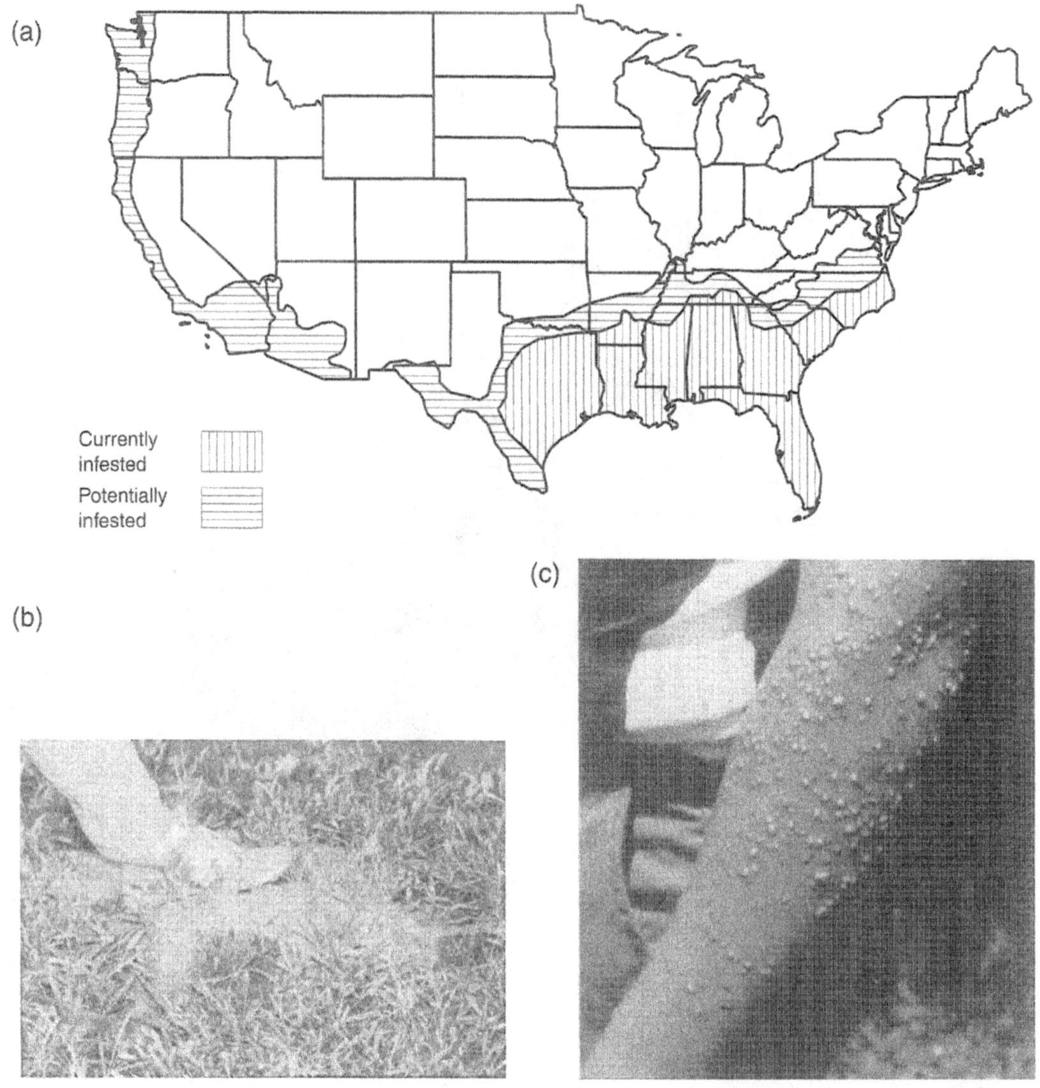

Figure 4.14. Red imported fire ants: (a) Map of USA actual and projected distribution (adapted from USDA data); (b) mounds surrounding nest (photograph by Les Greenberg); (c) blisters caused by fire ants (courtesy of John D. Edman).

are inhaled, they may cause breathing difficulty, and if ingested, can cause inflammation of the mouth and throat.

The body fluids of blister beetles (Meloidae) contain vesicating substances that cause skin blisters. These body fluids are an important source of cantharidin (the so-called Spanish fly), which is used as a counterirritant for skin blisters and erroneously as an aphrodisiac. Digestive tract irritation (colic) in livestock can result from the ingestion of crushed beetles in alfalfa and other forage. In Zimbabwe, the misuse of aphrodisiac preparations from crushed *Mylabris alterna* has resulted in death. Body fluids from larval beetles in the families Chrysomelidae and Carabidae are used as lethal arrow poisons in South Africa.

Millipedes (Diplopoda) are primarily herbivorous arthropods with 2 pairs of legs on each segment of their wormlike bodies. However, they possess defensive secretions in specialized glands on their body. Some species are able to expel these secretions several centimeters when threatened. Secretions include hydrogen cyanide, aldehydes, esters, phenols and quinonoids that can cause burning and blistering of skin and conjunctivitis when introduced into the eyes.

Airborne Arthropod Allergens

A variety of molecules derived from arthropods are potent allergens that induce a variety of respiratory symptoms when inhaled (Mathews 1989). The importance of cockroaches and free-living mites in the stimulation of house dust allergy is well established, and specific allergens have been characterized (Thien et al. 1994). Mites of the genus *Dermatophagoides* are linked to IgE-mediated respiratory illness. Species commonly responsible for the allergens of house dust allergy include *Dermatophagoides pteronyssinus*, *Dermatophagoides farinae* and *Euroglyphus maynei* (Arlian 1989a).

House dust mites (Fig. 4.17) are ubiquitous in dwellings. The highest populations are associated with high humidity in areas such as bedrooms, where sloughed human skin scales are available as food. The number of mites varies considerably within homes and are found most commonly in mattresses, overstuffed furniture and loose pile carpets. Low humidity, especially during the winter, will reduce mite populations. Species of house dust mites differ in their prevalence in geographic regions and among homes within a region (Arlian 1989a).

An important advance that facilitated the characterization of *Dermatophagoides* allergens was the colonization of these mites. This allowed the isolation of cDNA that coded for Der p I, a major allergen of *D. pteronyssinus*. *Dermatophagoides farinae* contains a cysteine protease allergen, designated Der f I (Heymann et al. 1986). Der p I and Der f I are known as Group I allergens. Group II allergens do not appear to possess enzymatic activity, and their nature remains to be fully characterized. Group III allergens are serine proteases, whereas Group IV allergens are amylases. The Group I, III and IV allergens of both *D. pteronyssinus* and *D. farinae* are secreted by the mite digestive tract, resulting in high levels of allergen in mite fecal pellets. Due to their size, mite fecal pellets can be dispersed readily into the air and inhaled. Fecal pellets as well as mite bodies contain Group II allergens (Heymann et al. 1989). Immunotherapy with house dust mite allergens is effective in selected patients (Thien et al. 1994).

In addition to mites, house dust allergy frequently is associated with sensitization to cockroach allergens (Kang 1976). The American cockroach (Fig. 4.18), *Periplaneta americana*, and the German cockroach, *Blattella germanica*, are the species to which individuals are sensitized most commonly in the USA (Thien et al. 1994). A number of other insect species have been linked to respiratory allergies (Mathews 1989). Sensitization often results from occupational

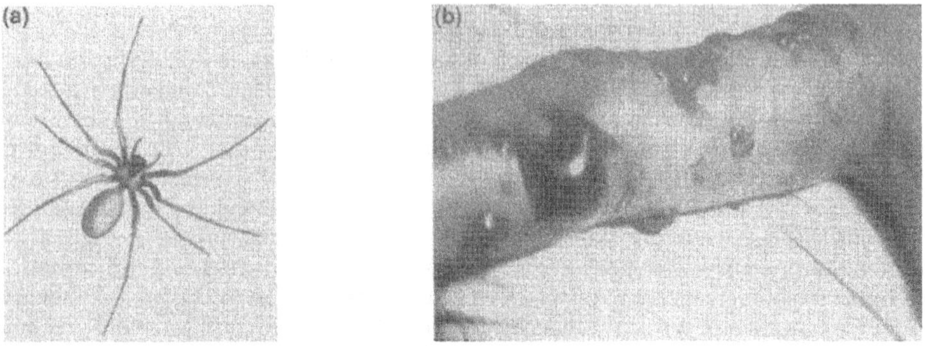

Figure 4.15. The brown recluse spider, *Loxoceles reclusa:* (a) Adult female; (b) lesions on human arm resulting from bite (from Tipton 1974).

exposure to insects in either large numbers or for long periods.

Hypersensitivity to cockroach allergens is the most significant cause of house dust induced respiratory disease in many settings (Kang 1976). The likelihood of sensitization to cockroach antigens is greatest in multifamily, crowded dwellings and lower socioeconomic settings. Cockroach feces, saliva and shed cuticle contain allergens (Platts-Mills 1995). The major allergens identified are Bla g I and Bla g II from the German cockroach, and Per a I from the American cockroach. Allergen Bla g II is a heat labile, 36 kdal protein to which twice as many humans have specific IgE antibodies, when compared with sensitivity to Bla g I. Immunotherapy using aqueous cockroach extracts is effective, but avoidance of the allergens is the best management strategy (Kang et al. 1988).

Mathews (1989) reviewed respiratory allergies induced by inhalation of insect allergens. Specific sensitivities are not rare and reinforce the importance of occupational exposures in insect allergy, particularly among entomologists. Selected examples of occupational related hypersensitivities include the development of rhinitis and/or asthma from exposure to the grain weevil, *Sitophilus granarius*, the fruit fly, *Drosophila melanogaster* and locusts, *Locusta migratoria* and *Schistocerca gregaria*. The bed bug, *Cimex lectularis*, also has been implicated as a causative agent of asthma attacks (Sternberg 1929). Additional insects to which respiratory hypersensitivities have been reported include crickets, grasshoppers, mealworms, silk moths, butterflies, screwworm flies, house flies, sewer flies, box elder bugs, mosquitoes and aphids (Mathews 1989).

Immune Responses to Ectoparasitic Arthropods

Feingold et al. (1968) proposed that immunological reactions to arthropod bites are caused by antigenic substances contained in the arthropod's oral secretions. They suggested that repeated challenge by oral antigens in sequential bites produces an effect characterized by 5 distinct stages: 1) an initial bite, causing no obvious reaction, 2) a delayed skin reaction, 3) an immediate skin reaction followed by delayed reactions, 4) an immediate reaction only and 5) no reaction. Many people who have long-term exposure to bites by specific arthropod species achieve stage 5 and no longer react to the bites of those species.

Significant advances have been made in recent years in understanding the spectrum of

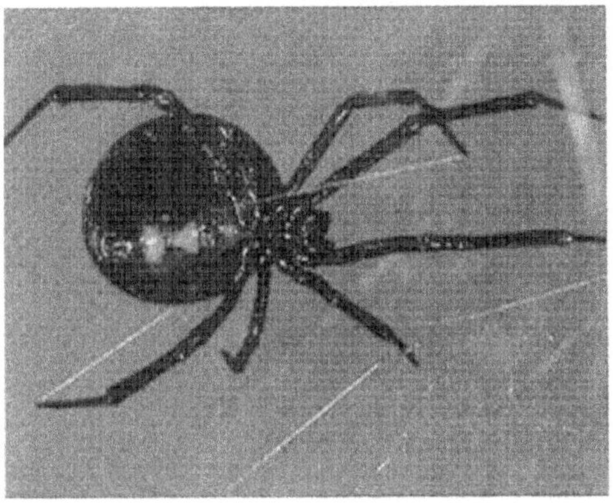

Figure 4.16. The black widow spider, *Latrodectus mactans* (courtesy of Maricopa County Cooperative Extension).

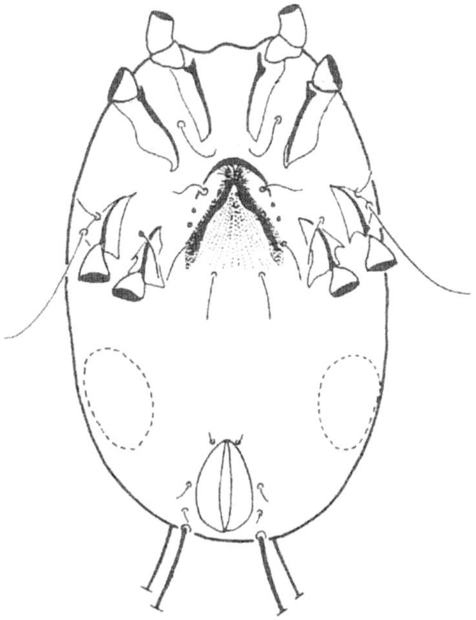

Figure 4.17. The house dust mite, *Dermatophagoides pteronyssinus* (from Krantz 1978).

immune responses induced by ectoparasitic arthropods, particularly bloodfeeders (Wikel 1996). The complexity of host-arthropod immune interactions is increased by the fact that many ectoparasitic arthropod species are capable of modulating host immune defenses (Wikel et al. 1996b). Furthermore, immunosuppression of the host can be an important factor in pathogen transmission, such as in the case of ticks (Zeidner et al. 1996, Wikel and Bergman 1997). In addition to immunosuppressive molecules, ectoparasitic arthropod saliva contains an array of pharmacologically active molecules including anticoagulants, inhibitors of platelet aggregation, vasodilators and pain inhibitors (Ribeiro 1995, Champagne and Valenzuela 1996). Bloodfeeding has arisen independently multiple times among the ectoparasites, so it is not surprising that a diverse array of pharmacologically active molecules are found in the saliva of these species. These molecules contribute to successful feeding and the ability of vector species to transmit disease pathogens. Immune responses associated with individual groups of arthropods are reviewed below.

Lice

Host immune responses to sucking lice have been studied for many years. Repeated infestations of the human body louse results in intense cutaneous irritation and development of systemic symptoms suggestive of anaphylaxis. Immediate and delayed hypersensitivity reactions in sensitive individuals are elicited by skin tests with louse heads and feces. Skin test sensitivity is correlated directly with the intensity of cutaneous reactivity to infestation.

Polyplax serrata, a sucking louse of mice, is a useful experimental model for studying the immunology of host-louse interactions. Mice that were prevented from grooming developed

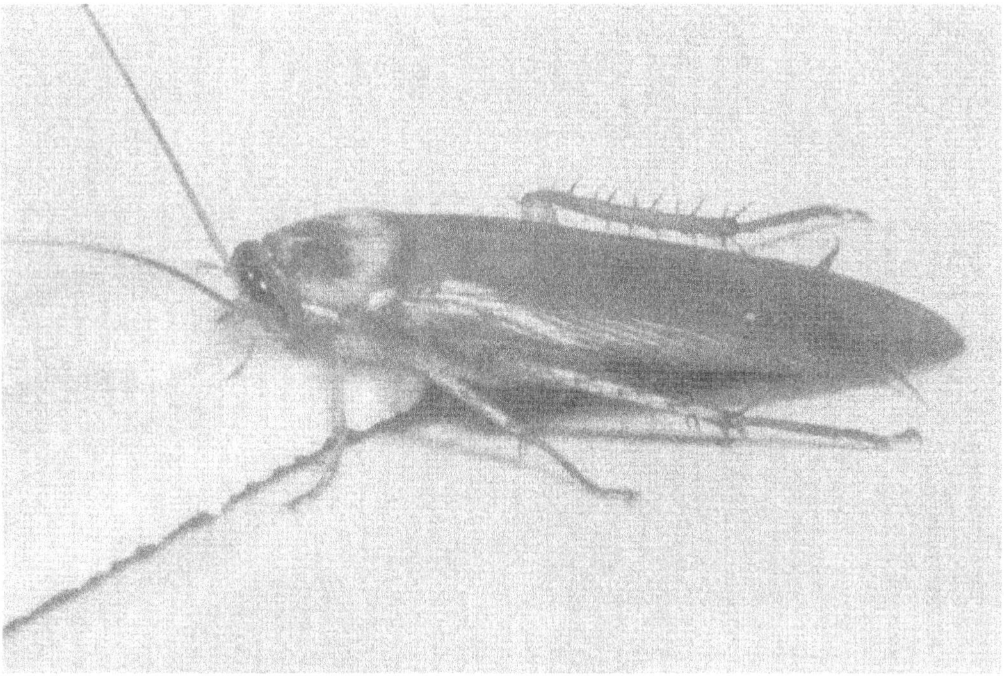

Figure 4.18. The American cockroach, *Periplaneta americanum* (photograph by Daniel R. Suiter).

peak louse burdens by the 4th week of infestation, after which the number of lice declined to few or none (Bell et al. 1966). Development of resistance to infestation appeared to be correlated directly with the length and intensity of infestation. Resistance to reinfestation with *P. serrata* was limited to the site of prior louse feeding, because naive sites on previously infested mice were susceptible to engorgement. Lice allowed to feed on previously uninfested cutaneous areas on resistant mice were reduced by 59% when compared to susceptible controls. Systemic acquired immunity to *P. serrata* was developed by mice within 50 days of infestation (Ratzlaff and Wikel 1990). Systemic anti-louse immune responses were responsible for acquired resistance. One week after initial infestation, lymphocytes of infested mice proliferated *in vitro* when cultured in the presence of *P. serrata* extract, indicating immune recognition of louse immunogens. A distinct sequence of skin test reactivity to louse antigens developed during the course of a primary infestation. Delayed reactions were elicited by day 19 of infestation and both immediate and delayed reactions developed to skin testing by day 34. Continued infestation resulted in a loss of skin test reactivity.

True bugs

Hemipteran saliva has been characterized extensively for anti-platelet, vasodilatory and anticoagulant molecules (Ribeiro 1995), but not for immunogens. Cimicid bug bites cause the sequential development of the 5 stages of cutaneous reactivity previously described (Feingold et al. 1968).

Fleas

Host-flea immunological interactions were among the first studied in detail. A progressive pattern of skin responsiveness was described for guinea pigs repeatedly exposed to bites from cat fleas, *Ctenocephalides felis* (Benjamini et al. 1961). The sequence of host reactions is similar to that described by Feingold et al. (1968). In animals exposed to flea bites daily for 10 days and then twice weekly for several months, immediate hypersensitivity only (Stage 4) occurred after 60–90 days of exposure. Fleas sensitize and elicit host immune reactions involving antibodies and specifically sensitized T-lymphocytes.

Characterization of flea immunogens is of practical importance, not only in understanding flea allergy, but also for developing strategies for desensitizing individuals who are hypersensitive to flea bites. Cat flea allergens are hypothesized to be haptens that complex with host components, especially collagen (Benjamini et al. 1963, Michaeli et al. 1965). The biological activities of these allergens and their potential usefulness in control of flea hypersensitivity remain to be determined.

Mosquitoes

The role of mosquito saliva as the source of allergens first was established by cutting the main salivary duct and then allowing mosquitoes to feed on sensitive individuals (Hudson et al. 1960). Bites of females with severed ducts did not elicit cutaneous hypersensitivity, whereas bites of mosquitoes with intact salivary ducts did elicit skin responses. Mellanby (1946) described 5 stages of cutaneous hypersensitivity to *Aedes aegypti* bites, similar to those described for bed bugs and fleas. Immediate and delayed type hypersensitivity reactions were observed after mosquito bites (Wilson and Clements 1965). These cutaneous reactions indicate the involvement of mosquito specific IgE for immediate hypersensitivity, circulating IgM and/or IgG for the Arthus reaction, and specifically sensitized T-cells mediating delayed type hypersensitivity responses. Individuals with immediate hypersensitivity to *Aedes albopictus* had serum antibodies that reacted intensely on immunoblots (Shan et al. 1995). Severe systemic anaphylaxis and even death can occur from mosquito bites (Ohtaki and Oka 1994). Cutaneous and serological cross-reactivity among mosquito species have been observed for bite sensitive individuals (Peng and Simons 1997).

Other Diptera

Cutaneous hypersensitivity is caused by bites of the midge, *Culicoides imicola* (Ungar-Waron et al. 1990), the tsetse, *Glossina morsitans centralis* (Ellis et al. 1986), the black fly, *Simulium vittatum* (Cross et al. 1993b) and the horn fly, *Haematobia irritans* (Kerlin and Allingham 1992). Antibody and cell-mediated immune responses are stimulated by feeding flies. An important feature of host immune reactivity to dipterans is that the responses elicited do not protect the host from further bites. Molecules essential for the feeding success of the fly are poorly immunogenic, perhaps to assure continued fly survival. Genetic composition of the animal being attacked is also an important factor in determining the nature of the immune response to fly bites, and most flies feed on a wide variety of animal hosts.

Ticks

Host immune responses to ticks have been studied more than any other arthropod ectoparasite-host relationship. The initial stimulus for investigating tick-host immune interactions was the observation that vertebrate hosts acquired resistance to infestation (Willadsen 1980, Wikel 1982a). A practical implication is that resistance to ticks might be inducible by immunization. Acquired resistance to tick infestation is characterized by reduced engorgement, prolonged feeding, diminished egg production, inhibition of molting and death (Wikel et al. 1996a). Many experiments focusing on host immune responses to ticks have utilized unnatural vertebrate-tick associations. This has been a source of criticism and concern. However, hosts also acquire immunologically-based resistance to ticks infesting them in nature. Laboratory mice respond to *Ixodes ricinus* infestation in a manner similar to mice in nature (Mbow et al. 1994). The black vole (*Clethrionomys glareolus*) is a natural host of *Ix. ricinus* and still acquires resistance to infestation (Dzij and Kurtenbach 1995). An advantage of laboratory animal studies of tick-host immune interactions is the availability of reagents to study specific mediators and cell populations, inbred lines of animals and the availability of "knockout" mice lacking genes encoding specific elements of the immune system.

Early investigations of host immunity to ticks focused on the histology of bite sites on susceptible and resistant hosts. Guinea pigs and cattle expressing acquired resistance developed intense basophil accumulations at tick attachment sites (Allen 1973, Allen et al. 1977), which had the characteristics of cutaneous basophil hypersensitivity (CBH) responses. Eosinophils are a prominent component of the cellular influx surrounding the imbedded mouthparts of ticks attached to vertebrate animals with acquired resistance (Allen 1973). Resident mast cells and infiltrating basophils possess Fc receptors for IgE and other classes of antibodies. These are capable of interacting with specific immunogens introduced into the bite site in tick saliva, resulting in the release of bioactive mediators and the influx of eosinophils (Brossard and Fivaz 1982). Passive administration of anti-basophil or anti-eosinophil serum results in reduced resistance to infestation with the lone star tick, *Amblyomma americanum*, indicating a role for both cell types in acquired resistance (Brown et al. 1982). Histamine, an important component of basophil and mast cell granules, causes the disruption of tick feeding (Paine et al. 1983).

Circulating antibodies in vertebrate animal hosts contribute to acquired resistance. Antibodies for acquired resistance to infestation by a number of tick species, including *Dermacentor variabilis, Ixodes holocyclus, Boophilus microplus* and *A. americanum*, can be transferred passively. Guinea pigs develop increasing levels of salivary gland-specific antibodies during the course of repeated infestations with *Dermacentor andersoni* (Whelen et al. 1986). These antibodies are important in resistance expression (Whelen and Wikel 1993).

Host immunity to ticks is not entirely dependent on antibodies. T-lymphocytes have regulatory and effector functions in acquired resistance. Resistance has been adoptively transferred with viable lymphocytes (Wikel and Allen 1976). The CBH reactions elicited by ticks feeding on resistant hosts are a form of delayed hypersensitivity, and are mediated by Th1 lymphocytes (Mosmann and Coffman 1989). Furthermore, *in vitro* cultures of peripheral blood or splenic lymphocytes proliferate in the presence of tick salivary gland molecules (Brossard et al. 1991). Use of the T-lymphocyte specific immunosuppressant, Cyclosporin A, resulted in a significant reduction of immunity to *Ix. ricinus*. (Girardin and Brossard 1989).

The contribution of T-lymphocytes to the regulation and expression of acquired resistance to ticks is complex and dynamic, changing during the course of infestation. Delayed type hypersensitivity responses and the production of IgM and IgG are orchestrated in part by IL-2, IFN-G and other cytokines elaborated by the Th1 subpopulation of T-helper cells. Th2 cytokine IL-4 contributes to the development of the homocytotropic antibody response induced by tick feeding. Significant gaps remain in our understanding of acquired resistance to ticks, particularly the identity and characteristics of immunogens responsible for antibody and cell-mediated responses. Construction of salivary gland cDNA libraries and application of the tools of molecular biology should result in significant advances in this area.

Mites

Several ectoparasitic mite species have been studied in relation to their interactions with host immune systems. These include: *Psoroptes ovis*, *Psoroptes cuniculi*, and *Sarcoptes scabiei*. A comprehensive review of the literature on host responses to mange mites (other than *S. scabiei*) and chiggers has been prepared by Wrenn (1996).

Psoroptes *spp.*

Sheep scab, caused by *P. ovis*, was recognized as a serious disease of sheep prior to 180 BC (Kirkwood 1986). *Psoroptes ovis* cause great economic loss, and extensive efforts have been made to eradicate this mite. *Psoroptes ovis* and *P. cuniculi* (rabbit ear mite) are not reproductively isolated (Wright et al. 1983). Host erythrocytes are ingested by both *P. ovis* and *P. cuniculi* (Wright and DeLoach 1980), but their mechanisms of feeding are not well defined. Host sensitization arises from exposure of host skin to mite antigens introduced during feeding (Kirkwood 1986). Rabbits infested with *P. cuniculi* develop mite specific antibodies that are thought to be responsible for the skin lesions. Cattle infested with *P. ovis* develop circulating antibodies that bind components of extracts of both *P. ovis* and *P. cuniculi*, suggesting the presence of shared epitopes (Fisher and Wilson 1977). Development of dermatitis caused by *P. ovis* by Hereford heifers restricted from grooming was correlated to the presence of circulating antibodies (Purett et al. 1986). Cattle previously infested with *P. ovis* develop resistance to a second infestation, during which mite populations are greatly reduced (Stromberg and Fisher 1986). Host grooming is an important factor limiting the number of mites infesting cattle (Guillot 1981). The development of immunologically-mediated cutaneous hypersensitivity would contribute significantly to the cutaneous sensations that stimulate grooming.

Sarcoptes scabiei

The most extensively studied mite-host immunological relationship is scabies and humans (Arlian 1989b). Scabies mites reside in the epidermis and burrow from the bottom of skin creases by the action of the mouthparts and the blade-like claws of the first and second legs (Burgess 1994). Mouthparts of adult female *S. scabiei* can penetrate the spinose layer of the epidermis. The constant growth of the epidermis results in the burrow and larvae are found in the cornified layers of the epidermis. The burrow eventually is lost as the keratinized layer of the epidermis is sloughed off. Mites deposit immunogens that stimulate and elicit host immune responses in the burrow. Immunogens are found in mite saliva, feces, cuticle, eggs, feces and dead mites.

Arlian (1996) reviewed the immunology of scabies. The sequence of cutaneous immune reactions occurring during the course of scabies is similar to that described for bites by bed bugs, mosquitoes and fleas. The induction, or sensitization, phase lasts 4–8 weeks, depending on the intensity of infestation. During this time there is an absence of cutaneous hypersensitivity. The delayed hypersensitivity response to infestation involves antigen presenting cells, specifically sensitized T-cells and inflammatory cells.

Immediate hypersensitivity responses are mediated by mite specific IgE, or other immunoglobulin isotypes, that are bound to mast cells situated around blood vessels in the dermis. With the onset of immediate hypersensitivity, the characterization of persistent or delayed reactions becomes more complicated due to the phenomenon of the late phase allergic reaction, which can persist for more than 24 hr (Lemanske and Kaliner 1993). Late phase allergic reactions involve the infiltration of granulocytes and mononuclear cells over a period of hours into the site of an immediate hypersensitivity response. Eventually, individuals become specifically unresponsive to *S. scabiei* immunogens. Unresponsiveness (anergy) to mite antigens could be occurring in individuals with crusted (Norwegian) scabies. This would explain the lack of cutaneous irritation and the large numbers of mites experienced by these patients. Circulating IgM and IgG immunoglobulins are induced by *S. scabiei* infestation (Arlian 1996). Circulating antibodies are not dependent upon an IgE response. During infestation, serum IgA levels were found to be lower than after successful treatment (Hancock and Ward 1974). IgE, IgM and IgG levels are increased during infestation, while IgA is lowered (Morsy et al. 1993) and complement levels are not changed. Antibodies and complement levels appear to contribute to the cutaneous pathology of scabies.

The first *S. scabiei* infestation induces resistance to reinfestation. Primary infestation of human volunteers was characterized by development of peak mite populations after 8–15 days, followed by a rapid decline in mite numbers (Mellanby 1944). A second infestation resulted in a rapid onset of cutaneous irritation, and the mite population was only a fraction of that reached during the first exposure. Subsequent investigations of acquired immunity to scabies established the importance of cell-mediated immunity in protection, i.e., the predominance of Th1 responses in resistant hosts, and the dominant Th2 reactivity of susceptible individuals (Arlian 1996). These results indicate that a cell-mediated, pro-inflammatory response is associated with resistance to scabies, while a predominantly immediate hypersensitivity response is associated with susceptibility to infestation.

Arthropod Modulation of Host Immunity

The survival of bloodfeeding arthropods is threatened by the hemostatic and immune defenses of vertebrate hosts. Problems of host immune defenses are particularly important for arthropods such as ticks that feed for long periods of time. Short-term bloodfeeders also are faced with host immunity stimulated by repeated feedings over time. Therefore, it is not surprising that ectoparasitic arthropods can modulate host defenses. Disease pathogens possess a variety of mechanisms for circumventing host defenses (Marrack and Kapler 1994, Kotwal 1996). Many of the immunosuppression strategies employed by pathogens also are used by ectoparasitic arthropods.

A dynamic balance occurs between the immune defenses of the host and arthropod-mediated immunosuppression that likely evolved to facilitate feeding. Saliva of hematophagous arthropods contains anticoagulants, anti-platelets and vasodilatory molecules, all of which are essential for a continuous supply of blood while the arthropod is feeding (Ribeiro 1995, Champagne and Valenzuela 1996). Pathogens appear to have taken advantage of this situation to facilitate their own transmission. Inhibiting immunity to the point where hosts are overwhelmed by commonly encountered microorganisms would deprive the arthropod of essential food and/or habitat. Thus, a balance is achieved between the parasite and host defenses. Arthropod modulation of host defenses is a topic of current research interest and will enhance understanding of host-arthropod relationships and vectorborne pathogen transmission.

Diptera are capable of suppressing host innate and specific acquired immune responses. These reductions in host defenses facilitate successful bloodfeeding or migration through host tissues for myiasis larvae. Furthermore, the immune defenses suppressed by the arthropod are those that would be involved in the host response to arthropodborne disease agents. Therefore, arthropod suppression of host immunity provides a favorable environment for transmission and establishment of vectorborne pathogens (Wikel 1999).

Salivary gland extracts of the black fly, *S. vittatum*, modulate host immunity (Cross et al. 1993a). These extracts reduce the expression of molecules essential for effective presentation of antigens to T-lymphocytes, decrease the ability of T- and B-lymphocytes to proliferate and inhibit production of specific cytokines. In essence, *S. vittatum* has the ability to suppress key host defense mechanisms.

Infectivity of *Leishmania major* in the vertebrate host is enhanced by saliva of the sand fly, *Lutzomyia longipalpis* (Titus and Ribeiro 1990). This enhanced infectivity can be attributed in part to the fact that sand fly saliva inhibits the ability of macrophages to effectively participate in the development of a specific T-cell-mediated immune response to *Le. major* (Theodos and Titus 1993). Furthermore, saliva of the Old World sand fly, *Phlebotomus papatasi*, inhibits the production of macrophage molecules that are capable of killing intracellular pathogens (Hall et al. 1995). The importance of these observations is that disease agents transmitted by these flies are deposited into a bite site where host defenses against them are reduced.

Mosquitoes also inhibit host innate and specific immune defenses. Salivary gland extracts of female, but not male, *Ae. aegypti* reduces the release of pro-inflammatory cytokines from mast cells (Bissonnette et al. 1993). The importance of this observation is that mast cells are found around small blood vessels from which female mosquitoes will be attempting to obtain a meal. The presence of biologically active molecules released from mast cells has the potential to disrupt the normal events associated with bloodfeeding. By blocking mast cell mediator release, the female mosquito provides a more favorable environment in which to dine. Furthermore, salivary extracts of female *Ae. aegypti* reduce pro-inflammatory cytokine production by T-lymphocytes (Cross et al. 1994). Reduced inflammatory/immune responses are to the advantage of the female mosquito as she attempts to feed.

Tissue-migrating fly larvae are in intimate contact with host immune defenses so that any reduction in host immunity would provide a more favorable environment for their development within host tissues. Larvae of the warble fly, *Hypoderma lineatum*, degrade host complement levels (Boulard 1989), inhibit T-lymphocyte proliferation (Chabaudie and Boulard 1992) and reduce T-cell production of the cytokine IL-2 (Nicolas-Gaulard et al. 1995). *Phaenicia cuprina* larvae, which infest sheep, have the ability to reduce host antibody responses (Kerlin and East 1992), degrade host antibodies of the IgG class (Sandeman et al. 1995) and inactivate complement (O'Meara et al. 1992). Clearly, these myiasis larvae have the ability to counteract elements of the host defenses that would impair their ability to survive.

Tick modulation of host defenses has been investigated more thoroughly than other ectoparasitic arthropod-host relationships. Readers interested in an in-depth discussion of this topic should consult the reviews by Wikel and Bergman (1997) and Wikel (1999). Because ticks feed over a period of days, host immune defenses are a particular threat to their survival. Ticks have developed countermeasures to avoid these host defenses. Saliva of *Ix. scapularis* inhibits activation and biological activity of specific components of the complement system (Ribeiro and Spielman 1986, Ribeiro 1987). Tick infestation reduces the ability of infested animals to develop an antibody response to a foreign antigen (Wikel 1985). Inhibiting the ability of the host to make antibodies and the inhibition of complement action can directly facilitate tick feeding and reduce the host defensive response to any infectious agent introduced by the engorging tick.

In addition to reducing the antibody arm of the immune response, infestation with *D. andersoni* (Wikel 1982b) or *Ix. ricinus* (Ganapamo et al. 1996) reduces T-lymphocyte responses. Furthermore, infestation with these tick species or *in vitro* exposure of cells to salivary gland extracts, modulates host T-lymphocyte cytokine responses (Ramachandra and Wikel 1992, Ganapamo et al. 1996). Tick suppression of these specific aspects of host cell-mediated immunity is induced by specific proteins produced by tick salivary glands (Urioste et al. 1994, Bergman et al. 1995).

Tick immunosuppression of the host influences transmission and/or establishment of the tickborne pathogen *Borrelia burgdorferi*. Cytokines passively administered to mice suppressed transmission by feeding *Ix. scapularis* nymphs infected with *B. burgdorferi* whereas control mice were not protected (Zeidner et al. 1996). In addition, mice repeatedly infested with pathogen-free *Ix. scapularis* nymphs were resistant to the transmission of *B. burgdorferi* when subsequently infested with infected ticks (Wikel et al. 1997). The basis for resistance to tick-transmitted infection following repeated infestation with pathogen-free ticks appears to be linked to development of a host response that reduces the intensity of tick-induced host immunosuppression.

Anti-Arthropod Vaccines

Anti-arthropod vaccines are based in part on the observations that hosts acquire natural resistance to infestation by some arthropods. Furthermore, arthropod-derived molecules not introduced into hosts during feeding can be used to stimulate anti-arthropod immunity. The most striking success to date is development of a vaccine to control *B. microplus* ticks. This vaccine consists of a recombinant digestive tract surface membrane glycoprotein, Bm86 (Tellam et al. 1992). Significant advances are being made toward the development of a vaccine to control myiasis by the sheep blowfly, *P. cuprina* (Sandeman 1996). Although attempts to develop anti-insect vaccines have been variable, advances being made in vaccine research, molecular biology and biotechnology certainly will further efforts to develop commercially viable anti-arthropod vaccines in the future.

MYIASIS

Myiasis is the invasion of vertebrates by dipteran larvae that feed on living or dead tissue and fluids or on the ingested food of the host. Such invasions may be asymptomatic; however, they can be severe and result in death if left untreated. Myiasis may be **obligatory** when the parasite is dependent on the host to complete its development, or **facultative** when the larva, though normally free-living, can under certain circumstances adapt itself to a parasitic existence. **Accidental** myiasis results from inadvertent ingestion of fly larvae or eggs. Tables 4.2, 4.3 and 4.4 present common species of Diptera associated with each of these types of myiasis.

Zumpt (1965) hypothesized that Diptera causing myiasis evolved independently along 2 routes: the saprophagous and the sanguinivorous. Modern species that followed the saprophagous route developed through 3 steps: (1) carrion- or excrement-feeding larvae invaded hosts through diseased or ill-smelling wounds; (2) healthy tissues, usually those in contact with the necrotic ones, were invaded; (3) finally, the parasite became obligatory and malign, requiring healthy tissues for development. In some species, the final step still has not been bridged thoroughly. For example, the sarcophagid fly *Wohlfahrtia vigil* is an obligatory parasite in its earlier stages and usually throughout its larval development. However, it can survive as a final instar after its host has died.

Diptera evolving along the sanguinivorous route began as predaceous larvae that pierced the cuticle and sucked the hemolymph of other insects that shared their habitat in excrement or carrion. Maggots then proceeded to attack vertebrates as bloodsuckers (e.g., the Congo floor maggot, *Auchmeromyia senegalensis*) and finally became warble-forming or body-invading parasites.

James (1969) discussed the probable evolution of the 3 obligatory screwworms: the primary screwworm (*Cochliomyia hominivorax*), the Old World screwworm (*Chrysomya bezziana*) and *Wohlfahrtia magnifica*, compared to their scavenger-facultative counterparts *Cochliomyia macellaria* (the secondary screwworm), *Chrysomya megacephala* and *Wohlfahrtia meigenii*. Strains of the common green-bottle fly, *Phaenicia sericata*, include one harmless enough to be used safely in maggot therapy of wounds and osteomyelitis, while another is economically important as a sheep wool maggot. More in depth reviews of myiasis can be found in Harwood and James (1979), James (1947), Zumpt (1965), Morikawa (1958), Scott (1964), Lee (1968), (Broce 1985) and Kettle (1990). The last-named author also provided an economic analysis.

Accidental Enteric Myiasis

About 50 species of fly larvae that may invade the human digestive track and cause enteric myiasis belong to the genera *Musca*, *Calliphora*, *Sarcophaga*, *Fannia*, *Eristalis* and *Drosophila*. Usually no development of these larvae occurs within the human host. Infestation is caused by ingestion of food or liquids contaminated with larvae or eggs. Enteric myiasis as a pathological condition in humans is discussed in detail by Riley (1939), James (1947) and West (1951). They contend that true enteric myiasis may occur occasionally when chemical and physical conditions within the host alimentary tract favor parasite survival. Leclercq (1974) presented a concise account of enteric myiasis and its associated dipterans. Zumpt (1965) does not consider enteric infestation in humans to be true myiasis because the ingested larvae normally do not feed while they complete their development in the human digestive tract. He sees this as the ability of the fly maggots to resist an extremely unfavorable environment, rather than an adaptation to a facultative form of parasitism. An obligatory type of enteric myiasis does occur in the Gasterophilidae that naturally invade the digestive tract of herbivorous mammals.

Accidental enteric myiasis (Table 4.2) sometimes may cause severe clinical symptoms in humans. Symptoms vary with the larval species, number or location within the digestive tract. Severe infestations may cause depression and malaise in the patient. Vomiting, nausea, vertigo, abdominal pain and diarrhea with bloody discharge may occur as a result of injury to the intestinal mucosa. Diptera involved in human enteric myiasis include the house fly (*Musca domestica*), the little house fly (*Fannia canicularis*), the latrine fly (*Fannia scalaris*), the false stable fly (*Muscina stabulans*), the cheese skipper (*Piophila casei*), the black soldier fly (*Hermetia illucens*) and the drone fly (*Eristalis tenax*).

Some fly larvae invade the intestine via the anus, which may explain reported cases of intestinal myiasis caused by *E. tenax*, *F. scalaris*, *F. canicularis*, *M. stabulans* and certain species of *Sarcophaga*. These larvae are excrement feeders and may complete their development in the host's rectum or terminal part of the intestine. Parasitism of this type may occur in humans, usually small children or bedridden adults, living under filthy conditions, or in domestic animals that are partially paralyzed or otherwise helpless.

Facultative myiasis may occur when a species that normally is a saprophage or carrion feeder becomes adapted to a parasitic existence (Table 4.3). *Phaenicia cuprina* is a facultative parasite that is benign in many parts of the world but has become an important wool maggot in Australia since the introduction of sheep.

Urinary Myiasis

Urinary myiasis occurs when the bladder and urinary passages are invaded by Diptera larvae. This may cause obstruction, pain, pus, mucus, blood in the urine and a frequent desire to urinate. *Fannia canicularis* is the species most frequently involved, although *F. scalaris*, *M. domestica*, *M. stabulans*, the ephydrid *Teichomyza fusca* and other species have been associated with urinary myiasis. Infestation probably occurs at night in warm weather when people (usually females) sleep without covering. Discharges from diseased reproductive organs, soiled or unbathed pubic areas or use of unsanitary toilets may stimulate oviposition and infestation of the host.

Cutaneous Myiasis

Larvae that normally develop in carrion or processed meats may become involved in traumatic cutaneous myiasis. Blow flies (Calliphoridae), including species such as *Calliphora vicina*, *Phaenicia sericata*, *P. cuprina*, *Lucilia illustris*, *Lucilia caesar*, *Phormia regina*, *C. macellaria* and several species of *Chrysomya* are involved most commonly. Other related families of flies that may produce this type of myiasis include sarcophagid species such as *Sarcophaga haemorrhoidalis* and the phorid *Megaselia scalaris*.

Species such as *P. regina* and *P. sericata* may be attracted to neglected, suppurating, malodorous wounds, especially if the patient is incapacitated. Considerable pain may accompany the invasion of these maggots. Hall et al. (1986) described a case of human myiasis caused by *P. regina*. *Phaenicia sericata* is often a benign human wound parasite, feeding primarily on necrotic tissues; however, healthy tissues may be invaded. Greenberg (1984) described 2 cases of human myiasis caused by *P. sericata* in Chicago hospitals. There is evidence that *Chrysomya rufifacies*, while having larvae that usually are predatory, may be an important pest in some areas. Richard and Gerrish (1983) recorded the first confirmed human case of myiasis produced by *C. rufifacies* in the continental USA; Baumgartner (1993) discussed its taxonomy, distribution, bionomics and medical/veterinary importance. He stated that late instars are beneficial as predators of the maggots of pathogen-transmitting and myiasis-producing flies, but strains from Australia, India and Hawaii can cause secondary myiasis. In Hawaii, *C. rufifacies* produces an unusual myiasis in newborn calves in which larvae eat the epidermis, causing death by dehydration. Other *Chrysomya* species may be either primary or secondary parasites.

The secondary screwworm of the Americas, *Cochliomyia macellaria*, may be a secondary wound invader of some consequence, particularly in domestic animals. However, much of the damage attributed to this fly in early literature was caused by its close relative, the primary screwworm, *Cochliomyia hominivorax*. A similar situation occurs in the Old World, where *Chrysomya bezziana* was responsible for much of the traumatic myiasis previously attributed to *Chrysomya megacephala*.

Wool Maggots

Although wool maggot, or fleeceworm, infestation is most serious in Australia, it also is important in certain other parts of the world. Frogatt (1922) suggested that prior to the introduction of cattle and sheep into Australia, these blow flies existed as simple scavengers. Losses from wool maggots in Australia, due primarily to *P. cuprina*, were estimated at $28 million during the 1969-1970 season. *Phaenicia cuprina* is also the most important species in South Africa. *Phaenicia sericata* replaces *P. cuprina* in importance in some areas, such as in Scotland. In the USA, *P. sericata*, *P. regina* and *C. macellaria*

Table 4.2. Some common species of Diptera associated with accidental myiasis in humans (adapted mainly from James 1947).

Family and species	Type of myiasis	Distribution
Syrphidae		
Eristalis tenax	Enteric (anal)	Cosmopolitan
Ephydridae		
Teichomyza fusca	Urinary	Palearctic, Neotropical
Stratiomyidae		
Hermetia illuscens	Enteric	Nearctic, Neotropical, Palearctic, Australian
Piophilidae		
Piophila casei	Enteric	Cosmopolitan
Muscidae		
Fannia scalaris	Enteric (anal), urinary, auricular	Palearctic, Nearctic, Ethiopian
Fannia canicularis	Enteric (anal), urinary, auricular	Palearctic, Nearctic, Neotropical, Ethiopian, Australian
Musca domestica	Enteric, urinary, nasal, auricular, general tissue	Cosmopolitan
Muscina stabulans	Enteric (anal), urinary	Palearctic, Nearctic, Neotropical, Ethiopian, Australian
Stomoxys calcitrans	Enteric, general tissue	Cosmopolitan

are responsible for most attacks on sheep (known as sheep strike. Another important species is *Calliphora stygia* in New Zealand and to a lesser extent in Tasmania and Australia. A sometimes important species native to Australia is *Calliphora augur*. *Ophyra rostrata* can be of secondary importance in South America.

Surgical Maggots

The role of maggots in healing was discovered as early as 1799 and was utilized first during the US Civil War (Miller 1997). Although now mainly of historical interest, the use of sterile maggots in wound therapy was practiced commonly from 1931 until the advent of sulfa drugs and antibiotics a decade later. One of the authors (SEK) continues to receive inquiries about the availability of *Phaenicia* maggots for surgical purposes and they still appear to be used on a limited, but not well advertised, basis to treat certain bone and other deep injuries. One of the first known natural antibiotics, allantoin, was isolated from blow fly larvae invading human wounds. This discovery gave credence to observations during the Civil War that injured soldiers with maggots in their wounds seemed to suffer less gangrenous bac-

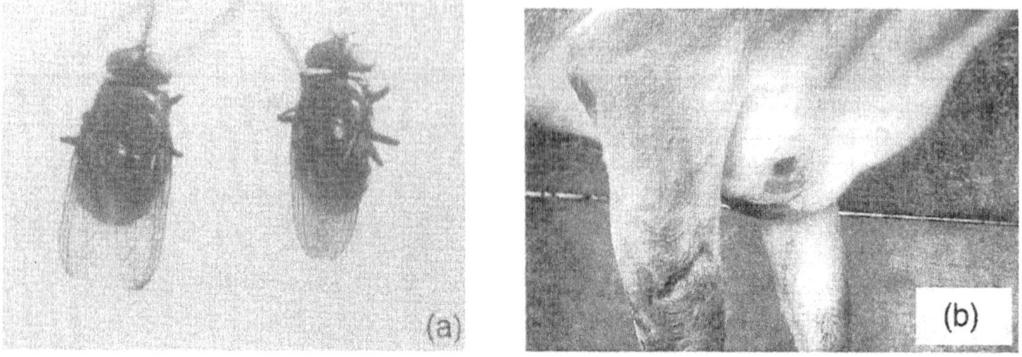

Figure 4.19. (a) The New World primary screwworm, *Cochliomyia hominivorax*; (b) cattle damage from screwworm (courtesy of USDA Livestock Insects Laboratory).

terial infections. The subject of maggot therapy is summarized by Greenberg (1973).

Obligatory Myiasis

Domestic and wild mammals are subject to attack primarily by dipteran larvae that are obligatory flesh parasites (Table 4.4). However, birds, reptiles and amphibians occasionally are invaded by this group of parasites. These fly larvae can develop only in living host tissue. Flies of several families may be involved, usually Calliphoridae, Sarcophagidae, Oestridae, Gasterophilidae and Cuterebridae.

The primary screwworm

Cochliomyia hominivorax (Fig. 4.19a) can be an important human parasite in tropical areas where humans are associated closely with livestock (Fig. 4.19b). The last reported human case was a soldier returning from duty in Panama (Mehr et al. 1991). The primary screwworm is a major pest of livestock, and has resulted in significant economic losses.

Cochliomyia hominivorax is an obligate parasite requiring living tissue to complete its development. However, a cut, abrasion or other injury to the skin is generally required for the larvae to gain entry to host tissues. Larvae also can enter through body orifices, but it is believed that the mucous membranes still must be broken or abraded. Many livestock husbandry practices such as castration, dehorning, branding and shearing, along with cuts, scratches and insect bites, often create wound stimuli that attract ovipositing female screwworm flies.

Adult flies are approximately 3 times the size of a house fly and characteristically are metallic blue or blue-green in color. Females lay batches of 200–300 eggs in compact masses on the skin around fresh or necrotic wounds. Eggs hatch in 12–24 hr and larvae are gregarious, feeding in a characteristic head-down position with the posterior spiracles toward the opening of the wound. Larvae continue development for 4–10 days and grow to a length of approximately 17 mm before dropping from the wound to pupate in the top 2.5 cm of soil. The pupal stage lasts from about 1 week during summer to about 3 months during winter. Females typically mate only once and lay their first egg batch 5–10 days after eclosion. The summer life cycle averages approximately 24 days. The screwworm has no true hibernating life stage and thus cannot overwinter in cold climates.

Beginning in the early 1950s, entomologists with the US Department of Agriculture dem-

Table 4.3. Some common species of Diptera associated with facultative myiasis in humans (adapted mainly from James 1947).

Family and species	Type of myiasis	Distribution
Sarcophagidae		
Wohlfahrtia meigenii	Traumatic	Nearctic
Sarcophaga haemorrhoidalis	Traumatic, urinary, enteric	Nearly Cosmopolitan; rare in Oriental and Australian
Sarcophaga bullata	Dermal	Nearctic
Sarcophaga barbata	Traumatic	Nearctic, Palearctic, Australian
Sarcophaga carnaria	Traumatic	Palearctic
Calliphoridae		
Cochliomyia macellaria	Traumatic	Nearctic, Neotropical
Chrysomya megacephala	Traumatic	Nearctic, Neotropical, Ethiopian, Australian
Chrysomya rufifacies	Traumatic	Palearctic, Oriental, Australian, Nearctic
Phormia regina	Traumatic	Palearctic, Nearctic
Protophormia terrae-novae	Traumatic	Palearctic, Nearctic
Lucilia caesar	Traumatic, auricular, enteric	Palearctic
Lucilia illustris	Traumatic	Palearctic, Nearctic, Neotropical, Australian
Phaenicia cuprina	Traumatic	Oriental, Australian, Ethiopian
Phaenicia sericata	Traumatic, dermal, auricular	Nearly Cosmopolitan
Calliphora vicina	Associated with various diseased tissues	Nearctic, Palearctic, Australian
Calliphora vomitoria	Traumatic	Nearctic, Palearctic, Australian

onstrated that the exposure of screwworm pupae to ionizing radiation resulted in sterile adult males. Matings between sterile males and wild-type females produced eggs that did not hatch. Because of the single-mating tendency of the female screwworm, it was theorized that mass release of sterile flies into a natural population would produce significant sterility among wild-type females. In testing this autocidal technique, mass-reared and sterilized flies were used to eradicate natural screwworm populations from the islands of Curacao and Puerto Rico and from the southeastern USA in the late 1950s. That program was successfully extended throughout the USA and Mexico. Currently, the screwworm has been eradicated from most of Central America except Costa Rica and Panama. Efforts are underway to eradicate the screwworm from all of Central America and the Caribbean. A recent accidental introduction of *C. hominivorax* into northern Africa (Libya) was eradicated successfully by the release of sterile males being mass produced for the Central American program.

Old World screwworms

In Africa, India and the nearby islands of the Pacific and Indian Oceans (e.g., Indonesia, New Guinea and the Philippine Islands), another calliphorid fly (*Chrysomya bezziana*) occupies a position similar to that of *Cochliomyia hominivorax* in the Americas. It also differs from the other members of the genus (including the widespread *Chrysomya megacephala*, which it closely resembles) in being an obligatory parasite.

Wohlfahrtia traumatic myiasis

A third screwworm fly, *Wohlfahrtia magnifica* (Sarcophagidae), is widespread over the warmer parts of the Palearctic Region. The female does not lay eggs, but rather deposits active first-stage maggots. Skin lesions and injuries in prospective hosts may be the site of larval entry. Mucous membranes, eyes, ears and genital openings also may be used as entry points. Fatal cases in humans have been reported. Nearly all warmblooded animals are attacked.

Gasterophilus enteric myiasis

Obligatory enteric myiasis occurs in horses, donkeys, mules, zebras, elephants and rhinoceroses (James 1947). Zumpt (1965) listed 6 *Gasterophilus* species that attack domestic equids (*G. intestinalis, G. haemorrhoidalis, G. nasalis, G. inermis, G. pecuorum* and *G. nigricornis*). The first 3 of these species have been introduced into the Americas from Europe. All are widely distributed in the Old World. Broce (1985) reviewed the biology of the 3 species found in North America.

Adults of *G. intestinalis, G. nasalis*, and *G. haemorrhoidalis* possess nonfunctional mouthparts, are short lived and similar in size to honey bees, which they superficially resemble. Adults mate shortly after emergence and females begin ovipositing almost immediately by attaching eggs to hairs on the host's body (Fig. 4.20). The particular site of oviposition varies with the species, but all newly hatched larvae penetrate the soft tissues of the mouth (lips, gums and tongue) where they spend approximately 3 weeks. Larvae then migrate to the stomach or small intestine where they attach to the lining (Fig. 4.21). After several months, full grown larvae (3rd instars) detach and pass out with the feces. Pupation takes place in the soil under the dung pile. Adults emerge 2–8 weeks later, depending on climatic conditions.

The horse bot fly, *G. intestinalis*, mates in the vicinity of horses, and females may lay 500–1,000 eggs. The pale yellow or grayish eggs usually are attached to the inside of the forelegs, but also can be found on the outside of the legs, the mane and the flanks. Females oviposit while in flight, hovering about the host and occasionally darting in to deposit an egg. The incubation period for horse bot fly eggs is about 5 days, after which they can be stimulated to hatch by the warmth and moisture of the host's tongue during grooming. After a short period in the mouth, larvae attach to the mucosa of the left sac or esophageal area of the stomach and remain there for 7–10 months. Mature larvae pass out with the feces and pupate in the soil. Adults are active during the early summer.

The throat bot fly, *G. nasalis*, attaches its whitish-yellow eggs to the host's hair beneath the

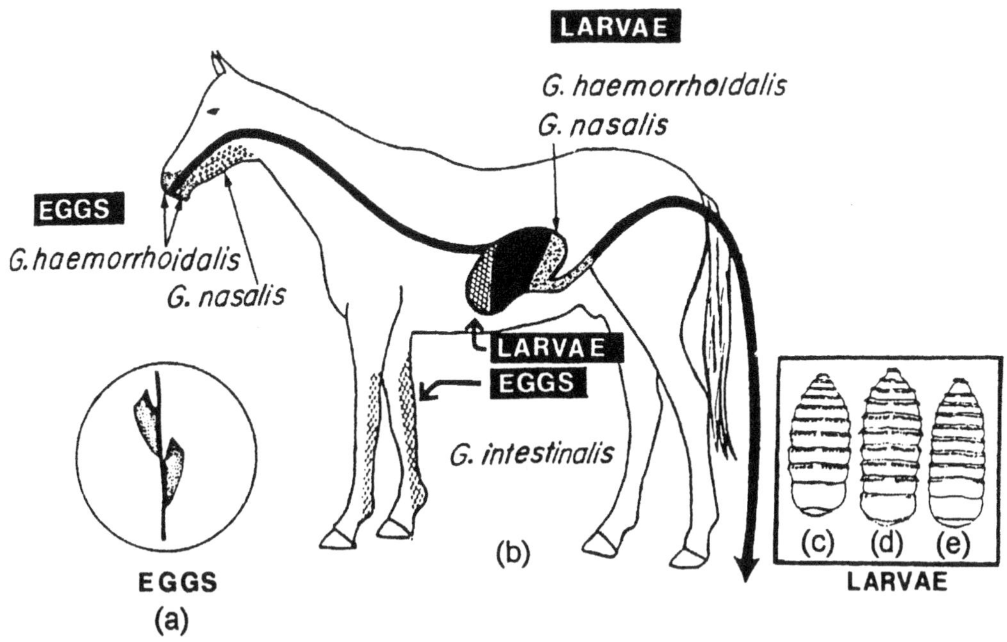

Figure 4.20. Horse bot life cycles (*Gasterophilus* spp.): (a) Eggs attached to hair of host; (b) oviposition and attachment sites; (c), (d) and (e) larvae of *G. nasalis*, *G. intestinalis* and *G. haemorrhoidalis*, respectively (from Broce 1985).

jaw. Each female can produce 450–500 eggs that may hatch within 4–5 days. These eggs do not require an external stimulus to hatch. Newly hatched larvae crawl along the skin to the horse's mouth where they penetrate the soft tissue. After ~20 days, they leave the mouth and attach to the pyloric region of the stomach or the upper portion of the duodenum. Mature larvae pass out with the feces. Most adult activity occurs during late spring or early summer.

The nose bot fly, *G. haemorrhoidalis*, is a rapid flier. Females attach their blackish eggs to hairs on the upper and lower lips of horses. Females produce an average of 160 eggs. Hatching, stimulated by moisture, occurs in about 2 days. Young larvae burrow into the tongue or lips, where they remain for about 6 weeks. After migration down the esophagus, they attach to the pyloric region of the stomach or the duodenum in a manner similar to *G. nasalis*. Mature larvae do not pass directly out with the feces, but rather, reattach to the wall of the rectum near the anus for 2–3 days before leaving the host. Adults are active during the summer months.

Damage caused by horse bot flies falls into several categories. Reactions of horses to ovipositing females may be so violent that they injure themselves or others. First-stage larvae cause irritation and pus pockets when burrowing in the oral tissue. The attachment of larvae to the stomach and duodenum walls probably interferes with digestion and gives rise to the characteristic "stretching out" behavior. Infested animals gain weight at a slower rate than uninfested ones. Surveys conducted in the USA and other countries indicate that nearly all pastured horses are affected by bot flies. Of the 3 horse bot species in the USA, *G. intestinalis* is the most prevalent.

Table 4.4. Some common species of Diptera associated with obligate myiasis in humans and livestock (adapted mainly from James 1947).

Family and species	Type of myiasis	Distribution
Sarcophagidae		
Wohlfahrtia magnifica	Traumatic, ocular, auricular, anal, etc.	Palearctic
Wohlfahrtia vigil	Furuncular	Nearctic
Calliphoridae		
Cochliomyia hominivorax	Nearly all types	Nearctic, Neotropical
Chrysomya bezziana	Nearly all types	Oriental, Ethiopian
Auchmeromyia senegalensis	Sucks blood	Ethiopian
Cordylobia anthopophaga	Furuncular	Ethiopian
Cordylobia rodhaini	Furuncular	Ethiopian
Gasterophilidaae		
Gasterophilus intestinalis	Creeping	Nearly Cosmopolitan
Gasterophilus haemorrhoidalis	Creeping	Nearctic, Neotropical, Oriental, Australian, Ethiopian
Gasterophilus nasalis	Creeping	Nearctic, Neotropical, Oriental, Australian, Ethiopian
Oestridae		
Cuterebra sp.	Nasal, oral, auricular, furuncular	Nearctic, Neotropical
Dermatobia hominis	Furuncular	Neotropical
Hypoderma bovis	Creeping, furuncular	Nearctic, Palearctic
Hypoderma lineatum	Creeping, furuncular	Nearctic, Palearctic, Neotropical, Ethiopian
Oestrus ovis	Nasal, auricular, ocular	Cosmopolitan

Control of bot flies has involved removal or killing of eggs, as well as chemical control of larvae in the digestive tract. Clipping of the eggs from the hair will prevent larvae from entering the host, but generally, this is not practical. Sponging the eggs of *G. intestinalis* with warm water induces hatching, after which larvae may be washed off the animal.

Other genera of parasitic Gasterophilidae include *Gyrostigma* in rhinoceroses, and *Cobboldia*, *Platycobboldia* and *Rodhainomyia* in elephants.

Fig. 4..21. Horse bots attached to intestinal wall (courtesy of John D. Edman).

Cattle grubs

Larvae of 2 species of the genus *Hypoderma* (the heel or ox warble flies) parasitize cattle and Old World deer but occasionally they invade horses and humans as well. *Hypoderma lineatum*, the common cattle grub (Fig. 4.22), is distributed widely in the USA, Canada, Europe and Asia. *Hypoderma bovis*, the northern cattle grub, is found in the northern USA, Canada and Europe. Literature on the cattle grub is voluminous; Gebauer (1958) and Scholl (1993) are comprehensive works.

Adults of both cattle grub species superficially resemble honey bees and are termed "heel flies" because of their habit of ovipositing on the lower legs of cattle. Their mouthparts are nonfunctional, and adults usually live for only 2–3 days. Activity of the adult stage of the 2 species is separated chronologically, the northern cattle grub usually appearing about a month later than the common cattle grub. There is only one generation per year. Eggs of the common cattle grub typically are attached in rows of 3–10 on a single hair, while those of the northern species generally are attached singly. Females are estimated to produce 500–800 eggs. Cattle may be severely agitated by ovipositing heel flies, especially the northern species, termed the "bomb fly." This gadding behavior is typified by running and stampeding, but affected cattle also are known to stand in ponds or streams or to seek dense shade to avoid ovipositing females.

Eggs of both cattle grub species usually hatch in about 4 days and the resultant larvae crawl down the hairs and enter the skin, usually through a hair follicle (Fig. 4.23). Such penetration is probably irritating, with the point of entry marked by a clear scab of dried serum. Mortality of first-stage larvae is estimated to be greater than 50%. Little is known about the subsequent migration of the larvae through the host's body, but in 1–2 months they are about the size of a grain of rice and often are observed in the submucosa of the gullet (common cattle grub) or in the tissue surrounding the spinal column (northern cattle grub). Larvae feed at these sites for variable lengths of time, increasing in size about 6-fold before migrating to the back of the host.

Upon reaching the upper back, the grubs cut a small opening in the hide, reverse their position, and form a small pouch called a warble between the layers of skin. Feeding on secretions from the tissue, larvae grow rapidly for about 2 additional months, then exit through the hole and drop to the ground and pupate. In North America, grubs appear in the backs of animals between September and February, depending on the latitude. They appear in the early autumn in central and southern Texas, but not until midwinter in the northern USA. This typical pattern may be disrupted by transport of infested cattle between locations. Duration of the pupal stage varies with climate, ranging from 2–8 weeks.

Emerging females are ready to mate almost immediately with males gathering at aggregation sites. Females are ready to oviposit within 20 minutes after copulation. There are no effective mechanical or cultural procedures available for the control of cattle grubs in pastured or range cattle. Confinement of animals in barns

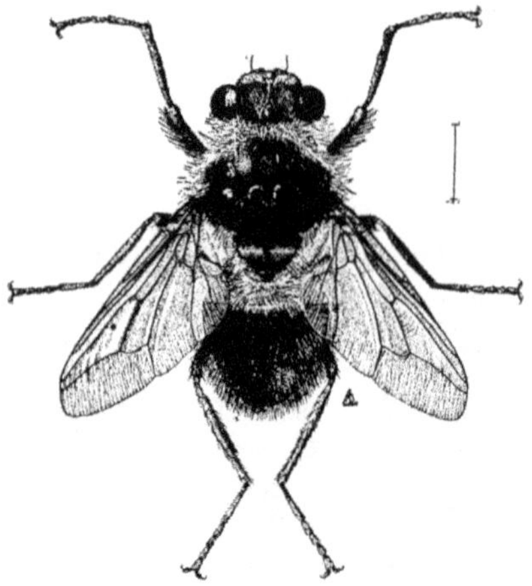

Fig. 4.22. The cattle grub, *Hypoderma lineatum* (from James 1947).

to prevent oviposition by adult heel flies may be effective, but is not always practical. Control of cattle grubs generally is effected by insecticide application. Although systemic insecticides have been effective in controlling cattle grubs, they remain a serious pest and can rebound quickly if insecticidal treatment is not rigorously followed.

Various systemic compounds have excellent therapeutic value by killing the small migrating larvae, but because of the short residual life of these materials, it is important to delay treatment until after the end of all adult fly activity. However, there are geographically variable "cutoff" dates after which systemic treatments should not be applied to cattle because the presence of large numbers of dead grubs following late treatment may cause potentially fatal anaphylactic shock.

Hypoderma as a human parasite

Numerous reports of attacks on humans by *H. lineatum*, *H. bovis*, and *Hypoderma diana* have been published. Unlike *Gasterophilus*, *Hypoderma* can complete its larval development in humans, often with serious consequences. In 1996, a case was reported to one of the authors (SEK) from a hospital in central Texas.

Caribou flies

The caribou, or reindeer, warble fly, *Hypoderma tarandi*, is distributed widely over the range of its host in northern Eurasia and northern North America. Its life history is similar to that of the warble flies of cattle. This parasite may cause heavy mortality among young animals. Warbles tend to form mainly on the rump. Lung involvement may occur and death from suffocation may result. As with other *Hypoderma*, partial immunity develops, older animals being more resistant than younger ones. In addition to death or damage to the host, this fly may be involved in the spread of brucellosis.

Sheep bot fly

The sheep bot, *Oestrus ovis*, is widely distributed. Females are slightly smaller than a honey bee and deposit live larvae in the nostrils of domestic sheep, goats and related wild hosts from early summer to autumn. Larvae move up the nasal passages into the nasal and frontal sinuses and reach maturity the following spring. There usually is only one generation per year. Parasitized animals have a purulent discharge from the nostrils, loss of appetite and grating of the teeth. The majority of cases are not fatal, but death may result after the appearance of aggravated symptoms.

Head maggots

Rhinocephalus purpureus is an important head maggot of horses in parts of Europe, Asia and Africa. Its habits are similar to those of *O. ovis*. Deer, elk, caribou and other related animals are commonly infested with head maggots. European species include *Cephenemyia stimulator* in roe deer, *Cephenemyia auribarbis* in the red deer,

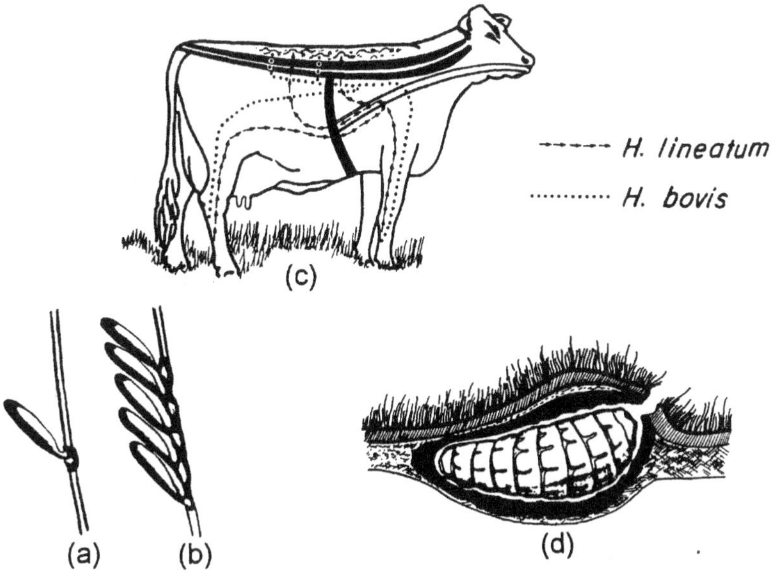

Figure 4.23. Cattle grubs (*Hypoderma* spp.): (a),(b) Eggs of *H. bovis* and *H. lineatum*, respectively; (c) routes followed by larvae within host from the legs to the back; (d) mature grub in warble (after Broce 1985).

Cephenemyia ulrichii in the European elk and *C. trompe* in the reindeer. *Cephenemyia trompe* also occurs in reindeer and caribou in the New World. American species include *Cephenemyia phobifer* in white-tailed deer, *Cephenemyia pratti* in mule deer, *Cephenemyia jellisoni* in Pacific black-tailed deer and white-tailed deer, and *Cephenemyia apicata* primarily in California deer.

Cuterebra

Larvae of the genus *Cuterebra* (Family Cuterebridae) are common parasites in the surface tissues of rodents, and wild and domestic rabbits and hares in the New World (Fig. 4.24a). Occasionally animals other than rodents or lagomorphs, including dogs, cats, monkeys and humans, are parasitized by *Cuterebra*. Fatal cases have been recorded in cats and dogs, but human cases are rare. One of the authors (JDE) identified larvae from 2 cases in recent years, one in the cheek of a young male shade-leaf tobacco worker and the other in the breast of an elderly woman with no pets and little outdoor exposure. Both individuals were living in Massachusetts and reported that the invasion (i.e., skin boil) was quite painful. Adult flies are large, dark, tabanid-like flies without functional mouthparts. Eggs are laid in and around the burrows or nests of their host. Egg hatch is stimulated by the presence of the host and hatched larvae crawl onto the host and burrow into the skin. The cutaneous bots usually develop in places where the hosts cannot groom, i.e., on the neck or hind region (Fig. 4.24b).

Dermatobia

Dermatobia hominis is common in parts of Mexico and Central and South America and has been called the human bot fly or tórsalo. It parasitizes a wide range of hosts: cattle, swine, cats, dogs, horses, mules, sheep, goats, humans,

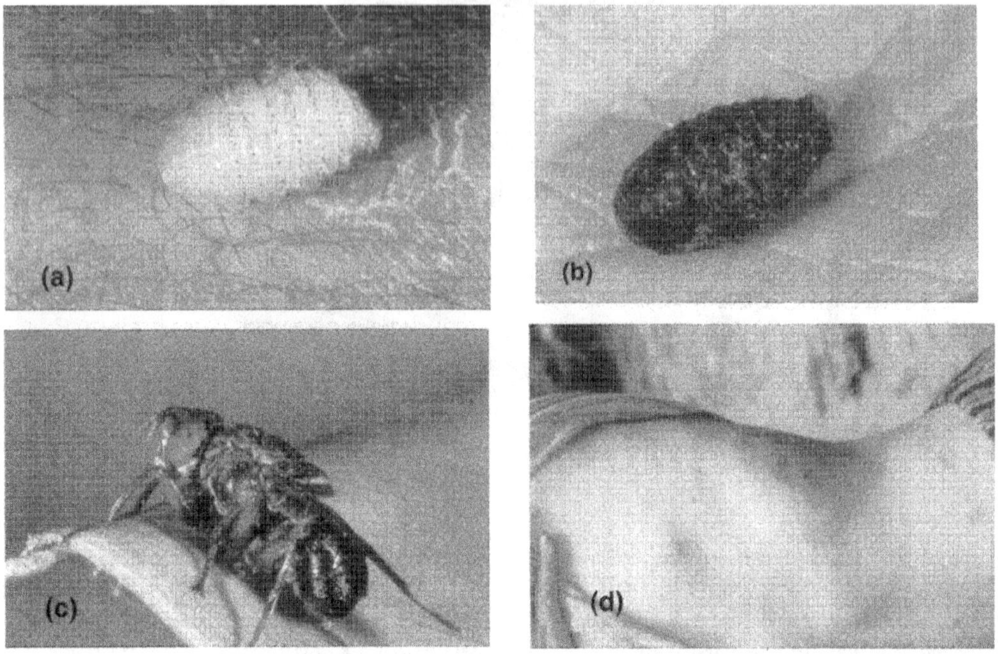

Figure 4.24. Bot flies: (a) *Dermatobia hominis* larva emerged from the arm of Darrell Ubick, California Academy of Sciences, in Costa Rica, 53 days after egg hatched on skin; (b) pupa after 65 days; (c) adult fly after 110 days; (d) wounds in arm of girl in Ecuador after removal of human bot fly larvae (a, b, and c, photographs by Tom Davies; d, photograph by J. Gordon Edwards).

monkeys and certain other wild mammals. It is a serious pest of cattle in Latin America.

This fly inhabits forested areas and it is claimed that its life history includes a bizarre adaptation for gaining access to its host. Females do not deposit eggs directly on the host, rather they capture other host-frequenting Diptera and glue their eggs along the ventral or lateral side of the carrier's body with a quick-drying adhesive. When the carrier lands on its host to feed, the *Dermatobia* eggs hatch rapidly and young larvae transfer to the host. Guimarães and Papavero (1966) listed as known carriers 48 species of flies and one tick, *A. cajennense*. The flies include 24 species of mosquitoes as well as black flies, species of *Chrysops* and *Fannia*, *M. domestica*, *S. calcitrans* and others.

Larvae produce warble-like lesions and swelling at the point where they enter the skin. They display no wandering behavior within the host. Development in the body of the host requires about 6 weeks. Mature larvae drop to the soil for pupation. The entire life cycle requires 3–4 months.

Dermatobia hominis myiasis has been diagnosed on several occasions in people who have acquired the parasite in tropical America and then returned to their homes in North America and Europe before the parasite completed its development. As with *Cuterebra*, humans experiencing *Dermatobia* report painful symptoms. Guimarães and Papavero (1966) reviewed the biology, systematics, pathogenesis, economic importance and control of *D. hominis*.

REFERENCES

Abbitt, B. and Abbitt, L.G. 1981. Fatal exsanguination of cattle attributed to an attack of salt marsh mosquitoes (*Aedes sollicitans*). J. Am. Vet. Med. Assoc. **179:**1397-1400.

Allen, J.R. 1973. Tick resistance: basophils in skin reactions of resistant guinea pigs. Intern. J. Parasitol. **3:**195-200.

Allen, J.R., Doube, B.M. and Kemp, D.H. 1977. Histology of bovine skin reactions to *Ixodes holocyclus*, Neuman. Canad. J. Compar. Med. Vet. Sci. **41:**26-35.

Anderson, J.R. and Nilssen. 1998. Do reindeer aggregate on snow patches to reduce harrassment by parasitic flies or to thermoregulate? Rangifer **18:**3-17.

Arlian, L.G. 1989a. Biology and ecology of house dust mites, *Dermatophagoides* spp. and *Euroglyphus* spp. Immunol. Allerg. Clinics N.A. **9:**339-356.

Arlian, L.G. 1989b. Biology, host relations, and epidemiology of *Sarcoptes scabiei*. Annu. Rev. Entomol. **34:**139-161.

Arlian, L.G. 1996. Immunology of scabies. Pp. 232-258 *in* S.K. Wikel (ed.), The Immunology of Host-Ectoparasitic Arthropod Relationships. CAB International, Wallingford, Oxon, UK.

Bänziger, H. and Buttiker, W. 1969. Records of eye-frequenting Lepidoptera from man. J. Med. Entomol. **6:**53-58.

Baumgartner, D.L. 1993. Review of *Chrysomya rufifacies* (Diptera: Calliphoridae). J. Med. Entomol. **30:**338-352.

Bell, J.F., Clifford, C.M., Moore, G.J. and Raymond, G. 1966. Effects of limb disability on lousiness in mice. III. Gross aspects of acquired resistance. Exp. Parasitol. **18:**49-60.

Benjamini, E., Feingold, B.F. and Kartman, L. 1961. Skin reactivity in guinea pigs sensitized to flea bites. The sequence of reactions. Proc. Soc. Exp. Biol. Med. **108:**700-702.

Benjamini, E., Feingold, B.F., Young, J.D., Kartman, L. and Shimizu, M. 1963. Allergy to flea bites. IV. *In vitro* collection and antigenic properties of the oral secretion of the cat flea, *Ctenocephalides felis felis* (Bouche). Exp. Parasitol. **13:**143-154.

Berenbaum, M. 1998. Are insects getting a bad rap? Pp. 14-15, Parade Magazine.

Bergman, D.K., Ramachandra, R.N. and Wikel, S.K. 1995. *Dermacentor andersoni*: salivary gland proteins suppressing T-lymphocyte responses to concanavalin A in vitro. Exp. Parasitol. **81:**262-271.

Bissonnette, E.Y., Rossignol, P.A. and Befus, A.D. 1993. Extracts of mosquito salivary gland inhibit tumor necrosis factor alpha. Parasite Immunol. **15:**27-33.

Boulard, C. 1989. Degradation of bovine C3 by serine proteases from parasite *Hypoderma lineatum* (Diptera, Oestridae). Vet. Immunol. Immunopathol. **20:**387-398.

Broce, A.B. 1985. Myiasis producing flies. Pp. 83-100 *in* R.E. Williams, R.D. Hall, A.B. Broce and P.J. Scholl (eds.), Livestock Entomology. John Wiley and Sons, New York.

Brossard, M. and Fivaz, V. 1982. *Ixodes ricinus* L.: mast cells, basophils and eosinophils in the sequence of cellular events in the skin of infested and reinfested rabbits. Parasitol. **85:**583-592.

Brossard, M., Rutti, B. and Huang, T. 1991. Immunological relationships between host and ixodid ticks. Pp. 177-200 *in* C.A. Toft, A. Aeschliman and L. Bolic (eds.), Parasite-host associations: coexistence or conflict. Oxford University Press, Oxford, UK.

Brown, S.J., Galli, S.J., Gleich, G.J. and Askenase, P.W. 1982. Ablation of immunity to *Amblyomma americanum* by anti-basophil serum: cooperation between basophils and eosinophils in expression of immunity to ectoparasites (ticks) in guinea pigs. J. Immunol. **129:**790-796.

Burgess, I. 1994. *Sarcoptes scabiei* and scabies. Adv. Parasitol. **33:**235-292.

Campbell, J.B. 1988. Arthropod-induced stress in livestock. Veterinary Clinics of North America. Food Animal Practice **4:**551-555.

Chabaudie, N. and Boulard, C. 1992. Effect of hypodermin A, an enzyme secreted by *Hypoderma lineatum* (Insecta, Oestridae) on the bovine immune system. Vet. Immunol. Immunopathol. **31:**167-177.

Champagne, D.E. and Valenzuela, J.G. 1996. Pharmacology of haematophagous arthropod saliva. Pp. 85-106 *in* S.K. Wikel (ed.), The immunology of host-ectoparasitic arthropod relationships. CAB International, Wallingford, Oxon, UK.

Cordoves, C.O., Farinas, J.L., Fleites, R., Garrido, P., Garcia, J. and Hernandez, S. 1986. The economic damage due to ticks in a dairy herd. Vet. Med. Rev. **1:**46-49.

Cotgreave, P. and Clayton, D.H. 1994. Comparative analysis of time spent grooming by birds in relation to parasite load. Behaviour **134:**171-187.

Cross, M.L., Cupp, M.S., Cupp, E.W., Galloway, A.L. and Enriquez, F.J. 1993a. Modulation of murine immunological responses by salivary gland extract of *Simulium vittatum* (Diptera: Simuliidae). J. Med. Entomol. **30:**928-935.

Cross, M.L., Cupp, M.S., Cupp, E.W., Ramberg, F.B. and Enriquez, F.j. 1993b. Antibody responses of BALB/c mice to salivary antigens of haematophagous black flies (Diptera: Simuliidae). J. Med. Entomol. **30**:725-734.

Cross, M.L., Cupp, E.W. and Enriquez, F.J. 1994. Differential modulation of murine cellular responses by salivary gland extract of *Aedes aegypti*. Am. J. Trop. Med. Hyg. **51**:690-696.

Downes, C.M., Theberge, J.B. and Smith, S.M. 1986. The influence of insects on the distribution, microhabitat choice, and behavior of the Burwash caribou herd. Canad. J. Zool. **64**:622-629.

Dzij, A. and Kurtenbach, K. 1995. *Clethrionomys glareolus*, but not *Apodermus flavicollis* acquires resistance to *Ixodes ricinus* L., the main European vector of *Borrelia burgdorferi*. Parasite Immunol. **17**:177-183.

Ellis, J.A., Shapiro, S.Z., ole Moi-Yoi, O. and Moloo, S.K. 1986. Lesions and saliva-specific antibody responses in rabbits with immediate and delayed hypersensitivity reactions to the bites of *Glossina morsitans centralis*. Vet. Pathol. **23**:661-667.

FAO. 1998. Insect and Pest Control Newsletter 58. Food and Agriculture Organization.

Feingold, B.F., Benjamini, E. and Michaeli, D. 1968. The allergic responses to insect bites. Annu. Rev. Entomol. **13**:137-158.

Fisher, W.F. and Wilson, G.I. 1977. Precipitating antibodies in cattle infested by *Psoroptes ovis* (Acarina: Psoroptidae). J. Med. Entomol. **14**:146-151.

Frances, A. and Munro, A. 1989. Treating a woman who believes she has bugs under her skin. Hospital and Community Psychiatry **40**:1113-1114.

Freeman, T.M. 1997. Insect and fire ant hypersensitivity: what the primary physician needs to know. Comprehensive Therapy **23**:38-43.

Frogatt, W.W. 1922. Sheep-maggot flies. N.S.W. Dept. Agric. Farmers' Bull. **144**:32.

Ganapamo, F., Rutti, B. and Brossard, M. 1996. Immunosuppression and cytokine production in mice infested with *Ixodes ricinus* ticks: a possible role of laminin and interleukin-10 on the in vitro responsiveness of lymphocytes to mitogens. Immunology **87**:259-263.

Gebauer, O. 1958. Die Dasselfliegen des Rindes und ihre Bekampfung. Parasitol. Schriftenreihe. Heft. **9**:97.

Gieler, U. and Knoll, M. 1990. Delusional parasitosis as 'folie a trois.' Dermatologica **181**:122-125.

Girardin, P. and Brossard, M. 1989. Effects of Cyclosporin A on humoral immunity to ticks and on cutaneous immediate and delayed hypersensitivity reactions to *Ixodes ricinus* L. salivary gland antigens in re-infested rabbits. Parasitol. Res. **75**:657-662.

Grace, J.K. and Wood, D.L. 1987. Delusory cleptoparasitosis: delusions of arthropod infestation in the home. Pan-Pacific Entomol. **63**:1-4.

Graft, D.F. 1996. Stinging insect hypersensitivity in children. Curr. Opin. Pediatr. **8**:597-600.

Gray, E.W., Adler, P.H. and Noblet, R. 1996. Economic impact of black flies (Diptera: Simuliidae) in South Carolina and development of a localized suppression program. J. Am. Mosq. Contr. Assoc. **12**:676-678.

Greenberg, B. 1973. Flies and disease. Vol. 2, Biology and disease transmission. Princeton University Press, Princeton, New Jersey. 447 pp.

Greenberg, B. 1984. Two cases of human myiasis caused by *Phaenicia sericata* (Diptera: Calliphoridae) in Chicago area hospitals. J. Med. Entomol. **21**:615.

Guillot, F.S. 1981. Population increase of *Psoroptes ovis* (Acari: Psoroptidae) on stanchioned cattle during summer. J. Med. Entomol. **18**:44-47.

Guimarães, J.H. and Papavero, N. 1966. A tentative annotated bibliography of *Dermatobia hominis* (Linnaeus Jr., 1781) (Diptera, Cuterebridae). Arq. Zool. **14**:223-294.

Habermann, E. 1972. Bee and wasp venoms. Science **177**:315-322.

Hall, R.D., Anderson, P.C. and Clark, D.P. 1986. A case of human myiasis caused by *Phormia regina* (Diptera: Calliphoridae) in Missouri, USA. J. Med. Entomol. **23**:578-579.

Hall, R.A., Broom, A.K., Hartnett, A.C., Howard, M.J. and Mackenzie, J.S. 1995. Immunodominant epitopes on the NS1 protein of MVE and KUN viruses serve as targets for a blocking ELISA to detect virus-specific antibodies in sentinel animal serum. J. Virol. Meth. **51**:201-210.

Hancock, B.W. and Ward, A.M. 1974. Serum immunoglobulins in scabies. J. Investig. Dermatol. **63**:484-484.

Harwood, R.F. and James, M.T. 1979. Entomology in human and animal health, 7th ed. Macmillan, New York. 548 pp.

Heymann, P.W., Chapman, M.D., Aalberase, R.C., Fox, J.W. and Platts-Mills, T.A.E. 1989. Antigenic and structural analysis of group II allergens (Der f II and Der p II) from house dust mites (*Dermatophagoides* spp.). J. Allerg. Clin. Immunol. **83**:1055-1067.

Heymann, P.W., Chapman, M.D. and Platts-Mills, T.A.E. 1986. Antigen Der f 1 from the dust mite *Dermatophagoides farinae*: structural comparison a with Der p 1 from *D. pteronyssinus* and epitope specificity of murine IgG and human IgE antibodies. J. Immunol. **137**:2841-2847.

Hollander, A.L. and Wright, R.E. 1980. Impact of tabanids on cattle: blood meal size and preferred feeding sites. J. Econ. Entomol. **73**:431-433.

Hudson, A., Bowman, L. and Orr, C.W.M. 1960. Effects of absence of saliva on blood feeding by mosquitoes. Science **131**:1730-1731.

James, M.T. 1947. The flies that cause myiasis in man. US Dept. Agric. Misc. Publ. 631, Washington, DC. 175 pp.

James, M.T. 1969. A study of the origin of parasitism. Bull. Entomol. Soc. Am. **15:**251-253.

Kang, B. 1976. Study on cockroach antigen as a probable causative agent in bronchial asthma. J. Allerg. Clin. Immunol. **58:**357-365.

Kang, B., Johnson, J., Morgan, C. and Chang, J.L. 1988. The role of immunotherapy in cockroach asthma. J. Asthma **25:**205-218.

Kerlin, R.L. and Allingham, P.G. 1992. Acquired immune response of cattle exposed to buffalo fly (*Haematobia irritans exigua*). Vet. Parasitol. **43:**115-129.

Kerlin, R.L. and East, I.J. 1992. Potent immunosuppression by secretory/excretory products of larvae from the sheep blowfly *Lucilia cuprina*. Parasite Immunol. **14:**595-604.

Kettle, D.S. 1990. Medical and veterinary entomology. CAB International, Wallingford, Oxon, UK. 658 pp.

Kirkwood, A.C. 1986. History, biology and control of sheep scab. Parasitol. Today **2:**302-307.

Kotwal, G.J. 1996. The great escape. Immune evasion by pathogens. The Immunologist **4:**157-164.

Krantz, G.W. 1978. A manual of acarology, 2nd ed. Oregon State University, Corvallis. 509 pp.

Lane, R.P. and Crosskey, R.W. 1993. Medical insects and arachnids. Chapman and Hall, London. 723 pp.

Leclercq, M. 1974. Myiases of the human digestive tract. Med. Chirurg. Digest. **3:**147-152.

Lee, D.J. 1968. Human myiasis in Australia. Med. J. Austral. **1:**170-173.

Lemanske, R.F., Jr. and Kaliner, M.A. 1993. Late-phase allergic reactions. Pp. 321-361 *in* E. Middleton, Jr., C.E. Read, E.F. Ellis, N.F. Adkinson, J.W. Yunginger and W.W. Busse (eds.), Allergy Principles and Practice, 4th ed. C.V. Mosby Co., St. Louis, Missouri.

Linley, J.R. and Davies, J.B. 1971. Sandflies and tourism in Florida and the Bahamas and Caribbean area. J. Econ. Entomol. **64:**264-287.

Marrack, P. and Kapler, J. 1994. Subversion of the immune system by pathogens. Cell **76:**323-332.

Marschall, M.A., Dolezal, R.F., Gohen, M. and Marschall, S.F. 1991. Chronic wounds and delusions of parasitosis in the drug abuser. Plastic and Reconstructive Surgery **88:**328-330.

Mathews, K.P. 1989. Inhalant insect-derived allergens. Immunol. Allerg. Clinics N.A. **9:**321-338.

Mbow, M.L., Rutti, B. and Brossard, M. 1994. Infiltration of CD4+, CD8+ T cells, and expression of ICAM-1, Ia antigens, IL-1 alpha and TNF-alpha in the skin lesion of BALB/c mice undergoing repeated infestations with nymphal *Ixodes ricinus* ticks. Immunology **82:**596-602.

McKenna, W.R. 1994. Characteristics of multiple-massive honey bee sting cases. J. Allerg. Clin. Immunol. **93 (part 2):**244, abstract #368.

Mehr, Z., Powers, N.R. and Konkol, K.A. 1991. Myiasis in a wounded soldier returning from Panama. J. Med. Entomol. **28:**553-554.

Mellanby, K. 1944. The development of symptoms, parasitic infection and immunity in human scabies. Parasitol. **35:**197-206.

Mellanby, K. 1946. Man's reaction to mosquito bites. Nature **158:**554-556.

Mellanby, K. 1972. Scabies, 2nd ed. E.W. Classey, Hampton, UK. 81 pp.

Michaeli, D., Benjamini, E., deBuren, F.P., Larrivee, D.H. and Feingold, B.F. 1965. The role of collagen in the induction of flea bite hypersensitivity. J. Immunol. **95:**162-170.

Miller, G.L. 1997. Historical natural history: insects and the Civil War. Am. Entomol. **43:**227-245.

Morikawa, T. 1958. Studies on myiasis. I. A revision of human myiasis reported in Japan. Ochanomizu Egaku Zasshi **6:**1451-1466.

Morsy, T.A., Kenawi, M.Z., Zohdy, H.A., Abdalla, K.F. and Fakahany, A.F.E. 1993. Serum immunoglobulin and complement values in scabietic patients. J. Egypt. Soc. Parasitol. **23:**221-228.

Mosmann, T.R. and Coffman, R.L. 1989. TH1 and TH2 cells: different patterns of lymphokine secretion lead to different functional properties. Annu. Rev. Immunol. **7:**145-173.

Mumcuoglu, K.Y. 1996. Control of human lice (Anoplura: Pediculidae) infestations: past and present. Am. Entomol. **42:**175-178.

Mumford, J. 1982. Entomophobia: the fear of arthropods. Antenna **6:**156-157.

Muradov, S.M., Davetklychev, A.A. and Berdyev, B. 1980. The effect of blood sucking flies on the productivity of animals in the Turkmenian SSR. Parazitologiya **14:**520-522.

Nicolas-Gaulard, I., Moire, N. and C. Boulard, C. 1995. Effect of the parasite enzyme hypodermin A, on bovine lymphocyte proliferation and interleukin-2 production via the prostaglandin pathway. Immunology **84:**160-165.

Noirtin, C. and Boiteux, P. 1979. Death of 25 farm animals (including 24 bovines) as a result of simuliid bites in the Vosges. Bull. Mens. Soc. Veterin. Pratique **63:**41-54.

Ohtaki, N. and Oka, K. 1994. A quantitative study of specific immunoglobulins to mosquito salivary gland antigen in hypersensitive and common types of mosquito bite reaction. J. Dermatol. **21:**639-644.

O'Meara, T.J., Nesa, M., Raadsma, H.W., Saville, D.G. and Sandeman, R.M. 1992. Variation in skin inflammatory responses between sheep bred for resistance or susceptibility to fleece rot and blowfly strike. Res. Vet. Sci. **52:**205-220.

Orkin, M. and Maibach, H.I. 1978. This scabies pandemic. New England J. Med. **298**:496-498.

Overskaug, K. 1994. Behavioural changes in free-ranging red foxes (*Vulpes vulpes*) due to sarcoptic mange. Acta Vet. Scand. **35**:457-459.

Paholpak, S. 1990. Delusion of parasitosis: a report of ten cases at Srinagarind hospital. J. Med. Assoc. Thailand **73**:111-114.

Paine, S.H., Kemp, D.H. and Allen, J.R. 1983. In vitro feeding of *Dermacentor andersoni* (Stiles): effects of histamine and other mediators. Parasitol. **86**:419-428.

Peng, Z. and Simons, F.E.R. 1997. Cross-reactivity of skin and serum specific IgE responses and allergen analysis for three mosquito species with worldwide distribution. J. Allerg. Clin. Immunol. **100**:192-198.

Platts-Mills, T.A.E. 1995. Allergens. Pp. 1231-1256 *in* M.M. Frank, K.F. Austen, H.N. Claman and E.R. Unanue (eds.), Samter's Immunologic Diseases, 5th ed. Little, Brown and Company, Boston.

Purett, J.H., Guillott, F.S. and Fisher, W.F. 1986. Humoral and cellular immunoresponsiveness of stanchioned cattle infested with *Psoroptes ovis*. Vet. Parasitol. **22**:121-133.

Ramachandra, R.N. and Wikel, S.K. 1992. Modulation of host immune responses by ticks (Acari: Ixodidae): effect of salivary gland extracts on host macrophages and lymphocyte cytokine production. J. Med. Entomol. **29**:818-826.

Ratzlaff, R.E. and Wikel, S.K. 1990. Murine immune responses and immunization against *Polyplax serrata* (Anoplura: Polyplacidae). J. Med. Entomol. **27**:1002-1007.

Ribeiro, J.M.C. 1987. *Ixodes dammini*: salivary anti-complement activity. Exp. Parasitol. **64**:347-353.

Ribeiro, J.M.C. 1995. Blood-feeding arthropods: live syringes or invertebrate pharmacologists? Infec. Agents Dis. **4**:143-152.

Ribeiro, J.M.C. and Spielman, A. 1986. *Ixodes dammini*: salivary anaphylatoxin inactivating activity. Exp. Parasitol. **62**:292-297.

Richard, R.D. and Gerrish, R.R. 1983. The first confirmed field case of myiasis produced by *Chysomya* sp. (Diptera: Calliphoridae) in the continental United States. J. Med. Entomol. **20**:685.

Riley, W.A. 1939. The possibility of intestinal myiasis in man. J. Econ. Entomol. **32**:875-876.

Sagan, C. 1977. The dragons of Eden: speculations on the evolution of human intelligence. Random House, New York. 263 pp.

Sandeman, R.M. 1996. Immune responses to mosquitoes and flies. Pp. 175-203 *in* S.K. Wikel (ed.), The Immunology of Host-Ectoparasitic Arthropod Relationships. CAB International, Wallingford, Oxon, UK.

Sandeman, R.M., Chandler.R.A. and Seaton, D.S. 1995. Antibody degradation in blowfly strike. Intern. J. Parasitol. **25**:621-628.

Schofield, C.J. 1994. Triatominae: biology and control. Eurocommunica Publ., West Sussex, UK. 80 pp.

Scholl, P.J. 1993. Biology and control of cattle grubs. Annu. Rev. Entomol. **38**:53-70.

Scott, H.G. 1964. Human myiasis in North America (1952-1962, inclusive). Fla. Entomol. **47**:255-261.

Shan, E., Yoshiki, T., Shimizu, M., Ando, K., Chinzei, Y., Suto, C., Ohtaki, T. and Ohtaki, N. 1995. Immunoglobulins specific to mosquito salivary gland proteins in the sera of persons with common or hypersensitive reactions to mosquito bites. J. Dermatol. **22**:411-418.

Steelman, C.D. 1976. Effects of external and internal arthropod parasites on domestic livestock production. Annu. Rev. Entomol. **21**:155-178.

Sternberg, L. 1929. A case of asthma caused by the *Cimex lectularis* (bed bug). J. Allerg. Clin. Immunol. **1**:83.

Stromberg, P.C. and Fisher, W.F. 1986. Determatopathology and immunity in experimental *Psoroptes ovis* (Acari: Psoroptidae) infestation of naive and previously exposed Hereford cattle. Amer. J. Vet. Res. **47**:1551-1560.

Tellam, R.L., Smith, D., Kemp, D.H. and Willadsen, P. 1992. Vaccination against ticks. Pp. 301-331 *in* W.K. Young (ed.), Animal parasite control using technology. CRC Press, Boca Raton, Florida.

Theodos, C. and Titus, R.G. 1993. Salivary gland material from the sand fly *Lutzomyia longipalpis* has an inhibitory effect on macrophage function in vitro. Parasite Immunol. **15**:481-487.

Thien, F.C.K., Leung, R.C.C., Czarny, D. and Walters, E.H. 1994. Indoor allergens and IgE-mediated respiratory illness. Immunol. Allerg. Clinics N.A. **14**:567-590.

Thorne, E.T., Kingston, N., Jolley, W.R. and Bergstrom, R.C. (eds). 1982. Diseases of wildlife, 2nd ed. Wyoming Fish and Game Department, Laramie.

Tipton, V.J. (ed.). 1974. Medical entomology, mediated instructional kit. Entomological Society of America and Brigham Young University. College Park, Maryland and Provo, Utah.

Titus, R.G. and Ribeiro, J.M.C. 1990. The role of vector saliva in transmission of arthropod-borne disease. Parasitol. Today **6**:157-160.

Ungar-Waron, H., Braverman, Y., Gluckman, A. and Trainin, Z. 1990. Immunogenicity and allergenicity of *Culicoides imicola* (Diptera: Ceratopogonidae) extracts. J. Vet. Med., Ser. B **37**:64-72.

Urioste, S., Hall, L.R., Telford, S.R., III and Titus, R.G. 1994. Saliva of the Lyme disease vector, *Ixodes dammini*, blocks cell activation by a non-prostaglandin E2-dependent mechanism. J. Exp. Med. **180**:1077-1085.

US Army. 1957. Ixodid ticks of Japan, Korea and the Ryukyu Islands. 406th Medical General Laboratory, Camp Zama, Japan. 37 pp.

US Public Health Service. 1969. Pictorial keys to arthropods, reptiles, birds and mammals. Publication No. 1955, US Department of Health, Education and Welfare, Communicable Disease Center, Atlanta, Georgia.

Usinger, R.L. 1944. The Triatominae of North and Central America and the West Indies and their public health significance. Public Health Bull. **228**: 1-83.

Usinger, R.L. 1966. Monograph of Cimicidae (Hemiptera - Heteroptera). Vol. 7. Entomological Society of America, Thomas Say Foundation, College Park, Maryland. 585 pp.

Valentine, M.D. 1995. Insect venom allergy. Pp. 1367-1375 *in* M.M. Frank, K.F. Austen, H.N. Claman and E.R. Unanue (eds.), Samter's Immunologic Diseases, 5th ed. Little, Brown and Company, Boston.

West, L.S. 1951. The housefly, its natural history, medical importance, and control. Comstock, Ithaca, New York. 584 pp.

Whelen, A.C., Richardson, L.K. and Wikel, S.K. 1986. Dot-ELISA assessment of guinea pig antibody responses to repeated *Dermacentor andersoni* infestations. J. Parasitol. **72**:155-162.

Whelen, A.C. and Wikel, S.K. 1993. Acquired resistance of guinea pigs to *Dermacentor andersoni* mediated by humoral factors. J. Parasitol. **79**:908-912.

Wikel, S.K. 1982a. Immune responses to arthropods and their products. Annu. Rev. Entomol. **27**:21-48.

Wikel, S.K. 1982b. Influence of *Dermacentor andersoni* infestation on lymphocyte responsiveness to mitogens. Ann. Trop. Med. Parasitol. **79**:513-518.

Wikel, S.K. 1985. Effects of tick infestation on the plaque-forming cell response to a thymic dependent antigen. Ann. Trop. Med. Parasitol. **79**:195-198.

Wikel, S.K. 1996. Immunology of the tick-host interface. Pp. 204-231 *in* S.K. Wikel (ed.), The immunology of host-ectoparasitic arthropod relationships. CAB International, Wallingford,Oxon,UK.

Wikel, S.K. 1999. Modulation of the host immune system by ectoparasitic arthropods. Biosci. **49**:311-320.

Wikel, S.K. and Allen, J.R. 1976. Acquired resistance to ticks. I. Passive transfer of resistance. Immunology **30**:311-316.

Wikel, S.K. and Bergman, D.K. 1997. Tick-host immunology: significant advances and challenging opportunities. Parasitol. Today **13**:383-389.

Wikel, S.K., Bergman, D.K. and Ramachandra, R.N. 1996a. Immunological-based control of blood-feeding arthropods. Pp. 290-315 *in* S.K. Wikel (ed.), The immunology of host-ectoparasitic arthropod relationships. CAB International, Wallingford, Oxon, UK.

Wikel, S.K., Ramachandra, R.N. and Bergman, D.K. 1996b. Arthropod modulation of host immune responses. Pp. 107-130 *in* S.K. Wikel (ed.), The immunology of host-ectoparasitic arthropod relationships. CAB International, Wallingford, Oxon, UK.

Wikel, S.K., Ramachandra, R.N., Bergman, D.K., Burkot, T.R. and Piesman, J. 1997. Infestation with pathogen-free nymphs of the tick *Ixodes scapularis* induces host resistance to transmission of *Borrelia burgdorferi* by ticks. Infect. Immun. **65**:335-338.

Willadsen, P. 1980. Immunity to ticks. Adv. Parasitol. **18**:293-313.

Wilson, A.B. and Clements, A.N. 1965. The nature of the skin reaction to mosquito bites in laboratory animals. Intern. Arch. Allerg. **26**:294-314.

Wrenn, W.J. 1996. Immune responses to mange mites and chiggers. Pp. 259-289 *in* S.K. Wikel (ed.), The immunology of host-ectoparasitic arthropod relationships. CAB International, Wallingford, Oxon, UK.

Wright, F.C. and DeLoach, J.R. 1980. Ingestion of erythrocytes containing 51Cr-labeled hemoglobin by *Psoroptes cuniculi* (Acari: Psoroptidae). J. Med. Entomol. **17**:186-187.

Wright, F.C., Riner, J.C. and Guillott, F.S. 1983. Cross-mating studies with *Psoroptes ovis* (Hering) and *Psoroptes cuniculi* Delafond (Acarina: Psoroptidae). J. Parasitol. **69**:696-700.

Zeidner, N., Dreitz, M., Belasco, D. and Fish, D. 1996. Suppression of acute *Ixodes scapularis*-induced *Borrelia burgdorferi* infection using tumor necrosis factor-alpha, interleukin-2, and interferon-gamma. J. Infect. Dis. **173**:187-195.

Zumpt, F. 1965. Myiasis in man and animals in the Old World. Butterworths, London. 267 pp.

Chapter 5

Arthropod Transmission of Vertebrate Parasites

JOHN D. EDMAN
University of California, Davis

ROUTES OF TRANSMISSION

Survival and reproduction of vertebrate parasites depend on a reliable mechanism for their movement from one host animal to another. Transmission is accomplished more easily by vertebrate ectoparasites that live on or near the surface of the host than by endoparasites that must penetrate exterior defenses and get inside their host. There are three basic channels for infectious agents to get on or inside their vertebrate hosts: (1) direct contact, (2) ingestion of contaminated water or food and (3) inhalation of contaminated air. Direct contact includes physical contact with contaminated surfaces or infected individuals (including sexual contact), active penetration (usually in water) by invasive stages of parasites, and contact via infected vectors as a result of their feeding or defecation behavior. Therefore, vectors are agents of parasite infection and represent a unique route of direct transmission.

TRANSMISSION BY VECTORS

Parasites often can reproduce in multiple host species, but in vectorborne agents this host range is unusually diverse phylogenetically. These parasites must alternate between vertebrates and invertebrates. In addition to hosting the development and/or reproduction of the pathogen/parasite, the invertebrate also serves as the vehicle for the transfer of the infectious agent between infected and noninfected vertebrates. This role defines the word **vector** as normally used by medical entomologists and disease epidemiologists. In recent years, other vehicles such as rodent urine or feces (e.g., in hantavirus transmission) also have been termed vectors but this is an expansion of the classical definition. Transfer via an arthropod vector usually takes place while arthropods are bloodfeeding on the vertebrate host of the parasite. Thus, the vertebrate host experiences double parasitism: first by the feeding arthropod (e.g., tick, flea or biting fly), then by the much smaller infectious parasite transmitted (= vectored) during arthropod feeding.

B.F. Eldridge and J.D. Edman (eds.), Medical Entomology, Revised Edition, 151–163.

ORIGIN AND EVOLUTION OF ARTHROPOD-TRANSMITTED INFECTIONS

Arthropods and microbes existed, and undoubtedly interacted, millions of years before vertebrates evolved. Bloodfeeding arthropods appeared early in the history of land vertebrates and are thought to have arisen via 2 quite different pathways (Waage 1979). In one case, insects with piercing-sucking mouthparts adapted for feeding on plant juices (e.g., mosquitoes) or prey hemolymph (e.g., reduviid bugs) appear to have made the transition to a more nutritious liquid, vertebrate blood. In the other pathway to bloodfeeding, arthropods with chewing mouthparts and adapted to living in the nests of vertebrates and feeding on nest detritus such as fur/skin surfaces of the nest occupants, gradually acquired mouthparts for piercing the skin and feeding on vertebrate blood.

Helminth, protozoan and bacterial parasites of vertebrates have a long history as well. When and how bloodfeeding arthropods began serving as dual hosts and vehicles for the transmission of these parasites is less clear. Some modern day vertebrate parasites, such as the rickettsiae, appear to have originated with arthropods and later were introduced into vertebrates after bloodfeeding arthropods evolved. In the case of other parasites, arthropod involvement may have come after they were already firmly established in their vertebrate hosts. Obviously, reliable interaction between certain arthropods and their vertebrate hosts had to occur before these relationships could be exploited by parasites as a more efficient mechanism for invading new hosts. At some transitional phase, many parasites undoubtedly utilized multiple mechanisms for their transfer, arthropods being only one of them. Some modern insect-vectored parasites still do not rely on a single mechanism for their transfer and represent such a transitional phase.

Zoonotic infections illustrate how changing animal abundances can influence the host-feeding patterns of vectors and thereby the types of new parasites to which expanding host populations such as humans might then be exposed. With time and further specialization in vector feeding behavior, such relationships may be transformed from zoonotic to anthroponotic ones.

Protozoan and filarial parasites and vectors associated primarily with humans are modified variants of older parasite-vector relationships involving primates, rodents or other vertebrates that lived in close proximity with early human populations. Destruction of wild hosts and their habitat, along with the expansion of stable human populations, undoubtedly provided the backdrop for the selection and evolution of these newer relationships in concert with the changing host-feeding behavior of the insect vector. The development of agrarian societies about 40,000 years ago and the domestication of certain wild animals, beginning about 14,000 years ago, greatly influenced the evolution of many of the disease agents that human societies are battling today (Edman 1988). Hunter-gatherer societies appear to have low levels of parasitic infections. The long-term maintenance of many acute disease agents that stimulate lifelong immunity (e.g., many viruses and bacteria) seems to require concentrated settlements of ~200,000 or more individuals (Anderson and May 1979, 1991).

Many Old World parasites apparently were introduced into the new world during early European exploration or through the slave trade that occurred within the last 500 years. Thus, the New World vectors of these transplanted parasites represent comparatively recent parasite adaptations. In addition, the parasites themselves have undergone limited, but recognizable, genetic modifications during the short period since their introduction into the New World. This process is still continuing, as new pathogens and vectors are being discovered or rediscovered in new regions of the world annually. These often are referred to as "emerging" or "reemerging" diseases. The 1999 outbreak of West Nile virus in the New York City area is an example of such a new introduction (Centers for Disease Control and Prevention 1999). The mechanism by which this virus entered the USA remains a mystery. Sequencing the microbial genome can help to determine the origins and strain relationships of recently introduced or geographically isolated parasites/pathogens (Enserink 1999).

Transmission Efficiency

Among vertebrate parasites with well defined, multiple developmental stages (e.g., helminths), transfer between hosts usually is accomplished by the egg or early immature stage. The transfer may be active, involving movement and penetration by the juvenile stage of the parasite, or it may be entirely passive, e.g., relying on the host or vector to ingest material containing infectious forms of the parasite. Eukaryotic parasites are capable of little movement on their own and often appear to depend entirely on chance for their transmission.

Although changing hosts is precarious for parasites, the probability of successful transfer can be enhanced or reduced by many events. The success rate or efficiency of transmission is a major determinant of: (1) the geographic or host distribution of parasites and (2) the incidence of any given parasite and the disease it brings to its hosts. Parasites often enhance the probability of their own transfer. Hosts, through their own behavioral patterns, unwittingly encourage transmission as well. Thus, parasite transmission usually involves a combination of parasite-induced and parasite-exploited behavior on the part of both the vertebrate and invertebrate host. The life cycles of some vectorborne parasites seem so complex and inefficient that the biggest unsolved mystery is why these parasites have not become extinct. Clearly, there are elements of transmission efficiency that remain unrecognized or under appreciated.

Parasites that employ vectors for their transmission are often the most complex because their success involves adaptation to both a vertebrate and invertebrate host. Moreover, parasite success relies on a close relationship between these two very different but equally important hosts. Some parasites actively promote this relationship, a phenomenon that will be discussed later in this chapter.

Transmission Frequency

The shorter the life cycle of a parasite, the more frequently it must be transferred between hosts and the more efficient its transfer mechanism must be. If parasitism causes acute infection that can kill the host, rapid cycling to new, noninfected hosts takes on added urgency. It has been argued that vectorborne parasites generally cause more severe pathology and mortality in their vertebrate hosts than do non-vectorborne parasites. The logic for this lies in the assertion that these parasites cannot afford to become too adapted to their vertebrate host and still live successfully within their invertebrate host. Of course, some parasites spend their time in each host while in a different developmental stage, so stage-specific adaptations may occur.

Vectorborne parasites tend to be transmitted frequently because the life span of the vector often is short, and frequent transmission favors greater pathogenicity (Ewald 1993, 1996). Following this logic, one would expect that parasites with long-lived vectors (e.g., ticks) that also serve as reservoir hosts might be less pathogenic than those with vectors that only live a few days or weeks.

Both transmission frequency and efficiency are related to the bloodfeeding frequency and efficiency of the vector. These are key parameters in **vectorial capacity,** discussed in Chapter 6.

ACQUISITION OF INFECTIOUS AGENTS AND TRANSFER TO NEW VERTEBRATE HOSTS

Vector Acquisition of Parasites From Vertebrates

The transfer of infectious agents from vertebrate hosts to invertebrate vectors is as tenuous as the reverse process, but it is less variable. Bloodfeeding vectors always acquire the infectious agents during the uptake of blood, although the parasites may be in lymph as well as in blood. When parasites are acquired from the lymphatic system rather than from blood, the vectors are pool-feeding insects rather than capillary feeders. Stages of parasites that are infectious to the vector often are present in blood for only short periods of time, i.e., a few days in the case of most viruses. This often is referred to as the parasitemic (or viremic in the case of viruses) period.

Ingestion of viable parasites during feeding does not always result in the infection of the vector. One factor that can influence the success of the transfer is the number (= dose) of parasites ingested by the vector. If the dose is too small, it may be overcome by the arthropod's immune system so that insufficient parasites are available to invoke an infection in the arthropod. After ingestion, parasites must find their way to target organs in the vector where they can develop or replicate. The road has many curves that may lead to the demise of the parasite. Understanding the mechanisms and genetics of parasite resistance and susceptibility in vector arthropods is an important topic of contemporary research. One strategy under consideration for interrupting the transmission of vectorborne agents is to engineer vectors that can no longer successfully transmit these pathogens and replace wild vector populations with these resistant forms.

Vector-Mediated Mechanical Transmission to Vertebrates

In bloodborne mechanical transmission, arthropod mouthparts act as hypodermic needles, transferring microorganisms between hosts in the same fashion as the needles shared by IV drug users. This is an inefficient process for several reasons, so no infectious agent relies solely on arthropods to guarantee its transfer in this manner. It is no coincidence that the most ill-adapted bloodfeeding insects make the best mechanical vectors. Horse flies and deer flies (Diptera: Tabanidae) mechanically transmit the virus causing equine infectious anemia (EIA) among horses and the bacteria causing tularemia among rabbits. Nearly all tabanids are autogenous (i.e., able to develop eggs without a bloodmeal) because their ability to successfully obtain blood from vertebrates is severely limited by the painful way in which they attempt to extract blood from potential hosts. Tabanids lack an effective salivary anesthetic or anticoagulant and have no ability to capillary feed. Thus, these flies must rely on repeated slashing movements with their serrated mouthparts to open surface wounds from which they can sponge up limited amounts of pooled blood. The natural clotting process soon forces them to create new wounds in order to continue feeding. The annoyance associated with their feeding leads to frequent interruption by the host and such disturbances often results in flies relanding on a different host (their hosts are generally herd animals) where they immediately reattempt to feed. This primitive bloodfeeding scenario is well suited for the quick transfer of bloodborne pathogens that can adhere to the mouthparts of these rather large, robust flies. Still, both EIA and tularemia have alternate transmission mechanisms because mechanical transmission by flies is too tenuous for the parasite to rely on.

Public concern over insect transmission of human immunodeficiency virus (HIV) focused renewed attention on both the possibilities and the inefficiencies of mechanical transmission by insects. Lice, bed bugs, mosquitoes, black flies and biting midges that, unlike tabanids, often feed on human hosts, are small by comparison to tabanids and are much less likely to exhibit the sequential biting and host switching behavior seen among tabanids. HIV is fragile once externally exposed; it is present only in low titers in blood and cannot replicate within any known insect. Thus, the chance for this virus to be mechanically transmitted is extremely remote. Epidemiological patterns of HIV infection provide further evidence that insect transmission of this virus, even if possible, is inconsequential (Miike 1987).

The most common form of mechanical transmission occurs among bacterial infections of the eye and skin. Certain wound- and eye-frequenting Diptera such as face flies (*Musca autumnalis*), house flies (*Musca domestica*) and eye gnats (*Hippelates* spp.) are attracted to the very sites where these bacteria (e.g., those causing conjunctivitis and yaws) abound. Bacteria are picked up readily by the sponging mouthparts of flies feeding on lachrymal secretions and oozing sores on the skin. Like tabanids, these flies have a tendency to visit several hosts in the course of their feeding and thereby can quickly pass on these bacterial agents before the harsh external environment destroys them. Nonetheless, these microbes also are not solely dependent on the feeding of these flies for their transmission (for details see Chapter 12).

Vector-Mediated Biological Transmission to Vertebrates

The most successful arthropodborne parasites are those that depend on their arthropod vector to support part of their life cycle. This support may take 1 of 3 forms: propagative transmission, cyclo-developmental transmission and cyclo-propagative transmission

Propagative transmission

In the case of viruses, rickettsiae, and bacteria, parasites reproduce (i.e., propagate) in the vector, but do not pass through developmental stages. Transmission cannot occur until the parasite population has increased manyfold and been disseminated, e.g., to the salivary glands.

Cyclo-developmental transmission

Insectborne filarial parasites are the only known examples of this type of transmission. Microfilarial parasites ingested by the vector with the bloodmeal undergo development to 3rd-stage larvae (L3) before being transferred back to another vertebrate host. There is no replication and not all ingested microfilaria survive to L3. Thus, the number of parasites leaving the vector always is less than the number entering.

Cyclo-propagative transmission

Protozoan parasites undergo both development and reproduction within the vector. These parasites both increase in number and transform to a different life stage before transmission can take place. However, the vertebrate stages of these parasites sometimes can invoke a new infection if they are directly transferred to another vertebrate host via blood transfusion or across the placenta (e.g., malaria parasites) or through sexual intercourse (e.g., trypanosomes).

The time from when an arthropod ingests an infected bloodmeal until it becomes infective to other vertebrate hosts when it again feeds on blood is termed the extrinsic incubation period. This period varies from a few days up to 2 weeks or more. It generally is longer at cooler ambient temperatures and shorter at higher ambient temperatures. In contrast, the intrinsic incubation period is the time in the vertebrate host from parasite delivery by the vector until the vertebrate host is able to infect new vectors. In most infectious diseases, the intrinsic incubation period is defined as the time from exposure until clinical symptoms appear. This definition does not fit many vectorborne infections because the vertebrate host may become infectious prior to, or even without, the appearance of clinical symptoms (e.g., in some arboviruses and filariae). Moreover, some malaria parasites are not infectious to vectors until after clinical symptoms have abated.

Mechanisms of Parasite Transfer By Vectors

Transmission of pathogens by arthropods may be either horizontal or vertical (generational). It also may be from vector to vertebrate, from one life cycle stage to another within vectors or between vectors.

Vertical transmission (vector to vector)

Complete vertical transmission (i.e., generational transmission within the vector) occurs when some arboviral and rickettsial pathogens are passed from the female vector via her eggs to the next generation. This also can occur when infected male sperm fertilizes eggs laid by noninfected females (Table 5.1). Partial vertical transmission occurs in some arthropods in which the immature stages also bloodfeed (e.g., ticks and kissing bugs). Parasites are transferred from one immature stage to the next or to the adult stage (= transstadial transmission) but not to the next generation (= transovarial or genetic transmission).

Horizontal transmission (vector to vector)

Venereal transmission

Venereal transmission is a unique form of horizontal transmission. It occurs with certain arboviruses that also are transovarially transmitted when infected male mosquitoes (which do not bloodfeed) infect females during mating.

Co-feeding

Another recently reported vector to vector transfer method is co-feeding. It involves 2 or more ticks feeding in close proximity on a naïve host and occurs when the salivary secretions of one infected tick pass the infection to the noninfected tick feeding nearby, the host tissues acting as a conduit (Randolph et al. 1996, Randolph 1998).

Horizontal transmission (vector to vertebrate)

Horizontal transmission from vector to vertebrate host can occur through any of 6 different mechanisms (Table 5.1).

Salivarian transmission

The most common and efficient method of transmission of pathogens by arthropods is through salivary secretions that are injected during feeding or while probing for blood prior to feeding. This type of transmission is not possible for parasites that remain within the arthropod gut or hemocoel. These parasites escape their arthropod host by more tenuous routes. The most active escape occurs with filariae. After development, the infective stages of these parasites congregate in the head and mouthparts of infective vectors and actively crawl out while the vector is bloodfeeding. Within a very short time (before desiccation occurs) these small worms must locate the feeding puncture of the insect and use it to gain entrance to the vertebrate host.

Stercorarian transmission

Some parasites (e.g., *Trypanosoma cruzi* and *Rickettsia typhi*) exit posteriorly with arthropod feces that are deposited on the host during or just after feeding. Defecated parasites still must gain entrance to the host. This generally is ac-

Table 5.1. Mechanisms of parasite transfer involving arthropod vectors.

General category	Specific route of transmission	Parasite/vector example
Vertical[1]	Transovarial (mother to offspring)	*Babesia bigemina*/tick
Vertical[1]	Transstadial (stage to stage)	*Borrelia burgdorferi*/tick
Horizontal[1]	Venereal (male to female)	La Crosse virus/mosquito
Horizontal[1]	Co-feeding	*Borrelia burgdorferi*/tick
Horizontal[2]	Salivation	*Plasmodium* spp./mosquito
Horizontal[2]	Stercorarian	*Trypanosoma cruzi*/triatomid
Horizontal[2]	Regurgitation	*Yersinia pestis*/flea
Horizontal[2]	Assisted escape / passive transfer	*Borrelia recurrentis*/louse
Horizontal[2]	Active escape / active invasion	*Onchocerca* spp./black fly
Horizontal[2]	Ingestion by host	*Dipylidium caninum*/flea

[1]From vector to vector.
[2]From vector to vertebrate.

complished by the host itself when it scratches its skin or rubs its eyes. Defecation behavior becomes an important aspect of the transmission efficiency of potential vectors of these parasites.

Regurgitation

Other gut-inhabiting parasites (e.g., *Leishmania* and *Yersinia*) exit anteriorly during regurgitation by the arthropod while attempting to bloodfeed. This unnatural response is stimulated by the presence of masses of these parasites in the arthropod gut to the point that they interfere with the vector's ability to bloodfeed successfully. Thus, the parasite is responsible for the repeated unsuccessful feeding attempts of the vector, which in turn greatly augments its own chances for transmission to a new host.

Fleas transmit plague bacilli (*Yersinia pestis*) by regurgitation of infected blood. Regurgitation results from blockage of the alimentary canal by masses of multiplying bacteria that adhere to the spines of the proventriculus. Eventually, this prevents normal bloodfeeding by the flea, and regurgitation occurs when the flea attempts to clear the obstruction. The vector competence of individual fleas is related to the degree of blockage by bacteria.

Assisted escape and passive transfer of pathogens

A few bacterial parasites rely on the vertebrate host to macerate the body of the annoying arthropod vector (e.g., body lice), thereby releasing the parasites onto the surface of the host. These parasites (e.g., *Rickettsia recurrentis*) also rely on host behavior to gain entrance to their vertebrate host (Table 5.1).

Parasites that are deposited on vertebrate skin via the crushed body parts or feces of the vector require the behavioral assistance of the host to assure successful invasion of a new host. Itching associated with the immediate allergic reaction to salivary antigens injected into subcutaneous tissues during probing and feeding by vectors helps expedite this process. Vertebrate hosts inadvertently abrade contaminated areas of skin because these areas also are where the localized itching occurs. Resulting skin breaks allow the parasites to enter. The natural tendency of hosts to rub their eyes provides another point of entry for the parasites if the fingers are contaminated. The "Eye of Romaña," a characteristic early symptom of Chagas disease (*Trypanosoma cruzi*), is an example of this method of parasite entry. The feeding wound of the vector also may provide the skin break that is utilized by the parasites to invade their new host (e.g., 3rd-stage filarial larvae).

Active escape and active transfer of pathogens

The transmission of most filariae occurs during bloodfeeding, but not by injection of saliva. In this type of transmission, L3 migrate to the mouthparts of the vector and escape from the mouthparts at the tip, sometimes through openings torn by the larvae. The L3 then enter the feeding wound by serpentine movement.

Ingestion of vector

In a few instances, transmission to zoonotic reservoirs occurs through grooming or foraging when the animal eats the infected arthropod. Examples of this method are rodents eating reduviid bugs infected with *T. cruzi* and dogs eating fleas infected with the dog tapeworm. Other infections such as rodent malaria have been shown experimentally to be transmissible by this same oral route, but the role of this mechanism in natural transmission remains unknown.

Barriers to Transmission

When parasites leave the protected environment of one host and are provided a brief opportunity to invade a new host, they may encounter potentially fatal hazards. Most parasites can survive only for short periods away from one or the other of their hosts. Using an arthropod vector may help to assure that the transfer will take place quickly and safely. Parasites that do not utilize a vector often have a specialized resistant stage to carry them through the environmentally harsh and protracted time spent away from a living host. Desiccation and adverse temperatures are major hazards parasites face during transfer between living hosts. This is true even for some vectorborne parasites, but the degree of risk depends on the efficiency of the transmission provided by the vector. Parasites that are transmitted in vector salivary fluids are not exposed to the external environment because they are injected directly into the vertebrate host during feeding. In contrast, mechanical transmission is high-risk and so parasites do not rely solely on this method for their transfer to new vertebrate hosts.

Once the parasite successfully invades the host, whether vertebrate or invertebrate, the host's defensive armament confronts the parasite. Parasite transmission to resistant or immune hosts is wasted transmission. Thus, there is selective pressure on parasites to develop mechanisms for invading susceptible hosts and for neutralizing or avoiding immune or other defensive responses. This conflict has been characterized as a coevolutionary battle between parasites and their hosts. When mosquitoes ingest virus in a bloodmeal, the barriers the virus must negotiate if it is to successfully pass to a new host have been put into five categories: the peritrophic membrane barrier, the midgut barrier, the hemolymph barrier, the salivary gland barrier and the salivary gland escape barrier (Hardy et al. 1983, Hardy 1988). Parasites also face the insect's immune system and may fall victim to encapsulation or melanization responses or antibacterial peptides such as defensins and cercropins. It now appears that the latter also may help protect resistant insects against invasion by certain protozoans such as malaria parasites (Paskewicz and Christensen 1996).

Parasite Maintenance and Amplification

Not all hosts that become infected are essential to perpetuating or maintaining parasites. Because some parasites are transmitted inefficiently, they frequently end up in hosts that either cannot support development and/or reproduction or that are unlikely to serve as reservoir hosts of infection for the future transmission of the parasites. Transmission that is vital to the long-term perpetuation of parasites is termed **primary transmission**, while nonessential transmission that may occur sporadically or for limited periods, such as only during epizootics or epidemics, is referred to as **secondary transmission**. Transmission to hosts that cannot serve as reservoirs of future infection is referred to as **dead-end transmission** and the hosts involved are called **dead-end hosts**. Transmission of all zoonotic infections to humans fall into either the secondary or dead-end categories. There also are dead-end arthropod hosts that become infected but cannot serve as vectors. This is why finding a parasite in an arthropod species is not sufficient proof that it serves as a vector for that parasite. Vector incrimination involves several other criteria as well (see Chapter 6).

Hosts that serve as reservoirs provide for the long-term maintenance of the parasite and can be either the vertebrate or invertebrate host or both. Stable transmission between reservoir hosts and vectors is referred to as **enzootic transmission** (or endemic transmission if humans are the reservoir) and the hosts and vectors involved are termed the **enzootic hosts** and **enzootic vectors** of the parasites. By definition, endemic and epidemic transmission involves human hosts. However, some human outbreaks of zoonotic infections such as certain arboviruses also are referred to often as epidemics even though all the human transmission is dead-end. For that matter, even horse outbreaks of certain arboviral diseases have been referred to as epidemics because, as with humans, horses are not natural enzootic hosts.

Parasite amplification, or rapid spread, can occur for limited periods when ideal environmental and biological conditions for transmission exist. These episodes of accelerated transmission lead to an increased incidence of parasitism and disease, i.e., epizootics (increased incidence among nonhuman animals) or epidemics (when humans are involved). They also widen the distribution of parasites into hosts and geographic regions where they do not normally occur. Amplification among avian hosts is sometimes referred to as an epiornitic.

Parasite Enhancement of Transmission

Parasites are not merely spectators to their own transmission. Many infectious agents participate actively by changing the behavior or physiology of both their vertebrate and invertebrate hosts. This increases the probability of successful transmission.

Arthropod-transmitted parasites must be circulating in the vertebrate bloodstream (or lymph and surface tissues) to be ingested by bloodfeeding vectors. Blood provides the humoral defense system that protects vertebrates from infection. It may also, after repeated exposure, provide some protective immunity against protozoan and helminth parasites. To limit exposure of their surface antigens to the host's immune system, it is advantageous for parasites to be present in blood only when the probability of transmission is high. The infective stages of many filarial worms have circadian or seasonal patterns of appearance in host blood that mirrors the daily and seasonal biting cycle of their mosquito vector (Sasa and Tanaka 1972). Still other parasites have evolved to congregate at locations where the vector prefers to feed. For example, *Onchocerca bovis* parasites congregate in the umbilicus where black flies that feed on the underbelly are likely to become infected. Of course, feeding on selected regions of the body also may lead to regional concentrations of the parasite so that it naturally remains high in the region where flies are most likely to feed.

Phlebotomine sand flies are attracted to and selectively feed around the skin lesions associated with cutaneous leishmaniasis and so their chances of becoming infected with highly localized leishmanial parasites are enhanced (Coleman and Edman 1988).

As already described, microbes that mechanically interfere with bloodfeeding can enhance their transmission. Certain parasites also enhance transmission at the biochemical level. For example, malarial parasites in the salivary glands of mosquitoes disrupt the production of apyrase, a component of the salivary secretion that aids the probing mosquito in locating a blood vessel. Lack of apyrase prolongs probing time and thereby increases the probability of transmission of the malaria parasites (Rossignol et al. 1984, Ribeiro et al. 1985, Ribeiro 1989). Parasites also manipulate their vertebrate hosts physiologically and behaviorally to increase the feeding success of vectors that attack infective hosts (Day and Edman 1984, Rossignol et al. 1985).

As discussed in Chapter 4, salivary secretions injected by vectors during feeding also may modify the vertebrate host milieu in ways that assist the parasite in establishing an infection. However, this is an example of parasite exploitation rather than parasite induction.

PARASITES AND PATHOGENS WITH ARTHROPODS AS HOSTS OR VECTORS

Some members of most major groups of parasitic helminths, protozoans and microbes are transmitted to their vertebrate hosts by bloodfeeding arthropods. For many helminths, non-bloodfeeding arthropods serve as intermediate hosts, but do not directly vector the vertebrate parasites they harbor through their feeding behavior. Instead, the parasitic stages found in the arthropods enter their vertebrate host in 1 of 2 ways: some are ingested by the vertebrate host while eating or drinking; others have motile, invasive stages that leave their invertebrate host (usually in water) and actively penetrate the vertebrate host on their own. These parasites will be covered briefly here.

Arboviruses

As detailed in Chapter 11, many different viral taxa in at least 5 different families (Togaviridae, Flaviviridae, Bunyaviridae, Reoviridae and Rhabdoviridae) are vectored by arthropods (Monath 1988, 1989). These are collectively referred to as the arthropodborne viruses or arboviruses. By definition, arboviruses are viruses that infect vertebrates and are transmitted in some way by hematophagous arthropods.

Over 500 arboviruses are recognized and well over 100 of these have been associated with naturally acquired disease in humans and domestic animals. They generally have febrile, hemorrhagic or neuropathic effects on their vertebrate hosts. They cause acute infections with rapid onset following transmission and generally result in lifelong immunity among individuals that survive exposure. Disease severity ranges from inapparent infection or mild febrile illness to severe encephalitis or hemorrhaging and death. Arboviruses are transmitted biologically by various species of mosquitoes, sand flies, biting midges and ticks. Much of the literature on transmission dynamics and vector-parasite interaction has been generated from research with arboviruses and mosquitoes.

Following viral replication and dissemination within the arthropod host, transmission is effected through the salivary secretions when infective arthropods bloodfeed on vertebrate hosts. All human arboviral diseases, except for dengue fever, normally cycle in nonhuman hosts (i.e., are zoonotic infections) but human to mosquito to human transmission sometimes may occur during epidemics of other arboviruses as well (e.g., yellow fever). Because only yellow fever and Japanese encephalitis vaccines are available generally to prevent arboviral diseases and there are no curative drugs for treating these infections, interruption of transmission through vector control has been widely promoted.

Rickettsiae

Nearly all rickettsiae are associated with hematophagous arthropods with which they appear to have coevolved. The family Rickettsiaceae consists of small coccoid to rod-shaped, gram negative bacteria, 0.3-0.7 μ by 1.5-2.0 μ in size. These tiny bacteria once were considered intermediate between bacteria and viruses. They contain both RNA and DNA and are bounded by a 3-layered cell wall. They are obligate intracellular parasites capable of independent metabolic activity and reproduce, as do other bacteria, by binary fission.

Many rickettsial infections are maintained transovarially in the arthropod vector that serves as the reservoir host. Horizontal transmission to vertebrate hosts, including humans, takes place during bloodfeeding via the infected salivary secretions or feces deposited on the skin by mites, fleas, lice and, in particular, ticks. The genera *Coxiella* (agent of Q-fever) *Rickettsia* and *Orientia* contain the important species that are transmitted by arthropods to humans. *Rickettsia* and *Orientia* are grouped according to their antigenic and clinical features into 3 subgroups: (1) Spotted Fever, (2) Typhus and (3) Scrub Typhus. All human rickettsial diseases, with the possible exception of epidemic typhus, are zoonotic infections. Rickettsial infections often are lethal but fortunately they are treatable with certain antibiotics. Few vaccines for prevention exist, and those that do (e.g., Rocky Mountain Spotted Fever) rarely are used. Interruption of transmission by controlling the vector is therefore a common practice whenever these mostly zoonotic infections become epidemic.

Bacteria

Only a few of the many bacterial diseases of humans and other animals involve arthropods. Some of these may be transmitted mechanically on the mouthparts of bloodfeeding or wound- and eye-frequenting flies. Those that are biologically-transmitted by bloodfeeding arthropods (e.g., fleaborne plague bacilli and louse- and tickborne *Borrelia* spirochetes) are zoonotic infections (except for louseborne relapsing fever) caused by non-spore-forming bacteria. Transmission from arthropods to vertebrate hosts (usually rodents) is via infected saliva, crushed body parts or regurgitant associated with failed bloodfeeding attempts. Bacterial infections that are treated quickly respond well to antibiotic therapy. No vaccines existed to prevent any arthropod-transmitted bacterial infection until the recent release of 2 different vaccines. Their benefits are still being evaluated.

Protozoa

Flagellates (*Trypanosoma* and *Leishmania*) and blood-invading sporozoans (*Babesia*, *Theileria*, *Leucocytozoon* and *Plasmodium*) utilize arthropods as vectors and as hosts in which they undergo critical development. They produce the most serious and widespread arthropodborne diseases of humans: malaria, trypanosomiasis and leishmaniasis. Although single-celled animals, protozoans have complex life cycles and reproduce both sexually and asexually. As with all other prokaryotes, they have been intractable to vaccine development thus far despite major efforts in recent years. These parasites undergo replication and transformation in or around the insect's gut and then make their way to the salivary glands from where they are transmitted during feeding. The exception is *T. cruzi*, which develops in the gut of its kissing bug vector but is transmitted via fecal contamination of host skin. Most human protozoan infections are zoonotic, but 3 of the 4 species of human malaria have no wild animal reservoir. Drugs, often too harsh for long-term use, are available to treat protozoan infections, but drug resistant strains of some parasites have developed and spread rapidly in recent years. Vector control is a common strategy for interrupting transmission, except in the poorer developing countries where such preventive measures are often too costly.

Helminths

About 50 species of filarial worms (Nematoda) infect the lymphatic and circulatory systems of humans and domestic animals. All reproduce sexually and are transmitted biologically by species of bloodfeeding flies. These parasites develop but do not reproduce within their arthropod vector/host. Circulating microfilariae, the motile embryonic stage of the worm, are continuously produced by the long-lived female worms living (with males) in the vertebrate host. Microfilaria are ingested when vector flies bloodfeed and then develop into the infective stage (L3) within the flight muscles or Malpighian tubules of the vector. Infective stage larvae migrate to the head and mouthparts of the fly where they escape and penetrate the host skin when flies bloodfeed again. Chronic infection by nematodes is debilitating but usually not fatal to their human host. The 3 recognized species of human filariae do not infect naturally any other animal. Drugs traditionally used to treat adult worms are harsher than those used to prevent or treat microfilariae. The new ivermectins seem effective against both stages with few adverse side effects. No vaccine exists for any stage of any human nematode parasite, but research to develop such vaccines, especially for animal helminths, continues.

TRANSFER OF HELMINTHS THROUGH ARTHROPOD INTERMEDIATE HOSTS

Many Trematoda (flukes) and Cestoda (tapeworms) use insects (especially scarab beetle larvae), oribatid mites, isopods, crustaceans or mollusks as their intermediate hosts. However, these invertebrates do not actively transmit the infective stage of these parasites to their definitive, vertebrate host. Nonetheless, they are critical hosts in the life cycle of these parasites and often serve as the vehicle by which the parasites are transmitted indirectly to their vertebrate host (e.g., by ingesting the infected invertebrate host).

EPIDEMIOLOGICAL PATTERNS OF TRANSMISSION

Transmission involving vectors is not random and periods of accelerated transmission (epidemics and epizootics) are particularly disruptive to human and animal populations. These patterns often relate to changes in vector or vertebrate populations that may in turn be related to changes in weather, climate or the biotic environment. In temperate zones, where the transmission of most vectorborne diseases ceases during the colder months of the year, a summer pattern of disease often is used to indicate that the disease may be vectorborne. Transmission patterns also may be seasonal and repetitive from year to year even in the tropics, as with accelerated malaria and dengue transmission, which often are associated with the rainy season.

Some arbovirus epidemics occur in clusters of years separated by several years without transmission. The basis for such unusual patterns is not clear; it could be related to herd immunity. Epidemic patterns also may be associated with unusual weather patterns such as *El Niño* events or to catastrophic events such as those brought on by war, earthquakes, floods or drought. Epidemics of louse and fleaborne infections generally follow catastrophic events due to the disruption of host populations rather than increases in vector populations. Many of these features of transmission are discussed for specific diseases in the chapters that follow.

REFERENCES

Anderson, R.M. and May, R.M. 1979. The population biology of infectious diseases: Part I. Nature **280:**361-367.

Anderson, R.M. and May, R.M. 1991. Infectious diseases of humans: dynamics and control. Oxford University Press, New York and Oxford. 757 pp.

Centers for Disease Control and Prevention. 1999. Outbreak of West Nile-like viral encephalitis – New York, 1999. MMWR **48:**845-849.

Coleman, R.E. and Edman, J.D. 1988. Feeding-site selection of *Lutzomyia longipalpis* (Diptera: Psychodidae) on mice infected with *Leishmania mexicana amazonensis*. J. Med. Entomol. **25:**229-233.

Day, J.F. and Edman, J.D. 1984. The importance of disease induced changes in mammalian body temperature to mosquito blood feeding. Comp. Biochem. Physiol. **77A:**447-452.

Edman, J.D. 1988. Disease control through manipulation of vector-host interaction: some historical and evolutionary perspectives. Pp. 43-50 *in* T.W. Scott and J. Gumstrup-Scott (eds.), The role of vector-host interactions in disease transmission. Entomological Society of America, Landover, Maryland.

Enserink, M. 1999. Groups race to sequence and identify New York virus. Science **286:**206-207.

Ewald, P.W. 1993. Evolution of infectious diseases. Oxford University Press, New York and Oxford. 298 pp.

Ewald, P.W. 1996. Guarding against the most dangerous emerging pathogens: insights from evolutionary biology. Emerg. Infect. Dis. **2:**245-257.

Hardy, J.L. 1988. Susceptibility and resistance of vector mosquitoes. Pp. 87-126 *in* T.P. Monath (ed.), The arboviruses: epidemiology and ecology, Vol. 1. CRC Press, Boca Raton, Florida.

Hardy, J.L., Houk, E.J., Kramer, L.D. and Reeves, W.C. 1983. Intrinsic factors affecting vector competence of mosquitoes for arboviruses. Annu. Rev. Entomol. **28:**229-262.

Miike, L. 1987. Do insects transmit AIDS? Staff Paper 1. Office of Technology Assessment, US Congress, Washington, DC 43 pp.

Monath, T.P. (ed.) 1988, 1989. The arboviruses: epidemiology and ecology. 5 vol. CRC Press, Boca Raton, Florida. 1,319 pp.

Paskewicz, S.M. and Christensen, B.A. 1996. Immune responses of vectors. Pp. 371-392 *in* B.J. Beaty and W.C. Marquardt (eds.), The biology of disease vectors. University Press of Colorado, Niwot.

Randolph, S.E. 1998. Ticks are not insects: consequences of contrasting vector biology for transmission potential. Parasitol. Today **14:**186-192.

Randolph, S.E., Gern, L. and Nutthall, P.A. 1996. Co-feeding ticks: epidemiological significance for tick-borne pathogen transmission. Parasitol. Today **12:**472-479.

Ribeiro, J.M.C. 1989. Vector saliva and its role in parasite transmission. Exp. Parasitol. **69:**104-106.

Ribeiro, J.M.C., Rossignol, P.A. and Spielman, A. 1985. *Aedes aegypti* model for blood finding strategy and prediction of parasite manipulation. Parasitol. Today **60:**118-132.

Rossignol, P.A., Ribeiro, J.M.C., Jungery, M., Turell, M.J., Spielman, A. and Bailey, C.L. 1985. Enhanced mosquito blood-finding success on parasitemic hosts: evidence for vector-parasite mutualism. Proc. Nat. Acad. Sci. **82:**7725-7727.

Rossignol, P.A., Ribeiro, J.M.C. and Spielman, A. 1984. Increased intradermal probing time in sporozoite-infected mosquitoes. Am. J. Trop. Med. Hyg. **33:**17-20.

Sasa, M. and Tanaka, H. 1972. Studies on the methods for statistical analysis of the microfilarial periodicity survey data. S.E. Asian J. Trop. Med. Publ. Hlth. **4:**518-536.

Waage, J.K. 1979. The evolution of insect/vertebrate associations. Biol. J. Linnean Soc. **12:**187-224.

Chapter 6

The Epidemiology of Arthropodborne Diseases

BRUCE F. ELDRIDGE
University of California, Davis

Epidemiology is the study of disease ecology. A **disease** is any departure from health. Infectious diseases result from infections by pathogens or parasites. However, as Chapter 1 explained, there are many other kinds of diseases (e.g., vitamin deficiencies, metabolic disorders, psychoses, etc.). **Ecology** studies the relationships between individuals or populations and their biological and physical environment. Understanding arthropodborne disease ecology is fundamental to medical entomology. Thorough knowledge of ecological relationships among host, parasite and vector may permit accurate prediction of the risk of arthropodborne diseases to humans or other vertebrate animals. Such knowledge also is vital to optimize strategies for disease prevention. Lastly, understanding pathogen-host interactions of a disease may help develop therapeutic measures.

EPIDEMIOLOGICAL CONCEPTS AND TERMS

Historically, epidemiology studied only human diseases. It now studies diseases of animals other than humans, and even those of plants. In this chapter, epidemiology will be discussed solely in relation to diseases of humans and other animals. An **epidemic** is an unusually large number of cases of a disease (Barr 1979). Epidemics encompass human disease outbreaks, whereas the term **epizootic** is used for diseases in other animal species, including invertebrate animals. **Incidence** of a disease means the number of new cases in a defined population occurring during some time interval. This is commonly expressed as an **incidence rate**, or the number of new cases per unit of time in a population, e.g., 100 cases per year per 100,000 people. **Prevalence** refers to the number of cases of a disease present in a population at any given time.

When the incidence of a human disease is localized and stable, and new cases are balanced by increases in disease-free hosts so that prevalence remains relatively constant, the disease is said to be **endemic**. The term **enzootic** refers to the same situation in other animals.

A **zoonosis** is an infectious disease of nonhuman vertebrate animals that is secondarily transmissible to humans. This concept is especially important in medical entomology because most arthropodborne infectious diseases are zoonoses, and many zoonoses are arthropod-

B.F. Eldridge and J.D. Edman (eds.), Medical Entomology, Revised Edition, 165–185.

Effects on Species 1 (columns); **Effects on Species 2** (rows)

	+	O	—
+	**Mutualism**	**Commensalism**	**Parasitism** Predation
O	**Commensalism**	No interaction Neutrality	Amensalism
—	**Parasitism** Predation	Amensalism	Competition

Figure 6.1. Various kinds of symbiotic relationships based on effects of species involved. "+" = benefit received; "O" = no effect; "–" = harm. Terms in boldfaced type discussed in text (courtesy of Richard Karban).

transmitted. There are notable exceptions: rabies is a zoonotic disease that is not arthropodborne, but rather spreads by direct contact. Malaria may be the most important arthropodborne human disease of all, yet it is not a zoonosis. Malaria also occurs in other primates, as well as rodents, birds and lizards, but other species of plasmodial parasites cause diseases in these hosts. Diseases such as malaria that occur only in humans are known as **anthroponoses**. Dengue is another anthroponosis. Claims that nonhuman vertebrate hosts for dengue occur in nature have never been confirmed, and they are not essential for the maintenance of the disease. Epidemic (louseborne) typhus, another arthropodborne disease, was long thought to lack a nonhuman reservoir. However, recent studies have demonstrated the causal agent of epidemic typhus in North American flying squirrels, raising the possibility of other vectors and vertebrate hosts (Sonenshine et al. 1978).

When individuals of 2 different species of organisms live intimately, the relationship is called **symbiosis**. This intimacy may have evolved to the point where the 2 species cannot exist without one another. There are many different kinds of symbiotic relationships (Fig. 6.1). When both species derive some benefit, the relationship is termed **mutualism**. **Commensal-**

ism means that one species benefits and the other does not. A special case of commensalism is **phoresy**, in which one species transports another. In **parasitism**, one of the species (the parasite) benefits, while the other (the **host**) is harmed. The parasite usually depends on the host for nutrition, as well as a place to live. Parasitology is a field of study closely related to medical entomology, and is concerned with parasitic organisms, transmitted by both arthropods and non-arthropods. Arthropods themselves can be parasites. When they occur at or near the surface of the skin of the parasitized hosts, they are called **ectoparasites**. Common examples of ectoparasites are sucking lice and fleas. Parasites that usually spend most of their life cycles inside the body of their hosts are called **endoparasites**. A number of arthropods, such as lung mites and bot flies, fit this classification because they invade body cavities and tissues of vertebrate animals. The invasion of vertebrate tissue by Diptera larvae is called **myiasis**.

Arthropodborne infectious diseases have 3 components in their natural transmission cycles. A **pathogen** is an organism whose presence in or on the host may cause disease (i.e., a pathogenic condition). The host is a vertebrate animal, although arthropod vectors may be considered hosts for the pathogens they transmit. The establishment of a pathogen in a host is termed an **infection**. For a variety of reasons (e.g., low pathogenicity, host immune status) infections do not always display signs of disease. The **vector** is the organism that transmits the pathogen from host to host. For all diseases capsulated within medical entomology, vectors are arthropods. The terms pathogen and parasite have overlapping definitions. However, as explained in Chapter 1, microorganisms that are not classified as animals (e.g., viruses, rickettsiae, bacteria) are referred to as pathogens, whereas organisms (both microorganisms and larger animals) classified as animals (e.g., protozoans, helminthes, arthropods) are termed parasites. The relationship between pathogens and vectors, and the various mechanisms of pathogen transmission, were covered in Chapter 5.

The relationships between pathogens and their hosts are complex. Although beyond the scope of this book, **pathology** and **immunology** are important in understanding the epidemiology of arthropodborne diseases. According to Barr (1979), these disciplines consider questions such as the **pathogenicity** of an etiological agent (the proportion of hosts that develop symptoms among those infected by a given agent) and **virulence** (the proportion of infected individuals that develop severe symptoms). An important consideration in medical entomology is the occurrence of infectious stages of pathogens within the peripheral bloodstream of infected hosts (called **parasitemia** for parasites, **viremia** for viruses, etc.).

VECTOR INCRIMINATION

A fundamental activity of medical entomologists is to establish the role that a particular arthropod species or population plays in the transmission of a particular infectious disease agent. A group of arthropods may have no relationship to the transmission of a given pathogen, or may range in importance from involvement in rare, occasional or secondary transmission (**secondary vector**) to the most important means of transmission (**primary vector**). In the case of some zoonoses, arthropods may not necessarily be the primary means of transmission to humans, but may be responsible for transmission in enzootic cycles (**enzootic vector**).

The role of vectors is quantitative as well as qualitative. In other words, it is not enough simply to record that a given population of arthropods is or is not a vector of a given pathogen; rather, the epidemiological role of the arthropods as vectors must be described. Establishment of the relationship of arthropod popula-

Table 6.1. Criteria for incrimination of arthropods as vectors of pathogens of humans and other animals (after Barnett 1962).

Criteria
1. Demonstration that members of the suspected arthropod population commonly feed upon vertebrate hosts of the pathogen, or otherwise make effective contact with the hosts under natural conditions.
2. Demonstration of a convincing biological association in time and space between the suspected vectors and clinical or subclinical infections in vertebrate hosts.
3. Repeated demonstration that suspected vectors, collected under natural conditions, harbor the identifiable, infective stage of the pathogen.
4. Demonstration of efficient transmission of the identifiable pathogen by the suspected vectors under controlled experimental conditions.

tions to transmission of a particular disease agent is called **vector incrimination**. It is a critical activity because disease prevention and control often depend upon vector abatement. Under many circumstances, effective control is impossible unless the arthropod vector has been determined accurately. Vector incrimination must be carried out at the population level, not the species level, because there can be considerable intraspecific variation among arthropod population ecology and vector competence that can affect their importance as vectors (DeFoliart et al. 1987).

Medical entomologists have long recognized the significance of vector incrimination. Early investigators, such as Sir Patrick Manson (1844-1922) and Sir Ronald Ross (1857-1932), realized that species of arthropods varied in their ability to transmit pathogens, but the methodology used to incriminate arthropods systematically evolved much later. Barnett (1962) provided a detailed description of methods used to incriminate vectors of human disease pathogens, based on the well-known postulates of Robert Koch (1843-1910), a German bacteriologist. An adaptation of Barnett's methods for incriminating arthropods as causes of human diseases is presented in Table 6.1.

It was by the application of these methods in the early 1950s that Barnett and his fellow workers proved that the mosquito *Culex tritaeniorhynchus* was the primary vector of the virus causing Japanese encephalitis, a serious human viral disease in the Far East (Fig. 6.2). They found that a curve plotted for human cases of encephalitis closely followed a curve plotted for relative abundance of *Cx. tritaeniorhynchus*, with a lag equal to the average intrinsic incubation period of the disease in humans (Barnett 1962). Recently, the failure of researchers to satisfy these criteria in exploring the possibility that mosquitoes might transmit the human immunodeficiency virus (HIV) to humans led to the conclusion that such transmission was extremely unlikely (Miike 1987).

The methods used in vector incrimination range from classical methods of sampling organisms in nature to modern molecular studies of organismal gene sequences. Estimates of vectorial capacity (described below) are central in vector incrimination. However, vector incrimination can never be carried out without regard to the other components of transmission cycles (hosts and pathogens), nor without consideration of other biotic and abiotic factors in the environment.

VECTORIAL CAPACITY

The term **vectorial capacity** describes the dynamic relationship between vectors of infectious disease agents and vertebrate hosts. It combines the physiological attributes of vectors that determine their susceptibility to infection and their ability to transmit pathogens (**vector competence**), with relevant ecological and behavioral traits of vectors such as longevity, host preference and abundance. By definition, vectorial capacity represents the number of new infections disseminated per case per day by a vector (Fine 1981). The concept of vectorial capacity grew out of early attempts by Ross (1910) to quantify the transmission of malarial parasites by mosquitoes, and later models developed by Macdonald (1952). Garrett-Jones (1964), who first used the term, simplified the basic Macdonald model to include only entomological factors. There have been many modifications of this formula and many criticisms of the basic approach. Reisen (1989) and Dye (1992) provide excellent reviews of the subject. Black and Moore (1996) summarized the relationship between vectorial capacity and a general equation for the probability of infectious disease transmission (the Reed-Frost equation). A formula for vectorial capacity is shown in Fig. 6.3, as modified by Reisen (1989) to include a term (V) for vector competence. This modification illustrates the concept that vector competence (i.e., susceptibility plus transmissibility) is a component of vectorial capacity. In the Macdonald (1957) malaria models, the concept of vector competence was represented by a factor "b," the proportion of anopheline vectors with infective parasites in their salivary glands (i.e., are infective). This factor then was used to develop an **entomological inoculation rate**.

The sections that follow discuss some of the important components of vectorial capacity and methods used to estimate them. For a detailed discussion of all of the methods used to estimate vectorial capacity in mosquitoes, one should consult Service (1993).

Density of Vectors in Relation to Density of Hosts

Density means the number of individuals per unit of space. **Abundance** is a general term that addresses the question "how many?" **Absolute abundance**, or absolute number, describes the total number of individuals in a population. If the unit of space used to define density is large enough to encompass an entire population, density and absolute abundance will be equal. An example of an expression of density would be the number of ticks per hectare. There are instances when it is not feasible to relate numbers of individuals to units of space. In these cases, numbers can be expressed as indices of **relative abundance** (Odum 1983). The number of insects per trap, the number of insect larvae per dip and the number of ticks attaching to a host per hour are all expressions of relative abundance. Density and absolute abundance of populations are almost never known with certainty, and sampling is used to produce estimates. In most cases, expressions of relative abundance are meaningful only in comparison with other data collected in the same manner.

The density of vectors in relation to the density of hosts is an important component of vectorial capacity, because it bears directly upon the probability of effective vector-host contact. In Fig. 6.3, this factor is represented by "m." The element "ma" (host biting rate in bites per host per day) is derived by combining "m" with "a," the proportion of vectors feeding on a host multiplied by the host blood index. The density of potential vectors can be high at times, but if these periods do not coincide with periods of host and pathogen availability, transmission will not occur. Vector density is influenced not only by population factors (e.g., natality and mortality), but also by weather factors such as precipitation and temperature, and by competition from other organisms.

In nature, a number of physiological and ecological factors influences the rate of growth of a population as well as maximum absolute abundance. The level beyond which no major increases can occur is called the **carrying capacity** for that population (Odum 1983). Carrying capacity is determined by food, growth and

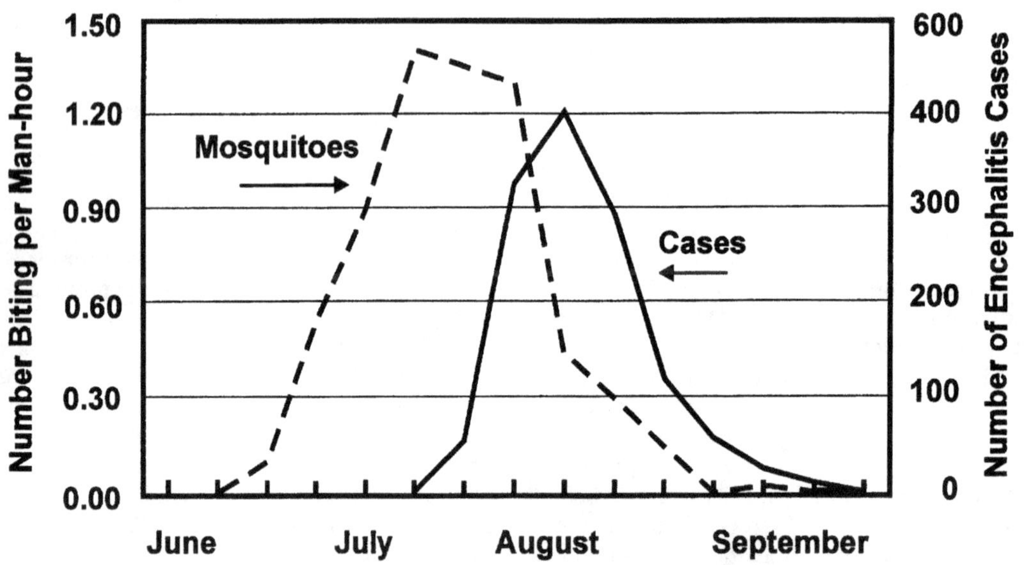

Figure 6.2. Relative abundance (number of biting mosquitoes per man-hour) for *Culex tritaeniorhynchus* superimposed upon human cases of Japanese encephalitis during the Tokyo epidemic of 1950 (after Barnett 1962).

mortality factors specific for the environment in which the population exists.

A population characteristic that bears directly on density is its **reproductive capacity**. This is the ability of populations to replace their numbers lost by mortality. It can be expressed quantitatively by the formula $R_o = B - D$, where R_o = reproductive capacity, B = births and D = deaths. If a population has not reached its carrying capacity, and $R_o > 0$, the population will increase. If $R_o = 0$, it will remain stable in size, and if $R_o < 0$, it will decrease in size. Estimating reproductive capacity requires the analysis of survivorship and natality in populations, either in the laboratory or in the field.

Population dispersal supplements natality and mortality in the determination of population density (Odum 1965). The relative ability of vectors to disperse (e.g., from areas where immature stages develop to areas occupied by vertebrate hosts) is related to the factor "m" in Fig. 6.3. Dispersal tends to be greater in flying insects than in non-flying insects or other arthropods. If the site where immature insects develop is located far from concentrations of human or other vertebrate hosts, and if those insects can fly only limited distances, then the chance of vector-host contact is small. If, on the other hand, insects fly great distances from their immature development sites in search of bloodmeals, then the chance of vector-host contact is increased. **Flight range** is an expression used in medical entomology to describe the farthest distance bloodfeeding flying insects will travel in search of bloodmeals. Of course, there may be other reasons for flight, such as searching for mates and oviposition sites, or searching for carbohydrate sources for energy, but these kinds of flight have received little attention. There are reports in the literature of flying insects being collected long distances from their nearest possible source. A classic example is a large collection of female *Aedes sollicitans* mosquitoes on a ship that had been no closer than 175 km from the North Carolina coast (Curry 1939). Such extreme examples undoubtedly represent the movement of insects in strong wind currents, and do not represent typical flight ranges.

Flight range can be estimated in 2 general ways. In the laboratory, insects are tested in devices known as flight mills (Rowley et al. 1968). These tests usually consist of tethering the insects on a pivoting arm of some kind, then recording the number of revolutions of the arm under various experimental conditions. Typically, flight mills record the speed and distance of the pivot arm, and store the resulting data in a computer. Data obtained by this approach probably bear little relationship to natural flight behavior. However, they can be useful in comparing the flight potential of different species of insects, and individuals of the same species that differ in some ways (e.g., male vs. female, pathogen-infected vs. noninfected, and starved vs. fed).

A method known as **mark-release-recapture** (MRR) may be used to estimate flight range in nature. In addition to estimating flight range, MRR also can be used to estimate other vector characteristics, including abundance and daily survival (see below). To estimate flight range, a sample of arthropods, either reared in the laboratory or captured in nature, is marked in some way, often with a nontoxic fluorescent dust or paint, and released into the environment. Traps for recapture are placed in concentric circles at various distances from the release point. Arthropods collected in these traps are examined (with ultraviolet illumination in the case of fluorescent dusts) and the number of marked individuals recorded. The distance traveled from the release point to the traps then can be known exactly. Usually, the mean distance traveled by recaptured individuals is more interesting than the farthest distance traveled by any one individual. For practical reasons, traps for recapture will be spaced further apart as the distance from the release point increases. To compensate for differences in trap density, a correction equation should be used to estimate mean distance traveled (Brenner et al. 1984).

Other methods have been used to mark released insects. Fay and Craig (1969) used genetic markers for female *Aedes aegypti*. Reiter et al. (1995) used radioactive rubidium to mark females of this same species, and then were able to detect labeled eggs in ovitraps.

The distance non-flying arthropods can disperse also is important, and some of these same methods can be used to estimate the distance traveled in search of bloodmeals (in this case, the host-seeking range).

Host Preference and Host-Feeding Patterns

As shown above, the component "ma" in the equation for vectorial capacity in Fig. 6.3 is derived from the product of vector density relative to hosts (m) and host biting habit (a). In turn, the host biting habit is the product of the frequency of biting and a host blood index. Underlying these factors is the concept of **host preference**. This concept is based on the observation that bloodfeeding arthropods frequently will feed on certain species of vertebrate hosts irrespective of the availability of other vertebrate species in an area. Estimates of host preference usually are based on laboratory choice tests of some kind that observe behavioral patterns in an olfactometer or field tests that record mosquitoes attracted to caged animals. An olfactometer records the behavior of animals in response to olfactory stimuli. The relative availability of hosts in time and space influences patterns of bloodfeeding by many arthropods. Some entomologists use the term **host-feeding**

pattern to describe the pattern of host selection at a given time and place. In epidemiological investigations, host-feeding patterns usually are estimated by sampling arthropods that have taken a recent bloodmeal, and then identifying the blood source. Identification usually is done using some kind of immunological procedure. Precipitin tests (e.g., capillary tube, ring or gel diffusion) have been the standard for years, but more recently, other methods such as the enzyme-linked immunosorbent assay (ELISA) and gel immunoelectrophoresis have become popular. These tests vary considerably in ease of use, cost, cross-reactivity and sensitivity, but all can provide satisfactory results if properly administered and interpreted.

It is not always easy to obtain large, unbiased samples of bloodfed arthropods for identification of bloodmeals. Within a given arthropod population, a significant portion of individuals usually will be seeking bloodmeals at a given time. Thus, collections of arthropods made with traps using attractants such as carbon dioxide are apt to attract individuals that are blood-seeking, but few individuals that recently have ingested blood. The most effective traps for blood-engorged arthropods are those designed to capture inactive arthropods (e.g., resting boxes or vacuum aspirators for mosquitoes). These traps are less biased toward any particular type of bloodmeal, and can result in relatively high proportions of bloodfed individuals, i.e., 10–30%.

For ticks and other ectoparasites, host preferences can be estimated by recording the proportions of various species on different vertebrate hosts.

Sampling bias can affect markedly estimates of host-feeding patterns. Local concentrations of a single species of vertebrate host may bias the estimate. Several statistical approaches have been proposed to account for variation in the relative occurrence of vertebrate hosts in the study area. One such approach is the **forage ratio** of Hess et al. (1968). This ratio is the percentage of arthropods in a sample that have fed on a particular host divided by the percentage that host comprises of the total number of potential hosts in the area. As pointed out by Edman (1971), there are problems with this approach, not the least of which is the difficulty in determining the total number of potential hosts in an area. To avoid some of these problems, Kay et al. (1979) devised the **feeding index**. In this approach, a comparison is made between proportions of bloodmeals observed and expected on 2 different hosts, in effect testing for differences that cannot be explained on the basis of chance. To derive realistic values for expected bloodmeals, 3 factors are taken into consideration: (1) the temporal and spatial concurrence of hosts and vectors, (2) the feeding success of vectors and (3) the relative body size of hosts.

The phenomenon of multiple bloodmeals, if blood from 2 or more vertebrate species is present in the arthropod tested, can complicate the application of results of bloodfeeding tests. It was thought at one time that certain arthropod species (e.g., mosquitoes) feed only once during the course of a reproductive cycle. However, several studies have shown that multiple bloodfeeding does occur in nature, in some species frequently. The explanation for multiple feeding is that sequential bloodmeals are taken during a single gonotrophic cycle or to a lesser extent, feeding is interrupted on the initial host, then continued on one or more subsequent hosts. Multiple bloodfeeding appears to be especially common in mosquitoes of the genus *Anopheles* (Briegel and Horler 1993), but it also has been recorded for the yellow fever mosquito, *Ae. aegypti* (Gubler 1988). This habit may contribute significantly to the ability of populations of this species to transmit dengue viruses even at low population levels (Scott et al. 1997).

Under most circumstances, a bloodfeeding arthropod must take at least 2 bloodmeals in order to transmit a pathogen to a vertebrate host; the first bloodmeal to acquire the pathogen from an infected (= donor) host; the second for transmission of the pathogen to a susceptible noninfected (= recipient) host. This is why the factor "a" in the formula in Fig. 6.3 is squared. However, situations involving **vertical transmission** of pathogens within vectors, i.e., the transmission of pathogens from parent to offspring, or from one life stage to another (e.g., from a larval to a nymphal tick), may permit transmission at the time of the first bloodmeal. If an arthropod adult lives sufficiently long to take many bloodmeals, its vectorial capacity is enhanced. Some arthropods take multiple bloodmeals during their life span. Both nymphs and adults, and both sexes, of human body lice take bloodmeals. Further, each louse may take several bloodmeals per day (Service 1996).

$$C = \frac{ma^2VP^n}{-\text{Log}_eP}$$

C = Vectorial capacity, the number of new infections disseminated per case per day by a vector
m = Density of vectors in relation to density of hosts
a = Proportion of vectors feeding on a host divided by the length of the gonotrophic cycle of the vector in days
V = Vector competence
P = Daily survival of vectors
n = Extrinsic incubation period

Figure 6.3. A formula for vectorial capacity, modified from Reisen (1989).

Daily Survival and Longevity

Vector survivorship is a critical determinant of vectorial capacity. It is directly related to the size of the infective vector population as well as the duration of infective life. The methods used to estimate daily survival in vectors generally fall under one of 2 types: horizontal or vertical (Milby and Reisen 1989).

Horizontal methods involve estimating daily survivorship in a single cohort of vectors over time. For these approaches some method of distinguishing the cohort is necessary, especially in populations having overlapping generations. Mark-release-recapture can be used for this purpose. To estimate daily survival, traps for recapture are placed at various locations surrounding the point of release (randomly, if possible) for a period of days or weeks. The number of marked recaptures will decay over time, and a decay line can be fitted by regression that will permit an estimate of the daily survival of the marked and released population (Milby and Reisen 1989). One disadvantage of MRR is that it confounds loss of individuals from a cohort because of death with loss due to emigration.

For purposes of estimating vectorial capacity, daily survivorship usually is estimated only for the instar responsible for transmission, typically the adult. However, there are times when it is useful to know which life stages are most vulnerable to various mortality factors (e.g., for designing control strategies). The **life table** method can be used for this study. This horizontal approach tabulates various mortality factors (predation, disease, accidents, senescence) for various age classes in a cohort (Kendeigh 1961, Southwood 1978). For arthropods, life stages (i.e., egg, first-stage larva, second-stage larva, etc.) usually are used instead of age classes. For experimental purposes, data are gathered in a variety of ways. Natural mortality factors often are estimated by comparing survival in cages that exclude predators with those that permit predation.

A laboratory approach to estimating daily survivorship is to place a group of arthropods from a single cohort in a cage of some kind, and then simply remove and record any dead

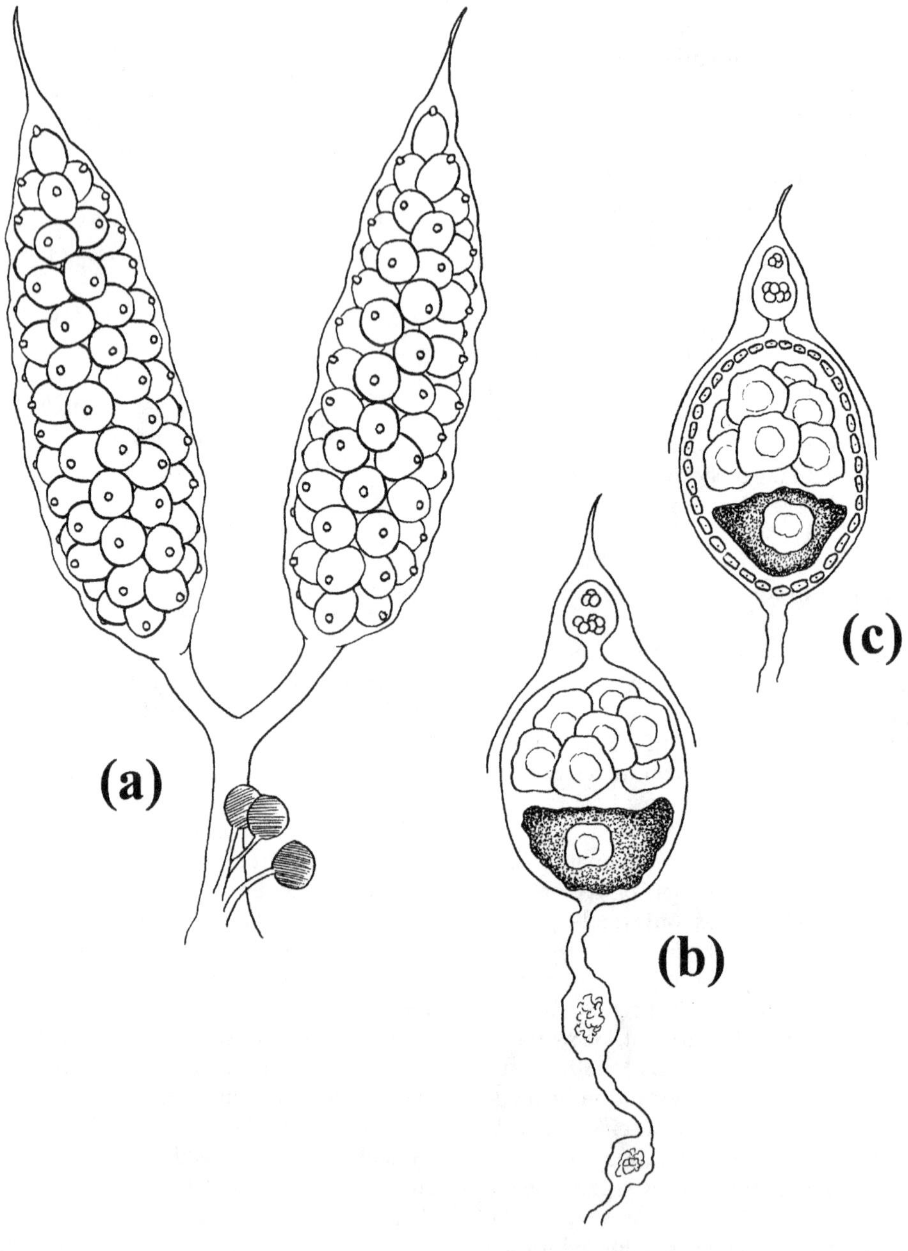

Figure 6.4. Female reproductive organs of a culicine mosquito: (a) Paired ovaries and associated structures; (b) details of individual ovarian follicle of a female that has never developed a batch of eggs; (c) details of individual ovarian follicle of a female that has been through two previous gonotrophic cycles (drawings by Carol S. Barnett).

individuals daily. The disadvantage of this approach is that many natural mortality factors, such as predation and infection, are excluded, and longevity nearly always is overestimated.

Vertical methods of estimating daily survivorship require examining samples of females in a population to determine the proportion that have passed some point in their life cycle, e.g., oviposition (Milby and Reisen 1989). This determination is done by dissecting samples of females drawn from the population, and examining the irreversible morphological changes in female reproductive organs that occur in the course of egg development and egg-laying. To describe the females at different points in their reproductive history, the term **nulliparous** is used for individuals that have never reproduced, **parous** for those that have. The latter group can be subdivided further into females that have developed a single batch of eggs (**uniparous**), or more than one batch of eggs (**multiparous**).

Another vertical method for estimating survival in arthropod populations falls under the term **age grading**. Many methods have been used. One observes the amount of wear on wings of flying insects and another counts the number of aquatic hydrachnid mites present on the bodies of adult mosquitoes. Infestations of these mites take place at the time of adult eclosion, and many drop off at the time of the first oviposition. Consequently, female mosquitoes with large numbers of hydrachnid mites ordinarily are nulliparous (Gillett 1971).

A relatively easy way of differentiating parous from nulliparous adult dipterans involves examination of the tracheae of ovaries (Detinova 1945). This method is based on the fact that in adult females that have never developed a batch of eggs, the ends of ovarian tracheae terminate in tightly coiled skeins of tracheoles (Fig. 6.5a). In contrast, after development of the first batch of eggs the coils become irreversibly stretched (Fig. 6.5b).

A tedious and controversial technique of age grading that has been applied to Diptera is the inspection of individual ovarioles for dilatations representing relicts of previously laid eggs (Polovodova 1949). With this technique, it is possible under some circumstances to differentiate between parous and nulliparous females, and to determine the number of egg batches laid. The Polovodova technique can be difficult to master because individual ovarioles must be dissected from the ovaries (Fig. 6.4), and because the application and interpretation of the method frequently have been questioned. Improvements in technique, such as the injection of paraffin oil into ovaries to increase the visibility of internal structures (Lange et al. 1981), have permitted more accurate determinations of reproductive history. An analysis by Sokolova (1994) of the morphological changes that occur during vitellogenesis (egg development) has clarified some of the confusion about the process.

Other age grading methods, including examination for the presence of a meconium (yellowish or greenish waste product present in mosquito guts), measuring apodemal growth rings and the assay of insect eyes for pteridines (fluorescent compounds) are described in Service (1993). Recently, observing temporal changes in the concentration of selected hydrocarbons on the epicuticle has shown promise for more accurate aging of mosquitoes (Desena et al. 1999).

Extrinsic Incubation Period

In vectorial capacity formulas, the **extrinsic incubation period** (EIP) is an exponent of the daily survival rate of vectors. EIP is the time from the uptake of an infectious bloodmeal by a vector to the time the vector is capable of transmitting the pathogen. The relationship between survival and EIP is obvious. In the case of very short EIPs, vectorial capacity may be high even if daily survivorship is relatively low. Because

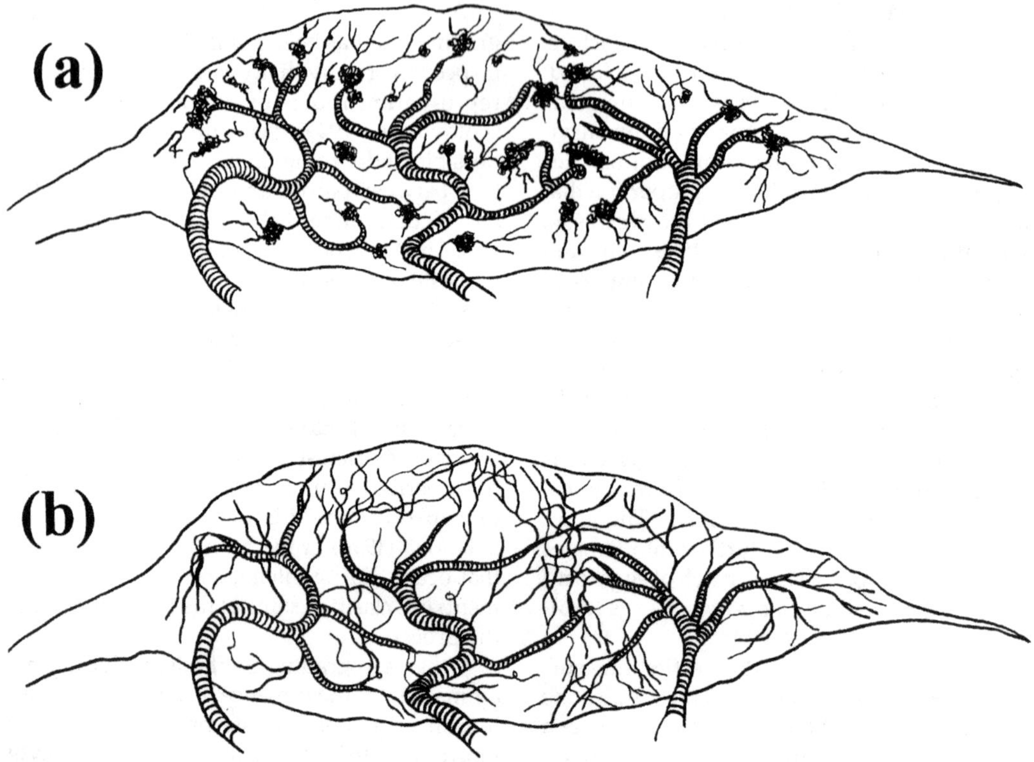

Figure 6.5. Diagram showing an ovary from a mosquito that has (a) never developed a batch of eggs and (b) one that has developed at least one batch of eggs (drawings by Carol S. Barnett).

EIP is influenced by temperature, estimates of EIP to determine vectorial capacity often are based on day-degree models for ambient temperature (Reisen 1989).

Vector Competence

Previous sections have dealt with various ecological and behavioral attributes of individuals as well as population parameters of arthropods that determine their vectorial capacity. Not all arthropods that feed upon an infective host become infected. Further, not all the arthropods that become infected will become infective. Vector competence is the ability of individuals in a population of arthropods to become infected by, and to transmit, a given strain of pathogen. Vectorial competence is an important component of vectorial capacity, and is one of the underlying requirements for vector incrimination. Estimates of vector competence are based on infection and transmission experiments carried out in the laboratory.

Natural infections usually are detected by dissection and direct microscopic observation, or by inoculation of ground arthropod suspensions into experimental animals, cell cultures or artificial media. To enhance the probability of pathogen detection in situations such as arboviruses in mosquitoes, where infection rates are very low, arthropods are tested in pools of 25–100. In these cases, the proportion of positive pools usually is reported as a **minimum infection rate** (MIR) of infected over noninfected individuals tested. The term "minimum" derives from the principle that there is no way to determine how many infected arthropods were present in a positive pool, but that there must have been at least one. Because infection rates usually are low, MIR frequently is expressed as the number of infected arthropods per 1,000 individuals tested. Identification of pathogens isolated from vectors is tedious and expensive. Historically, immunological methods such as the hemagglutination-inhibition test or the cross-neutralization test were used to identify pathogens recovered from arthropod pools. Current tools can provide more rapid and less expensive identifications. Enzyme-linked immunosorbent assays (ELISAs) have been developed to detect malarial parasites by species in anopheline mosquitoes (Wirtz et al. 1987). Recently, a combination of electrophoresis and DNA-specific probes identified leishmanial parasites isolated from phlebotomine sand flies (Vasconcelos et al. 1994).

A method known as **polymerase chain reaction** (PCR) has extended the detection and identification of a number of arthropodborne pathogens. This 3-step procedure involves (1) the denaturing of DNA from unknown samples, (2) the annealing of synthetic primers developed from known antigens and (3) the repetitive synthesis of DNA. If the annealing of the primers is successful (i.e., a positive test), the DNA in the sample is amplified many times, and then can be detected by electrophoresis. Primers presently are available for a number of pathogens.

Pathogens also can be isolated and identified from vertebrate hosts. This is the usual procedure for malaria, where the frequency of occurrence of parasites in mosquitoes may be low, and where closely related nonhuman parasites may be present.

Arthropods themselves may be used to detect some pathogens, either from vertebrates, or from other arthropods. **Xenodiagnosis** uses noninfected (clean) arthropods to detect the presence of pathogens in vertebrates. It has successfully detected trypanosomes causing Chagas disease using triatomid bugs. Noninfected arthropods also have been used for detection and propagation of arboviruses by injecting mosquito tissue extracts into the large-bodied adults of *Toxorhynchites* mosquitoes (Rosen 1981).

Traditionally, laboratory transmission experiments have involved infection of laboratory animals (mice, guinea pigs, etc.), followed by permitting noninfected arthropods to feed upon these parasitemic animals after a suitable incubation period (**intrinsic incubation period**). Bloodfed arthropods then are held under suitable temperature conditions for a time sufficient to become infective (extrinsic incubation period), and then permitted to feed on another group of noninfected laboratory animals. Finally, these recipient laboratory animals are maintained during an intrinsic incubation period and tested for evidence of infection. Using live animals for experimental transmission studies has a number of drawbacks. One is that the level of infection within the donor animal is impossible to control, and thus subject to considerable within-subject variation. Also, the purchase, care and feeding of laboratory animals is expensive and highly regulated. Moreover, the use of live animals now is discouraged in most laboratories if adequate alternatives are available. Consequently, most laboratories are now shifting to *in vitro* methods for such studies. These *in vitro* systems usually involve artificial feeding systems such as membrane feeders (Rutledge et al. 1964) or blood-soaked cotton pledgets for infection of arthropods, and cell culture systems in place of vertebrate hosts. An innovative method of capturing pathogens expelled in arthropod saliva utilizes a capillary tube feeder (Aitken 1971). These systems are less natural than live animal systems, but cost less and provide better control over pathogen dose, while avoiding the problems associated with the care and handing of experimental animals. An excellent discussion of the methods used to estimate vector competence of mosquitoes for arboviruses is found in Reeves (1990).

HOST AND PATHOGEN FACTORS

The preceding discussion approached epidemiology from the standpoint of the vector; however, host and pathogen factors also are critical to understanding vectorborne diseases and to preventing and controlling arthropodborne diseases. Studies that lead to the development of vaccines or therapeutic drugs are examples of pathogen-oriented or host-oriented approaches. Host factors to consider include immune status of hosts to given strains of pathogens, the innate susceptibility of hosts to pathogens and pathogen virulence for given hosts. These factors collectively may influence the level of parasitemia, and therefore, the level of pathogens available for infection of bloodfeeding arthropods.

Immunity of Hosts

Host mechanisms for immunity to arthropodborne pathogens can influence the ecology of a particular disease. Not only will immune status affect the proportion of infected hosts that show signs of diseases, but also it will affect the level and duration of pathogens in the peripheral bloodstream (viremia, bacteremia, etc.). Herd immunity is important as a population phenomenon. The proportion of individual hosts in a population immune to infection by a given pathogen can significantly affect the activity for that pathogen during a given period of time, especially if the hosts amplify the pathogen in nature. Vertebrate amplification is most often the case where pathogens have seasonal cycles, and fall to low levels during periods when little or no transmission takes place.

Susceptibility of Hosts

The innate susceptibility of hosts to infection by different pathogens varies greatly. It can range from nearly complete insusceptibility (e.g., rats and mice to human malarial parasites) to nearly complete susceptibility (e.g., horses for African horse sickness virus). Hosts infected with pathogens without showing signs of disease are called silent hosts. The ratio of silent infections to those causing symptoms in the host is referred to as the ratio of inapparent to apparent infections. Host susceptibility and host immune status are important for the same reasons. Milby and Reeves (1990) enumerated the characteristics of a vertebrate host as a source of arboviruses that infect mosquito vectors. The following criteria illustrate the importance of host characteristics in disease ecology:

1. High abundance in the area.
2. Lack of apparent signs of infection.
3. Development of a high and long lasting viremia after dermal inoculation of a small virus dose.
4. Lack of transfer of first season's protection to offspring from transfer of maternal antibodies.
5. Service as a preferred host for the mosquito vector.

Pathogen Characteristics

Virulence and tissue tropisms are significant considerations. Certain viruses infect the nervous system of vertebrate animals (e.g., western equine encephalomyelitis virus), whereas others infect visceral tissues (e.g., yellow fever virus). Not only does this difference affect the course of disease in the host, it also may influence the role the vertebrate host plays in the overall ecology of the disease. If viral infections do not produce high viremias of relatively long duration in a particular host, then that vertebrate may be a **dead-end host** and not serve as a source of additional infections to arthropods. Humans are dead-end hosts for a number of zoonotic pathogens (e.g., filarial worms responsible for dog heartworm).

LANDSCAPE EPIDEMIOLOGY

The study of the geographical distribution of living organisms is known as **biogeography**. For an examination of the relationship of environmental factors to the distribution of animals the reader should refer to Andrewartha and Birch (1984). Because the broad geographic distribution of arthropod-associated diseases is determined primarily by the geographic distribution of vertebrate hosts, disease pathogens and arthropod vectors, biogeography is important to medical entomology. The Russian scientist E. N. Pavlosky, who wrote extensively on what he termed the nidality of disease, emphasized the relationship between biogeography and arthropodborne diseases (Pavlovsky 1966). A **nidus (focus)** is defined as an area within the range of a disease where transmission is possible or more probable than elsewhere (McClelland 1987). Nidality uses the fact that within large geographical areas, defined by climatic factors (i.e., temperature and precipitation), numerous smaller regions vary because of **edaphic** (soil related) or other local factors such as slope, proximity of oceans, bays or rivers, or human land use activities such as urbanization, deforestation or agriculture. These combined local factors produce a variety of ecological zones or **landscapes** within geographic regions, and some arthropodborne diseases are associated with certain landscapes. A classic example is the occurrence of miteborne rickettsiosis (scrub typhus) in areas of Southeast Asia where forests have been cleared and replaced by grasslands (Traub and Wisseman 1974). Some of these disturbed areas are marked by the simultaneous occurrence of the etiologic agent of scrub typhus, *Orientia (= Rickettsia) tsutsugamushi*, wild rats in the genus *Rattus*, and mite vectors in the genus *Leptotrombidium*. The study of the relationship of geographic features of areas to the incidence of various diseases is called **landscape epidemiology**. Whereas other approaches in epidemiology focus on human activity and vector ecology, landscape epidemiology seeks to integrate biological and environmental factors to predict long-term disease distribution trends.

Weather and Climate

The basis for the distribution of plants and animals on the earth is the pattern of weather and climate. Climatologists classify climates mainly on the basis of long-term temperature and precipitation patterns. From the standpoint of medical entomology, the tropical rainy and the humid mesothermal climates (Trewartha and Horn 1980) are the most important. Dry climates are associated with arthropodborne diseases under special circumstances (e.g., when irrigation is used extensively) as are microthermal climates, but to a lesser extent. Tropical rainy climates concern medical entomologists for a number of reasons. Warm temperatures cause short life cycles for arthropods as well as short extrinsic incubation periods for some pathogens, and ample precipitation provides developmental sites for vectors with aquatic larvae. More species of organisms of all kinds, including vectors and pathogens, live in the tropics than elsewhere. Human densities are high in many tropical countries, and arthropod contact is frequent. Climate, along with socioeconomic factors, is a predominant reason for the large number of arthropod-associated diseases in South and Central America, sub-Saharan Africa and tropical Asia.

Vegetation

Because plants are more stationary than animals, they often indicate climatic patterns known as climatic associations, or biomes. Many schemes have classified these vegetative patterns. One of the best is the classification of vegetation by Holdridge (1947). Based on evapotranspiration determined from temperature and precipitation data for a given region, it predicts the dominant type of vegetation that will occur in each region under ideal (undisturbed) conditions. Differences in vegetation that can be correlated with the incidence of vectorborne diseases form the basis for geographic information systems (GIS) in assessment and prediction (Washino and Wood 1994). Chapter 13 contains a more detailed description of these approaches.

Human Culture and Behavior

Human activities influence the incidence of arthropod-associated diseases of all kinds, including those of livestock. The study of human populations (**demography**) and their relationship to arthropodborne diseases were covered extensively by Service (1989). This book includes topics such as human migration, urbanization, irrigation development and resettlement schemes.

A basic tenant of medical entomology is that arthropodborne disease problems rarely exist in undisturbed areas of the world. This is true even in tropical rain forests. Many epidemics occur in disturbed areas where forests have been removed and land has been converted to agricultural or urban use. In undisturbed tropical forests, diversity is high and host density is low. When such environments are converted to agricultural use, diversity decreases and host density (humans or domestic animals) increases. This combination increases disease transmission. In North America, the opposite may be true. The reversion of agricultural land to secondary forests and the construction of housing developments in forested areas may increase disease incidence by the increase in human-vector contact (e.g., Lyme disease).

Land use planners have not appreciated the effects of human activities on the incidence of arthropodborne diseases. Many disease outbreaks might have been avoided with proper planning. Medical entomologists must play an expanded role in alerting public officials to the dangers of arthropodborne diseases, and help plan major projects such as housing developments, seasonal wetlands and water redistribution projects.

ASSESSMENT AND PREDICTION OF DISEASE RISK

Medical entomologists spend much of their time preventing outbreaks of arthropodborne diseases. This is especially vital when the probability of an outbreak is high because of unusual weather, or after a widespread disaster such as a flood or an earthquake. Medical entomologists analyze patterns of occurrence of vectors, hosts and pathogens to predict the probability of arthropodborne disease for a given area or year. Such analysis requires many principles discussed earlier in this chapter.

For some particularly serious arthropodborne diseases, systematic disease surveillance programs have been designed and implemented. Disease surveillance can be passive (the recording of cases reported by caregivers) or active (the search for cases in connection with epidemiological studies). Passive surveillance usually underreports disease cases. For diseases where the ratio of apparent to inapparent infections is low (e.g., western equine encephalomyelitis), even fewer infections will be reported. Reasons for underreporting of arthropodborne diseases include insufficient resources for diagnosis and reporting, poorly defined case definitions and lack of adequate medical care.

The World Health Organization (WHO) operates a worldwide disease surveillance and reporting system. Much of this surveillance records the cases of certain diseases reported by member nations of WHO, but the reporting is often uneven and unreliable. Nevertheless, these reports are the best gauge of the worldwide incidence of certain reportable arthropodborne diseases currently available. In the USA, diseases such as Lyme disease, malaria, plague and mosquitoborne viral diseases are the subject of state-sponsored surveillance programs. Chapter 13 discusses the methods used for surveillance and prediction of arthropodborne diseases.

MODERN APPROACHES TO EPIDEMIOLOGICAL STUDIES

Although the principles of epidemiology have changed little over the past few decades, the approaches used to collect and process information for analysis have improved dramatically. One of the most significant changes has been the development of inexpensive, powerful computers. A number of new applications for computers have helped epidemiological studies. They include **remote sensing** (Washino and Wood 1994) and the use of **geographic information systems** (Kitron et al. 1994). These methods hold great promise because they integrate large amounts of diverse information into predictive models (see below). Still, their incorporation into surveillance programs and similar operations is in the future.

Biochemical methods have vastly improved techniques for the identification and genetic characterization of arthropods and pathogens. Consider how such methods have recently studied leishmanial parasites transmitted by phlebotomine sand flies. Thirty years ago, leishmanial parasites were characterized and named only on the basis of the human symptoms of leishmaniasis associated with the various geographic areas where they were found, and to some extent on their associations with certain species of sand flies. Now, isozyme analyses are standard for identification of *Leishmania*, and newer methods, such as random amplified polymorphic DNA-polymerase chain reaction (RAPD-PCR) are being used (Noyes et al. 1996). Interestingly, taxonomic studies using isozymes have confirmed the validity of the classification of *Leishmania* based on the earlier methods (Kreutzer et al. 1987).

Improved methodology has not lessened the importance of biosystematic studies of vectors, pathogens and hosts in arthropodborne disease epidemiology. Rather, new approaches have focused attention on the importance of areas of biosystematics such as taxonomy, genetics and evolution, by providing new and better ways to identify precisely genetic relationships within and between populations of the components of disease systems.

Genetic Analyses

Many studies have demonstrated genetic variation within populations of vectors, pathogens and hosts. This variation extends to a number of traits of epidemiological significance, including vector competence in arthropods and virulence in pathogens.

For many years, a standard tool for studying population genetics has been **electrophoresis**. This method is based on variation that occurs in soluble enzymes present in all arthropods. Because these enzymes vary in detectable ways, yet still carry out identical physiological functions, they are called isoenzymes, or just **isozymes**. These isozymes are not associated in any known way with vector competence. However, they still are useful to estimate the amount of gene flow within and between arthropod populations, and thus to estimate geographic, ecological and behavioral boundaries for species and populations. Electrophoretic analysis requires fresh or frozen material, but

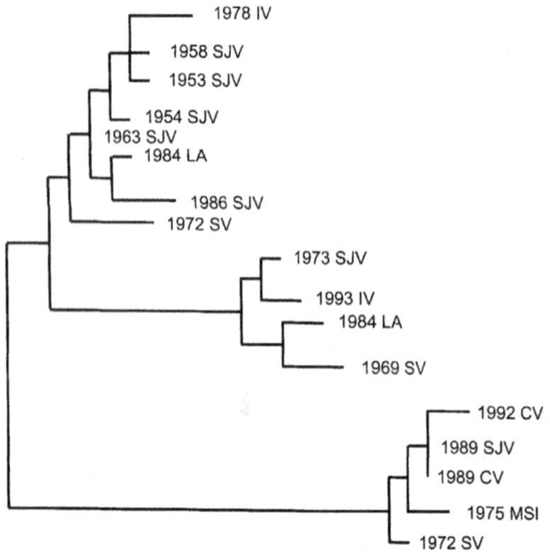

Figure 6.6. Phylogram generated by parsimony analysis of some nucleotide sequences of St. Louis encephalitis viral strains from California. Strains are designated by year and geographic source; CV = Coachella Valley, IV = Imperial Valley, LA = Los Angeles, SV = Sacramento Valley, SJV = San Joaquin Valley, MSI = comparison strain from Mississippi (data courtesy of Laura D. Kramer).

variation in the composition of cuticular hydrocarbons can be studied in samples stored in alcohol, or dried.

Still newer approaches in epidemiological studies involve molecular genetics. These approaches afford a more precise way to examine genetic variation, and involve methods such as the determination of the actual sequence of nucleotides that make up the nuclear DNA of vectors. New preparative techniques, such as PCR, have made it possible to study minute samples of arthropod material, including dried specimens. Molecular markers known as microsatellites also are useful in population genetic studies (Bruford and Wayne 1993). Microsatellites are small, heritable repeating DNA sequences that vary among and within species of many organisms, including arthropods.

For a review of the various experimental methods that study the population genetics and systematics of vectors the reader is referred to Tabachnick and Black (1996) and Black and Munstermann (1996).

Evolutionary Studies

The study of the evolution of parasitic relationships among biological organisms is fascinating. Diseases such as AIDS require knowledge of the recent evolutionary history of HIV viruses, especially their host ranges before the appearance of human diseases. The genetic research tools described above can be combined with modern methods of **cladistic analysis** (a mathematical procedure permitting the construction of evolutionary branching diagrams, or cladograms, from analysis of character state data) to study evolution. The advent of methods that detect DNA sequences in mitochondria or chromosomes have made possible highly refined and presumably highly accurate phylogenetic reconstructions, especially when molecular data are combined with morphological and isozyme data. Even phylogenetic reconstructions for submicroscopic organisms such as viruses now are possible. A **phylogram** (Fig. 6.6) also reflects phylogenetic relationships, but differs from a cladogram by portraying the degree of divergence by the length of the branches.

Modeling of Disease Systems

Scientists have attempted to model various infectious disease systems for almost as long as such systems have been known. Sir Ronald Ross was one of the first to become interested in constructing models for insectborne diseases. Most of the early models were conceptual, but later workers developed mathematical models, i.e., models in which various natural phenomena can be simulated by mathematical formulas. The malaria model proposed by Macdonald

(1957) represents such a model. Modern, desktop computers can run complex simulation models in a few minutes that would have taken hours to run just a few years ago. Models that include random variables are called **stochastic**; their outputs are in the form of probabilities. Models that include no random variables are called **deterministic**. The output of these models always will be the same for a given set of input data. Models that use sequentially different approaches to the solution of a problem, evaluating each approach in turn, are called **heuristic**.

Simulation models have been developed for the life cycles of a number of medically-important arthropods, including *Aedes aegypti*, the yellow fever mosquito (Focks et al. 1993). Life cycle simulations for *Amblyomma americanum*, the lone star tick, have been developed (Mount et al. 1993) as well as models that evaluate various integrated strategies for its control (Barnard et al. 1994).

CONCLUSION

This chapter has presented the study of arthropodborne diseases from an ecological perspective. Although the general relationships among vectors, pathogens and hosts now are understood for the most important diseases, new pathogens continue to be discovered, and new disease relationships detected. Understanding of new diseases and a fuller knowledge of old ones will require application of a combination of many of the concepts and approaches discussed. Furthermore, the future undoubtedly will bring still newer tools for the study of the epidemiology of arthropodborne diseases.

REFERENCES

Aitken, T.H.G. 1971. An *in vitro* feeding technique for artificially demonstrating virus transmission by mosquitoes. Mosq. News **37**:130–132.

Andrewartha, H.G. and Birch, L.C. 1984. The ecological web. The University of Chicago Press, Chicago. 506 pp.

Barnard, D.R., Mount, G.A., Haile, D.G. and Daniels, E. 1994. Integrated management strategies for *Amblyomma americanum* (Acari: Ixodidae) on pastured beef cattle. J. Med. Entomol. **31**:571–585.

Barnett, H.C. 1962. The incrimination of arthropods as vectors of disease. Proc. 11th Intern. Congr. Entomol. Wein (1960) **2**:341–345.

Barr, A.R. 1979. Epidemiological concepts for entomologists. Bull. Entomol. Soc. Am. **25**:129–130.

Black, W.C., IV. and Moore, C.G. 1996. Population biology as a tool for studying vector-borne diseases. Pp. 393–416 *in* B.J. Beaty and W.C. Marquardt (eds.), The biology of disease vectors. University Press of Colorado, Niwot.

Black, W.C., IV and Munstermann, L.E. 1996. Molecular taxonomy and systematics of arthropod vectors. Pp. 438–470 *in* B.J. Beaty and W.C. Marquardt (eds.), The biology of disease vectors. University Press of Colorado, Niwot.

Brenner, R.J., Wargo, M.J., Stains, G.S. and Mulla, M.S. 1984. The dispersal of *Culicoides mohave* (Diptera: Ceratopogonidae) in the desert of Southern California. Mosq. News **44**:343–350.

Briegel, H. and Horler, E. 1993. Multiple blood meals as a reproductive strategy in *Anopheles* (Diptera: Culicidae). J. Med. Entomol. **30**:975–985.

Bruford, M.W. and Wayne, R.K. 1993. Microsatellites and their application to population genetic studies. Curr. Opin. Genet. Dev. **3**:939–943.

Curry, D.P. 1939. A documented record of a long flight of *Aedes sollicitans*. Proc. N.J. Mosq. Contr. Assoc. **26**:36–39.

DeFoliart, G.R., Grimstad, P.R. and Watts, D.M. 1987. Advances in mosquito-borne arbovirus/vector research. Annu. Rev. Entomol. **32**:479–505.

Desena, M.L., Clark, J.M., Edman, J.D., Symington, S.B., Scott, T.W., Clark, G.G. and Peters, T.M. 1999. Potential for aging adult *Aedes aegypti* (Diptera: Culicidae) by gas chromatographic analysis of cuticular hydrocarbons, including a field analysis.. J. Med. Entomol. **36**:811–823.

Detinova, T.S. 1945. The determination of the physiological age of females of *Anopheles* by changes in the tracheal system of the ovaries. Med. Parazitol. Parazit. Bolezni **14**:45–49.

Dye, C. 1992. The analysis of parasite transmission of bloodsucking insects. Annu. Rev. Entomol. **37**:1–19.

Edman, J.D. 1971. Host-feeding patterns of Florida mosquitoes. 1. *Aedes, Anopheles, Coquillettidia, Mansonia* and *Psorophora*. J. Med. Entomol. **8**:687–695.

Fay, R.W. and Craig, G.B., Jr. 1969. Genetically marked *Aedes aegypti* in studies of field populations. Mosq. News **29**:121–127.

Fine, P.E.F. 1981. Epidemiological principles of vector mediated transmission. Pp. 77–91 *in* J.J. McKelvey, Jr., B.F. Eldridge and K. Maramorosch (eds.), Vectors of disease agents. Praeger Scientific, New York.

Focks, D.A., Haile, D.G., Daniels, E. and Mount, G.A. 1993. Dynamic life table model for *Aedes aegypti* (Diptera: Culicidae): analysis of the literature and model development. J. Med. Entomol. **30**:1003–1017.

Garrett-Jones, C. 1964. The human blood index of malaria vectors in relation to epidemiological assessment. Bull. Wld. Hlth. Organiz. **30**:241–261.

Gillett, J.D. 1971. Mosquitos. Weidenfeld and Nicolson, London. 274 pp.

Gubler, D.J. 1988. Dengue. Pp. 223–260 *in* T.P. Monath (ed.), The arboviruses: epidemiology and ecology, Vol. 2. CRC Press, Boca Raton, Florida.

Hess, A.D., Hayes, R.O. and Tempelis, C.H. 1968. The use of the forage ratio technique in mosquito host preference studies. Mosq. News **28**:386–389.

Holdridge, L.R. 1947. Determination of world plant formations from simple climatic data. Science **107**:367.

Kay, B.H., Boreham, P.F. and Edman, J.D. 1979. Application of the "feeding index" concept to studies of mosquito host-feeding patterns. Mosq. News **39**:68–72.

Kendeigh, S.C. 1961. Animal ecology. Prentice-Hall, Edgewood Cliffs, New Jersey. 468 pp.

Kitron, U., Pener, H., Costin, C., Orshan, L., Greenberg, Z. and Shalom, U. 1994. Geographic information system in malaria surveillance: mosquito breeding and imported cases in Israel, 1992. Am. J. Trop. Med. Hyg. **50**:550–556.

Kreutzer, R.D., Souraty, N. and Semko, M.E. 1987. Biochemical identities and differences among *Leishmania* species and subspecies. Am. J. Trop. Med. Hyg. **36**:22–32.

Lange, A.B., Khok, C.K. and Sokolova, M.I. 1981. The method of intraovarial oil injection and its use in the determination of the physiological age of females of blood-sucking mosquitoes (Diptera, Culicidae). Med. Parazitol. Parazit. Bolezni **50**:51–53.

Macdonald, G. 1952. The analysis of the sporozoite rate. Trop. Dis. Bull. **49**:569–586.

Macdonald, G. 1957. The epidemiology and control of malaria. Oxford University Press, New York. 201 pp.

McClelland, G.A.H. 1987. Medical entomology, an ecological perspective, 11th ed. University of California, Davis. 433 pp.

Miike, L. 1987. Do insects transmit AIDS? Staff Paper 1. Office of Technology Assessment, US Congress, Washington, DC 43 pp.

Milby, M.M. and Reeves, W.C. 1990. Natural infection in vertebrate hosts other than man. Pp. 26–65 *in* W.C. Reeves (ed.), Epidemiology and control of mosquito-borne arboviruses in California, 1943–1987. California Mosquito and Vector Control Association, Sacramento.

Milby, M.M. and Reisen, W.K. 1989. Estimation of vectorial capacity: vector survivorship. Bull. Soc. Vect. Ecol. **14**:47–54.

Mount, G.A., Haile, D.G., Barnard, D.R. and Daniels, E. 1993. A new version of LSTSIM for computer simulation of *Amblyomma americanum* population dynamics (Acari: Ixodidae). J. Econ. Entomol. **30**:843–857.

Noyes, H.A., Belli, A.A. and Maingon, R. 1996. Appraisal of various random amplified polymorphic DNA-polymerase chain reaction primers for *Leishmania* identification. Am. J. Trop. Med. Hyg. **55**:98–105.

Odum, E.P. 1965. Fundamentals of ecology. W.B. Saunders Co., Philadelphia. 546 pp.

Odum, E.P. 1983. Basic Ecology. Saunders College Publishing, Philadelphia. 613 pp.

Pavlovsky, E.N. 1966. Natural nidality of transmissible diseases (English translation). University of Illinois Press, Urbana. 261 pp.

Polovodova, V.P. 1949. The determination of the physiological age of female *Anopheles* by the number of gonotrophic cycles completed. Med. Parazitol. Parazit. Bolezni **18**:352–355.

Reeves, W.C. 1990. Epidemiology and control of mosquito-borne arboviruses in California, 1943–1987. California Mosquito and Vector Control Association, Sacramento. 508 pp.

Reisen, W.K. 1989. Estimation of vector capacity: introduction. Bull. Soc. Vect. Ecol. **14**:39–40.

Reiter, P., Amador, M.A., Anderson, R.A. and Clark, G.A. 1995. Short report: dispersal of *Aedes aegypti* in an urban area after blood feeding as demonstrated by rubidium-marked eggs. Am. J. Trop. Med. Hyg. **52**:177–179.

Rosen, L. 1981. The use of *Toxorhynchites* mosquitoes to detect and propogate dengue and other arboviruses. Am. J. Trop. Med. Hyg. **30**:177–183.

Ross, R. 1910. The prevention of malaria. E.P. Dutton & Co., New York. 669 pp.

Rowley, W.A., Graham, C.L. and Williams, R.E. 1968. A flight mill system for the laboratory study of mosquito flight. Ann. Entomol. Soc. Am. **61**:1507–1514.

Rutledge, L.C., Ward, R.A. and Gould, D.J. 1964. Studies on the feeding response of mosquitoes to nutritive solutions in a new membrane feeder. Mosq. News **24**:407–419.

Scott, T.W., Naksathit, A., Day, J.F., Kittayapong, P. and Edman, J.D. 1997. A fitness advantage for *Aedes aegypti* and the viruses it transmits when females feed only on human blood. Am. J. Trop. Med. Hyg. **57**:235–239.

Service, M.W. (ed.) 1989. Demography and vector-borne diseases. CRC Press, Boca Raton, Florida. 402 pp.

Service, M.W. 1993. Mosquito ecology. Field sampling methods, 2nd ed. Chapman and Hall, London. 988 pp.

Service, M.W. 1996. Medical entomology for students. Chapman and Hall, London. 278 pp.

Sokolova, M.I. 1994. A redescription of the morphology of mosquito (Diptera: Culicidae) ovarioles during vitellogenesis. Bull. Soc. Vect. Ecol. **19**:53–68.

Sonenshine, D.E., Bozeman, F.M., Williams, M.S., Masiello, S.A., Chadwick, D.P., Stocks, N.I., Lauer, D.M. and Elisberg, D.L. 1978. Epizootiology of epidemic typhus (*Rickettsia prowazekii*) in flying squirrels. Am. J. Trop. Med. Hyg. **27**:339–349.

Southwood, T.R.E. 1978. Ecological methods with particular reference to the study of insect populations, 2nd ed. Chapman and Hall, London. 524 pp.

Tabachnick, W.J. and Black, W.C., IV. 1996. Population genetics in vector biology. Pp. 417–437 *in* B.J. Beaty and W.C. Marquardt (eds.), The biology of disease vectors. The University Press of Colorado, Niwot.

Traub, R. and Wisseman, J., C.L. 1974. The ecology of chigger-borne rickettsiosis (scrub typhus). J. Med. Entomol. **11**:237–303.

Trewartha, G.T. and Horn, L.H. 1980. An introduction to climate, 5th ed. McGraw-Hill, New York. 416 pp.

Vasconcelos, I.A.B., Vasconcelos, A.W., Fe Filho, N.M., Queiroz, R.G., Santana, E.W., Bozza, M., Sallenave, S.M., Valim, C., David, J.R. and Lopes, U.G. 1994. The identity of *Leishmania* isolated from sand flies and vertebrate hosts in a major focus of cutaneous leishmaniasis in Baturite, northeastern Brazil. Am. J. Trop. Med. Hyg. **50**:158–164.

Washino, R.K. and Wood, B.L. 1994. Application of remote sensing to arthropod vector surveillance and control. Am. J. Trop. Med. Hyg. **50**:134–144.

Wirtz, R.A., Burkot, T.R., Graves, P.M. and Andre, R.G. 1987. Field evaluation of enzyme-linked immunosorbent assays for *Plasmodium falciparum* and *Plasmodium vivax* sporozoites in mosquitoes (Diptera: Culicidae) from Papua, New Guinea. J. Med. Entomol. **24**:433–437.

Chapter 7

Malaria, Babesiosis, Theileriosis and Related Diseases

THOMAS R. BURKOT[1] AND PATRICIA M. GRAVES[2]
[1]*Centers for Disease Control and Prevention, Fort Collins, Colorado and* [2]*Department of Preventive Medicine and Biometrics, University of Colorado Health Sciences Center, Denver*

INTRODUCTION

Descriptions and Definitions

The diseases malaria, babesiosis and theileriosis are caused by parasitic protozoa in the genera *Plasmodium*, *Babesia* and *Theileria*, respectively. All have complex life cycles in which both vertebrates and bloodsucking arthropods serve as hosts. The vectors of *Plasmodium* species are mosquitoes (of the genus *Anopheles* for human malarias), while ticks in a wide range of genera transmit babesial and theilerial parasites. These important human and veterinary pathogens share characteristics in their structure, antigenicity and the diseases they cause. However, malaria is a greater source of morbidity in humans than babesiosis, and theileriosis is strictly a disease of veterinary importance. Insects transmit a number of species in 3 related genera (*Leucocytozoon*, *Hepatocystis* and *Hemoproteus*) to birds and mammals. In some cases, these infections can cause serious diseases in domestic animals, especially fowl.

Classification

The protozoa that cause the diseases discussed in this chapter are in the Phylum Apicomplexa (Levine 1988). Their life cycles vary considerably. There are 2 groups of parasites of interest here: the class Haemosporidea, Order Haemosporida (which includes *Plasmodium*, *Leucocytozoon*, *Hepatocystis* and *Hemoproteus*), and the class Piroplasmea, Order Piroplasmida (which includes *Babesia* and *Theileria*). In both of these orders, the life cycle includes 3 phases of cell division and development known, respectively, as **merogony**, **gametogony** and **sporogony**. Merogony (often referred to as **schizogony**) is a type of asexual reproduction. It begins with the invasion of a host cell by a **merozoite**, followed by developmental changes and nuclear divisions that result in the budding and nearly simultaneous release of numerous daughter merozoites. In *Plasmodium*, the growing parasite within the erythrocyte is termed a **trophozoite**. Gametogony begins with the differentiation of some merozoites into **gametocytes**, thus beginning a phase of sexual reproduction. Gametocytes undergo cell division to produce male (**microgametes**) and female (**macrogametes**) sex cells that join

B.F. Eldridge and J.D. Edman (eds.), Medical Entomology, Revised Edition, 187–230.

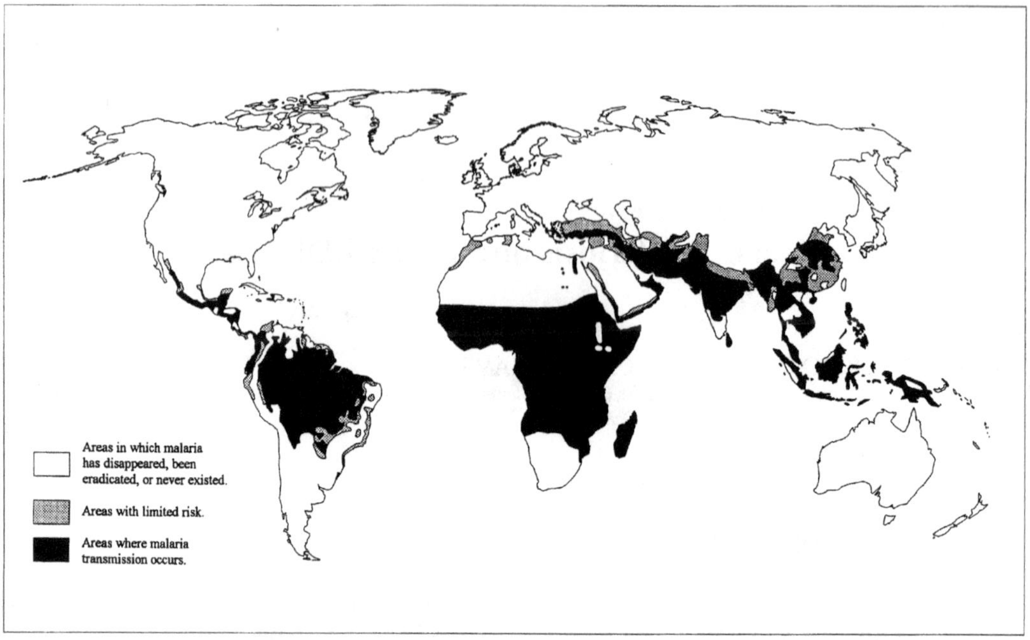

Figure 7.1. Worldwide distribution of human malaria (courtesy US Centers for Disease Control and Prevention).

to form a **zygote** within the insect midgut. Sporogony, another asexual phase, immediately follows fertilization and involves cell division within spores (**oocysts** in most species) resulting in the production of **sporozoites.**

All members of the class Haemosporida undergo merogony within erythrocytes of vertebrates and fertilization/sporogony in blood-sucking insect vectors. Other characteristics of the Haemosporida are that the **macrogametocytes** and **microgametocytes** develop independently in the vertebrate erythrocytes, the microgametocyte produces 8 microgametes, and there is a motile zygote (**ookinete**). Within the Haemosporida, *Plasmodium* is classified in the family Plasmodiidae; the closely related genera that infect birds and mammals, but not humans, are placed in the families Haemoproteidae and Leucocytozoidae (Bennett et al. 1994).

Protozoa in the Order Piroplasmida (such as *Babesia* and *Theileria*) also undergo merogony, gametogony and sporogony. All known vectors are ticks. Merogony is not restricted to erythrocytes, but occurs also in lymphocytes in *Theileria* spp. The taxonomic distinction between *Theileria* and *Babesia* is based on the lack of exoerythrocytic schizonts in *Babesia*. However, *Babesia equi* is an exception to this.

Distribution and Significance

Human malaria, a serious and prevalent disease, infects up to 500 million people and causes 2.5 million fatalities annually, primarily in the tropics (Fig. 7.1, World Health Organization 1994). Malaria is most serious in young children and pregnant women; World Health Organization (1993) attributes the deaths of at least 1 million children a year to it. In Africa, malaria is one of the 3 infectious diseases that contributes to the burden of disease as estimated by disability adjusted life years (World Bank 1993).

Babesiosis and theileriosis are economically important diseases in cattle and horses, particularly in tropical and subtropical areas (Kakoma and Mehlhorn 1994, Mehlhorn et al. 1994). However, babesiosis pathogens of rodents and bovines also have been responsible for about 200 human cases (Kreier and Ristic 1989), primarily in North America and Europe. Serosurveys suggest that infection of humans with babesiosis may be much more common in some tropical areas (reviewed by Telford et al. 1993).

Diseases of domestic fowl caused by apicomplexan parasites occur throughout the world. Infections of *Leucocytozoon caulleryi* in chickens represent a serious problem in certain parts of Asia (Morii 1992, Nakamura et al. 1997).

Historical Aspects

In 1888, Victor Babes characterized an intraerythrocytic protozoan that eventually became known as *Babesia bigemina*, the pathogen responsible for Texas cattle fever. Later, Smith and Kilbourne (1893) demonstrated that the tick, *Boophilus annulatus*, transmitted the pathogen. This was the first demonstration of transmission by a bloodfeeding arthropod of any microbial pathogen. It was also the first demonstration of vertical transmission of a pathogen in an arthropod. Subsequently, Ronald Ross, at the urging of Patrick Manson (who first observed the infection of mosquitoes by the parasites causing lymphatic filariasis), described the transmission of malarial parasites by mosquitoes, for which he was awarded the Nobel Prize (Harrison 1978).

Recent Disease Outbreaks and Current Status

In the USA, malaria was essentially eradicated in the 1950s, but the vectors are still present and sporadic outbreaks continue to occur. Since the late 1980s, several episodes have occurred in California associated with returning military personnel or immigrant populations from Central America. In the 1990s, clusters of cases apparently due to local transmission were reported from New York, New Jersey and Texas. In Europe, cases occasionally occur in the summer after persons living near airports are bitten by mosquitoes brought in by flights from endemic areas.

After the partial success achieved in eradicating malaria by the DDT indoor spray campaigns of the 1960s, malaria has returned to many parts of the world (e.g., India, Sri Lanka). Major epidemics have been aggravated by limited control activities, reduced immunity of the human population, the emergence of drug resistant parasite strains and physiologic and behavioral resistance of the vector to insecticides.

MALARIA

The Parasite

Description

The genus *Plasmodium* is classified into several subgenera. Of the 4 human malaria species, *Plasmodium vivax*, *Plasmodium malariae* and *Plasmodium ovale* are in the subgenus *Plasmodium*, whereas the subgenus *Laverania* contains *Plasmodium falciparum* together with some malaria parasites of nonhuman primates (Kreier and Baker 1987). The subgenus *Vinckeia* comprises species infecting lemurs and lower mammals. Mosquitoes of the genus *Anopheles* transmit the species in the subgenera *Plasmodium*, *Laverania* and *Vinckeia*. The 450 species of avian malaria parasites are classified in 4 other subgenera of *Plasmodium*, and are transmitted by both anopheline and culicine mosquitoes. The reptilian malarias are divided into 3 other subgenera. While the vectors of most reptilian malarias are still unknown, some are transmitted by phlebotomine sand flies and *Culicoides* biting midges. Recent analyses based on small subunit ribosomal RNA sequences suggest that *P. falciparum*, the most lethal of the human malarias, is more closely related to the rodent and avian malarias than to the primate malarias (McCutchan et al. 1984, Waters et al. 1993).

Life cycle

General

Developmental and propagative changes occur in both the vertebrate and mosquito host. All stages in the vertebrate are haploid. This life cycle is best known for the human malarias and they will be used as an example (Fig. 7.2).

Vertebrate stages

The vertebrate portion of the malaria life cycle begins when an infected mosquito inoculates sporozoites from the salivary glands while probing or taking a bloodmeal on a susceptible human host. Within 5 min, sporozoites invade the liver where they form hepatic **trophozoites**. Internal divisions (hepatic schizogony) result in a multinucleated hepatic **schizont** that will rupture to release thousands of **merozoites**. Within minutes of their release, the merozoites will invade red blood cells. The time from the introduction of sporozoites into the host until blood stage parasites can be detected is known as the **prepatent** period. However, not all the sporozoites of *P. vivax* and *P. ovale* that invade hepatocytes immediately form trophozoites. Some remain latent in the hepatocytes as **hypnozoites** for months or years before resuming development. These hypnozoites initiate the blood stage infections of relapses that occur in *P. vivax* and *P. ovale* infections. A relapse should be distinguished from a recrudescence, which is the detection of blood stage malaria parasites in an individual after an apparent aparasitemic period. Unlike the relapse, a recrudescence is not initiated by a fresh release of merozoites from the liver.

Further developmental changes occur in the erythrocyte: the merozoite first becomes a ring, then a trophozoite before dividing (erythrocytic schizogony) to form the erythrocytic schizont (Fig. 7.3). In *P. falciparum*, the schizont-infected red blood cells attach to the endothelial cells lining the venules, a process called **sequestration**. Intraerythrocytic stages of the parasite are contained within a parasitophorous vacuole. Merozoites produced by merogony (or schizogony) are released into the bloodstream and penetrate noninfected erythrocytes, thus initiating further rounds of parasite multiplication.

Erythrocytic trophozoites also can develop into gametocytes. *Plasmodium vivax* gametocytes may be seen simultaneously with the first appearance of merozoites in the blood, whereas the mature *P. falciparum* gametocytes generally are not seen until at least 7–10 days after merozoites appear, because immature gametocytes are sequestered. Like the rings, trophozoites and schizonts, the gametocytes remain intracellular in the erythrocyte unless ingested by a mosquito. The blood stages of *P. falciparum*, including mature gametocytes, are maintained in culture routinely (Ifediba and Vandenberg 1981, Trager and Jensen 1976).

Invertebrate stages

The vertebrate and invertebrate hosts of malaria offer different environments for the parasite, which has distinct ribosomes in the blood and mosquito stages (Waters et al. 1989). After ingestion into the mosquito gut, the mature gametocytes are released from the parasitophorous vacuole and erythrocyte membrane (Fig. 7.4). The drop in temperature from the vertebrate to the mosquito gut and xanthurenic acid induces gametogenesis (Garcia et al. 1998).

The female gametocyte (macrogametocyte) matures to become a macrogamete. The male gametocyte (microgametocyte) rapidly undergoes division into 8 cells that migrate to the surface of the microgametocyte and into finger-like projections. This process is called **exflagellation** (Fig. 7.5). The projections detach to become individual motile microgametes that can fertilize a macrogamete to form the diploid zygote. Within 16–30 hr of blood ingestion, exflagellation, fertilization, zygote formation and transformation into a motile ookinete have taken place inside the mosquito gut. To cross the peritrophic membrane, the ookinete has an enzyme with chitinolytic activity that is activated by the presence of mosquito midgut proteases (Shahabuddin and Kaslow 1994). After passage through the gut wall, the ookinete rounds up to form an oocyst underneath the basement membrane (Fig. 7.5).

Malarial infection may harm the mosquito host. Oocyst infections >10 resulted in decreased survivorship in *Plasmodium cynomolgi*-infected *Anopheles dirus* (Klein et al. 1986), but studies with *Anopheles gambiae* infected with *P. falciparum* by feeding on gametocytemic individuals found no effect of malaria infection on mosquito survivorship (Chege and Beier 1990, Robert et al. 1990). Of a sample of 620 infected mosquitoes in the *Anopheles punctulatus* complex in Papua New Guinea, 79% had <10 oocysts, with 46% of infected mosquitoes having only 1 or 2 oocysts (Fig. 7.6). The median number of oocysts of human malaria in naturally infected mosquitoes usually is <4.

Meiosis occurs rapidly after fertilization, followed by sporogony (asexual division), so the developing sporozoites in the oocyst are haploid. During fertilization in the mosquito, there is opportunity for cross-fertilization if parasites of more than one clone are present in an individual or possibly if the mosquito takes a mixed bloodmeal from 2 individuals with gametocytes in their blood. The extent to which self- or cross-mating occurs probably depends on the local intensity of transmission (Paul et al. 1995, Walliker 1989). The time from gametocyte ingestion by a mosquito to release of sporozoites from the salivary glands is known as the **extrinsic incubation period (EIP)**. The EIP varies with temperature and the species of malaria. Because the EIP usually is at least 8–12 days, the daily survival rate is an important component of their vectorial capacity.

The sporozoite stage has several unique molecular attributes. Chief among them is the presence of an abundantly expressed surface molecule with a region of species-specific amino acid repeats, the **circumsporozoite (CS) protein**. This molecule, which begins to be expressed while the sporozoites are developing in the oocyst, is the target antigen for species-specific immunologic detection and quantitation of the sporozoite stage. Eventually, the oocyst excysts, releasing the sporozoites that migrate to the salivary glands. Estimates of sporozoite loads in mosquitoes based on detecting CS protein and on counting sporozoites are given in Table 7.1. Perhaps only 20% of *P. vivax* sporozoites released from oocysts into the hemocoel reach the salivary glands from which they can be discharged when the mosquito takes a subsequent bloodmeal (Rosenberg and Rungsiwongse 1991).

Sporozoite invasion of salivary glands and bloodfeeding

Successful invasion of salivary glands by sporozoites induces pathological changes that influence mosquito bloodfeeding behavior (Ribeiro et al. 1985). Work with the bird malaria *Plasmodium gallinaceum* revealed that sporozoites will interfere with the apyrase secretions of the salivary glands. The reduced apyrase activity limits the mosquito's ability to inhibit vertebrate platelet formation. This interference with the mosquito's ability to take a normal bloodmeal leads to increased probing, which presumably would increase numbers of sporozoites inoculated, interrupted bloodmeals and multiple feedings within the same gonotrophic cycle. The result would be an increased inoculation rate and more efficient transmission.

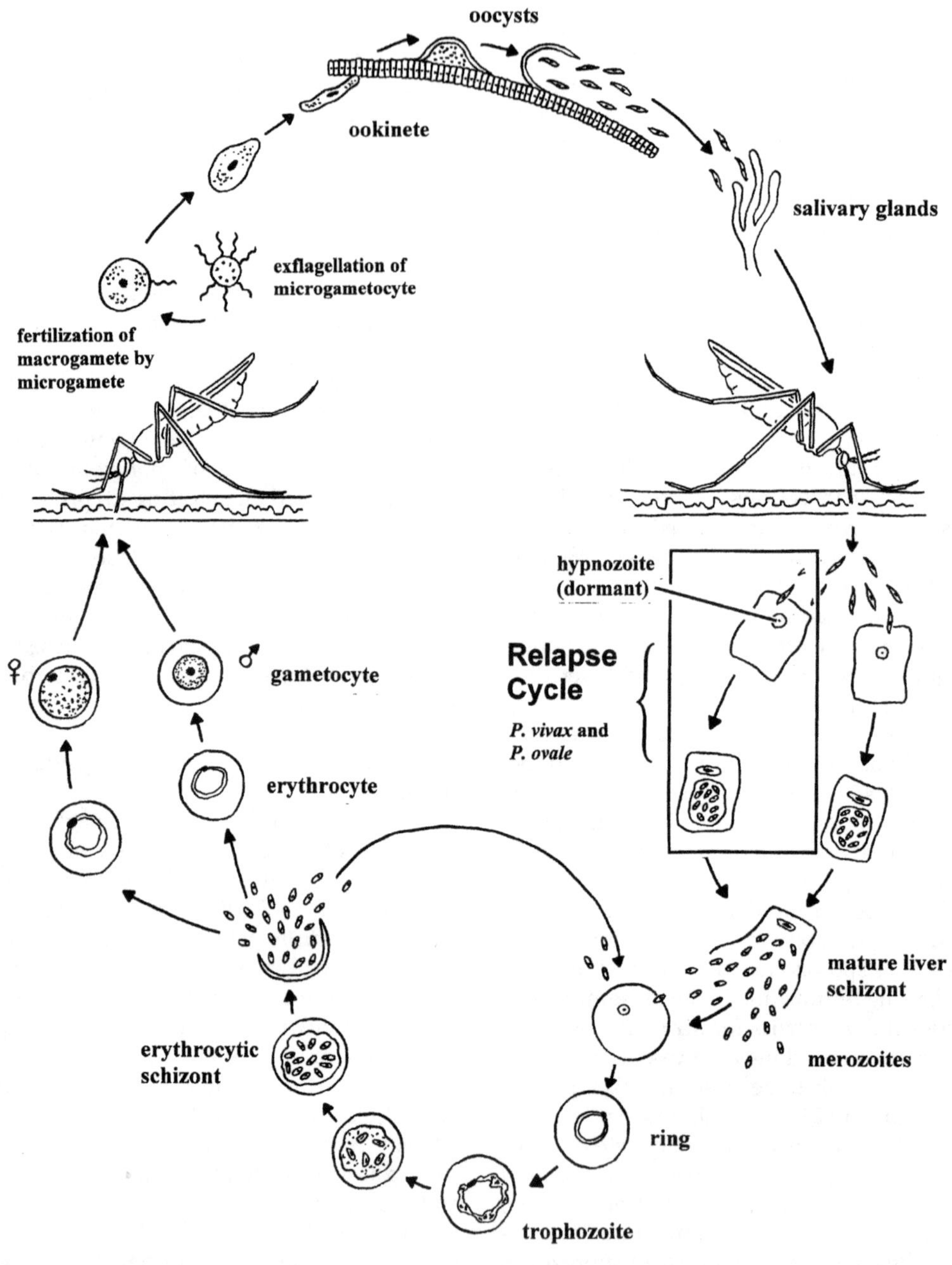

Figure 7.2. Generalized malaria life cycle: (a) Parasite development in mosquito host; (b) parasite development in human host (drawing by Carol S. Barnett).

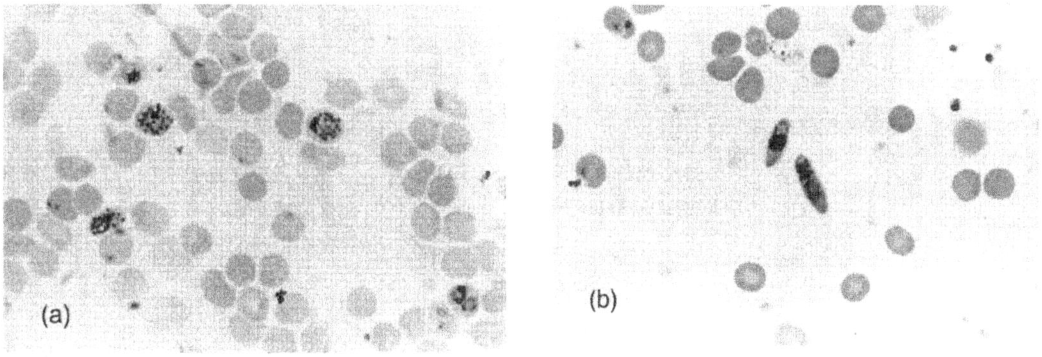

Figure 7.3. Blood stages of the human malaria parasite, *Plasmodium falciparum*: (a) Asexual blood stages as seen in culture; (b) mature male and female gametocytes (courtesy of Walter Reed Army Institute of Research).

Human Disease

Clinical aspects, including pathology

The classic clinical symptoms of human malaria in nonimmune individuals are chills followed by fevers and sweating (Table 7.2). The fevers are associated with the synchronous release of merozoites and associated toxins from the erythrocytic schizont. These **paroxysms** repeat every 48 hours for the tertian malarias, *P. vivax* and *P. ovale*, and every 72 hours for quartan malaria caused by *P. malariae*. *Plasmodium falciparum* may present an irregular tertian cycle. However, in immune individuals, **asymptomatic malaria** is 4- to 20-fold more common than **symptomatic malaria** (Covell 1960). Associated with repeated malaria attacks are pathological changes to a number of internal organs (Conner et al. 1976). Best known is the enlargement of the spleen, which traditionally has been used to define endemic areas.

Plasmodium falciparum is the species responsible for most of the mortality and severe morbidity due to malaria, the most severe effects appearing in young children and pregnant women. Factors that affect the severity of the malarial attack include the immune status of the individual, use of chemotherapeutics and host genetic factors, which will be discussed in subsequent sections. *Plasmodium falciparum* can rise rapidly to high densities and cause severe and potentially fatal complications (e.g., cerebral malaria, severe anemia and renal failure). **Cerebral malaria** occurs because of the ability of this species to undergo sequestration in brain capillaries. Infected erythrocytes thus would avoid passage through the spleen where they could be removed from circulation. In order to sequestrate, the parasite must insert certain antigens (known as PfEMP for *P. falciparum* erythrocyte membrane protein) onto the surface of the erythrocyte in structures known as knobs. There is a large family of PfEMP1 genes and it appears that antigenic variation resulting from differential expression of these genes enables the parasite to escape the host's immune system response (Borst et al. 1995).

Sporozoites and severity of infection

The size of the sporozoite inoculum may be related to the severity of the resulting clinical infection (Marsh 1992). In theory, the number of hepatic merozoites released from the liver will be proportional to the number of sporozoites inoculated by an infectious anopheline. The release of merozoites resulting from the bite of

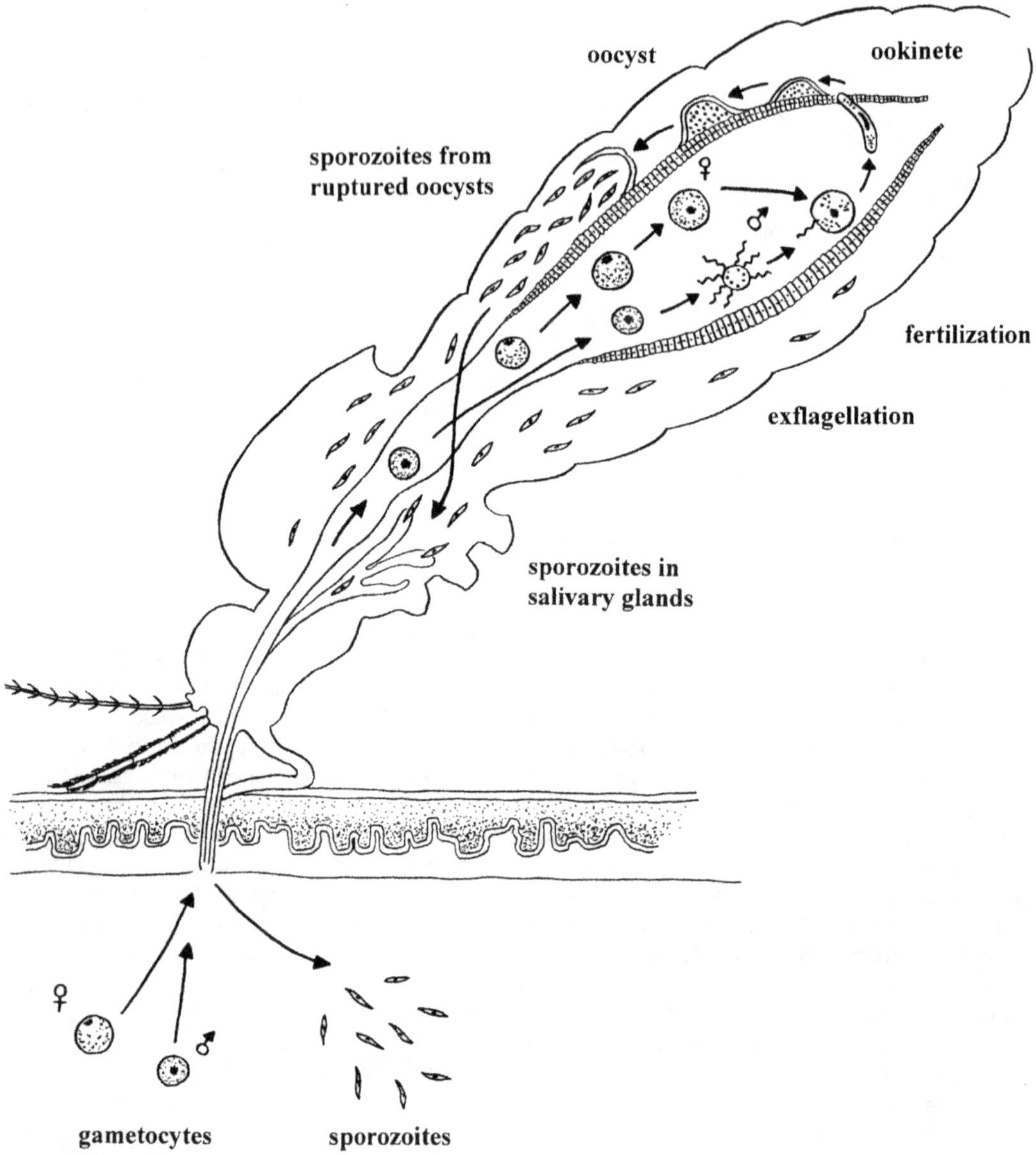

Figure 7.4. Malaria life cycle in the mosquito host (vector). The malaria parasite undergoes both developmental and multiplicative changes in the mosquito. Following ingestion of mature male and female gametocytes while bloodfeeding, exflagellation of the microgametocyte in the mosquito gut releases microgametes (male cells) that fertilize macrogametes (female cells) to form a zygote. Further developmental changes result in a mobile ookinete that traverses the mosquito gut, where an oocyst develops. In the oocyst, multiplication of the parasite eventually releases sporozoites that subsequently invade the salivary glands (drawing by Carol S. Barnett).

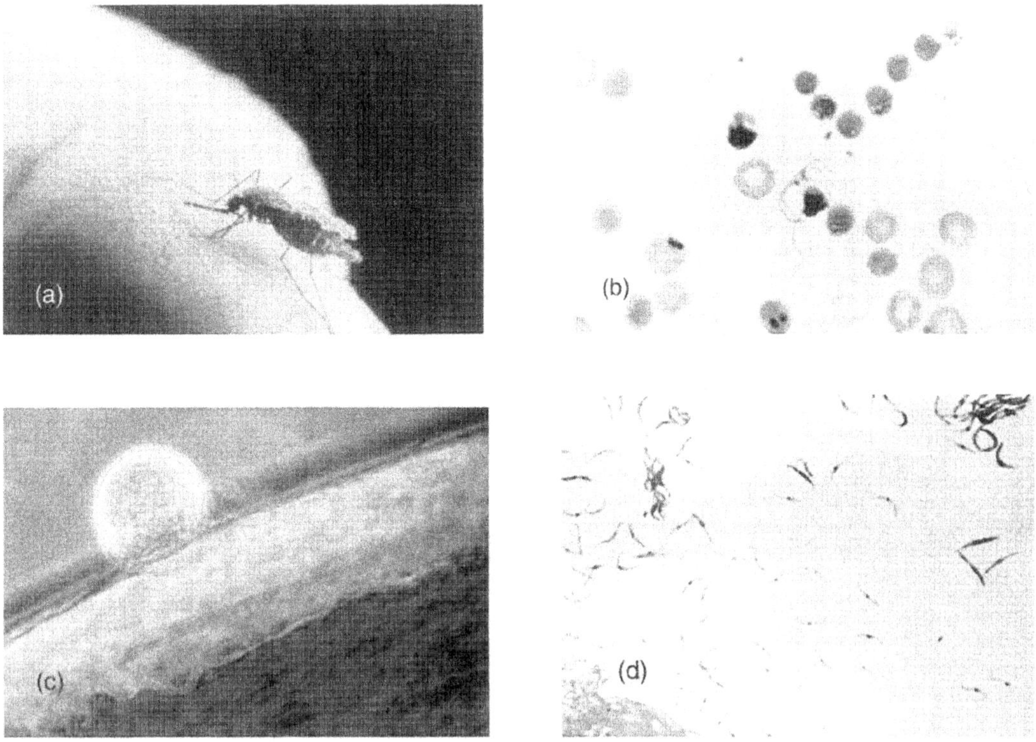

Figure 7.5. (a) Female *Anopheles gambiae* feeding on a human arm (courtesy of Wellcome Trust Tropical Medicine Resource); (b) exflagellation of microgametes occurs in the mosquito gut (courtesy of Walter Reed Army Institute of Research); (c) oocyst on the mosquito gut — note blood in mosquito gut from a previous bloodmeal (photograph by Patricia M. Graves); (d) malaria sporozoites (courtesy of Walter Reed Army Institute of Research).

a single infectious mosquito should be synchronous. A larger inoculum theoretically would diminish the prepatent period, reducing the time for the immune system to react to the infection, thereby minimizing the ability of the host's immune system to suppress the infection. In this scenario, larger inocula would result in a more severe clinical infection.

Unfortunately, experiments with both human, monkey and bird malarias that might have confirmed the relationship between size of inocula and clinical outcome used inocula that were not epidemiologically relevant (Burkot and Graves 1994). Sporozoite inocula, whether delivered by needle or the bite of infected mosquitoes, far exceeded those which might reasonably be expected to be delivered in a malaria endemic area.

Diagnosis

Diagnosis usually is based on clinical criteria, with parasite detection by examination of Giemsa-stained thick blood smears for confir-

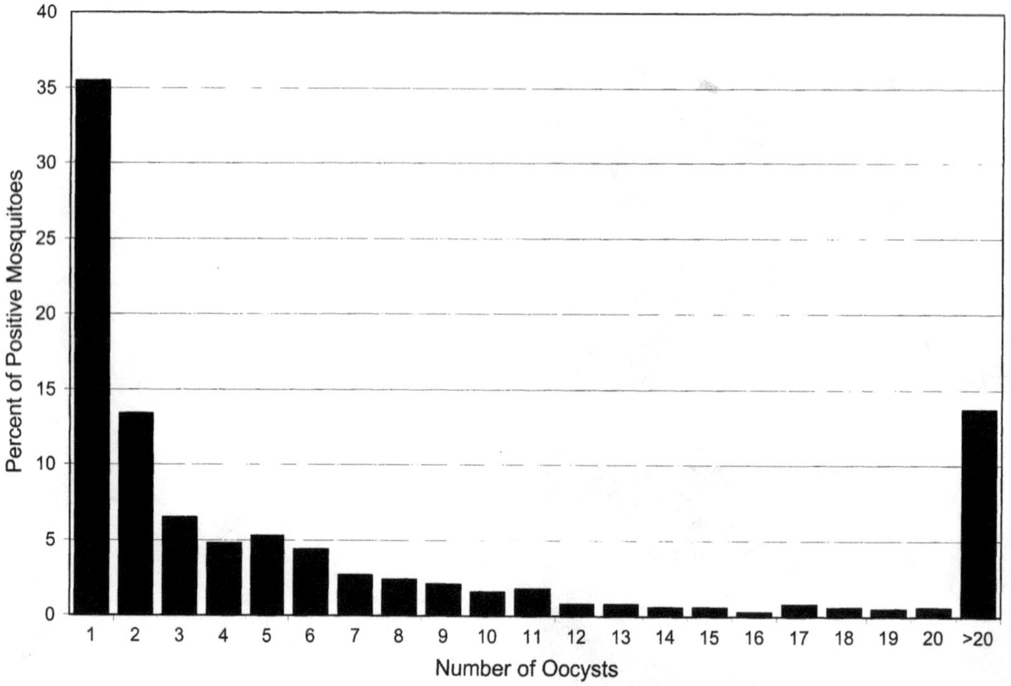

Figure 7.6. Oocyst distribution in naturally-infected mosquitoes in the *Anopheles punctulatus* complex in Papua New Guinea (data from Burkot et al. 1988).

mation if facilities permit. The reliability of clinical diagnosis varies widely: in one study, 84–99% of physician diagnoses were predictive of parasitemia while in another, only 26–28% of diagnosed cases were parasite-positive (reviewed in Gomes et al. 1994). The difficulty stems in part from the preponderance of asymptomatic parasitemias in highly endemic areas and the frequency of other causes of fever.

While parasitemias in humans as high as 10% have been reported, the typical parasite density is much lower (<0.01%). Consequently, failure to find malaria parasites in blood smears generally does not preclude presumptive treatment for malaria in an endemic area. When the number of thick blood films examined for malaria parasites was doubled from 200 to 400 fields, the relative prevalence of *P. falciparum* and *P. malariae* increased by 10% and 24%, respectively (Molineaux and Gramiccia 1980). Two new types of diagnostic tests for malaria based on detection of parasite antigens recently have become available (Makler et al. 1998, World Health Organization 1996). These "dipstick" tests have sensitivity and specificity equal to conventional microscopy.

Immunity

Natural

The vertebrate host can affect the development of the malaria parasite in a number of ways (Miller 1988). First, the parasite may be unable to invade erythrocytes of a particular type. An example is the inability of *P. vivax* to invade erythrocytes lacking the human Duffy blood group antigen (as found in people of West Africa and many Americans of African descent), or the increased refractoriness to invasion caused by increased rigidity of the ovalocytic erythrocytes found in certain Melanesian populations. Genetic factors also may influence the ability of parasites to grow in the erythrocyte. For example, development of *P. falciparum* within the cell is hindered in individuals with sickle cell anemia, lending individuals heterozygous for this trait an advantage in areas where malaria is endemic. Finally, genetic factors such as HLA (human leucocyte antigen) type may influence the severity of malaria (Hill et al. 1991).

Acquired

Exposure to malaria slowly builds up immunity to all stages of the parasite, based on both antibody and cellular responses. Thus, adults in endemic areas have reduced parasitemias and reduced symptomatology. However, it is not clear which immune responses (i.e., to which antigens) are critical in the development of protective immunity. Some rodent models display protective immunity in the absence of a detectable antibody response while others possess immunity without a demonstrable cellular immune response (Long 1993). Natural infections of *P. vivax* (Mendis et al. 1987) and *P. falciparum* (Graves et al. 1988b) induce antibodies that block transmission of gametocytes from humans to mosquitoes.

Epidemiology of Malaria

Epidemiological patterns

Plasmodium falciparum and *P. vivax* are the two most common human malaria parasites worldwide, with *P. vivax* having the most widespread geographic distribution and being found in both tropical and temperate climates. Malaria endemicity classically was not defined by detection of parasites. Rather, endemicity was defined by the proportion of children (2–9 years of age) and adults with enlarged spleens. A spleen-based index can screen individuals more rapidly than classification systems based on parasite rates. In this system, a **hypoendemic** area is one in which spleen rates in children are <10%. The **mesoendemic** area is characterized by spleen rates in children from 11 – 50%. When spleen rates in children are >50% and spleen rates in adults >25%, the area is described as **hyperendemic. Holoendemic** areas have spleen rates > 75% in children but low spleen rates in adults.

Prevalence now often defines endemicity. However, there are no clear-cut distinctions between levels of endemicity. Protective measures such as chemotherapy may reduce transmission even in areas with high inoculation rates. Age-specific prevalence data are the most informative (Fig. 7.7). The higher the transmission level the lower the age at which peak prevalence occurs.

The above terms (hypo-, meso-, hyper- and holoendemic) generally are used to describe large geographic areas. Cattani et al. (1986) described the importance of small area variations in malaria endemicity where consistent differences in prevalence of parasites or transmission exist within an endemic area in a limited geographic area.

Malaria also may be defined as being **stable** or **unstable**. Stable malaria means that the parasite is endemic (i.e., constantly present in a given area). There may be seasonal variation in transmission, but it remains constant from year to year. The infection is most severe in children because adults have acquired some immunity; young babies acquire some level of immunity from their mothers (maternal immunity). Stable malaria occurs in areas which have high vector populations, sufficiently high mean temperatures for the completion of the extrinsic cycle within the vector's lifespan, and opportunity for contact between host and vector (preference

Table 7.1. Sporozoite loads in anopheline mosquitoes from Africa, Southeast Asia and the Pacific. Sporozoite densities can be quantified by counting in a hemocytometer or after Giemsa-staining, or estimated in an ELISA that detects the circumsporozoite protein. Sporozoite densities are stated as the arithmetic (A) or geometric (G) mean.

Vector	Parasite	Sporozoite no.	Quantitative method	Gametocyte source
An. dirus[1]	*P. vivax*	3,688/oocyst	Hemocytometer	Human
An. dirus[1]	*P. falciparum*	3,385/oocyst	Hemocytometer	Human
An. punctulatus[2]	*P. vivax*	1,050 (A) 350 (G)	CS-assay	Wild-caught mosquitoes
An. punctulatus[2]	*P. falciparum*	6,400 (A) 2,300 (G)	CS-assay	Wild-caught mosquitoes
An. farauti[2]	*P. vivax*	330 (A) 150 (G)	CS-assay	Wild-caught mosquitoes
An. farauti[2]	*P. falciparum*	6,010 (A) 2,140 (G)	CS-assay	Wild-caught mosquitoes
An. koliensis[2]	*P. vivax*	250 (A) 160 (G)	CS-assay	Wild-caught mosquitoes
An. koliensis[2]	*P. falciparum*	6,250 (A) 2,420 (G)	CS-assay	Wild-caught mosquitoes
An. gambiae[3]	*P. falciparum*	4,845 (A) 962 (G)	Giemsa-stained slides	Wild-caught mosquitoes
An gambiae[4]	*P. falciparum*	17,030 (A) 6,380 (G)	Giemsa-stained slides	Wild-caught mosquitoes
An. funestus[3]	*P. falciparum*	2,843 (A) 812 (G)	Giemsa-stained slides	Wild-caught mosquitoes
An. funestus[4]	*P. falciparum*	8,670 (A) 4,540 (G)	Giemsa-stained slides	Wild-caught mosquitoes

[1]Rosenberg and Rungsiwonge (1991).
[2]Burkot et al. (1988).
[3]Beier et al. (1991).
[4]Pringle (1966).

for human biting combined with housing conditions that enables contact between humans and mosquitoes between dusk and dawn). Examples of areas with stable malaria include Africa south of the Sahara, the lowlands of New Guinea and the forested areas of Latin America and Southeast Asia.

In unstable malaria, epidemics periodically occur that affect people of all ages. Malaria epidemics may occur in areas where vectors are present but transmission is marginal because

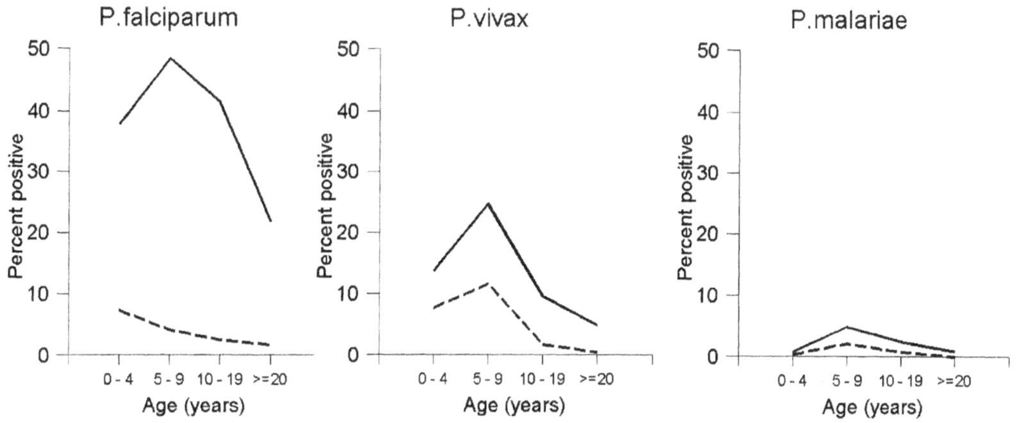

Figure 7.7. Age-specific prevalence of 3 species of malaria in residences of coastal villages near Madang, Papua New Guinea in 1983-85. Solid lines = asexual stages, dashed lines = gametocytes (data from Graves et al. 1988a).

of low average temperature due to elevation (e.g., the highlands of New Guinea) or distance from the equator (Middle East and southern Europe), or because of improved living conditions (southern USA). Epidemics also occur because control programs fail or large numbers of infected people move to areas where environmental and housing conditions favor transmission even though malaria previously was controlled (e.g., Sri Lanka, Madagascar).

Plasmodium vivax predominates where transmission is unstable due to its longer duration of infection in the human host, its ability to relapse and its shorter extrinsic incubation period.

Vectors

Culicine mosquitoes, phebotomine sand flies and culicoid biting midges are vectors of many of the avian, rodent and reptilian malarias; only *Anopheles* mosquitoes are vectors of human malarias (Fig. 7.8, Table 7.3). Larvae of *Anopheles* mosquitoes may develop in a variety of aquatic habitats, including water with salinity ranging from brackish to fresh. Breeding sites include transient water sources (puddles, bromeliads and water-filled footprints) as well as large permanent ponds, swamps and stream margins. Generally, *Anopheles* do not tolerate polluted water sources. Eggs are laid singly on the water surface and are distinguished from those of other mosquito genera by the presence of floats. Like all mosquitoes, they have 4 larval stages and a pupal stage. Anopheline larvae differ from culicines in that anophelines lack a siphon and therefore lay parallel to the water surface and have distinctive palmate hairs (Fig. 7.9b-ph). Developmental times vary widely with conditions in the breeding site, but generally take longer than culicine mosquitoes. Adults often may be distinguished from other mosquitoes by the presence of wing spots, maxillary palps nearly equal in length to the proboscis in both male and females, and by the raised posture of the abdomen and hind legs during bloodfeeding (Fig. 7.9d, d-w and d-p, respectively).

Vector identification technology

Vectors of human malaria often occur as sibling species (i.e., closely related species that are nearly indistinguishable morphologically) in a species complex. Species complexes include some of the most important malaria vectors.

Table 7.2. Clinical features of human malaria.

Clinical feature	Species			
	P. falciparum	*P. vivax*	*P. malariae*	*P. ovale*
Incubation period (days)	9–14	12–17	18–40	16–18
Erythrocytic cycle (hours)	48	48	72	49–50
Sequestration	Yes	No	No	No
Relapses	No	Yes	No (?)	Yes
Duration of infection (years)	1–2	1.5–3	3–50	1.5–3 (?)
Case fatality rate*	>10%	0%	0%	0%

* In untreated nonimmune patients.

Members of the *Anopheles quadrimaculatus* (Reinert et al. 1997) and *Anopheles albimanus* complexes are vectors of malaria in North and South America, respectively. In Asia, the *An. dirus, Anopheles culicifacies, Anopheles maculatus* and *Anopheles minimus* complexes are responsible for malaria transmission, while the members of the *Anopheles maculipennis* complex were major vectors of malaria in Europe. The *An. punctulatus* complex transmits much of the malaria and filariasis in the southwest Pacific while the *An. gambiae* complex dominates malaria transmission in Africa.

The technology for identifying members of species complexes has evolved so that the number of identified species in the complexes has increased. Morphologically-based identifications, while still the most commonly used, may not distinguish species as adults; for example, the 7 members of the *An. maculipennis* complex were identified from differences in egg morphology. Classic methods for identifying isomorphic *Anopheles* species include examining the offspring of cross-matings between standard colony mosquitoes and the offspring of unidentified parents, or examining the banding patterns of polytene chromosomes. These techniques are labor intensive and require live mosquitoes, thus limiting the number of mosquitoes that can be used for vector incrimination and transmission studies. Isozyme analysis has been used widely but is restricted by the need for freshly killed mosquitoes that must be analyzed immediately or stored at –70°C.

To circumvent the restrictions of the above techniques, nucleic acid-based techniques have been developed for the species in the *An. gambiae, An. dirus, An. punctulatus* and *Anopheles freeborni* complexes. The stability of DNA allows field-collected mosquitoes to be stored frozen, dried or preserved in a variety of alcohols until analyzed. The sensitivity of the techniques allows identifications to be made with only a small portion of the mosquito, leaving the remainder available for additional analyses (e.g., malaria parasite identification, host blood source identification, age grading).

There now are 2 basic approaches to developing DNA probes for vector identification. In the first, genomic libraries constructed with the DNA of one species are made and screened with the DNA of the homologous and heterologous species in order to identify species-specific sequences. Typically, the resulting probe contains short repetitive high copy number sequences. Such probes lend themselves to the

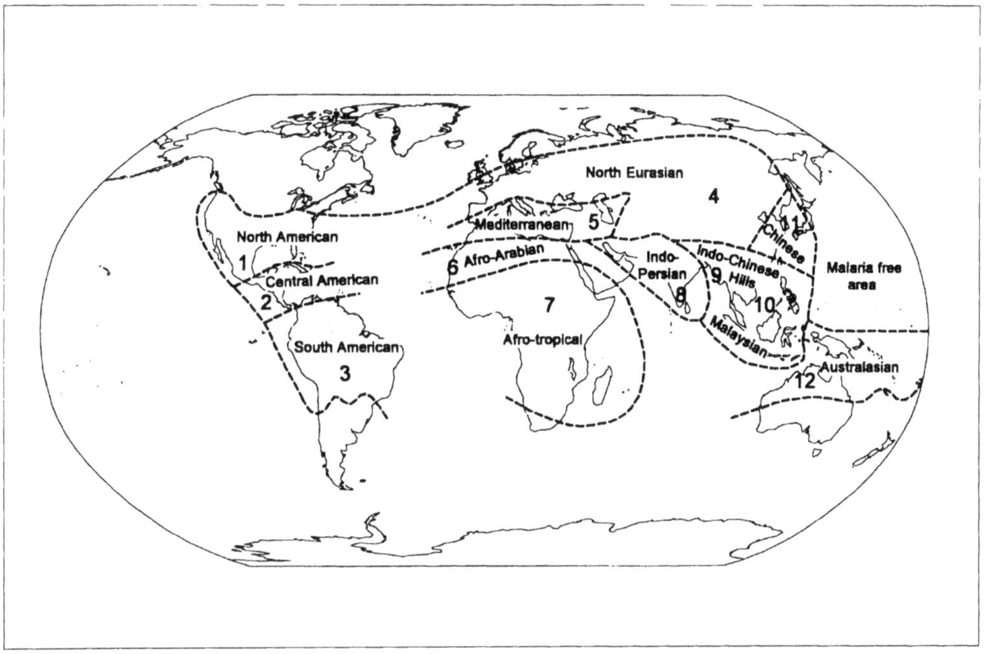

Figure 7.8. Map of epidemiological zones for anopheline mosquitoes (adapted from MacDonald 1957 and other sources).

preparation of synthetic oligonucleotide probes that are compatible with probing squash-blots. While labor-intensive to prepare, the simplicity of squash-blot analysis allows large numbers of mosquitoes to be identified simultaneously.

The second nucleic acid approach to vector identification exploits differences in the restriction fragment sizes of ribosomal DNA which then are hybridized with a single probe. Although this approach rapidly develops methods for vector discrimination, it is more labor intensive for processing field-collected anophelines, requiring DNA extraction from each mosquito followed by restriction digests, electrophoresis and Southern blotting before probing. Use of the polymerase chain reaction to amplify the diagnostic variant regions eliminates the need for Southern blotting and probing because identification is based on fragment size alone (Fig. 7.10). This approach has a further advantage: it allows systematists to determine taxonomic relationships among species based on ribosomal DNA sequence analyses. The development of DNA-based taxonomic methods for anopheline species complexes should facilitate studies on the biological characteristics and vectorial capacity of the mosquitoes in these complexes.

Although morphologically similar, members of species complexes may differ significantly in their importance as vectors of malaria. For example, some may be more **anthropophilic** (attracted to humans). Alternatively, some members of the complex may have characteristics that would make them more amenable than others to control strategies. Permethrin-impregnated bednets may be more effective against an

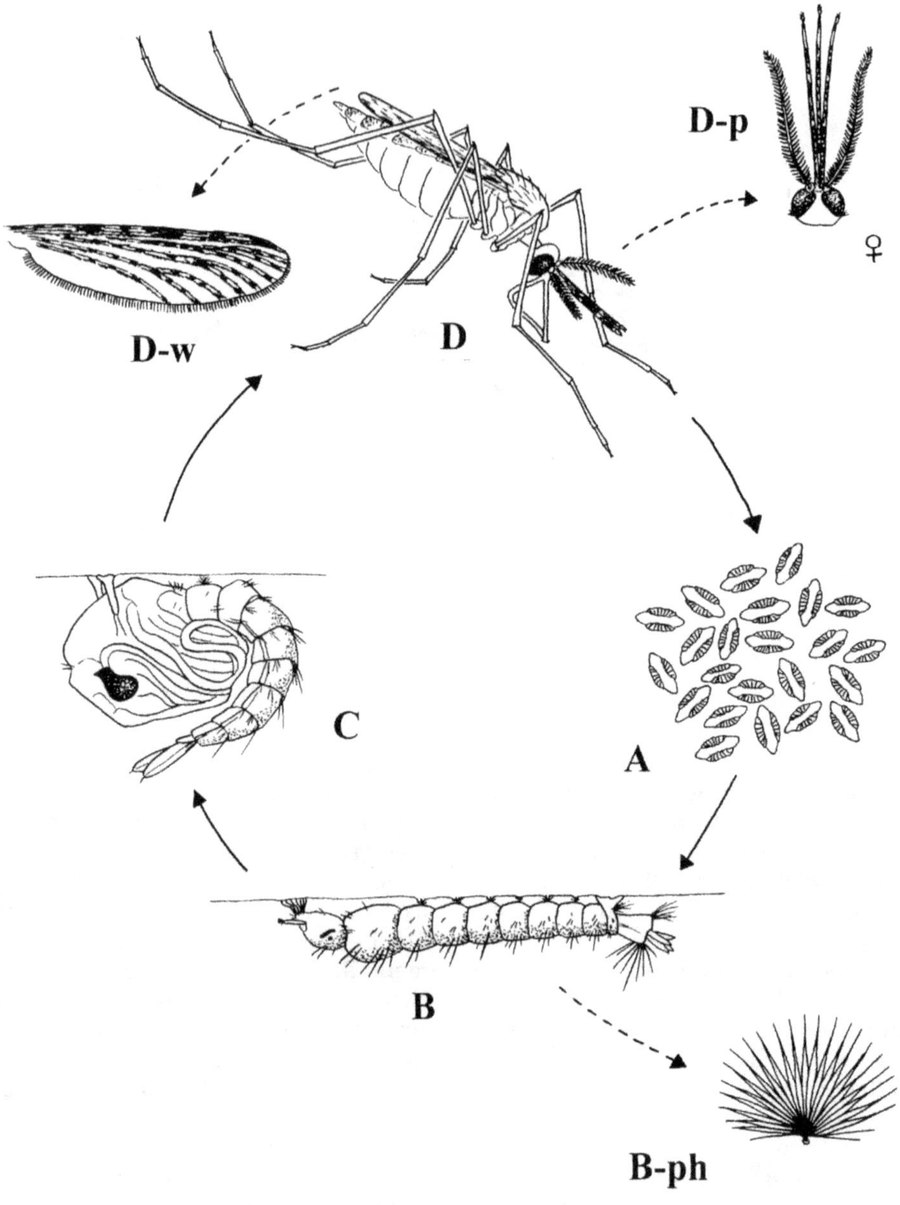

Figure 7.9. Important diagnostic characters of *Anopheles*. The eggs (a) are laid singly and have floats. Larvae (b) lie flat on the water surface and have palmate hairs (b-ph). Pupae (c) are similar to those of other Culicidae. Adult females (d) bloodfeed with hind legs raised, have spotted wings (d-w), and palps (d-p) equal in length to the proboscis (drawing by Carol S. Barnett).

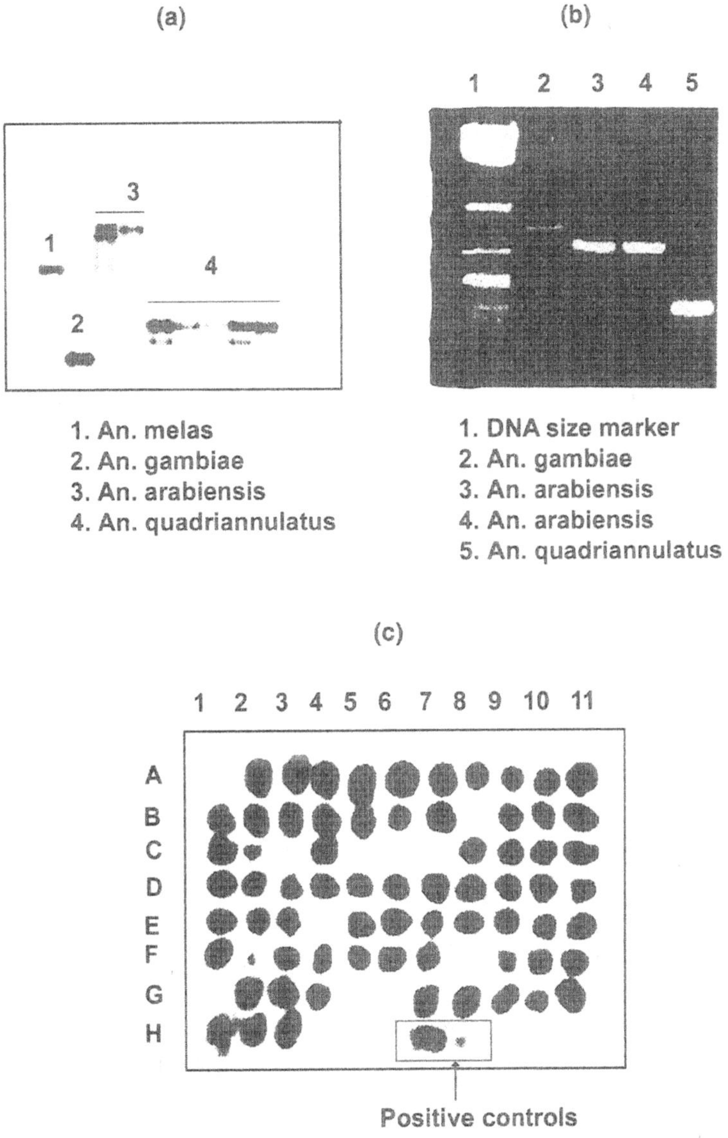

Figure 7.10. DNA-based anopheline identification techniques: (a) Hybridization of the anopheline rDNA probe pAGr12A to EcoR1 digests of DNA from single female mosquitoes identifies mosquitoes based on species-specific differences in the length of a ribosomal intergenic spacer (Collins et al. 1988); (b) ethidium bromide-stained PCR gel of mosquito DNA amplified using primers derived from mosquito ribosomal sequences (Paskewitz et al. 1993); (c) squash blots of field-collected mosquito heads probed with a genomic probe specific for *An. farauti* No. 2 (Burkot and Cooper unpublished data).

endophagic (feeds indoors) vector with peak biting activity in the middle of the night compared to an **exophagic** (feeds outdoors) vector that feeds early in the evening).

Usually, the preference for human blood determines the efficacy of a mosquito as a vector, although differences in susceptibility or abundance also could be involved. For example, only 4 of the 7 members of the *An. maculipennis* complex were incriminated in the transmission of human malaria in Europe. Similarly, *Anopheles quadriannulatus* in the *An. gambiae* complex is not considered to be a malaria vector, while in India, Pakistan and Arabia, *An. culicifacies* A is considered the primary malaria vector. *Anopheles culicifacies* B is a poor vector in the Indian subcontinent (Subbarao 1988). Biological differences (e.g., endophagy) of these species also may make some members of a complex more susceptible than others to control measures.

Up to 10 species now have been identified in the *An. punctulatus* complex, but studies on their biological characteristics and their importance as vectors of human disease in the southwestern Pacific have been performed only on the morphologically distinguishable members of this complex, *Anopheles farauti, Anopheles koliensis* and *An. punctulatus.* While found sympatrically over most of their range, particularly in coastal areas, they differ in a number of biological characteristics. *Anopheles farauti* will breed in brackish water pools, whereas *An. koliensis* and *An. punctulatus* are found only in fresh water. All 3 species are somewhat catholic in their bloodmeal sources, but *An. farauti* is decidedly more **zoophilic** (attracted to nonhuman animals) (Burkot 1988). All the members exhibit both endo- and exophagic behavior, yet the peak biting time for *An. farauti* tends to be earlier (10 PM) than that of either *An. koliensis* or *An. punctulatus* (midnight to 2 AM). Due to the more anthropophilic nature of *An. koliensis* and *An. punctulatus*, they have higher sporozoite and inoculation rates and therefore are more important vectors of human malaria (Burkot et al. 1988, Burkot and Graves 1995).

More is known about the biological characteristics of the 6 members of the *An. gambiae* complex: *An. gambiae, Anopheles arabiensis, An. quadriannulatus, Anopheles merus, Anopheles melas* and *Anopheles bwambae* (Coluzzi 1984, 1992). *Anopheles gambiae* and *An. arabiensis* are the most anthropophilic members of the complex and *An. quadriannulatus* is the most zoophilic. *Anopheles merus, An. melas* and *An. bwambae* exhibit partial zoophagy and are more exophilic than *An. gambiae* and *An. arabiensis. Anopheles merus* and *An. melas* are saltwater breeders; *An. bwambae* breeds only in mineral pools. The other 3 members of the complex breed in open, shallow, sunlit freshwater pools. *Anopheles gambiae* and *An. arabiensis* have the widest distribution, often occurring sympatrically. *Anopheles bwambae* has the most restricted range, limited to the Semliki Forest of Uganda. *Anopheles melas* is found on the western coast of Africa while *An. merus* is restricted to the eastern coast of Africa. *Anopheles quadriannulatus* is found in southeastern Africa and Ethiopia.

Studies on the **polytene chromosomes** of the members of the *An. gambiae* complex revealed **inversion polymorphisms**. These inversions may confer selective advantages in ecological zones and initiate the speciation process by preserving associations of genes in geographic or ecologically marginal zones. According to this controversial theory, inversions have adaptive benefits for the ecotypic area in which they originated (Coluzzi 1984). Indeed, clinal geographic variations in inversion frequencies have been described for *An. gambiae* and *An. arabiensis.* Inversions associated with humid areas increase in frequency during the rainy season. Similar inversions have been found in association with the dry season. Such inversions may allow expansion of the species into ecologically marginal areas.

Polytene chromosome differentiation has its highest complexity in *An. gambiae* in Mali where

Table 7.3. Malaria vectors associated with epidemiological zones. Major vectors are in boldfaced type. Locations of epidemiological zones are shown in Fig. 7.8 (Table continued on next page).

	Subgenera			
Epidemiological zone	*Anopheles*	*Nyssorhynchus*	*Kerteszia*	*Cellia*
1. North America	*freeborni*[1] *quadrimaculatus*[2]	*albimanus*		
2. Central America	*aztecus* *pseudopuncti-pennis* *punctimacula*	***albimanus*** *albitarsus* ***aquasalis*** ***argyritarsis*** ***darlingi***		
3. South America	***pseudopuncti-pennis*** ***punctimacula***	***albimanus*** ***albitarsus*** ***aquasalis*** *argyritarsis* *braziliensis* ***darlingi*** *nuneztovari* *triannulatus*	*bellator* *cruzii* *neivai*	
4. North Eurasia	***atroparvus***[3] *messae*[3] *sacharovi*[3] *sinensis*			*pattoni*
5. Mediterranean	***atroparvus***[3] *claviger* ***labranchiae*** *messae*[3] ***sacharovi***[3]			*hispaniola* ***superpictus***
6. Afro-Arabian				***culicifacies***[5] *fluviatilis* *hispaniola* *multicolor* ***pharoensis*** ***sergentii***
7. Afro-tropical				**arabiensis**[6] *funestus* ***gambiae***[6] *melas*[6] *merus*[6] *moucheti* *nili* *pharoensis*

Table 7.3. (Continued).

	Subgenera			
Epidemiological zone	*Anopheles*	*Nyssorhynchus*	*Kerteszia*	*Cellia*
8. Indo-Iranian	*sacharovi*[3]			*aconitis* *annularis* ***culicifacies***[5] ***fluvilatilis*** *jeyporensis* *minimus*[7] *philippinensis* *pulcherrimus* *stephensi* *sundiacus* *superpictus* *tessellatus* *varuna*
9. Indo-Chinese	*nigerrimus*			*annularis* ***culicifacies***[5] ***dirus***[9] *fluvilatilis* *jeyporensis* *maculatus*[8] *minimus*[7]
10. Malaysian	***campestris*** ***donaldi*** ***letifer*** ***nigerrimus*** *whartoni*			***aconitus*** ***balabacencis***[10] ***dirus***[9] ***flavirostris*** *jeyporensis* ***leucosphyrus*** ***ludlowae*** ***maculatus***[8] *mangyanus* ***minimus***[7] *philippinensis* ***subpictus***[11] ***sundiacus***
11. Chinese	***anthropophagus*** ***sinensis***			*balabacencis*[10] *pattoni* *jeyporensis*
12. Australasian	*bancroftii*[4]			***farauti***[12] *hili* *karwari* ***koliensis***[12] ***punctulatus***[12] *subpictus*[11]

[1]*An. freeborni* complex; [2]*An. quadrimaculatus* complex; [3]*An. maculipennis* complex; [4]*An. bancrofti* complex; [5]*An. culicifacies* complex; [6]*An. gambiae* complex; [7]*An. minimus* complex; [8]*An. maculatus* complex; [9]*An. dirus* complex; [10]*An. leucosphyrus* complex; [11]*An. subpictus* complex; [12]*An. punctulatus* complex. Adaped from Gillies and Warrell (1993), with permission from E. Arnold, and Lane and Crosskey (1993).

distinctive chromosomal forms have been used to split this species into 3 incipient species: Mopti, Bamako and Savanna (Coluzzi 1984). The Mopti chromosomal form predominates along the middle Niger River and other riverine-irrigated areas that provide suitable dry season larval breeding areas. The Bamako form is found along the upper Niger River. Breeding habitat for this form occurs late in the rainy season. The Savanna form is associated with rain-dependent breeding sites. Localities where at least 2 of the forms are sympatric revealed incomplete intergradation between Savanna and Mopti, and between Bamako and Savanna. However, no hybrids were found between Mopti and Bamako. These 2 forms are reproductively isolated, with Savanna acting as a bridge for gene flow.

Each of these incipient species exhibits characteristic behavioral differences much as those described for the other species in the complex (Coluzzi 1992). For example, behavioral variations that involved resting habits and biting preferences within *An. melas* were associated with 2Rn inversion karyotypes in The Gambia. *Anopheles gambiae*, with different chromosome-2 karyotypes, displayed a 2-fold difference in *Plasmodium* infection rates. A nonuniform distribution took place in bovine- and human-fed *An. arabiensis* with different 2Rb inversion karyotypes found resting indoors (Petrarca and Beier 1992).

Vector susceptibility to malaria

Some evidence indicates geographic incompatibility between strains of malaria parasites and *Anopheles* spp. European *Anopheles* competent for transmitting European isolates of human malarial parasites cannot become infected with African strains of *P. falciparum* (Curtis and Graves 1983). In laboratory experiments, anophelines may support the development of oocysts to maturation by releasing sporozoites, but the sporozoites may be unable to penetrate the salivary glands (Rosenberg 1985). The F_1 progeny of 7 species of field-collected *Anopheles* in Brazil showed a wide range in susceptibility to both oocyst infection and numbers of *P. vivax* sporozoites in the salivary glands (Klein et al. 1991). In susceptible species, the speed of bloodmeal digestion and rate of peritrophic membrane formation relate to the variation in susceptibility: individual *An. gambiae* and *Anopheles stephensi* with slower rates of digestion had significantly higher *P. falciparum* oocyst infections (Ponnudurai et al. 1988).

Malaria susceptible and refractory lines of *An. gambiae* and *An. stephensi* have been selected (Collins et al. 1986, Feldmann and Ponnudurai 1989). Although the *An. gambiae* colony was selected for resistance against a monkey malaria, the colony was refractory to rodent, bird, human and other simian malarias. Two unlinked genetic loci appear to control expression of the susceptible and refractory phenotypes through physiologically independent mechanisms (Vernick et al. 1989).

Refractoriness to malarial parasites was expressed as the ability of a mosquito line to encapsulate ookinetes via melanin (Collins et al. 1986). Other mechanisms of resistance to malaria parasites include nitric acid and possibly insect defensins such as serine proteases.

Detection of parasites in vectors

Malaria infections in mosquitoes may be determined by dissections followed by microscopic examinations for the presence of oocysts on the gut wall or sporozoites in the salivary glands. Oocysts of the different *Plasmodium* spp. are not distinguishable morphologically, so immunologic or molecular methods must ascertain species.

Determining the sporozoite rate by the dissection of salivary glands is particularly tedious. Like oocysts, sporozoites cannot be microscopically differentiated by morphology (e.g., sporozoites of the human malarias could not be distinguished from each other, nor could they be distinguished from sporozoites of monkey, rodent and bird malarias). Recently, Wirtz and Burkot (1991) developed tests, both antibody and nucleic acid-based, to detect and identify sporozoites.

The most popular and widely used of these tests are the enzyme-linked immunosorbent assays (ELISAs) based on detecting circumsporozoite (CS) proteins (Burkot and Wirtz 1986). The CS proteins are abundantly expressed proteins found on the surface of sporozoites. They are highly immunogenic with a central domain of amino acid repeats. The composition of the repeats is unique for a given malaria species. For example, *P. falciparum* has a 4 amino acid sequence (asparagine-alanine-asparagine-proline) repeated up to 40 times while the CS repeat region of *P. vivax* is composed of a 9 amino acid sequence repeated up to 20 times. Two variants of *P. vivax* have been found with the composition of the amino acids in the repeat region differing (Rosenberg et al. 1989). The tandem repeats allow a single monoclonal antibody to be used for antigen capture as well as enzyme-labeled for signal generation. In addition to allowing both sporozoite detection and identification, the CS ELISAs have further advantages over traditional dissections. They are compatible with storage of mosquitoes under a variety of conditions (e.g., dried or frozen). With a sensitivity of 10–25 sporozoites, the CS ELISAs are capable of detecting a single infected mosquito in pools of up to 50–100 noninfected mosquitoes. Because CS proteins begin to be expressed while sporozoites are developing in the oocyst, a researcher must assay only heads and thoraces of mosquitoes to avoid overestimating the sporozoite rate by CS antigen detection. This focus will distinguish sporozoite rates determined by the dissection of salivary glands from CS antigen positivity rates based on ELISA detection.

Studying transmission

It is important to understand the transmission of malaria parasites from the human host to the anopheline mosquito as well as from the mosquito vector to susceptible human hosts. Consequently, these 2 types of transmission will be considered separately. In each case, transmission may be measured either directly or indirectly.

Estimating transmission to humans

Direct estimates based on human infections. Malaria transmission to humans may be estimated directly in a human population by determining the rate at which aparasitemic individuals acquire infections. To determine the incidence in the absence of immunity in highly endemic areas, transmission can be estimated by the infant conversion rate (i.e., the proportion of infants in a time interval that convert from aparasitemic to parasitemic). In areas of lower endemicity with minimal immunity to malaria, the study population may include older individuals. Close monitoring for new malaria episodes over time can determine incidence in immune individuals. In highly endemic areas, some studies have cleared current parasitemias with chemotherapy to facilitate detection of new infections by increasing the proportion of aparasitemic individuals at the start of the study.

Indirect estimates based on levels in humans. Transmission may be estimated indirectly by determining the rate at which individuals shift from being antibody negative to antibody positive for a particular malaria antigen, such as the CS proteins. The prevalence of antibodies recognizing the *P. falciparum* CS protein successfully estimated malaria transmission in areas of seasonal transmission (Esposito et al. 1988) or during epidemics as individuals seroconvert to the CS protein following their first exposure (Webster et al. 1987).

Indirect estimates of transmission potential based on entomological parameters. Risk of potential malaria transmission may be estimated from the **vectorial capacity** of anopheline populations. As introduced in Chapter 6, the concept of vectorial capacity originated in studies of malaria, and was defined as an estimate of the daily rate at which future inoculations arise from a currently infective malaria case (Reisen 1989). Traditionally, vectorial capacity estimates are independent of parasitologic measurements. Five variables determine vectorial capacity: the **human biting rate** (bites per person per night), the **human blood index** (proportion of bloodmeals taken from humans), the **vector competence** of the mosquito for the parasite, the **daily survival rate of the vector**, and the extrinsic incubation period (EIP) for the malarial parasite in the vector. The EIP is the period from ingestion of an infectious bloodmeal to the time of pathogen transmission capability.

Vectorial capacity may be expressed mathematically by a number of formulas, including the one shown in Fig. 7.11. In the context of malariology, vectorial capacity means the number of new infections disseminated per day by each mosquito that has fed on a person infectious to mosquitoes.

The human biting rate is estimated most frequently by landing rates or by exit and bednet traps. Estimates by landing catches generally overestimate the true mosquito attack rate because this collection method minimizes defensive behavior and cultural methods that prevent mosquito bites. The scale of the overestimate may be as much as 20-fold (Saul 1987). At the other extreme, estimates based on bednet and exit traps underestimate the true biting rate due to the inefficiency of the trapping method. Estimates derived from these methods assume, probably falsely, that vector and host populations randomly mix. According to Burkot (1988), most studies showed adults to be more attractive to mosquitoes than children, but all studies demonstrated a wide range in the attractancy of individuals to mosquitoes (e.g., some individuals, regardless of size, will be bitten by more mosquitoes than others).

The human blood index is the proportion of bloodfed resting mosquitoes containing human blood (Table 7.4). A variety of immunologic or molecular-based methods may identify bloodmeal source. Determination of the human blood index is subject to sample collection bias and it is difficult to accumulate sufficient numbers of engorged mosquitoes from which bloodmeal source identifications can be made. In areas where engorged resting mosquitoes cannot be found easily, the human blood index often is estimated by comparative collections from humans and livestock. The proportion of anopheline bloodmeals on humans may vary widely in some species within a small geographic area. The human blood index affects vectorial capacity in a linear manner; interrupted bloodmeals will increase vectorial capacity estimates.

Mosquito survivorship can be estimated as the proportion of a cohort of adult female mosquitoes surviving per day, per gonotrophic cycle, per feeding cycle or per EIP (Table 7.5). In malaria, the EIP varies from 10–30 days, depending on temperature and the species of malaria. If the average daily temperature is too low (e.g., usually at elevations above 3,000 m near the equator, lower elevations in cooler climates), the EIP exceeds the vector life span and transmission cannot occur. Measurement of parity and methods of age grading estimate the probability of surviving through a gonotrophic cycle, while mark-release-recapture (MRR) results and mosquito infection rates estimate survival through the feeding cycle. This estimate must be raised to a power equal to the number of feeding cycles that comprise the EIP to estimate the probability of surviving the EIP. The length of the feeding cycle can be estimated through MRR experiments and the length of the gonotrophic cycle by cross-correlations of time-series parous rate data. An educated guess

$$C = \frac{ma^2VP^n}{-\mathrm{Log}_e P}$$

C = Vectorial capacity, the number of new infections disseminated per case per day by a vector
m = Density of vectors in relation to density of hosts
a = Proportion of vectors feeding on a host divided by the length of the gonotrophic cycle of the vector in days
V = Vector competence
P = Daily survival of vectors
n = Extrinsic incubation period

Figure 7.11. Formula for vectorial capacity (Reisen 1989).

estimates the length of the EIP. Obviously, this approach is imprecise.

Alternatively, the probability of survival through the EIP can be estimated directly from mosquito infection rates (sporozoite rates or sporozoite and oocyst rates) in landing and resting collections (Burkot et al. 1990, Graves et al. 1990, Saul et al. 1990). The approach is feasible only in endemic areas where mosquito infection rates and vector population levels are high enough to produce adequate sample sizes. When applied to anopheline collections in Papua New Guinea, no significant differences in probability of survival through the feeding cycle were found among the morphologically identified members of the *An. punctulatus* complex in different villages (Graves et al. 1990) nor at different times of the year (Burkot et al. 1990). However, significant differences in probability of survival through the EIP appeared both in the different villages and at different times of the year. The results suggest that the length of the feeding cycle, and therefore the number of feeding cycles per EIP, varies with time and location in that hyperendemic area. In contrast, large seasonal differences in probability of survival through the feeding cycle were found in The Gambia (Lindsay et al. 1991).

Estimates of vectorial capacity frequently are inaccurate. Consequently, key components of vectorial capacity in a given epidemiological situation should be identified to improve the value of entomological transmission estimates that can predict parasitologic changes in human hosts (Dye 1986). Various studies have identified components of vectorial capacity with better correlations to parasitologic changes in humans than overall estimates of vectorial capacity. In many cases, the key component is the human biting rate, but this may differ among epidemiological situations.

The imprecision associated with estimates of the components of vectorial capacity does not necessarily eliminate their usefulness as measures of changes in transmission (Dye 1990, Dye 1992). However, for a component of vectorial capacity to remain valid as an indicator of changes in transmission during interventions, the bias associated with the parameter estimate must remain constant during the study.

Direct estimates based on entomological parameters. An alternative to estimating transmission potential from vectorial capacity is to measure transmission directly through inoculation rate estimates. Determining the risk of being bitten by an infectious mosquito requires measuring both the proportion of mosquitoes with sporozoites in their salivary glands (**sporozoite rate**) and the rate of contact between humans and vectors (the human biting rate). The product of these 2 factors is known as the **inoculation rate**. Today, sporozoite rates generally are estimated by the CS antigen positivity rate. As discussed above, care must be taken to minimize the effect of detecting CS antigens in the oocyst.

The human biting rate usually is estimated by landing rates or bednet and exit traps. The inaccuracy of these methods was discussed pre-

Table 7.4. Human blood index values for members of the *Anopheles punctulatus* complex in villages near Madang, Papua New Guinea. The catholic nature of these mosquitoes can be seen in the wide anthropophilic range found in this limited geographic area.

Village	*An. farauti* HBI ± se (n)	*An. koliensis* HBI ± se (n)	*An. punctulatus* HBI ± se (n)
Buksak	—	—	0.86±0.03 (136)
Butelgut	—	—	0.71±0.12 (14)
Dogia	0.11±0.02 (222)	0.29±0.17 (7)	1.00±0.00 (3)
Erima	0.25±0.03 (159)	0.76±0.06 (49)	0.60±0.22 (5)
Hudini	—	—	0.20±0.18 (5)
Maraga	0.08±0.03 (686)	0.36±0.10 (22)	—
Mebat	0.45±0.11 (22)	0.82±0.06 (38)	0.50±0.12 (16)
Panim	—	—	0.50±0.35 (2)
Sah	—	—	0.75±0.22 (4)
Umun	—	0.85±0.10 (13)	0.26±0.17 (7)

viously. Because these methods yield inaccurate estimates, sporozoite inoculation rates must be considered relative estimates of the risk of being bitten by an infectious mosquito, and estimates of inoculation rates from different epidemiological areas must be compared with caution.

Despite the magnitude of the error associated with estimates of biting rates and therefore inoculation rates, estimates of inoculation rates are approximately 3-fold more precise than estimates of transmission based on vectorial capacity (Burkot and Graves 1995). Inoculation rates offer further advantages over vectorial capacity for estimating the relative risk of transmission, because inoculation rates require less technical expertise to perform and CS antigen positivity rates may be determined using dried or frozen mosquitoes. Field studies therefore can collect mosquitoes during the transmission season and analyze for CS proteins later.

Just as the sporozoite rate varies, the numbers of sporozoites released per bloodmeal also will vary. Studies have shown that the number of sporozoites released are surprisingly small. *Anopheles stephensi* that were infected by feeding on tissue-cultured *P. falciparum* gametocytes released an average of 15 sporozoites (range: 0–978) when allowed subsequently to salivate into an oil droplet (Rosenberg et al. 1990). The number of sporozoites released was proportional to the total sporozoite load in the salivary glands. Similar estimates were found for *An. gambiae*; a geometric mean of <6.1 sporozoites per bloodmeal were released from the salivary glands per bloodmeal (Beier et al. 1991).

Table 7.5. Mosquito daily survivorship estimates using different methods in Papua New Guinea. Vectorial capacity theory implies that changes in survivorship will have the greatest impact on transmission.

Village	Mosquito	Method of estimation of survival rate			
		Multiparous age-grading	Mark-release-recapture	Parity rates	Infection rates
Bilbil and Maraga	*An. farauti*	0.42–0.49	0.29–0.41	0.29–0.71	0.61
Mebat	*An. koliensis*	—	0.30–0.41	0.70	0.60
Butelgut	*An. punctulatus*	0.34	0.46	0.65	0.58

Sporozoites released, whether directly into a capillary or into tissue, are capable of initiating an infection. Furthermore, an inoculation of 10 sporozoites is sufficient to establish a blood stage infection (Ungureanu et al. 1976).

Transmission to mosquitoes

When a noninfected mosquito feeds on a human with malaria, many factors affect the transmission of parasites to the mosquito. These factors include the proportion of persons with gametocytes in their blood, the density of those gametocytes, the presence or absence of antibodies that affect gametocyte infectivity and the susceptibility of the mosquito to malaria infections. Some estimates of the probability of mosquito infection include an estimate of the proportion of humans who are infectious, sometimes called the **reservoir of infection** (Gamage-Mendis et al. 1991). However, an infectious person may not necessarily infect all mosquitoes that feed on that individual. The chance that a mosquito will become infected following a bloodmeal from a human may be termed the 'mosquito infection probability.' There are 3 ways to estimate this probability. The first involves feeding mosquitoes on people at random, while the second involves feeding mosquitoes on gametocyte carriers and extrapolating to the whole population using separate data on the frequency of gametocyte carriers. The third method measures infection rates in wild mosquito populations to estimate the infection probability (Graves et al. 1990, Lines et al. 1991).

Table 7.6 summarizes the estimates of mosquito infection probability made by all 3 methods. Feeding studies in endemic areas have shown that approximately 40–60% of gametocyte carriers of both *P. falciparum* and *P. vivax* were infectious (Graves et al. 1988a). However, after allowing for the prevalence of gametocyte carriers and the proportion of mosquitoes infected per batch, the infection probability is actually about 0.6–1.9% (Table 7.6). Using infection rates in wild-caught mosquitoes resulted in much higher estimates, ranging from 2.8–21% (Table 7.6). The reason for this discrepancy has not been resolved, but might include low susceptibility of mosquito colonies, nonrandom feeding behavior, or both.

Treatment

A wide variety of materials have been used to treat human malaria cases. Current antimalarial drugs can be classified into the following groups (World Health Organization 1995):

1. 4-aminoquinolines (e.g., chloroquine)
2. Arylamino alcohols (e.g., quinine, mefloquine)

3. Phenanthrene methanols (e.g., halofantrine)

4. Artemisinin derivatives

5. Antimetabolites (e.g., proguanil, pyrimethamine, sulfadoxine)

6. Antibiotics (e.g., doxycycline, tetracycline)

7. 8-aminoquinolines (e.g., primaquine)

All of the first 6 groups are active against blood stages, while 8-aminoquinilines are active only against gametocytes and liver stages (hypnozoites). In addition, proguanil and the antibiotics are active against hypnozoites as well as blood stages.

Drugs in the first 4 groups act the most rapidly against malarial parasites, and are the mainstay of treatment. Serious side effects have been reported for some drugs (e.g., halofantrine).

Resistance to antimalarial compounds by *P. falciparum*, particularly to chloroquine, is now widespread. Chloroquine resistance by *P. vivax* has been demonstrated in some areas. Resistance to Fansidar® (pyrimethamine/sulfadoxine), quinine and mefloquine also has been documented, though not as widely as for chloroquine. Drug resistance can be determined through *in vitro* tests (for *P. falciparum*) or by *in vivo* tests. In the *in vivo* test, an antimalarial is administered to a patient. If the parasitemia is cleared within 7 days and does not return, the parasite is deemed drug-sensitive. Resistant parasites are categorized as R1, R2 or R3 if after treatment blood stage parasites reappear after 7 days, if parasitemia drops but does not become subpatent or if it has little effect, respectively.

Prevention and Control

Following the insight of Macdonald (1957) that the probability of survival of the adult mosquito is a vulnerable point in the transmission cycle, malaria control traditionally relied on house-spraying with residual insecticides such as DDT, propoxur and malathion to kill adult mosquitoes. While this approach initially was effective and eradicated malaria in some areas, it has fallen out of favor because of high cost, insecticide resistance in the vectors, ineffectiveness against outdoor transmission and lack of acceptance by local populations. The current global strategy for malaria control recommended by the World Health Organization (1993) has 4 components:

1. Early diagnosis and prompt treatment of malaria cases;

2. Application of selective and sustainable vector control methods;

3. Early detection to contain or prevent epidemics;

4. Regular reassessment of a country's malaria situation, in particular the ecological, social and economic determinants of the disease.

The first component of this strategy requires strengthening of primary health care systems in endemic countries. The second strategy most concerns medical entomologists. Vector control is crucial to sustainable malaria control. Selective and sustainable methods do not exclude the application of indoor residual insecticides, but such use is likely to be much more restricted than in the past. They should be used only in areas where the behavior of the vector (i.e., tendency to rest indoors), the intensity of transmission and the acceptance of adulticiding by people justify the high cost of this method. An alternative effective, community-based and sustainable vector control strategy is to use mosquito bednets or house curtains impregnated with pyrethroid insecticides such as permethrin, deltamethrin or lambda-cyhalothrin. Lengeler and Snow (1996) showed that mosquito net impregnation effectively reduces human-vector contact and lowers malaria incidence in several areas with different levels of endemicity (western and eastern Africa, Papua New Guinea, Thailand, Sri Lanka). Large trials in Africa significantly reduced childhood mortality in populations using impregnated bednets (e.g., Alonso et al. 1991, Binka et al. 1996, Nevill et al. 1996). This strategy is being promoted worldwide.

Table 7.6. Estimates of mosquito infection probability (probability of a mosquito acquiring malarial parasites after feeding upon a parasitemic human).

Place	Method	Mosquito species	Malaria species	Infection probability	Reference
Liberia	Feeding studies	*An. gambiae*	*P. falciparum*	1.9%	Muirhead-Thompson 1957
Madang, Papua New Guinea	Feeding studies	*An. farauti*	*P. falciparum/ P. vivax*	1.3–1.5%	Graves et al. 1988a
Kano, Kenya	Feeding studies	*An. gambiae*	*P. falciparum*	0.8%	Githeko et al. 1992
Yaoundé, Cameroon	Feeding studies	*An. gambiae*	*P. falciparum*	0.6%	Tchuinkam et al. 1993
Muheza, Tanzania	Infection rates and age-grading in wild-caught mosquitoes	*An. gambiae* *An. funestus*	*P. falciparum*	6.9% 6.1%	Gilies and Wilkes 1965
Muheza, Tanzania	Infection rates and age-grading in wild-caught mosquitoes	*An. gambiae*	*P. falciparum*	21%	Lines et al. 1991
Nigeria	Infection rates and age-grading in wild-caught mosquitoes	*An. gambiae* *An. funestus*	*P. falciparum*	5.5% 4.3%	Service 1965
3 villages near Madang, Papua New Guinea	Infection rates in wild-caught mosquitoes	*An. farauti* *An. koliensis* *An. punctulatus*	*P. falciparum/ P. vivax*	7.4–10.5% 4.9–17.4% 7.4%	Graves et al. 1990
Buksak, near Madang, Papua New Guinea	Feeding studies and infection rates in wild-caught mosquitoes	*An. punctulatus*	*P. falciparum/ P. vivax*	5.4% (feeds) 4.5–5.9%	Burkot et al. 1990
Numawala, Tanzania	Infection rates in declining population of wild-caught mosquitoes	*An arabiensis*	*P. falciparum*	1.8%	Charlwood et al. 1995
Kisegese, Tanzania	Oocyst rates and sizes in wild-caught mosquitoes	*An. gambiae* *An. funestus*	*P. falciparum*	2.8% 4.2%	Haji et al. 1996

Other vector control measures include selective larviciding of breeding sites with insecticides such as temephos, and environmental management (filling or periodic draining of breeding sites). These methods are difficult to implement effectively because regulation of larval population size often is density dependent. Also, many anophelines breed in numerous temporary water sources, making complete coverage with larvicides difficult, or in large bodies of water that cannot be eliminated because of the loss of drinking and washing sources and because of the elimination of valuable aquatic habitats. Several vectors with different breeding preferences also may mediate transmission, and elimination of breeding sites for one vector may have little effect on intensity of malaria transmission. Nevertheless, selective larval or environmental control may have an important role, particularly in the containment of potential epidemics, but careful evaluations of these methods have not been performed.

Surveillance

Medical entomologists also have a crucial role to play in the strategy of early detection, containment or prevention of epidemics. If epidemics are to be prevented, effective entomological surveillance should monitor the entomological inoculation rate (product of sporozoite rate and the bites per person per night). When this rate exceeds a certain value defined for each monitoring station, selective control measures should be implemented. Monitoring must employ the most efficient and accurate way to sample the human biting anopheline population. Entomological surveillance should be integrated with monitoring of parasite incidence through health care systems, but reliance should not be on clinical indicators alone, since they react to epidemics rather than proactively prevent them.

Vaccines

An international effort currently is underway to develop vaccines against human malaria. Molecules on the sporozoite, exoerythrocytic stages, erythrocytic stages and sexual stages have been targeted. In order for anti-sporozoite and anti-exoerythrocytic vaccines to prevent malaria, they must be completely effective. It is argued that vaccines against the blood stages, even if they do not confer sterile immunity, could greatly reduce mortality and severe clinical disease. The antisexual stage vaccines aim to prevent transmission. These vaccines would not protect the individual receiving the vaccine, except as a result of overall reduction of transmission due to prevention of the vaccinated individual from acting as a reservoir to infect anophelines. Thus far, none of the vaccines have demonstrated sufficient efficacy in field trials to warrant large-scale production and dissemination (Graves et al. 1998).

BABESIOSIS

Introduction

Babesiosis is an important economic disease in cattle and equines. At least 100 species of *Babesia* have been described. In Ireland alone, annual loses are estimated at $15 million while Australia's losses in 1974 exceeded $40 million. In addition, hosts for these protozoans include canines, cervids, rodents and humans. Canine babesiosis affects dogs throughout the world, with clinical cases in Europe corresponding to the activity patterns of the questing tick vectors. Hundreds of cases of babesiosis have been reported in humans, caused primarily by rodent and cattle *Babesia* spp.

The Parasite

The life cycle of *Babesia* spp. is characterized by schizogony in the vertebrate host and a sexual cycle (gametogony) in the tick vector followed by asexual sporogony (Fig. 7.12). Sporozoites inoculated by the bite of a *Babesia*-infected tick usually will invade erythrocytes directly, although sporozoites of *B. equi* will develop first in lymphocytes prior to erythrocyte invasion. Unlike malaria, there is no invasion of tissues outside of the circulatory system. The parasite develops into trophozoites (or **meronts**) by binary fission. However, the small *Babesia* spp. (e.g., *B. equi* and *Babesia microti*) produce 4 merozoites, termed a schizont by some. These merozoites, when released, invade other erythrocytes (Fig. 7.13). Unlike malaria parasites, the erythrocytic stages are not enclosed in a parasitophorous vacuole. Like *Plasmodium* spp., differentiation of some of the merozoites into the sexual stage (gametocytes) will occur, although in *Babesia* this differentiation occurs in the tick rather than the vertebrate host.

After ingestion by the tick vector, most parasites are destroyed. However, some gametocyte-forming cells survive and develop into **gamonts** (strahlenkorper or "ray bodies"). The gamont then divides into 4 gametes. Male and female gametes fuse in the tick to form a diploid elongated zygote. An arrowhead organelle allows penetration of the peritrophic membrane and invasion of a midgut cell by the ookinete or mobile zygote. Division occurs in the gut cell, and kinetes (sporokinetes or vermicules) are released into the hemolymph. Other tick cells then are invaded to begin new cycles of kinete development. Ovaries of adult ticks invaded in this way result in transovarial transmission, which is not seen in malaria. The salivary glands in the tick embryos are invaded; here a final round of development produces the sporozoites. This developmental cycle allows one-host ticks, such as *Boophilus* spp., to act as vectors. In addition, transovarial transmission enables adult 3-host ticks, such as *Ixodes* spp., to be *Babesia* vectors via their progeny.

In *Babesia*-infected larvae and nymphs, ookinetes from the tick gut invade the salivary glands as well as other organs where a second generation of ookinetes reinvade the nymph's salivary glands which are reformed following the molt. These secondary ookinetes produce the sporozoites.

Vectors and Transmission

Ticks in the genera *Boophilus, Rhipicephalus, Dermacentor, Haemaphysalis, Ixodes, Hyalomma* and *Anocentor* transmit *Babesia* parasites. A summary of the important vectors of 20 species of *Babesia* is given in Table 7.7, together with the geographic areas where they occur.

Animal Babesiosis

At least 20 parasites may cause animal babesiosis, an important economic disease transmitted by 8 genera of ticks (Table 7.7). After a prepatent period of 8–16 days following exposure to infectious ticks, the increasing erythrocytic infection raises the body temperature. In addition to hematologic changes associated with babesiosis, pathology may involve hemoglobinuria, jaundice, anemia and renal failure. The coagulation cascade is disrupted, with coagulation time reduced 33–66%. Pathology may be associated with either high parasitemias or a shock syndrome that occurs at apparent low parasitemias. Shock syndrome is associated with parasites that exhibit cytoadherence, the binding of parasitized erythrocytes to the endothelium resulting in the accumulation of infected erythrocytes in the capillaries in a manner similar to that seen in *P. falciparum*-infected erythrocytes. Because of cytoadherence, the density of circulating parasites may be low in heavily parasitized animals. Severity of disease is related to both the animal breed as well as the strain and species of parasite (*Babesia bovis*

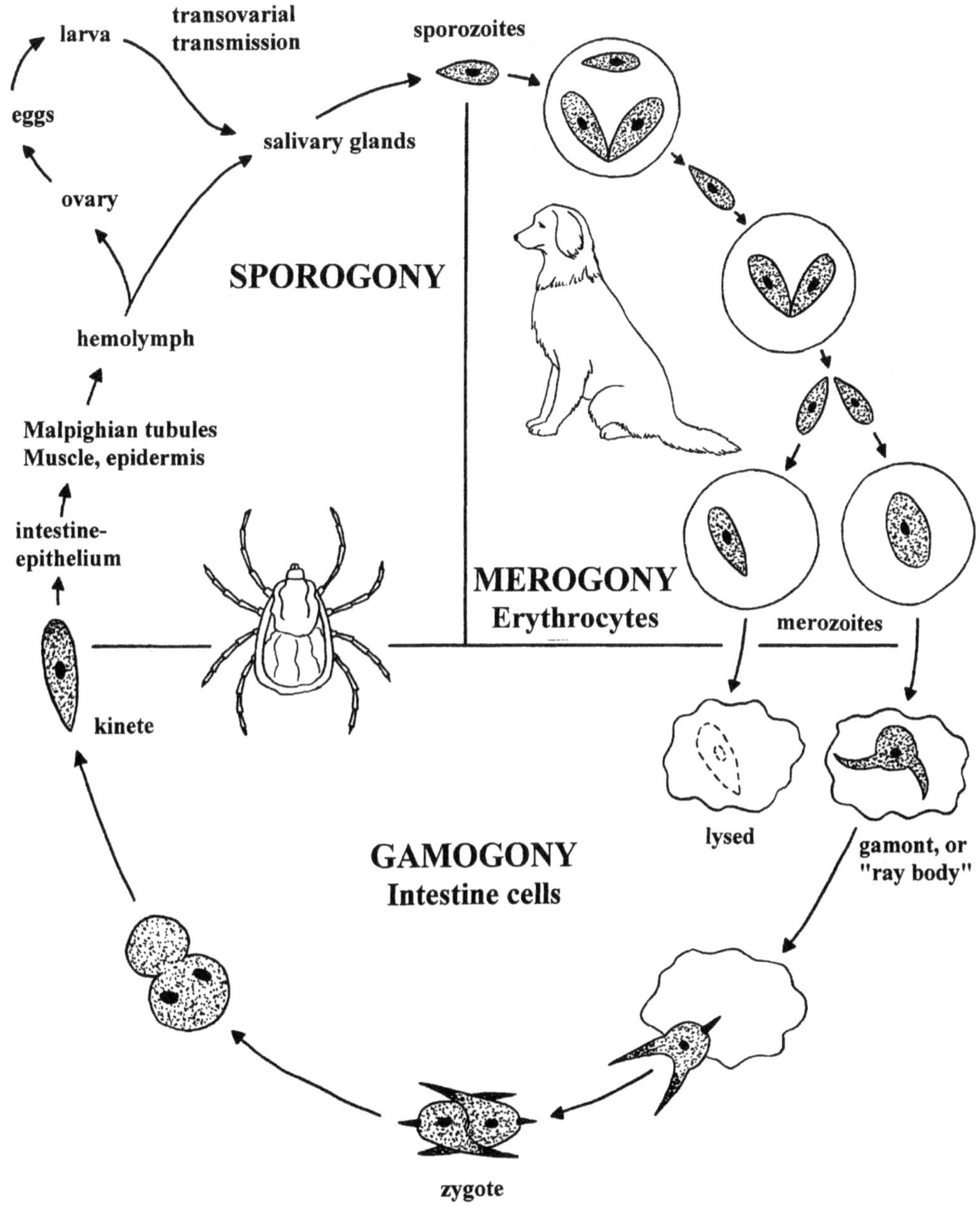

Figure 7.12. *Babesia* life cycle (drawing by Carol S. Barnett).

often causes more acute disease in cattle than *B. bigemina*). Maternal antibodies passed in the colostrum from immune cows offer protection to nursing calves, as does innate immunity.

Like bovine babesiosis, equine babesiosis, caused by *B. equi* and *Babesia caballi*, causes fever, anemia, hemoglobinuria and mortality rates up to 50%. Canine babesiosis due to *Babesia canis* and *Babesia gibsoni* is found throughout the world. In contrast to bovine babesiosis, young dogs are more susceptible to severe illness than are older animals because they lack previous exposure and have partial immunity. The disease induces high fevers, anemia and increasing hypoxia. In Europe, cases occur primarily in the spring, with a smaller number of cases in the fall, corresponding to the biting activity of the primary vector, *Dermacentor reticulatus*. Cervid babesiosis, an asymptomatic infection caused by *Babesia odocoilei* in white-tailed deer, is transmitted by *Ixodes scapularis* in North America.

Human Babesiosis

In Europe, tens of splenectomized individuals have been infected with the cattle parasites *Babesia major* or *Babesia divergens*, presumably transmitted by *Ixodes ricinus*. Hundreds of cases of babesiosis in humans have been reported to date in the USA, predominately in New England. Here, human infection with *B. microti* by *Ix. scapularis* occurs in both splenectomized and nonsplenectomized elderly individuals (Anderson et al. 1991). *Babesia microti* normally is a parasite of mice and voles. In humans, *Babesia* induces a malaria-like disease after an incubation period of 1–4 weeks. Severe disease may occur in splenectomized patients infected with bovine or equine *Babesia* spp. as well as in older individuals with intact spleens. Symptoms begin with fever, chills, sweating, headaches and general myalgia and may lead to arthralgia, nausea, vomiting and prostration. Infected splenectomized individuals often die. Chronic symptoms include cirrhosis, anemia, icterus and hemoglobinuria.

Recently, an unusual babesia-like piroplasm, designated WA1, was identified in 5 patients from Washington State and California (Quick et al. 1993, Thomford et al. 1994, Persing et al. 1995). Based on small-subunit ribosomal DNA, WA1 was related most closely to the canine pathogen, *B. gibsoni*. Clinical symptoms ranged from asymptomatic to influenza-like to fulminant, fatal disease.

Immunity

Infection with *Babesia* spp. elicits a strong humoral and cellular immune response. Antibodies against both merozoites and exoantigens (by-products of parasite metabolism) are detectable, but these antibodies do not necessarily protect against further erythrocyte invasions by the parasites. The antibody response will vary due to the strain of parasite and host and the stage of infection. Persistent subclinical infections may result in immunity. Furthermore, calves receive maternal antibodies that persist for several months. If infected during this period, a strong immune response develops which persists for >4 years.

Diagnosis

Diagnosis may involve clinical observations, microscopic examination of blood smears and serological tests. Morphological similarities to *P. falciparum* make examination of Giemsa-stained thick blood smears difficult. However, unlike *Plasmodium*, *Babesia* does not produce pigment in parasitized erythrocytes. The most commonly used serological test, the indirect immunofluorescent antibody test, has high sensitivity. As with all serological tests for babesiosis, it lacks specificity.

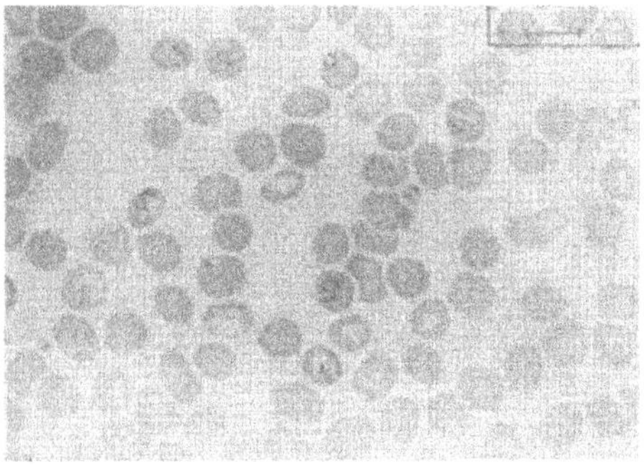

Fig. 7.13. Blood stage of *Babesia* parasite (courtesy of M.L. Eberhard, Centers for Disease Control and Prevention).

Treatment and Control

When required, dipping and spraying livestock with acaricides usually controls the tick vectors. Vaccination with live attenuated parasites, killed organisms and purified antigens or recombinant proteins may control the parasite. While most vaccines offer protection only against homologous parasites, vaccines derived from extracts of infected erythrocytes induce some protection against heterologous challenge (Wright 1991). A variety of babesiacidal drugs, including diminazene and imidocarb, may treat infected canines. Without treatment, the infection may persist in asymptomatic carriers. Environmental management, including fencing, vegetative selection (to prevent ascent of questing ticks) and quarantine measures, is difficult to implement over large areas (Smith and Kakoma 1989).

In humans, *B. microti* infections, while generally self-limiting, may be treated with quinine and clindamycin. In older or asplenic patients, severe infections also may require exchange transfusions (White and Breman 1998). In addition, the antimalarial drugs chloroquine and pyrimethamine alleviate the symptoms associated with babesiosis. None of these drugs have unequivocal babesiacidal activity, as parasitemia often persists, albeit with a reduction of symptoms.

Implementing control strategies against the vectors in an area requires knowledge of the biology of tick vectors. The incidence of clinical cases often can be related to the activity patterns of the vectors. For example, *B. divergens* cases in Europe correspond to the bimodal host-seeking behavior of *Ix. ricinus*. *Boophilus annulatus*, an important vector of both *B. bovis* and *B. bigemina*, may be maintained in the USA on white-tailed deer. Although the white-tailed deer is resistant to most *Babesia* spp., the deer may help to disseminate the tick.

Endemic stability is required to minimize the impact of babesiosis in an area. The goal is to achieve herd immunity by balancing the inoculation rate (derived from the proportion of infected ticks, the daily mean number of ticks attaching to hosts and the proportion of parasitized ticks that infect nonimmune hosts) with asymptomatic carriers and maternal immunity, among other factors. In an endemic stable situation, low level continuous tick transmission acquired from asymptomatic carriers maintains a threshold inoculation rate. Calves infected by such ticks, while under the influence of maternally derived antibodies, develop mild or asymptomatic infections. Calves born to unexposed mothers develop more resistance as they age due to repeated infections. Animals surviving mild illness become subclinical carriers experiencing recrudescences that infect ticks. To maintain this balance with a threshold inoculation rate, the tick population should be suppressed, but not eradicated. Upsetting the balance (e.g., by overly efficient tick control or the introduction of highly susceptible cattle breeds) may result in epidemics.

THEILERIOSIS

Introduction

Theileria spp. are important economic parasites of cattle, sheep, goats and wild artiodactyls. Seven species of *Theileria* infect cattle in Africa (Table 7.8). Disease caused by *Theileria parva* is known as East Coast disease, Corridor fever or January disease, depending on the severity of the disease. January disease is relatively mild; Corridor fever refers to the most serious form. *Theileria annulata*, which causes tropical theileriosis, infects livestock in the area encompassing North Africa, the middle east, Turkey and southern Europe. The bites of infected ticks of the genera *Rhipicephalus, Amblyomma, Hyalomma* and *Haemaphysalis* spread theileriosis.

The Parasite

Like *Plasmodium* spp., *Theileria* spp. are erythrocytic parasites. The differences in the life cycles among these parasites will be emphasized (Fig. 7.14). Unlike malaria parasites, *Theileria*-infected erythrocytes are not contained within a parsitophorous vacuole and differentiation into the sexual stages does not occur until the erythrocyte is lysed after ingestion by the tick vector.

Conditions initiating gametocytogenesis are uncertain, but the temperature drop in the replete tick together with conditions in the gut lumen are believed to initiate gametogony, possibly from the large ring forms. Male and female gametes fuse to form the zygote in the tick gut within a week. The zygote then invades a gut epithelial cell to develop into a single kinete. The kinete can invade the salivary glands only after redevelopment of the salivary glands following the molt. Estimates of the number of sporozoites produced per cell during sporogony is highly variable, ranging from 5,000 in *Theileria mutans* to 30,000 for *T. parva* in *Rhipicephalus appendiculatus* to 140,000 for *Theileria taurotragi*.

While salivary glands have not been demonstrated to release chemoattractants, recognition and invasion of salivary gland cells appears to be specific, because other tick organs and cell types, including the ovum, are not penetrated. Thus, while transstadial transmission is essential to transmission since the kinete cannot invade the salivary gland until after the molt, transovarial transmission does not occur.

Unlike malaria, the duration of *Theileria* infection in the tick is not indefinite and has been related to elevation (probably reflecting temperature) and stage. At 2,100 m, adult *R. appendiculatus* remain infected for life (up to 2 yr) whereas at sea level, *Theileria* infections persist in ticks for only 9 months. Infections in nymphs are of shorter duration than in adults.

Sporozoites are believed to be released slowly from the individual salivary gland cells after several days of bloodfeeding. Once in the bovine host, the sporozoite invades lymphocytes to initiate the vertebrate infection. The infected lymphocyte is transformed by the schizont stage to become lymphoblastoid, producing infected daughter cells, a process termed reversible parasite-induced transformation. After schizogony, merozoites invade erythrocytes (Fig. 7.14).

Giemsa-stained smears can detect *Theileria* spp. in host lymph nodes, peripheral blood smears and tick vectors. Morphological similarities among the species make this original identification method only as reliable as the experience of the observer. Molecular-based detection and identification methods including DNA probes, PCR and restriction fragment length polymorphism analysis now have been developed as well as serological tests using specific antigens and monoclonal antibodies.

Table 7.7. Host-vector relationships for the most common *Babesia* species[1].

Species	Vertebrate hosts	Principal tick vectors	Geographic distribution
B. bigemina	Bovines	*Boophilus annulatus, B. microplus, B. decoloratus, B. geigyi*	Central and South America, southern USA, Europe, Africa, Asia, Australia
B. bovis[2]	Bovines	*B. annulatus, B. microplus, B. geigyi, Ixodes ricinus*	Same as *B. bigemina*
B. major	Bovines	*Haemaphysalis punctata*	Europe and northern Africa
B. divergens	Bovines, humans[3]	*Ix. ricinus*	Northern Europe
B. jakimovi	Bovines	*Ix. ricinus*	Northern former USSR
B. ovata	Bovines	*Haemaphysalis longicornis*	Japan
B. occulatus	Bovines	*Hyalomma marginatum rufipes*	Southern Africa
B. caballi	Equines	*Anocentor nitens, Hyalomma* spp., *Rhipicephalus* spp.	Worldwide
B. equi	Equines	*A. nitens, Hyalomma* spp., *Rhipicephalus* spp.	Worldwide
B. canis	Canines	*Rhipicephalus* spp., *Dermacentor* spp.	Southern Europe, Africa, Asia, North America
B. gibsoni	Canines	*Rhipicephalus sanguineus, Haemaphysalis bispinosa*	Asia, Europe, North America
B. motasi	Ovines	*Haemaphysalis* spp., *Rhipicephalus* spp.	Europe, Asia, Africa
B. ovis	Caprines	*Rhipicephalus* spp.	Southern Europe, Middle East, Africa
B. trautmani	Porcines	Unknown	Southern Europe, former USSR, Africa
B. perroncitoi	Porcines	Unknown	Southern Europe, Africa
B. herpailuri	Felines	Unknown	Africa
B. felis	Felines	Unknown	Africa, southern Asia
B. microti	Rodents, humans	*Ixodes scapularis, Ix. ricinus, Ix. trianguliceps*	USA, Europe
B. lotori[4]	Racoons	*Ixodes texanus* (?)	USA
B. odocoilei[5]	Cervids	*Ix. scapularis, Amblyomma americanum* (?)	USA

[1]Modified after Kuttler (1988), with permission of CRC Press, Inc.; [2]Synonyms include *B. argentina* and *B. berbera;* [3]Most human infections in splenectomized individuals; [4]Anderson et al. (1981); [5]Waldrup et al. (1990).

Hosts

A wide range of ungulates is capable of maintaining both the parasite and the vector. *Syncerus caffer*, the African buffalo, has been hypothesized to be the original host for *T. parva*, *T. mutans* and *Theileria velifera* with *T. taurotragi* being capable of infecting a wide range of wild ungulates. *Theileria annulata* and *Theileria buffeli* infect both Asian buffalo and cattle.

The parasite-induced transformation of lymphocytes and their subsequent destruction, both during merogony and by cytolytic T-lymphocytes, causes the pathology seen in several host species. Destruction of erythrocytes by both the host immune system and by release of merozoites causes anemia in other *Theileria* spp. infections. Reductions in productivity in domestic animals due to chronic *Theileria* infections are difficult to ascertain due to the economic impacts of tick infestations alone. Acute *Theileria* infections with associated high mortality and severe disease are serious problems in Africa and Asia, particularly in areas where cattle are being upgraded for increased productivity. In *T. annuata* endemic areas, the local cattle breeds (Sanga, Zebu and their crosses) and domestic buffalo resist both *Theileria* infection and tick infestation. The introduction of imported breeds results in problems with theileriosis.

Vectors

Major vectors of *Theileria* spp., their tick vectors and distribution are presented in Table 7.8. A complete discussion of the population dynamics of each of these vectors is beyond the scope of this chapter. However, a brief discussion of the diversity of ecological zones in which *R. appendiculatus* is found will illustrate the potential effects on the transmission of *Theileria*. In the presence of year-around rainfall and favorable temperatures, *R. appendiculatus* has 3 generations per year (trivoltine). In drier areas, the species may be only univoltine, with behavioral diapause in the adult tick controlling the seasonality of the instars and their appearance during the more humid wet season. Heavy infestations of *R. appendiculatus* may be found in areas with both high rainfall and dense host populations. High tick densities are facilitated by the wide host range of the species and by the lack of resistance of many of the host species. However, overgrazing by excessively large ungulate populations will destroy the microenvironment needed for survival of the flat instars, resulting in the disappearance of this tick species in the area.

As with different species of ticks, different species of vertebrate hosts vary in their susceptibility to infection with *Theileria*, just as populations of the same tick species vary in their vector competency. *Rhipicephalus zambeziensis* is a more efficient vector of *T. parva* in buffalo than is *R. appendiculatus*. Nondiapausing Kenyan *R. appendiculatus* are more efficient at *T. parva* transmission than diapausing Zimbabwean *R. appendiculatus*.

Treatment and Control

There presently are 4 approaches to controlling theileriosis in cattle: restriction of cattle movement, control of vector ticks, chemotherapy of infected cattle and immunization of cattle at risk in endemic areas. The effectiveness of restricting the movement of *Theileria*-infected cattle from disease-free areas depends on the cultural practices in an area. In eastern and central Africa, the social customs of cattle due for dowry and funereal rites, together with cattle rustling and movement of people with their cattle, makes restrictions on cattle movements as a control option difficult to enforce, thereby limiting its effectiveness.

Tick control generally is attempted through the application of acaricides, especially arsenicals, organochlorines, organophosphates and pyrethroids. Besides the development of acaricide resistance by the ticks, effectiveness is de-

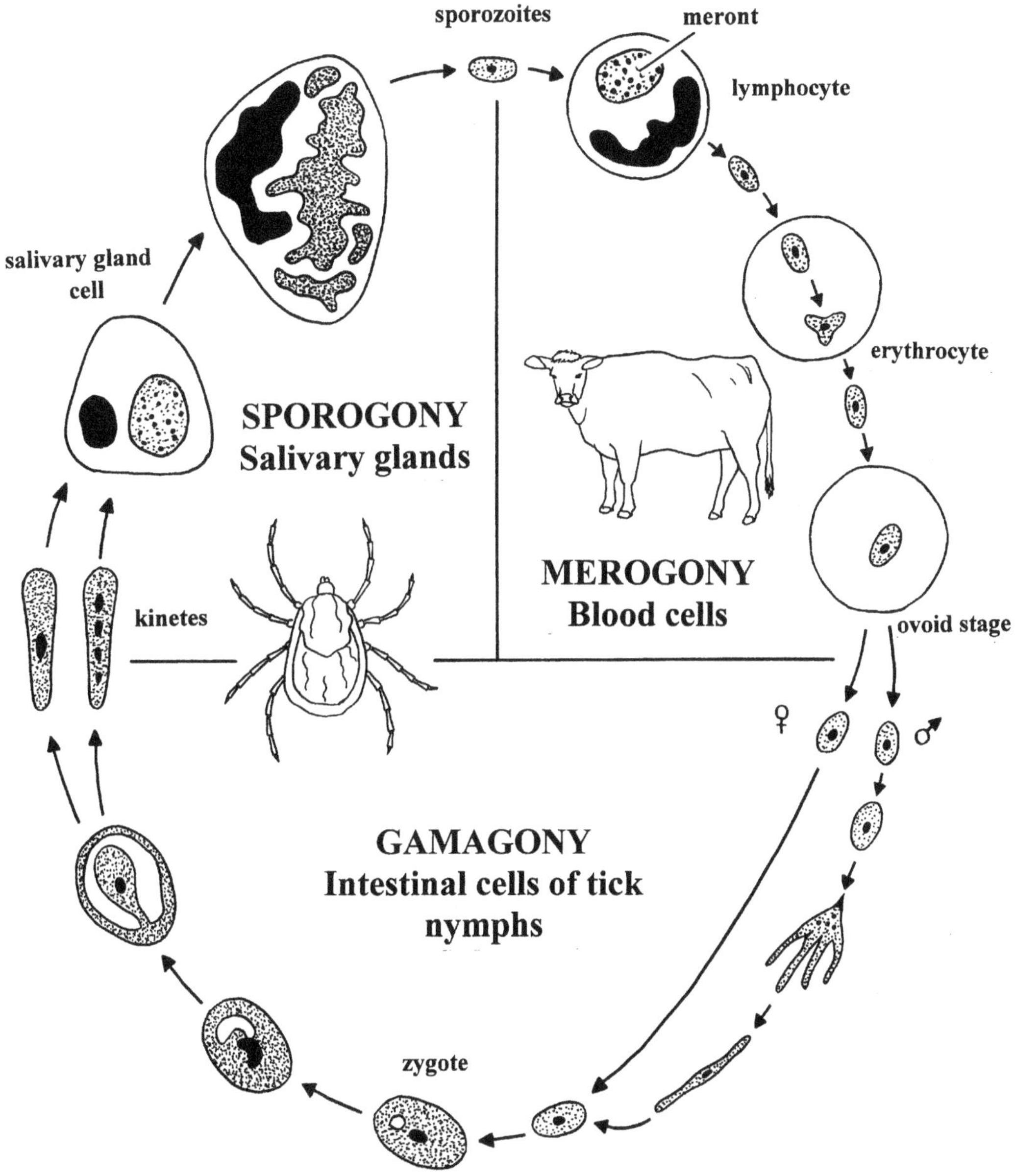

Figure 7.14. *Theileria* life cycle (drawing by Carol S. Barnett).

pendent on strict adherence to the recommended regimen of the acaricide. Acaracide applied rigorously to cattle for an extended period can control theileriosis. However, under such conditions, the cattle become highly susceptible to tickborne diseases, compounding the problems when coverage breaks down or resistance to the acaracide develops.

Three drugs effectively treat *Theileria*-infected cattle, if used early. In addition to destroying the schizont stage, halofuginone lactate and parvaquone reduce or eliminate fevers. Buparvaquone is even more effective than the other 2 drugs, but has a narrower safety margin. Two problems are associated with chemotherapy. The first is that relapses in recovering animals may lead to death of the animal. The second problem is that the drugs, like the acaracides, are manufactured outside the endemic countries and require hard currency to purchase.

Immunization is the fourth control method in use today. While efforts are being directed towards the development of bioengineered subunit vaccines, schizont vaccines and sporozoite stabilate vaccines, an infection and treatment immunization method has effectively controlled *T. parva*. In this approach, cattle are inoculated with *Theileria* at the same time that antibiotics (tetracyclines or buparvaquone) are administered. Due to the costs involved, infection and treatment immunization is cost-effective only in improved strains of cattle and must be implemented by specialized veterinary personnel to ensure proper quality controls to avoid the transmission of other pathogens.

RELATED ARTHROPODBORNE PATHOGENS

There are many other arthropodborne parasites of vertebrates in the Phylum Apicomplexa. Some of them, such as *Leucocytozoon caulleryi* (a parasite of chickens) have considerable economic significance (Morii 1992). These parasites are contained in the Order Haemosporida, along with parasites of the genus *Plasmodium*, discussed above.

Leucocytozoon

Several species of *Leucocytozoon* infect domestic fowl. The primary vectors of these parasites are species of black flies (Simuliidae) and biting midges of the genus *Culicoides* (Ceratopogonidae). Avian species in at least 15 orders are susceptible, but most infections produce only mild symptoms (Fallis et al. 1974). However, severe symptoms can result from infections of *Leucocytozoon simondi* in ducks, *Leucocytozoon smithi* in turkeys and *L. caulleryi* in chickens. This disease in chickens significantly decreases poultry production in the Far East and Southeast Asia. Symptoms in domestic fowl include loss of appetite, emaciation, drowsiness, enlarged spleen, damaged liver, congested lungs and heart, reduced egg production and thin eggshells (Nakamura et al. 1997). *L. caulleryi* in chickens is the most important parasite transmitted by *Culicoides*; as far as is known, black flies of the genus *Simulium* transmit the other species of *Leucocytozoon*.

Hepatocystis

Parasites of the genus *Hepatocystis* infect several Old World mammals, especially arboreal animals such as monkeys. These parasites, which cause only mild symptoms in mammals, presumably, are all transmitted by *Culicoides* midges (Desser et al. 1993).

Haemoproteus

Another group of sporozoan parasites is placed in the genus *Haemoproteus*. *Culicoides* midges transmit many species, and a few are transmitted by other arthropods such as horse and deer flies (Tabanidae) and louse flies (Hip-

Table 7.8. The most common *Theileria* species, their hosts, principal tick vectors and geographic distribution[1].

Species	Vertebrate hosts	Principal tick vectors	Geographic distribution
T. annulata	Cattle, buffalo	*Hyalomma anatolicum, H. asiaticum, H. detrium, Hyalomma* spp.	Northern Africa, Sudan, western and central Asia
T. mutans	Cattle, buffalo	*Amblyomma variegatum, A. gemma, A. hebraeum, A. lepidum, A. cohaerens, Amblyomma* spp.	Sub-saharan Africa and some Caribbean islands
T. ovis	Sheep	*Rhipicephalus* spp., *Dermacentor* spp., *Haemaphysalis* spp., *Hyalomma* spp., *Ornithodorus* spp.	Worldwide
T. parva	Cattle, buffalo	*Rhipicephalis appendiculatus, R. zambeziensis, R. duttoni, R. nitens*	Eastern, southern and central Africa
T. sergenti[1]	Cattle	*Haemaphysalis* spp.	Worldwide except North America
T. taurotragi	Eland, cattle	*R. appendiculatus, R. zambeziensis, R. pulchellus*	Eastern, central and southern Africa
T. velifera	Cape buffalo, cattle	*Amblyomma variegatum, A. hebraeum, A. lipidum*	Sub-saharan Africa
T. buffeli	Cattle	*Haemaphysalis* spp.	Europe, eastern Africa, Asia, including Japan and Australia, probably USA
T. hirci	Sheep, goats	*Hyalomma anatolicum*	Southern and eastern Europe, northern Africa
T. separata	Sheep	*Rhipicephalus evertsi*	Tanzania
T. cervi[2]	White-tailed deer	*Dermacentor* spp. (?), *Amblyomma americanum*	USA

[1]Based on Norval et al. (1992). Synonyms include *T. orientalis* and *T. buffeli*. With permission of Academic Press, Inc.; [2]Based on Mitema et al. (1991).

poboscidae). Both wild and domestic birds are affected, including ducks, geese and other waterfowl. Wild birds probably serve as the reservoir for epizootics among domestic birds (Garnham 1966). Pigeons are infected by *Haemoproteus columbiae* transmitted by the pigeon fly (Hippoboscidae). Other species of hippoboscids (*Lynchia* and *Stilometopa*) transmit similar parasites to other birds, including quail.

Haemoproteus metchnikovi, a parasite of turtles, is the only haemosporidian protozoan known to be transmitted by tabanid flies. In this case, transmission is cyclopropagative.

Control of Parasites in Domestic Birds

There have been few attempts to control infections by these parasites in birds. Aerial applications of pesticides have been applied to control vectors of turkey *Leucocytozoon* (Kissam et al. 1975). A killed vaccine has been tested to protect chickens against *L. caulleryi* (Morii et al. 1990).

REFERENCES

Alonso, P.L., Lindsay, S.W., Armstrong, J.R.M., Conteh, M., Hill, A.G., David, P.H., Fegorn, G., de Francisco, A., Hall, A.J., Shenton, F.C., Chan, K. and Greenwood, B.M. 1991. The effect of insecticide-treated bednets on mortality of Gambian children. Lancet **337**:1499–1502.

Anderson, J.F., Magnarelli, L.A. and Sulzer, A.J. 1981. Raccoon babesiosis in Connecticut, USA. *Babesia lotori* Sp.N. J. Parasitol. **67**:417–425.

Anderson, J.F., Mintz, E.D., Gadbaw, J.J. and Magnarelli, A.A. 1991. *Babesia microti*, human babesiosis and *Borrelia burgdorferi* in Connecticut. J. Clin. Microbiol. **29**:2779–2783.

Beier, J.C., Onyango, F.K., Koros, J.K., Ramadan, M., Ogwang, R., Wirtz, R.A., Koech, D.K. and Roberts, C.R. 1991. Quantitation of malaria sporozoites transmitted by wild Afrotropical *Anopheles*. Med. Vet. Entomol. **5**:71–79.

Bennett, G.F., Pierce, M.A. and Earle, R.A. 1994. An annotated checklist of the valid avian species of *Haemoproteus, Leucocytozoon* (Apicomplexa, Haemosporida) and *Hepatozoon* (Apicomplexa, Haemogregarinidae). Syst. Parasitol. **29**:61–73.

Binka, F.N., Kubaje, A., Adjuik, M., Williams, L.A., Lengeler, C., Maude, G.H., Armah, G.E., Kajihara, B., Adiamah, J.H. and Smith, P.G. 1996. Impact of permethrin impregnated bednets on child mortality in Kassena-Nankana district, Ghana: a randomized controlled trial. Trop. Med. Inter. Health **1**: 147-154.

Borst, P., Bitter, W., McCulloch, R., Van Leeuwen, F. and Rudenko, G. 1995. Antigenic variation in malaria (minireview). Cell **82**:1–4.

Burkot, T.R. 1988. Non-random host selection by anopheline mosquitoes. Parasitol. Today **4**:156–162.

Burkot, T.R. and Graves, P.M. 1994. Human malaria transmission: reconciling field and laboratory data. Pp. 149–182 *in* K.F. Harris (ed.), Advances in Disease Vector Research, vol. 11. Springer-Verlag, New York.

Burkot, T.R. and Graves, P.M. 1995. The value of vector-based estimates of malaria transmission. Ann. Trop. Med. Parasitol. **89**:125–134.

Burkot, T.R., Graves, P.M., Paru, R., Battistutta, D., Barnes, A. and Saul, A. 1990. Variations in malaria transmission rates are unrelated to the anopheline survivorship per feeding cycle. Am. J. Trop. Med. Hyg. **43**:321–327.

Burkot, T.R., Graves, P.M., Paru, R., Wirtz, R.A. and Heywood, P.F. 1988. Human malaria transmission studies in the *Anopheles punctulatus* complex in Papua New Guinea: sporozoite rates, inoculation rates and sporozoite densities. Am. J. Trop. Med. Hyg. **39**:135–144.

Burkot, T.R. and Wirtz, R.A. 1986. Immunoassays of malaria sporozoites in mosquitoes. Parasitol. Today **2**:155–157.

Cattani, J.A., Tulloch, J.L., Vrbova, H., Jolley, D., Gibson, F.D., Moir, J.S., Heywood, P.F., Alpers, M.P., Stevenson, A. and Clancy, R. 1986. The epidemiology of malaria in a population surrounding Madang, Papua New Guinea. Am. J. Trop. Med. Hyg. **35**:3–15.

Charlwood, J.D., Kihonda, J., Sama, S., Billingsley, P.F., Hadji, H., Verhave, J.P., Lyimo, E., Luttikhuisen, P.C. and Smith, T. 1995. The rise and fall of *Anopheles arabiensis* (Diptera: Culicidae) in a Tanzanian village. Bull. Entomol. Res. **85**:37–44.

Chege, G.M.M. and Beier, J.C. 1990. Effect of *Plasmodium falciparum* on the survival of naturally infected Afrotropical *Anopheles* (Diptera: Culicidae). J. Med. Entomol. **27**:454–458.

Collins, F.H., Petrarca, V., Mpofu, S., Brandling-Bennett, A.D., Were, J.B.O., Rasmussen, M.O. and Finnerty, V. 1988. Comparison of DNA probe and cytogenetic methods for identifying field collected *Anopheles gambiae* complex mosquitoes. Am. J. Trop. Med. Hyg. **39**:545-550.

Collins, F.H., Sakai, R.K., Vernick, K.D., Paskewitz, S., Seeley, D.C., Miller, L.H., Collins, W.E., Campbell, C.C. and Gwadz, R.W. 1986. G enetic selection of a *Plasmodium*-refractory strain of the malaria vector *Anopheles gambiae*. Science **234**:607–609.

Coluzzi, M. 1984. Heterogeneities of the malaria vectorial system in tropical Africa and their significance in malaria epidemiology and control. Bull. Wld. Hlth. Organiz. **62**:107–113.

Coluzzi, M. 1992. Malaria vector analysis and control. Parasitol. Today **3**:113–118.

Conner, D.H., Neafie, R.C. and Hockmeyer, W.T. 1976. Malaria. Pp. 273–283 *in* C.H. Binford and D.H. Conner (eds.), Pathology of tropical and extraordinary diseases. Armed Forces Institute of Pathology, Washington, DC

Covell, G. 1960. Relationship between malarial parasitaemia and symptoms of the disease. Bull. Wld. Hlth. Organiz. **22:**605–619.

Curtis, C.F. and Graves, P.M. 1983. Genetic variation in the abililty of insects to transmit filariae, trypanosomes and malarial parasites. Pp. 31–62 *in* K.F. Harris (ed.), Current Topics in Vector Research, vol. 1. Praeger, New York.

Desser, S.S., Bennett, G.F. and Kreier, J.P. 1993. The genera *Leucocytozoon, Haemoproteus* and *Hepatocystis*. Pp. 273–307 *in* J.P. Kreier (ed.), Parasitic Protozoa, 2nd ed, vol. 4. Academic Press, New York.

Dye, C. 1986. Vectorial capacity: must we measure all its components? Parasitol. Today **2:**203–209.

Dye, C. 1990. Epidemiological significance of vector-parasite interactions. Parasitol. **101:**409–415.

Dye, C. 1992. The analysis of parasite transmission of bloodsucking insects. Annu. Rev. Entomol. **37:**1–19.

Esposito, F., Lombardi, S., Modiano, D., Zavala, F., Reeme, J., Lamizana, L., Coluzzi, M. and Nussenzweig, R.S. 1988. Prevalence and levels of antibodies to the circumsporozoite protein of *Plasmodium falciparum* in an endemic area and their relationship to resistance against malaria infection. Trans. Royal Soc. Trop. Med. Hyg. **82:**827–832.

Fallis, A.M., Desser, S.S. and Khan, R.A. 1974. On species of *Leucocytozoon*. Pp. 1–67 *in* B. Dawes (ed.), Advances in parasitology, vol. 12. Academic Press, New York.

Feldmann, A.M. and Ponnudurai, T. 1989. Selection of *Anopheles stephensi* for refractoriness and susceptibility to *Plasmodium falciparum*. Med. Vet. Entomol. **3:**41–52.

Gamage-Mendis, A.C., Rajalaruna, J., Carter, R. and Mendis, K.N. 1991. Infectious reservoir of *Plasmodium vivax* and *Plasmodium falciparum* malaria in an endemic region of Sri Lanka. Am. J. Trop. Med. Hyg. **45:**479–487.

Garcia, G.E., Wirtz, R.A., Barr, J.R., Woolfitt, A. and Rosenberg, R. 1998. Xanthurenic acid induces gametogenesis in *Plasmodium*, the malaria parasite. J. Biol. Chem. **273:**12003–12005.

Garnham, P.C.C. 1966. Malaria parasites and other Haemosporida. Oxford Blackwell Scientific, Philadelphia. 1114 pp.

Gilles, H.M. and Warrell, D.A. 1993. Bruce-Chwatt's essential malariology, 3rd ed. E. Arnold, London. 340 pp.

Gillies, M. and Wilkes, T.J. 1965. A study of the age composition of populations of *Anopheles gambiae* Giles and *A. funestus* Giles in north-eastern Tanzania. Bull. Entomol. Res. **56:**237–262.

Githeko, A.K., Brandling-Bennett, A.D., Beier, M., Atieli, F., Owaga, M. and Collins, F.H. 1992. The reservoir of *Plasmodium falciparum* malaria in a holoendemic area of Western Kenya. Trans. Royal Soc. Trop. Med. Hyg. **86:**355–358.

Gomes, M., Espino, F.E., Abaquin, J., Realon, C. and Salazar, N.P. 1994. Symptomatic identification of malaria in the home and primary health care clinic. Bull. Wld. Hlth. Organiz. **72:**383–390.

Graves, P.M., Burkot, T.R., Carter, R., Cattani, J.A., Lagog, M., Parker, J., Brabin, B.J., Gibson, F.D. and Alpers, M.P. 1988a. Measurement of malarial infectivity of human populations to mosquitoes in the Madang area, Papua New Guinea. Parasitol. **96:**251–263.

Graves, P.M., Burkot, T.R., Saul, A., Hayes, R. and Carter, R. 1990. Estimation of anopheline survival rate, vectorial capacity, and mosquito infection probability from malaria vector infection rates in villages near Madang, Papua New Guinea. J. Appl. Ecol. **27:**134–147.

Graves, P.M., Carter, R., Burkot, T.R., Quakyi, I.A. and Kumar, N. 1988b. Antibodies to *Plasmodium falciparum* gamete surface antigens in Papua New Guinea sera. Parasite Immunol. **10:**209–218.

Graves, P.M., Gelband, H.M. and Garner, P. 1998. Spf66 malaria vaccines. What is the evidence for efficacy? Parasitol. Today **14:** 218-220.

Haji, H., Smith, T., Meuwissen, J.T., Sauerwein, R. and Charlwood, J. 1996. Estimation of the infectious reservoir of *Plasmodium falciparum* in natural vector populations based on oocyst size. Trans. Royal Soc. Trop. Med. Hyg. **90:**494–497.

Harrison, G. 1978. Mosquitoes, malaria and man: a history of the hostilities since 1880. E.P. Dutton, New York. 314 pp.

Hill, A.V.S., Allsopp, C.E.M., Kwiatkowski, D., Anstey, N.M., Twumasi, P., Rowe, P.A., Bennett, S., Brewster, D., McMichael, A.J. and Greenwood, B.M. 1991. Common West African HLA antigens are associated with protection from severe malaria. Nature **352:**595–600.

Ifediba, T. and Vandenberg, J.P. 1981. Complete *in vitro* maturation of *Plasmodium falciparum* gametocytes. Nature **294:**364–366.

Kakoma, I. and Mehlhorn, H. 1994. *Babesia* of domestic animals. Pp. 141–216 *in* J.P. Kreier and J.R. Baker (eds.), Parasitic Protozoa, vol. 7. Academic Press, New York.

Kissam, J.B., Noblet, R. and Garris, G.I. 1975. Large-scale aerial treatment of an endemic area with Abate granular larvicide to control black flies (Diptera: Simuliidae) and suppress *Leucocytozoon smithi* in turkeys. J. Med. Entomol. **12:**359–362.

Klein, T.A., Harrison, B.A., Grove, J.S., Dixon, S.V. and Andre, R.G. 1986. Correlation of survival rates of *Anopheles dirus* A (Diptera: Culicidae) with different infection densities of *Plasmodium cynomolgi*. Bull. Wld. Hlth. Organiz. **64:**901–907.

Klein, T.A., Lima, J.B.P., Tada, M.S. and Miller, R. 1991. Comparative susceptibility of anopheline mosquitoes in Rondonia, Brazil to infection by *Plasmodium vivax*. J. Trop. Med. Hyg. **45:**463–470.

Kreier, J.P. and Baker, J.R. 1987. Parasitic Protozoa. Allen and Unwin, Boston. 241 pp.

Kreier, J.P. and Ristic, M. 1989. Introduction. P. 1 *in* J.P. Krier, M. Ristic and J.R. Baker (eds.), Malaria and babesiosis. 3rd Intern. Congr. Malaria and Babesiosis, Annecy, France, Vol. 83(Suppl.). Transactions of the Royal Society of Tropical Medicine and Hygiene, London.

Kuttler, K.L. 1988. Babesiosis of domestic animals and man. Pp. 1–22 *in* M. Ristic (ed.), World-wide impact of babesiosis. CRC Press, Boca Raton, Florida.

Lane, R.P. and Crosskey, R.W. (eds). 1993. Medical insects and arachnids. Chapman and Hall, London. 723 pp.

Lengeler, C. and Snow, R.W. 1996. From efficacy to effectiveness: insecticide-treated bednets in Africa. Bull. Wld. Hlth. Organiz. **74:**325–332.

Levine, N.D. 1988. Progress in taxonomy of the apicomplexan protozoa. J. Protozool. **35:**518–520.

Lindsay, S.W., Wilkins, H.A., Zidler, H.A., Daly, R.J., Petrarca, V. and Byasi, P. 1991. Ability of *Anopheles gambiae* to transmit malaria during the dry and wet seasons in an area of irrigated rice cultivation in the Gambia. J. Trop. Med. Hyg. **94:**313–314.

Lines, J.D., Wilkes, T.J. and Lysimo, E.O. 1991. Human malaria infectiousness measured by age-specific sporozoite rates in *Anopheles gambiae* in Tanzania. Parasitol. **102:**167–177.

Long, C.A. 1993. Immunity to blood stages of malaria. Curr. Opinions Immunol. **5:**548–556.

Macdonald, G. 1957. The epidemiology and control of malaria. Oxford University Press, New York. 201 pp.

Makler, M.T., Palmer, G., J, and Ager, A.L. 1998. A review of practical techniques for the diagnosis of malaria. Ann. Trop. Med. Parasitol. **92:**419–433.

Marsh, K. 1992. Malaria—a neglected disease? Parasitol. **104:**S53–S69.

McCutchan, T.F., Dame, J.D., Miller, L.H. and Barnwell, J. 1984. Evolutionary relatedness of *Plasmodium* species as determined by the structure of DNA. Science **225:**808–811.

Mehlhorn, H., Schein, E. and Ahmed, J.S. 1994. *Theileria*. Pp. 217–304 *in* J.P. Kreier and J.R. Baker (eds.), Parasitic Protozoa, vol. 7. Academic Press, New York.

Mendis, K.N., Munesinghe, Y.D., de Silva, Y.N.Y., Keragalla, I. and Carter, R. 1987. Malaria transmission-blocking immunity induced by natural infections of *Plasmodium vivax* in humans. Infect. Immun. **55:**369–372.

Miller, L.H. 1988. Malaria. P. 493 *in* W.H. Wernsdorfer and I.A. McGregor (eds.), Principles and practice of malariology. Churchill Livingston, Edinburgh.

Mitema, E.S., Kocan, A.A., Mukolwe, S.W., Sangiah, S. and Sherban, D. 1991. Activity of buparvaquone against *Theileria cervi* in white-tailed deer. Vet. Parasitol. **38:**49–53.

Molineaux, L. and Gramiccia, G. 1980. The Garki Project: research on the epidemiology and control of malaria in the Sudan savanna of West Africa. World Health Organization, Geneva. 311 pp.

Morii, T. 1992. A review of *Leucocytozoon caulleryi* infection in chickens. J. Protozool. Res. **2:**128–133.

Morii, T., Fujita, J., Akiba, K., Isobe, T., Nakamoto, K., Masubuchi and Ishahara, H. 1990. Protective immunity to *Leucocytozoon caulleryi* in chickens by a killed vaccine. Parasitol. Res. **76:**630–632.

Muirhead-Thompson, R.C. 1957. The malaria infectivity of an African village population to mosquitoes (*Anopheles gambiae*). Am. J. Trop. Med. Hyg. **6:**971–979.

Nakamura, K., Mitarai, Y., Tanimura, N., Hara, H., Ikeda, A., Shimada, J. and Isobe, T. 1997. Pathogenesis of reduced egg production and soft-shelled eggs in laying hens associated with *Leucocytozoon caulleryi* infection. J. Parasitol. **83:**325–327.

Nevill, C.G., Some, E.S., Mung'ala, V.O., Mutemi, W., New, L., Marsh, K., Lengeler, C. and Snow, R.W. 1996. Insecticide-treated bednets reduce mortality and severe morbidity from malaria among children on the Kenya coast. Trop. Med. Intern. Hlth. **1:**139–146.

Norval, R.A.I., Perry, B.D. and Young, A.S. 1992. The epidemiology of theileriosis in Africa. Academic Press, Inc., London. 481 pp.

Paskewitz, S.M., Ng, K., Coetzee, M. and Hunt, R.H. 1993. Evaluation of the polymerase chain reaction for identifying members of the *Anopheles gambiae* (Diptera: Culicidae) complex in southern Africa. J. Med. Entomol. **30:**953-957.

Paul, R.E.L., Packer, M.J., Walmsley, M., Lagog, M., Ranford-Cartwright, L.C., Paru, R. and Day, K.P. 1995. Mating patterns in malaria parasite populations of Papua New Guinea. Science **269:**1709–1711.

Persing, D.H., Herwaldt, B.L., Glaser, C., Lane, R.S., Thomford, J.W., Mathiesen, D., Krause, P.J., Phillip, D.F. and Conrad, R.A. 1995. Infection with a babesia-like organism in northern California. New England J. Med. **332:**298–303.

Petrarca, V. and Beier, J.C. 1992. Intraspecific chromosomal polymorphism in the *Anopheles gambiae* complex as a factor affecting malaria transmission in the Kisumu area of Kenya. Am. J. Trop. Med. Hyg. **42:**229–237.

Ponnudurai, T., Billingsley, P.F. and Rudin, W. 1988. Differential infectivity of *Plasmodium* for mosquitoes. Parasitol. Today **4:**319–321.

Pringle, G. 1966. A quantitative study of naturally-acquired malarial infections in *Anopheles gambiae* and *Anopheles funestus* in a highly malarious area of East Africa. Trans. Royal Soc. Trop. Med. Hyg. **60**:626–632.

Quick, R.E., Herwaldt, B.L., Thomford, J.W., Garnett, M.E., Eberhard, M.L., Wilson, M., Spach, D.H., Dickerson, J.W., Telford, S.R., III, Steingart, K.R., Pollock, R., Persing, D.H., Kobayashi, J.M., Juranek, D.D. and Conrad, P.A. 1993. Babesiosis in Washington state: a new species of *Babesia*? Ann. Intern. Med. **119**:284–290.

Reinert, J.F., Kaiser, P.E. and Seawright, J.A. 1997. Analysis of the *Anopheles (Anopheles) quadrimaculatus* complex of sibling species (Diptera: Culicidae) using morphological, cytological, molecular, genetic, biochemical and ecological techniques in an integrated approach. J. Am. Mosq. Contr. Assoc. **13(suppl.)**:1–102.

Reisen, W.K. 1989. Estimation of vector capacity: introduction. Bull. Soc. Vect. Ecol. **14**:39–40.

Ribeiro, J.M.C., Rossignol, P.A. and Spielman, A. 1985. Salivary gland apyrase determines probing time in anopheline mosquitoes. J. Insect Physiol. **31**:689–692.

Robert, V., Verhave, J.P. and Carnevale, P. 1990. *Plasmodium falciparum* infection does not increase the precocious mortality rate of *Anopheles gambiae*. Trans. Royal Soc. Trop. Med. Hyg. **84**:346–347.

Rosenberg, R. 1985. Inability of *Plasmodium knowlesi* sporozoites to invade *Anopheles freeborni* salivary glands. Am. J. Trop. Med. Hyg. **34**:687–691.

Rosenberg, R. and Rungsiwongse, J. 1991. The number of sporozoites produced by individual malaria oocysts. Am. J. Trop. Med. Hyg. **45**:574–577.

Rosenberg, R.A., Wirtz, R., Lanar, D.E., Sattabongkot, J., Hall, T., Waters, A.P. and Prasittisuk, C. 1989. Circumsporozoite protein heterogeneity in the human malaria parasite *Plasmodium vivax*. Science **245**:973–976.

Rosenberg, R., Wirtz, R.A., Schneider, I. and Burge, R. 1990. An estimation of the number of malaria sporozoites ejected by a feeding mosquito. Trans. Royal Soc. Trop. Med. Hyg. **84**:209–212.

Saul, A. 1987. Estimation of survival rates and population size from mark-release-capture experiments of bait-caught haematophagous insects. Bull. Entomol. Res. **77**:589–602.

Saul, A., Graves, P.M. and Kay, B.H. 1990. A cyclical feeding model for pathogen transmission and its application to determine vectorial capacity from vector infection rates. J. Appl. Ecol. **27**:123–133.

Service, M.W. 1965. Some basic entomological factors concerned with the transmission and control of malaria in Northern Nigeria. Trans. Royal Soc. Trop. Med. Hyg. **59**:291–296.

Shahabuddin, M. and Kaslow, D.C. 1994. *Plasmodium*: parasite chitinase and its role in malaria transmission. Exp. Parasitol. **79**:85–88.

Smith, R.D. and Kakoma, I. 1989. A reappraisal of vector control strategies for babesiosis. Trans. Royal Soc. Trop. Med. Hyg. **83(1, Suppl.)**:43–52.

Smith, T. and Kilbourne, F.L. 1893. Investigations into the nature, causation and prevention of Texas or southern cattle fever. US Dept. Agric. Bull. No. 1, Washington, DC 301 pp.

Subbarao, S.K. 1988. The *Anopheles culicifacies* complex and control of malaria. Parasitol. Today **4**:72–75.

Tchuinkam, T., Mulder, B., Dechering, K., Stoffels, H., Verhave, J.P., Cot, M., Carnevale, P., Meuwissen, J.H.E.T. and Robert, V. 1993. Experimental infections of *Anopheles gambiae* with *Plasmodium falciparum* of naturally infected gametocyte carriers in Cameroon: factors influencing the infectivity to mosquitoes. Trop. Med. Parasitol. **44**:271–276.

Telford, S.R., III, Gorenflot, A., Brasseur, P. and Spielman, A. 1993. Babesial infections in humans and wildlife. Pp. 1–47 *in* J.P. Kreier (ed.), Parasitic Protozoa, vol. 5. Academic Press, New York.

Thomford, J.W., Conrad, P.A., Telford, S.R., III, Mathiesen, D., Bowman, B.H., Spielman, A., Eberhard, M.L., Herwaldt, B.L., Quick, R.E. and Persing, D.H. 1994. Cultivation and phylogenetic characterization of a newly recognized human pathogenic protozoan. J. Infect. Dis. **169**:1050–1056.

Trager, W. and Jensen, J.B. 1976. Human malaria parasites in continuous culture. Science **193**:673–675.

Ungureanu, E., Killick-Kendrick, R., Garnham, P.C.C., Branzei, P., Romanescu, C. and Shute, P.G. 1976. Prepatent periods of a tropical strain of *Plasmodium vivax* after inoculations of tenfold dilutions of sporozoites. Trans. Royal Soc. Trop. Med. Hyg. **70**:482–483.

Vernick, K.D., Collins, F.H. and Gwadz, R.W. 1989. A general system of resistance to malaria infection in *Anopheles gambiae* controlled by two main genetic loci. Am. J. Trop. Med. Hyg. **40**:585–592.

Waldrup, K.A., Kocan, A.A., Barker, R.W. and Wagner, G.G. 1990. Transmission of *Babesia odocoilei* in white-tailed deer (*Odocoileus virginianus*) by *Ixodes scapularis* (Acari: Ixodidae). J. Wildlife Dis. **26**:390–391.

Walliker, D. 1989. Genetic recombination in malaria parasites. Exp. Parasitol. **69**:303–309.

Waters, A.P., Higgins, D.G. and McCutchan, T.F. 1993. The phylogeny of malaria: a useful study. Parasitol. Today **9**:246–250.

Waters, A.P., Syin, C. and McCutchan, T.F. 1989. Developmental regulation of stage-specific ribosome populations in *Plasmodium*. Nature **342**:438–440.

Webster, H.K., Boudreau, E.F., Pang, L.W., Permpanich, B., Sookto, P. and Wirtz, R.A. 1987. Development of immunity in natural *Plasmodium falciparum* malaria: antibodies to the falciparum sporozoite vaccine 1 antigen (R32tet32). J. Clin. Microbiol. **25**:1002–1008.

White, N.J. and Breman, J.G. 1998. Malaria and other diseases caused by red blood cell parasites. Pp. 1180–1189 *in* A.S. Fauci, E. Braunwald, K.J. Isselbacher, J.D. Wilson, J.B. Martin, D.L. Kasper, S.L. Hauser and D.L. Longo (eds.), Harrison's principles of internal medicine, 14th ed. McGraw-Hill, New York.

Wirtz, R.A. and Burkot, T.R. 1991. Detection of malarial parasites in mosquitoes. Pp. 77–106 *in* K.F. Harris (ed.), Advances in Disease Vector Research, vol. 8. Springer-Verlag, New York.

World Health Organization. 1993. Global malaria control. Bull. Wld. Hlth. Organiz. **71:**281–284.

World Health Organization. 1994. World malaria situation in 1992. Weekly Epidemiol. Rec. **69:**309–314.

World Health Organization. 1995. WHO model prescribing information. Drugs used in parasitic diseases, 2nd ed. World Health Organization, Geneva. 146 pp.

World Health Organization. 1996. A rapid dipstick antigen capture assay for the diagnosis of falciparum malaria. Bull. Wld. Hlth. Organiz. **74:**47–54.

World Bank. 1993. World development report 1993: investing in health. Oxford University Press, New York. 329 pp.

Wright, I.G. 1991. Towards a synthetic *Babesia* vaccine. J. Parasitol. **21:**156–159.

Chapter 8

Leishmaniasis and Trypanosomiasis

PHILLIP G. LAWYER[1] AND PETER V. PERKINS[2]
[1]Uniformed Services University of the Health Sciences, Bethesda, Maryland and [2]US Army Medical Service Corps, retired

INTRODUCTION

Leishmaniasis and trypanosomiasis are separate groups of arthropodborne diseases of humans and other animals caused by infection with protozoan hemoflagellates of the genus *Leishmania* and *Trypanosoma*, respectively. Both genera are included in the family Trypanosomatidae, order Kinetoplastida (Fig. 8.1).

The order Kinetoplastida includes parasites that possess a **kinetoplast**, a deeply staining structure giving rise to a **flagellum**. A single locomotory flagellum, either free or attached to a **pellicle** as an **undulating membrane**, arises from the kinetoplast (Fig. 8.2).

All species in the genera *Leishmania* and *Trypanosoma* alternate between vertebrate and invertebrate hosts. Within both hosts, these parasites undergo changes in morphology that have lead to misunderstandings in identification and nomenclature. Originally, each morphological form was referred to by the genus in which the form predominated. Thus, the "crithidial" form occurred mainly in the genus *Crithidia* and "leishmanial" forms were found mainly in *Leishmania*. Subsequently, a new nomenclature used names for the morphological forms according to the position of the flagellum.

Kingdom	Protista
Subkingdom	Protozoa
Phylum	Sarcomastigophora
Subphylum	Mastigophora
Class	Zoomastigophora
Order	Kinetoplastida
Suborder	Trypanomatina
Family	Trypanosomatidae
Genera	*Leishmania* *Trypanosoma*

Figure 8.1. Classification of *Leishmania* and *Trypanosoma* (after Molyneaux and Ashford 1983).

B.F. Eldridge and J.D. Edman (eds.), Medical Entomology, Revised Edition, 231–298.

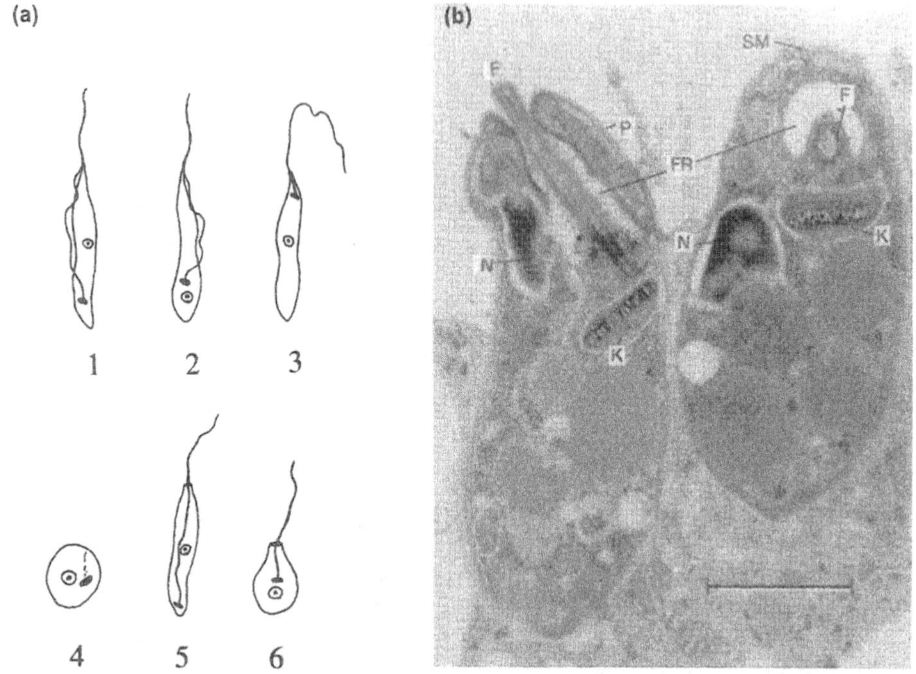

Figure 8.2. Morphology of the Trypanosomatidae, order Kinetoplastida: (a) Diagram of the developmental stages and genera (in parentheses); 1. Trypomastigote (*Trypanosoma*); 2. Epimastigote (*Trypanosoma, Blastocrithidia, Endotrypanum*); 3. Promastigote (*Trypanosoma, Blastocrithidia, Endotrypanum, Phytomonas, Leishmania, Leptomonas*); 4. Amastigote (*Trypanosoma, Blastocrithidia, Endotrypanum, Phytomonas, Leishmania, Leptomonas, Herpetemonas, Crithidia*); 5. Opisthomastigote (*Herpetomonas*); 6. Choanomastigote (*Crithidia*)[modified from Hoare (1972)]; (b) electron micrograph, saggital section of *Leishmania mexicana* amastigotes in hamster macrophage cell showing characteristic features of the Kinetoplastida. F = flagellum, FR = flagellar reservoir, K = kinetoplast, N = nucleus, P = pellicle, SM = subpellicular microtubules. Approximate magnification 33,440x; bar = 1 µm.

Leishmania and *Trypanosoma* undergo obligatory development and propagation in their respective invertebrate hosts. Both groups of parasites have unique mechanisms for circumventing the immune systems of their vertebrate hosts. The *Leishmania* are intracellular parasites that invade and proliferate in mononuclear phagocytic cells, the blood cells largely responsible for vertebrate immune defenses. The *Trypanosoma*, on the other hand, are extracellular parasites. While they do occur in the blood, many species sequester in tissues. Here, they are protected from the host's immune response during periods of reproductive activity and it is here where new variants may arise.

The leishmaniases and trypanosomiases occur primarily in the tropics and subtropics. They cause suffering, disability, permanent disfigurement and death in millions of people and their livestock annually. In 1990, the World Health Organization (WHO) estimated that 350 million people were at risk of leishmaniasis in 21 different countries, with 12 million people infected at any one time (World Health Organization 1990). The current estimate of the annual global incidence of leishmaniasis is 1.5–2 million new cases per year in 88 countries, with 1–1.5 million cases of cutaneous disease and 500,000 cases of visceral disease (Desjeux 1996). An estimated 350 million people are at risk of contracting this disease worldwide, with a prevalence of 12 million (Desjeux 2001). The WHO estimates that 50 million people living in endemic areas are at risk of African trypanosomiasis (sleeping sickness), with an estimated 10,000–20,000 newly reported cases per year (World Health Organization 1998). American trypanosomiasis (Chagas disease) places 35 million people at risk in the Americas, with an estimated 16–18 million infected and 50,000 deaths per year (World Health Organization 1991). In Latin America, of 24 million seropositive people, 10–30% will experience clinical disease, and about 70,000 will die each year (Kettle 1995).

THE LEISHMANIASES

Probably only malaria is more important than the leishmaniases among the protozoan diseases in terms of human suffering and economics. At least 23 *Leishmania* species cause these diseases, occurring mainly in the tropics and subtropics of the Old and New Worlds in a wide range of physical environments and with great epidemiological diversity (Tables 8.1, 8.2). Killick-Kendrick (1990a) observed that a wider range of parasites cause leishmaniasis than any other parasitic disease. The public health impact of the leishmaniases has been underestimated, due mainly to lack of public awareness. Only the last 2 decades have shown that leishmaniasis is more prevalent than had been realized. The leishmaniases are endemic in 22 New World and 66 Old World countries on 4 continents. Major epidemics of visceral leishmaniasis have occurred recently in Brazil, India and the Sudan. In Bihar State in eastern India, it is believed that 200,000–250,000 people contracted the disease in 1993. Reports from the southern Sudan confirm 15,000 cases treated and 100,000 deaths in a 5-year period (1991–1996). Cutaneous leishmaniasis is on the rise in many countries, with epidemics ongoing in newly settled areas of the Amazon Basin, North Africa and the Middle East (Desjeux 1996, 2001, Magill 1995).

Epidemiological studies of leishmaniasis often are complicated because 2 or more leishmanial taxa and morphologically similar nonleishmanial trypanosomatids coexist in many localities (Lainson 1983). In addition, the number of taxa is increasing as more foci are discovered and studied. Identifying these organisms and associating them with specific vectors remains a challenge even with state-of-the-art biochemical analyses.

Scientists continue to discover new foci of leishmaniasis, sometimes well beyond the known geographic range of a particular *Leishmania* species. Such discoveries may represent long established but undetected foci, or they may indicate recent introductions of the disease.

Sand flies in the genera *Phlebotomus* (Old World; Fig. 8.3, Table 8.1) and *Lutzomyia* (New World; Table 8.2) are the only known vectors of *Leishmania*. However, specific vectors and reservoir mammals are unknown in many foci. About 80 of the approximately 700 known sand fly species worldwide have been incriminated as vectors. The existence of sympatric morphospecies (sibling species), particularly among the *Lutzomyia* in the New World, often complicate vector incrimination studies. For example, specimens of *Lutzomyia wellcomei*, a proven vector of *Leishmania braziliensis* complex parasites, are difficult to distinguish by morphological characters alone from those of *Lutzomyia complexa*, a nonvector, in Brazil (Ready and da Silva 1984, Lane and Ready 1985).

Leishmania transmission takes place in a variety of habitats in both the Old and New Worlds, but epidemiological details of foci may vary, even within similar habitats in the same

Table 8.1. Old World *Leishmania* and proven and suspected vectors (Young and Lawyer 1987; Killick-Kendrick 1990, 1999; Lane 1993. Updated by S.A. Kamhawi and P.G. Lawyer).

Parasite (disease)[1]	Sand fly species[2]	Sand fly distribution[3]
Leishmania aethiopica (CL)	***Phlebotomus longipes***	Kenya, Ethiopia
	P. pedifer	Kenya, Ethiopia
Le. donovani s.l. (VL, PKDL)	***P. alexandri***	North Africa to western China
	P. argentipes	Bangladesh, Nepal, India
	P. celiae	Kenya, southern Ethiopia
	P. martini	East Africa, Ethiopia
(CL, MCL, VL)	***P. mongolensis***	Central Asia
	P. orientalis	Sudan, Ethiopia, Saudi Arabia, Yemen
Le. infantum (VL, CL)	***P. ariasi***	Western Mediterranean
	P. brevis	Northern Iran to Caucasus
	P. chinensis	Northern, Central China
	P. halepensis	Jordan, Lebanon, Iraq
	P. kandelakii	Iran, Afghanistan
	P. langeroni	Egypt, Tunisia
	P. longicuspis	North Africa
	P. longiductus	North Africa, Central Asia
	P. neglectus	Eastern Mediterranean
	P. perfiliewi	Mediterranean Basin, Algeria
	P. perniciosus	Western Mediterranean
	P. sichuanensis	China
	P. smirnovi	Central Asia
	P. tobbi	Eastern Mediterranean
	P. transcaucasicus	Caucasus
Le. killicki (CL)	*P. papatasi*	Central Tunisia
	P. alexandri	Central Tunisia
	P. chaubaudi	Central Tunisia
Le. major (CL)	*P. alexdandri*	North Africa to Western China
	P. ansarii	Iran
	P. caucasicus	Iran
	P. duboscqi	Sahelian Africa, Kenya
	P. papatasi	North Africa, Middle East
	P. salehi	Iran, Pakistan
Le. tropica (CL, LR, VL)	*P. aculeatus*	Kenya
	P. guggisbergi	Kenya
	P. sergenti	Middle East, North Africa

[1]For disease abbreviations, see text.

[2]Species in boldfaced type are proven vectors on the basis of isolation of fully identified parasites from sand flies, demonstration that parasites develop to metacyclic stage in sand fly gut, and demonstration of human-vector and reservoir-vector contact. Experimental transmission by bite shown in some cases. Other species are suspected vectors on the basis of human contact, but with only a few isolations of fully identified parasites from sand flies, or with many isolations of parasites that have not been fully identified.

[3]Geographic distribution of sand fly species in areas where transmission of indicated parasite may occur.

Table 8.2. New World *Leishmania* and proven and suspected vectors (information sources as for Table 8.1). (Table continued on next page).

Parasite (disease)[1]	Sand fly species[2]	Sand fly distribution[3]
Leishmania amazonensis (CL)	***Lutzomyia flaviscutellata***	Northern South America
	Lu. olmeca nociva	Amazon basin
Le. braziliensis (ML, MCL)	*Lu. amazonensis*	Northern Amazon basin
	Lu. ayrozai	Southeastern Brazil
	Lu. carrerrai	Western Amazon basin
	Lu. complexa	Para, Brazil
	Lu. intermedia	Southern Brazil
	Lu. llanomartinsi	Brazil
	Lu. migonei	Brazil, Venezuela
	Lu. ovallesi	Guatemala, Venezuela
	Lu. panamensis	Central, northern South America
	Lu. paraensis	Northern South America
	Lu. pessoai	Southern Brazil
	Lu. spinicrassa	Colombia
	Lu. trinidadensis	Venezuela
	Lu. wellcomei	Para, Brazil
	Lu. whitmani	Eastern Brazil
	Lu. yucumensis	Bolivia
Le. chagasi (VL)	*Lu. evansi*	Colombia
	Lu. longipalpis	Central and South America
Le. colombiensis (CL)	*Lu. hartmanni*	Colombia
	Lu. gomezi	Panama
	Lu. panamensis	Panama
Le. garnhami (CL)	*Lu. youngi*	Venezuela
Le. guayanensis (CL, MCL)	***Lu. anduzei***	Northern South America
	Lu. umbratilis	Amazon basin
Le. lainsoni (CL, MCL)	*Lu. ubiquitalis*	Amazon basin
Le. mexicana (CL)	***Lu. anthophora***	Southern Texas, USA
	Lu. ayacuchensis	Ecuador
	Lu. diabolica	Southern Texas, USA
	Lu. olmeca olmeca	Central America
	Lu. ylephiletor	Guatemala
Le. mexicana ?? (DCL)	***Lu. christophei***	Dominican Republic

[1]For disease abbreviations, see text.
[2]Species in boldfaced are proven vectors; others are suspected (see explanations of criteria for vector incrimination below Table 8.1).
[3]Geographic distribution of sand fly species in areas where transmission of indicated parasite may occur.

Table 8.2 (continued). New World *Leishmania* and proven and suspected vectors (information sources as for Table 8.1).

Parasite (disease)[1]	Sand fly species[2]	Sand fly distribution[3]
Le. naiffi (CL)	*Lu. squamiventris*	Brazil
Le. panamensis (CL, MCL)	*Lu. gomezi*	Central, n. South America
	Lu. panamensis	Central, n. South America
	Lu. trapidoi	Central America
	Lu. ylephiletor	Central America
Le. peruviana (CL)	*Lu. peruensis*	Northern Andes
	Lu. verrucarum	Northern Andes
Le. pifanoi (CL)	***Lu. flaviscutellata***	Northern South America
Le. shawi (CL)	*Lu. whitmani*	Brazil
Le. venezuelensis (CL)	*Lu. olmeca*	Northern South America

[1]For disease abbreviations, see text.
[2]Species in boldfaced are proven vectors; others are suspected (see explanations of criteria for vector incrimination below Table 8.1).
[3]Geographic distribution of sand fly species in areas where transmission of indicated parasite may occur.

area. Changes in the landscape, usually resulting from human activities, affect species diversity and population densities of vectors and reservoirs (Killick-Kendrick and Ward 1981). The leishmaniases are largely zoonotic diseases. The many mammal reservoir species may or may not show signs of infection. Humans usually are incidental hosts, becoming infected when their activities bring them into contact with natural transmission cycles. However, anthroponotic forms of the disease occur for which there are no known nonhuman reservoirs, or for which humans are the primary reservoir. These include kala-azar, caused by *Leishmania donovani*, and urban cutaneous leishmaniasis, caused by *Leishmania tropica*.

Forms of the Disease

The leishmaniases in humans manifest themselves in various clinical forms: cutaneous (CL), mucocutaneous (MCL) and visceral (VL). These forms vary considerably, depending on the species of *Leishmania*, immunological responses of the individual and other factors. Certain species of *Leishmania* traditionally are associated with a particular clinical syndrome, but correlation between *Leishmania* species, as characterized by isoenzyme analysis and clinical diseases, is not absolute (Barral et al. 1991).

Cutaneous leishmaniasis (CL)

Cutaneous leishmaniasis occurs in the Old World throughout the Mediterranean Basin of North Africa, the Middle East and southern Europe, parts of sub-Saharan Africa, southern Asia, the western part of the Indian subcontinent, and China (Fig. 8.4). In the New World, it is found from Texas in North America to Argentina in South America (Fig. 8.5). It typically appears as a nonhealing ulcer, referred to as localized cutaneous leishmaniasis (LCL) (Fig. 8.6a,b,c). The lesion usually develops within weeks or months after a sand fly bite and slowly evolves from a papule to a nodule to an ulcer.

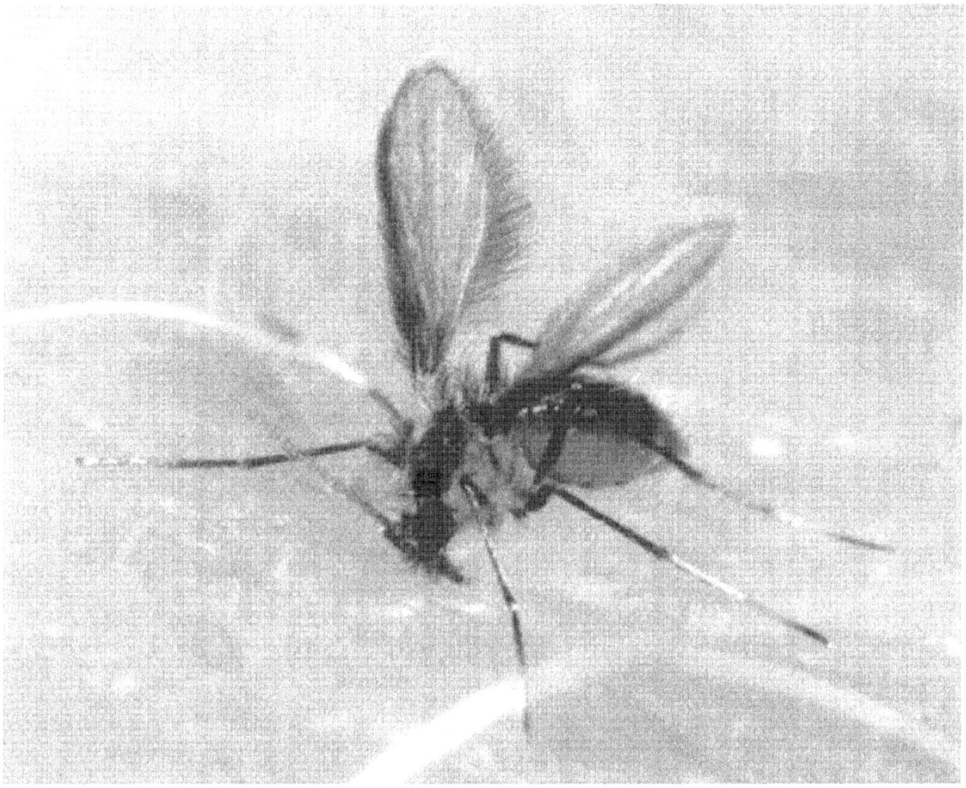

Figure 8.3. Female sand fly, *Phlebotomus argentipes* (photograph by Edward D. Rowton, all rights reserved).

Cutaneous lesions may resolve quickly (2–3 months) without treatment or they may become chronic (lasting many months to years) and seldom will heal without treatment. Scarring accompanies healing. *Leishmania* species most commonly associated with CL are *Leishmania mexicana* and *Le. braziliensis* complex parasites in the New World, and *Leishmania major* and *Le. tropica* in the Old World.

Diffuse cutaneous leishmaniasis (DCL), an uncommon disease resembling lepromatous leprosy, causes disseminated nonulcerative lesions, including soft, fleshy nodules filled with parasitized macrophages (Magill 1995; Fig. 8.6d). This condition is recognized only in patients with abnormal immune responses. In most cases, DCL occurs in areas endemic for CL and represents a small portion of the total number of human cases. However, this is the only form of leishmaniasis that occurs in the Dominican Republic (Walton 1987).

Leishmaniasis recidivans (LR) occurs in North and East Africa and the Middle East, characterized by chronic, localized cutaneous lesions, and usually caused by *Le. tropica*. (Fig. 8.6e). The primary lesion, frequently crescent-shaped and on the cheek below the eye, heals slowly over a period of years, with significant scarring. After the primary lesion appears to have healed, satellite lesions may erupt on the

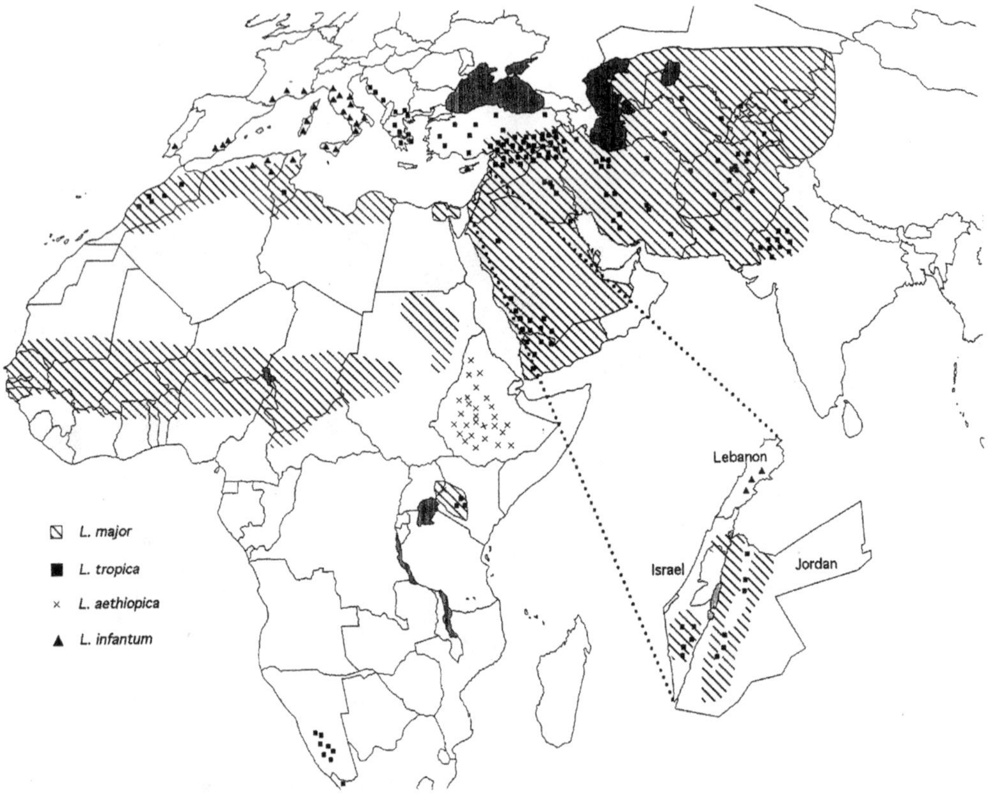

Figure 8.4. Geographic distribution of Old World cutaneous leishmaniasis (Magill 1995, with permission of W.B. Saunders Publishers).

periphery of the primary scar, from which parasites can be isolated (Mebrahtu et al. 1992).

Another form of CL is post-kala-azar dermal leishmaniasis (PKDL), seen in some individuals following treatment for kala-azar (Fig. 8.6f). These individuals develop plaques, small papules or nodules that may persist for many years, and may serve as a source of infection of susceptible vectors. In India, perhaps up to 20% of treated kala-azar patients develop PKDL (Rees and Kager 1987) and in the Sudan up to 56% of patients develop PKDL following treatment of African VL (Zijlstra et al. 1994).

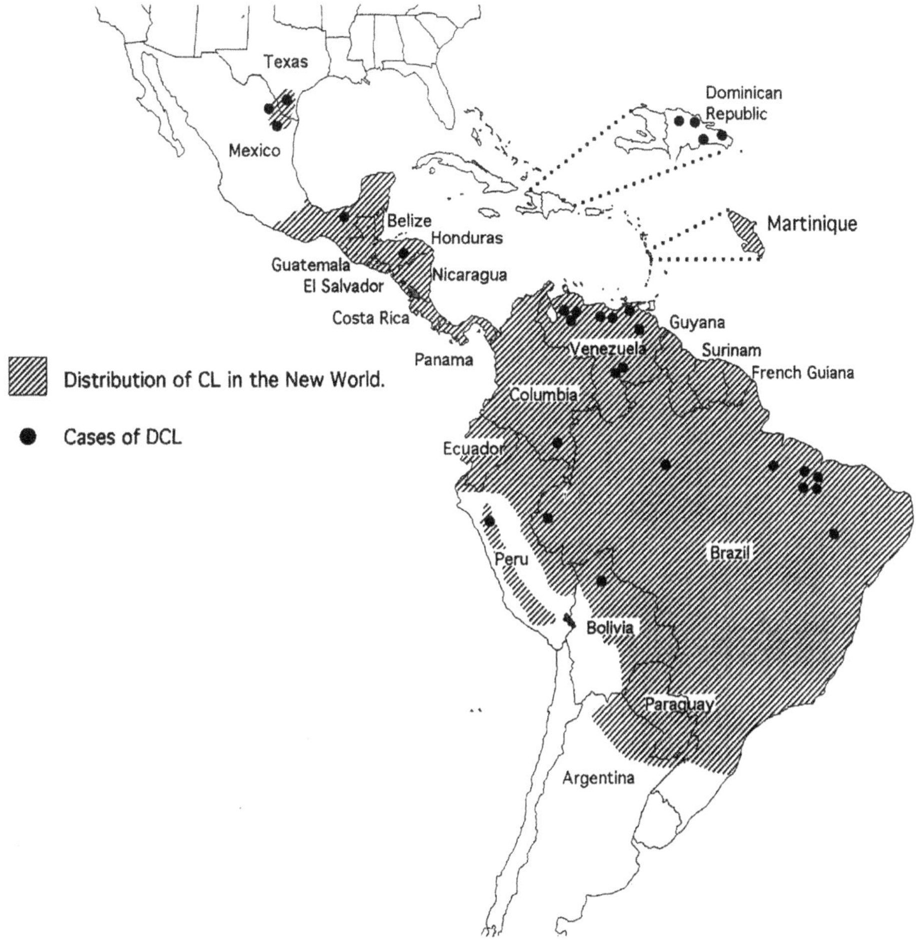

Figure 8.5. Geographic distribution of New World cutaneous leishmaniasis (Magill 1995, with permission of W.B. Saunders Publishers).

Mucocutaneous leishmaniasis (MCL)

Mucocutaneous leishmaniasis occurs most frequently in the New World (Brazil and Central America) and develops in about 5% of patients months to years following LCL caused by *Leishmania braziliensis* or, occasionally, *Leishmania panamensis* and *Leishmania guyanensis*. It sometimes will appear in patients with no history of a primary lesion, or it may be concurrent with a primary lesion. MCL patients develop ulcerative or granulomatous lesions of the nasal, oral and pharyngeal mucosa (Fig. 8.7). The disease develops and progresses despite a vigorous cell-mediated response. The inflammatory response limits parasite replication, but

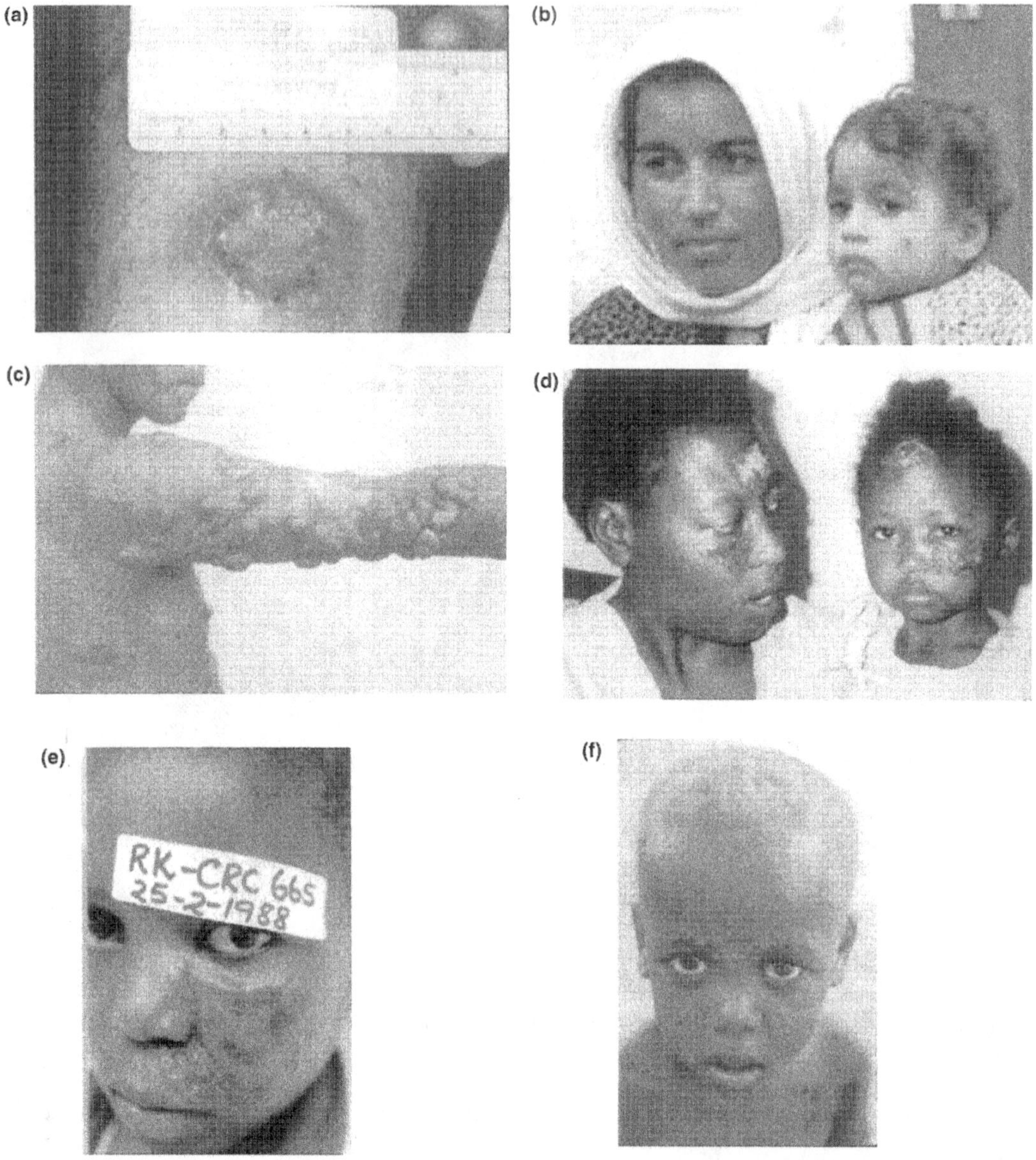

Figure 8.6. Typical clinical manifestations of leishmaniasis: (a) Cutaneous lesion due to *Leishmania panamensis*; (b) cutaneous lesion due to *Le. major*, Jordan; (c) diffuse cutaneous leishmaniasis; (d) cutaneous lesions due to *Le. tropica*, Kenya; (e) leishmaniasis recidivans due to *Le. tropica*, Kenya; (f) post-kala-azar dermal leishmaniasis, Kenya (a, courtesy of Walter Reed Army Institute of Research; b, d and e, photographs by Phillip G. Lawyer; c, courtesy of Phillippe Desjeux; f, photograph by Charles Oster).

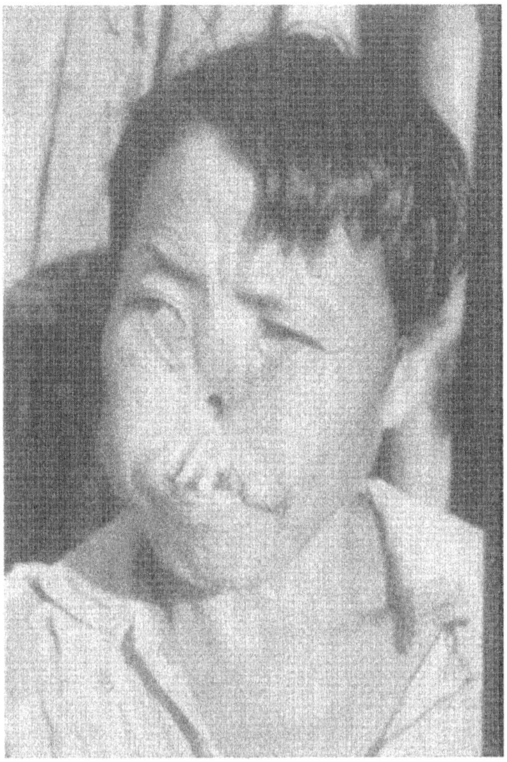

Figure 8.7. Mucocutaneous leishmaniasis due to *Leishmania braziliensis* (courtesy of Philippe Desjeux).

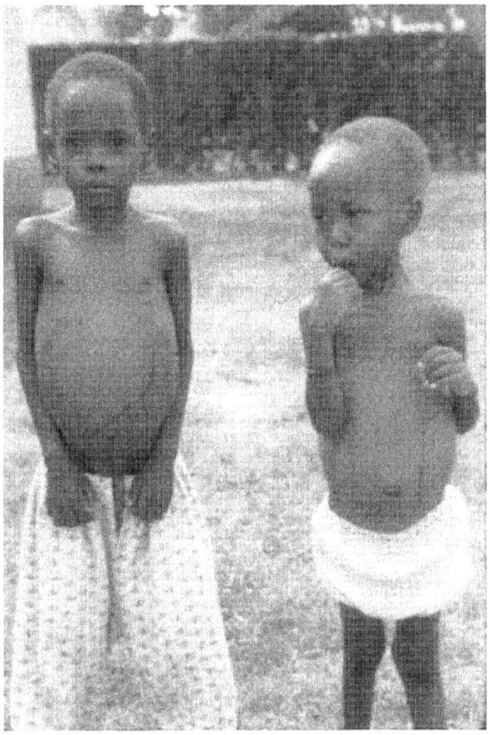

Figure 8.8. Children with visceral leishmaniasis (kala-azar) due to *Le. donovani*, Kenya (photograph by Charles Oster).

results in disfiguring tissue destruction (Magill 1995).

Visceral leishmaniasis (VL)

Visceral leishmaniasis is the most severe form of leishmaniasis, with mortality as high as 95% in untreated cases (Fig. 8.8). It is a chronic disease characterized by fever (2 daily peaks), lymphadenopathy, hepatosplenomegaly, anemia and progressive emaciation and weakness caused by parasite proliferation in organs associated with the reticuloendothelial system (Lainson 1982). In the Old World, VL usually is attributed to *Le. donovani* or *Leishmania infantum* and occurs in the Mediterranean Basin countries of North Africa, the Middle East and southern Europe, and in East Africa, South Central Asia and China (Fig. 8.9). Viscerotropic *Le. tropica* also has been reported (Mebrahtu et al. 1989) and was discovered in veterans of the Persian Gulf war (Magill et al. 1993). In the New World, VL is attributed to *Leishmania chagasi* and occurs in South America, especially in Brazil (Fig. 8.10).

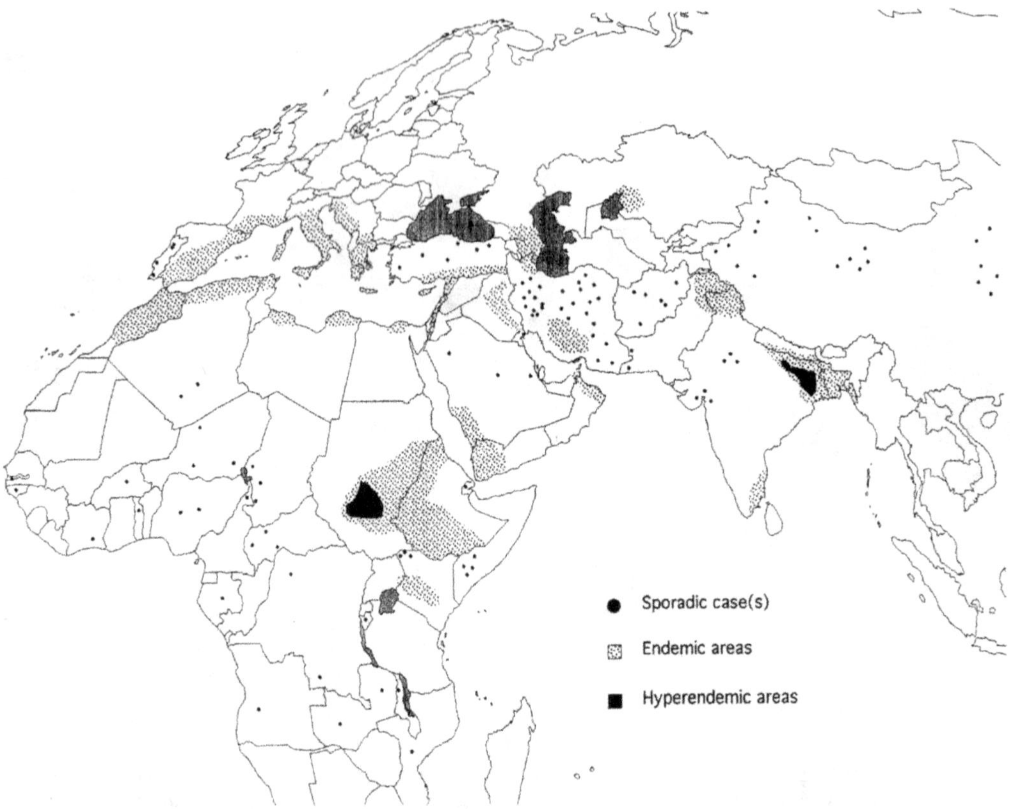

Figure 8.9. Geographic distribution of Old World visceral leishmaniases (Magill 1995, with permission of W.B. Saunders Publishers).

Sand Fly Vectors

Sand fly incrimination criteria

Applying the principles of vector incrimination to sand flies and the leishmaniases, Killick-Kendrick and Ward (1981) outlined 5 criteria that must be fulfilled to declare with reasonable certainty that a sand fly species is a vector of human leishmaniasis:

1. The suspected sand fly vector must bite humans and be present in a place where humans become infected. Exclusive anthropophily is never the case; most sand flies feed on a range of hosts. In Nepal, *Phlebotomus argentipes* is the known vector of *Le. donovani* and this species feeds predominantly on cattle and only opportunistically on humans and other animals. Biting behavior may be determined by direct observation, as in landing/biting collections using humans as bait, or by analysis of sand fly

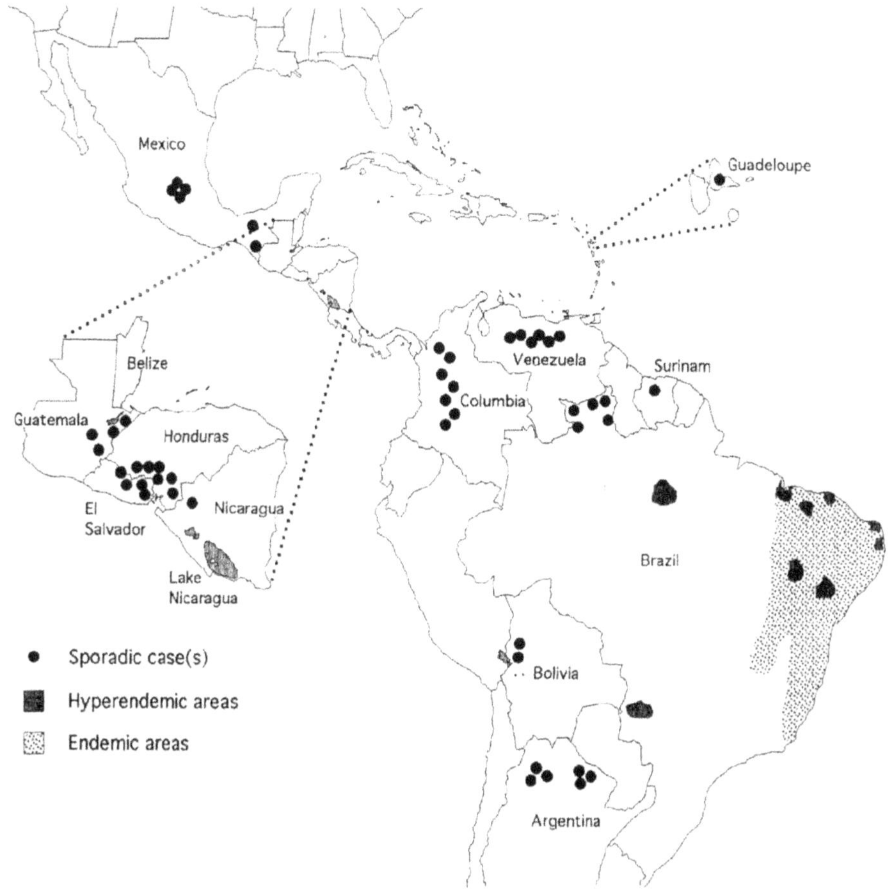

Figure 8.10. Distribution of New World visceral leishmaniasis (Magill 1995, with permission of W.B. Saunders Publishers).

bloodmeals. Young and Lawyer (1987) reported that 90 of 354 *Lutzomyia* species from the New World feed on humans. In the Old World, at least 39 species of the genus *Phlebotomus* bite humans (Killick-Kendrick 1990a). In a few localities, leishmaniasis has been reported in humans or dogs where no mammal-feeding sand flies have been discovered (Anderson et al. 1980). These occurrences may be due to undiscovered vectors or the occurrence of imported cases erroneously considered autochthonous. In foci where only one anthropophilic species is known, that species must be suspected as the likely vector. Where more than one anthropophilic species exists, the most common one is not necessarily responsible for transmission. For example, of 2 anthropophilic species, *Phlebotomus papatasi* and *P. argentipes*, collected in the Terai Plain of southeastern Nepal, the former outnumbers the latter in human-biting collections 4:1, yet only *P. argentipes* transmits kala-azar parasites.

2. The geographic distribution of the suspected vector should encompass the distribution of the disease in humans, and the sand fly must be sufficiently abundant to maintain parasite transmission. *Phlebotomus argentipes* first was suspected as the vector of kala-azar in India because its distribution coincided with the distribution of human disease (Sinton 1922). In Italy, *Phlebotomus perfiliewi* is the suspected vector of CL because its distribution coincides with that of the disease, it is abundant and anthropophilic (Killick-Kendrick et al. 1977). On the other hand, another anthropophilic species, *Phlebotomus mascitti*, is not considered to be a vector of leishmanial parasites because its distribution does not coincide with that of leishmaniasis in Italy and nowhere is it abundant (Rioux and Golvan 1969).

3. *Leishmania* should be isolated from wild-caught sand flies and be the same species of parasite causing disease in humans in the same area. There is a tendency to suspect as a vector any sand fly species containing flagellates. However, all *Leishmania* species occurring in sand flies are morphologically indistinguishable, and nonhuman trypanosomatids (e.g., *Endotrypanum* in porcupines or sauroleishmanias in reptiles) also may exist sympatrically with human leishmanial species. In Panama, approximately 8% of anthropophilic sand flies were naturally infected with flagellates, but fewer than one in 1,000 thousand (0.06%) were infected with human parasites (Christensen et al. 1983).

4. It should be demonstrated that naturally or experimentally infected flies can maintain the infection through the complete extrinsic life cycle of the parasite. Sand flies may feed on infected hosts and ingest amastigotes that may persist for several days in the midgut, but unless parasites develop to the infective stage, transmission will not occur. In infection experiments with *Phlebotomus duboscqi* and *Sergentomyia schwetzi*, *Le. major* promastigotes were isolated from the guts of females of both species during the first 72 hr post-feeding (Lawyer et al. 1990). However, in *P. duboscqi*, a known vector of *Le. major*, the parasites developed through the complete extrinsic life cycle, whereas in *S. schwetzi*, they did not complete development, rarely were found after 48 hr, and never were found after 90 hr post-feeding. The presence of early developmental forms of *Leishmania* from nonvectors may confound vector incrimination, and detecting leishmanial DNA in wild-caught sand flies (Mukherjee et al. 1997) is not sufficient for vector incrimination.

5. If other criteria are satisfied, experimental transmission of *Leishmania* by bite generally is considered conclusive proof that a sand fly is a vector of a given parasite. There is a close host-parasite relationship between *Leishmania* and their phlebotomine sand fly vectors. Attempts to transmit *Leishmania* experimentally using unnatural parasite-sand fly combinations have been unsuccessful, largely because refractory barriers limit parasite colonization, differentiation, migration and attachment (Adler 1947, Adler and Theodor 1939, Lawyer et al. 1990, Walters et al. 1992, Sacks and Kamhawi 2001). Experimental evidence and field observations have shown that particular phlebotomine vectors transmit only certain species of *Leishmania* (Pimenta et al. 1994). This limit further suggests a close evolutionary relationship between *Leishmania* species and their vectors.

Sand fly morphology

Sand fly taxonomy is based primarily on morphology. Several authors have published general descriptions of sand fly anatomy, with some variation in the structural terminology used in their identification (Abonnenc 1972, Fairchild and Hertig 1947, Forattini 1973, Quate and Vockeroth 1981, Young and Duncan 1994). Figs. 8.11 and 8.12 provide the general morphology of adult female and male sand flies, using the terminology of Young and Duncan (1994). Superb drawings of sand fly morphology may be found in Jobling's "Anatomical drawings of biting flies" (Jobling and Lewis 1987). Like all true Diptera, sand flies are holometabolous, having 4 life stages: egg, larva, pupa and adult (Fig. 8.13). Although each life cycle stage possesses distinctive morphological features that are useful taxonomically, practical sand fly taxonomy relies primarily on adult structures for species diagnosis.

Adults

Compared to other Diptera, sand flies are small (<5 mm long) and fragile. Seldom recognized in the field by nonspecialists, these tiny, densely-haired flies characteristically rest with their wings held upward and outward above the body so that the costal margins form an angle of approximately 60° (Fig. 8.3). This resting stance, coupled with a dark-gray to dusty-brown color, relatively large eyes, mouthparts at least as long as the head, long (16-segmented) antennae, long palpi and a fuzzy appearance constitute an unmistakable gestalt. The sand fly head is suboval when viewed anteriorly and bears daggerlike piercing mouthparts flanked by long, often recurved, 5-segmented maxillary palpi. Adapted for bloodfeeding, the female mouthparts comprise a fascicle of 6 blade-like stylets (labrum, paired mandibles and maxillae, and hypopharynx) within a labial sheath (Fig. 8.14). Males lack mandibles and are not hematophagous. The large dark eyes vary in size and shape and help to separate species groups (Young 1979). Internally, the head contains a cibarium armed with teeth and other sclerotized structures that help identify and classify.

Females (Fig. 8.15a) have a plump, cylindrical abdomen, somewhat tapered distally, that appears to end with 2 rounded, paddle-like segments (cerci). However, internally there is a well-sclerotized genital fork (shaped like an inverted "Y"), common or individual sperm ducts and paired spermathecae. The actual and relative lengths of these ducts and size and shape of the spermathecae vary considerably and are diagnostic at the species and subgeneric level. The 2 wings, with numerous fine hairs and distinctive venation, are elongate and pointed distally.

Males (Figs. 8.12, 8.15b) have elaborate, bilaterally symmetrical external clasping structures on the distal end of their abdomen, which, along with conspicuously sclerotized internal organs of their genitalia, are diagnostic.

The body color of males and females ranges from light brown to black, depending on the species, and reflects differences in setation and cuticular color patterns. Because of their small size and taxonomically important internal structures, museum specimens are mounted on glass slides in Canada balsam or other resinous media and require microscopic examination to determine species.

Eggs

At oviposition, sand fly eggs are white to light gray or light brown in color, but depending on the species, turn darker brown, gray or black within a few hours. Somewhat banana-shaped and almost microscopic in size, they measure 0.3–0.5 mm long by 0.1–0.15 mm wide. The chorion is sculptured with ridges, polygons or protuberances that form species-specific patterns (Fig. 8.16). Electron microscopy has enabled researchers to describe the eggs of many Old and New World sand fly species (e.g., Endris et al. 1987, Zimmerman et al. 1977).

Larvae

Larvae (Fig. 8.17a–c) are caterpillar-like, usually with a grayish white body and a dark, well-developed head capsule. They may measure up to 4 mm long at maturity. Unlike most sand fly larvae, *Lutzomyia panamensis* has a charcoal gray body color and a head capsule possessing 2 hornlike protuberances (Fig. 8.17b). Sand fly larval antennae are small and leaflike. The thorax is not differentiated from the abdomen, although the abdominal segments bear ventral

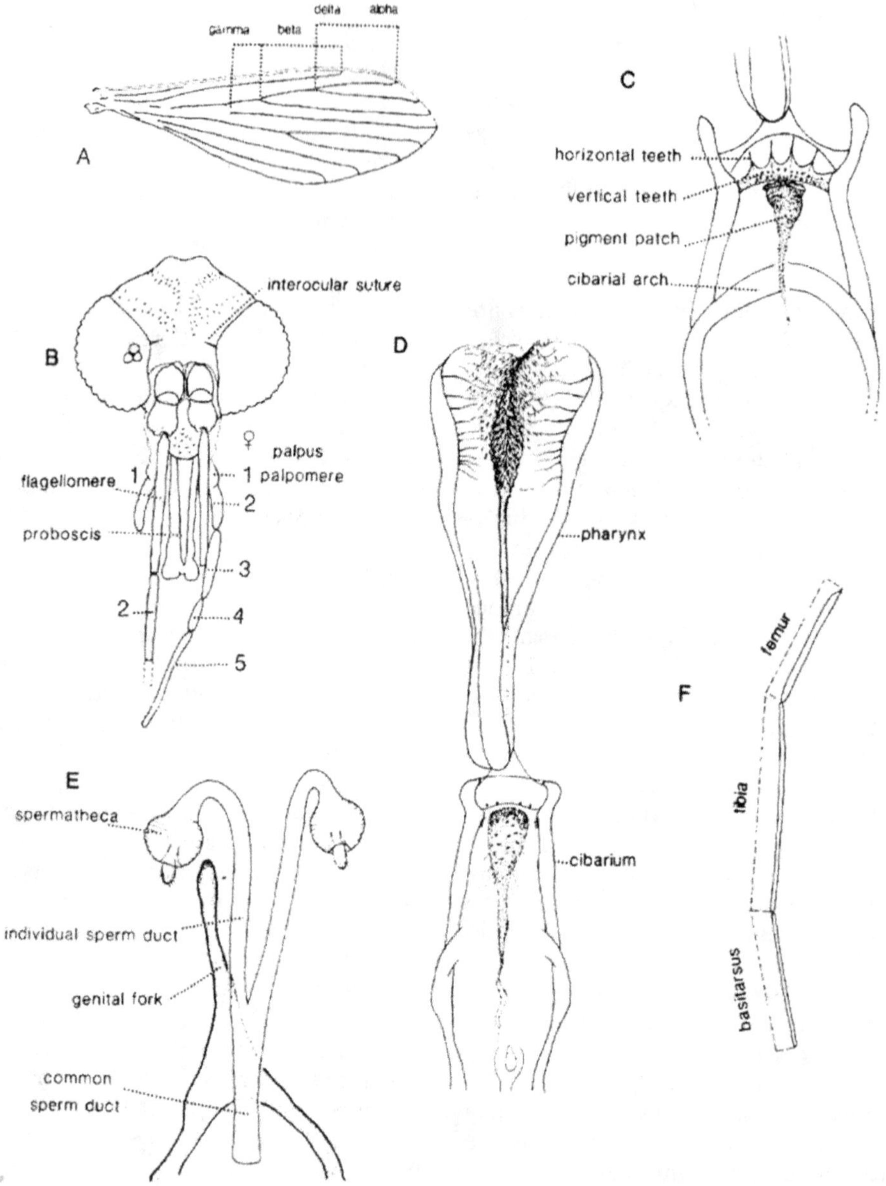

Figure 8.11. Female sand fly structures: A, Wing; B, head; C, cibarium; D, cibarium and pharynx; E, spermathecae and genital fork (furca); (F) leg, in part [Young and Duncan (1994), with permission of Associated Publishers, American Entomological Institute].

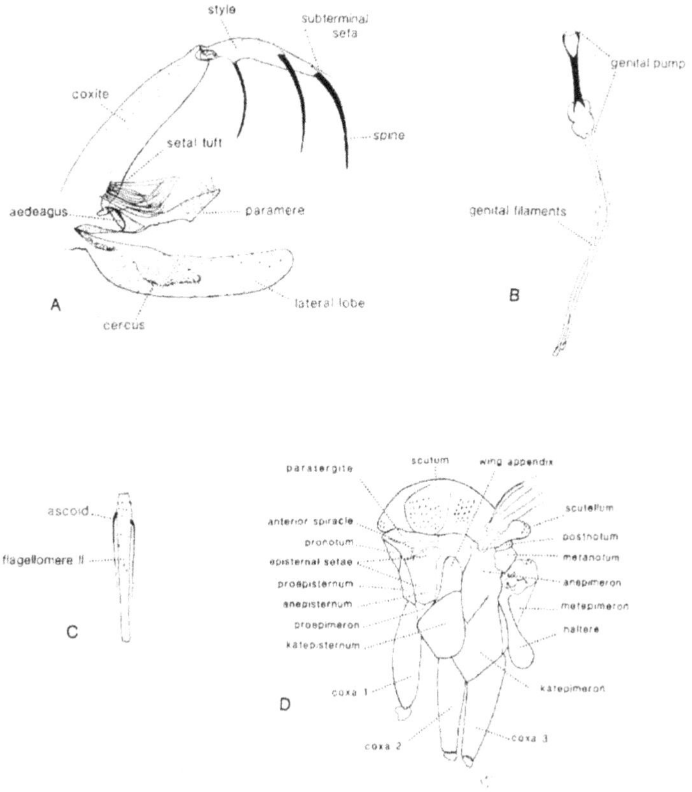

Figure 8.12. Male sand fly structures: A, Terminalia, lateral view; B, genital pump and filaments; C, flagellomere II, showing paired ascoids; D, thorax. [Young and Duncan (1994), with permission of Associated Publishers, American Entomological Institute].

pseudopods, i.e., unjointed evaginations of the body that are used for progression (Kettle 1994). On the body segments are many short, pinnate, lateral and dorsal hairs and 2 or 4 caudal setae on the terminal segment. The length of the caudal setae may be related to the habits of the species; surface feeders generally have longer setae than those that burrow (Hanson 1961). The dark head capsule and the caudal setae are diagnostic features of phlebotomine larvae. However, larvae are not often used in taxonomy because so few species have been reared and fewer still collected in nature.

There are 4 larval instars, each of which can be distinguished with the aid of a dissecting microscope. Newly hatched first instars are about 0.5 mm long by 0.1 mm wide and grow to about twice their original length and 2–3 times their original width before molting. They are characterized by a conspicuous egg burster on the vertex of the head capsule, and 2 long caudal bristles on the last abdominal segment. Lateral and dorsal segmental setae are small and inconspicuous. Prior to molting, the first larval instar inflates, causing the skin to stretch and take on a shiny appearance. Molting always is

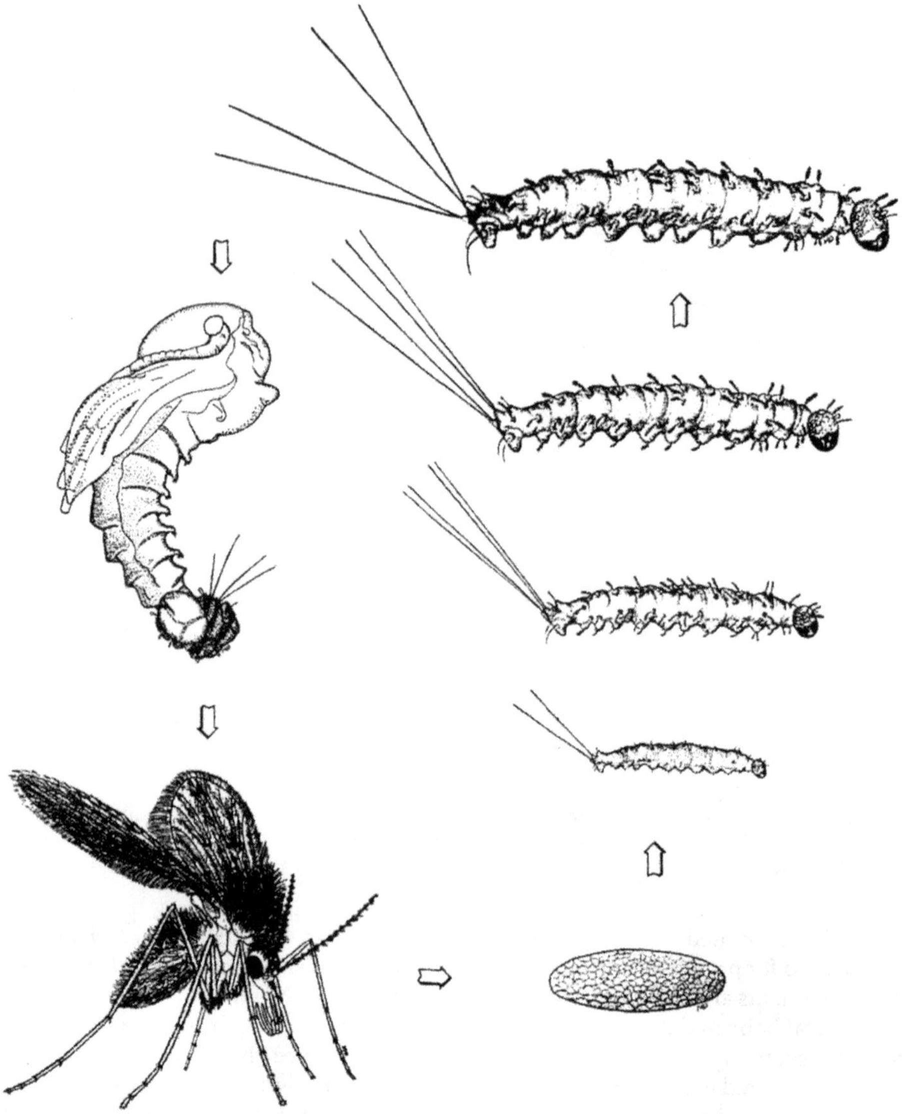

Figure 8.13. Sand fly life cycle showing egg, 4 larval instars, pupa and adult (drawing of immature stages by Margo Duncan; drawing of adult by Hilda Muñoz).

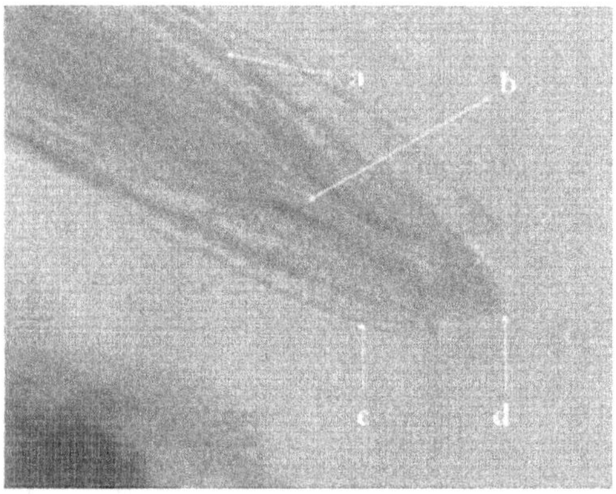

Figure 8.14. Female sand fly mouthparts: (a), lacinia; (b), labrum; (c), mandible; (d), hypopharynx (photomicrograph by Phillip G. Lawyer).

preceded by evacuation of the gut contents and cessation of movement. At the initiation of the molt, the old skin, which is glued posteriorly to the substrate by the excretion (secretion) of an adhesive substance, splits in the region of the head capsule and the new instar pulls itself out by anteriorly directed peristaltic movements. The egg burster is absent in the 2nd instar, having been shed with the old skin, and usually 4 (instead of 2) caudal setae are present. The 2nd instar grows to about twice the maximum length of the first before molting. On the 3rd instar, the lateral and dorsal segmental setae are stouter and conspicuously more spatulate than on the 2nd instar. The 3rd instar grows to about twice the maximum length of the 2nd instar before molting. The 4th instar is robust and easily distinguishable from the others with the unaided eye (Fig. 8.17c). Unlike the other instars, it bears a heavily sclerotized, saddle-shaped dorsal anal plate. Its maximum dimensions just prior to pupation are about 3.2 mm long and 0.6 mm wide. Prior to pupation, the 4th instar turns white, a result of evacuation of the gut.

Pupae

Pupae (Fig. 8.17d) closely resemble a butterfly chrysalis, with developing eyes, wings and legs evident externally. The collapsed 4th-stage larval exuvium can be seen at the caudal end, attached to some solid surface, thus holding the pupa upright. The pinnate hairs and caudal setae of the last larval exuvium easily identify phlebotomine pupae. Newly formed pupae are whitish in color but soon turn varying shades of golden brown or orange, depending on the species, and darken further just prior to eclosion.

Sand fly classification and taxonomy

Lewis et al. (1977) proposed a conservative classification for the phlebotomine (literally "vein cutter") sand flies as the subfamily Phlebotominae, family Psychodidae, recognizing only 5 genera but a large number of subgenera, species groups and lesser categories. Most sand fly workers have accepted this approach. The genus *Chinius* was added later, for a total of 6 genera (Leng 1987), 3 in the Old World (*Phlebotomus*, *Sergentomyia* and *Chinius*), and 3 in the New World (*Lutzomyia*, *Brumptomyia* and *Warileya*). The Old World genera can be differentiated by their having the 5th palpal segment longest, no post-spiracular setae and no posterior bulge of the cibarium (Lane 1993). In the New World genera, palpal segment 3 usually is the longest, post-spiracular setae are present and there is a bulge to the cibarium. As a general rule, Old World phlebotomines are savanna and desert species, whereas most New World phlebotomine species inhabit wet tropical forests. However, several species in North and South America also inhabit drier, savanna-like regions. Species in the genus *Phlebotomus* are mammal-biters and include all the known Old World vectors of leishmaniasis. The genus is characterized by the absence of cibarial teeth in females and by the presence of erect hairs on the hind borders of the abdominal tergites.

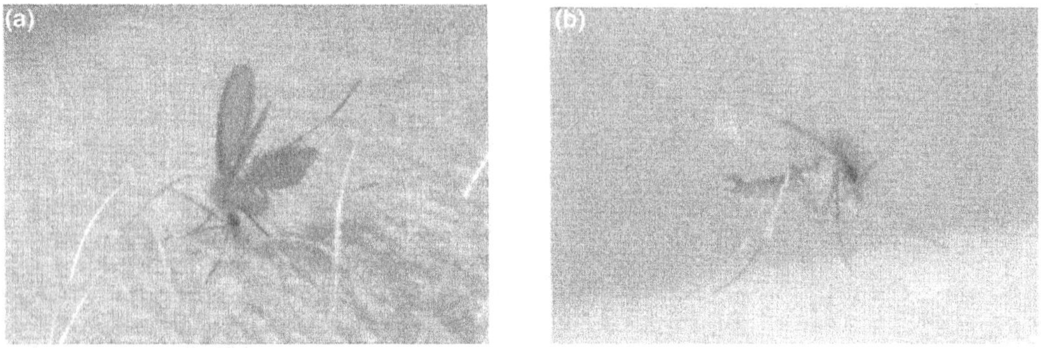

Figure 8.15. Adult sand flies: (a) Female *Phlebotomus papatasi*; (b) male *Lutzomyia* sp. (a, photograph by Edgar D. Rowton, all rights reserved; b, photograph by Peter V. Perkins).

Sergentomyia is the dominant genus of the Old World tropics of Africa, India and Australia. Members of this genus feed primarily on reptiles and amphibians, although some, including *Sergentomyia clydei*, *Sergentomyia garnhami*, *S. schwetzi*, *Sergentomyia bedfordi* and *Sergentomyia africanus* (Quate 1964) of Kenya, and *Sergentomyia babu* and *Sergentomyia salehi* in India and Nepal, feed on mammals, including humans (Lawyer et al. 1990, Mukherjee et al. 1997). This genus contains no known vectors. A posterior transverse row of cibarial teeth in females and recumbent setae on the abdominal tergites characterize it.

Lutzomyia is the largest sand fly genus and contains the majority of the New World Phlebotominae, including all New World *Leishmania* vectors. *Lutzomyia* feed on both mammals and reptiles. Members of this genus are characterized by a transverse row of posterior teeth and one or more rows of anterior teeth on the female cibarium.

Lane (1993) provides an excellent key to the genera and subgenera of Phlebotominae (excluding *Lutzomyia* subgenera). Lewis (1982) reviewed the genus *Phlebotomus* and provided keys to 11 subgenera, 96 species and 17 subspecies. Young and Duncan (1994) produced a comprehensive monograph on the Nearctic and Neotropical genera, which includes keys to nearly 400 species of *Lutzomyia* in 14 subgenera, 22 species of *Brumptomyia*, and 6 species of *Warileya* in 2 subgenera.

Sand fly biology and ecology

Immature stages

Female sand flies deposit eggs singly or in small batches in a variety of sites that provide moisture, relatively constant temperatures of 15.6°–26.7° C, protection from desiccation, and proximity to a reliable source of decaying organic matter, such as animal dung or forest detritus. Immature stages have been recovered from organically enriched soil in a wide variety of microhabitats.

Eggs usually hatch within 10 days post-oviposition, but some eggs may have a prolonged incubation period, hatching 30 or more days after others in the same batch (Lawyer and Young 1991). Delayed hatching (obligatory diapause) of a few eggs in a batch spreads risk and insures survival of at least some offspring. Some sand fly species, such as *Lutzomyia diabolica* in south Texas, undergo diapause in response to hot, dry summer temperatures, or to cooler temperatures and shorter photoperiods (Lawyer and Young 1991).

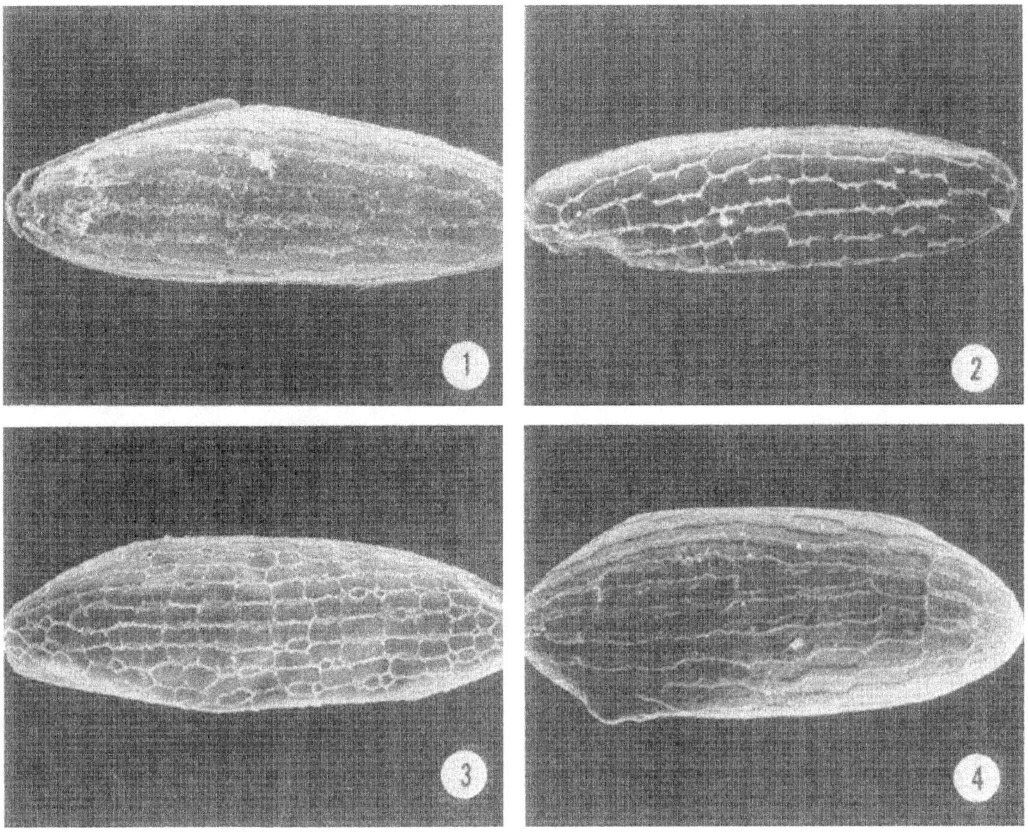

Figure 8.16. Sand fly eggs: Scanning electron photomicrographs showing chorionic sculpturing of (1) *Lutzomyia diabolica*; (2) *Lutzomyia shannoni*; (3) *Lutzomyia vexator*; (4) *Lutzomyia cruciata* (from Endris et al. 1987, with permission of Entomological Society of America).

Upon hatching, first-stage larvae begin feeding on available dead organic matter, including the egg shell, with which they may ingest oocysts of gregarine parasites (Protozoa). Sand fly larvae in the laboratory develop in 3–8 weeks, depending on the species. *Lutzomyia diabolica* larvae in laboratory colonies average ~31 days at 24ºC (range 21–49 days) and 25 days at 27ºC (range 18–30 days) (Lawyer 1984). Diapause or quiescence triggered by environmental extremes such as heat, cold, drought or reduced photoperiod may prolong development time for months (Killick-Kendrick and Killick-Kendrick 1987b, Lawyer and Young 1991).

When mature, 4th-stage larvae stop feeding and seek a drier place to pupate. Pupae usually attach to an object, such as a dead leaf or a stone. The compressed larval skin can be seen at the posterior end of the pupa (attached to the substrate). The pupal stage normally lasts 7–12 days.

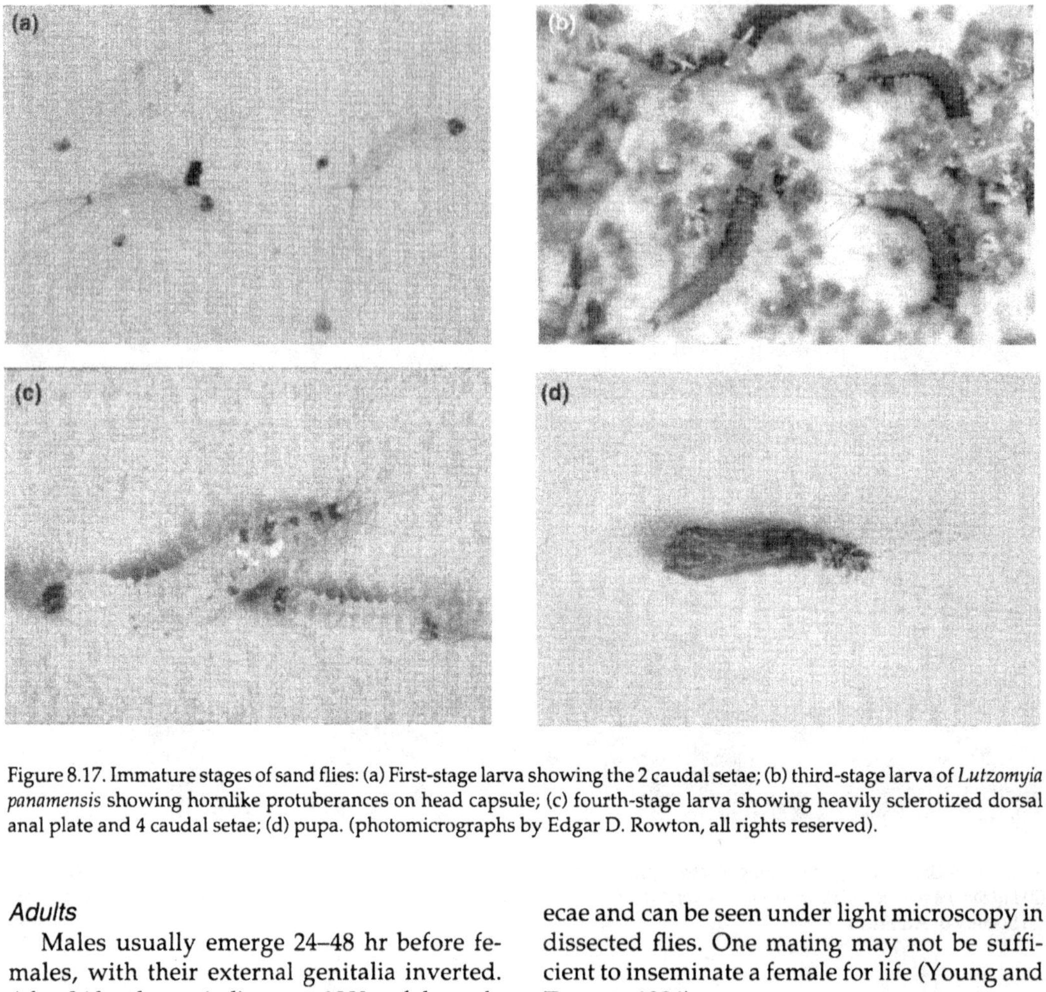

Figure 8.17. Immature stages of sand flies: (a) First-stage larva showing the 2 caudal setae; (b) third-stage larva of *Lutzomyia panamensis* showing hornlike protuberances on head capsule; (c) fourth-stage larva showing heavily sclerotized dorsal anal plate and 4 caudal setae; (d) pupa. (photomicrographs by Edgar D. Rowton, all rights reserved).

Adults

Males usually emerge 24–48 hr before females, with their external genitalia inverted. After 24 hr, the genitalia rotate 180° and the male is sexually mature.

Courtship and mating. Pheromones, and locating a resting site or vertebrate host where females are present, aid the male's search for a receptive female. Premating courtship behavior, including rapid wing beating of males and other lek-like behavior, has been observed when both sexes are close together (within several centimeters). In colonies, copulation has been observed before, during and after a bloodmeal. Spermatozoa are held in the female spermathecae and can be seen under light microscopy in dissected flies. One mating may not be sufficient to inseminate a female for life (Young and Duncan 1994).

Feeding habits. Both male and female adult sand flies require carbohydrates for energy and longevity. The presence of fructose, glucose and sucrose has been detected in the crops of Old and New World phlebotomines (Lewis and Domoney 1966, Killick-Kendrick 1979). Suggested natural sources of such sugars include floral and extra-floral nectars, ripe fruits, other plant juices and aphid honeydew (Schlein and Warburg 1986, Killick-Kendrick and Killick-Kendrick 1987a). Tesh et al. (1992) showed that the aphid alarm pheromone, trans-beta-farnesene, stimulated feeding of male and female *Lutzomyia longipalpis* under laboratory conditions.

In addition to sugars, females of most sand fly species require one or more bloodmeals for each batch of eggs. Blood hunger in females in laboratory colonies does not develop until after about 72 hr, whereas a sugar meal is taken on the first day post-emergence. Suitable combinations of light intensity, ambient temperature, relative humidity, air movement and other physical conditions stimulate hungry sand fly females to search for bloodmeals (Young and Lawyer 1987). Some observations indicate that the presence of ingested sugar in the crop of the female sand fly may stimulate host-seeking behavior (Williams 1970).

Sand fly bites usually are quite painful. Sand fly females often probe the exposed skin surface several times to locate a suitable site before inserting their mouthparts. Insertion of the tiny mouthparts disrupts cells and capillaries, and a potent vasodilatory component of sand fly saliva induces blood and lymph to flow into the wound, forming a pool from which it sucks blood. This also triggers several repair reactions in the host that may be antagonized or modified by pharmacological salivary components in the sand fly's saliva, thus facilitating feeding (Ribeiro et al. 1989). Feeding activity is influenced by environmental factors such as temperature, humidity and air movement. Minimal air movement, along with other optimal conditions, may increase biting activity. In some foci, biting rates may be >1,000/hr, but <1/hr in others. Most anthropophilic species feed at dusk and during the evening, when temperatures drop and relative humidity rises, but *Lu. wellcomei, Lutzomyia carrerai, Lutzomyia pesoana* and other forest dwellers will attack during the daytime if their habitat is disturbed. Bloodfeeding females may release aggregation pheromones that attract other females to the same host. Sand flies may be endophilic (*P. papatasi*) or exophilic (*Lutzomyia trapidoi*).

Oviposition. In laboratory cultures, sand flies often will deposit eggs in grooves, pits or bubble holes on the surface of the plaster of Paris bottoms of rearing jars, on pieces of larval food or on the remains of dead adults. A combination of environmental factors, of which little is known, probably controls the selection of suitable oviposition sites. Laboratory experiments to improve insectary breeding efficiency have shed some light on the subject. Studies by workers in Ethiopia and Panama focused on regulating physical factors, such as temperature and humidity (Chaniotis 1986, Foster et al. 1970). Schlein et al. (1989) found that cow manure attracted non-gravid *P. papatasi* females. Schlein et al. (1990) observed that oviposition by *P. papatasi* was subject to seasonal variation, despite relatively uniform insectary conditions. The low rate of oviposition in October–November could be increased by short exposure to UV sources of 254 and 312 nm, but not by white light, suggesting that levels of UV irradiation set the seasonal oviposition cycle of *P. papatasi*. They also observed that cutting furrows in the plaster of Paris lining the bottom of cages and placing cow manure in the cages promoted oviposition. Oviposition decreased when larvae were present or when the cage size was reduced by half. Elnaiem and Ward (1990, 1991) provided evidence for an oviposition attractant-stimulant pheromone associated with the eggs of *Lu. longipalpis* and later showed that rabbit feces possess both chemical oviposition attractants and stimulants.

Female sand flies usually lay 30–70 eggs during a single gonotrophic cycle, depending on species, size and nature of the previous bloodmeal, larval diet and other factors. Anautogenous females require vertebrate blood for the maturation of their eggs. However, in some sand fly species (e.g., *P. papatasi, Lutzomyia beltrani, Lutzomyia shannoni*) a portion of the females are facultatively autogenous only during their first gonotrophic cycle. The general rule in laboratory colonies of sand flies is that one bloodmeal results in the development of one batch of eggs. However, in some species (e.g., *Lu. shannoni, Lu. longipalpis*), females may take multiple bloodmeals prior to oviposition (Buescher et al. 1984). This extra nutrition obviously would enhance vector potential, but the extent to which it happens in nature is unknown. Autogenous egg batches tend to be smaller than anautogenous batches (Johnson 1961). As the eggs pass through the oviduct, they are coated with a mucilaginous fluid produced by the accessory glands. When this fluid dries or hardens, it glues the eggs in place.

Dispersal behavior. Sand flies are weak fliers, traveling in short hops rather than sustained flight in search of a host, a sugar meal, a mate, or resting and developmental sites. Although slight air movement aids the detection of hosts along odor plumes, wind speeds >1.5 m/sec inhibit flight, and it ceases altogether in winds >4–5 m/sec (Lane 1993). Most sand fly species fly horizontally near ground level where they feed on ground-dwelling rodents and other mammals. Some forest-inhabiting sand flies also move vertically between the forest floor and canopy (Chaniotis et al. 1974, Ready et al. 1986). Forest-inhabiting sand flies, which are mostly Neotropical, rarely fly more than 200 m from their resting site (Chaniotis et al. 1974, Chippaux et al. 1984, Alexander 1987). In savanna and semi-desert habitats typical of the Old World, flight ranges may be longer. Dispersal behavior of *Lu. longipalpis*, found in arid to semiarid habitats in the New World, resembles more closely that of Old World sand flies in similar habitats than the forest-inhabiting Neotropical sand flies (Morrison et al. 1993).

Resting sites. Various factors, including availability of suitable microhabitats, seasonal conditions and moisture, affect the sites used by sand flies for resting (Fig. 8.18). Tropical forests offer the greatest variety of resting sites, and in the New World that is where most sand fly species occur. Forest microhabitats include tree buttresses, tree holes, fissures in tree bark and leaf litter. The forest floor itself is the most extensive microhabitat used by many resting sand flies. *Lutzomyia longipalpis* in the New World, and *Phlebotomus guggisbergi* and related species in the Old World, may be found resting in rock crevices and in caves. Many species inhabit rodent burrows, which afford protection from the elements and ready access to a bloodmeal. In Texas and elsewhere in the southwestern United States and northern Mexico, *Lutzomyia anthophora* inhabits the inner chambers of woodrat (*Neotoma micropus*) nests; *Lutzomyia texana* and *Lutzomyia vexator* often are found in animal burrows, such as those occupied by rodents or armadillos. Many peridomestic sand flies rest in dark, cool and humid corners in human dwellings or animal shelters. Lawyer (1984) routinely collected *Lu. diabolica* from the cool tile walls of public restrooms in Garner State Park, near Uvalde, Texas.

Geographic and seasonal distribution

Sand flies occur mainly in the tropics and subtropics, with a few species ranging into temperate zones of the northern (to 50°N) and southern (to 40°S) hemispheres. Distribution is limited to areas that have temperatures above 15.6°C for at least 3 months of the year. There are no sand flies in New Zealand or on Pacific Islands (Lane 1993). Anthropophilic sand flies of the Old World are distributed mostly in the subtropics, with a few human-biters south of the Sahara and none in Southeast Asia. In the New World, they are limited mainly to the tropics.

Sand flies occur in a wide variety of habitats distributed in elevation from below sea level in areas surrounding the Dead Sea in Israel and Jordan to 2,800 m or more above sea level in the Andes and Ethiopia. Most species have specific ecological requirements that sometimes co-

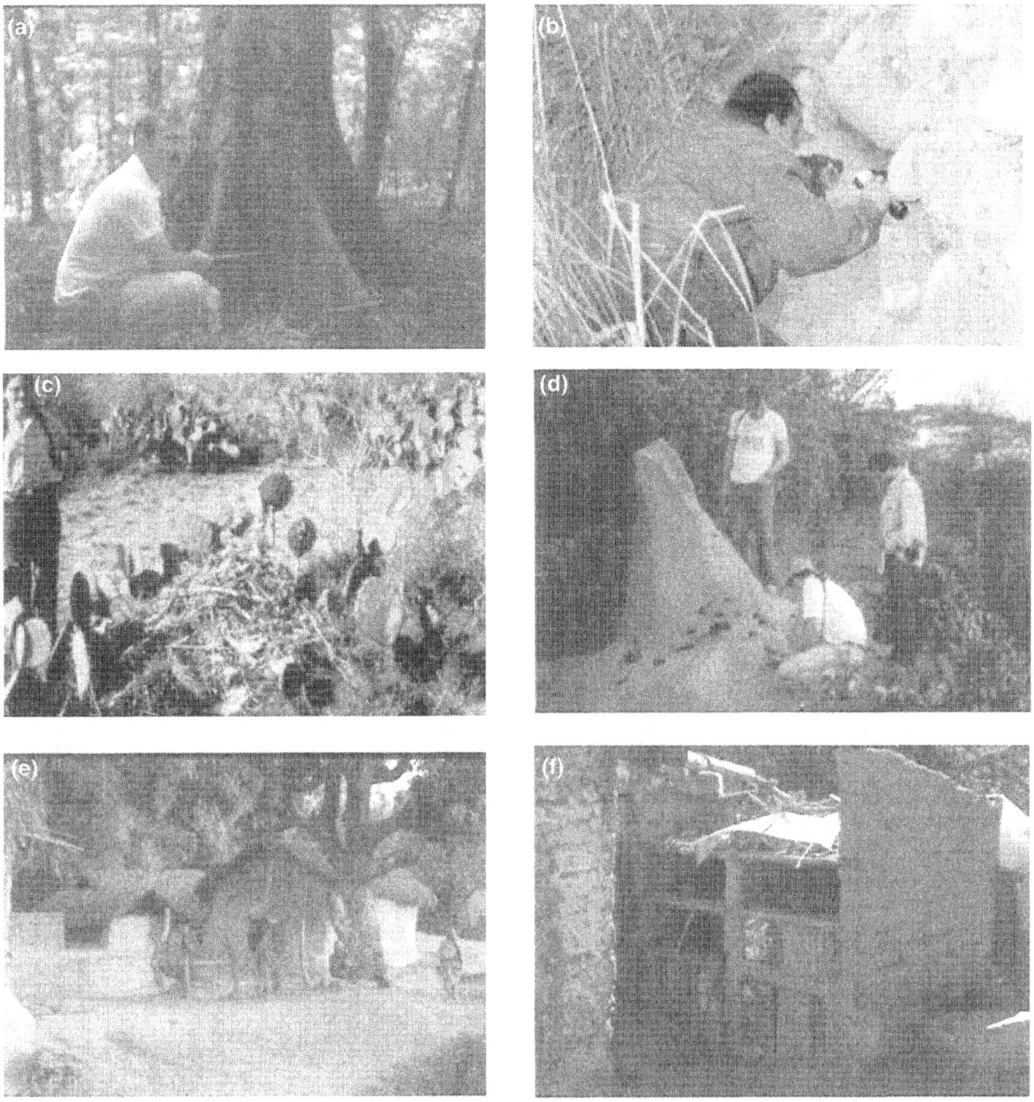

Figure 8.18. Sand fly resting sites: (a) Tree buttresses from which *Lutzomyia shannoni* were collected; (b) collecting *Lutzomyia verrucarum* from rock crevices in Caraz, Peru; (c) nest of the plains woodrat (*Neotoma micropus*), favored resting site of *Lutzomyia anthophora*; (d) collecting *Phlebotomus martini* from a termitary in Baringo, Kenya; (e) human and animal dwellings constructed of mud, dung and wattle — typical daytime resting sites of *Phlebotomus argentipes* in Nepal; (f) chicken coop resting site of *Lutzomyia verrucarum*, Caraz, Peru (a, d, courtesy of Peter V. Perkins; b, f, photographs by Richard G. Andre; c, courtesy of Chad McHugh; e, photograph by Phillip G. Lawyer).

incide with conditions around human and domestic animal dwellings. Depending on the locality, sand fly population densities may or may not vary greatly throughout the year. Marked seasonal changes in sand fly populations occur where there are significant seasonal fluctuations in temperature and precipitation. In East African savannas, and in many other arid or semiarid habitats, sand fly populations are highest toward the end of the rainy seasons and lowest toward the end of the dry seasons. In hot, dry deserts or in dry temperate climes with hot summers and cold winters, adults of some species may disappear entirely during the driest and/ or coldest seasons of the year.

The Ecology of *Leishmania* Transmission

Leishmania transmission is cyclo-propagative and involves interactions among the parasite, the vector and a susceptible vertebrate host. In the following paragraphs we discuss parasite-vector and vector-host interactions. A comprehensive review of parasite-host interactions is provided by Peters and Killick-Kendrick (1987).

Parasite-vector interactions

When feeding on infected hosts, female sand flies may ingest amastigotes with the bloodmeal, some of which may divide and transform into promastigotes. Leishmanias may be grouped into 3 sections based on recognized patterns of development in the gut of the sand fly (Fig. 8.19). Parasites in Section Hypopylaria are primitive flagellates that develop posteriorly in the pylorus, ileum and rectum. Parasites in this section formerly were classified as *Leishmania*, but now are placed in a separate genus, *Sauroleishmania;* none are human pathogens. Their only known hosts are Old World lizards. Because infection is limited to the hindgut of the sand fly, transmission probably occurs when the lizard eats infected sand flies (Lainson 1982).

In the Section Peripylaria, the *Leishmania* establish an initial infection in the pylorus where rounded paramastigotes and promastigotes attach to the cuticular wall, migrating anteriorly during the course of development to the midgut and foregut. These include the *Le. braziliensis* complex of parasites, which are important Neotropical pathogens of humans and other mammals (Lainson 1982). Peripylarian leishmanias occur in mammals only in the amastigote form, and invade macrophages of the skin and viscera. Except for establishing infections in the hindgut of the vector, these parasites develop much the same as other mammalian leishmanias. Transmission is by bite of infected sand flies. Important species in this section include: *Le. braziliensis*, *Le. guyanensis*, *Le. panamensis* and *Leishmania peruviana*.

The vast majority of human leishmanias are in the Section Suprapylaria. Having lost the primitive attached forms in the hindgut of the sand fly host, their development is restricted to the midgut and foregut. As with the Peripylaria, transmission is by bite of infected sand flies. Vertebrate hosts are limited to wild and domestic animals, including humans; parasites are found only in the amastigote form in macrophages of the skin, viscera and blood (Lainson 1982). In this section are parasites of the *Le. mexicana* and *Leishmania hertigi* complexes of the New World, the *Le. tropica* and *Le. major* complexes of the Old World, and the *Le. donovani* complex of the Old and New World. Subspecies of the *Le. mexicana* complex are common parasites of rodents and marsupials; those of the *Le. hertigi* complex are restricted to Neotropical porcupines. Reservoirs of the *Le. tropica* complex are mostly rodents or, as in the case of *Leishmania aethiopica*, hyraxes. Humans usually are accidental hosts, except in the case of urban *Le. tropica*, for which a feral reservoir is not known. In humans, *Le. mexicana*, *Le. major* and *Le. tropica* usually produce simple skin lesions. However, *Le. tropica* may produce "leishmaniasis recidivans," a condition of long-lasting,

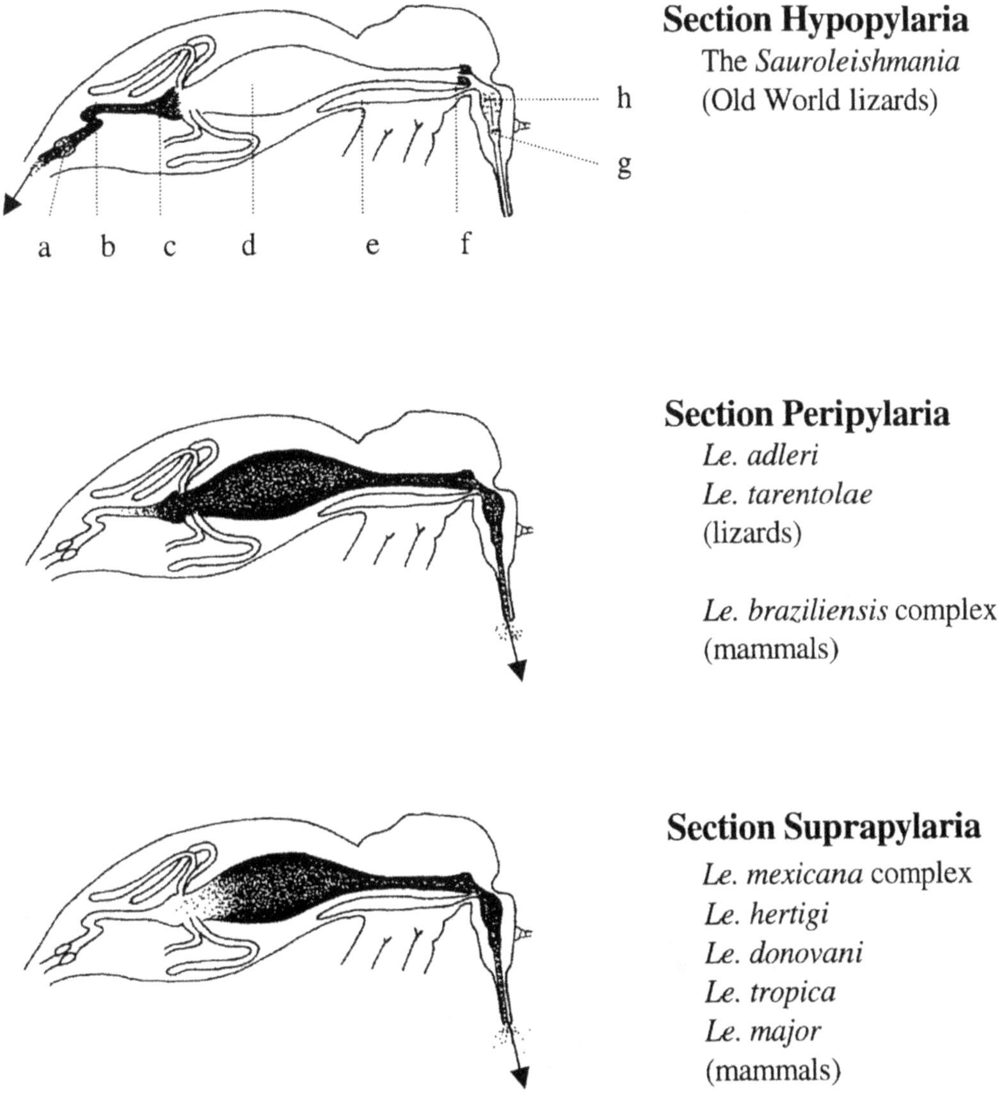

Figure 8.19. Classification of leishmanial parasites according to their development in the vector (after Lainson and Shaw 1987). a, rectum; b, ileum; c, hindgut (pylorus); d, abdominal midgut; e, diverticulum; f, stomodeal valve; g, cibarium; h, pharynx.

nonhealing ulcers that may persist for as long as 20 years, or "viscerotropic leishmaniasis," such as reported from Kenya, Israel, India and the Persian Gulf region (Schnur et al. 1981, Mebrahtu et al. 1989, Magill et al. 1993).

The extrinsic (within the sand fly) life cycle of *Leishmania* can be divided into 3 stages. In

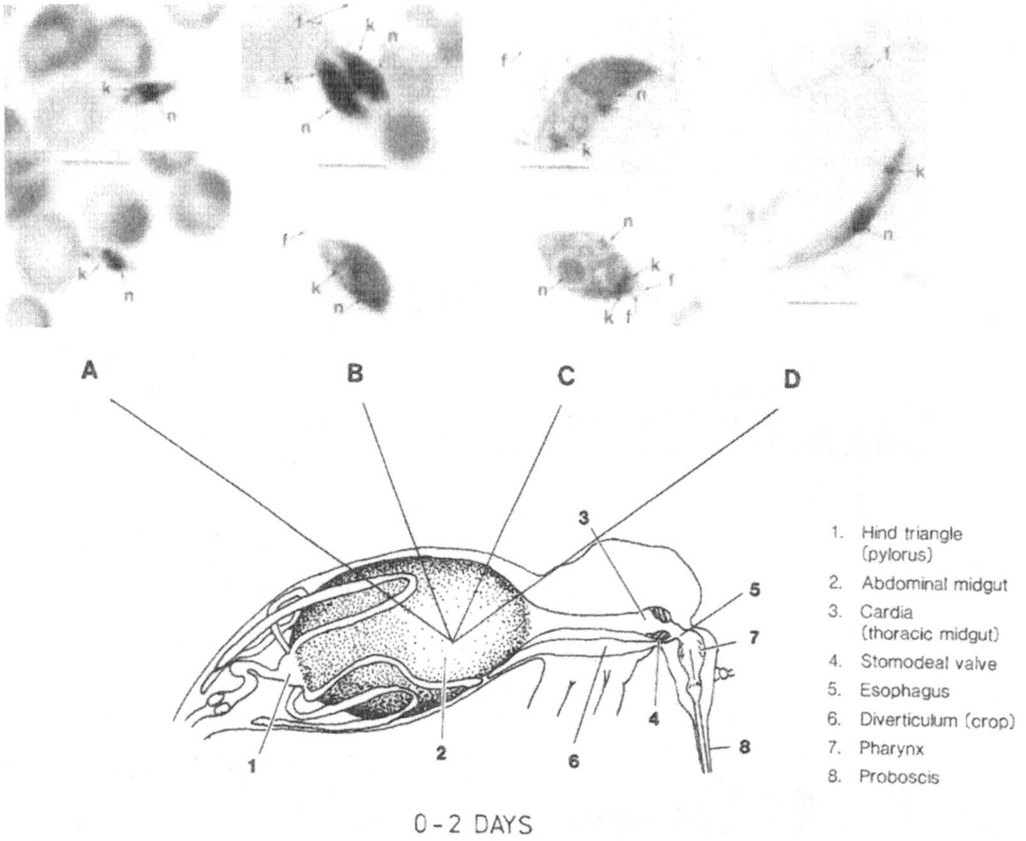

Figure 8.20. Development of *Leishmania major* in the alimentary tract of *Phlebotomus duboscqi*, 0–2 days after infectious bloodmeal. Bars = ~ 5 µm; n=nucleus, k=kinetoplast, f=flagellum. (a) Amastigotes <1 hr postfeeding; (b) "procyclic" promastigotes dividing (top) and swimming freely in bloodmeal (bottom), within interperitrophic space of abdominal midgut, 12–24 hr postfeeding; (c) longer, broader "procyclic" promastigotes swimming freely in bloodmeal (top) and dividing (bottom), 24–36 hr postfeeding; (d) nectomonad, swimming in bloodmeal within interperitrophic space of midgut, 36–48 hr postfeeding (with permission of American Journal of Tropical Medicine and Hygiene).

the first stage, 0–2 days postfeeding (Fig. 8.20), all parasite development occurs in the bloodmeal within the peritrophic membrane in the midgut. During this time, procyclic promastigotes develop from amastigotes and undergo at least 2 divisions. Some workers have reported division of amastigotes prior to transformation to procyclic promastigotes (Killick-Kendrick et al. 1979). Transformation of ingested amastigotes to promastigotes begins 6–12 hr postfeeding, while the host bloodmeal, which occupies the entire abdominal midgut, is still red in color and enveloped within the peritrophic membrane (Fig. 8.20a). Brun et al. (1976) suggested that a mammalian bloodmeal may contain a transformation blocking factor

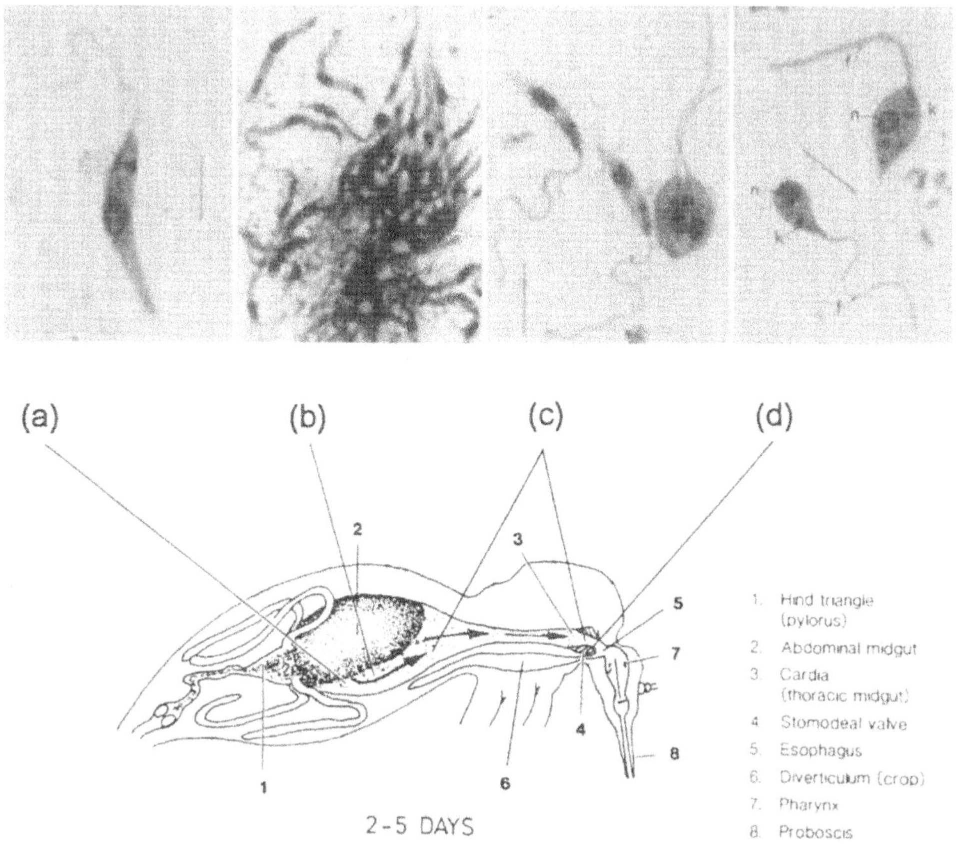

Figure 8.21. Development of *Leishmania major* in the alimentary tract of *Phlebotomus duboscqi*, 2–5 days after infectious bloodmeal. Bars = ~ 5 μm; n = nucleus, k = kinetoplast, f = flagellum. (a) nectomondad >2 days postfeeding; (b) cluster (rosette) of nectomonads in lumen of midgut following deterioration of peritrophic membrane, 2.5–5 days postfeeding; (c) dividing haptomonads in lumen of anterior midgut, thoracic midgut and cardia, and at stomodeal valve, 4–5 days postfeeding; (d) paramastigotes with juxtaposed nucleus and kinetoplast (top) and with nucleus to the side and and posterior to kinetoplast (bottom) at cardia and attached to stomodeal valve, 4–5 days postfeeding (with permission of American Journal of Tropical Medicine and Hygiene).

and that the change from amastigote to promastigote is inhibited until this factor is degraded by the sand fly's digestive enzymes. Procyclics are short, ovoid, slightly motile forms with large, diffuse nuclei and short flagella, typically seen swimming about the periphery of the bloodmeal; some may be seen dividing by binary fission (Fig. 8.20b, c). Between 36 and 48 hr, these early promastigotes begin to transform into long, slender, highly motile forms with long flagella (nectomonads of Killick-Kendrick et al. 1974 and others; Fig. 8.20d).

During the 2nd stage, 2–5 days postfeeding (Fig. 8.21), the peritrophic membrane ruptures

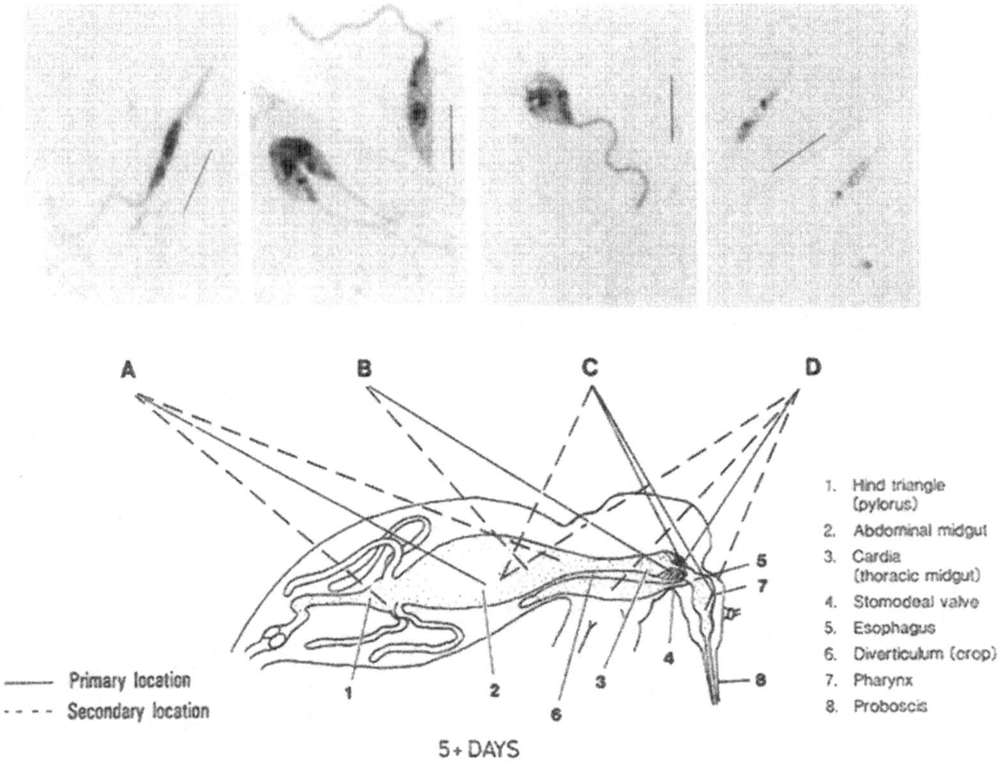

Figure 8.22. Development of *Leishmania major* in the alimentary tract of *Phlebotomus duboscqi*, >5 days. Bars = ~ 5 µ; n = nucleus, k = kinetoplast, f = flagellum. (a) Nectomondad in anterior and abdominal and thoracic midgut >5 days postfeeding; (b) dividing haptomonads in anterior thoracic at cardia and attached to stomodeal valve >5 days postfeeding; (c) paramastigote in anterior thoracic midgut, cardia and attached to stomodeal valve, esophagus and posterior pharynx >6 days postfeeding; (d) metacyclic promastigotes, usually anterior to stomodeal valve, >6 days postfeeding (with permission of American Journal of Tropical Medicine and Hygiene).

and the partially digested gut contents begin their passage through the pylorus and into the ileum. There is strong evidence that deterioration of the peritrophic membrane, which consists of a chitin framework and a protein-carbohydrate matrix, is mediated by chitinases secreted by the *Leishmania* promastigotes (Schlein et al. 1991). This event is accompanied by intense multiplication of procyclic promastigotes and transformation into nectomonads, which fill the anterior abdominal midgut (Fig. 8.21a). Many attach themselves by interdigitating their flagella with the microvillar lining of the gut wall, and in this way avoid being passed out with the digested bloodmeal. As the bloodmeal diminishes, parasites migrate forward into the

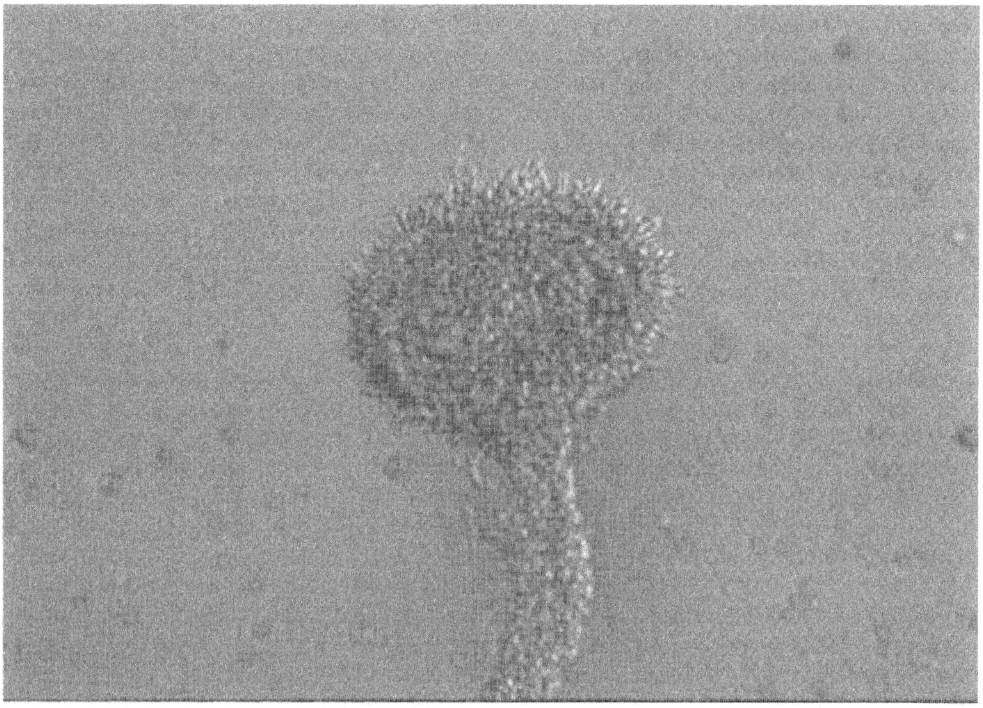

Figure 8.23. Stomodeal valve infection in *Lutzomyia diabolica* by *Leishmania mexicana* (photograph by Phillip G. Lawyer).

thoracic midgut (Fig. 8.21b) and transform into actively dividing haptomonads and paramastigotes (Fig. 8.21c,d). These parasites attach themselves to the cuticular intima of the cardia and stomodeal valve by foot-like enlargements of their flagella called hemidesmosomes (Killick-Kendrick et al. 1974), or they bond to each other by a gel-like matrix (Lawyer et al. 1987). At this point the sand fly's eggs are nearly mature.

During the 3rd and final stage, 5+ days postfeeding (Fig. 8.22), the gut is cleared of the bloodmeal, egg development is completed and oviposition begins. Concomitantly, a massive infection of dividing haptomonads, paramastigotes and nectomonads erupts at the stomodeal valve, forming a darkened lens- or ball-shaped mass, effectively impeding the flow of blood into the midgut (Fig. 8.22a). Haptomonads continue to divide and anterior migration of haptomonads and paramastigotes progresses into the esophagus and posterior pharynx (Fig. 8.22b,c). This advance is accompanied by the development of metacyclic (infective) promastigotes that spread forward into the pharynx, cibarium and proboscis, and rearward into the midgut (Figs. 8.22d, 8.23). By now, the fly is ready to seek another bloodmeal and may inject metacyclic promastigotes into a susceptible vertebrate host.

Metacyclogenesis

The most important recent advance in the study of leishmaniae in sand flies has confirmed the production of metacyclic promastigotes, or infective-stage promastigotes. The advent of the metacyclic promastigote coincident with the vector's search for another bloodmeal is a key factor in *Leishmania* transmission (Sacks and Perkins 1984). Metacyclogenesis has been described as a "preadaptation to survival" within the vertebrate host (Sacks 1989). Unlike other developmental forms that are immobilized because of attachment to or entrapment around the stomodeal valve, metacyclic promastigotes are unattached and highly motile, and are able to migrate anteriorly past the stomodeal valve into the pharynx, cibarium and proboscis, making it the ideal form for transmission. This is the only form that has been observed in sand fly mouthparts or on the skin at the site of the bite (Killick-Kendrick et al. 1979).

Mechanisms of transmission

Several authors have described stomodeal valve infections of adherent haptomonads and paramastigotes as blocks, plugs or balls similar to those seen in blocked fleas infected with plague organisms. Reports of experimental transmission by infected flies that appear to have difficulty in engorging (Molyneux 1977, Beach et al. 1985) support a blockage theory. Killick-Kendrick et al. (1974) disagreed with this interpretation and suggested that parasites in sand flies may interfere directly with sensilla that control probing and engorgement or alter the flow rate of blood, disrupting coordination of pumping of the cibarium and pharynx. This interference would result in egestion of parasites onto the skin. Schlein et al. (1992) observed deterioration of the cuticular lining of the stomodeal valve of *P. papatasi*, presumably caused by chitinolytic enzymes secreted by parasites attached *en masse* to the valve. They suggested a mechanism of infection based on the dysfunction of the stomodeal valve in infected flies. Sand flies ingest food by the action of the pharyngeal and cibarial pumps located in the head (Fig. 8.11d). These pumps are separated by the cibarial valve, and the main sphincter, the stomodeal valve, is posterior to the esophagus. In normal flies, when the cibarial and pharyngeal pumps are expanded in unison, the cibarial valve has to be open and the stomodeal valve closed to permit unilateral flow from the food canal into both pumps. In infected flies, the damaged stomodeal valve remains open; therefore, the pump sucks in both directions and the midgut contents, including parasites, are drawn into the pumps and mixed with the ingested blood. Afterward, the pumps empty in a similar manner. The pulsation of the pumps only swirls their contents, and each beat draws and delivers fluid in both directions. Partial function of the stomodeal valve causes less regurgitation and allows for partial engorgement of the fly. During the bite, this regurgitation engulfs parasites from the midgut and deposits them in the host tissue. Since metacyclics are the only parasites swimming freely at this point, they are the ones most likely to be inoculated into the host.

Not all sand flies that imbibe blood containing amastigotes can harbor *Leishmania* through their complete extrinsic life cycle and then transmit them to a susceptible host. Differences in the innate ability of various sand fly species to support the growth of *Leishmania* are well documented (Killick-Kendrick 1985, Lawyer et al. 1990). As described above, the microecology of *Leishmania* in the vector presents a number of barriers to complete development. Success depends on the ability of the parasites to overcome these barriers, which are: (1) the peritrophic membrane, which prevents migration out of the abdominal midgut, overcome by the action of parasite-derived chitinase (Schlein et al. 1992); (2) the proteolytic enzymes responsible for digestion that may inhibit early development and survival (Killick-Kendrick 1990b); and (3) excretion with the digested bloodmeal following deterioration of the peritrophic matrix, avoided by attaching to the microvillar lining of the midgut. Schlein and Romano (1986) and Borovsky and Schlein (1987) studied the effect of *Le. major* and *Le. donovani* on the digestive enzymes of *P. papatasi*, a competent vector of *Le. major* that is essentially refractory to *Le. donovani*. They showed that *Le. major* depresses digestive enzyme activity but that *Le. donovani* enhances it. Thus, *Le. major* thrives, but growth of *Le. donovani* is inhibited. Pimenta et al. (1992) determined that midgut attachment of promastigotes is controlled by the expression of terminally exposed galactose residues on lipophosphoglycan (LPG), the major promastigote surface molecule. During metacyclogenesis, these residues are down-regulated in favor of oligosaccharides terminating in arabinose (McConville and Homans 1992). There is evidence that gut lectins or lectin-like molecules serve as parasite attachment sites, and the stage-specific expression of the complementary sugars on LPG provides an underlying mechanism controlling the retention of attached, dividing parasites during bloodmeal digestion and excretion. This expression of sugars also would control the release of infective-stage promastigotes, permitting their anterior movement so as to be positioned for transmission by bite. In some cases, promastigotes of different *Leishmania* species and their LPGs display inherently different binding capacities for the midguts of different phlebotomine vectors. The extent of binding in each species is an important factor in vector competence. Pimenta et al. (1994) found that species of *Leishmania* other than *Le. major* lack the appropriate ligands for binding to midguts of *P. papatasi*, explaining why this fly, the natural vector of *Le. major*, is not involved in transmission of any other *Leishmania* species. Likewise, *P. sergenti*, a natural vector of *Le. tropica*, supports experimental infection with *Le. tropica* but not with *Le. infantum*. In contrast, *P. argentipes* midguts possess receptor sites lacking in *P. papatasi*, and may serve as vectors for transmission of available sympatric parasites (e.g., *Le. major*, *Le. tropica* and *Le. donovani*). Thus, species with only one type of receptor in their midguts would be expected to vector only leishmanias with complementary ligands on their LPG surface molecules. It follows that species with multiple midgut receptor types are capable of transmitting more than one species of *Leishmania*.

Vector-Host Interactions

Sand fly-host contact depends on a combination of factors. For the sand fly, these are habitat preferences, aggregation or lekking behavior, horizontal and vertical distribution, seasonal distribution and feeding behavior. For the vertebrate host, including humans, these factors are exposure (dependent on habitat preferences), activity cycles, defensive behavior and type and location of the nest (house).

Habitat relationships

For a competent vector to transmit infective parasites to a susceptible host, there must be at least some temporary overlap in their respective habitats so that the vector has the opportunity to feed upon the host. Lainson and Shaw (1987) described several scenarios to show how vector and host habitat relationships determine where transmission will occur (Fig. 8.24); if vector and host (reservoir) habitats do not overlap, there will be no transmission (Fig. 8.24a). Where some overlap exists (Fig. 8.24b-g), so does the potential for transmission. Fig. 8.24b, c, and d represent strictly enzootic cycles involving only one reservoir host. Enzootic VL caused by *Le. chagasi* in the Amazon region of Brazil probably follows closely the cycle described in Fig. 8.24b, while in northeast Brazil, Fig. 8.24c is a more accurate description. These scenarios also may apply to anthroponotic VL caused by *Le. donovani* in Kenya, the Sudan, India and Nepal, and to anthroponotic CL caused by *Le. tropica* in some towns and villages in the Middle East (Lysenko and Beljaev 1987, Vioukuv 1987), where humans are the only or main reservoir

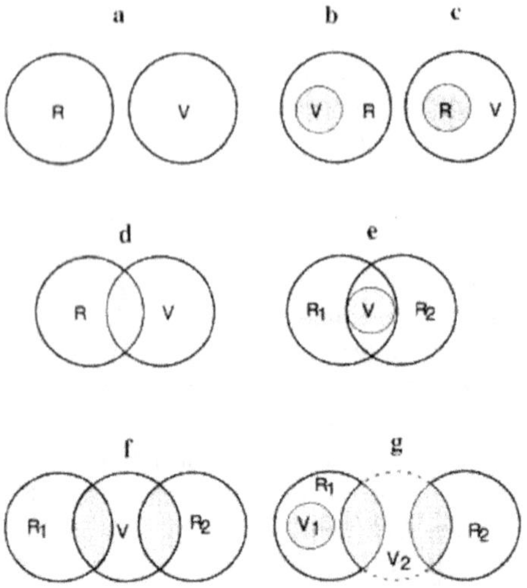

Figure 8.24. Vector-host habitat relationships and *Leishmania* transmission: (a) No overlap, no transmission; (b) transmission throughout vector habitat; (c) transmission throughout reservoir habitat; (d) transmission limited to small area of habitat overlap; (e) transmission to 2 hosts with different preferences throughout the habitat of the vector; (f) transmission to 2 different hosts in different habitats by the same vector; (g) transmission by a secondary vector to a secondary host outside the habitat of of the primary host and vector. R = reservoir; V = vector; R1 = primary reservoir; R2 = secondary reservoir; V1 = primary vector; V2 = secondary vector. (a–d represent enzootic cycles, e–g represent possible zoonotic cycles in which humans are secondary or accidental hosts; from Lainson and Shaw 1987).

and source of infection. Fig. 8.24e, f and g represent probable zoonotic cycles in which humans are accidental or secondary hosts (R2). In Fig. 8.24e, transmission may occur to 2 hosts in the same habitat, even though neither has exactly the same ecological distribution. This scenario applies in the transmission of *Le. major* by *P. papatasi* in the Mediterranean Basin and South and Central Asia, and *P. duboscqi* in sub-Saharan Africa, respectively. In these foci, humans become infected accidentally when their habitations and activities encroach upon those of the vector and the primary reservoir (R1). Fig. 8.24f illustrates transmission outside the habitat of R1 to R2. In Jordan, along the northern coast of the Dead Sea, it appears that human CL caused by *Le. major* occurs in villages far removed from colonies of R1, *Psammomys obesus*. This distance suggests either that the vector, *P. papatasi* is capable of flight between *Psammomys* (R1) and human (R2) habitats, or another animal such as a commensal rodent serves as R1. Fig. 8.24g illustrates transmission by a secondary vector (V2) to R2 outside the habitat of R1 and a primary vector (V1). This scenario may apply in the transmission of *Le. mexicana* to humans in south central Texas (Fig. 8.25). In this focus, V1, *Lu. anthophora*, lives in the nest of the woodrat, *N. micropus*, R1. A sympatric sand fly, *Lu. diabolica* (V2), is an opportunistic feeder and will take blood from woodrats as well as humans, thus transmitting *Le. mexicana* from the rodent population (R1) to humans living in the area (R2).

Transmission season

The degree of habitat overlap between vector and host may depend on seasonal climatic changes. In many leishmaniasis foci, sand fly

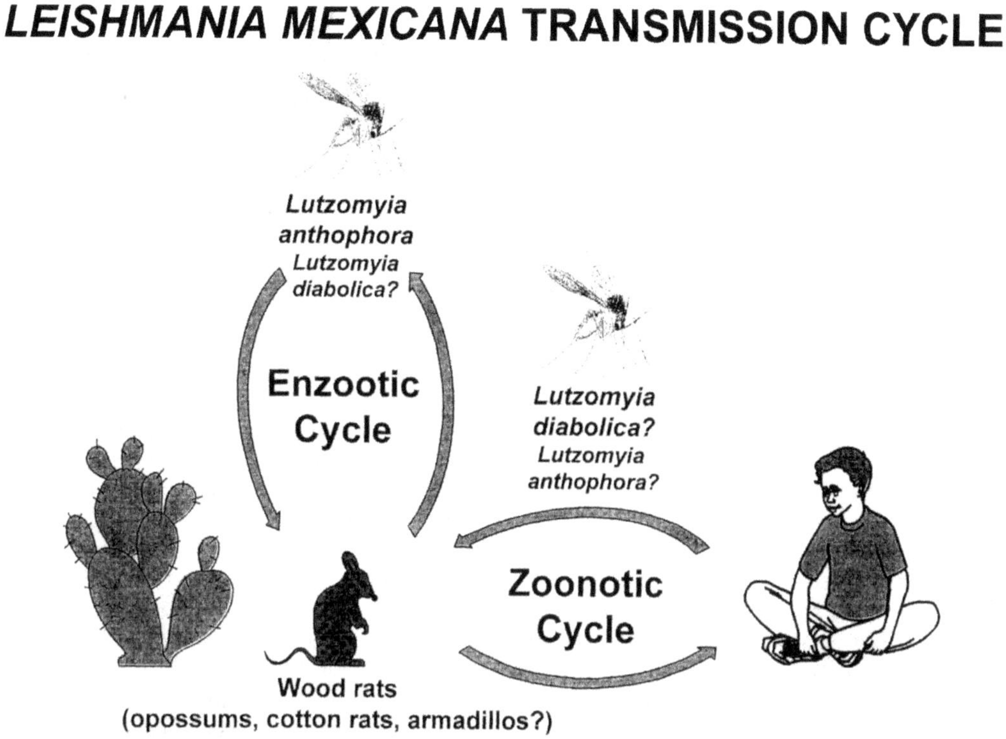

Figure 8.25. Transmission cycle of *Leishmania mexicana* in Texas (courtesy of Phillip G. Lawyer).

populations fluctuate in density and age as temperature and humidity become more or less favorable for their existence. In Kenya, populations of *Phlebotomus martini* (the vector of *Le. donovani*) and *P. duboscqi* (the vector of *Le. major*) reach their peaks at the ends of the rainy seasons, but almost disappear by the end of the dry seasons. Seasonal climatic changes also may cause changes in the reservoir habitat, resulting in die-off or migration. As the primary reservoirs become less numerous, vector sand flies may seek an alternative host such as humans. If the incubation period for a form of leishmaniasis in a given focus is known, the season of greatest transmission risk can be calculated. This time may not necessarily correspond to periods when environmental conditions favor sand fly growth and the highest populations occur. Periods of peak transmission often correspond to times when the vector population is relatively old and waning, and not when it is at its highest density. At the latter stage, a large proportion of the population is made up of freshly emerged females that never have engorged and therefore are noninfected.

Surveillance Methods

Sand fly collection methods

Selection of sand fly collection methods (Fig. 8.26) will depend on the objectives of the study, the degree of familiarity with the local fauna and the ecology of the study site. If the intent is to collect live sand flies for colonization or experimentation, specific techniques will be required. Where little is known about sand fly populations, a combination of sampling methods is desirable in order to determine the species spectrum. If the habits and ecology of the target sand fly population are well known, one or 2 methods tailored to the situation may suffice.

Immatures

Searches for immature sand flies should be directed to protected habitats with high humidity and soil moisture and an abundance of decomposing organic matter, such as leaf litter or animal feces. Laboratory observations of ovipositing females indicate that they disperse eggs individually in the environment. Therefore, larvae usually are not clumped in nature. Adult emergence traps set on the floor of forests where large numbers of adult flies occur have confirmed this separation. Except for desert species, in which larval habitats are restricted to the humid lower portions of animal burrows, few adults occur per square meter of soil. Tremendous collecting efforts involving sifting through literally tons of soil generally have yielded few larvae or pupae. Methods for extracting immature stages from soil samples include sugar floatation and filtration (Hanson 1961, Rutledge and Mosser 1972). Some workers prefer using a modified Berlese funnel apparatus (Killick-Kendrick 1987).

Adults

Adult sand flies usually are collected from their resting sites during the early morning hours, or during their dusk to dawn activity cycles. Their small size and secretive habits require a variety of specialized methods, including direct searches, human- or animal-baited collections, and trapping.

Direct searches. Searches of resting sites, such as inside houses, animal shelters, animal burrows, tree trunks/buttresses/treeholes, caves, rock crevices and cracks in walls and wells, are especially useful for collecting live bloodfed or gravid females. The simplest way to collect resting sand flies is with an aspirator fitted with a fine enough mesh to exclude these tiny insects (Fig. 8.26a,b). Mechanical, battery-operated aspirators (Fig. 8.26c) are available commercially and are used by some workers, but they often are cumbersome and noisy. A flashlight illuminates dark areas and disturbs the flies so that they are more easily spotted and captured. Collectors commonly disturb resting flies by moving a small branch or stick over suspected resting surfaces, or by poking sticks or throwing small objects into holes to achieve the same result. Aspirators may be used to collect resting sand flies at the bases of large trees in well-canopied forests. A Damasceno trap is a tent-like structure resembling a circular shower curtain made of bed sheeting. The trap is suspended from the end of a pole over an animal burrow, or over the space enclosed by the buttresses at the base of large trees. Flies, flushed from hiding when the burrow entrance or the leaf litter between the buttresses is disturbed, land on the fabric sides of the trap and can be collected with an aspirator. Aspirator collections can be made from the walls and ceilings of small caves and similar natural resting places, or from the interiors of animal or human dwellings that are insulated from the heat of the day.

Human- or animal-baited collections. Collections using humans as bait remain the most reliable method for determining which species commonly bite humans (Fig. 8.26b). However, there is the obvious risk of becoming infected with leishmaniasis while using this technique

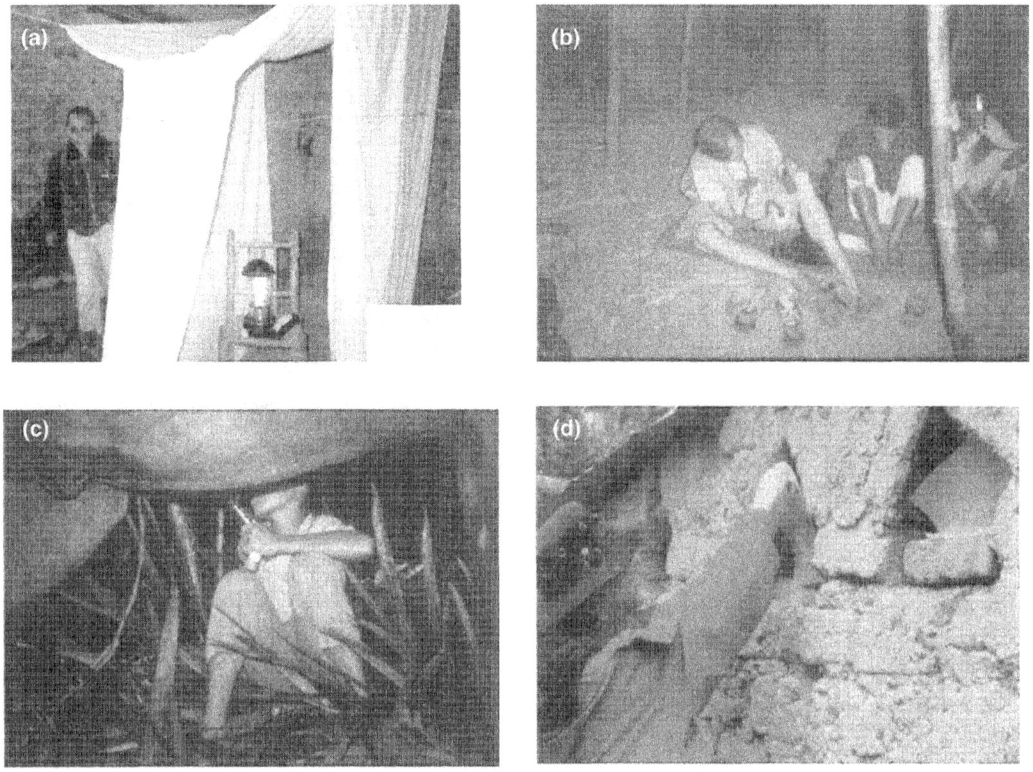

Figure 8.26. Sand fly collection methods: (a) a Shannon trap; (b) using a mouth aspirator to collect sand flies attracted to human bait; (c) using a powered aspirator to collect sand flies from beneath a horse; (d) setting out castor oil-coated traps to sample sand flies adjacent to a human dwelling (a, photograph by Lyman Roberts; b, photograph by Shredhar Pradhan; c, courtesy of Robert Killick-Kendrick; d, courtesy of Richard G. Andre). Figure continued on next page.

Figure 8.26 (continued). Sand fly collection methods: (e) a CDC miniature light trap with a fine-mesh screen; (f) using a mouth aspirator to collect resting sand flies from interior walls of a home (e, photograph by Phillip G. Lawyer; f, photograph by Richard G. Andre).

in endemic areas. The risk of infection can be reduced by wearing protective clothing and collecting the flies as they land on shirt sleeves or trouser legs. A large, tethered domestic animal, such as a horse, cow or goat, often will serve as an acceptable alternative to human bait.

Traps. There are 2 types of traps: interceptive and attractive. Interceptive traps capture sand flies as they fly in search of sugar or a bloodmeal, a mate, or a resting or oviposition site. In dry, open habitats, sticky paper traps are useful for sampling adult activity (Fig. 8.26d). These consist of pieces of photocopy paper soaked in castor oil and placed in potential resting places or mounted on sticks in areas where sand flies might be expected to fly. Traps are set out in late afternoon and collected in the morning. Flies then are removed with a fine paint brush and rinsed in 70% ethanol if they are to be mounted for identification, or rinsed in a dilute detergent if they are to be dissected.

Exit-entrance traps can capture sand flies as they fly in or out of animal burrows, tree holes, termitaria or other small cavities that provide suitable harborage. This type of trap consists of a plastic cylinder, open at both ends and divided in the middle by a fine mesh partition. Paper sheets coated with castor oil line the inside surface of each half of the trap to provide a trapping surface. When inserted into a rodent burrow, the trap seals off the entrance. Flies emerging from the burrow are caught on the sticky surface of the inner half of the trap; flies coming from outside and attempting to enter the burrow are trapped on the sticky surface of the outer half of the trap (Yuval and Schlein 1986).

Emergence traps have been used to collect newly-emerged adults from larval breeding sites. These can consist of an inverted plastic washtub fitted with a collecting bottle, or a wood or metal frame covered with fine mesh screen that then is placed on the ground at a suspected breeding site (Rutledge and Ellenwood 1975).

CDC miniature light traps, fitted with a tungsten (not UV) bulb and fine-mesh collecting nets (20/cm), frequently are used at night to collect phototropic species (Fig. 8.26e). This trap is particularly useful in wet, forested environments where unbaited sticky traps usually are unproductive. The traps should be hung close to the ground (within 1 m) or near the entrance of rodent burrows or termitaria. Carbon dioxide or a small, caged animal may be used to increase the trap's attractiveness to non-phototropic species. CDC traps are least efficient in open semidesert areas, but their effectiveness increases as they are placed closer to sand fly resting sites and host habitats.

In Neotropical forest environments, a Shannon trap frequently is used as a nocturnal collecting station (Shannon 1939). This trap consists of a rectangular tent-like structure made of bed sheeting or other cloth that is suspended above the ground (Fig. 8.26f). A light source hung inside attracts the sand flies so collectors standing nearby can gather the flies as they land on the fabric.

Traps also capture sand flies that are attracted to animal bait. The Disney trap (Disney 1966) consists of a shallow metal pan coated with castor oil. A small animal is placed in a cage near the center of the pan, close to the sticky surface. Sand flies stick in the oil as they attempt to reach or leave the animal. In Brazil, Ward (1977) used a funnel-type trap, baited inside with a small mammal, to collect blood-engorged sand flies for laboratory colonization.

Sand fly dissection and examination for parasites

Detection, isolation and typing of *Leishmania* from wild-caught flies is an important step in the incrimination of vectors. This study involves dissection and examination of the fly for parasites and if parasites are found, attempted isolation of the parasite for cell culture. Sterile handling techniques are essential, because leishmanial promastigotes do not compete well in culture with bacteria or fungi in growth media. Antibiotics and antifungal agents may be used when dissecting field-collected flies to minimize contamination. Preserving field-collected specimens in liquid nitrogen is an excellent but little-used method for holding them until they can be dissected in a laboratory.

Any of the collecting methods described above can produce suitable specimens for dissection. Flies collected on sticky papers frequently still are alive when the papers are collected, and if removed, washed and chilled quickly, remain fresh for dissection. An aspirator removes flies collected live by other methods from the collecting bags; then they are blown into a weak detergent solution to immobilize them and to remove dust and setae. The flies are next surface sterilized in antibiotic solutions prior to dissection. Johnson et al. (1963), Rioux et al. (1984) and Perkins et al. (1988) describe methods to dissect sand flies and to culture *Leishmania*. Characterization or typing of *Leishmania* parasites isolated from sand flies involves a variety of biochemical, biomolecular and immunological methods, such as isoenzyme analysis, nucleic acid-based techniques (DNA probes, PCR), enzyme-linked immunosorbent assays (ELISA) and immunofluorescent assays (IFA). A discussion of these methods is beyond the scope of this chapter.

Bloodmeal analysis

Where risk of contracting leishmaniasis is high and human-bait collections cannot be used, anthropophilic species can be determined by identifying the bloodmeals of wild-caught sand flies. This is done by precipitin tests, countercurrent gel electrophoresis, ELISA or other means. Bloodmeals can be preserved by crushing engorged flies on filter paper and desiccating them in sealed vials containing silica gel.

Preservation of sand fly specimens for identification

Field-collected flies can be killed by various means, such as freezing on dry ice, heating in direct sunlight or exposure to poisons (ethyl acetate) in killing bottles. Because of their small size and the absence of obvious external diagnostic characters, most sand flies must be cleared and mounted on microscope slides for identification under a compound microscope. Field-collected material may be stored dry between layers of tissue paper in small containers (e.g., pill boxes), or in vials containing 80% alcohol. Dry storage is preferred to storage in alcohol because the latter discolors the specimens and causes the muscles to harden with age (6+ months), making them difficult to clear and macerate. Young and Duncan (1994) give detailed instructions for mounting phlebotomine sand flies for identification.

Leishmaniasis Prevention and Control

Environmental approaches

Environmental changes

Natural or human environmental changes or disruptions (such as war) may cause humans to move into vector habitats where they become exposed to infected populations of sand flies. For example, in 1984–1986, as a result of drought and famine, a clan of Turkana nomads migrated from less favorable areas of Turkana District, in Kenya's Rift Valley Province, to a more favorable area in Nadume, at the southern tip of the district. Apparently, this migration was into habitat that favored *Leishmania* vectors as well. In late 1986 and 1987, the clan suffered from an acute kala-azar epidemic that killed many of its members. A similar phenomenon occurred in the late 1980s and early 1990s in Sudan's western Upper Nile region, where inhabitants of kala-azar endemic villages were driven by war and famine into nearby *Acacia / Balanites* forests; there they encountered large populations of *P. orientalis*, a competent vector of *Le. donovani* (Schorscher 1991). The increased exposure to sand fly bites caused an epidemic that, over 5 years, took the lives of more than 100,000 individuals in a population of just over 1 million. The emergence of zoonotic visceral leishmaniasis (ZVL) as an increasingly important public health problem in tropical America provides another example. The massive destruction of primary forests in this region, together with rapid human population growth and concomitant development of new farmland and rural settlements, have led to conditions that now support large populations of the vector *Lu. longipalpis*, as well as dogs and foxes, the major reservoir hosts. Consequently, ZVL now occurs in many regions of Latin America where it was not found previously (Tesh 1995).

Occupational and recreational hazards

Humans may engage in occupational or recreational activities that expose them to bites of infected flies. Military personnel who operate in tropical forested or semi-desert areas where sand flies and their reservoir hosts occur are at risk, as are surveyors, game wardens, park rangers and others active in such areas. Human-fly contact is increased by sleeping outdoors and by visiting areas with infected flies during times of peak biting activity.

Location and construction of housing

People who live in sand fly-infested areas risk infection when housing is substandard or located close to habitats supporting sand fly populations and reservoir hosts, as is the case in most leishmaniasis foci in developing countries. Settlements in Neotropical forests, in places infested by infected rodents and in semi-arid foci on the periphery of certain towns increase the possibility of transmission. In several foci in the Middle East and North and East Africa, houses built near rock cliffs or caves inhabited by hyraxes or other reservoir animals increase human-fly contact. Stone walls with crevices, hollow trees or rodent burrows/nests serve as sand fly developmental or resting places near houses and also are important in rural VL, ACL and zoonotic cutaneous leishmaniasis (ZCL). In kala-azar foci in India and Nepal, typical human dwellings are constructed of wattle (a framework of interwoven sticks) plastered with mud and animal dung, with thatch or ceramic tile roofs. Such houses rarely have glass-paned windows or tight-fitting doors, and almost invariably have gaps between the tops of walls and the roof through which flying insects can enter. Animal shelters and storage sheds of similar construction, usually within the same compound, afford ideal habitat for the kala-azar vector, *P. argentipes*. This highly zoophilic species feeds primarily on cattle and other domestic livestock but is an opportunistic human biter. The proximity of livestock and human dwellings in these foci increases human-fly contact. Furthermore, people in such rural villages often sleep in the same shelters as their domestic animals, exposing themselves to greater risk of being bitten by infected sand flies.

A well-designed leishmaniasis control program must address all 3 elements of the epidemiological triad and capitalize on those points of interface between parasite, vector and host that can be blocked or interrupted. Efforts should control the parasite pool (treatment of human cases, especially in anthroponotic disease), the reservoir and the vector. Thorough knowledge of the life cycle and behavior of the reservoir(s) and vector(s) will provide pivotal information on which to base an integrated control program.

Treatment of humans

Antimony compounds are available for treatment of human cases, but they are expensive and difficult to administer routinely in rural health clinics. Antifungal agents, such as paromomycin and amphoterycin, also have been tested in clinical trials against both cutaneous and visceral leishmaniasis with mixed success (Lane 1993). Treatment of active cases reduces morbidity and mortality, yet it does little to reduce the incidence of new cases, particularly in the case of zoonotic diseases (Tesh 1995). There are no effective vaccines against leishmaniasis.

Reservoir control

Domestic dogs that serve as reservoirs of ZVL can be examined and the infected animals destroyed. According to G.B. Fairchild, elimination of domestic dogs and installation of fly-proof screening on breeding facilities for dogs used for food all but eliminated human leishmaniasis from a large area of China in the early 20th Century. However, in some areas prevention of the disease in dogs appears to be the best approach for interrupting the domestic cycle (Tesh 1995). Killick-Kendrick et al. (1997) investigated deltamethrin-impregnated collars as a low-cost, humane intervention for prevention of sand fly bites on dogs. If successful, this would eliminate the prime source of infection in foci of human visceral leishmaniasis with a canine reservoir. In laboratory experiments in which sedated dogs were exposed to the bites of *Phlebotomus perniciosus* (a widespread vector in the western Mediterranean Basin), the anti-feeding effect of deltamethrin-impregnated dog collars on the sand flies was shown to protect the dogs from 94% of bites for up to 34 weeks (Killick-Kendrick et al. 1997). The collars act as a depot, gradually releasing the deltamethrin into the lipids of the skin. Controlled field trials are in progress to see if this intervention can reduce the circulation of the parasite in nature, thereby diminishing the risk of infection to the human population. Where reservoirs are the colonial murid rodents *P. obesus* (giant sand rat, Middle East), *Rhombomys*, *Meriones* and *Arvicanthus* (African grass rats, North and East Africa), some success has been achieved with rodenticides or by deep plowing to destroy rodent habitat. Reservoir control in forests or mountainous areas has not been effective (Lane 1993).

Sand fly control

Applying insecticides, sometimes in conjunction with environmental management, commonly has controlled sand flies. The emphasis has been residual insecticide applications in peridomestic environments, especially in houses and in animal shelters to control anthroponotic leishmaniasis (Vioukuv 1987). In many cases, sand fly control was a collateral benefit of insecticide spraying to control malaria vectors. Such was the case in India and Nepal, where kala-azar all but disappeared during the DDT era (Bista et al. 1993). A resurgence of the disease in 1980 was attributed to the cessation of routine spraying of houses with DDT for malaria vector control (Joshi et al. 1990). The more environmentally friendly residual pesticides available today for use in peridomestic settings also are very efficacious. However, they are more expensive and less enduring than DDT, requiring more frequent application and a whole new strategy for their use, i.e., precision targeting rather than blanket application.

Vioukov (1987) cited a 1941 example of successful elimination of a ZCL focus in Turkmenia accomplished by treating with chloropicrine all great gerbil burrows within a 1.5 km radius of human settlements. However, he further stated that apart from this one example, the effects of sand fly control in foci of ZCL have been minimal and usually of short duration. More recently, in a dual focus of kala-azar and ZCL in Baringo District, Kenya, Robert and Perich (1995) achieved up to 90% control of adult sand flies for 2 weeks by treating termitaria and animal burrows with a mixture of 1.5% cyfluthrin in corn oil. In a densely forested focus in Guatemala, Perich et al. (1995) achieved up to 78% reduction in vector sand fly populations for 12 days with ULV applications of cyfluthrin in a 100 m barrier around a mock bivouac site.

Robert et al. (1997) investigated the efficacy of the biological control agent *Bacillus sphaericus* incorporated in sugar solutions. These workers sprayed vegetation near rodent burrows and termitaria with a sugar solution containing this larval toxicant. Adult sand flies that take such bait apparently become *B. sphaericus* carriers and later return to their resting and breeding places inside rodent burrows and die, where they presumably are eaten by their larvae, resulting in a significant decrease in adult populations of some sand fly species.

Ecological or environmental modification has been used with some success in limited areas to destroy or reduce sand fly breeding sites and control *Leishmania* vectors. Sand fly breeding site distribution falls into 3 categories: 1) peridomestic situations, such as in and around human and animal shelters and out buildings; 2) areas with hot, dry climates (deserts, steppes and savannas), where breeding sites are concentrated in animal habitats that provide the conditions essential for sand fly development and survival; and 3) humid areas (forests), where breeding sites may be scattered and not limited to specific animal habitats.

In peridomestic settings, improvements in sanitation can eliminate many breeding sites and cases of leishmaniasis. Resurfacing mud walls and areas close to human habitations to cover holes and cracks, demolition and removal of uninhabited buildings, removal of rubble and organic waste and removal of unwanted vegetation to reduce plant-sugar sources will discourage sand fly and rodent breeding. Where economically feasible, house construction with cement floors and masonry walls, tight-fitting doors, windows and ceilings, and the use of bed nets and window screens will deny sand flies access to indoor habitats and reduce human-fly contact.

In arid regions of the Middle East, Central Asia and North Africa that are inhabited by colonial rodents (such as *Psammomys*, *Rhombomys* or *Meriones*), deep plowing over concentrations of rodent burrows has been used successfully to eradicate *Le. major* reservoir and vector habitats. By contrast, habitat reduction in humid, Neotropical forests has shown very little promise.

Where habitat reduction or modification is not feasible (e.g., in Neotropical forests), conscientious use of personal protective measures can reduce human-fly contact. In these areas, people who are outdoors during twilight or darkness should wear protective clothing and apply repellents to all exposed skin. Treatment of outer garments with permethrin will further reduce contact with sand flies. When sleeping in the open or in buildings from which sand flies are not excluded, the use of pyrethroid-impregnated bednets provides effective protection.

Sustainable management

Methods to control *Leishmania* vectors will vary in different circumstances. The most successful control programs employ integrated control of reservoirs and vectors. Control measures must be specifically designed with the local epidemiology of the disease in mind, and they must be locally sustainable or they will be abandoned due to expense when outside-agency funding ceases. Development, production and distribution of affordable vaccines for the hundreds of millions of people in endemic areas is a future challenge to a world besieged with devastation caused by nature and mankind and where starvation is an all too likely outcome. The challenge to medical entomology is to develop prevention and control strategies based on a firm understanding of parasite-vector-host interactions to break the chain of infection.

THE TRYPANOSOMIASES

Trypanosomal Diseases of Humans and Livestock

Parasites in the genus *Trypanosoma* cause several serious human and livestock diseases in Africa and in the New World tropics. In Africa, a human disease known as African sleeping sickness is caused by 2 subspecies of *Trypanosoma*: *Trypanosoma brucei rhodesiense* and *Trypanosoma brucei gambesiense*. Both parasites are transmitted by muscoid flies in the genus *Glossina*, commonly called tsetse. African diseases of livestock are caused by other *Trypanosoma* that are transmitted by tsetse as well as other Diptera.

In the New World tropics, a disease known as Chagas disease is caused by *Trypanosoma cruzi* and is transmitted by triatomid bugs.

In 1995, WHO estimated that 300,000 people were infected annually with some form of *Trypanosoma* worldwide (World Health Organization 1995).

African Trypanosomiasis in Humans

The disease

The best known of the vectorborne trypanosomes are the agents of "African sleeping sickness." This disease has been known since the 1400s, when an Arab writer described a long

Table 8.3. The species and subspecies of *Glossina* (from World Health Organization 1998).

Subgenus	Species	Subspecies
Glossina (*morsitans* group)	*G. morsitans*	*G. m. morsitans** *G. m. centralis** *G. m. submorsitans*
	*G. swynnertoni** *G. longipalpis* *G. pallidipes** *G. austeni*	
Nemorhina (*palpalis* group)	*G. palpalis*	*G. p. palpalis** *G. p. gambiensis**
	G. fuscipes	*G. f. fuscipes** *G. f. quanzensis* *G. f. martinii*
	G. pallicera	*G. p. pallicera* *G. p. newsteadi*
	G. caliginea *G. tachnioides**	
Austenina (*fusca* group)	*G. fusca*	*G. fusca fusca* *G. f. congolensis*
	G. nigrofusca	*G. n. nigrofusca* *G. n. hopkinsi*
	G. fuscipleuris *G. haningtoni* *G. schwetzi* *G. tabaniformis* *G. nashi* *G. vanhoofi* *G. medicorum* *G. frezili* *G. severini* *G. brevipalpis* *G. longipennis*	

*Major vectors of African sleeping sickness.

illness that killed a Sultan of the Mali Kingdom in terms that many authors agree describes African sleeping sickness. Widespread across Africa, human trypanosomiasis is reported by WHO from around 200 discrete foci from 36 of the sub-Saharan African countries within areas where tsetse occur. An estimated 60 million people are at risk, with over 300,000 new cases of the disease each year (World Health Organization 1998). Since surveillance often is inadequate due to a shortage of skilled personnel and the remoteness of vast areas where the disease is endemic, these should be considered minimum estimates. The movement of people

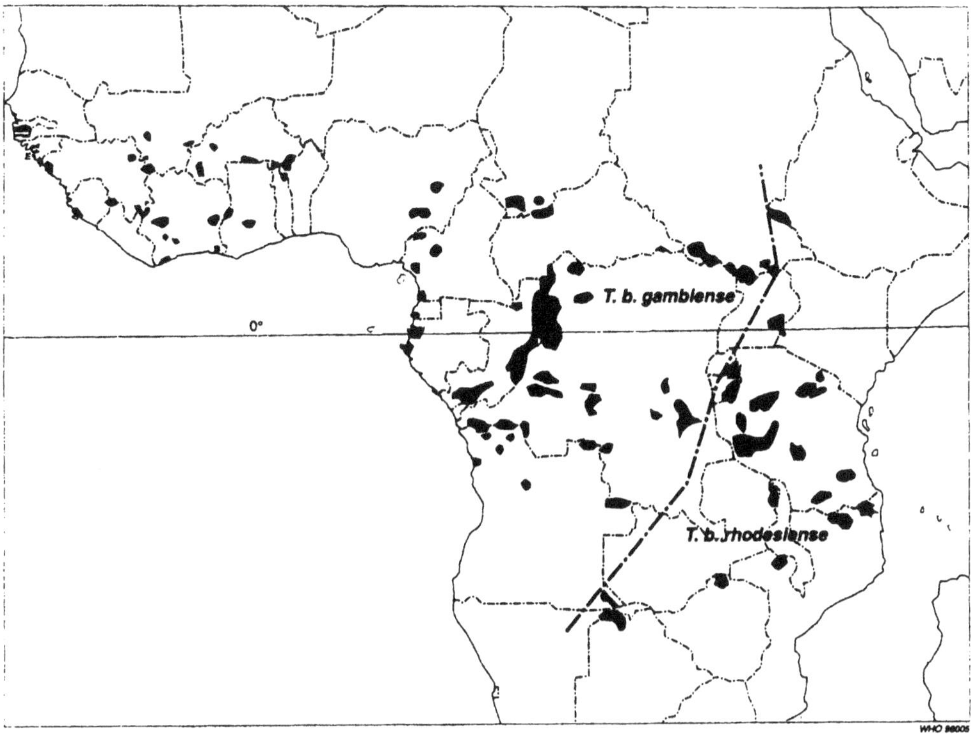

Figure 8.27. Geographic distribution of major epidemic foci of sleeping sickness in Africa, 1995 (World Health Organization 1998).

carrying the parasites has spread the disease to nearly all areas where tsetse vectors exist. Civil wars and other conflicts have resulted in large numbers of refugees in parts of Africa in the past 20 years, and during this time disease has increased in some areas where control methods previously had kept it in check. This disease nearly always is fatal if untreated.

There are 2 forms of sleeping sickness: Gambian, caused by *T. brucei gambiense*, and Rhodesian, caused by *T. brucei rhodesiense*. These organisms are morphologically indistinguishable from the nonhuman pathogen *Trypanosoma brucei brucei*. However, the metacyclic trypanosomes of *T. brucei brucei* are lysed by a nonspecific factor in human serum that is unable to lyse the metacyclic trypanosomes of *T. brucei gambiense* and *T. brucei rhodesiense*.

The 2 human diseases are separated geographically and ecologically, with Gambian sleeping sickness occurring in western and north central Africa and Rhodesian sleeping sickness occurring in central and eastern Africa (Fig. 8.27). The clinical manifestations of African sleeping sickness have both similarities and differences. Gambian sleeping sickness usually is a chronic disease, with death occurring after years of infection. Rhodesian sleeping sickness tends to be more acute. The course of the illness in humans in both diseases follows 2 or 3 stages.

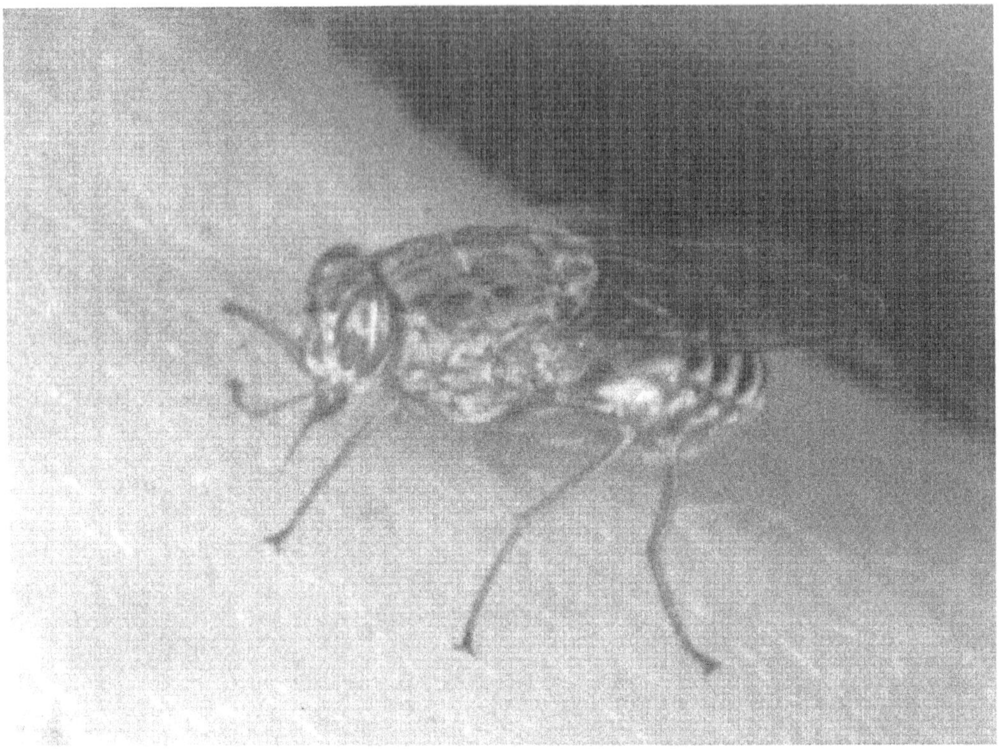

Figure 8.28. Adult female tsetse, *Glossina morsitans* (photograph by Edgar D. Rowton, all rights reserved).

The first stage occurs in only about 30% of the cases of both Gambian and Rhodesian sleeping sickness cases. This stage is characterized by inflammation, called a **trypanosomal chancre,** at the bite site of the tsetse vector. When present, the lesion occurs within 5–15 days of the infectious bite as a raised sore that is warm, painful and somewhat tender. Resolution of the chancre usually occurs spontaneously within 3 weeks. Lymph glands in the region of the bite may become swollen as drainage occurs from the bite site.

The 2nd stage is termed the **hemolymphatic stage** (sometimes the first symptomatic stage). It occurs within 1–3 weeks after the infectious bite and is characterized by periodic fever, general fatigue, joint pain and headache. These symptoms resemble malaria or influenza, with muscular and bone aches, fevers and generalized weakness. This stage lasts about 3 months, with symptoms occurring a week at a time and interspersed with asymptomatic periods. Swelling of the posterior cervical lymph nodes is termed **Winterbottom's sign,** and is characteristic of *T. b. gambiense*, while general swelling of lymph nodes is more characteristic of *T. b. rhodesiense*.

The hemolymphatic stage is followed by a 3rd stage (sometimes the 2nd symptomatic stage) called the **meningoencephalitic stage**. This stage results from the invasion of the central nervous system within weeks to months after infection with *T. b. rhodesiense* or in months to years with *T. b. gambiense*. Early in this stage,

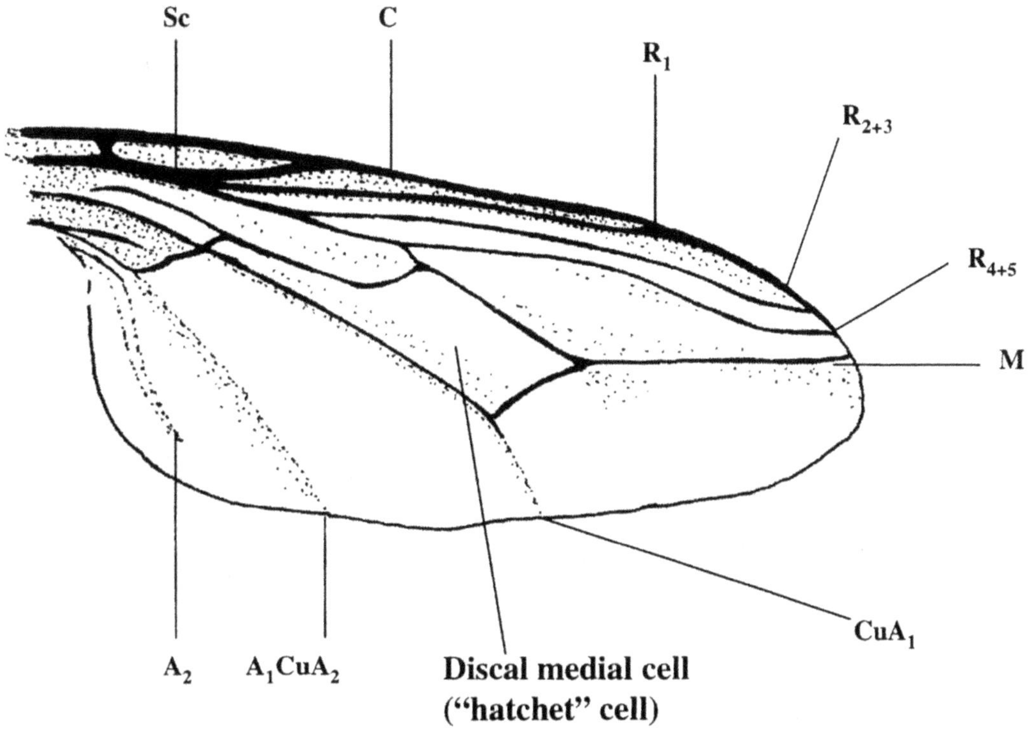

Figure 8.29. Tsetse wing showing cleaver-shaped cell (drawing by Phillip G. Lawyer).

symptoms usually are absent, but the infected host still can serve as an asymptomatic carrier who may travel and spread the disease. Eventually, the victim experiences increasing and persistent headaches, loss of sleep, disorientation and progressive loss of consciousness, abnormal behavior and finally, coma and death. Death may occur in as short a period as 9 months (Rhodesian form) or as long as 4 years (Gambian form) after the infective bite of the tsetse vector.

Vectors

As stated previously, muscoid flies known as tsetse transmit the typanosomes that cause African sleeping sickness. There are about 30 species and subspecies of tsetse, all in the genus *Glossina,* family Glossinidae (Table 8.3). The word "tsetse" is thought to have been derived from the African word "setswana," meaning destroyer of cattle. Many of the species and subspecies of tsetse do not transmit African sleeping sickness. *Trypanosoma brucei gambiense* is transmitted by tsetse in the *palpalis* group (e.g., *Glossina palpalis palpalis, Glossina tachinoides* and *Glossina fuscipes fuscipes*); *T. b. rhodesiense* is

Figure 8.30. Mature third-stage larva (white) and puparia (dark brown) of *Glossina morsitans* (photographs by Edward D. Rowton, all rights reserved).

transmitted mostly by flies in the *morsitans* group (*Glossina morsitans morsitans, Glossina morsitans centralis*).

Although some taxonomists place tsetse in the family Muscidae, most include them in a separate family (Glossinidae) because of their distinctive features. Adult flies are medium to large-sized (6–15 mm), brownish or grayish in color, with sword-shaped biting mouth parts directed forward well beyond the front of the head, and with palps nearly as long as the proboscis. These narrow-bodied, strong fliers have long wings that extend beyond the tip of the abdomen and fold scissors-like across the back when resting (Fig. 8.28). Wings have a characteristic venation with the discal medial cell shaped like a hatchet or cleaver (Fig. 8.29). Tsetse have distinctive antennae with bilaterally branched hairs arising from a single, elongated arista.

Distribution

The genus *Glossina* is restricted almost completely to continental Africa. Tsetse occur on some small offshore islands such as Zanzibar (Minter 1989), and there are records for 2 species in southwestern Saudi Arabia (Elsen et al. 1990). Fossil records from the Oligocene period in Colorado shale suggest there are extinct New World species. The genus is separable into 3 species groups, each associated with different habitats and hosts. Detailed maps of the distribution of tsetse in Africa were published by Ford and Katondo (1977).

Species of the *Glossina palpalis* group are found primarily in wet lowland forests, especially along the shores of lakes and rivers, closely associated with riverine vegetation. Their distribution extends into adjacent savanna beyond forest boundaries along rivers and streams that flow into the Atlantic Ocean and the Mediterranean Sea, but not those that flow into the Indian Ocean. Some species of this group have become adapted to peridomestic habitats in transitional savanna and forest zones in lieu of their native forest environments. Important vectors of human trypanosomes include *G. palpalis palpalis*, *Glossina palpalis gambiensis*, *G. fuscipes fuscipes* and *G. tachinoides*.

The *Glossina morsitans* group is the most widespread, ranging through a large sub-Saharan nonlinear belt across Africa from west to east. The cold winter temperatures of South Africa and the dry deserts of North Africa define the limits of this group. Flies of this complex are found in bushy thickets and are the usual savanna or game vectors of trypanosomes causing both Rhodesian sleeping sickness (*T. b. rhodesiense*) and nagana (*T. brucei brucei*). In savanna areas, these flies are scarce because they lack sufficient woody vegetation or wild animals. Minter (1989) listed the recognized species and subspecies important to transmission of *T. b. rhodesiense* as *Glossina morsitans submorsitans*, *G. morsitans centralis*, *G. morsitans morsitans* and *Glossina swynnertoni*. Of animal trypanosomes, he listed *Glossina longipalpis* and *Glossina pallidipes*, with only occasional human transmission.

The *Glossina fusca* group is found in gallery forests or lowland rainforests. This group contains no members that transmit parasites causing human trypanosomiasis, although some may be minor vectors of livestock parasites.

Life history

Female flies sequentially produce single larvae that are nourished within the female abdomen (Fig. 8.30). With the exception of the prepupal stage (the mature 3rd-stage larva that leaves the female fly just prior to pupation), there is no free-living larval stage. Each full-grown larva requires several bloodmeals 2–3 days apart as nourishment through its 8–10 day developmental period. The intrauterine developing larva feeds on fluids from special milk glands inside the female fly. The female deposits the mature 3rd-stage larva on loose soil in a shaded area and the larva burrows into the soil for protection from predators and desiccation. Within an hour, the integument of the 3rd-stage larva becomes a darkened, barrel-shaped **puparium** with 2 prominent lobes on the posterior end (Fig. 8.30). Within the puparium, the 4th-stage larva pupates. The pupal stage lasts 2–4 weeks or longer, depending on soil temperature. Each female has the potential, if sufficiently long lived, to produce 8–10 larvae.

Behavior

Both sexes of all species of tsetse are obligate blood feeders and can transmit parasites equally. Adults of the most important vectors of human trypanosomiases prefer moving targets as bloodmeal sources and are attracted to wild animals, domestic animals and humans. Male tsetse feed less frequently so there is less tsetse-human contact. Tsetse use both movement and olfactory cues to find suitable hosts. Carbon dioxide, octanol and certain urine and breath components attract female tsetse (Vale and Hall 1985).

Adults have a diurnal activity pattern and their habits conserve energy. They spend long periods resting on vegetation interspersed with short hunting periods. Although they will disperse long distances (up to 21 km), their normal flight activity is limited to the shortest distance to a suitable blood source and back to the vegetation favored by the adults for resting and deposition of prepupae (Fig. 8.31).

Adults of most species rest on the topside of leaves of trees at night and on the woody parts of vegetation during the day (Jordan 1993). The distance of resting sites from the ground appears to depend on microclimatic conditions.

Figure 8.31. Trapping tsetse with a biconical trap in a typical tsetse habitat in Africa (courtesy of Peter V. Perkins).

Larvipositing females search for low-level vegetation or fallen logs that provide shade and larviposit on loose soil. This positioning assures survival of the pupa.

The role of adult tsetse as vectors of trypanosomes to vertebrate hosts directly relates to their feeding habits. Tsetse are daytime feeders that seek moving targets and bite humans easily through thin clothing. With Rhodesian sleeping sickness, humans are most often only an opportunistic source of blood for the *Glossina* vectors. However, sleeping sickness caused by *T. brucei gambiense* usually is regarded as an interhuman disease because the vector flies feed predominantly on human hosts, once thought to be the only reservoir for the parasite. Humans are relatively unattractive to most tsetse species; most feed preferentially on animals such as swine or cattle. Some species appear to feed on humans only in the absence of other hosts. In such cases, the amount of human-tsetse contact may depend on the relative abundance of humans compared to other hosts. Tsetse do not have exceptionally long flight ranges and usually choose the closest bloodmeal source they can find. Consequently, one of the best predictors of disease risk is the occupation of human hosts.

Those tsetse that prefer riverine habitats may become aggressive biters during periods of drought when water holes dry up and few such sites are available. Game animals may become scarce due to migration in search of water, and humans will be bitten more often as they travel to these limited water sources.

Parasitic infections affect diurnal activity of *Glossina*. Large parasite populations within the tsetse result in rapid depletion of energy, stimu-

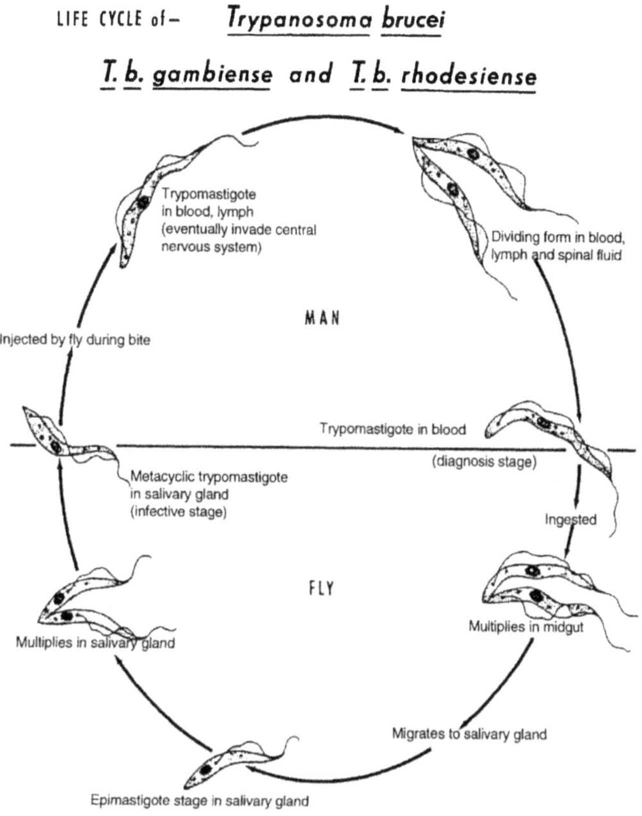

Figure 8.32. Life cycle of trypanosome parasites causing African sleeping sickness (courtesy of Centers for Disease Control and Prevention).

lating the fly to feed more frequently than non-infected tsetse (World Health Organization 1986).

Parasite life cycle

Trypanosomes causing Gambian sleeping sickness normally alternate between human and tsetse hosts, although they also may infect pigs. Multiplication and change in form occur in both (Fig. 8.32). Tsetse become infected when they feed on human blood containing shortened forms called **stumpy promastigotes**. These forms survive in the midgut by circumventing the peritrophic membrane and not succumbing to the low pH of the posterior midgut. Within about an hour these stumpy forms elongate to forms called **slender trypomastigotes** that migrate to the cardia and posterior region of the midgut. Here the parasites multiply by binary fission and migrate via the esophagus, pharynx and hypopharynx to the salivary glands. In the salivary glands, the parasites change again to **metacyclic trypomastigotes** (metacyclics for short). It is believed that some parasites pass as trypomastigotes through the gut wall and go through a sexual cycle resulting in trypomastigotes that ultimately find their way to the salivary glands.

The discovery of electrophoretically variant forms of *T. brucei brucei* suggested the existence of a sexual cycle, and it was confirmed by feeding a mixture of variants to *G. morsitans centralis* and allowing full cyclical development of the trypanosomes in the flies without any vertebrate involvement (Jenni et al. 1986). Genetic exchange may produce hybrid or heterozygous offspring and may be responsible for the great genetic variation found within species of trypanosomes. Random genetic exchange and the existence of diploid trypanosomes are unmistakable in biochemical data from the study of the parasites (Jenni et al. 1986). To occur in nature requires the individual tsetse vector to receive mixed infections of trypanosomes from a single host or feed on 2 separate hosts harboring differing strains of parasites. Since tsetse require 4 bloodmeals per larva and often will feed more frequently when infected with parasites, random genetic exchange could occur frequently in nature. Mixed infections of trypanosomes in a single host rarely have been found, suggesting that multiple hosts with differing strains is the more plausible route for hybrid development. This method would represent a unique, but undemonstrated, form of cyclopropagative transmission.

Metacyclics are the only parasite forms that infect vertebrates. Although trypanosomes may be spread by blood transfusion and congenitally, the usual method of transmission to humans is by tsetse. Tsetse transmission is cyclopropagative and salivarian (anterior station). Although stable flies and horse flies are involved in mechanical (and perhaps occasionally developmental) transmission of trypanosomes to other animals, there is little or no evidence of transmission by these vectors to humans.

After entering the vertebrate bloodstream, metacyclics change in form to slender trypomastigotes. These forms then multiply and change to stumpy trypomastigotes to complete the cycle.

Natural infections in wild-caught tsetse range from <1% to 80%. Infections in vectors are controlled by a number of factors, including the age of individual tsetses when they become infected. Young flies are more susceptible to *T. brucei brucei* infection than are older flies. Even when fed 24–48 hr after eclosion, the infection rate of some *Glossina* is only 10%. The different infection rates by a given trypanosome species of a given *Glossina* species relate to peritrophic membrane development. Younger flies have a less completely developed peritrophic membrane that reaches only a short way into the midgut. Dipeolu and Adam (1974) found that 50% of young flies developed infections immediately after taking a bloodmeal through an artificial membrane. However, infections of flies fed 3 days later had dropped to 9%. Bursell et al. (1973) reported that older flies had an elongated peritrophic membrane and that trypanosomes that had to travel far down the midgut to circumvent the membrane were exposed to lethal pH levels. These and other studies show that younger flies have high infection rates and are able to maintain those infections for several months.

Strains of *T. brucei* complex parasites characteristically change their antigenic nature, especially their surface coat. Various authors have described the waves of primary antigenic expression found in endemic areas of the tsetse-vectored trypanosomes. This change keeps the parasites ahead of the host's immune system that eventually provides protection against the most recent wave of antigen expression, while the current wave moves through the host population. Strategies for vaccination of people and domestic animals have been unsuccessful largely because of this antigenic change.

Epidemiologic considerations

Humans are important reservoirs for *T. b. gambiense* because during the long early stage of the disease infected people can continue their normal activities. Contact between humans and tsetse often is most intense at the end of prolonged hot dry spells, since both tsetse and humans aggregate in the vicinity of the remaining water sources. Droughts are not necessarily seasonal, but are related more often to changes in climatic conditions over longer periods of time, so the increase in fly activity and subsequent disease frequently is linked to migration of humans and their domestic animals rather than to cyclic seasonal variations in weather. An outbreak of Rhodesian sleeping sickness in the Lambwe valley of Kenya was reported following the heaviest April/May rainfall in 10 years (Wellde 1989). This deluge was followed by the driest June in 10 years. Under these conditions, large numbers of tsetse dispersed from the thickets in the national park into areas adjacent to the park where human and domestic animal populations had increased markedly during the previous decade.

Surveillance

Tsetse surveillance

Host-seeking behavior studies have led to the development of a number of surveillance methods that capture tsetse in proportion to their density. Recently developed insecticide-impregnated traps and cloth targets bear little resemblance to hosts of tsetse. Tsetse activity normally exhibits a V-shaped pattern with highest activity in the morning and afternoon of each day. This pattern is not, as previously thought, an avoidance of the hottest part of each day but rather an endogenous circadian rhythm much like that described in *Phlebotomus* by Mebrathu et al. (1996). Odors of ox urine, acetone and octenol have been used along with carbon dioxide to lure tsetse to land on target materials. The odor clues provide tsetse information about, and provide directional cues to, stationary hosts. These hints are not as important in open terrain as in highly vegetated areas. The synthetic ox odors made from the chemicals prevalent in ox urine still are less effective than actual urine (Vale and Hall 1985). The availability of urine for baiting traps in remote areas is greater than the availability of artificial lures. This supply provides less expensive tsetse surveillance and control. When attracting tsetse to baits, shape and color are important; dark blue is the most attractive for some species. The attraction of *G. palpalis palpalis* to UV reflective traps might be based on their natural attraction to the often white underbellies of their hosts, which are highly UV reflective. However, *G. pallidipes* and *G. morsitans morsitans* preferred landing on black, then blue and less frequently on white. Colvin and Gibson (1992) list shape, orientation, brightness, contrast, movement and color as important visual cues in attracting tsetse. They found the biconical (Fig. 8.31) trap to be effective, even though it bears no obvious resemblance to any animal. Movement can be important; mobile traps caught 16 times more *G. morsitans morsitans* than stationary baits in experiments where angular velocity of the moving object seemed to be the critical feature (Brady 1972).

Parasite surveillance

Parasite assays are the definitive diagnosis, but discovering parasites in the blood, cerebrospinal fluid or aspirates of lymph nodes is difficult. Parasite numbers often are low, even in very ill people, and complications of concurrent infections of other diseases such as malaria and leishmaniasis further cloud the diagnosis. Most blood tests require detailed microscopic examination of at least 100 fields. Examination of buffy coats and the centrifuged sediment of cerebrospinal fluid helps. When direct examination fails to reveal the organism, *in vitro* culture and inoculation into test animals may be necessary. Most serological tests are unreliable because cross-reactions provide false positives. An ELISA (Enzyme Linked Immunosorbent Assay) is useful if antigens are purified. However, due to the antigenic variation of the parasite within hosts, these tests may be so specific that some parasite variants go undetected. Therefore, the "gold standard" of diagnosis remains the microscopic detection of parasites from patient fluids.

Nonhuman trypanosomes in tsetse, and even tabanids, in endemic areas complicate detection of parasites in vectors. Positive identification of parasites requires culturing the isolated organism and comparative specific ELISA tests, or the DNA-based assays now available. These culturing procedures and tests require substantial resources, especially of laboratory supplies and trained technicians, that often are beyond the means of many countries where sleeping sickness is endemic. Here, limited medical resources are most often spent for other kinds of health support.

Prevention and control

Chemotherapy and chemoprophylaxis

The options for therapeutic and prophylactic drugs are limited. These options, and the possibility of the development of an antitrypanosomal vaccine, are discussed in Chapter 15. Selection and administration of therapeutic drugs depends on several factors, including the stage of infection.

Tsetse suppression

Because larval tsetse are protected for so much of their life cycle, most tsetse control schemes are directed against adults. In fact, only about 20% of tsetse populations are adults, and 5–6 pesticide applications may be required for lasting control. Jordan (1986) reviewed methods to control tsetse. The susceptibility of tsetse to pesticides has offered hope of eliminating them by the judicious use of ULV sprays of insecticides specifically timed to the life cycle of the flies. The expense and temporary nature of control has limited such techniques in most areas of Africa. Control of nonhuman tsetse hosts has been practiced in the past, including wholesale slaughter of wildlife. This approach is no longer regarded as an environmentally acceptable approach.

Control based on habitat modification, or destruction of natural vegetation and wild hosts, has been tried in some areas with success, but it drastically disrupts the environment. Such techniques are not popular in this age of environmental enlightenment. The search for more specific and less environmentally damaging methods is ongoing.

The concept of using imitation and actual hosts as bait to attract and kill tsetse is more than 100 years old; it successfully eradicated tsetse from the island of Principe (Colvin and Gibson 1992). Over the years, a number of traps have been developed and tested.

Stationary traps with a mixture of color patterns have yielded good long term "trap-out" results. Traps with 2 white and 2 blue panels have been highly effective. Dark colors and odor sources together attract tsetse, especially a mixture of ox urine and dark blue or black panels (Vale 1993). Such traps may cause tsetse populations to decline or even collapse. These types of traps may even be used as protective barriers against tsetse invasion of cleared zones. Traps and insecticide-impregnated targets employing artificial female-mimicking decoys coupled with combinations of other forms of control have had success in limited areas (Colvin and Gibson 1992). Community-based control programs in which local people do the trapping and see the results firsthand have proven to be the most successful with species responsive to traps. In the end, these low-cost programs provide effective surveillance linked with control that can eliminate or greatly reduce tsetse populations that have flourished in 40% of Africa for hundreds of years.

Because of the low reproductive capacity of tsetse, genetic manipulation and sterile male techniques have attracted interest. Although some of these methods have been effective in field trials on geographically isolated populations, they have yet to prove themselves at large.

American Trypanosomiasis of Humans

The disease

American trypanosomiasis, or Chagas disease (after its discoverer Carlos Chagas, who in 1909 showed that 2 species of triatomid bugs transmitted the parasite, *T. cruzi*), is a zoonotic disease of Central and South America, with a few cases each year reported from South Texas and southern California (Benenson 1990). It is one of the most important and widespread insect-vectored diseases in the New World. It is estimated that there are 16-18 million people infected with the disease, and 90 million people at risk of acquiring the disease in 21 countries in the Americas (World Health Organization 1990). Its acute condition usually is limited to young children and may be fatal. However, the chronic condition occurs approximately 20 years or more after acquisition and involves adults who suffer from smooth muscle degeneration of the heart, digestive and other systems, resulting in a variety of debilitating conditions and death.

Chagas disease tends to be a disease of the poor and rural areas of Central and South America, where it often goes undetected and builds in human reservoir populations. Construction of dwellings with thatched roofs, adobe or mud over wood walls and dirt floors enhances vector populations.

The parasite causing Chagas disease, *T. cruzi*, resembles the one causing African sleeping sickness, but with different insect vectors, animal reservoirs and mode of transmission.

Chagas disease resembles African sleeping sickness in having several stages. In the case of Chagas disease, the first stage is rare; it is marked by a local lesion called a **primary chagoma** at the site of parasite inoculation. When inoculation occurs at the mucous membranes of the eye, a swelling known as **Romaña's sign** results from the partial closing of the eye due to the enlargement of surrounding tissues (Fig. 8.33). There usually is spontaneous healing of the entry site of the parasite.

The 2nd stage of infection (also called the acute stage) is manifested by daily fever, swelling of lymph nodes, liver and spleen; rash and acute heart conditions, most often in young children; occasionally it can be fatal. These symptoms generally subside in several months and are replaced by the chronic phase. The acute stage of Chagas disease usually is either mild or asymptomatic, producing chronic cases with no recollection by the patient of an acute phase infection.

The 3rd stage of Chagas disease is the chronic stage. Its most serious and frequent clinical signs occur generally 5–15 years after infection. Enlargement of the heart caused by destruction of heart muscle tissue by invading parasites and enlargement of the esophagus and colon are the most common manifestations of this lingering disease. Debilitation and death occur most often as a result of complications involving the affected heart or digestive tract.

Vectors

Insect vectors of Chagas disease parasites are true bugs in the family Reduviidae, subfamily Triatominae (Table 8.4). For a thorough taxonomic treatment of this group, the reader should consult Lent and Wygodzynski (1979). A more recent summary of triatomid systematics and ecology is that of Carcavallo (1987).

All stages of both sexes are obligate bloodfeeders. They feed primarily at night and hide during daylight hours. All species of Triatominae should be regarded as potential vectors of American trypanosomiasis. However, the most important species are those adapted to the ecology of human habitations. The 5 most important vector species are *Triatoma brasiliensis*, *Triatoma dimidiata*, *Triatoma infestans*, *Rhodnius prolixus* and *Panstrongylus megistus*, but some 115 species have been found naturally infected or have been experimentally infected with *T. cruzi* (Schofield and Dolling 1993). The bite of these large bugs is painless and normally occurs when the host is asleep. Triatomids acquire *T. cruzi* parasites when they feed on infected hosts, but transmit the parasites to susceptible new hosts through the infected feces they deposit on the host. The parasites enter the host through the skin with the assistance of the host (scratching).

Triatomid bugs have a distinctive shape (Fig. 8.34) that allows them to hide in crevices. Their color is most often brownish or blackish. The prominent head is elongate with large eyes and a neck-like narrowing between the head and thorax. As with all true bugs, there are 2 sets of wings. The forewings (hemelytra) are thickened basally and are thin apically. The hind wings are entirely membranous. Wings seldom are used for locomotion except for short nocturnal flights.

Distribution

Triatomid bugs are distributed throughout the New World. Some potential vectors of *T. cruzi* also occur in Asia, Africa and Australia, but the parasite still is unreported from the Old World. In the Western Hemisphere, triatomid bugs are found from the southwestern United States, through Mexico and Central America, and into South America as far as northern Argentina. The presence of triatomid bugs in local habitats depends upon availability of warm-blooded hosts. The distribution of trypanosomiasis is not concordant with the distribution of triatomid vectors in the New World. *Trypanosoma cruzi* is unknown in the heavily forested areas of Central and South America, although there are one or more vectors present in these areas.

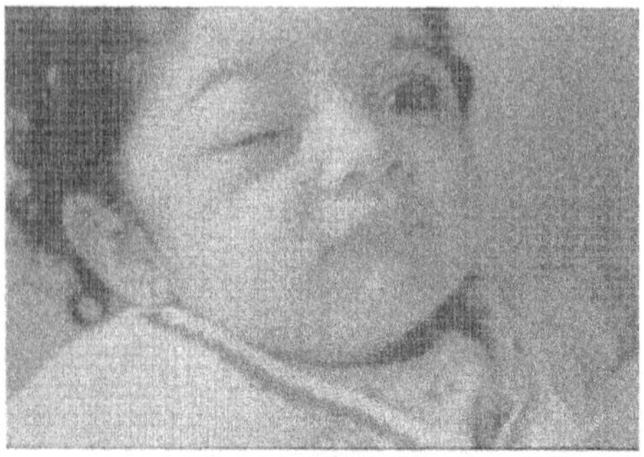

Figure 8.33. Child with Romaña's sign resulting from the bite of a triatomid (from collection of Walter Reed Army Institute of Research).

Life history

Triatomid females deposit small operculate eggs singly in secluded areas of their habitat. After 8–10 days, depending upon temperature, a tiny nymph emerges. Nymphs of both sexes are bloodfeeders and feed on small mammals in their nests. They are able to fit into tiny cracks when unfed, but swell greatly after a bloodmeal. Triatomids have 5 nymphal stages; a full bloodmeal is required before molting to the next stage. Even when provided frequent full bloodmeals in the laboratory, a complete life cycle from egg to adult may require as long as a year, and in nature as long as 2 years. Longevity of adults in nature is not known; in the laboratory they may live up to a year. The long maturation period plus the longevity of adults means that a single infected bug may serve as a vector for up to 3 years.

Triatomid bugs occur in a variety of habitats that provide shelter from harmful environmental conditions, access to vertebrate bloodmeals and protection from natural enemies. Habitats that provide these conditions include rodent nests and houses that furnish easy entry for bugs and places to hide. Houses with thatched roofs are particularly suitable.

Triatomids frequently hitchhike on people or their belongings.

Table 8.4. Selected species of triatomid vectors of Chagas disease.

Vector species	Presence in human habitats	Geographic distribution
Triatoma brasiliensis	Highly adapted	Brazil
Triatoma dimidiata	Highly adapted	Ecuador, Central America, Mexico
Triatoma infestans	Highly adapted	Argentina, Uruguay,Brazil, Bolivia, Paraguay, Chile, Peru
Panstrongylus megistus	Highly adapted	Argentina, Brazil
Rhodnius prolixus	Highly adapted	Venezuela, Colombia, Central America
Rhodnius pallescens	With sylvatic ecotypes, but commonly found	Panama, Colombia
Triatoma barberi	With sylvatic ecotypes, but commonly found	Mexico
Triatoma guasayana	With sylvatic ecotypes, but commonly found	Argentina
Triatoma herreri	With sylvatic ecotypes, but commonly found	Peru
Triatoma maculata	With sylvatic ecotypes, but commonly found	Venezuela
Triatoma pseudomaculata	With sylvatic ecotypes, but commonly found	Brazil
Triatoma sordida	With sylvatic ecotypes, but commonly found	South America
Triatoma spinolai	With sylvatic ecotypes, but commonly found	Chile

Behavior

Triatomids generally come out only at night to feed and defecate on a sleeping host. Rodents in houses can serve as reservoirs for parasites. Rodents may acquire trypanosome infections by exposure to feces in association with triatomid bites, or by eating infected triatomids.

Another factor in the transmission of trypanosomes by triatomid vectors is the peculiar habit these bugs have of taking vertebrate bloodmeals from one another (a form of hyperparasitism). This practice, coupled with the longevity of these insects and their adaptability to human dwellings, provides for sustained endemic foci of the disease. A behavior critical to transmission is the tendency of bugs to defecate on the vertebrate host during and just after feeding. The stealthy nighttime visits of these bloodfeeding bugs to humans often may go completely unnoticed as compared to the noisy

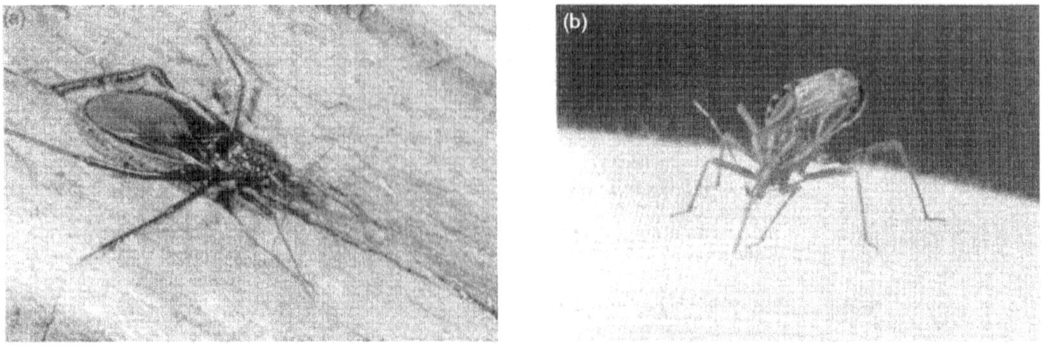

Figure 8.34. Triatomid vectors of Chagas disease: (a) *Panstrongylus* sp.; (b) *Triatoma* sp. (photographs by Edgar D. Rowton, all rights reserved).

flights of various species of mosquitoes and other flying insects.

Three factors determine the vectorial capacity of a given species of triatomid bug for *T. cruzi*. First, the degree of susceptibility of the triatomid species to the trypanosome parasite (vector competence); some triatomids show low infection rates (Sousa 1988). Second is the time interval between feeding and defecation, because the bug that defecates while feeding, or soon after, is more apt to leave trypanosome parasites on the skin of the host. Third, the degree of contact with human hosts is important, and species that cohabit with humans most efficiently obtain human bloodmeals and transmit parasites (Minter 1989).

Parasite life cycle

As with the parasites causing African sleeping sickness, the parasites causing Chagas disease alternate between human and arthropod hosts (Fig. 8.35). Triatomid bugs become infected by picking up trypomastigotes during ingestion of vertebrate blood. Parasites enter the midgut and hindgut of the bugs, where they multiply and change form to become **amastigotes** and **epimastigotes**. After 1–2 weeks, depending upon ambient temperature, metacyclic trypomastigotes appear in the hindgut. They also appear in the rectum of the triatomid, where they may be attached to the inner cuticular layer of that organ. Only metacyclics infect vertebrates.

Transmission to humans occurs primarily by stercorarian (posterior station) cyclopropagative transmission. *Trypanosoma cruzi* infections occur by the entry of compacted blood or liquid bug feces containing metacyclics into feeding lesions caused by the bite of the bug. The parasites also may enter the host body through mucous membranes and conjunctivae or through skin abrasions caused by scratching in response to the bite (Minter 1989).

After entering the bloodstream of the vertebrate host, parasites enter different cell types. Trypomastigotes then differentiate into nonflagellated amastigotes that undergo binary fission in a pseudocyst within the infected cell. The pseudocyst eventually ruptures, releasing parasites into interstitial spaces, where they pass through 2 additional stages (promastigote and epimastigote). They then change once more before moving into the bloodstream as trypomastiogotes that can infect other cells or be ingested by feeding triatomid vectors.

Trypanosoma cruzi is an asexual parasite, at least in the majority of its range in Central and

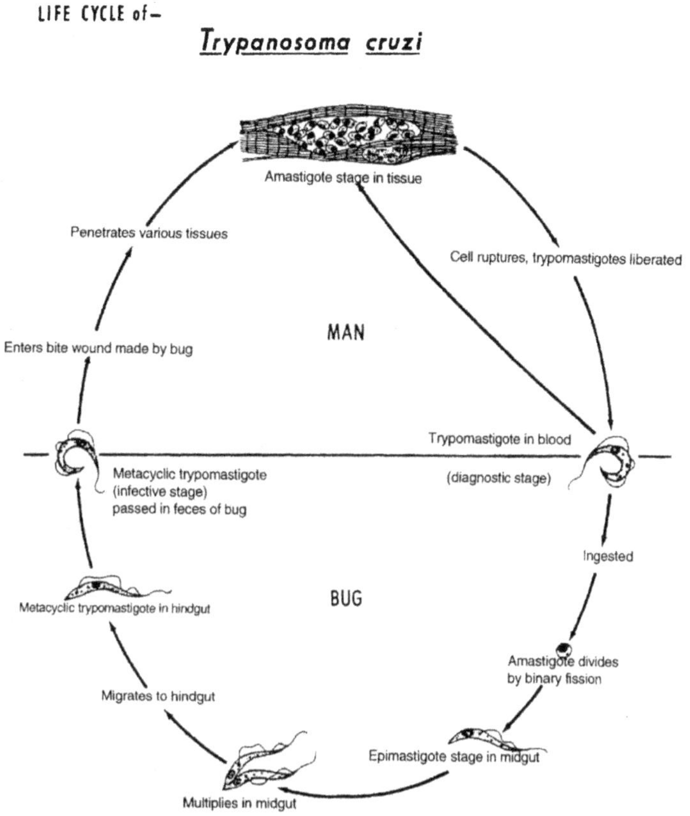

Figure 8.35. Life cycle of *Trypanosoma cruzi*, causitive agent of Chagas disease (courtesy of Centers for Disease Control and Prevention).

South America. However, recent results using isozyme and random amplified polymorphic DNA (RAPD) analysis of strains from multiple locations, including sylvatic mammals, suggest that genetic exchange occurs during sylvatic transmission (Carrasco et al. 1996). This genetic cycle contributes to the generation of phenotypic and genotypic diversity of *T. cruzi*.

An important difference between the salivarian trypanosomes and the stercorarian trypanosomes is their respective vector infection rates. In contrast to stercorarian vectors, where 100% infection rates following feeding on infectious hosts often occurs, salivarian vector infection rates among tsetse often are low, even when done in the laboratory under optimum conditions (Kettle 1995). When this rate is coupled with the relative longer life of stercorarian triatomid vectors, it is easy to understand the greater tenacity of Chagas disease in endemic areas. Many species of triatomid bugs do not feed on humans, but still maintain the disease in reservoir animals.

Epidemiologic considerations

American trypanosomiasis is a zoonotic disease. More than 100 vertebrates have been found naturally infected with *T. cruzi*. More than half of the 115 known triatomid species can transmit the parasite, but only a few species are involved in transmission to humans. The most important vector species (discussed earlier) are those found in close association with human dwellings. These species possess the attributes necessary to be efficient vectors: they readily feed on human hosts, occupy human dwellings, defecate during and after taking bloodmeals and exist close to possible nonhuman reservoir hosts. Although a number of species of triatomids invade human dwellings, only the few named previously have established colonies in such dwellings and traveled from dwelling to dwelling in furniture and other household articles moved by people.

Surveillance

Specific surveillance techniques vary from species to species depending upon behavior of the bugs in and around human dwellings. Most surveillance techniques involve hand-capture of bugs in dwellings with the aid of a flashlight, sometimes supplemented by application of flushing agents and knockdown insecticides (Gurtler et al. 1993).

Prevention and control

Chemotherapy and chemoprophylaxis

Chapter 15 reviews chemotherapeutic agents for Chagas disease. Several agents effectively treat acute stage infections, but not chronic infections.

Triatomid suppression

Residual pesticide sprays inside and outside houses have reduced the incidence of Chagas disease in the past, but triatomid bugs have developed resistance to many conventional insecticides, including DDT, lindane and dieldrin. Organophosphates and synthetic pyrethroids now replace them. Space sprays may kill bugs in houses, but this approach is not sustainable.

Recently, Mlot (1997) reported efforts to genetically alter the bacterium *Rhodococcus rhodnii*, which lives symbiotically in the gut of the reduviid bug *R. prolixis*. The altered bacterium expresses a peptide that kills the insect stages of *T. cruzi* and the release of bugs infected with the altered bacterium into the wild may spread this trypanosome-killing bacteria. The bacteria normally are passed from mature bugs to young triatomids when the young probe the fecal droplets of the adults as a food source. Other triatomids have a different gut flora, so this approach will not work for species other than *R. prolixus*.

Some government sponsored and locally enacted conventional triatomid control projects have made excellent progress. These programs provide pesticides and application equipment and educate people on the relationship between triatomid bugs and Chagas disease.

Other approaches

Sustainable control of Chagas disease is based on preventing contact between infected vectors and humans. The incidence of the disease is highly correlated with construction standards of dwellings. Homes with mud walls over stick-wood supports and with thatched roofs provide ideal bug habitats. Eliminating such housing is a socioeconomic problem of vast proportions in countries endemic for Chagas disease. Corrugated steel roofing makes indoor temperatures uninhabitable for bugs during midday, and replacement of thatched roofs eliminates harborage for peridomestic rodents, such as the roof rat, that can act as disease reservoirs. Unfortunately, steel roofing also makes indoor temperatures hotter for human inhabitants and increases noise levels during rain storms. Maintaining domestic animals away from places where humans sleep may reduce contact with bugs, but may require more land area than is available in many places.

In some countries, plastering of walls to seal all cracks and crevices that provide harborage to triatomids has reduced transmission of Chagas disease.

Screening of blood used in transfusions and collecting vectors from houses by school children taught about the vector help interrupt transmission of Chagas disease.

Trypanosomal Diseases of Livestock

Diseases

Several trypanosome species and subspecies affect domestic animals and limit cattle and swine production in various areas of the world. In Africa, livestock infections by *Trypanosoma* affect an estimated 10 million km^2. Within the range of *Trypanosoma* transmission, loss of meat production severely depresses human dietary protein. The progressive emaciation of beasts of burden such as horses, donkeys and camels has further limited agriculture by requiring that humans be the major beasts of burden in vast areas where mechanized farming is still only a distant dream.

Certain trypanosome strains causing diseases in animals are highly virulent, while others cause chronic infections that allow survival while weakening the host and making it susceptible to other diseases. Chronically infected livestock maintain reservoir populations where wild animals are scarce.

In tropical Africa, these diseases in livestock are called **nagana** (a Zulu word for the infection in cattle and horses) and **dourine**. The diseases, caused by a complex of parasite species that includes *T. brucei brucei, Trypanosoma vivax vivax, Trypanosoma suis,* and *Trypanosoma congolense,* are transmitted primarily by tsetse (Table 8.5). Trypanosomes also infect domestic animals in areas such as northern Africa where tsetse are absent. These diseases have common names like **Surra**, Gufar and El-debab and are caused by the parasites *Trypanosoma evansi* and *T. vivax vivax*. In tsetse-free areas, parasites are transmitted mechanically by horse flies (*Tabanus*) and stable flies (*Stomoxys*).

Surra, an important, frequently fatal disease of animals such as horses, camels and dogs, occurs in cattle and buffalo as a benign infection. Although widespread in Africa, where it is transmitted by tsetse, it also is present north of the tsetse zone and in Yemen, southern Asia, including the Philippines and Indonesia, and in Central and South America. In these areas, tabanid flies are mechanical vectors, and dogs, foxes and other carnivores may acquire the parasite by eating the flesh of infected animals. Flies in other genera of Tabanidae (e.g., *Chrysops*) are of little or no importance in transmission. Stable flies have been incriminated as secondary vectors. Vectoring by vampire bats has been demonstrated, but in these instances, the main vector in the area still has proven to be a tabanid.

Both tabanids and stable flies spread trypanosomes within herds of domestic animals kept in close proximity (such as feed lots), promoting unexpected epidemics that cause great loss to producers in areas where the disease has been unknown previously. This possibility of infection has limited the movement of uninspected cattle from one country to another as well as within certain areas of countries in Africa.

Several other trypanosomes have been implicated in animal diseases. *Trypanosoma theileri* of domestic cattle, oxen and wild antelope occurs throughout the world. *Trypanosoma simiae* causes a fatal disease of domestic swine, with warthog reservoirs. Present in tropical Africa, it is transmitted by tsetse. *Trypanosoma vivax viennei* causes disease in cattle in Mauritania, West Indies, and Central and South America and is transmitted by tabanids.

Vectors

In Africa within the range of tsetse, approximately 20 species of tsetse transmit trypano-

Table 8.5. Some trypanosomal diseases of livestock.

Tsetse species	Vertebrate hosts	Parasite species
G. swynnertoni	Suids	*T. simiae,* *T. suis*
G. fuscipleuris	Suids	*T. simiae,* *T. suis*
G. pallidipes	Bovids	*T. uniformis*
G. longipalpis	Bovids	*T. uniformis*
G. fusca fusca	Bovids	*T. uniformis*
G. morsitans morsitans	Bovids, suids, occasionally humans	*T. congolense,* *T. vivax*
G. morsitans centralis	Bovids, suids, occasionally humans	*T. congolense,* *T. vivax*
G. morsitans submorsitans	Bovids, suids, occasionally humans	*T. congolense,* *T. vivax*
G. brevipalpis	Other than suids and bovids	*T. brucei brucei*
G. longipennis	Other than suids and bovids	*T. brucei brucei*
G. palpalis gambesiensis	Opportunistic; nearly any mammal	*T. brucei gambesiense,* *T. b. rhodesiense*
G. palpalis palpalis	Opportunistic; nearly any mammal	*T. brucei gambesiense,* *T. b. rhodesiense*
G. tachinoides	Opportunistic; nearly any mammal	*T. brucei gambesiense,* *T. b. rhodesiense*
G. fuscipes fuscipes	Opportunistic; nearly any mammal	*T. brucei gambesiense,* *T. b. rhodesiense*

some parasites to wildlife and domestic livestock. Of the 6 most important species, 3 are riverine and 3 are savanna dwellers. More than 100 African mammals, mainly wildlife species, serve as reservoirs. In some tsetse areas, horse flies and stable flies are developmental or mechanical vectors but their role as vectors is minor compared to cyclo-propagative transmission by tsetse.

The ability of tabanids to transmit trypanosomes of animal diseases such as nagana and surra was discovered by noting that cattle contracted *T. theileri,* horses *T. evansi,* and swine *T. vivax viennei* in areas devoid of tsetse. Researchers found the flies transmitted trypanosomes mechanically by interrupted feeding. There is some evidence that transmission of *T. vivax viennei, T. theileri* and *T. simiae* is cyclo-propagative by tabanids. This observation is based on the low numbers found circulating in animal blood as compared to the large numbers in vector midguts (Hoare 1972).

Ticks also may transmit *T. theileri* in certain parts of East Africa, notably Ethiopia. Burgdorfer et al. (1973) noted that species of both *Rhipicephalus pulchellus* (often found on warthogs, cattle and humans) and *Boophilus decoloratus* (of cattle) contained developmental stages of these parasites.

Vector-pathogen interactions

Because *T. evansi* does not multiply within insect vectors, it may be transmitted only mechanically (see Chapter 12). The parasite may change forms within the insect vector mouthparts; in these cases may it be considered a type of developmental transmission.

Some researchers believe that trypanosomes such as *T. evansi*, *T. vivax viennei* and *T. theileri* are able to multiply by binary fission on the mouthparts of horse flies and possibly even stable flies. However, most believe that simple mechanical (non-propagative) transmission is the only mechanism involved. The question is often asked: if no multiplication occurs, why do flies caught in nature have large numbers of parasites on their mouthparts when there are few parasites in host blood? Most species tested in the laboratory pick up trypanosomes during bloodfeeding and transmit them to other animals mechanically. Some authors believe that cyclic changes take place within the mouthparts (Greenberg 1973). Another possibility: parasites aggregate at the feeding site, as is the case with insect-transmitted filariae.

Prevention and control

Controlling trypanosomal diseases of livestock in tropical Africa is based on the same principles of tsetse elimination practiced to prevent human disease. Chemotherapy and chemoprophylaxis have some value in areas of the periphery of the range of tsetse (Murray and Urquhart 1977). As discussed above, practical vaccines have not yet been developed because of the problem of antigenic variation in parasites.

There are fewer options for trypanosomes mechanically transmitted by tabanids and stable flies. Stable flies in developing countries are controlled primarily by sanitation, specifically control of livestock manure and other organic media that may serve for larval development. There has been limited success in controlling tabanids through the use of insecticide-baited traps. Without highly focused and well funded programs, neither of these insects is likely to be controlled in areas where trypanosome transmission is present.

REFERENCES

Abonnenc, E. 1972. Le phlebotomes de la Region Ethiopienne (Diptera: Psychodidae). Mem. ORSTOM **55**:1–289.

Adler, S. 1947. The behavior of a Sudan strain of *Leishmania donovani* in *Phlebotomus papatasii*. A comparison of strains of *Leishmania*. Trans. Royal Soc. Trop. Med. Hyg. **40**:701–712.

Adler, S. and Theodor, O. 1939. The behavior of *Leishmania chagasi* in *Phlebotomus papatasii*. Ann. Trop. Med. Parasitol. **33**:45–47.

Alexander, J.B. 1987. Dispersal of phlebotomine sand flies (Diptera: Psychodidae) in a Colombian coffee plantation. J. Med. Entomol. **24**:552–558.

Anderson, D.C., Buckner, R.G., Glenn, B.L. and MacVean, D.W. 1980. Endemic canine leishmaniasis. Vet. Pathol. **17**:94–96.

Barral, A., Pedral-Sampaio, D., Grimaldi, G., Jr., Momen, H., McMahon-Pratt, D., DeJesus, A.R., Almeida, R., Badaro, R., Barral-Netto, M., Carvalho, E.M. and Johnson, W.D., Jr. 1991. Leishmaniasis in Bahia, Brazil: evidence that *Leishmania amazonensis* produces a wide spectrum of clinical disease. Am. J. Trop. Med. Hyg. **44**:536–546.

Beach, R., Kiilu, G. and Leeuwenburg, J. 1985. Modification of sand fly biting behavior by *Leishmania* leads to increased parasitic transmission. Am. J. Trop. Med. Hyg. **34**:278–282.

Benenson, A.S. 1990. Control of communicable diseases in man, 15th ed, Washington, DC 532 pp.

Bista, M.B., Shrestha, K. and Devotka, U.N. 1993. Gasteroenteritis, encephalitis, meningitis and kala-azar: an epidemiological review. Epidemiology Division, Ministry of Health, Kathmandu, Nepal. 70 pp.

Borovsky, D. and Schlein, Y. 1987. Trypsin and chymotrypsin-like enzymes of the sandfly *Phlebotomus papatasi* infected with *Leishmania* and their possible role in vector competence. Med. Vet. Entomol. **1:**235–242.

Brady, J. 1972. The visual responsiveness of the tsetse fly *Glossina morsitans* West. (Glossinidae) to moving objects: the effects of hunger, sex, host odour and stimulus characteristics. Bull. Entomol. Res. **62:**257–279.

Brun, R., Berens, R.L. and Krassner, S.M. 1976. Inhibition of *Leishmania donovani* transformation by hamster spleen homogenates and active human lymphocytes. Nature **262:**689–691.

Buescher, M.D., Rutledge, L.C., Roberts, J. and Nelson, J.H. 1984. Observations of multiple feedings of *Lutzomyia longipalpis* in the laboratory (Diptera: Psychodidae). Mosq. News **44:**76–77.

Burgdorfer, W., Schmidt, M.L. and Hoogstraal, H. 1973. Detection of *Trypanosoma theileri* in Ethopian cattle ticks. Acta Trop. **30:**340–346.

Bursell, E., Berridge, M.J. and Freeman, J.C. 1973. The penetration of the peritrophic membrane of the tsetse flies by trypanosomes. Acta Trop. **30:**347–355.

Carcavallo, R.U. 1987. The subfamily Triatominae (Hemiptera: Reduviidae): systematics and some ecological factors. Pp. 1–39 *in* R.R. Brenner (ed.), Chagas' disease vectors, Vol. 1. CRC Press, Boca Raton, Florida.

Carrasco, H.J., Frame, I.A., Valente, S.A. and Miles, M.A. 1996. Genetic exchange as a possible source of genomic diversity in sylvatic populations of *Trypanosoma cruzi*. Am. J. Trop. Med. Hyg. **54:**418–424.

Chaniotis, B.H. 1986. Successful colonization of the sandfly *Lutzomyia trapidoi* (Diptera: Psychodidae) with enhancement of its gonotrophic activity. J. Med. Entomol. **23:**163–166.

Chaniotis, B.H., Correa, M.A., Tesh, R.B. and Johnson, K.J. 1974. Horizontal and vertical movements of phlebotomine sandflies in a Panamanian rain forest. J. Med. Entomol. **11:**369–375.

Chippaux, J.P., Pajot, F.X. and Barbier, D. 1984. Leishmaniasis in French Guinea. 5. Fuller data on the ecology of the vector in the cacao forest village. Cah. ORSTOM Ser. Entomol. Med. Parasitol. **22:**313–318.

Christensen, H.A., Fairchild, G.B., Herrer, A., Johnson, C.M., Young, D.G. and de Vasques, A.M. 1983. The ecology of cutaneous leishmaniasis in the Republic of Panama. J. Med. Entomol. **20:**463–484.

Colvin, J. and Gibson, G. 1992. Host-seeking behavior and management of tsetse. Annu. Rev. Entomol. **37:**21–40.

Desjeux, P. 1996. Leishmaniasis, public health aspects and control. Clinic. Dermatol. **14:**417–423.

Desjeux, P. 2001. The increase in risk factors worldwide. Trans. R. Soc. Trop. Med. Hyg. **95:**239-243.

Dipeolu, O.O. and Adam, K.M.G. 1974. On the use of membrane feeding to study the development of *Trypanosoma brucei* in *Glossina*. Acta Trop. **31:**185–201.

Disney, R.H.L. 1966. A trap for phlebotomine sandflies attracted to rats. Bull. Entomol. Res. **56:**445–451.

Elnaiem, D.A. and Ward, R.D. 1990. An oviposition pheromone on the eggs of sandflies (Diptera: Psychodidae). Trans. Royal Soc. Trop. Med. Hyg. **84:**456–457.

Elnaiem, D.A. and Ward, R.D. 1991. Response of sandfly *Lu. longipalpis* to an oviposition pheromone associated with conspecific eggs. Med. Vet. Entomol. **5:**87–91.

Elsen, P., Amoudi, M.A. and Leclerqc, M. 1990. First record of *Glossina fuscipes fuscipes* Newstead, 1910 and *Glossina morsitans submorsitans* Newstead 1910 in southwestern Saudi Arabia. Ann. Soc. Belg. Med. Trop. **70:**281–287.

Endris, R.G., Young, D.G. and Perkins, P.V. 1987. Ultrastructural comparison of egg surface morphology of five *Lutzomyia* species (Diptera: Psychodidae). J. Med. Entomol. **24:**412–414.

Fairchild, G.B. and Hertig, M. 1947. Notes on the *Phlebotomus* of Panama (Diptera: Psychodidae). II. Descriptions of three species. Ann. Entomol. Soc. Am. **40:**617–622.

Forattini, O.P. 1973. Entomologia medica. Vol. 4, Psychodidae, Phlebotominae, Leishmanoses, Bartonelose. Edgard Blucher, Sao Paulo. 658 pp.

Ford, J. and Katondo, K.M. 1977. Maps of tsetse fly (*Glossina*) distribution in Africa, 1973, according to subgeneric groups on scale 1:5,000,000. Bull. Anim. Hlth. Prod. Africa **15:**187–193.

Foster, W.A., Tesfa-Yohannes, T.M. and Teckle, T. 1970. Studies on leishmaniasis in Ethiopia. II. Laboratory culture and biology of *Phlebotomus longipes* (Diptera: Psychodidae). Ann. Trop. Med. Parasitol. **64:**403–409.

Greenberg, B. 1973. Flies and disease. Vol. 2, Biology and disease transmission. Princeton University Press, Princeton, New Jersey. 447 pp.

Gurtler, R.E., Schweigmann, N.J., Cecere, M.C., Chuit, R. and Wisnevesky-Colli, C. 1993. Comparison of two sampling methods for domestic populations of *Triatoma infestans* in north-west Argentina. Med. Vet. Entomol. **7:**238–242.

Hanson, W.J. 1961. The breeding places of *Phlebotomus* in Panama (Diptera: Psychodidae). Ann. Entomol. Soc. Am. **54:**317–322.

Hoare, C.A. 1972. The trypanosomes of mammals: a zoological monograph. Blackwell Scientific Publications, Oxford, UK. 749 pp.

Jenni, L., Marti, S., Schweizer, J., Betschart, B., LePage, R.W.F., Wells, J.M., Tait, A., Paindavoine, P., Pays, E. and Steinert, M. 1986. Hybrid formation between African trypanosomes during cyclical transmission. Nature **322**:173–175.

Jobling, B. and Lewis, D.J. 1987. Anatomical drawings of biting flies. British Museum of Natural History and Wellcome Foundation, London, UK. 119 pp.

Johnson, P.T. 1961. Autogeny in Panamanian *Phlebotomus* sandflies (Diptera: Psychodidae). Ann. Entomol. Soc. Am. **54**:116–118.

Johnson, P.T., McConnell, E. and Hertig, M. 1963. Natural infections of leptomonad flagellates in Panamanian *Phlebotomus* sandflies. Exp. Parasitol. **14**:107–122.

Jordan, A.M. 1986. Trypanosomiasis control and African rural development. Longman, London and New York. 357 pp.

Jordan, A.M. 1993. Tsetse-flies (Glossinidae). Pp. 333–388 *in* R.P. Lane and R.W. Crosskey (eds.), Medical insects and arachnids. Chapman and Hall, London.

Joshi, D.D., Shrestha, J.D., Pradhan, S.P. and Joshi, A.B. 1990. Kala-azar in Morang District: epidemiological situation. J. Instit. Med. (India) **12**:205–209.

Kettle, D.S. 1994. Medical and veterinary entomology, 2nd ed. Cambridge University Press, Cambridge, UK. 725 pp.

Kettle, D.S. 1995. Trypanosomiasis and leishmaniasis in medical and veterinary entomology, 2nd ed. CAB International, Wellingford, UK. 725 pp.

Killick-Kendrick, R. 1979. Biology of *Leishmania* in phlebotomine sandflies. Pp. 395–460 *in* W.H.R. Lumsden and D.A. Evans (eds.), Biology of the Kinetoplastida, Vol. 2. Academic Press, London, UK.

Killick-Kendrick, R. 1985. Some epidemiological consequences of the evolutionary fit between leishmaniae and their phlebotomine vectors. Bull. Soc. Pathol. Exot. **78**:747–755.

Killick-Kendrick, R. 1987. Appendix II: methods for the study of phlebotomine sandflies. Pp. 473–497 *in* W. Peters and R. Killick-Kendrick (eds.), The leishmaniases in biology and medicine, Vol. 1, Biology and epidemiology. Academic Press, New York.

Killick-Kendrick, R. 1990a. Phlebotomine vectors of leishmaniasis: a review. Med. Vet. Entomol. **4**:1–24.

Killick-Kendrick, R. 1990b. The life-cycle of *Leishmania* in the sandfly with special reference to the form infective to the vertebrate host. Ann. Parasitiol. Hum. Comp. **65**:37–42.

Killick-Kendrick, R. 1999. The biology and control of phlebotomine sand flies. Clinics in Dermatology. **17**: 279-289.

Killick-Kendrick, R. and Killick-Kendrick, M. 1987a. Honeydew of aphids as a source of sugar for *Phlebotomus ariasi*. Med. Vet. Entomol. **1**:297–302.

Killick-Kendrick, R. and Killick-Kendrick, M. 1987b. The laboratory colonization of *Phlebotomus ariasi* (Diptera: Psychodidae). Ann. Parasitiol. Hum. Comp. **62**:354–356.

Killick-Kendrick, R., Killick-Kendrick, M., Focheux, C., Dereure, J., Puech, M-P., and Cadiergues, M.C. 1997. Protection of dogs from bites of phlebotomine sand flies by deltamethrin collars for control of canine leishmaniasis. Med. Vet. Entomol. **11**: 105-111.

Killick-Kendrick, R., Molyneux, D.H. and Ashford, R.W. 1974. *Leishmania* in phlebotomid sandflies. I. Modifications of the flagellum associated with attachment to the mid-gut and oesophageal valve of the sandfly. Proc. Roy. Soc. Entomol. (B) **187**:409–419.

Killick-Kendrick, R., Molyneux, D.H., Leaney, A.J. and Rioux, J.A. 1979. Aspects of the life-cycle of *Leishmania* in the sandfly. Pp. 89–95 *in* Association of Yugoslav Parasitologists (ed.), Proceedings of the Second European Multicolloquy of Parasitology. Prosveta, Belgrade, Yugoslavia.

Killick-Kendrick, R., Ready, P.D. and Pampiglione, S. 1977. Notes on the prevalence and host preferences of *Phlebotomus perfiliewi*. Ecologie des Leishmanioses, Colloques Internationaux du C.N.R.S. **239**:169–175.

Killick-Kendrick, R. and Ward, R.D. 1981. Ecology of the *Leishmania*. Parasitol. **82**:143–152.

Lainson, R. 1982. Leishmaniasis. Pp. 41–103 *in* J.H. Steele (ed.), CRC Handbook Series in Zoonoses, Vol. 1. CRC Press, Boca Raton, Florida.

Lainson, R. 1983. The American leishmaniases: some observations on their ecology and epidemiology. Trans. Royal Soc. Trop. Med. Hyg. **77**:569–596.

Lainson, R. and Shaw, J.J. 1987. Evolution, classification and geographic distribution. Pp. 1–120 *in* W. Peters and R. Killick-Kendrick (eds.), The leishmaniases in biology and medicine, Vol. 1, Biology and epidemiology. Academic Press, New York.

Lane, R.P. 1993. Sandflies (Phlebotominae). Pp. 78–119 *in* R.P. Lane and R.W. Crosskey (eds.), Medical insects and arachnids. Chapman and Hall, London.

Lane, R.P. and Crosskey, R.W. (eds.) 1993. Medical insects and arachnids. Chapman and Hall, London. 723 pp.

Lane, R.P. and Ready, P.D. 1985. Multivariate discrimination between *Lutzomyia welcomei*, a vector of mucocutaneous leishmaniasis, and *Lu. complexus* (Diptera: Phlebotominae). Ann. Trop. Med. Parasitol. **79**:225–229.

Lawyer, P.G. 1984. Biology and colonization of the sand fly *Lutzomyia diabolica* (Hall) (Diptera: Psychodidae) with notes on its potential relationship to human cutaneous leishmaniasis in Texas, USA. Ph.D. thesis. University of Florida, Gainesville.

Lawyer, P.G., Ngumbi, P.N., Anjili, C.O., Odongo, S.O., Mebrahtu, Y.B., Githure, J.I., Koech, D.K. and Roberts, C.R. 1990. Development of *Leishmania major* in *Phlebotomus duboscqi* and *Sergentomyia schwetzi* (Diptera: Psychodidae). Am. J. Trop. Med. Hyg. **43:**31–43.

Lawyer, P.G. and Young, D.G. 1991. Diapause and quiescence in *Lutzomyia diabolica* (Diptera: Psychodidae). Pp. 353–360 *in* M. Maroli (ed.), First International Symposium on Phlebotomine Sandflies, Vol. 33(1, Suppl.). Parrasitologia, Rome.

Lawyer, P.G., Young, D.G., Butler, J.F. and Akin, D.E. 1987. Development of *Leishmania mexianca* in *Lutzomyia diabolica* and *Lutzomyia shannoni* (Diptera: Psychodidae). J. Med. Entomol. **24:**347–355.

Leng, Y.J. 1987. A preliminary survey of phlebotomine sandflies in limestone caves of Sichuan and Guizhou Provinces, south-west China, and description and discussion of a primitive new genus *Chinius*. Ann. Trop. Med. Parasitol. **81:**311–317.

Lent, H. and Wygodzinsky, P. 1979. Revision of the Triatominae (Hemiptera, Reduviidae), and their significance as vectors of Chagas' disease. Bull. Am. Mus. Nat. Hist. **163:**123–520.

Lewis, D.J. 1982. A taxonomic review of the genus *Phlebotomus* (Diptera: Psychodidae). Bull. Brit. Mus. Nat. Hist. (Ent.) **45:**121–209.

Lewis, D.J. and Domoney, C.R. 1966. Sugar meals in Phlebotominae and Simuliidae. Proc. Roy. Soc. Entomol. (B) **41:**175–179.

Lewis, D.J., Young, D.G., Fairchild, G.B. and Minter, D.M. 1977. Proposals for a stable classification of phlebotomine sandflies (Diptera: Psychodidae). Syst. Entomol. **2:**319–332.

Lysenko, A.J. and Beljaev, A.E. 1987. Quantitative approaches to epidemiology. Pp. 263–290 *in* W. Peters and R. Killick-Kendrick (eds.), The leishmaniases in biology and medicine, Vol. 1, Biology and epidemiology. Academic Press, New York.

Magill, A.J. 1995. Epidemiology of the leishmaniases. Clinic. Dermatol. **13:**505–523.

Magill, M.D., Grogl, M., Gasser, R.A., Jr., Sun, W. and Oster, C.N. 1993. Viscerotropic leishmaniassis caused by *Leishmania tropica* in soldiers returning from Operation Desert Storm. New England J. Med. **328:**1383–1387.

McConville, M.J. and Homans, S.W. 1992. Identification of the defect in lipophosphoglycan biosynthesis in a non-pathogenic strain of *Leishmania major*. J. Biol. Chem. **267:**5855–5861.

Mebrahtu, Y., Lawyer, P., Githure, J., Were, J.B., Muigai, R., Hendricks, L., Leeuwenburg, J., Koech, D. and Roberts, C. 1989. Visceral leishmaniasis unresponsive to Pentostam caused by *Leishmania tropica* in Kenya. Am. J. Trop. Med. Hyg. **41:**289–294.

Mebrahtu, Y.B., Beach, R.F., Lawyer, P.G. and Perkins, P.V. 1996. The blood-feeding behavior of *Phlebotomus martini* (Diptera: Psychodidae): is it a question of photoperiodism or circadian rhythm? Ann. Trop. Med. Parasitol. **90:**665–668.

Mebrahtu, Y.B., Lawyer, P.G., Ngumbi, P.M., Kirigi, G., Mbugua, J.K., Gachihi, G., Wasunna, K., Pamba, H., Sherwood, J.A., Koech, D.K. and Roberts, C.R. 1992. A new rural focus of cutaneous leishmaniasis caused by *Leishmania tropica* in Kenya. Trans. Royal Soc. Trop. Med. Hyg. **86:**381–387.

Minter, D.M. 1989. Geographical distribution of arthropod-borne diseases and their principal vectors. 134 pp.

Mlot, C. 1997. Insect-borne disease: curing the carrier. Science News **151:**223.

Molyneux, D.H. 1977. Vector relationships in the Trypanosomatidae. Adv. Parasitol. **15:**1–82.

Molyneux, D.H. and Ashford, R.W. 1983. The biology of the *Trypanosoma* and *Leishmania*, parasites of man and domestic animals. Taylor and Francis, Inc., New York. 294 pp.

Morrison, A.C., Ferro, C., Morales, A., Tesh, R.B. and Wilson, M.L. 1993. Dispersal of the sand fly *Lutzomyia longipalpis* (Diptera: Psychodidae) at an endemic focus of visceral leishmaniasis in Colombia. J. Med. Entomol. **30:**427–435.

Mukherjee, S., Hassan, M.Q., Ghosh, A., Ghosh, K.N., Bhattacharya, A. and Adhya, S. 1997. Short Report: *Leishmania* DNA in *Phlebotomus* and *Sergentomyia* species during a kala-azar epidemic. Am. J. Trop. Med. Hyg. **57:**423–425.

Murray, M. and Urquhart, G.M. 1977. Immunoprophylaxis against African trypanosomiasis. Pp. 209–241 *in* L.H. Miller, J.A. Pino and J.J. McKelvey, Jr. (eds.), Immunity to blood parasites of animals and man. Plenum Press, New York.

Perich, M.J., Hoch, A.L., Rizzo, N. and Rowton, E.D. 1995. Insecticide barrier spraying for the control of sand fly vectors of cutaneous leishmaniasis in rural Guatemala. Am. J. Trop. Med. Hyg. **52:**485–488.

Perkins, P.V., Githure, J.I., Mebrahtu, Y., Kiilu, G., Ngumbi, P.S., Nzovu, J., Oster, C.N., Whitmire, R.E., Leeuwenburg, J., Hendricks, L.S. and Koech, D.K. 1988. Isolation of *Leishmania donovani* from *Phlebotomus maritini* in Baringo District, Kenya. Trans. Royal Soc. Trop. Med. Hyg. **82:**695–700.

Peters, W. and Killick-Kendrick, R. (eds). 1987. The leishmaniases in biology and medicine, Vol. 2, Clinical aspects and control. Academic Press, New York. 941 pp.

Pimenta, P., Saraiva, E., Rowton, E., Modi, G., Cilmi, S., Beverly, S. and Sacks, D. 1994. The vectorial competence of phlebotomine sand flies for different species of *Leishmania* is controlled by structural polymorphisms in the surface lipophosphoglycan. Proc. Nat. Acad. Sci. **91:**9155–9156.

Pimenta, P.F.P., Turco, S.J., McConville, M.J., Lawyer, P.G., Perkins, P.V. and Sacks, D.L. 1992. Stage-specific adhesion of *Leishmania* promastigotes to sandfly midgut. Science **256:**1813–1815.

Quate, L.W. 1964. *Phlebotomus* sandflies of the Paloich Area in the Sudan (Diptera, Psychodidae). J. Med. Entomol. **1:**213–268.

Quate, L.W. and Vockeroth, J.R. 1981. Psychodidae. Pp. 293–300 *in* J.F. McAlpine, B.V. Peterson, G.E. Shewell, H.J. Teskey, J.R. Vockeroth and D.M. Wood (eds.), Manual of Nearctic Diptera, Vol. 1. Agriculture Canada, Monograph 27, Ottawa, Ontario.

Ready, P.D. and da Silva, R.M.R. 1984. An alloenzymic comparison of *Psychodopygyus wellcomei*—an incriminated vector of *Leishmania braziliensis* in Para State, Brazil, and the sympatric morphospecies *Ps. complexus* (Diptera, Psychodidae). Cah. ORSTOM Entomol. Med. Parasitol. **22:**3–8.

Ready, P.D., Lainson, R., Shaw, J.J. and Souza, A.A. 1986. The ecology of *Lutzomyia umbratilis* Ward & Fraiha (Diptera; Psychodidae), the major vector to man of *Leishmania braziliensis guyanensis* in north-eastern Amazonian Brazil. Bull. Entomol. Soc. Am. **76:**21–40.

Rees, P.H. and Kager, P.A. 1987. Visceral leishmaniasis and post-kala-azar dermal leishmaniasis. Pp. 584–615 *in* W. Peters and R. Killick-Kendrick (eds.), The leishmaniases in biology and medicine, Vol. 2, Clinical aspects and control. Academic Press, New York.

Ribeiro, J.M.C., Vachereau, A., Modi, G.B. and Tesh, R.B. 1989. A novel vasodilatory peptide from the salivary glands of the sand fly *Lutzomyia longipalpis*. Science **243:**212–214.

Rioux, J.A. and Golvan, Y.A. 1969. Epidemiologie des leishmanioses dans le sud de la France, Monograph 37. Insitut National de la Sante et de la, Recherche Medicale, Paris. 221 pp.

Rioux, J.A., Jarry, D.M., Lanotte, G., Maazoun, R. and Killick-Kendrick, R. 1984. Ecologie des leishmanioses dans le sud de la France. 18. Identification enzymatique de Leishmania infantum Nicolle, 1908, isole de Phlebotomus ariasi Tonnoir, 1921 spontanement infeste en Cevennes. Ann. Parasitiol. Hum. Comp. **59:**331–333.

Robert, L.L. and Perich, M.J. 1995. Phlebotomine sand fly (Diptera: Psychodidae) control using a residual pyrethroid insecticide. J. Am. Mosq. Contr. Assoc. **11:**195–199.

Robert, L.L., Perich, M.J., Schlein, Y., Jacobson, R.L., Wirtz, R.A., Lawyer, P.G. and Githure, J.I. 1997. Phlebotomine sand fly control using bait-fed adults to carry the larvicide *Bacillus sphaericus* to the larval habitat. J. Am. Mosq. Control Assoc. **13:**140–144.

Rutledge, L.C. and Ellenwood, D.A. 1975. Production of phlebotomine sandflies [*Lutzomyia, Brumptomyia*] on the open forest floor in Panama: the species complement [leishmaniasis, insect vectors, control]. Environ. Entomol. **4:**71–77.

Rutledge, L.C. and Mosser, H.L. 1972. Biology of immature sandflies (Diptera: Psychodidae) at the bases of trees in Panama. Environ. Entomol. **4:**71–77.

Sacks, D.L. 1989. Minireview: metacyclogenesis in *Leishmania* promastigotes. Folia Entomol. Mexicana **69:**100–103.

Sacks, D.L. and Perkins, P.V. 1984. Identification of an infective stage of *Leishmania* promastigotes. Science **223:**1417–1419.

Sacks, D.L. and Kamhawi, S. 2001. Molecular aspects of parasite-vector and vector-host interactions in leishmaniasis. Annu. Rev. Microbiol. **55:**453-483.

Schlein, Y., Borut, S. and Jacobson, R.L. 1990. Oviposition diapause and other factors affecting the egg laying of *Phlebotomus papatasi* in the laboratory. Med. Vet. Entomol. **4:**69–78.

Schlein, Y., Jacobson, R.L. and Messer, G. 1992. *Leishmania* infections damage the feeding mechanism of the sandfly vector and implement paarasite transmission by bite. Proc. Nat. Acad. Sci. **89.**

Schlein, Y., Jacobson, R.L. and Shlomai, J. 1991. Chitinase secreted by *Leishmania* functions in the sandfly vector. Proc. Roy. Soc. Entomol. (B) **245:**121–126.

Schlein, Y. and Romano, H. 1986. *Leishmania major* and *L. donovani*: effects on proteolytic enzymes of *Phlebotomus papatasi* (Diptera: Psychodidae) and their possible role in adaptation to the vector. Exp. Parasitol. **62:**376–380.

Schlein, Y. and Warburg, A. 1986. Phytophagy and the feeding cycle of *Phlebotomus papatasi* (Diptera: Psychodidae) under experimental conditions. J. Med. Entomol. **23:**11–15.

Schlein, Y., Yuval, B. and Jacobson, R.L. 1989. Leishmaniasis in the Jordan valley: differential attraction of dispersing and breeding site populations of *Phlebotomus papatasi* (Diptera: Psychodidae) to manure and water. J. Med. Entomol. **26:**411–413.

Schnur, L.F., Chance, M.L., Ebert, F., Thomas, S.C. and Peters, W. 1981. The biochemical and serological taxonomy of visceralizing *Leishmania*. Ann. Trop. Med. Parasitol. **75:**131–144.

Schofield, C.J. and Dolling, W.R. 1993. Bedbugs and kissing-bugs (bloodsucking Hemiptera). Pp. 483–516 *in* R.P. Lane and R.W. Crosskey (eds.), Medical insects and arachnids. Chapman and Hall, London.

Shannon, R. 1939. Methods for collecting and feeding mosquitoes in jungle yellow fever studies. Am. J. Trop. Med. **19:**131–140.

Sinton, J.A. 1922. Entomological notes on field service in Waziristan. Indian J. Med. Res. **9:**575.

Sousa, O.E. 1988. Relationship between vector species and their vectorial capacity for certain strains of *T. cruzi*. Revista Argentina de Microviologia **20(Suppl.):**63–70.

Tesh, R.B. 1995. Control of zoonotic visceral leishmaniasis: is it time to change strategies? Am. J. Trop. Med. Hyg. **52:**287–292.

Tesh, R.B., Guzman, H. and Wilson, M.L. 1992. Trans-beta-farnesene as a feeding stimulant for the sand fly *Lutzomyia longipalpis* (Diptera: Psychodidae). J. Med. Entomol. **29:**226–236.

Vale, G.A. 1993. Visual responses of tsetse flies (Diptera: Glossinidae) to odour-baited traps. Bull. Entomol. Res. **83:**277–289.

Vale, G.A. and Hall. D.R. 1985. The role of 1-octen-3-ol, acetone and carbon dioxide in the attraction of tsetse flies, *Glossina* spp. (Diptera: Glossinidae), to ox odour. Bull. Entomol. Res. **79:**209–217.

Vioukuv, V.N. 1987. Control of transmission. Pp. 909–928 *in* W. Peters and R. Killick-Kendrick (eds.), The leishmaniases in biology and medicine, Vol. 2, Clinical aspects and control. Academic Press, London.

Walters, L.L., Irons, K.P., Modi, G.B. and Tesh, R.B. 1992. Refractory barriers in the sand fly *Phlebotomus papatasi* (Diptera: Psychodidae) to infection with *Leishmania panamensis*. Am. J. Trop. Med. Hyg. **46:**221–228.

Walton, B.C. 1987. American cutaneous and mucocutaneous leishmaniasis. Pp. 637–664 *in* E. Peters and R. Killick-Kendrick (eds.), The leishmaniases in biology and medicine, Vol. 2, Clinical aspects and control. Academic Press, New York.

Ward, R.D. 1977. The colonization of *Lutzomyia flaviscutellata* (Diptera: Psychodidae), a vector of *Leishmania mexicana amazonensis* in Brazil. J. Med. Entomol. **14:**469–476.

Wellde, B.T. 1989. Trypanosomiasis in the Lambwe Valley, Kenya. Ann. Trop. Med. Parasitol. **83 (Suppl.):**220.

World Health Organization. 1986. Epidemiology and control of African trypanosomiasis. Report of a WHO expert committee. WHO Tech. Rep. Ser. No. 739. 127 pp.

World Health Organization. 1990. World report on tropical diseases. Trop. Med. Hyg. News **39:**73.

World Health Organization. 1991. Control of Chagas' disease: report of a WHO expert committee. WHO Tech. Rep. Ser. No. 811.

World Health Organization. 1995. Planning overview of tropical diseases. Division of Control of Tropical Diseases, WHO, Geneva.

World Health Organization. 1998.Control and surveillance of African trypanosomiasis. Report of a WHO expert committee. WHO Tech. Rep. Ser. No. 881. 113 pp.

Williams, P. 1970. Phlebotomine sandflies and leishmaniasis in British Honduras (Belize). Trans. Royal Soc. Trop. Med. Hyg. **64:**317–368.

Young, D.G. 1979. A review of the bloodsucking psychodid flies of Colombia (Diptera: Phlebotominae and Sycoracinae). University of Florida Agricultural Experiment Station Bulletin 806, Gainesville. 266 pp.

Young, D.G. and Duncan, M.A. 1994. Guide to the identification and geographic distribution of *Lutzomyia* sand flies in Mexico, the West Indies, Central and South America (Diptera: Psychodidae). Associated Publishers, American Entomological Institute, Gainesville, Florida. 881 pp.

Young, D.G. and Lawyer, P.G. 1987. New World Vectors of the Leishmaniases. Pp. 29–71 *in* K.F. Harris (ed.), Current Topics in Vector Research, Vol. 4. Springer-Verlag, New York.

Yuval, B. and Schlein, Y. 1986. Leishmaniasis in the Jordan Valley. III. Nocturnal activity of *Phlebotomus papatasi* (Diptera: Psychodidae) in relation to nutrition and ovarian development. J. Med. Entomol. **23:**411–415.

Zijlstra, E., El-Hassan, A., Ismael, A. and Ghalib, H.W. 1994. Endemic kala-azar in eastern Sudan: a longitudinal study on the incidence of clinical and subclinical infection and post-kala-azar dermal leishmaniasis. Am. J. Trop. Med. Hyg. **51:**826–836.

Zimmerman, J.H., Newson, H.D., Hooper, G.R. and Christensen, H.A. 1977. A comparison of egg surface structure of six anthropophilic phlebotomine sand flies (*Lutzomyia*) with scanning electron microscope. J. Med. Entomol. **13:**574–579.

Chapter 9

Filariasis

JAMES B. LOK[1], EDWARD D. WALKER[2] AND GLEN A. SCOLES[3]
[1]University of Pennsylvannia, Philadelphia, [2]Michigan State University, East Lansing and [3]Yale University, New Haven, Connecticut

INSECT-TRANSMITTED FILARIAL PARASITES

Description and Classification

Parasitic round worms (nematodes) in the superfamily Filarioidea, family Onchocercidae, are the only metazoon disease agents of vertebrates that undergo true biological transmission by arthropod vectors. These worms parasitize the tissues and tissue spaces of all classes of vertebrates except fish (Anderson 1992). The insect-transmitted parasites are elongate, slender worms and differ from other Filarioidea in that the adults bear live young as highly specialized larval forms called **microfilariae**. Microfilariae migrate to the peripheral blood circulation or skin where they are accessible to bloodfeeding arthropod vectors. These arthropod vectors are obligate, intermediate hosts for the parasite, necessary for development from the microfilaria to the **third-stage** or **infective larva** (commonly abbreviated "**L3**"). This is the stage infectious to vertebrates. None of the nematode larvae propagates in the vector.

The terms **filaria** (plural, **filariae**) or **filarial nematode** refer to any member of the family Onchocercidae, superfamily Filarioidea (Anderson 1992). Nematodes of the Filarioidea are classified within the phylum Nematoda, order Spirurida, suborder Spirurina; this suborder includes a diverse array of parasitic nematodes that have arthropod intermediate hosts. Adults of this suborder are superficially nondescript with external taxonomic characters limited to rows of caudal, submedial papillae and cuticular ridges or spines in some species. External features such as the pseudolabia and other cephalic structures, so conspicuous in the gut-dwelling ascarids and strongylids, are absent in the filarioids. The buccal cavity itself is highly reduced. In females, the vulva characteristically is positioned on the anterior third of the body, and the copulatory spicules of males are subequal in length (Anderson 1992). Insofar as they lack an alimentary tract, microfilariae might be characterized morphologically as motile, vermiform embryos. However, in other respects, such as the presence of a nerve ring, and sensory amphids and phasmids, the internal anatomy of microfilariae resembles that of a fully formed first-stage larval nematode. In some species, microfilariae retain the chitinous egg membrane after extrusion from the gravid female as a stretched, tightly fitting sheath. In other species, microfilariae shed the egg mem-

B.F. Eldridge and J.D. Edman (eds.), Medical Entomology, Revised Edition, 299–375.

brane within the reproductive tract of the gravid female and thus are seen unsheathed in peripheral blood or skin.

There are 2 families within the Filarioidea: the Filariidae, containing 5 genera among 2 subfamilies; and the Onchocercidae, containing 78 genera among 8 subfamilies. Several species of veterinary significance occur in the Filariidae, especially in the genera *Stephanofilaria* and *Parafilaria*. The vertebrate hosts of filariids include rodents, carnivores, swine, African antelope, horses, cattle, elephants and rhinoceroses. The female filariids release ova into the surrounding tissues and do not bear live microfilariae as do the onchocercids. Eggs and larvae of the filariids remain in cutaneous sites in close proximity to adult parasites, often where cutaneous lesions develop. These lesions attract muscoid flies (the biological vectors) that ingest the ova and larvae.

The family Onchocercidae contains the filarial species of public health importance. Table 9.1 summarizes the biological relationships among filarial worms in this family and their vertebrate and invertebrate hosts. Within the family Onchocercidae, 34 genera have been described biologically and taxonomically. There is a greater breadth of vertebrate hosts and arthropod vectors in the Onchocercidae than in the Filariidae. All major groups of hematophagous arthropods, including Diptera (mosquitoes, biting midges, black flies, sand flies, horse flies, deer flies and louse flies), fleas, lice, ticks and mites can transmit species of Onchocercidae. Reptiles, amphibians, birds and mammals can serve as vertebrate hosts.

Historical Aspects

The following historical summary of early research on lymphatic filariasis is based on Grove (1990). The reader also should consult the compilation of classic papers on the subject by Kean et al. (1978). The disease known as elephantiasis has been known since antiquity and probably is depicted in artifacts from ancient Egypt. Elephantiasis is unmistakably described in the writings of European explorers and colonists in the 16th through 18th Centuries. However, it was not until the latter half of the 19th Century that the filariases and their etiologic agents and vectors were systematically studied. The first stages to be studied were microfilariae of the lymphatic dwelling parasites, being readily accessible in blood, urine and tissue-abscess fluids. The French surgeon Jean-Nicolas Demarquay published the first account of microfilariae of what is now believed to have been *Wuchereria bancrofti* in hydrocele fluid from a Cuban expatriate (Demarquay 1863). Wucherer (1868) observed these same microfilariae in the urine of hematuric patients in Bahia, Brazil. In a landmark paper, Lewis (1872) reported finding microfilariae in the blood of patients in Calcutta, India, and presented what, in retrospect, appears to be the first evidence that numbers of these forms in the peripheral blood fluctuate in a nocturnally periodic cycle (see below).

The Australian physician Joseph Bancroft first described the adult stages of a filaria, presumably *W. bancrofti*, from a lymphatic abscess of the arm in a paper communicated to *The Lancet* by Cobbold (1877). Descriptions of the reproductive tracts of female worms by Lewis (1877) demonstrated formation of the microfilarial sheath (see below) from the attenuated egg membrane.

Patrick Manson, a British physician, was stimulated by Lewis' 1872 paper to test the possibility that hematophagous arthropods might transmit the filariae. Working in Amoy, China, Manson dissected mosquitoes that had fed on a microfilaremic donor (his own gardener) and discovered that not only were the parasites ingested by the mosquitoes but also that they underwent a marked physical transformation over the ensuing days. Grove (1990) highlighted Manson's diary entry summarizing what many today regard as the founding discovery of medical entomology:

"I found that my idea was correct, and that the haematozoon that entered the mosquito a simple, structureless animal, left it, after passing through a series of highly interesting metamorphoses, much increased in size, possessing an alimentary canal and being otherwise suited for an independent existence."

The initial hypotheses of Manson and others were that bathers acquired larval filariae in water contaminated by the carcasses of infected mosquitoes. This misconception persisted until 1899 when Thomas Bancroft, son of Joseph, perfected techniques that allowed him to rear adult mosquitoes in the laboratory for periods of up to 2 months. He infected mosquitoes by feeding them on microfilaremic volunteers and incubated them for the 16-day extrinsic incubation period. Although he did not know the true anatomical sites of filarial infection in mosquitoes, Bancroft first proposed in his 1899 paper that infective larvae might be transmitted from the insects in the act of biting. Low (1900) examined material histologically prepared by Bancroft and demonstrated migration of infective larvae into the labium of the vector. Low concluded that the bites of infected mosquitoes transmit parasites.

Associating filarial infection with the multiplicity of symptoms we now attribute to lymphatic filariasis was long in coming. Both Lewis and Demarquay suspected that the microfilariae they discovered in hydrocele fluid, urine and blood caused filariasis in their hosts, but they lacked proof. Manson, in his landmark paper of 1877, attributed elephantiasis, hydrocele and inguinal lymphadenopathy to adult filariae, but the role of these worms as disease pathogens remained controversial until O'Connor (1932) demonstrated that adult *W. bancrofti* in the lymphatic system, particularly the degenerating forms, elicited the inflammatory responses common to chronic human filariasis.

Nelson et al. (1991) summarized the early history of research on human onchocerciasis. They noted that skin lesions dubbed "craw-craw" in West African patients were associated first with dermal microfilariae, now believed to be *Onchocerca volvulus*, by Ship's Surgeon J. O'Neill working in present-day Ghana. Leuckart (1886) described the adult parasite from dermal nodules excised from African patients 2 decades later. This material also was laden with microfilariae that differed from the sheathed microfilariae (see below) found in patients infected with *Filaria noctua* (= *W. bancrofti*). The association between skin microfilariae and the nodular adults was unrecognized at first, but by the turn of the 20th Century a dermatitis known as "worm nodule disease" was the subject of increasing scrutiny (Prout 1901, Parsons 1909). Nelson et al. (1991) attribute the first incrimination of *O. volvulus* as the agent of ocular disease to a Guatemalan physician, Rodolfo Robles. Robles (1917) presented clear evidence linking not only the skin disease, known in Guatemala as *Erysipelas de la Costa*, but also anterior segment ocular lesions to onchocercal infection. Nelson et al. (1991) credit Strong et al. (1934), Hissette (1932), Bryant (1935) and Ridley (1945) with pioneering descriptions of ocular onchocerciasis in Africa.

Robles (1917) suspected early on that black flies (Simuliidae) were vectors of *O. volvulus*. However, Donald Blacklock of the Liverpool School of Tropical Medicine furnished proof from studies in Sierra Leone (Blacklock 1926). He described development and transmission of the parasite in *Simulium damnosum*. In his monograph, Dalmat (1955) provided the earliest scientific accounts of *Simulium ochraceum* and *Simulium metallicum* as vectors of *O. volvulus* in the Americas. This work remains one of the most comprehensive on this subject. For additional details on the classic studies in filariasis, including accounts of early studies of loiasis, the reader is referred to the works of Kean et al. (1978), Grove (1990) and Nelson et al. (1991).

Table 9.1. Summary of selected genera of onchocercid filarids. Adapted from Bain and Chabaud (1986) and Anderson (1992). (Table continued on next page).

Genus	No. of species	Distribution	Vector	Vertebrate hosts
Acanthocheilo-nema	6	Cosmopolitan	Hippoboscid flies, sucking lice, fleas, ticks	Many marine and terrestrial mammals
Aproctella	1	Neotropical	Mosquitoes	Blue-gray tanager
Breinlia	1	Oriental	Mosquitoes	Rodents, primates
Brugia	7	Ethiopian, Indomalayan, Nearctic	Mosquitoes	Various mammals, including humans
Cardiofilaria	1	Oriental	Mosquitoes	Chickens
Cercopithfilaria	4	Palearctic, Ethiopian	Ticks	Goats, domestic cats, dogs, porcupines
Chanderella	3	Cosmopolitan	Biting midges	Birds
Cherylia	1	Neotropical	Ticks	Unknown
Conispiculum	1	Oriental	Mosquitoes	Reptiles
Deraiophoro-nema	1	Oriental, Ethiopian, Australian	Mosquitoes	Camels
Dipetalonema	2	Neotropical	Biting midges	New World monkeys
Dirofilaria	7	Cosmopolitan	Mosquitoes, black flies	Carnivores, primates
Elaephora	1	Nearctic	Tabanid flies	Cervids
Eufilaria	5	Palearctic, Nearctic, Neotropical	Biting midges, biting lice	Birds
Foleyella	3	Palearctic, Ethiopian	Mosquitoes	Reptiles
Icosiella	1	Palearctic	Biting midges	Amphibians
Litomosoides	3	Neotropical, Nearctic	Mites	Rodents, oppossums

Table 9.1 (Continued).

Genus	No. of species	Distribution	Vector	Vertebrate hosts
Loa	1	Ethiopian	Tabanid flies	Humans
Loiana	1	Nearctic	Mosquitoes	Lagomorphs
Macdonaldius	1	Neotropical	Ticks	Snakes
Mansonella	7	Nearctic, Neotropical	Black flies, biting midges	Primates (including humans), raccoons
Molinema	3	Nearctic, Neotropical	Mosquitoes	Rodents
Monanema	3	Nearctic, Ethiopian	Ticks	Rodents
Onchocerca	8	Ethiopian, Indomalayan	Black flies, biting midges	Humans,cattle, horses, deer
Oswaldofilaria	5	Palearctic	Mosquitoes	Reptiles
Pelecitus	3	Australian, Nearctic, Indomalayan	Mosquitoes, tabanid flies	Lagomorphs, marsupials, birds
Saurositus	1	Ethiopian	Mosquitoes	Agamid lizards
Setaria	5	Cosmopolitan	Horn flies (*Haematobia*)	Ruminants
Skrjabinofilaria	1	Neotropical	Mosquitoes	Oppossums
Splendidofilaria	2	Nearctic	Black flies, biting midges	Ducks
Thamugadia	1	Palearctic	Phlebotomine sand flies	Geckoes
Waltonella	5	Nearctic, Neotropical, Palearctic	Mosquitoes	Amphibians
Wuchereria	2	Palearctic, Ethiopian, Neotropical, Indomalayan, formerly Australian and Nearctic	Mosquitoes	Primates, including humans
Yatesia	1	Neotropical	Ticks	Capybara

Distribution and Importance

Human filariases cause debilitating and disfiguring disease, resulting in much human suffering, especially in the tropics. In addition to their physical and psychosocial impact, they exact an enormous toll in terms of decreased economic production in nations where such decreases can be ill afforded. The filariases generally cause morbidity but not mortality in infected humans. Currently, some 146 million people are infected throughout tropical and subtropical regions of the world (Ottesen 1993, Michael and Bundy 1997), excluding people infected with species of *Mansonella*, for which infection is widespread but prevalence is unknown. Humans serve as definitive hosts for 8 species of filarial nematodes (Levine 1980, Anderson 1992). The global distribution of filarial infections in humans (Fig. 9.1) has been reviewed by several investigators (Hawking 1976, 1977, Hawking and Denham 1976, Sasa 1976, Michael et al. 1996, Michael and Bundy 1997).

Three mosquitoborne species of filariae cause **lymphatic filariasis**: *Wuchereria bancrofti* causes Bancroftian filariasis; *Brugia malayi* causes Brugian filariasis; and *Brugia timori* causes Timorian filariasis. *Wuchereria bancrofti* is the most widely distributed species, and its infections are the most prevalent. An estimated 115 million people are infected. It occurs in large sections of sub-Saharan Africa, parts of India, China, Bangladesh, Myanmar, Thailand, Malaysia, Laos, Vietnam, Indonesia, the Philippines, Papua New Guinea, and among island groups in the south Pacific Ocean. Foci existed in northern Australia, but have been eradicated. Bancroftian filariasis, introduced into the Western Hemisphere along with colonization by Europeans and the slave trade, formerly had a wider geographic distribution, including Trinidad and other islands of the Caribbean, and in coastal areas of Mexico and Central America. *Wuchereria bancrofti* infection also occurs in localized areas in Haiti, the Dominican Republic, Guyana, Surinam and the northeast coast of Brazil. Charleston, North Carolina, was endemic for Bancroftian filariasis until the early part of the 20th Century. However, the distribution of Bancroftian filariasis since has shrunk in the Western Hemisphere to its current status.

Brugia malayi has a more limited distribution than *W. bancrofti* and occurs in China, India, the Republic of Korea and Southeast Asia, including Indonesia and the Philippines. An estimated 13 million people are infected. *Brugia timori* is confined to certain southern islands of Indonesia in the Savu Sea: Timor, Flores, Alor, Sumba, Roti and Savu. The number of infected people has not been estimated, although humans in endemic areas have infection rates ranging to 30%.

Onchocerca volvulus, transmitted by black flies in the genus *Simulium*, causes the disease river blindness, Robles' disease or onchocerciasis. An estimated 17.7 million people are infected. The disease is associated with large watershed systems in parts of western, central, and eastern Africa where black fly vectors occur. Onchocerciasis was imported to the Western Hemisphere with the slave trade and became established in foci in southern Mexico, Guatemala, Colombia, Ecuador, Venezuela and Brazil.

Loa loa causes loiasis in humans. Loiasis occurs in parts of central and western Africa, encompassing the rain forest zones in a triangle-shaped region from eastern Nigeria, Equatorial Guinea and Cameroon; south to northern Angola; northeast to southern Sudan; then back again to eastern Nigeria. The number of people infected is unknown. Certain species of day-active, forest-dwelling tabanid flies in the genus *Chrysops* are vectors.

Mansonella ozzardi, *Mansonella perstans* and *Mansonella streptocerca* are the other 3 onchocercids that infect humans. The latter 2 species previously were classified within the genus *Dipetalonema*. *Mansonella ozzardi* causes

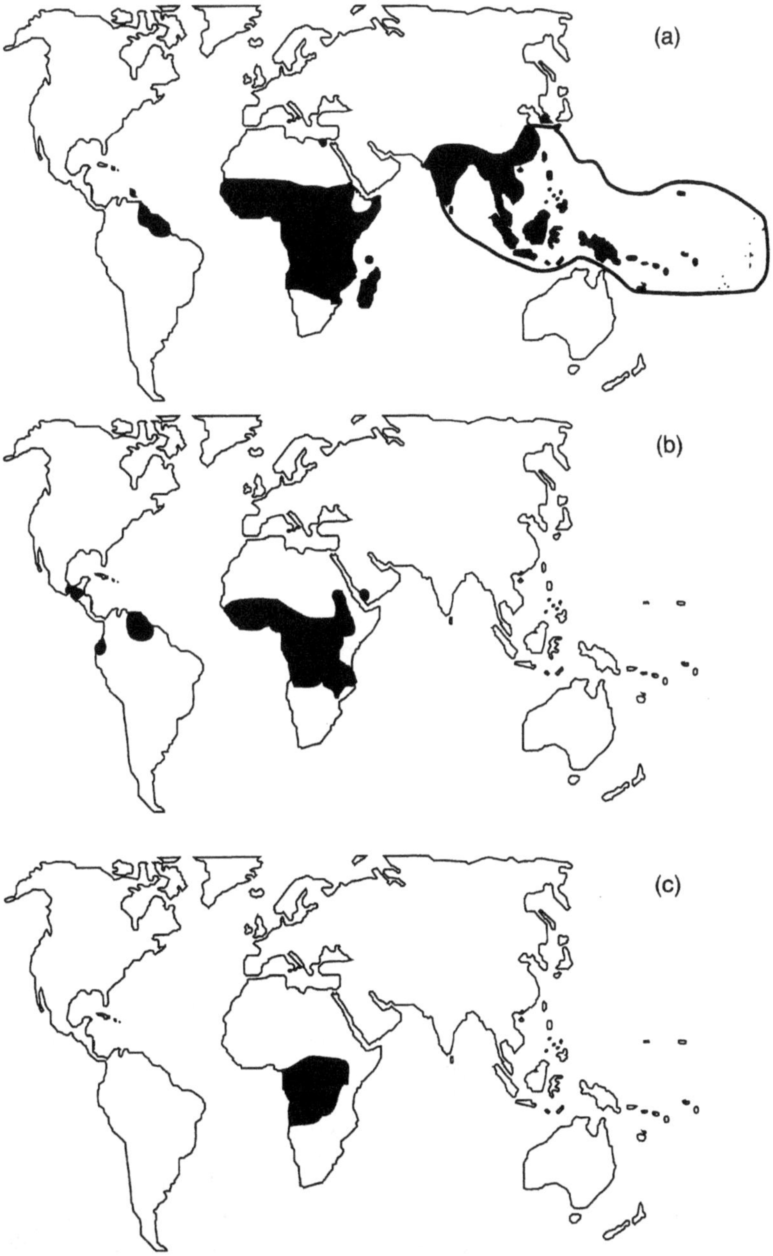

Figure 9.1. The geographic distribution of (a) lymphatic filariasis, (b) onchocerciasis and (c) loiasis in the world (courtesy World Health Organization).

widespread human infections in tropical areas of Central and South America and in some Caribbean islands. *Culicoides* biting midges and black flies of the *Simulium amazonicum* group of species are vectors. People infected with *M. ozzardi* usually show no overt symptoms, but some develop allergies to *M. ozzardi*-derived antigens, or may develop other symptoms such as chronic arthritis, headache and localized swelling at sites of infection. *Mansonella perstans* causes widespread human infection in western and central Africa and in certain parts of Central and South America. The vectors are *Culicoides* midges.

Mansonella streptocerca infections occur in parts of western and central Africa. The biting midge *Culicoides grahami* is the vector. This parasite often causes an itching dermatitis accompanied by skin lesions because the adult worms dwell in the skin. Rates of infection by *M. streptocerca* may be high in endemic areas, but the worldwide prevalence of infections is unknown because most individuals lack severe symptoms. Consequently, the estimated 146 million people currently infected with filarial nematodes does not include *M. streptocerca* infections.

Filarial infections are also of importance in veterinary medicine. *Dirofilaria immitis* causes a life threatening cardiopulmonary disease of increasing incidence in dogs and cats (Grieve et al. 1983, Boreham and Atwell 1988), and infections with certain *Onchocerca* spp. cause mild disease in horses, cattle, camels and other livestock. In some foci of lymphatic filariasis and onchocerciasis, the same vectors that transmit human parasites also transmit related livestock parasites. The resemblance of worms in livestock to those in humans confounds entomological transmission studies.

Many other species of onchocercids either cause clinically inapparent infections in domestic animals or are parasites of wildlife. The significance of wildlife diseases caused by filariae is not well known.

Parasite Life Cycles

Development in arthropods

The vector phase of filarial life cycles begins with the ingestion of microfilariae by a competent vector while bloodfeeding on an infected vertebrate host. The vertebrate tissue in which microfilariae dwell, whether skin or blood, is related to the mode of feeding of the particular arthropod vector. Filariae such as *Onchocerca* spp. have skin dwelling microfilariae and are transmitted exclusively by **telmophagic** or **pool feeding** vectors such as black flies (Simuliidae) or biting midges (Ceratopogonidae). In this case, microfilariae migrate from the dermis into the pool of blood and tissue fluid accumulating in the bite wound and are ingested by the feeding arthropod. Filarial species whose microfilariae occur in the peripheral blood may be transmitted either by pool feeders (e.g., the associations of *Mansonella* with ceratopogonids and *Loa* with tabanids) or by **solenophagic** or **vessel feeding** arthropods (e.g., mosquitoes in the cases of the lymphatic dwelling *Wuchereria* and *Brugia*).

Within a few hours of ingestion, microfilariae migrate from the alimentary tract of the vector to one of 4 possible sites of development: (1) the indirect flight muscles; (2) the Malpighian tubules; (3) the fat body; or (4) the hemocoel. Vector organ or tissue site specificity among the different species of filariae is marked (Table 9.2). At the organ site of development, microfilariae penetrate individual cells and remain as intracellular parasites for the majority of their development. Within cells of the vector organ site, microfilariae begin to differentiate into first stage larvae. Differentiation from microfilariae to first stage larvae does not involve a molt. During these changes, developing filarial larvae cease the vigorous motility characteristic of microfilariae. This quiescence is accompanied by shortening and widening of the body and by development of an alimentary tract. Further

Table 9.2. Sites of development in vectors for selected onchocercid filarids. Adapted from Bain and Chabaud (1986) and Anderson (1992).

Site of development in vector	Vector	Genera
Fat body	Mosquitoes	*Breinlia, Brugia, Molinema, Oswaldofilaria, Pelecitus, Waltonella*
	Black flies	*Splendidofilaria*
	Biting midges	*Eufilaria, Dipetalonema, Splendidofilaria*
	Tabanid flies	*Eleaophora, Loa*
	Horn flies	*Setaria*
	Hippoboscid flies	*Acanthocheilonema*
	Biting lice	*Eufilaria*
	Ticks	*Monanema*
Flight muscles	Mosquitoes	*Aproctella, Brugia, Conispiculum, Deraiophoronema, Saurositus, Skrjabinofilaria, Wuchereria*
	Black flies	*Mansonella, Onchocerca*
	Biting midges	*Chanderella, Eufilaria, Mansonella, Onchocerca*
	Phlebotomine sand flies	*Thamugadia*
	Horn flies	*Setaria*
	Biting lice	*Eufilaria*
Hemocoel	Mosquitoes	*Waltonella*
	Black flies	*Splendidofilaria*
	Biting midges	*Splendidofilaria*
	Mites	*Litomosoides*
Malpighian tubules	Mosquitoes	*Dirofilaria*
	Black flies	*Dirofilaria*
	Ticks	*Macdonaldius*
Muscles (in general)	Mosquitoes	*Waltonella*
	Biting midges	*Icosiella*
	Ticks	*Yatesia*

coalescence of the parasite digestive system and differentiation of the cellular precursors of the reproductive tract continues in the ensuing days (Schacher 1962). Near the end of the first stage (L1), larval filariae elongate and 2 molts to the second (L2) and third stages (L3) follow rapidly (Fig. 9.2). The parasite's oral cavity becomes functional during the second stage and the parasite begins to ingest organelles and other components of host cells. This rapid growth and onset of feeding disrupts the physical integrity of the cell, ultimately resulting in distention and lysis. These pathologic changes can reduce longevity or flight capability of the vector, depending upon the intensity of infection and the organ or tissue site in question. Following lysis of vertebrate host cells, L3 migrate through the hemocoel anteriorly to the hemocoelomic lumen of a mouthpart structure. Larvae then escape the mouthparts and invade the definitive host via the vector bite wound during bloodfeeding.

Microfilarial penetration and development in arthropods

Microfilariae traversing the foregut accumulate in the midgut in the first moments after ingestion, in most cases migrating posteriorly in the coagulating bloodmeal. Within 1–6 hr, the majority of microfilariae have left the midgut lumen and either penetrated the midgut wall and entered the hemocoel or migrated into the Malpighian tubules. In some microfilariae, penetration of the midgut may be facilitated by an array of cephalic armaments including a prominent, beak-like hook. It was believed that sheathed microfilariae (see above) cast off their sheaths within the midgut lumen and that this exsheathment is a prerequisite to penetration of the midgut epithelium and further migration. However, Agudelo-Silva and Spielman (1985) discovered that microfilariae of *B. malayi* begin to penetrate the midgut wall of *Aedes aegypti*, an experimental vector, prior to exsheathment. Exsheathment took place as the microfilariae penetrated the basement membrane and emerged into the hemocoel. Empty sheaths frequently were left protruding from ruptures in the gut. These cast sheaths may constitute "decoys" that assist the exsheathed worms to evade cellular and humoral factors in the hemolymph of the mosquitoes.

Physical disruption of the midgut wall by microfilariae may alter the course of a concomitant arboviral infection in vectors that ingest blood from dually infected hosts. The midgut wall is a potential barrier to arboviral infection and its permeability to virus particles is a primary determinant of vector competence for arboviruses. Even in highly susceptible arbovirus vectors, unassisted passage of the midgut wall is a complex process involving receptor-mediated invasion of epithelial cells, proliferation and egress to the hemocoel via the basement membrane. Damage to the midgut by migrating microfilariae apparently allows arboviruses to bypass this multistep route and enter the hemocoel directly, thereby decreasing the minimum level of virus uptake necessary for dissemination and accelerating its progress (Turell et al. 1984). However, the actual impact of this phenomenon on natural arboviral transmission cycles has not been demonstrated.

Within a few hours of ingestion, microfilariae that escape the midgut migrate either via the hemocoel to cells of the indirect flight muscles or fat body or to primary cells of the Malpighian tubules, presumably via their lumenal connection to the hindgut. The basis for the vector organ site specificity among the various species of filariae is not known. Until recently, the means by which microfilariae located their developmental site were obscure as well. However, evidence recently has come to light suggesting that this migration is directed by chemical attractants emanating from the target organs. Lehmann et al. (1995) demonstrated the existence of soluble factors within the thorax of *Simulium vittatum* that attract microfilariae of the thoracic muscle dwelling *Onchocerca lienalis.*

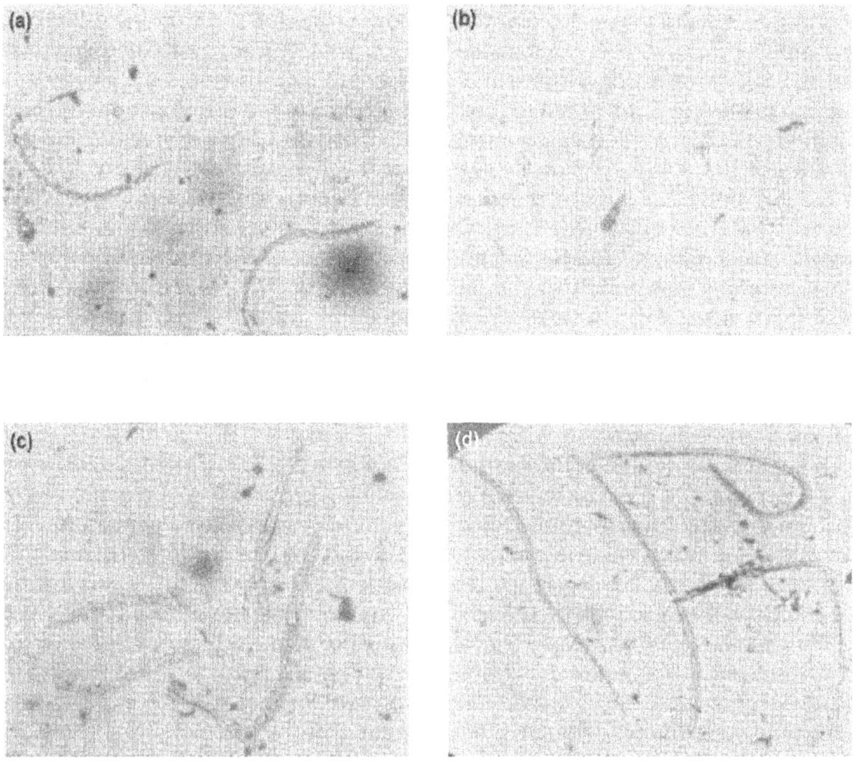

Figure 9.2. Developmental stages of *Wuchereria bancrofti* dissected from the mosquito *Aedes scutellaris* after having fed upon a volunteer in Polynesia: (a) Day 1 after feeding; (b) day 3; (c) day 8; (d) day 13. All photographs 115X (from a World War II training manual, courtesy of Michigan State University).

This material is diffusible and appears to be a protein exceeding 100 kDa in molecular weight.

Schacher (1962) described the development of *Brugia pahangi* in *Anopheles quadrimaculatus* and *Armigeres obturbans* mosquitoes and is one of the most detailed accounts of the vector phase of a filarial life cycle available. The following description summarizes this study:

After an infectious bloodmeal by the arthropod vector, microfilariae escaped from the midgut and migrated to the organ site of development. Vigorous serpentine movement by the microfilariae accompanied this activity. This movement ceased upon penetration of the development site and differentiation of the parasites began within a few hours. Cell division began first in the posterior of the parasite's body, in the first rectal, or R_1 cell (previously called the "G cell"). Division of these and other somatic cell precursors increased the width of the parasite, first at the level of the rectal cells and later throughout the length of the body. This increase in body width continued throughout the first larval stage, accompanied by a decrease in length. Eventually, a shortened, thickset worm resulted, bluntly rounded at both ends, but retaining a remnant of the pointed microfilarial tail. This form has been dubbed the "sausage-form" larva. The first development of the alimentary tract begins in this stage. Just prior to the first molt, cuticle loosens around both extremities of the worm. The first stage comprises roughly half of the developmental period in the vector in most instances.

The second larval stage in the filariae is characterized by rapid elongation and further organization of the alimentary tract. The gut lumen appears and the nerve ring becomes more prominent. During the second stage the genital rudiments, kidney-shaped clusters of cells giving rise to the reproductive tract, appear. The molting cycle begins near the end of the second stage with the loosening of the cuticle around both ends of the parasite. Molting to the third stage occurs late in vector phase development.

The third stage larva may be recognized by the absence of the spiked tail evident in the second stage. The esophagus of L3 clearly is differentiated into muscular and glandular regions and the rectum is functional. The L3 breaks out of its organ site of development shortly after molting and enters the hemocoel. Initially, migration through the body cavity appears random, but within a day or so of the second molt the majority of L3 converge upon the head and mouthparts of the vector. The worms ultimately invade some hollow structure of the feeding apparatus such as the labial sheath of mosquitoes or the labrum-epipharynx in black flies. Transmission occurs when the vector takes another bloodmeal.

McGreevy (1974b) described the escape of L3 from the mouthparts of *Ae. aegypti* infected experimentally with *D. immitis*. In this system, L3 within the labial sheath escape via the cuticle at the tips of labellar lobes (Fig. 9.3). Flexure of the labium during insertion of the fascicle into the host skin apparently stimulates invasive behavior in the L3 consisting of serpentine movement. Tearing of the vector's labellar cuticle by escaping L3 causes an outflow of hemolymph around the feeding site. McGreevy (1974b) postulated that the resulting pool of liquid constitutes a protective medium for the L3 as they emerge onto the host integument and invade via the feeding lesion. No comparably detailed description of L3 transmission from pool feeding vectors exists. However, the presence of *Onchocerca* L3 within the labrum-epipharynx of the feeding apparatus of *Simulium* spp. (Collins 1977) suggests that parasites either are placed directly into the bite wound during the initial stages of probing or escape into the pool of blood and tissue fluid during the engorgement phase of feeding.

Pathogenicity of larval filariae to their vectors

Filariae first exert pathogenic effects on their vectors by microfilarial penetration of the midgut wall. In heavily infected mosquitoes and black flies, high mortality occurs during the first 24 hr after infection. Presumably, this is a consequence of microfilarial penetration of the midgut. The exact cause of this mortality is unknown; possibly release of pathogenic microorganisms into the hemocoel causes a fatal infection.

During the first and early second stages, filarial larvae remain in cells of the vector organ site. However, lysis of these cells occurs as the parasites enter the rapid growth phase of the L2. Then, parasites may damage basement or other limiting membranes as parasites break out of their organ site of development and reenter the hemocoel as L3. This tissue damage may result in various pathological effects on the vectors ranging from impaired flight capacity in systems involving development in the thoracic musculature (e.g., *Wuchereria*/mosquito and *Onchocerca*/*Simulium* interactions) and excretory dysfunction caused by filariae such as *D. immitis* that develop in the Malpighian tubules. With heavy parasite burdens, these effects may kill the vector before transmission of L3 can occur. Sublethal effects of filariae may result in decreased dispersal or host-seeking capabilities

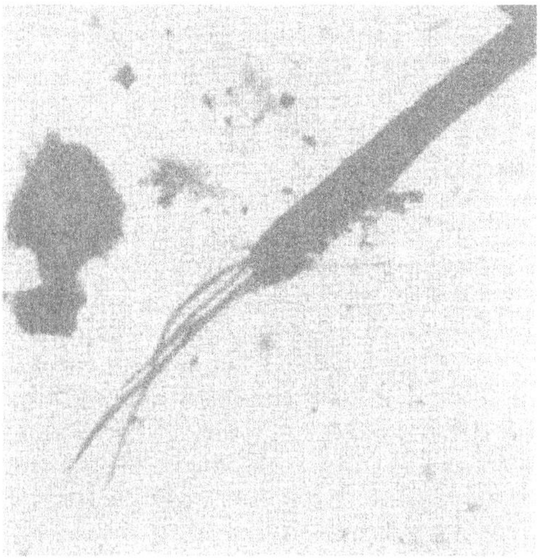

Figure 9.3. Escape of third-stage larvae (L3) of *Dirofilaria immitis* from the mouthparts of *Aedes aegypti* (courtesy of Edward D. Walker).

in infected arthropods, thus significantly limiting vector capacity. Pathogenesis by filariae in their vectors has a 2-fold effect. First, pathogenesis enhances the adaptive advantages provided by the various infection barriers and defensive systems that modulate infection intensity. Second, pathogenesis in vectors may exclude vertebrate hosts with extremely high microfilarial infections as reservoirs of infection.

Determinants of filarial susceptibility

As discussed in Chapter 6, vectorial capacity is a complex term combining vector competence and a number of population parameters and behavioral attributes that favor pathogen propagation and enhance vector-host contact. **Filarial susceptibility** means the ability of a vector to support development of ingested microfilariae to infective third-stage larvae. This ability is influenced by structural and physiological barriers to parasite migration, by active defensive responses on the part of the vector and by poorly characterized genetic factors.

As they pass down the alimentary tract of their vectors, microfilariae encounter the first of several physical barriers to infection in the form of **cibarial armature** (Fig. 9.4). This structure is a field of inwardly directed spines lining the posterior lumen of the cibarial pump within the buccopharyngeal apparatus. The cibarial armature may damage or "shred" the majority of microfilariae entering the foregut. The development of cibarial armature varies among vector species; when highly developed, it constitutes a significant modulating factor (Omar and Garms 1975). The cibarial barrier to microfilarial invasion may have 2 opposing effects on vector capacity. On the one hand, the cibarial armature may modulate output of L3 by reducing microfilarial intake. On the other hand, because of potential pathogenic effects of filariae upon their vectors, such modulation may be offset by increased survivorship among arthropods with low to moderate parasite burdens (Omar and Garms 1975). This survivorship may be especially true in foci of filariasis where prevailing microfilaremias in vertebrate hosts are high. In the latter instance, the cibarial armature, like all of the other infection barriers we will discuss, is a defense mechanism of the vector.

There also are physical barriers to microfilarial invasion in the midgut. The **peritrophic membrane** or **peritrophic matrix**, a noncellular, chitinous layer secreted around the bloodmeal by cells of the midgut epithelium, is an important barrier to microfilarial invasion in some vector species. In general, the time of formation of the peritrophic membrane relative to the time of penetration of the midgut wall determines the importance of this barrier. In most cases, microfilariae leave the midgut within 30 min of ingestion, well in advance of peritrophic membrane formation. However, rapid formation of the peritrophic membrane effectively traps a high proportion of *O. volvulus* microfilariae within the gut contents in some vectors, most notably rainforest dwelling members of

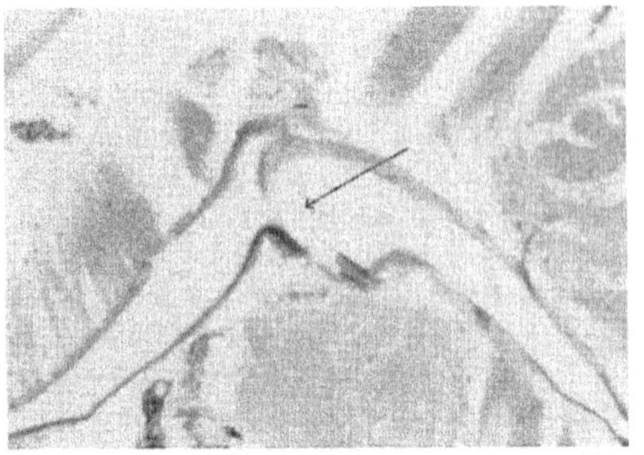

Figure 9.4. The buccopharyngeal armature (arrow) of *Simulium ochraceum* viewed in longitudinal section (courtesy of R.C. Collins, Centers for Disease Control and Prevention).

the *S. damnosum* complex, thereby counteracting the effects of the high prevailing levels of skin microfilariae in humans infected with forest form *O. volvulus* (Duke et al. 1966). Other factors inhibit filarial penetration through the midgut itself (Sutherland et al. 1986).

There also are physiological and biochemical modulators in the vector midgut that affect filarial movement, penetration and invasiveness. Recognizing that coagulation times for ingested bloodmeals varied widely by potential filarial vector species led to the hypothesis (Kartman 1953, Nayar and Sauerman 1975, Obiamiwe 1977, Buxton and Mullen 1980) that entrapment of microfilariae within clotted midgut contents might represent a significant mechanism for modulating infection intensity in inefficient vectors. Also, as is seen in *Glossina*/trypanosome interactions, lectin-like molecules within the midgut may modulate infection in potential filarial vectors by inhibiting outward migration of microfilariae. As evidence, Phiri and Ham (1990) showed that addition of N-acetyl-D-glucosamine (a lectin binding compound) to the infective bloodmeal resulted in an increased proportion of *B. pahangi* microfilariae migrating from the midgut of *Ae. aegypti* from a strain selected for refractoriness to infection. The effect was less pronounced in a susceptible mosquito strain.

As microfilariae migrate through the hemocoel, both cellular and humoral components of the vector's hemolymph-based defensive system attack them. These attacks may be categorized into 2 groups. The first response is phenol oxidase catalyzed formation of melanin from phenolic precursors such as tyrosine or dihydroxyphenylalanine (DOPA). This metabolite subsequently is involved in melanotic capsule formation with or without the participation of hemocytes. The second type of defensive response is the induction of soluble antimicrobial peptides and proteins such as cecropins, attacins and lysozyme. These molecules operate on microfilariae independently of melanotic encapsulation.

Melanotic encapsulation (Fig. 9.5) begins when phenoloxidase is cleaved from the proenzyme prophenoloxidase by serine proteases induced at the time of infection. Phenoloxidase, in turn, catalyzes the oxidation of phenols such as tyrosine, DOPA and perhaps other phenols and catechols to form melanin. Melanotic encapsulation may be humoral in nature involving the formation of a noncellular matrix around invading microfilariae. Conversely, melanin deposition may augment the cellular encapsulation response (Townson and Chaithong 1991, Ham et al. 1994, Li et al. 1994). Humoral melanization is not restricted to microfilariae within the hemocoel. Numerous investigators (Schacher 1962, Lehane and Laurence 1977, Nayar et al. 1989) have noted intracellular melanization of larvae developing within fibrils of the indirect flight muscles and primary cells of the Malpighian tubules. Although proteases responsible for cleavage/activation of prophenoloxidase are induced in response to injury or the presence of nonself entities such as infectious agents, the phenoloxidase-based defensive system is mainly a vector defensive mechanism against infection.

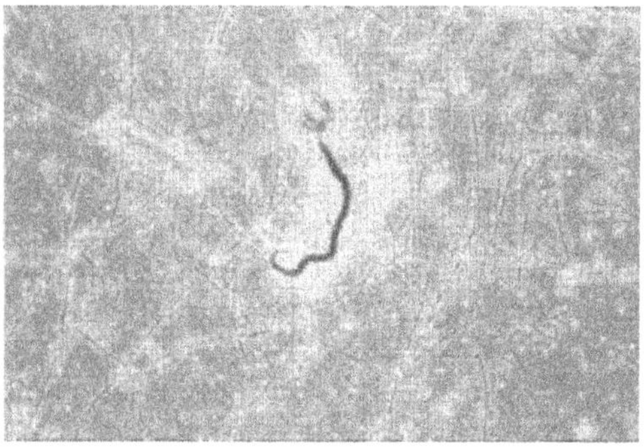

Figure 9.5. Melanized microfilaria of *Dirofilaria immitis* from the hemocoel of *Aedes aegypti* illustrating the result of a cell-mediated vector defense response (courtesy of Bruce M. Christensen).

Induction of lytic or antimicrobial proteins and peptides in the hemolymph in response to primary infection is a type of acquired resistance. Ham et al. (1995) reviewed the best-documented examples of these responses to filariae. In one example, hemolymph factors are induced that impart a degree of resistance to secondary infection during a primary onchocercal infection in susceptible black fly vectors. These factors may be transferred passively in hemolymph from resistant flies to confer resistance in näive recipients (Ham and Garms 1988) or to inhibit the motility of *Onchocerca* microfilariae *in vitro* (Ham and Garms 1988). Similar mechanisms may work in mosquito/filarial systems (Townson and Chaithong 1991). Here, transferable factors are induced in the hemolymph of *Ae. aegypti* in response to *B. pahangi* infection. These inducible responses in mosquitoes lack the specificity of humoral responses in vertebrates. Bacteria (*Escherichia coli*) or bacterial products such as lipopolysaccharides (LPS) bring about resistance to secondary filarial infection that is equivalent to that induced by microfilariae. The induction of resistance factors in black fly hemolymph coincides with that of specific proteases. The nature of their function is not certain, but Ham et al. (1995) hypothesized that they may activate components of the phenol oxidase system as well as lytic or toxic peptides and proteins.

Genetic basis for susceptibility to filarial infection in arthropod vectors

Roubaud et al. (1936) and Roubaud (1937) provided the first evidence that filarial susceptibility in vectors is an inherited characteristic, and showed that geographic populations of *Ae. aegypti* differed in their susceptibility to *D. immitis*. Ramachandran et al. (1960) obtained similar findings for *B. malayi* in *Ae. aegypti*. Kartman (1953) demonstrated that susceptibility to infection by *D. immitis* could be manipulated by artificial selection of *Ae. aegypti*. Macdonald (1962a) conducted the first detailed genetic analysis of filarial susceptibility. Working with *B. malayi* in *Ae. aegypti*, he increased susceptibility of a laboratory population of mosquitoes from 8–31% to 85% after 2 generations of selection. By crossing experiments between susceptible and refractory strains, he also showed that susceptibility was controlled by a single sex-linked recessive gene designated f^m. Selection of *Ae. aegypti* for susceptibility to *Brugia* automatically conferred susceptibility to another thoracic muscle-dwelling filaria, *W. bancrofti*, but did not affect susceptibility to *D. immitis* that develops in the Malpighian tubules (Macdonald and Ramachandran 1965).

Zielke (1973) and McGreevy et al. (1974a) selected populations of *Ae. aegypti* for susceptibility and refractoriness to *D. immitis*, and Terwedow and Craig (1977) selected this species for refractoriness to the fat body-developing filaria *Waltonella flexicauda*. As with f^m, back crossing experiments showed that susceptibility to *D. immitis* and *W. flexicauda* also was under the control of single sex-linked recessive genes. Selection for the gene currently designated f^i, controlling susceptibility to *Dirofilaria*, does not affect *Brugia* susceptibility, and development of *Dirofilaria* and *Brugia* are similarly unaffected by selection for and against f^{m2}, the gene conferring susceptibility to *W. flexicauda*. The modes of expression of these genes are unknown; they do produce melanization responses. Abortive parasite development, most commonly in the first larval stage, without signs of melanization, characterizes the course of infection in refractory mosquitoes.

More recent research has focused on the application of quantitative genetic methods to identify loci contributing to susceptibility and refractoriness to filarial development in *Ae. aegypti* (Severson et al. 1994, Beerntsen et al. 1995). Severson et al. (1993) constructed a complete restriction fragment length polymorphism (RFLP) map covering 75 loci among the 3 chromosomes of *Ae. aegypti* (Fig. 9.6). The RFLPs, when used as markers in segregation experiments, identified a quantitative trait fsb ("filarial susceptibility brugia") that mapped within a 10 centimorgan interval defined by the markers LF178 and LF198, and was called fsb{1,LF178}. This quantitative trait locus accounted for the most variance contributing to filarial development. A second locus also was identified on the second chromosome; it mapped along a 9 centimorgan interval marked by the LF282 and VCP loci, was named fsb{2,LF98} and had additive effects to those expressed at fsb{1,LF178} (Fig. 9.6) for filarial susceptibility. A third locus mapping to the same region called idb{2,LF181}(for "intensity determinant for *Brugia*") also was identified; it partly controlled intensity of infection, perhaps by affecting the number of filarial worms ingested and their capacity to penetrate the midgut (Beerntsen et al. 1995).

On the basis of independence of genetic control over susceptibility to filariae developing in different organ sites of *Ae. aegypti*, McGreevy et al. (1974a) proposed that the genes involved are tissue specific and that development of all filariae in a given organ site is controlled by a separate gene. However, a more complicated mode of action is suggested by Terwedow and Craig (1977), who found that f^{m2} and f^{m} are allelic. Despite inconclusive findings on the allelism of genes controlling filarial susceptibility in *Ae. aegypti*, other findings support the hypothesis of McGreevy et al. (1974a) that their action is not systemic but localized to the tissues in which the target filariae develop. Nayar et al. (1988) showed that larval *D. immitis* continue to develop in Malpighian tubules from susceptible females even when these tubules are transplanted to refractory individuals. Conversely, larvae in tubules from refractory individuals failed to develop after transplantation to susceptible recipients. Wattam and Christensen (1992) showed that unique thoracic polypeptides arising after bloodfeeding in strains of *Ae. aegypti* refractory to *Brugia* were not expressed in thoraces of mosquitoes from a strain refractory to *D. immitis*.

Despite these advances, progress towards elucidating the mode of expression of genetic factors that control filarial susceptibility in mosquitoes has been slow. Townson and Chaithong (1991) observed that some 3 decades after the discovery of f^{m}, "we are little nearer an understanding of the underlying processes by which such genes influence the development of the parasite." They correctly underscore that while *Ae. aegypti* is an excellent laboratory subject, it is not an important vector of any filaria in nature. Genetic factors affecting development of *W. bancrofti* in one of its natural vectors, *Culex quinquefasciatus*, have been identified (Zielke 1977), but their relationship to genes operating in *Ae. aegypti* remains obscure. Susceptibility to *W. bancrofti* in another natural vector, *Ae. scutellaris*, is inherited, yet factors involved are not genetic, but cytoplasmic, and correlated with the presence of symbiotic microorganisms (Trpis et al. 1981). Uncovering the molecular bases of filarial susceptibility in vector arthropods may be one of the primary challenges for medical entomologists in the 21st Century.

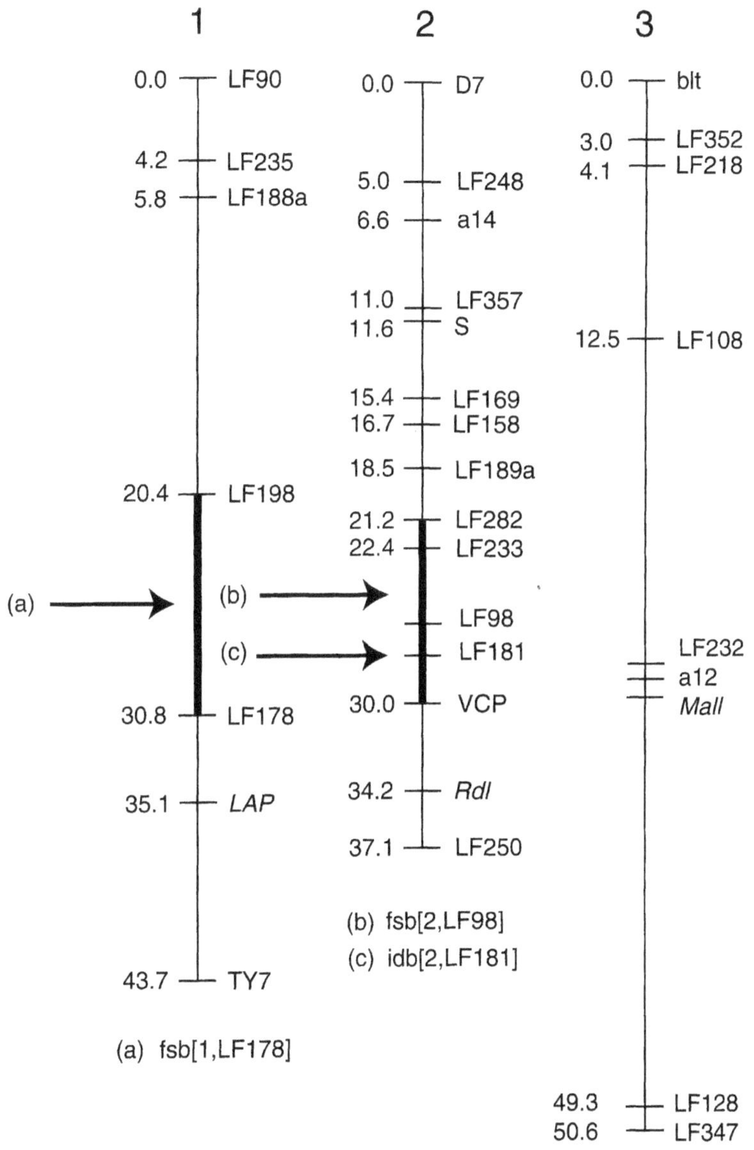

Figure 9.6. Linkage maps for filarial susceptibility in *Aedes aegypti* (compiled by Dr. David Severson, University of Notre Dame).

Methods for detecting filarial infection in field-collected vectors

Identification of L3 within field-collected vectors may be a problem in epidemiological studies in areas where more than one species of filaria is endemic or where filariae of domestic animals or wildlife share vectors with human parasites. Identification of filarial larvae in field-collected mosquito vectors involves dissection and microscopic observation of the caudal end of the L3. Nelson (1959) and Yen et al. (1982) furnish descriptions and keys. Length of worms can be used as an initial basis for determination; L3 of *Wuchereria*, *Brugia*, and *Setaria* spp. are substantially longer than L3 of *Dirofilaria* spp. The size, shape and arrangement of the caudal papillae and distance of the anus to caudal terminus identify genus and species. Generally, the caudal papillae of *W. bancrofti* are bubble-like, while those of *Brugia* are protuberant (Yen et al. 1982). The caudal papillae of *Dirofilaria* are knob-like, those of *Breinlia* are ear-like, and those of *Setaria* are thorn-like.

An alternative to microscopic identification of L3 is PCR-based detection using primers directed against appropriate target nucleotide sequences of *W. bancrofti* in mosquitoes (Siridewa et al. 1996, Furtado et al. 1997); *O. volvulus* in black flies (Katholi et al. 1995); and *D. immitis* in mosquitoes (Scoles and Kambhampati 1995; Fig. 9.7). Genomic DNA from individual or pooled, wild-caught vector insects is extracted using appropriate methods and subject to PCR assays where denaturation, primer hybridization and polymerase extension cycles have been optimized. Using this approach, Nicolas and Scoles (1997) detected both *D. immitis* and *W. bancrofti* in pools of *Aedes polynesiensis* collected in French Polynesia, an indication that this mosquito possibly serves as vector there of both filarial species.

Development in vertebrates

Vertebrates are the primary hosts for all filarial nematodes. They are the hosts for adult worms, and sexual reproduction occurs within these hosts. The tissues in which adult worms occur characterize the disease caused by the worms. Adult *W. bancrofti* remain in the lymphatic vessels and lymph nodes, and the disease caused by these and similar parasites is called lymphatic filariasis. The parasites causing river blindness, or onchocerciasis, lodge in subcutaneous tissues. The adults of parasites causing dog heartworm are found in the vena cava, right atrium, right ventricle and pulmonary artery. Some species of filarial parasites show patterns of periodicity in their vertebrate hosts, a trait that seems to have evolved to facilitate infection of their various arthropod vectors.

LYMPHATIC FILARIASIS

Parasites

Wuchereria bancrofti, *B. malayi* and *B. timori* cause human lymphatic filariasis. Adults of these parasites all inhabit the lymph nodes and lymphatic vessels of humans and, in the case of some populations of *B. malayi*, wild and domestic felids. Worms of the 3 species are similar in appearance and fit the general description of filarioids given earlier. They are slender and threadlike with bluntly tapering extremities, whitish in color and overtly featureless, lacking prominent cephalic armature. Adult *W. bancrofti* are the largest of the 3 species; females measure 80–100 mm in length and 0.24-0.3 mm in width, males measure 28–40 mm in length and 0.1 mm in width. Adult *B. malayi* are smaller; females measure 43–55 mm in length and 0.13–0.17 mm in width, males measure 14–23 mm by 0.8 mm. Adult *B. timori* are the smallest of the lymphatic dwelling filariae; females

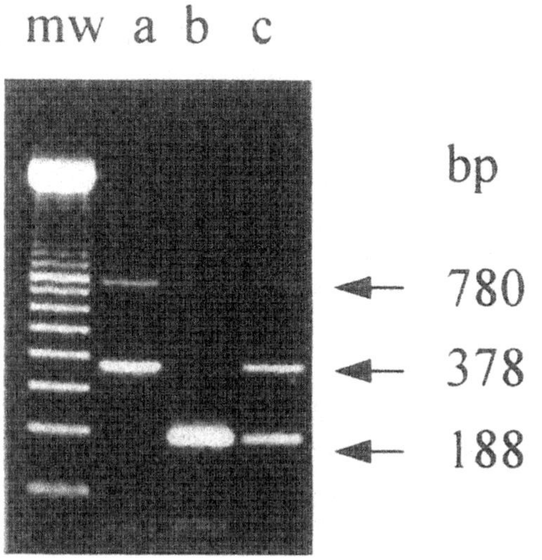

Figure 9.7. Gel showing PCR products of *Dirofilaria immitis* and *Wuchereria bancrofti* from *Aedes polynesiensis*.

measure 26.7 mm in length by 0.09 mm in width, males measure 16.9 mm by 0.07 mm (Ottesen 1990).

All of the lymphatic dwelling filariae of humans have sheathed bloodborne microfilariae. In stained blood films, these parasites can be distinguished from each other by the placement of nuclei in the cephalic and caudal extremities. *Wuchereria* is unique because it lacks caudal nuclei. Both *B. malayi* and *B. timori* have 2 terminal caudal nuclei but lack nuclei in their anterior extremities, creating the impression of a large caudal space. Unlike the sheath of *B. malayi*, that of *B. timori* does not stain with Giemsa. Microfilariae of *Brugia* spp. also can be identified on the basis of binding of species specific oligonucleotide probes to the Hha I repetitive DNA sequence (McReynolds et al. 1993).

The microfilariae of *W. bancrofti* and *B. malayi* show varying degrees of **periodicity** in the density of parasites in the peripheral blood during the day and night. The density fluctuates in a harmonic wave (Sasa and Tanaka 1972). The parasites are termed **periodic** when microfilariae disappear from the peripheral blood at some time during a 24-hr period. Filariae are described as **subperiodic** when densities of peripheral microfilariae fluctuate daily, but never fall to zero. Populations of filariae may be classified as either **nocturnally** or **diurnally periodic** or **subperiodic**, depending upon the time of day when densities of microfilariae peak in the peripheral circulation. In many cases, including all of the filariae of humans that exhibit periodicity, the peak of microfilariae in the peripheral circulation coincides with the time of peak biting activity of vector arthropods. Periodicity may vary within as well as between vector species, but the functional correlation with local vector feeding behavior persists. For example, urban populations of *W. bancrofti* are nocturnally periodic, with microfilariae peaking in the circulation at 1–2 AM. This periodicity enhances chances of uptake by the primary urban vector, the nocturnally feeding *Cx. quinquefasciatus*. On the other hand, populations of *W. bancrofti* from islands in the South Pacific are weakly diurnally periodic, corresponding with the diurnal biting habits of their aedine vectors. Circulating microfilariae of *L. loa* are most numerous around midday, which also is the time of peak biting activity of their *Chrysops* vectors. Adaptation of microfilariae to the feeding behaviors of their vectors is cited frequently as an example of host-parasite coevolution (Noble et al. 1989). Fig. 9.8 shows the distribution of the periodic forms of *W. bancrofti* and *B. malayi* in the world.

Several hypotheses have been advanced to explain physiological mechanisms of microfilarial periodicity; all suggest that host factors interact with a circadian rhythm in the parasite (Hawking 1967, Eberhard and Rabalais 1976, Zielke 1980). When absent from the peripheral circulation, microfilariae are sequestered in the precapillary arterioles of the lung. Sequestration of microfilariae is accompanied by subtle changes in motility, with increased amplitude of parasite undulations to the point where they become lodged in the downward tapering arterioles. Hawking (1967) hypothesized that this sequestering behavior depends on the sensitivity of microfilariae to the differential in oxygen tension between the pulmonary arterial and venous blood. The various manifestations of periodicity seen in different parasites are due to differences in the magnitude of this sensitivity and the degree to which it may be modified by other fluctuating parameters in the host, such as body temperature. Evidence for the participation of host factors in microfilarial periodicity varies, but in general involves manipulation of host physiology such as administration of oxygen (Hawking 1967) or varying activity patterns via shifts in the photoperiod (Grieve and Lauria 1983) to bring about corresponding changes in microfilarial behavior.

The basis for classification of intraspecific races of lymphatic filariae is the varying pattern of periodicity. Most populations of *W. bancrofti* show a strong nocturnal periodicity with numbers of parasites peaking in the circulation around 3 AM (Hawking and Gammage 1968). However, Polynesian and other island populations of *W. bancrofti* from the southern Pacific region either lack periodicity or show a slight diurnal periodicity. The significance of these differences to the biting habits of the local vector mosquitoes will be discussed in more detail below. An example is the mosquito *Aedes poicilius*, vector of *W. bancrofti* in the Philippines. This mosquito has a nighttime feeding behavior that coincides with the nocturnally periodic form of *W. bancrofti* occurring in that country. *Brugia malayi* comprises both a nocturnally periodic race in coastal estuary regions of Asia and a nocturnally subperiodic race distributed in regions of Asia with fresh water marshes and swamps. *Brugia timori* appears to be exclusively nocturnally periodic (Klion et al. 1991).

Life Cycle

Vertebrate hosts and distribution

In nature, *W. bancrofti* is restricted to human hosts. However, leaf monkeys, *Presbytis* spp., have been infected experimentally (Palmeri et al. 1983), potentially fulfilling the long-standing need for an animal model for Bancroftian filariasis. Similarly, no animal reservoir hosts for periodic *B. malayi* or for *B. timori* are known. On the other hand, wild and domestic felids, pangolins and leaf monkeys are naturally infected with the subperiodic race of *B. malayi*. The propensity of subperiodic *B. malayi* to infect nonprimate hosts has led to its adaptation to laboratory animals such as cats (Denham and Fletcher 1987) and gerbils (Klei et al. 1982), making it a widely used subject for research.

Diseases

Pathology

Lymphatic filariasis comprises a spectrum of acute and chronic clinical manifestations in humans (Ottesen 1990, Ottesen 1993). Many infected individuals apparently tolerate their infections well and show circulating microfilariae, but no clinical symptoms. This condition is termed **asymptomatic microfilaremia.** Such individuals constitute the primary reservoir of infection in non–zoonotic filariasis. In other infected individuals, lymphatic filariasis is manifested primarily as an **acute inflammatory lymphatic disease.** There is episodic pain and tenderness of the lymph nodes and occasionally of the lymphatic vessels (lymphangitis or **adenolymphangitis**, abbreviated **ADL**). Acutely affected individuals suffer from filarial fevers during episodes of ADL, particularly during the transmission season. Over time, deep abscesses may develop at the sites of inflammation. Dermal ulcers may form through the skin over these sites, and secondary bacte-

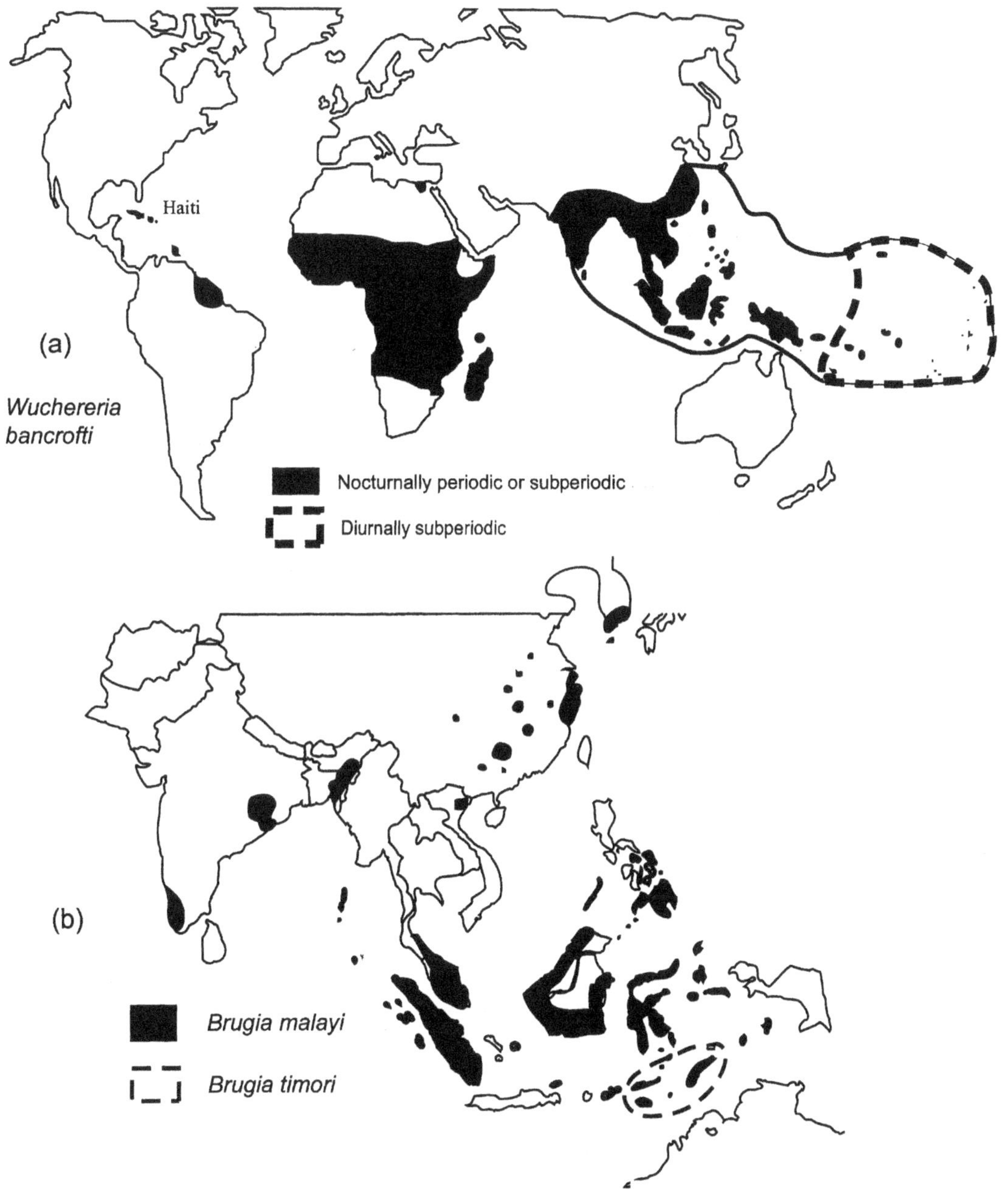

Figure 9.8. Distribution of *Wuchereria bancrofti* (a), *Brugia malayi* and *Brugia timori* (b), including periodic races and forms.

Figure 9.9. A Ghanian woman with lymphadenopathy of the leg and foot from *Wuchereria bancrofti* infection [courtesy of John M. Hunter, Michigan State University. Reprinted from Hunter (1992) with permission from Elsevier Press].

rial and fungal infection may develop and exacerbate inflammation. Circulating microfilariae usually are not observable in patients with the acute inflammatory disease. Another subset of lymphatic filariasis patients exhibit **chronic lymphatic pathology** indicative of compromised lymphatic function. These symptoms are associated most commonly with lymphatic filariasis and may include chyluria (a milky discoloration of urine because of contamination with lymphatic fluid), hydrocele (enlarged scrotum in men caused by fluid accumulation) and lymphadenopathies of breasts and / or extremities ranging from lymphoedema to elephantiasis in the most severely affected individuals (Fig. 9.9). As with acute lymphatic inflammation, microfilariae frequently are absent from the peripheral circulation in patients with chronic lymphatic pathology. In some patients, lymphatic filariasis may be a pulmonary disease not involving the lymphatic system. These patients have severe nocturnal coughing, diffuse lesions in the interstitial lung tissue, eosinophilia, elevated levels of IgE and specific antifilarial antibodies. This syndrome is called **tropical pulmonary eosinophilia** or **TPE.** Patients with TPE almost always lack circulating microfilariae and do not harbor adult filariae. Whether these different clinical manifestations represent a progression of disease is unknown. This uncertainty stems from the lack of an animal model recreating all aspects of the human infection and the great difficulty and ethical constraints associated with tracing the clinical course of disease in humans.

Immune responses

The different forms of disease described above are associated with different types of immune responses in patients (reviewed by Ottesen 1992). Historically, asymptomatic microfilaremics have been characterized as immunologic hyporesponsive or "immunosuppressed." When compared to asymptomatic microfilaremics, patients with elephantiasis or other chronic lymphatic pathology exhibit high lymphocyte proliferative responses to parasite antigen, elevated production of the inflammatory cytokine interferon gamma and increased production of total IgG antibodies in response to filarial antigens. However, Ottesen (1992) cautions that this hyporesponsiveness in asymptomatic microfilaremics pertains only to the pro-inflammatory type of cell-mediated immune response and that when anti-inflammatory responses are evaluated, these patients may be characterized as hyperresponsive. In recent years, it has become increasingly clear that humoral responses in chronic lymphadenopathy patients favor the production of pro-inflammatory IgG subclasses, e.g., IgG3, whereas asymptomatic microfilaremics produce high levels of IgG4 that inhibits allergic responses. Moreover, cytokines produced in response to filarial antigen by lymphocytes of patients with asymptomatic microfilaremia generally fit the model of an antiinflammatory Thl–type response, whereas cytokine profiles of lymphadenopathy or TPE are consistent with a pro–inflammatory Th2–type response (IL–4, IL–5).

Factors that predetermine the character of the immune response to lymphatic filarial infection, and consequently the nature of the disease, are under investigation but remain poorly understood. One possibility is that innate differences in antigen processing linked to the major histocompatibility complex or to blood group phenotypes may affect the outcome of the disease process (Ottesen 1992). Another possibility is that prenatal experience with filarial antigen may direct responses to postnatal infection. For example, there is a higher incidence of microfilaremia among individuals born to microfilaremic mothers than in the general population of filariasis patients (reviewed by Hightower et al. 1993).

In addition to the various disease manifestations discussed above, there is another group of individuals in areas of endemic lymphatic filariasis. This group includes persons who, despite ongoing exposure to the parasite, never become infected or develop disease signs. These individuals are called **endemic normals** and are regarded as immune.

Although immunopathology is important in the induction of chronic lymphatic pathology in filariasis patients, direct or non–immune mediated effects by the parasite can be involved as well. Providing evidence of this assumption, Nelson et al. 1991 and Vickery et al. 1991 found that congenitally immunodeficient mice exhibited lymphatic pathology in response to experimental infection with *B. pahangi* or *B. malayi*.

Diagnosis

Microfilaremic patients

Detecting microfilariae in peripheral blood samples is the most common method of diagnosing filarial infection in suspected lymphatic filariasis patients. Direct examination of 20 µl of blood prepared as a hematoxylin- or Giemsa-stained thin smear is a common screening procedure. Sensitivity is gained by using concentration methods that allow the examination of relatively large volumes of blood. One such method is Knott's concentration technique in which 1 cc of venous blood is lysed with an excess of 2% formalin, concentrated by centrifugation and the pellet containing lymphocytes and microfilariae examined microscopically. Another concentration method involves passage of a lysed, 1 cc blood sample through a polycarbonbate membrane filter of 3 µm pore size followed by staining and microscopic examination of the retained microfilariae. This method is easier to use than the Knott technique. It does not require a centrifuge and ensures that the entire proceeds of a 1 cc blood sample are confined to a 25 mm diameter field. Depending on the volume of pelleted material from a Knott test, larger areas may have to be scanned microscopically. The Knott concentration and stained direct smear methods allow morphological study of microfilariae, a critical factor where more than one filarial species is endemic and species must be differentiated. In addition to blood, microfilariae frequently may be demonstrated in urine or hydrocele fluid.

Table 9.3. Mosquito species incriminated as vectors of filarial nematodes with associations with nocturnally periodic (NP), nocturnally subperiodic (NS) or diurnally subperiodic (DS) forms. (Table continued on next page).

Geographic zone	Species	Parasite	Type of periodicity
Neotropical	*An. aquasalis*	*W. bancrofti*	NP
	An. bellator	*W. bancrofti*	NP
	An. darlingi	*W. bancrofti*	NP
	Ae. scapularis	*W. bancrofti*	NP
	Cx. quinquefasciatus	*W. bancrofti*	NP
	Ma. titillans	*W. bancrofti*	NP
Afrotropical	*An. funestus*	*W. bancrofti*	NP
	An. gambiae	*W. bancrofti*	NP
	An. arabiensis	*W. bancrofti*	NP
	An. bwambae	*W. bancrofti*	NP
	An. melas	*W. bancrofti*	NP
	An. meras	*W. bancrofti*	NP
	An. nili	*W. bancrofti*	NP
	An. pauliani	*W. bancrofti*	NP
	Cx. quinquefasciatus	*W. bancrofti*	NP
Middle Eastern	*Cx. pipiens molestus*	*W. bancrofti*	NP
Oriental	*An. barbirostris*	*B. malayi, B. timori*	NP
	An. campestris	*B. malayi*	NP
	An. donaldi	*B. malayi*	NP
	An. anthropophagus	*W. bancrofti, B. malayi*	NP
	An. kweiyangensis	*W. bancrofti, B. malayi*	NP
	An. nigerrimus	*W. bancrofti*	NP
	An. sinensis complex	*W. bancrofti*	NP
	An. letifer	*W. bancrofti*	NP
	An. whartoni	*W. bancrofti*	NP
	An. aconitis	*W. bancrofti*	NP
	An. flavirostris	*W. bancrofti*	NP
	An. minimus	*W. bancrofti*	NP
	An. candidiensis	*W. bancrofti*	NP
	An. balabacencis	*W. bancrofti*	NP
	An. leucosphyrus	*W. bancrofti*	NP
	An. maculatus	*W. bancrofti*	NP
	An. philippensis	*W. bancrofti*	NP
	An. subpictus	*W. bancrofti*	NP
	An. vagus	*W. bancrofti*	NP

Table 9.3. (Continued). Mosquito species incriminated as vectors of filarial nematodes with associations with nocturnally periodic (NP), nocturnally subperiodic (NS) or diurnally subperiodic (DS) forms.

Geographic zone	Species	Parasite	Type of periodicity
Oriental (continued)	*Ae. niveus*	*W. bancrofti*	NP
	Ae. harinasutai	*W. bancrofti*	NS
	Ae. togoi	*W. bancrofti, B. malayi*	NP
	Ae. poicilius	*W. bancrofti*	NP
	Cx. bitaeniorhynchus	*W. bancrofti*	NP
	Cx. sitiens complex	*W. bancrofti*	NP
	Cx. pipiens pallens	*W. bancrofti*	NP
	Cx. quinquefasciatus	*W. bancrofti*	NP
	Ma. uniformis	*W. bancrofti*	NP
		B. malayi	NP, NS
	Ma. bonnae	*B. malayi*	NP, NS
	Ma. annulata	*B. malayi*	NS
	Ma. indiana	*B. malayi*	NS
	Ma. dives	*B. malayi*	NP, NS
Western Pacific	*Cx. pipiens pallens*	*W. bancrofti*	NP
	Ae. togoi	*W. bancrofti, B. malayi*	NP, NS
Papuan	*An. bancrofti*	*W. bancrofti*	NP
	An. punctulatus	*W. bancrofti*	NP
	An. koliensis	*W. bancrofti*	NP
	Cx. annulirostris	*W. bancrofti*	NP
	Cx. bitaeniorhynchus	*W. bancrofti*	NP
	Ma. uniformis	*W. bancrofti*	NP
South Pacific	*Ae. fijiensis*	*W. bancrofti*	DS
	Ae. oceanicus	*W. bancrofti*	DS
		B. malayi	NP, NS
	Ae. samoanus	*W. bancrofti*	NS
	Ae. vigilax group	*W. bancrofti*	DS
	Ae. futunae	*W. bancrofti*	DS
	Ae. polynesiensis	*W. bancrofti*	DS
	Ae. pseudoscutellaris	*W. bancrofti*	DS
	Ae. tabu	*W. bancrofti*	DS
	Ae. tongae	*W. bancrofti*	DS
	Ae. upolensis	*W. bancrofti*	DS

Amicrofilaremic patients

As already discussed, lymphatic filariasis patients with acute or chronic lymphadenopathy or with pulmonary disease frequently do not have circulating microfilariae. Clinical presentation is the key to diagnosis in such cases. The presence of eosinophilia associated with radically elevated total serum IgE levels and filaria-specific IgE and IgG are diagnostic. Such findings in association with a history of prolonged or intense exposure and positive lymphangiographic results support the clinical diagnosis of filariasis. None of these clinical findings is sufficient for conclusive diagnosis of filariasis. Modern serodiagnostic tests measure specific parasite products in the serum by means of antigen capture immunoassays (Weil et al. 1987). The advantages of such assays are that they directly measure products of adult parasites rather than the nonuniform host's immunologic response to them. The assays do not rely on the presence of microfilariae which, as explained above, do not occur universally among filariasis patients, or may not be apparent in blood samples drawn during the daytime. Additionally, methods of detecting parasite nucleic acids by PCR have been developed, using primers directed against parts of the genome that generate species-specific gel phenotypes.

Vectors

Only mosquitoes transmit *W. bancrofti* and *Brugia* spp. The development of the lymphatic-dwelling filariae in their mosquito vectors was presented earlier under the generalized account of filarial life cycles. Mosquitoes can become infected by feeding on the peripheral blood of microfilaremic hosts. The numerous species of mosquitoes responsible for transmission of the 3 species of nematodes that cause lymphatic filariasis are reviewed below. In all cases, larval development takes place in the indirect flight muscles. As with all filariae, the duration of vector phase development in the lymphatic dwelling parasites is a function of ambient temperature and usually ranges from 8–16 days. Third-stage larvae (L3) are transferred to the host from the mouthparts of the vector during a subsequent bloodfeeding.

Table 9.3 lists the various vector species in relation to the species of filarial nematodes they transmit, the periodicity of the microfilariae, and the geographic occurrence of vectors and parasites. A wide range of mosquito species in the genera *Culex, Anopheles, Aedes* and *Mansonia* are competent vectors.

Vectors of periodic and subperiodic *W. bancrofti*

Culex quinquefasciatus is the principal vector of *W. bancrofti* in tropical urban environments (Fig. 9.10). *Culex quinquefasciatus* larvae occur commonly in domestic impoundments of water containing high levels of organic contamination, including cisterns, rain barrels, sewage pits, latrines and drains. *Culex quinquefasciatus* thrives in crowded urban environments with poor sanitation (Horsfall 1972). Adult *Cx. quinquefasciatus* are **endophilic** (their common name is the "southern house mosquito"). Although they readily enter dwellings to feed, they are timid bloodfeeders and react quickly to host defensive behavior. They bloodfeed at night on sleeping individuals. Their peak biting hours are from midnight to 4 AM, coinciding with the peak in microfilaremia in hosts of nocturnally periodic *W. bancrofti*, the most prevalent race of the parasite in urban ecosystems.

Other members of the *Culex pipiens* complex transmit nocturnally periodic *W. bancrofti* in subtropical areas. *Culex pipiens pallens* has been a primary vector in China and Japan, and *Culex pipiens molestus* in the eastern Mediterranean, including the Nile River delta. The larval breeding and biting habits of these mosquitoes resemble those of *Cx. quinquefasciatus*.

Zielke and Kuhlow (1977) investigated the genetics of *Wuchereria* susceptibility in *Cx.*

pipiens complex mosquitoes and found at least 2 genes to control development of the parasite. Neither the linkage relationships of these genes nor the degree of their homology to f^m, the gene controlling *Brugia* and *Wuchereria* susceptibility in *Ae. aegypti*, are known.

Mosquitoes of the genus *Anopheles* are important vectors of nocturnally periodic *W. bancrofti* in rural areas of Africa, Papua New Guinea and parts of Indonesia. In rural Africa, the principal vectors are members of the *Anopheles gambiae* species complex, chiefly *An. gambiae* s.str. and *Anopheles funestus*. Although *An. funestus* larvae may be found at the littoral zones of larger ponds and swamps, larval *An. gambiae* frequently inhabit small, temporary puddles such as foot or hoof prints, wheel ruts and other similar types of small, muddy habitats, nearly always in bright sunlight. Adults of *An. gambiae* and *An. funestus* frequent the dwellings of humans and domestic animals and tend to be endophilic and nocturnal in their biting habits. *Anopheles gambiae* is a relatively inefficient vector of *W. bancrofti* because of its highly developed cibarial armature. Anopheline vectors of periodic *W. bancrofti* in Irian Jaya, Indonesia, Papua New Guinea, and the Solomon Islands include *Anopheles farauti*, *Anopheles punctulatus* and *Anopheles koliensis*. Anopheline vectors of *W. bancrofti* in both Africa and Asia also are primary vectors of *Plasmodium* spp. within their distributions. *Anopheles darlingi* transmits nocturnally periodic *W. bancrofti* in localized endemic rural areas in South America.

Some aedine mosquitoes transmit nocturnally periodic *W. bancrofti* in tropical and subtropical Asia. In the Philippines, *Ae. poicilius* is a principal vector. Larvae of this mosquito develop in water contained in the leaf axils of plants such as abaca and banana. *Aedes togoi* is a vector of periodic *W. bancrofti* in areas of northern Asia, including China, coastal Japan and the Korean peninsula. Development of larval *Ae. togoi* occurs in brackish waters in rock pools or in a variety of domestic containers. This mosquito bloodfeeds at night and readily enters dwellings, but feeds outdoors as well (Horsfall 1972, World Health Organization 1987).

The primary vectors of diurnally subperiodic *W. bancrofti* throughout the islands of the eastern Pacific are members of the *Ae. scutellaris* species complex, the most important being *Ae. polynesiensis*. Larvae of this species develop in various small natural and domestic containers such as tree holes, hollow coconut shells, lacquered tin cans and discarded household utensils (Horsfall 1972). Females of *Ae. polynesiensis* are strongly **exophilic**, and they are active during daylight hours. The ability to concentrate microfilariae from the host circulation during bloodfeeding appears to enhance susceptibility of *Ae. polynesiensis* to infection with *W. bancrofti*. Although in different studies this concentration varies from 0.7 to as high as 12 (Bryan and Southgate 1976, Samarawickrema et al. 1985), the numbers of microfilariae ingested by *Ae. polynesiensis* are consistently higher than would be predicted on the basis of the microfilaremia. A slow rate of blood ingestion or a salivary chemoattractant may be the basis for this concentration phenomenon. In American Samoa and Samoa, *Aedes samoanus* is the primary vector of subperiodic *W. bancrofti*. Larvae of this mosquito develop in leaf axils and adult females bloodfeed and rest outdoors at night (World Health Organization 1987). Night biting by *Ae. samoanus* does not preclude transmission of subperiodic *W. bancrofti* because sufficient numbers of microfilariae remain in the peripheral circulation at night to produce optimum infections.

Vectors of *Brugia malayi* and *Brugia timori*

Like bancroftian filariasis, the nocturnally periodic and subperiodic races of *B. malayi* have different transmission cycles in different ecosystems, involving different genera of mosquito vectors. The transmission of *B. timori* is limited to isolated foci in Indonesia. *Brugia timori* is

Figure 9.10. Typical habitat of urban bancroftian filariasis, where *Culex quinquefasciatus* is the primary vector (photograph by Dr. Thomas Streit, University of Notre Dame).

native to areas in and around coastal wetlands (Ottesen 1990). This characterization aptly describes the larval breeding habitats of their principal vector in Indonesia, *Anopheles barbirostris*. Larvae of this mosquito occur in shallow impoundments with emergent vegetation such as swamps, rice fields, salt marshes and ponds. They also are present near algal blooms at the margins of flowing streams. Adults are exophilic and will feed at all hours, preferring shaded areas during the day. Although they will attack humans, they prefer larger hosts and the majority feed on cattle, if present (Horsfall 1972).

Anopheles sinensis transmits periodic *B. malayi* on the Asian mainland. The ecological niche of this species is similar to that of *An. barbirostris* in Indonesia, although *An. sinensis* is less tolerant of saline water. Adults are largely zoophilic, but a small proportion feeds on humans. In Malaysia, *Anopheles campestris* serves as the vector of periodic *B. malayi* (Horsfall 1972).

On the Indian subcontinent and in Sri Lanka, *Mansonia annulifera*, *Mansonia uniformis* and other *Mansonia* species transmit periodic *B. malayi*. *Mansonia uniformis* also transmits this parasite in Indonesia and northwestern Malaysia (World Health Organization 1987). Larvae of these mosquitoes are found in inland impoundments only in the presence of emergent vegetation. Larval *Mansonia* spp. penetrate air-filled cells in the roots of these plants with their respiratory siphons. *Mansonia annulifera* is found exclusively in association with water lettuce, *Pistia stratiotes*, while *M. uniformis* larvae associate both with *Pistia* and with other plants in the same types of habitats as water hyacinth (Horsfall 1972). Adults of *M. annulifera* are endophilic and anthropophilic. Bloodfed females leave dwellings of their hosts to rest and oviposit on leaves of the host plant of the larvae. Female *M. uniformis* prefer to bloodfeed outdoors, but also will enter houses.

In addition to periodic *W. bancrofti*, *Ae. togoi* also transmits nocturnally periodic *B. malayi* in China and Korea. The larval habitats and bloodfeeding tendencies of this species are discussed above.

Throughout its range in Malaysia, Indonesia and the Philippines, subperiodic *B. malayi* is associated with *Mansonia* spp., most importantly *Mansonia dives* and *Mansonia bonneae*. Chiang (1993) summarized the literature on these 2 species, noting that their larval biology resembles *M. uniformis* in that, unlike *M. annulifera*, both are associated with a wide range of nonrooted (e.g., *Pistia* spp.) and rooted aquatic plants. Both species are zoophilic but will feed on humans in the absence of preferred large animal hosts. Chiang (1993) cited an unpublished field study where the human biting index for *M. bonneae* was 0.04 in an area where cattle were present and 0.50 in an area where there were none. *Mansonia dives* and *M. bonneae* generally are exophilic and nocturnal in their bloodfeeding habits. They are strong fliers with maximum flight ranges of 3.2 km.

Epidemiology

Transmission of lymphatic filariae can be broken down into 4 components of the parasite life cycle. The first is the acquisition of microfilariae from the host by bloodfeeding mosquitoes. The second is the development of infective larvae in the vector. Efficiency of development of microfilariae to L3 was discussed previously under **Determinants of filarial susceptibility**. It varies considerably among species of mosquito vectors. The third component is the rate of transfer of L3 to human hosts during bloodfeeding by infective mosquitoes. The fourth component is the maturation of L3 to adult male and female worms and the immunopathological response of the human host to these parasites. Research on lymphatic filariasis has a long history of quantitative analyses of these different components, including the original studies by Hairston and DeMeillon (1968) and Hairston and Jachowski (1968), the compendium by Sasa (1976), and the studies by Southgate (1992b), Southgate and Bryan (1992), Michael (1993) and Kazura et al. (1997). A modeling exercise using an EXCEL spreadsheet program is available (Chan et al. 1998).

The efficiency of transmission of both microfilariae to mosquitoes and L3 to humans varies. Transmission is inefficient in endemic areas, and populations of filarial nematodes are strongly r-selected, i.e., very few microfilariae reach adulthood. However, precise estimates of their reproductive success are lacking. Since the studies by Hairston and DeMeillon (1968), many epidemiological analyses quantifying the inoculation rate of L3 to humans have been conducted. Consequently, the measurement of this rate has been standardized. The **annual biting rate (ABR)** is the number of bites a single person in an endemic area will receive from competent vectors during a single year. It is estimated by collecting biting mosquitoes on a daily basis to ensure coverage of the entire biting cycle of the mosquitoes, then extrapolating the number of mosquito bites for each species to the year, and calculating the ABR as follows:

ABR = Number mosquitoes caught x number of days in one year/number of sampling days.

The **annual infectious biting rate (AIBR)** equals:

Number of mosquitoes with L3s caught x number of days in one year/number of sampling days.

Then, the **annual transmission potential (ATP)** equals:

(Number of infective larvae received by one person in one year x number of days in one year/number of sampling days) x (number of mosquitoes caught/number of mosquitoes dissected for counts of L3).

The values can be calculated for other time units, e.g., month, week or day. Taken together, these 3 variables represent the parameters of the vector component of the epidemiology of filariasis. The basic human epidemiological variables are:

(1) the percentage of microfilaremic people, often called rate or prevalence of microfilaremia.

(2) the density (or intensity) of microfilaremia, i.e. the number of microfilaria per unit volume of blood.

(3) the age-specific prevalence of microfilaremia, usually calculated on a yearly basis; and

(4) the percentage and age-specific prevalence of acute and chronic symptoms, e.g., lymphadenopathy, lymphoedema and hydrocele.

Incidence data for these human variables are harder to obtain than prevalence data because of the length of time over which infections and symptoms become manifest in individuals.

Wuchereria bancrofti is the primary cause of lymphatic filariasis in the Philippines. There, 20 million people are at risk of exposure; *B. malayi* also occurs in some localized areas, particularly on southern islands. The Bicol region of southeastern Luzon is one of the most endemic regions for bancroftian filariasis in the country because abaca plants (*Musa textilis*) are cultivated to produce Manila hemp fiber. Larvae of the primary vector of *W. bancrofti* in the Philippines, *Ae. poicilius*, dwell in the water-filled leaf axils of these plants as well as in leaf axils of other plants such as banana and taro (Fig. 9.11). The vector capacity of *Ae. poicilius* for *W. bancrofti* is high because (1) female mosquitoes are competent vectors, nearly all females becoming infected experimentally with L3 after having fed on a microfilaremic person; (2) females bite both indoors and outdoors; (3) females have a nocturnal feeding rhythm synchronous with the periodic form of *W. bancrofti* (Fig. 9.12); (4) studies have shown a human blood index for female mosquitoes of about 0.70, indicating a predilection for human hosts; and (5) females have a daily survival rate of 0.85, providing sufficient time for the extrinsic incubation period of *W. bancrofti* (Walker et al. 1998).

Valeza and Grove (1979) developed a simple transmission model of *W. bancrofti* in the Bicol region of the Philippines based upon epidemiological and entomological data, where the infection rate of *Ae. poicilius* with L3 was 2.26% and a person was bitten an average of 19.5 times by *Ae. poicilius* per night. The product of the vector infection rate and the number of *Ae. pocilius* bites per person per night yields an average of 0.44 infective bites per person per night; this value is the nightly IBR. Assuming conservatively an annual transmission season of 6 months and an ABR of 3,559 (19.5 x 365/2), then each person in an endemic village would receive 80.3 infective bites per year (the AIBR; i.e., 0.0226 x 19.5 x 365/2). Given an average of 3.38 L3 per infectious bite (Valeza and Grove 1979), each person would be inoculated with ca. 271 larvae annually (the ATP). The age-specific microfilaremia rates increase at about 0.49% per year of age among villagers in abaca-growing regions of the Bicol region (Grove et al. 1978). That value (0.49%) divided by 80.3 infectious bites per person per year represents the efficiency of transmission, equal to 6.1×10^{-5}; i.e., only one patent microfilaremic human case of filariasis is produced for every 16,400 infectious bites (the reciprocal of 6.1×10^{-5}) and 55,432 infective stage larvae delivered per person per year. The former expression, 6.1 x 10^{-5}, is analogous to the parameter "b" in Macdonald's formula to estimate the malaria reproductive rate (Southgate 1992a). It represents the probability that a single infectious bite will yield a patent infection in a human.

The inefficiency of parasite transmission illustrated by the above example characterizes *W. bancrofti* and *B. malayi* (Hairston and deMeillon 1968, Gubler and Battacharya 1974, Southgate 1992a). Southgate (1992a) compiled data on transmission efficiencies relative to other epidemiological parameters (see Table 9.4). Efficiency of transmission varies geographically and among vector species. Transmission seems to be more efficient in Africa compared to Asia; transmission of *B. malayi* by *Mansonia* spp. is more efficient by 1-2 orders of magnitude than transmission of *W. bancrofti* by *Culex, Anopheles* or *Aedes* vectors. Transmission efficiency of *W. bancrofti* by *Cx. quinquefasciatus* varies from 10^{-3}–10^{-5}, depending upon the geographic area. The factors affecting efficiency are poorly known. One explanation, perhaps the most parsimonious one, is that there is a low probability of any one L3 finding its way to a suitable tissue site for development to the adult stage and then finding a mate at that site. It is well known that when näive human populations become exposed to infectious bites, such as has happened to American military personnel in the Pacific Islands in World War II and during the Vietnam War (Wartman 1947, Colwell et al. 1970), the number of resultant infections is high, suggesting that history of population exposure and immune factors are important. Another explanation is that the survival of L3 on human skin varies with temperature and humidity so that dry conditions reduce survival prior to the L3 penetrating the skin. Thus, many infectious bites may not result in L3 inoculations, reducing transmission efficiency.

The above analysis is done from the perspective of transmission from mosquito to human; it focuses on the inoculation rate of parasites and the rate at which infections become established. Another approach is to quantify transmission from human to mosquito. This approach relates the prevalence, intensity and periodicity of microfilaremia in the human population to the intensity and periodicity of mosquito biting on humans in order to predict the rate the vector population acquires infections (Sasa 1976, Webber and Southgate 1981). Webber and Southgate (1981) hypothesized a threshold vector biting density below which transmission cannot occur. They investigated changes in man-biting rate of *An. farauti* on Guadalcanal, Solomon Islands, in relation to acquisition of *W. bancrofti* microfilariae before and after residual house spraying with DDT (for malaria control) to test this hypothesis. Transmission ceased in areas where vector density was reduced below 3 bites per person per hr, and they concluded that a significant amount of vector biting could be tolerated while still achieving disease control. This observation may be valid only in areas where certain species of *Anopheles* are vectors.

There are at least 2 problems affecting the accuracy of the 2 analyses discussed above. First, both assume that vectors bite people randomly in space, and thus L3 would be transmitted to humans randomly regardless of their age, sex and location. Yet lymphatic filariasis is characterized by marked aggregation, or overdispersion, of infection, indicating that within any given endemic area, some humans are more likely to be exposed than others. These types of variations probably occur because of spatial heterogeneity in vector-human contact, and thus parasite transmission. For example, in the Philippines example given above, men tend to have higher rates of microfilaremia than do women, probably because annual abaca harvesting practices result in greater exposure of men (who work in temporary harvest camps

Table 9.4. Efficiency of transmission of *Wuchereria bancrofti* and *Brugia malayi* in different geographic areas. Note that when different vectors are present, efficiency is additive [data compiled from Southgate (1992), references therein and other sources]. (Table continued on next page).

Country and locale	Vector	ABR[1]	AIBR[2]	ATP[3]	TE[4]	Infectious bites required[5]
	Wuchereria bancrofti periodic strain					
Tanzania						
	Cx. quinquefasciatus				7.85×10^{-5}	12,674
	An. gambiae	1,090	8	8		
Tawalani	*An. gambiae*	10,846	152	368	3.30×10^{-4}	3,030
	An funestus	2,346	37	40		
Kwale	*An. gambiae*	2,648	24	82	4.33×10^{-4}	2,310
Machui	*Cx. quinquefasciatus*	2,311			3.72×10^{-3}	269
	An. gambiae	3,585	16	72		
Kenya						
Mambrui	*Cx. quinquefasciatus*	4,600	46	155	3.29×10^{-4}	3,040
Jaribuni	*An. gambiae*	818	9	22	1.36×10^{-4}	7,353
Liberia						
Marshall Terr.	*An. gambiae*	19,500	350	1,000	8.55×10^{-5}	11,696
Burkina Faso						
Tingrela	*An. gambiae*	3,049	56	135	1.60×10^{-4}	6,250
India						
Calcutta	*Cx. quinquefasciatus*	1.15×10^{-5}	1,850	5,904	5.80×10^{-5}	17,241
Pondicherry	*Cx. quinquefasciatus*		1,850		3.18×10^{-4}	3,145
Myanmar						
Rangoon	*Cx. quinquefasciatus*	82,873	298	1,353	3.15×10^{-5}	31,746
Indonesia						
Jakarta	*Cx. quinquefasciatus*	2.23×10^{-5}	647	1,941	1.73×10^{-5}	57,803
Philippines						
Sorsogon	*Ae. poicilius*	3,559	80	271	6.1×10^{-5}	16,400

Table 9.4 (Continued).

Country and locale	Vector	ABR[1]	AIBR[2]	ATP[3]	TE[4]	Infectious bites required[5]
	Wuchereria bancrofti subperiodic strain					
Fiji						
Koro	*Ae. polynesiensis*	$6.4x10^{-5}$	5,760	13,440	$1.48x10^{-5}$	67,568
	Brugia malayi subperiodic strain					
Malaysia						
East Pahang	*Ma. dives* *Ma. bonneae* *Ma. annulata*		31	155	$7.06x10^{-3}$	142

[1]ABR=annual biting rate.
[2]AIBR=annual infectious biting rate.
[3]ATP=annual transmission potential.
[4]TE=transmission efficiency.
[5]Estimated number of infectious bites per person required to generate a new human infection.

among abaca plants) to *Ae. poicilius*. The efficiency of transmission might be considerably higher because of spatial heterogeneity leading to aggregation in transmission, although as yet this hypothesis has not been rigorously studied for lymphatic filariasis.

The second problem is that the analyses discussed above are simplistic in that they necessarily discount the infectivity of people with very low, undetectable microfilaremias to biting mosquitoes. However, such people may be infective and the low doses of microfilaria ingested by the mosquitoes may facilitate survival and development of microfilariae to the L3 stage, or may be less injurious to the mosquitoes. Among several studies examining the infectiveness of apparently afilaremic humans for mosquitoes, Samarawickrema et al. (1985) is particularly useful. They compared the sensitivity of the finger prick, slide smear method of detecting *W. bancrofti* microfilaremia with the filter concentration method in American Samoans for a range of microfilaremia titers. This study also compared infections of *Ae. polynesiensis* and *Ae. samoanus* mosquitoes fed on these same individuals. The filtration method was more sensitive than the finger prick method; some people with low microfilaremic titers detected by filtration were classified as amicrofilaremic by the finger prick method. Further, although there was a dose dependent effect of microfilaremia titer on mosquitoes infection, many mosquitoes became infected despite having fed on people with very low or undetectable microfilaremias (Fig. 9.13). Thus, the importance of patients with low microfilaremias as sources of infection of *Ae. polynesiensis* by *W. bancrofti* may have been underestimated.

The concepts of **facilitation** and **limitation** were developed to explain the lack of agreement between the number of microfilariae ingested by different species of mosquitoes and the number of subsequently developing larvae in the mosquitoes' bodies (Pichon et al. 1974, Southgate and Bryan 1992, Wada et al. 1995).

Figure 9.11. Abaca leaf axil: larval development site of *Aedes poicilius* in the Bicol region of the Philippines.

In *Anopheles* mosquitoes, there is a positive correlation between the proportion of *W. bancrofti* microfilariae that successfully develop and the density of ingested microfilariae. A process termed "facilitation" describes this correlation. Facilitation may be mediated by density-dependent shredding of microfilariae by the cibarial armature of the vector during bloodfeeding. In other mosquitoes, particularly *Aedes*, there is a negative correlation between the proportion of microfilariae developing to L3 and the density of ingested microfilariae. This correlation is termed "limitation." Facilitation may be important because it suggests a low-density microfilaremia infection threshold for *Anopheles* vectors. This threshold could be a target for mass chemotherapy programs. Thus, facilitation has received considerable attention, especially from the World Health Organization (Zhong et al. 1996). However, Wada et al. (1995), through a review of experimental data and a modeling analysis, showed that the concept of facilitation was faulty and based upon erroneous interpretation of limited data. Their paper merits careful study, because their reanalysis suggests that natural reductions in microfilaremia prevalence and intensity could be explained not by facilitation, but rather by reductions in vector density due to vector control. The analysis of Webber and Southgate (1981) in Guadalcanal supports this concept.

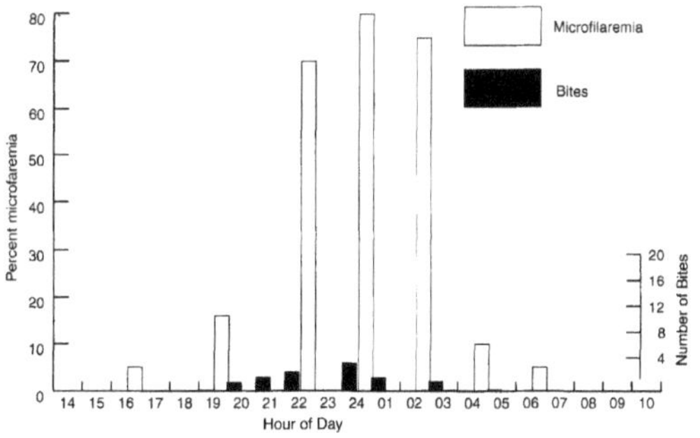

Figure 9.12. Diel periodicity of *Wuchererica bancrofti* in human blood and biting activity of *Aedes poicilius* in the Bicol region of the Philippines.

Transmission intensity appears to be a function of both the prevalence and intensity of microfilaremia in humans and the vectorial capacity of mosquito vectors. But how does intensity of transmission of parasites to humans influence the spectrum of disease in communities? This question has been addressed for both acute and chronic symptoms. Acute symptoms of lymphatic filariasis involve episodic inflammation and pain at the affected sites (termed adenolymphangitis, or ADL) that usually are accompanied by disabling fever. As transmission intensity increases, such as during the rainy season when vector populations burgeon, the increased exposure of people to antigens associated with L3 increase the number of ADL episodes. For example, in a study conducted in Ghana, West Africa (Gyapong et al. 1996), ADL episodes occurred at an annual incidence of 95.9 per 1,000 population. Episodes also were higher in frequency in the female population, lasted an average of 5 days with 3 days of incapacitation, and occurred more often during the rainy months of the year when *Anopheles* vector abundance was high (Fig. 9.14) and when peak agricultural work was required. Acutely affected individuals often were feverish and in pain, had difficulty working during the episodes and consequently suffered economic loss.

In addition to these studies linking transmission and acute disease, other studies have shown that transmission intensity, measured as annual transmission potential and annual infective biting rate, can affect prevalence of microfilaremia, intensity of microfilaremia and prevalence of edema of the leg. Kazura et al. (1997) studied these relationships in 5 villages in East Sepik Province, Papua New Guinea, an area well known to be endemic for bancroftian filariasis. These studies showed positive correlations between the village-specific ATP, expressed as a logarithm, and the percentage of people with microfilaremia, the intensity of microfilaremia and the percentage of people with leg edema, a chronic symptom. Further, there was a positive correlation between the ATP and the prevalence of hydroceles, although it was not statistically significant (Fig. 9.15). Aside from the epidemiological significance of these results in showing the relationship between clinical morbidity and transmission intensity, the data reveal a lower threshold of ATP of about 50 L3/person/year, or 1.7 logs below which microfilaremia and leg edema would not be predicted to occur.

Prevention and Control

Chemotherapy

The piperazine derivative diethylcarbamazine (DEC) has been the backbone of chemotherapy for lymphatic filariasis since its introduction in 1947. DEC affects microfilariae, with a single oral dose clearing up to 90% of these forms from the circulation within 1 hr. In contrast to its effect on *O. volvulus*, DEC also is adulticidal to *W. bancrofti* and *B. malayi*. Individuals treated for lymphatic filariasis remain microfilaria negative for several months following a regime of serial doses, but microfilaremia frequently reappears without follow-up treat-

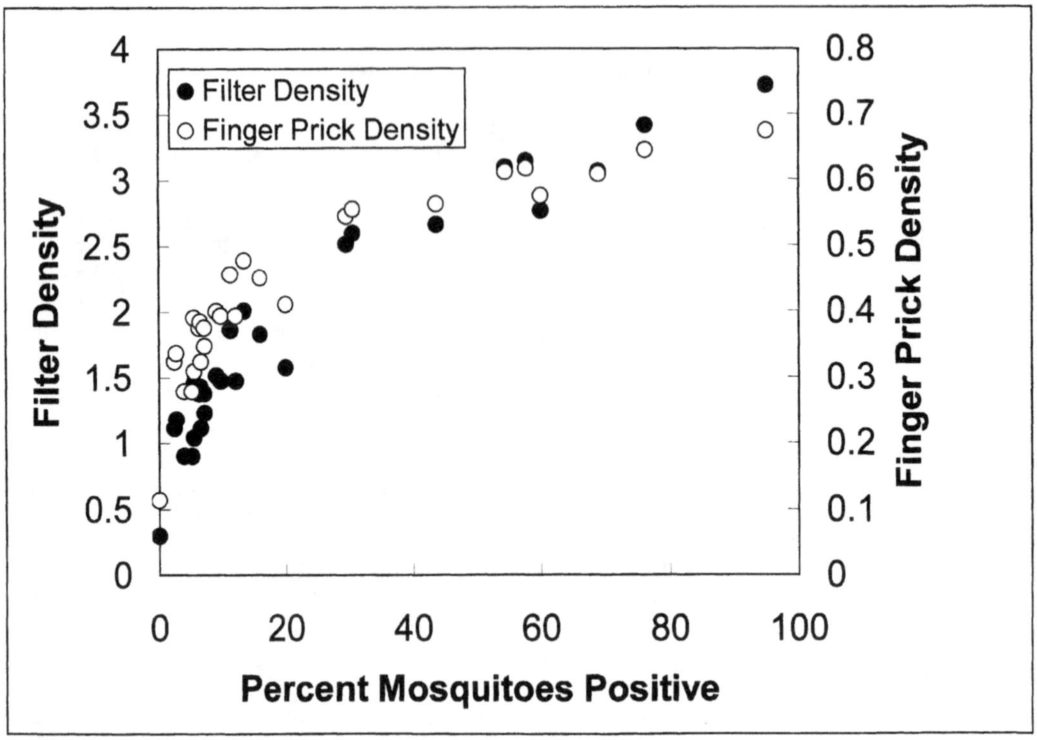

Figure 9.13. Infectivity of people with various microfilarial intensities to *Aedes polynesiensis* (from Samarawickrema et al. 1985).

ment, reflecting the failure of a single course of treatment to clear all adult worms. Recent studies indicate that longer courses of treatment (up to 18 months) with lower doses of DEC given at weekly intervals clear circulating microfilariae and also kill adults of both *W. bancrofti* and *B. malayi* (Ottesen 1990).

The mode of action of DEC against filariae remains a mystery. The fact that it exhibits no direct effects against microfilariae *in vitro* suggests that its action *in vivo* is mediated by the host's immune system. The drug may work via a 2-pronged mechanism involving opsonization of parasites and potentiation of host immune effector cells.

At normal doses, DEC is virtually free of toxic side effects. Adverse reactions commonly associated with the drug are actually responses of the host to dying microfilariae, and the severity of these side effects is proportional to the numbers of these parasites in the circulation. Clinical symptoms associated with these reactions include fever, headache, nausea, chills and dizziness, all generally occurring within 48 hr of treatment. Granulomatous reactions around adult worms killed by the drug can bring about apparent exacerbation of chronic lymphatic pathology (Ottesen 1990).

Microfilariae of the lymphatic dwelling filariae also are sensitive to the broad-spectrum parasiticide ivermectin. Effects of this compound on adults of *W. bancrofti* or *B. malayi* are unknown. While ivermectin causes side effects in microfilaremic patients similar to those seen with DEC, its advantage is that it effectively clears circulating microfilariae after a single oral dose rather than the repeated doses of DEC. In view of the logistical problems associated with drug delivery to patients in rural areas, the ability of ivermectin to control microfilaremia after a single dose may make it the drug of choice for treatment of human lymphatic filariasis patients (Ottesen 1990). The role of microfilaricidal drug treatment in controlling or even suppressing the transmission of the lymphatic dwelling filariae is discussed below. A third drug, albendazole, may kill both immature and adult worms in combination with microfilaricidal drugs.

Ministries of health in endemic countries often have the mission to control lymphatic filariasis. Some programs strive to reduce the burden of chronic disease through clinical treatment, while other programs encompass active case detection, drug treatment and vector control to reduce prevalence of acute and chronic disease. These programs depend upon political will and availability of funding. In some instances, vector control directed against malaria vectors has simultaneously resulted in control of filariasis.

Use of microfilaricidal drugs, namely DEC and ivermectin, help community-based filariasis control. Doses can be distributed through local and district health offices to the public to reduce microfilaremia and therefore reduce or block transmission. One of the more creative approaches has been the incorporation of DEC into ordinary table salt used in cooking in China (Jingyuan et al. 1992). In a population of over 5 million people among 20 counties and cities in Fujian Province, the microfilaremia prevalence for *W. bancrofti* declined from a mean of 3.09% to a mean of 0.07% after 6-9 months of treatment. Over 95% of the population complied with the treatment regime. The use of DEC-treated salt also has been adopted in other countries where the disease is endemic (Gelband 1994).

Vector control

Vector control is often an important component of mosquitoborne filariasis control, and usually is used as an adjunct to the treatment of chronic symptoms in people presenting to clinics, and to the distribution of drugs to clear microfilaremia and reduce transmission and disease. One approach is the elimination of aquatic habitats suitable for larval mosquito development, or the modification of aquatic habitats so that they no longer support larval development, a process is termed **source reduction**. The long-term effect of source reduction is to reduce the density of vectors and, consequently, the annual biting rate and annual transmission potential. In urban areas where *Cx. quinquefasciatus* is the primary vector of *W. bancrofti*, the larval stages occur in water polluted by organic material such as human waste, including drainage ditches, pit latrines and animal waste lagoons. These sites may be modified or eliminated by installation of waste water catchments and sewage treatment facilities as part of sanitation programs, using the principles of civil engineering. Obviously, to achieve this aim there must be well-organized and properly funded public works programs that can effect these environmental changes, and maintain and operate the sanitation systems after they have been installed. Environmental sanitation to eliminate the container habitats of *Ae. polynesiensis* has been effective in reducing the density of these vectors of *W. bancrofti* in the eastern Pacific. Containers such as tires and old cans can be collected for rubbish thus reducing the number of larval habitats. As with the example of *Cx. quinquefasciatus*, this kind of vector control requires organized public works programs such as municipal garbage collection and

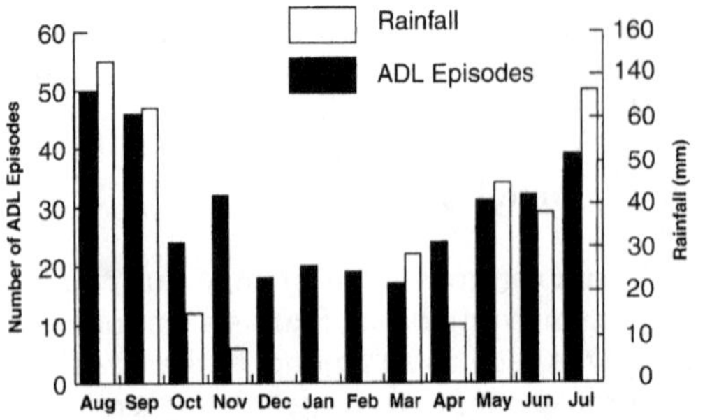

Figure 9.14. Frequency of ADL in relationship to rainfall throughout the season in Ghana (from Gyapong et al. 1996).

disposal. However, for other situations, larval habitat elimination is not practical or economical. For example, larvae of the vector of *W. bancrofti* in the Philippines, *Ae. poicilius*, dwell in the water-filled leaf axils of abaca plants. These plants are cultivated in large plantations to produce a strong fiber, manila hemp. Eliminating the larval habitats of *Ae. poicilius* also would eliminate the source of livelihood of residents in disease endemic areas. Other ways of modifying larval habitats to reduce mosquito production include removing vegetation that harbors larvae, varying water levels to strand larvae, planting fish that eat mosquito larvae or the vegetation that harbors larvae, and harvesting floating vegetation to eliminate the habitats of *Mansonia* and *Coquillettidia* larvae.

A second means of vector control is to reduce vector density by the application of chemical larvicides. This technology is well-developed, with a wide range of insecticides available such as oils forming surface films, organophosphate insecticides, insect growth regulators such as methoprene in sustained-release formulations, and formulations of bacterial-based insecticides such as *Bacillus thuringiensis israelensis* (Bti) and *Bacillus sphaericus*. Control of the larval stages of mosquito vectors requires an organized effort with trained crews of technicians to accomplish their tasks, working under the auspices of properly funded public health programs.

Adult vector mosquitoes also can be controlled to reduce transmission, generally by applying insecticides into their environment when the mosquitoes will be exposed. The use of residual insecticides in malaria control programs often has had the effect of reducing transmission of filarial nematodes as well, and in some cases this technology was adopted for filariasis control directly with considerable success, in particular in those instances where *Anopheles* species were the primary filariasis vectors.

Onchocerciasis

The Parasite

Adult females of the causative agent of onchocerciasis, *O. volvulus*, are long, slender worms with conspicuous cuticular annulations. They may be up to 500 mm in length and only 0.35 mm in width. Males are much smaller, measuring 19–42 mm in length and 0.17 mm in width. Adult worms coil in rounded bundles of 2–3 female worms and 1–2 males. These bundles, usually located in subcutaneous connective tissues, also may be found in deep intermuscular fascia. These aggregates of adult worms eventually may become encased in a translucent, fibrous capsule of host origin measuring up to 1 cm in diameter. These capsules are termed nodules or onchocercomata. Worms within nodules encounter host inflammatory responses including fibrin and various populations of inflammatory cells, and older nodular worms are often in varying stages of calcification and disintegration (Greene 1990).

Microfilariae of *O. volvulus* are unsheathed and occur in the skin of the host. They are distinguishable from other dermal microfilariae in stained preparations by their elongate cephalic

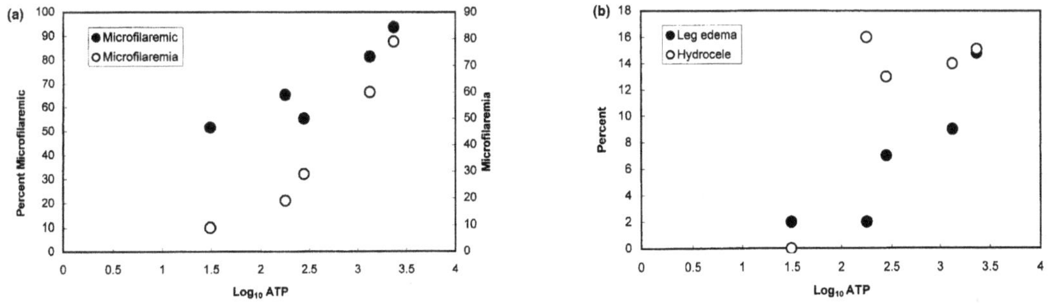

Figure 9.15. (a) Relationship between Log_{10} ATP and percent microfilaremic patients (left axis) and geometric mean of microfilaremia (right axis); (b) Relationship between Log_{10} ATP and percent of patients with leg edema and hydrocele (from Kazura et al. 1997).

space and sharply pointed tail. Molecular probes based on variant sequences within a highly repeated DNA element designated O-150 are available to identify different geographic strains of *O. volvulus* (Zimmerman et al. 1993) and are currently used in a PCR/hybridization assay to identify microfilariae from skin snips and larval stages within vector black flies.

The distribution of *O. volvulus* microfilariae in the skin of the host is nonuniform, and different dispersal patterns occur in different geographic populations of the parasite. Frequently, the highest concentration of parasites is seen in areas of the skin favored as feeding sites by local vectors. For example, in Guatemalan patients, the highest microfilarial densities occur in the skin of the upper body. Host-seeking females of *S. ochraceum* orient towards the upper torso and head. Similarly, in Africa, peak microfilarial densities in the skin of the lower torso, hips and legs correspond to the lower-body feeding sites preferred by *S. neavi* and *S. damnosum* s.l.

Life Cycle

The vectors of *O. volvulus* are black flies in the genus *Simulium*. Development of *O. volvulus* in the vector resembles the general life cycle of filarial parasites described above. The cycle begins when pool feeding black flies take a bloodmeal on a human host harboring skin microfilariae. During bloodfeeding, microfilariae in the dermis presumably migrate into, and are subsequently ingested with, the pool of blood and tissue fluid forming around the feeding lesion. Subsequent development is in the indirect flight muscles. As with other filariae, development of *O. volvulus* in vectors is temperature dependent; it is completed in approximately 7 days at 27°C. At the completion of development in black flies, infective L3 migrate to the mouthparts where they lodge in the labrum epipharynx (Collins 1977) and are transmitted to the host during a subsequent bloodfeeding.

Because of the lack of a small animal model, details of development and migration of *O. volvulus* in vertebrate hosts are unknown. However, data from experimental infections in chimpanzees indicate a highly variable prepatent period ranging from 7–34 months (Duke 1980,

Greene 1990). Male *O. volvulus* probably are not permanent residents of the worm nodules but rather invade them periodically to mate with females. Duke et al. (1991) aptly applies the term "itinerant" to the males to denote this behavior. Periodic reinsemination is required to sustain microfilarial production by female *O. volvulus*. Extensive studies of the uterine contents of worms from excised nodules (Schulz-Key et al. 1977, Duke et al. 1991) indicate a cyclic pattern of reproduction of approximately 3 months, each round consisting of insemination, embryogenesis and microfilarial release. Estimates of the life span of adult *O. volvulus* are as high as 18 yr, a fact that has dictated the very protracted vector control program used to control onchocerciasis in West Africa.

Vertebrate Hosts and Distribution

The geographic distribution of *Onchocerca volvulus* is presented in Fig. 9.16. This parasite is found in an equatorial belt in Africa from Senegal on the western coast eastward to the southern Arabian Peninsula. *O. volvulus* also is endemic in the northern highlands of Guatemala and the bordering states of Chiapas and Oaxaca in southern Mexico. More recently, endemic foci have been discovered in Ecuador, Colombia and Brazil. In nature, *O. volvulus* appears to be restricted to human hosts. There is no known animal reservoir. Patent experimental infections have been induced only in primates, including chimpanzees (Duke 1980) and Mangabey monkeys (Eberhard et al. 1991). Attempts to infect small laboratory animals have failed to date, and the lack of an experimental animal model for *O. volvulus* infection has been a major impediment to research on this parasite (Lok and Abraham 1992).

Human Disease

Geographical forms

Onchocerca volvulus may be divided into 3 forms based on pathogenicity in the host (Figs. 9.17, 9.18). Historically, these forms have been designated based on the regions where diseases associated with them are prevalent. These forms are the **African Sudan savanna** form, the **African forest** form and the **Guatemalan,** or **Central American** form. These forms diverge genetically; they may be reliably distinguished on the basis of variant sequences in the O-150 repetitive DNA element (Zimmerman et al. 1994). The vectors of these forms also differ. The ranges of the savanna and forest forms overlap significantly perhaps because of ecological changes occurring in the endemic zone of Africa and improved methods for field identification of parasite strains.

Skin disease

Pathology in human onchocerciasis is, except for nodule formation, caused by microfilariae. This contrasts with lymphatic filariasis, where chronic pathology is caused largely by adult parasites. Dermal microfilariae of *O. volvulus* cause several clinically significant changes in the skin. First, although the worms themselves do not appear to be associated with a tissue reaction, their presence results in localization of inflammatory cells around vessels in the dermis and an infiltration of fibroblasts and mast cells (Greene 1990). Subsequent physical changes in the upper dermis damage elastic fibers in the subcutis and chronically scar them. These changes eventually result in dermal atrophy with accompanying loss of skin elasticity and thinning of the epidermis. This condition manifests itself as premature wrinkling, loosening and enfolding of the skin. Changes in pigmentation occur in milder cases of dermal onchocerciasis, and some patients suffer depigmentation. This loss is more common in infections with the forest form of *O. volvulus* than with infections with the savanna form (Greene 1990). Thickening of the skin in the facial area leading to leonine facies in chronic infections and wrinkling of the skin at the superior aspect of the pinna of the ear are common dermatologic signs in American onchocerciasis patients. In addition to the signs accompanying dermal atrophy, a highly pruritic, localized, inflammatory dermatitis characterized by reddening, scaling and the formation of vesicular lesions also is seen. These lesions may be subject to secondary infection.

Although not associated with severe discomfort, patients dislike onchocercal nodules, which they consider to be disfiguring when they occur on exposed areas such as the head or face. Furthermore, nodules on the face or head indicate a heightened risk of ocular complications. Surgical nodulectomy is practiced systematically in Guatemala as a means of decreasing dermal and ocular microfilarial counts.

Ocular disease

The African form of human onchocerciasis derives its common name, river blindness, from the fact that visual impairment due to damage to ocular tissues by microfilariae is a major component of the pathology caused by *O. volvulus*. Microfilariae may damage all parts of the eye, but it is convenient to categorize pathogenic effects into anterior and posterior segment lesions.

Anterior segment lesions include conjunctivitis and photophobia, both reversible. Opacities of 2 types occur in response to invasion of the cornea by microfilariae. The more acute form consists of multiple punctate or "snowflake" opacities (Greene 1990). Histologically, these punctate opacities contain an infiltrate of inflammatory cells surrounding a degenerating microfilaria. Such loci typically resolve after microfilaricidal drug treatment. The more serious type of corneal lesion is a more generalized and progressive clouding known as sclerosing keratitis. Chronic inflammation, scarring and neovascularization characterizes this potentially blinding type of lesion. It occurs at higher frequencies among patients with the African savanna form of the parasite than among individuals with the forest form. Visual impairment due to chronic sclerosing keratitis is irreversible. Inflammatory responses to microfilariae invading the anterior uveal tract (the iris, ciliary body and choroid coat) may lead to pupillary deformity and constitute a major sight impairing complication of ocular onchocerciasis, particularly in the Americas (Greene 1990).

Posterior segment lesions damage the retinal pigment epithelium and the optic nerve. Chorioretinal lesions include atrophy and hyperpigmentation of the retinal pigment epithelium as well as atrophy of the underlying capillary bed. In contrast to sclerosing keratitis, they occur with roughly equal frequency in patients infected with forest and savanna form parasites and constitute the major cause of blindness among onchocerciasis patients within the forest zone of Africa. Invasion of the posterior ocular segment by microfilariae of *O. volvulus* also may result in optic nerve atrophy and resulting constriction of the visual field. Greene (1990) notes that in many cases this complication is either wholly or in part a side effect of drug treatment.

Lymphadenopathy

Lymphatic disease, mild in comparison to that seen in chronic Bancroftian or Brugian filariasis, is common in human onchocerciasis in Africa. The frequency of lymphatic disease is higher in the forest zone than in the savanna. The inguinal and femoral nodes are most frequently involved and their gross enlargement, along with fluid accumulation in surrounding tissues, gives rise to a condition called "hanging groin."

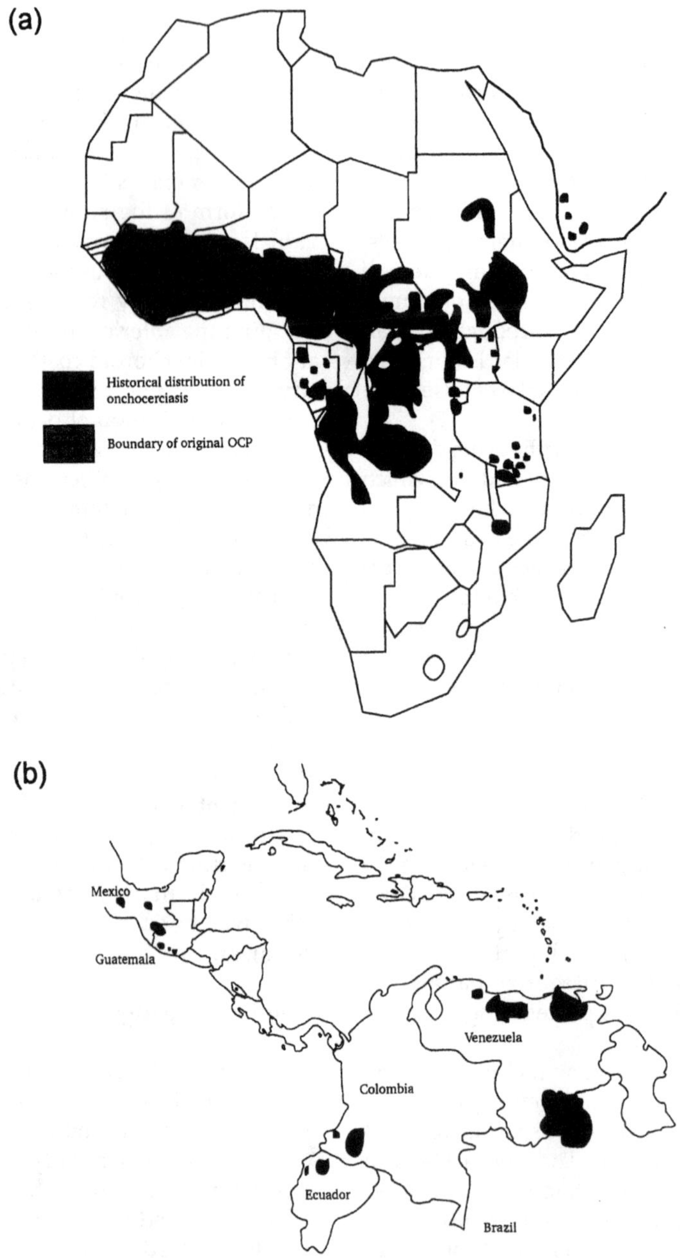

Figure 9.16. Geographic distribution of onchocerciasis: (a) In the Old World; (b) in the New World.

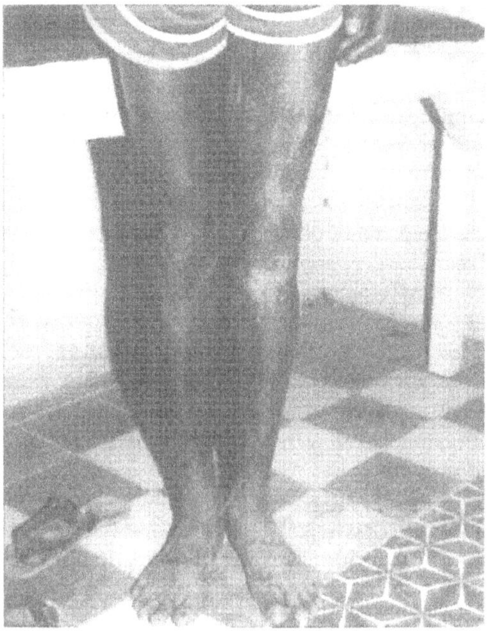

Figure 9.17. Onchocerciasis ("Sowda" variant) in a Sudanese man. Note unilateral inflamatory dermatitis on the left leg caused by immune-mediated killing of *Onchocerca volvulus* larvae in the skin (photograph by Michael A. Kron, reprinted by permission from Guzman, J.R., Jurando, H.M. and Kron, M.A. 1995. Infectious diseases in Ecuador. J. Travel Med. 2:89-95).

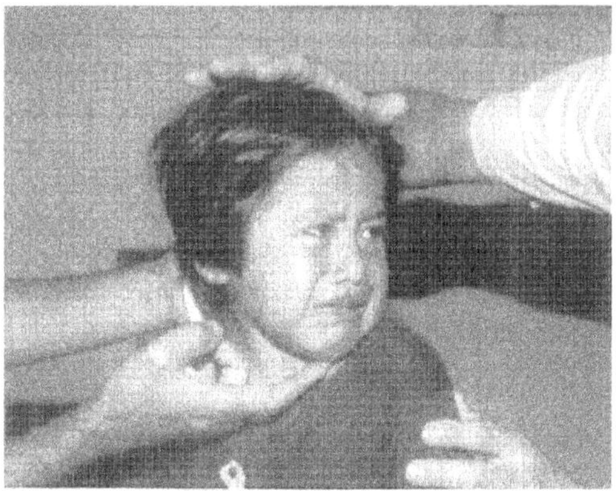

Figure 9.18. Subcutaneous nodule of *Onchocerca volvulus* adults in the forehead of a Chachi Indian girl from the Esmeraldas region of Ecuador (photograph by Michael A. Kron, by permission as for Fig. 9.17).

Diagnosis

Detection of microfilariae

As with lymphatic filariasis, the classic method of parasitologic diagnosis is to detect microfilariae in the peripheral tissues of the human host. Skin biopsies are collected by means of a corneoscleral punch or by using a scalpel to snip the top of a tiny piece of skin raised with a needle or straight pin. These biopsies should be bloodless to avoid contamination with bloodborne microfilariae. Samples are weighed and incubated for several hours in physiological saline or culture medium, typically in a multiwell culture plate, and checked for emergent microfilariae. Microfilariae may be seen occasionally in the anterior segment of the eye during slit lamp examination. Since the occurrence and severity of ocular and skin disease correlate with numbers of skin microfilariae, "skin-snip" diagnosis has the added advantage of allowing the microfilarial burden to be quantified, thereby providing a means of assessing the risk of severe disease in an individual patient.

Patients infected with *O. volvulus* microfilariae experience an adverse reaction to diethylcarbamazine (DEC). This reaction, termed the **Mazzotti reaction**, has been proposed as the basis of a diagnostic method in which symptoms are observed after a single provocative dose of the drug. Such a method has the advantage of greater diagnostic sensitivity than direct observation of microfilariae. However, the discomfort associated with the Mazzotti reaction and the risk of exacerbating ocular lesions argue against its use as a routine method.

Immune responses and serodiagnosis

The skin snip method is preferable to serologic methods for diagnosis, because immune responses to infection vary among infected people and with the range of symptoms. The primary immune response is cell-mediated and characterized by proliferation of eosinophils, in conjunction with production of IgG antibodies that mediate eosinophil function in recognizing antigens on the surface of microfilariae. As with lymphatic filariasis, the immune system also interacts with parasite antigens to produce disease, such as the action of eosinophils in causing microfilarial death, resulting in adverse dermatological conditions.

Detection and typing of parasite DNA

A PCR amplification/hybridization method involving strain- and species-specific plasmid and oligonucleotide probes of the O-150 repeat region of the *Onchocerca* genome recently has been developed and employed in the field to identify forest and savanna form parasites in human patients in West Africa (Toe et al. 1994). These analyses were performed on microfilariae isolated by conventional skin snip procedures. Repetitive parasite DNA sequences also can be amplified and identified from the blood of onchocerciasis patients. This "free DNA" appears to be present whether the patients have microfilariae in their skins or not and may represent the basis of a sensitive diagnostic method not requiring skin biopsy and isolation of microfilariae (Zhong et al. 1996).

Treatment

In the past, 2 drugs have been available for treatment of human onchocerciasis: DEC and suramin. However, both have severe side effects that now preclude routine clinical use. Although it does not kill adult worms, DEC is a highly effective microfilariacide against *O. volvulus*. It is administered in a series of twice- daily doses of 3-4 mg/kg/day for a period of 2 weeks. At the nominal dose, DEC is not intrinsically toxic; however, in most patients it precipitates a highly pruritic, erythematous skin reaction in response to dying parasites. This local reaction can be associated with multiple systemic symptoms including fever, headache, nausea, swelling of the lymph nodes and joint pain. More worrisome is the fact that the ocular complications associated with onchocercal infection, including both anterior and posterior segment lesions, may be exacerbated as a complication of DEC treatment. Because DEC leaves viable adult worms in place, patients may experience a recrudescence of skin microfilariae over a 6–18 month period after treatment. Periodic retreatment, which consequently is necessary to give long term suppression of skin microfilariae, may be associated with repeated adverse reactions. Although modification of the dosing schedule and concomitant corticosteroid therapy may dampen the response to DEC, the intense Mazzotti reaction, the potential for sight-impairing ocular complications and the requirement for periodic retreatment argue strongly against its use in mass, community level treatment programs.

Ivermectin, a macrocyclic lactone with broad-spectrum antiparasitic activity, was evaluated for its efficacy against *O. volvulus* infection in the mid-1980s. Like DEC, it was a fast-acting, potent microfilariacide. However, unlike DEC, ivermectin cleared microfilariae from the skins of onchocerciasis patients after a single dose. More importantly, side effects associated with the Mazzotti reaction to DEC treatment were relatively mild and infrequent with ivermectin (Greene et al. 1985). At the recommended dose of 150 mg/kg, ivermectin does not kill adult *O. volvulus*, and microfilariae may repopulate the skin in low numbers within 6–12 months, necessitating annual retreatment. Ivermectin recently has been licensed for use in humans for treatment of onchocerciasis and currently is the drug of choice for mass treatment campaigns.

In addition to alleviating the complications of ocular and dermal onchocerciasis, ivermectin treatment renders individual patients ineffective as donors of microfilariae for infection of black fly vectors (Cupp et al. 1986, Trpis et al. 1990). Long-term suppression of skin microfilariae with ivermectin in community level treatment campaigns eliminates a significant portion of the reservoir of infection and may decrease or block transmission altogether. The effect of mass chemotherapy on transmission of *O. volvulus* will be discussed below.

Suramin kills both adults and microfilariae of *O. volvulus*. However, this drug must be given intravenously, is highly nephrotoxic and precipitates a Mazzotti-like reaction to killing of microfilariae similar to that seen with DEC. Fatalities due to exfoliative dermatitis and renal complications have been reported in suramin-treated patients (Nelson 1972). Use of this drug generally is restricted to the very small minority of patients who cannot tolerate repeated treatments with ivermectin (Greene 1990). There currently is no safe and effective macrofilariacide available for treatment of human onchocerciasis patients.

Vectors

Distribution

Black flies (family Simuliidae) transmit *O. volvulus* worldwide (Fig. 9.19). In the Old World, vectors are restricted almost entirely to Africa (Table 9.5). In the New World, black fly vectors of *O. volvulus* occur in various parts of South and Central America (Table 9.6).

Africa

Simulium damnosum *s.l.*

Members of the *Simulium neavei* and *S. damnosum* species complexes are the primary vectors in Africa. *Simulium damnosum* s.l. occurs throughout sub-Saharan Africa and is a vector of *O. volvulus* in most of this region. Until the application of modern cytological methods in the mid-1970s, *S. damnosum* was regarded as a single, relatively uniform species (World Health Organization 1976). However, it now is known to be a complex of over 25 major cytotypes, many having specific status (Vajime and Dunbar 1975); see the review by Rothfels (1979). The vector status of species and cytotypes in the *S. damnosum* complex is determined primarily by their tendency to bloodfeed on human hosts. In East Africa, the majority of species in the complex are zoophilic and consequently not involved in transmission of *O. volvulus* to a significant degree. However, in West Africa, some 8 complex members either are partially or wholly anthropophilic and have been incriminated as vectors. Historically, these species are associated with either the savanna or rainforest ecosystems (Vajime and Dunbar 1975) and transmit either the savanna or forest forms of *O. volvulus* almost exclusively. But neither the virulent forms of *O. volvulus* nor the vectors within the *S. damnosum* complex are completely restricted to their classical geographic ranges (Garms et al. 1991, Toe et al. 1994). Nevertheless, the groupings of parasite forms with vector species within the *S. damnosum* complex is still generally valid.

Savanna members of S. damnosum *s.l. and the savanna form of* Onchocerca volvulus

Simulium sirbanum and *Simulium dieguerense* are the primary vectors of savanna form *O. volvulus* in the dry northern or Sudan-savanna belt of West Africa. Larval stages of these species develop in larger rivers of the region (Fig. 9.20). Adults generally are distributed longitudinally along rivers, spreading outward from these foci when protective cloud cover allows (World Health Organization 1976). To the south, in the wetter Guinea-savanna, *S. damnosum* s.str. and *Simulium sudanense* predominate as vectors of savanna form parasites.

Forest Simulium damnosum *s.l. and forest form* O. volvulus

Simulium yahense is the primary vector of *O. volvulus* in Liberia and Ivory Coast. Immature stages of these flies breed in small forest

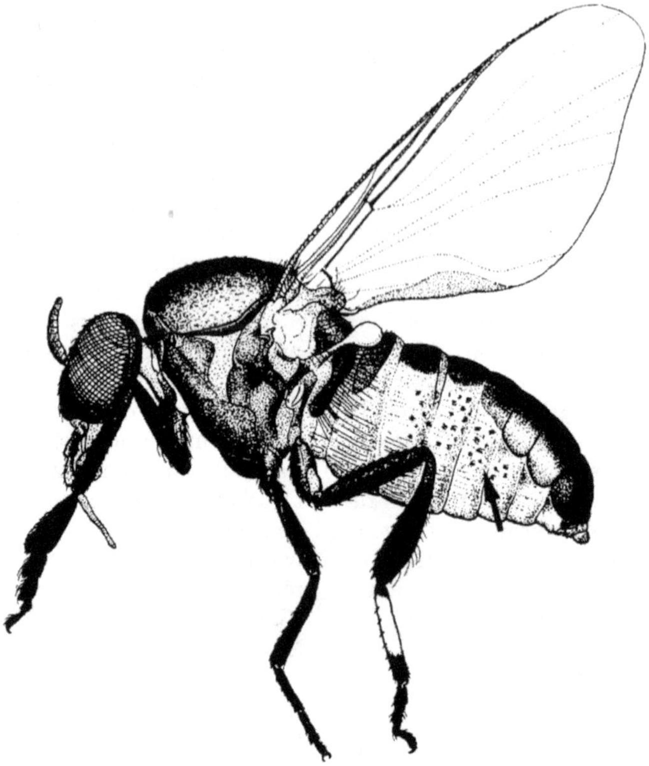

Figure 9.19. *Simulium damnosum* adult female (from Lane and Crosskey 1993).

streams, and canopy protection allows dispersal from these sites year-around. In areas surrounding larger streams and rivers of the forest zone, *Simulium sanctipauli* and *Simulium soubrense* are primary vectors.

The association of the forest and savanna black flies with respective forms of *O. volvulus* in West Africa not only is a function of geographic proximity but also is due to the intrinsic susceptibility patterns of black flies for the 2 parasite strains. Prior to the recognition of cryptic speciation in *S. damnosum* s.l., Duke et al. (1966) demonstrated that forest form *S. damnosum* s.l. were relatively insusceptible to infection by *O. volvulus* from the Sudan savanna zone while savanna form *S. damnosum* yielded relatively low numbers of infective larvae when fed on volunteers infected with the forest form of *O. volvulus*.

Simulium neavei s.l.

This group comprises another complex of black fly species that transmit *O. volvulus* in East Africa. Immature stages of these black flies are, almost without exception, found in phoretic association with fresh water crabs of the genus *Potamanautes*. They are confined to smaller streams in forested areas (Nelson 1972). *Simulium neavei* s.str., although eradicated now from the greater part of Kenya, persists in the border region with Uganda and then throughout that country. *Simulium woodi* is the primary vector in the complex in Malawi and Tanzania. *Simulium woodi ethiopense* has been incriminated as a vector in Ethiopia (World Health Organization 1976).

The Americas

Within the mountainous regions of northern Guatemala and neighboring Chiapas and Oaxaca states in Mexico, *S. ochraceum* is the primary vector of *O. volvulus,* with 2 other species, *S. metallicum* and *Simulium callidum,* playing a secondary role (Shelly 1988). Significant tracts in the Central American foci of onchocerciasis have been cleared of primary forest and given over to coffee plantations, and onchocerciasis has become an occupational disease of laborers on these plantations. Immature stages of *S. ochraceum* inhabit small rocky streams on densely forested slopes of the Sierra Madre. *Simulium metallicum* and *S. callidum* may occur in large creeks and rivers (Ramirez Perez 1988). The flight range of these 3 species of flies is 10 km or less (Shelly 1988). The uptake of microfilariae per unit volume of blood ingested by all these Central American vectors is high compared to members of the *S. damnosum* complex. This fact is attributed to concentration, mediated perhaps by a salivary attractant similar to *Aedes* vectors of *W. bancrofti.*

Four factors account for the primary vector status of *S. ochraceum.* First, this fly feeds on the upper body of humans where microfilarial densities are highest. *Simulium metallicum* and *S. callidum* feed on the lower extremities. Second, *S. ochraceum* has highly developed cibarial armature that modulates high infection levels, thus enhancing vector survival. *Simulium metallicum* and *S. callidum* lack these structures. Third, *S. ochraceum* is highly anthropophilic whereas its counterparts are zoophilic. Finally, in both *S. ochraceum* and *S. callidum,* virtually all microfilariae invading the thoracic muscles develop synchronously to the L3, and they do not cause significant mortality in the vector. However, in *S. metallicum* there is marked asynchrony of development by microfilariae in the muscles, a relatively small proportion of worms complete development to the L3, and worms cause significant mortality in vectors.

Like forest *S. damnosum* s.l., bloodfeeding by *S. ochraceum* females is markedly cyclical with most nulliparous (and therefore noninfected) flies feeding in the morning and most parous (potentially infected) flies feeding in the afternoon (Collins et al. 1981). Populations of bloodfeeding *S. ochraceum* peak during the dry season, coinciding with the seasonal harvest of coffee in the endemic zone, thus increasing the risk of transmission to field laborers (Shelly 1988).

Simulium metallicum is the primary vector in the densely populated northern Venezuelan focus of human onchocerciasis. Because of its greater zoophilism and its tendency to feed on areas of the host body with lower microfilarial densities, another indigenous black fly, *Simulium exiguum,* is a secondary vector. Both species lack cibarial armature, but show a low intrinsic susceptibility to infection relative to *S. ochraceum.* As already noted, parasite development in *S. metallicum* is asynchronous. However, development in *S. exiguum* from Venezuela is even less so, with a high frequency of stunting and other types of abortive development.

A diverse simuliid fauna occurs in the Amazonia focus of onchocerciasis in the savanna and rainforest of southern Venezuela and northern Brazil. Shelly (1988) implicates 5 species, *Simulium guianense, Simulium oyapockense, Simulium roraimense, Simulium yarzabali* and *S. exiguum,* as vectors in various ecosystems. *Simulium oyapockense* shares many vector attributes with *S. ochraceum* in Central America. It concentrates microfilariae of *O. volvulus* from the host skin, but its cibarial armature kills 98% of them. The remaining parasites complete development to the L3 in the thoracic muscles with-

Table 9.5. Old World black fly vectors of *Onchocerca volvulus* in endemic regions [from World Health Oganization (1995)].

Black fly species	Region	Habitat
Simulium damnosum complex		
Simulium damnosum subcomplex		
S. damnosum s.str.	West Africa to southern Sudan, Uganda	Savanna
S. sirbanum	West Africa to Sudan	Dry savanna
S. rasyani	Yemen	
S. dieguerense	West Africa	Savanna
S. sanctipauli subcomplex		
S. sanctipauli s.str.	West Africa	Forest
S. soubrense	West Africa	Forest
S. leonense	West Africa, esp. Guinea and Sierra Leone	Forest
S. konkourense	West Africa	Guinea highlands
S. squamosum subcomplex		
S. squamosum s.str.	West Africa	Forest, highlands
S. yahense	West Africa	Upland forest
S. megense	Cameroon	Forest
S. kilibanum	Zaire, Burundi, Uganda, Malawi, Tanzania	Montane areas
Simulium neavei complex		
S. neavei s.str.	Uganda, eastern Zaire	
S. woodi	Tanzania	
S. ethiopense	Southwestern Ethiopia	
S. albivirgulatum	'Cuvette central' Zaire	

Table 9.6. New World black fly vectors of *Onchocerca volvulus* in endemic regions [from World Health Organization (1995)].

Black fly species	Region	Habitat
S. ochraceum complex (at least 3 cytospecies)	Mexico, Guatemala	Highlands
S. metallicum complex (at least 11 cytospecies)	Mexico, Guatemala, Venezuela	Highlands
S. callidum	Mexico, Guatemala	Highlands
S. exiguum complex (at least 4 cytospecies)	Colombia, Ecuador, Venezuela	Forested foothills
S. guianense	Amazonian Brazil, Venezuela	Highlands, lowlands
S. incrustatum	Amazonian Brazil, Venezuela	Highlands
S. oyapockense complex (5 species)	Amazon and Orinoco basins	Lowland forest, savanna
S. quadrivittatum	Central and northern South America, Caribbean	Lowland and foothill forests
S. limbatum	Amazonian Brazil, Venezuela	Highlands

out stunting or rejection. *Simulium guianense* probably is the primary vector in the highlands of the Amazonia focus. A combination of factors accounts for the high natural infection rates occurring in this fly. It lacks cibarial armature, but unlike its "unarmed" counterparts in Central America, it allows synchronous development of a high proportion of parasites in the thoracic muscles. This lack of innate resistance mechanisms in *S. guianense* accounts for high mortality among flies fed upon hosts with extremely high microfilarial density. The secondary highland vector, *S. yarzabali*, has lower vector capacity than *S. guianense*, but may serve as a wet season vector when populations of the latter are low.

An endemic area of human onchocerciasis encompasses Esmeraldas, Ecuador, and San Antonio, Colombia (Shelly 1988). This area is largely a tropical forest on the western foothills of the Andes. It is unique among American foci of onchocerciasis in that the racial makeup of the affected human population is primarily black whereas other foci are predominantly Amerindian or Mestizo. Within this focus, *Simulium quadrivittatum* and *S. exiguum* are the principal vectors. Unlike the form in northern Venezuela, the Ecuadorian cytotype of *S. exiguum* is a highly efficient vector of *O. volvulus*; it is similar in efficiency to *S. guianense* in the Americas and *S. damnosum* s.l. in Africa. Like other members of the *S. exiguum* complex, it lacks cibarial armature, but in contrast to them it supports synchronous development of a high proportion of larvae penetrating the thoracic flight muscles. As in the other cases, high transmission efficiency is offset by high mortality among heavily infected flies. By virtue of its cibarial armature, the secondary vector, *S. quadrivittatum*, is a relatively inefficient vector for *O. volvulus*, on a par with *S. ochraceum* and *S. oyapockense*. However, as with *S. ochraceum*,

Fig. 9.20. Riverine black fly habitat in West Africa.

higher rates of survival among flies feeding on heavily infected hosts result in roughly equivalent rates of infectivity in populations of *S. quadrivittatum* and *S. exiguum*. Ultimately, the relative importance of the 2 species is a function of human attack rates, with landing rates for *S. exiguum* being over 20-fold greater than *S. quadrivittatum*.

Epidemiology

Intensity of transmission is determined by:

(1) Vectorial capacity of local black flies.

(2) Skin microfilarial densities in human hosts.

(3) Efficiency of microfilarial uptake by feeding flies.

(4) Efficiency of conversion of microfilariae to L3.

(5) Efficiency of transfer of L3 from black flies to susceptible hosts.

The parameters used in quantifying transmission of lymphatic filariasis, reviewed above, also apply to onchocerciasis. They include the annual biting rate (ABR, number of black fly bites per person per year), the annual infectious biting rate (AIBR, number of infectious black fly bites per person per year) and the annual transmission potential (ATP, number of L3 inoculations per person per year).

Sufficient data are available to characterize the transmission dynamics of all of the major foci of onchocerciasis. Selected endemic localities in the forest and Sudan-savanna zones of Africa and the Guatemalan sites in Central America have been the subjects of intensive scrutiny because of the control programs implemented there. Each vector/parasite system is different in terms of infectivity rates in vector populations, annual transmission potential, infective biting densities and disease presentation in the infected population. Each focus varies in the targets for reduction of annual transmission, prevailing microfilaria levels and infective biting densities. Thus, transmission dynamics are discussed below within the context of control as a means of elucidating facets of transmission.

In Guatemala, onchocerciasis occurs in departments (provinces) in certain mountainous areas in the southwestern part of the country along the Mexican border and to the southeast near the El Salvador border. Takaoka and Suzuki (1987) reviewed the transmission dynamics of *O. volvulus* in Guatemala. Rates of blindness due to onchocerciasis in endemic areas ranged from 0.5–4.6%, while the rate of skin-snip positives averaged 31%. Of 8 black fly species feeding on humans in these endemic areas, the primary vector of *O. volvulus* was *S. ochraceum*. Larvae of this species develop in small, cool streams at elevations of 500–1,500 m. Some of these streams are intermittent so that the number of streams with larvae varies between the dry (November to March) and wet (April to October) seasons. Because of the cool water temperatures, larval development requires at least 2 weeks before pupation. Adult *S. ochraceum* dispersed no more than 3–4 km. Their vectorial capacity was limited by the presence of cibarial armature that destroyed many microfilariae during bloodfeeding. The rate of infection with L3 was 0.02–3%, depending upon site of collection. *Simulium ochraceum* has low vector capacity because of poor vector competence. However, this is compensated by a high ABR (50,000–300,000 bites per person per year in the absence of black fly control) and an ATP of up to 9,000 L3 per person per year. Interestingly, the time of year with the most intense transmission was the dry season. This may be because then older adult female black flies persist from the wet season and the age structure of the black fly population shifts to an older, more infective population. Furthermore, elevated temperatures during the dry season enhance larval *O. volvulus* development in vectors.

Prevention and Control

A control program under the auspices of the Guatemala-Japan Cooperative Project on Onchocerciasis Research and Control was implemented in 1975 and continued to 1983. It involved locating streams with *S. ochraceum* larvae, followed by larviciding at regular intervals with temephos. Based upon a modeling analysis (Wada 1982), the target ABR for blocking transmission was 8,000 bites per person per year. As part of the control program, black fly adults were collected systematically inside and outside control areas from 1978 to 1983. Results (shown in Fig. 9.21) indicated that ABRs in 4 treated areas were reduced dramatically compared to 3 untreated areas and were brought well below the target levels.

The Onchocerciasis Control Program (OCP) is a multinational effort, administered by the WHO, to reduce morbidity caused by *O. volvulus* among people in several countries of West Africa. It began in 1974 and encompassed 7 countries; it now has expanded to include parts of 11 countries. The original focus was control of larval black flies, with the goal of a reduction in ABR and corresponding reduction in AIBR and ATP. With the availability of ivermectin (Mectizan) since 1988 as an intervention tool applied on a community basis and offered free by the manufacturer (Merck & Co.) through the Mectizan Donation Program, the OCP has adopted a dual strategy of vector control and community-based drug distribution. An expanded control program consisting primarily of ivermectin distribution in Africa has been named the African Program for Onchocerciasis Control (APOC); a similar program supporting elimination of onchocerciasis in the Americas is called the Onchocerciasis Elimination Program for the Americas (OEPA). However, the OCP remains a stellar example of the success of a multinational effort disease reduction. Davies (1994) and Boatin et al. (1987) have reviewed the voluminous literature on OCP. Molyneux and Davies (1997) reviewed the transition of onchocerciasis control from a vector control emphasis to an ivermectin-distribution emphasis. Hougard et al. (1997) reviewed the operational aspects and success of black fly control within the OCP (Table 9.7).

The basis of black fly control in the OCP is the reduction of black fly densities by larviciding to a level low enough to interrupt transmission. The program seeks to maintain the effort for the longevity of individual adult *O. volvulus* (approximately 14 yr). The program assumes transmission will be interrupted sufficiently so that, even though microfilariae are being produced, the number of biting black flies is too low to sustain transmission and result in new human cases. The OCP included entomological evaluations of larvicide efforts and adult black fly biting patterns and *O. volvulus* infection rates, studies of black fly cytotypes in relation to transmission intensity, migration of black flies from uncontrolled to controlled areas and epidemiological surveys for prevalence of microfilariae in skin-snip biopsies. Even prior to the introduction of ivermectin, the vector control-based OCP was successful. Initially, large scale application of insecticides to the large river systems where *S. damnosum* s.l. larvae were breeding began in 1975 and encompassed 654,000 km^2 in parts of Burkina Faso, Mali, Niger, Ivory Coast, Benin, Ghana and Togo. However, the effectiveness of the effort was diluted by invasion of adult *S. damnosum* s.l. from outside treated areas. Consequently, the areas treated were extended and during 1986-1991 they included 50,000 km of streams in 1 million km^2 of territory in 9 countries. Insecticide applications were adjusted to account for hydrology of the river systems to ensure optimal carry of active ingredient and kill of larvae. The insecticides included temephos, phoxim, pyraclofos, etofenprox, carbosulfan, permethrin and Bti. Applications were made from helicopters, boats and ground vehicles, usually weekly during the high water season. The target was a reduction of the ATP to below 100 infective larvae per person per year, because in the savanna areas of West Africa this level of inoculation was found to be below the intensity of ATP that resulted in ocular disease. This target was achieved over most of the control area, but a few foci remained at the edges.

The control of larval black flies through continuous insecticidal treatment of rivers has resulted in extensive developments in insect control technology, accompanied by environmental assessments and studies of insecticide resistance. Technological developments have included geographic information system-based mapping of larval habitats and satellite-linked hydrological monitoring of river flow and depth. Although resistance to organophosphate insecticides arose during the course of the OCP, rotation of insecticides prevented negative impacts on operations. Further, insecticide applications did not cause adverse environmental consequences to river systems. An ecological group component of the OCP was established early in the program (1975). Calamari et al. (1998) summarized their studies. The group found that weekly applications of larvicides over many years had no significant effect on riverine life. In addition, they noted that human resettlement and population increases resulting from disease reduction were of greater environmental concern than insecticide application in OCP areas.

LOIASIS

The Parasite

Loiasis is caused by the subcutaneous dwelling filaria *Loa loa*. Adult *L. loa* are migratory and Ottesen (1990) describes them as "wandering" through the subcutis of their human hosts. At times, these migrations involve the subconjunctiva and the worms become evident to both onlookers and the infected individual as they cross the eye. Because of this behavior, *L. loa* is called the "eyeworm." Females measure 50-70 mm by 0.5 mm and males 30-34 mm by 0.35 mm. Small cuticular protuberances give the adult worms a distinctive appearance in histologic sections and there is a cluster of papillae in the anal area of the worm (Ottesen 1990).

Microfilariae of *L. loa* are sheathed and circulate in the peripheral blood of the host. Stained specimens exhibit a continuous column of nuclei extending to the tip of the tail. Microfilariae of *L. loa* in human hosts are diurnally periodic whereas parasites in simian hosts show nocturnal periodicity.

Table 9.7. Outcomes of the Onchocerciasis Control Program (OCP) in terms of reductions in annual transmission potential (ATP) for the original OCP area and surrounding extension zones (compiled from Boatin et al. 1997).

Zone	Territory	Control modalities	Annual transmission potentials	
			Before control	1997
Original OCP	Benin, Burkina Faso, Ivory Coast, Ghana, Mali, Niger, Togo	Vector control and ivermectin 1974-present	1,800	0
Western extension 1	Senegal, western Mali, northern Guinea, Guinea Bissau	Ivermectin	590	Not available
Western extension 2	Northern Sierra Leone	Vector control (interrupted 1992-96), ivermectin	1,110	153
Western extension 3	Southern Sierra Leone	Ivermectin (therapeutic)	3,993	Not available
Western extension 4	Mali, southern Guinea	Vector control and ivermectin, 12 yr	1,326	22
Southern extension 1	Ivory Coast	Vector control and ivermectin, ongoing	4,500	51
Southern extension 2	Benin, Togo, Ghana	Vector control and ivermectin, 12 yr	9,900	344

Life Cycle

Tabanid flies in the genus *Chrysops* transmit *L. loa*. First- through third-stage larval development occurs in the fat body of the vector and is complete within 9–10 days at 27°C. Transmission occurs when L3 in the proboscis escape to the skin at the time of bloodfeeding and enter through the bite wound. Compared to the black fly and mosquito vectors discussed previously, tabanid vectors can support the development of exceedingly high numbers of L3 without appreciable effects on fly survival. Orihel and Lowrie (1975) report as many as 100 infective larvae from individual flies. This number may be due to the larger size of tabanid flies or because the fat body can sustain a greater degree of cell damage without compromising the physiology or mobility of the fly.

In the definitive host, parasites migrate subcutaneously. The L3–L4 molt occurs at 16–20 days of infection and the L4–juvenile molt at around 50 days. By day 90, females are inseminated; microfilariae appear in the peripheral circulation around day 150 of infection (Orihel et al. 1993).

Vertebrate Hosts and Distribution

Loa loa is found in the dense rainforests of western and central Africa and in the forested

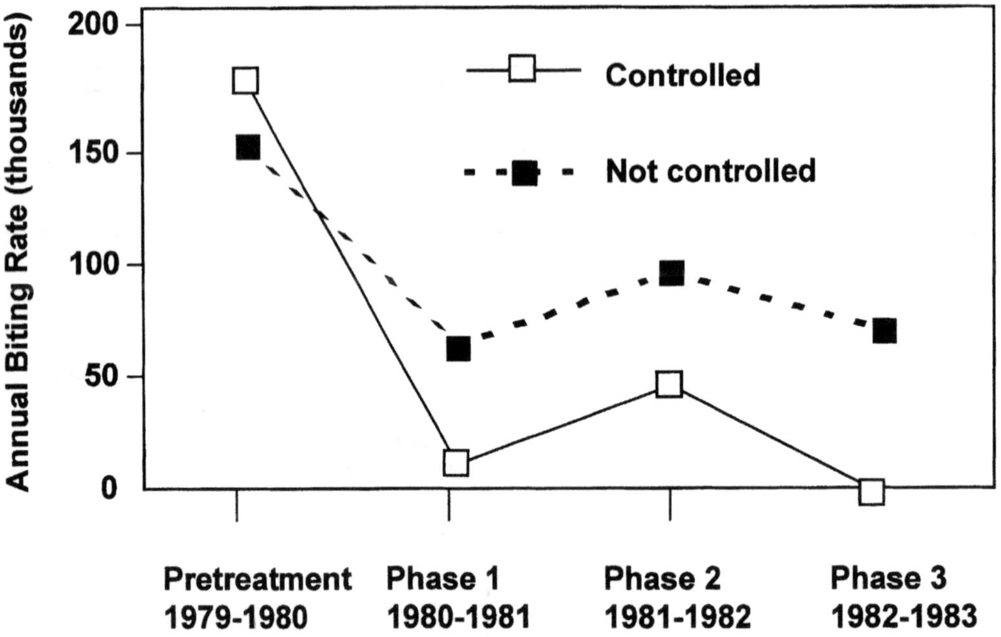

Figure 9.21. Reduction in ABR in Guatemala during larviciding program.

southern extreme of Sudan (Fig. 9.22). The parasite appears to comprise 2 physiologic races, one parasitizing humans and the other restricted to arboreal primates. There are no other known vertebrate hosts of this species. The lack of experimental animals hinders research on this parasite. However, significant strides have been made toward laboratory maintenance of *L. loa* outside the endemic zone. *Chrysops atlanticus* from the southern USA are susceptible to infection and yield large numbers of L3 after feeding on microfilaremic volunteers or experimentally infected primates (Orihel and Lowrie 1975). These infective larvae have been used to maintain the life cycle of the human parasite in a number of laboratory-reared primates, including patas monkeys (Eberhard and Orhiel 1981).

Human Disease

Loiasis takes 2 distinct forms, depending on the pattern of exposure. Asymptomatic microfilaremia predominates in persons native to the endemic zone and subject to continuous exposure. Occasional clinical symptoms in natives include migration of subconjunctival adult worms across the eye and episodes of transient angioedema usually localized to the extremities known as the "Calabar swelling." The Calabar swelling is thought to be a hypersensitivity response to migrating adult worms and sometimes is accompanied by a vermiform eruption or hives (Ottesen 1990).

Foreigners who acquire the infection while visiting an endemic area briefly are more likely to be amicrofilaremic and to exhibit allergic or hyperreactive disease than natives (Nutman 1988). Calabar swellings are more frequent and severe. Hypereosinophilia is common but not characteristic (Rakita et al. 1993). The latter condition has been tied to cardiomyopathies that sometimes accompany the infection in foreigners (Klion et al. 1991, Ottesen 1993).

Immune response

In certain respects, the immunologic relationship of loiasis in foreign and endemic populations resembles that of chronic inflammatory disease to asymptomatic microfilaremia in lymphatic filariasis. Although findings to date are insufficient to characterize the immunologic mechanisms underlying the 2 presentations as strictly pro- or anti-inflammatory (i.e., Th1 or Th2) in nature, key aspects suggest a hyporesponsive, nonallergic state in endemic patients. Eosinophil counts, total parasite-specific IgG levels and lymphoproliferative responses to parasite antigen are all low in these patients compared to infected foreigners (Klion et al. 1991). The role of prenatal exposure to antigen in determining the course of disease in loiasis has not been ascertained. Interestingly, recent studies reveal that levels of *L. loa*-specific IgG4 are equivalent in microfilaremic and amicrofilaremic patients (Akue et al. 1994).

Diagnosis

The presence of sheathed, diurnally periodic microfilariae in the peripheral blood suggests loiasis and diagnosis may be confirmed by examining fixed microfilariae for the presence of a continuous column of caudal nuclei. Diagnosis of loiasis in travelers returning after short periods of exposure in endemic areas is complicated by the fact that some 90% of these individuals will be amicrofilaremic. Although not uniformly present, migratory angioedema and eosinophilia contribute to a clinical diagnosis. Measurement of parasite-specific antibody may provide the best means of diagnosing amicrofilaremic or "occult" loiasis in foreigners. Levels of IgG4 recognizing low molecular weight *L. loa* antigens are not significantly different in microfilaremic and occult infections but are elevated compared to unexposed persons and to patients exposed to the sympatric filaria *M. perstans* (Akue et al. 1994).

Treatment

The drug of choice for treatment of *L. loa* infection is DEC. As with the lymphatic filariae, adults and microfilariae of *L. loa* are susceptible to the drug. One or 2 courses of DEC treatment may be required for complete clearance. Central nervous system complications, including encephalitis, may accompany DEC treatment in patients with extremely high microfilaremias. In these individuals, initial doses of the drug are lowered from the recommended 8–10 mg/kg/day to 1 mg/kg/day and corticosteroids are administered (Ottesen 1990). A single high dose of ivermectin (400 mg/kg) clears microfilariae of *L. loa*, but the effects of this drug on adult worms are unclear (Martin-Prevel et al. 1993).

Vectors

The vectors of *L. loa* are tabanid flies of the genus *Chrysops*. Day-biting flies transmit the human form of the parasite; crepuscular or nocturnally feeding *Chrysops* native to the forest canopy transmit the simian form. Pronounced host specificity among different species of vectors and spatial separation isolate the human and simian forms of *L. loa* so that microfilaremic primates do not constitute a reservoir for human infection (Duke and Wijers 1958, Ottesen 1990).

The vectors of *L. loa* include *Chrysops silacea* and *Chrysops dimidiata* (known as red flies), both day-biting flies that live in rain forests. Their daily peak biting time coincides with peak microfilaremia in humans. The larvae of these flies develop in sediments along the sides of shaded streams in the forest. The adult flies occur high in the forest canopy and descend to the forest floor to bite humans. Flies are attracted to smoke from wood fires and to vehicles. Rubber plantations favor transmission of *L. loa*. The high canopies of rubber trees provide good fly habitat, and the biting behavior of red flies results in effective host contact with day laborers at ground level. Other tabanids, in particular *Chrysops langi* and *Chrysops centurionis*, feed at night on primates such as mandrills and transmit a nocturally periodic form of *L. loa*.

Prevention and Control

Control aimed at vectors primarily is preventive, involving forest clearing to reduce habitat, and personal protective measures.

MANSONELLOSIS

The Parasites

Filariae of the genus *Mansonella* reside in the subcutis or peritoneal cavities of their vertebrate hosts and have unsheathed microfilariae. Three species are recognized as human parasites. *Mansonella perstans* adults reside in the peritoneal cavity of hosts. Female worms are 70 mm in length and 0.12 mm in width. Males measure 38 mm in length and 0.06 mm in width. Microfilariae circulate in the peripheral blood and may be recognized by their comparatively blunt tails with nuclei extending to the extremity. Adult *M. streptocerca* dwell in dermal collagen or in subcutaneous tissues of humans. Females measure 27 mm in length. Unlike *M. perstans*, microfilariae of *M. streptocerca* migrate through the skin of the host. Although the range of body lengths for *M. streptocerca* microfilariae overlaps that of *O. volvulus*, they may be distinguished from the latter because their nuclear column extends to the caudal extremity, whereas microfilariae of *O. volvulus* have a visible caudal space. Adults of *M. ozzardi*, like those of *M. perstans*, are found in the peritoneal cavity of their human host. Females measure 65–81 mm in length and 0.23–0.25 mm in width, males 38 mm in length and 0.2 mm in width. Unsheathed microfilariae circulate in the peripheral blood (Ottesen 1990).

Life Cycle

Development of *Mansonella* spp. commences when microfilariae are ingested from the peripheral blood or skin by a ceratopogonid or simuliid vector. Larval development conforms to the general outline presented for filarial life cycles. First- through third-stage development takes place in the indirect flight muscles of the vector and transmission occurs when infective L3 migrate from the mouthparts of vectors during bloodfeeding. The course of development by *Mansonella* spp. in humans is not known, but the prepatent period of *M. ozzardi* in patas monkeys is 163 days (Orihel et al. 1993).

Vertebrate Hosts and Distribution

Mansonella spp. are limited to western and central Africa and the American tropics. *Mansonella perstans* is found throughout equatorial Africa and along the Caribbean coast of South America. *Mansonella streptocerca* and *M. ozzardi* are found in western and central Africa and in central and northern South America, respectively. Gorillas and chimpanzees are natural hosts for *M. streptocerca*, but the importance of these animals as reservoirs of human infection is not known. Patas monkeys are susceptible to experimental infection with *M. ozzardi*, but natural infection with this parasite in primates has not been documented. The host range of *M. perstans* is not known to extend beyond humans.

Human Disease

Historically, *Mansonella* infections have been regarded of little medical importance or even as "nonpathogenic." Ottesen (1993), however, cautions that there has been little systematic clinical or epidemiological study of these infections and that even though the majority of infected individuals are asymptomatic, a significant minority present with microfilarial dermatitis (*M. streptocerca*), general constitutional illness, arthralgia, neurological signs and even fatal pericarditis (*M. perstans*). This fact has both positive and negative implications for chemotherapy. On the one hand, it implies that the highly effective microfilaricidal drug currently available for treatment of onchocerciasis patients (ivermectin) can bring about a resolution of clinical symptoms. However, the side effects of mass microfilarial killing can, in the case of onchocerciasis, exacerbate both acute and chronic pathology associated with the disease.

Vectors

Biting midges in the genus *Culicoides* transmit *Mansonella* parasites, particularly *Culicoides furens* in the new world for *M. ozzardi* and *C. grahami* for *M. streptocerca* in western and central Africa. *Culicoides austeni* is the primary vector of *M. perstans* in Africa. In Brazil and Guyana, black flies (*S. amazonicum*) are vectors of *M. ozzardi*.

Prevention and Control

Because *Mansonella* infections in humans usually are considered benign, few preventive or control measures have been developed against these parasites. Skin repellents probably can limit exposure to infective insect bites.

DIROFILARIASIS

The Parasite

The onchocercid nematode *Dirofilaria immitis* causes **canine cardiovascular dirofilariasis**, or **dog heartworm** disease, in dogs and other canids. The genus *Dirofilaria* is cosmopolitan in distribution, and *D. immitis* probably is the most widespread species because of its association with domestic dogs. Those species of *Dirofilaria* in the subgenus *Dirofilaria*, including *D. immitis*, inhabit the cardiovascular system of their definitive hosts. Species in the subgenus *Nochtiella* infect the subcutaneous tissues as adults. Among species in the subgenus *Nochtiella* are: *Dirofilaria corynodes*, a parasite of monkeys in Africa; *Dirofilaria magnilarvatum*, a parasite of macaques in Malaysia; *Dirofilaria repens*, a parasite of dogs and cats in Africa, southern Europe and parts of Asia; *Dirofilaria tenuis*, a parasite of raccoons in North America; *Dirofilaria ursi*, a parasite of bears in North America and Asia; *Dirofilaria subdermata*, a parasite of porcupines in North America; *Dirofilaria roemeri*, a parasite of kangaroos and wallabies in Australia; and *Dirofilaria striata*, a parasite of bobcats in North America.

Mosquitoes transmit all species of *Dirofilaria* in both subgenera except for *D. ursi*, which is transmitted by black flies.

Life Cycle

The life cycle of *D. immitis* alternates between canid vertebrate hosts and mosquito vectors. Mosquito vectors become infected by ingesting microfilariae with bloodmeals. The microfilariae remain in the midgut for about 1 day, then migrate to the Malpighian tubules, penetrating the

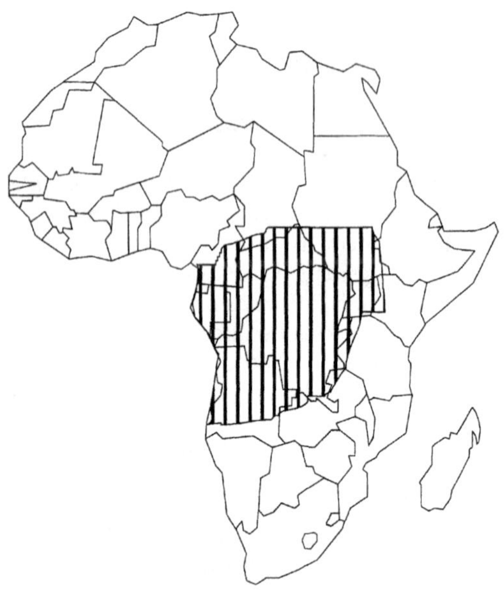

Fig. 9.22. Distribution of *Loa loa* in Africa.

primary cells of those organs. Within the primary cells, differentiation of the microfilariae occurs until by the third day post-ingestion they have become shorter and stouter (the "sausage stage"). Larvae move from the primary cells to the lumen of the Malpighian tubules 6–7 days after their ingestion by the mosquito. They molt to second stage larvae by day 10. A second molt, to the infective third larval stage, occurs by day 13. Infective larvae migrate anteriorly through the hemocoel of the mosquito and enter the proboscis. Dogs become infected when larvae exit the proboscis, generally from the tip of the labellum when the labium is bent during bloodfeeding, and enter the biting wound. Hemolymph deposited on the skin from the mouthparts may facilitate survival of the larvae on the skin prior to penetration.

Third stage *D. immitis* larvae remain in subcutaneous tissue close to the mosquito bite wound for about 3 days, whereafter they molt to fourth-stage larvae. The larvae then migrate to other subcutaneous, adipose or muscle tissues and molt again to a fifth-stage larva. These migrations and molts take place over a period of 70 days after infection of the vertebrate host. The worms, now approximately 18 mm in length, enter the venous circulation and become established in the heart and pulmonary arteries by day 115 post-infection. In the heart and pulmonary arteries, the fifth-stage parasites develop into sexually mature adults, mate, and the females release microfilariae into the circulation. At this time, the infected dog is infective for mosquitoes. Microfilaremia appears in an infected dog about 190-197 days after infection. Adult parasites may live in a dog for as long as 7.5 yr and microfilaremias may persist for 2.5 yr. The number of adult worms found in the hearts of infected dogs has ranged to as high as 15.

The concentration of microfilariae in the blood of dogs infected with adult worms varies from zero to 100,000 microfilariae per ml. Microfilaremia varies diurnally and seasonally; this variation results from movement of microfilariae between deep tissues and from peripheral circulation and probably is an adaptation to facilitate uptake by mosquitoes. Ambient temperature, light and oxygen tension in the small vessels of the lungs probably controls movement of microfilariae. Some dogs infected with *D. immitis* do not develop a microfilaremia, even though they may have heartworm disease. These dogs have **occult infections**. Occult infections may be caused by infection with parasites of only one sex, immune response resulting in reduced worm fertility, or the infection may be prepatent.

The factors controlling infectivity of microfilariae for mosquitoes, and the subsequent development of microfilariae to the L3, were reviewed above in the general discussion of filarial life cycles. *Dirofilaria immitis* is an important laboratory model for studying these interactions.

Vertebrate Hosts and Distribution

Dirofilaria immitis is primarily a parasite of canids, in particular those in the genus *Canis* such as the domestic dog (*Canis familiaris*) and coyote (*Canis latrans*). However, other canids have been infected with adult worms, and in some cases circulate microfilariae; they include the timber wolf, various species of foxes, the dingo, the maned wolf and the raccoon dog. Adult *D. immitis* also have infected hearts or the cardiopulmonary system of several species of felids, including the domestic cat, *Felis domesticus*. A wide range of other mammals likewise have been infected with adult *D. immitis*, including the California sea lion, common seal, wolverine, raccoon, ferret, otter, red panda, black bear, orangutan, gibbon and humans. Beaver, muskrat and European rabbit are other unusual hosts. Yet these unusual host records, largely based on postmortem examinations, generally were characterized by infections outside of the heart and were not accompanied by circulating microfilariae. Generally, the domestic dog and some closely related species such as coyote are thought to be the primary reservoir hosts in nature.

Pelecitus scapiceps and *Loaina uniformis*, both parasites of rabbits, previously were classified in the genus *Dirofilaria*.

The geographic distribution of *D. immitis* infections in dogs is widespread, involving all inhabitable continents, and it appears to be expanding. In North America, the distribution of *D. immitis* has expanded from the original focus in the southeastern USA in the 1950s to most of the eastern USA and parts of southern Canada, California and Arizona. Movement of infected dogs from one area to another probably facilitated this expansion.

Canine Disease

The literature of dog heartworm is voluminous and includes epidemiological, clinical and vector studies. The reader is referred to the compilation of papers edited by Boreham and Atwell (1988), the text by Rawlings (1986), and the review by Grieve et al. (1983), from which much of the information below was abstracted.

Adult *D. immitis* occupy the right ventricle of the canine heart and the pulmonary arteries extending from it (Fig. 9.23). The worms are 120-310 mm long and form aggregations of up to 50 or more individuals. In large aggregations, infection may extend to the right atrium. The pathobiology of heartworm disease in dogs is not a direct consequence of a heavy worm burden in the ventricle that impedes blood flow. Rather, infections result in deleterious changes in the endothelium and integrity of the walls of the pulmonary arteries, leading to pulmonary hypertension and right ventricular hypertrophy. These pathologic changes cause decreased cardiac output to the lungs, weakness, lethargy, chronic coughing and ultimately congestive heart failure. Dogs may die of infection if left untreated. Treatment of dog heartworm is discussed below under treatment and control.

Vectors

Lok (1988) listed 26 species of mosquitoes, incriminated on the basis of field or laboratory studies, as vectors of *D. immitis* in different geographic regions. Most vectors are in the genera *Aedes*, *Anopheles* or *Culex*. In some areas, native mosquito species became vectors after introduction of the parasite in infected dogs. In other instances, introduced mosquito species transmitted filariae after invasion of endemic areas. In California, the native treehole breeding species *Aedes sierrensis* became involved in the transmission of *D. immitis* after infected dogs appeared. The invasion of the exotic mosquito

Aedes albopictus into North America was accompanied by the involvement of this species in transmission of *D. immitis*, particularly in southern parts of the USA such as Louisiana and Florida.

The criteria for vector incrimination are described in Chapter 6. They apply to incrimination of mosquito species as vectors of *D. immitis*. Field collection of naturally infected mosquitoes is required. There must be evidence that the suspected mosquito feeds on dogs. The occurrence of infections in dogs and the presence of suspected vectors must coincide. It must be demonstrated in the laboratory that the parasite can develop to the infective-stage in the suspected mosquito, and that the mosquito can transmit infective larvae to susceptible hosts under realistic environmental conditions.

Table 9.8 lists the species that have been found to be naturally infected with filarial parasites presummed to be *D. immitis*. Table 9.9 summarizes 25 studies conducted in the USA in which naturally infected mosquitoes were collected and examined for *D. immitis*. Although not the sole studies of heartworm vectors, only these 25 collected naturally infected mosquitoes in the field. Studies in which mosquitoes were allowed to feed on infected dogs in dog-baited traps are excluded from this summary because in these instances vector stage development takes place in the laboratory rather than in the field, hence it does not represent natural survival of infected vectors. These studies, although inconclusive, provide important information on potential vectors of heartworm in a given region. A complete discussion of susceptibility studies is beyond the scope of this review; more than 70 species of mosquito have been shown to be suitable hosts for *D. immitis* in a large number of publications reviewed by Otto and Jachowski (1981) and Ludlam et al. (1970). Table 9.10 summarizes a few important susceptibility studies; it includes those already noted in Table 9.9.

Mapping studies have clarified the regional importance of different vector species (Fig. 9.24). Of the species listed, only *Aedes vexans* has been collected naturally infected in widespread parts of the USA. *Anopheles punctipennis* has been implicated everywhere but the western USA. Other species appear to be of only regional importance. Comparing studies in which a mosquito species has been collected infected with studies in which it harbored infective-stage larvae allows inferences about the vector capacity of the species. Although *Ae. vexans* has been collected with filariid nematodes indistinguishable from *D. immitis* in 14 different studies, it was collected with infective-stage larvae only in about half of them. This may indicate regional variation in susceptibility to infection or it may represent the failure of this species to survive long enough for completion of vector stage development of the parasite. The large number of collections of this species may have more to do with the ubiquity of *Ae. vexans* than with its vector capacity. On the other hand, infected *Aedes trivittatus* has been collected only in 6 studies, but in each it was collected with infective larvae, indicating not only that all populations sampled can support the development of the parasite, but also that the mosquito survives long enough in the field for the parasite to complete vector stage development.

Of the 22 North American species that have been collected naturally infected with third-stage larvae, only 8 have been shown to transmit naturally the parasite to noninfected dogs in the laboratory: *Ae. albopictus*, *Aedes canadensis*, *Aedes sollicitans*, *Aedes triseriatus*, *Ae. trivittatus*, *Ae. vexans*, *An. punctipennis* and *An. quadrimaculatus* (Newton 1957, Bickley et al. 1977, Christensen 1977, Hendrix et al. 1980, Roberts et al. 1985, Hendrix et al. 1986, Scoles 1997). Another species, *Ae. aegypti*, transmits filariae in the laboratory, although it never has been collected in the field with infective stage larvae (Hendrix et al. 1986). *Culex salinarius* has been collected with infective stage larvae, but failed to transmit larvae in the laboratory (Bickley et al. 1976); this failure demonstrates that susceptible individual mosquitoes do not always have the capacity to transmit. Two species exotic to the USA, *Aedes vigilax* from Australia and *Ae. togoi* from Japan, also transmit *D. immitis* in the laboratory (Kume and Itagaki 1955, Bemrick and Moorehouse 1968).

Representatives of several genera of filariid nematodes can be found in field collected mosquitoes; within the genus *Dirofilaria* alone there are several mosquito-transmitted species native to the USA. Arnott and Edman (1978) and Sauerman and Nayar (1983) discussed species of filariids in addition to *D. immitis* that might be encountered in mosquito surveys within the USA. Bain (1986) listed 21 North American filariid species for which the vector stage is known, 15 of which are transmitted by mosquitoes. His list included 5 species of *Dirofilaria*, 4 transmitted by mosquitoes: *D. immitis, D. repens, D. striata* and *D. tenuis*. *Dirofilaria lutrae* occurs in otters, but vector stages are unknown (Orihel 1965).

Epidemiology

Studies of geographical distribution

The movement of *D. immitis* into new areas of the USA has been widely discussed but little investigated (Noyes 1974, Otto 1975, Slocombe and McMillian 1978, 1981). In the 1960s, *D. immitis* was confined mostly to the Atlantic and Gulf coast regions (Otto 1975); since that time the parasite has spread throughout the USA. Although movement of infected dogs from regions of high endemicity in the south to regions formerly free of the parasite undoubtedly explains some of this range expansion, the distribution and spread of mosquito vectors also plays a critical role. Because of the essential role of mosquitoes in the life cycle of *D. immitis*, introduction of infected dogs into habitats without the presence of competent vectors will not result in the establishment of enzootic transmission. Consequently, the distribution of enzootic canine heartworm is linked closely to the distribution of competent vectors.

Identification of parasites

Distinguishing between closely related species is difficult. Several authors have prepared taxonomic keys to third–stage larvae of mosquito-transmitted filariid nematodes (Nelson 1959, 1960, Yen et al. 1982, Yen 1983, Bain and Chabaud 1986). Although these keys all include *D. immitis*, they often do not include the wide variety of other animal filariids that can be found in mosquitoes in the USA. For the most part, these keys were written for specific areas of the world where human filariasis is endemic; they are intended to distinguish the most common species of animal filariids (usually *D. immitis*) from filariid parasites of humans. Consequently they are not very useful for identifying vector stages of filariids of nonhuman animals in the USA.

Identification of filariids collected from mosquitoes naturally infected in the field has been a consistent problem in studies of the vectors of *D. immitis*. Different authors have used various characteristics for identification such as morphology (presence and arrangement of anal papillae, length of the tail and body size) and site of development in the vector and vertebrate hosts. None of these are definitive for species of *Dirofilaria*.

Third–stage larvae of filariids have few easily observable morphologic features that are useful for identification beyond genus. The characters that are present are variable (i.e., anal papillae) and thus not reliable. Furthermore, techniques for fixing and mounting L3 can affect the size and shape of the larvae, critical characters in most keys. Identifications based on morphology can be made reliably only in fresh, unfixed specimens.

Size often has been used as an identifying characteristic by comparing sizes of the worms collected with values published in the literature or to laboratory-reared specimens. Size is a useful character to separate some larger or smaller species (e.g., the large size of *Setaria* spp. allows these parasites to be separated from *D. immitis* with relative ease), but *Dirofilaria* species have overlapping size ranges. In addition, size of individual L3 can vary with vector mosquito size, mosquito species, the number of worms infecting the mosquito, as well as with the methods used to preserve the larvae. A wide range of sizes have been reported in the literature for *D. immitis* (Nelson 1959, Taylor 1960, Intermill 1973, Christensen 1977, Yen et al. 1982, Hendrix et al. 1984). Thus, definitive identification by size alone is not reliable.

The vector development site of the parasite may be an important identifying character. Most filariids develop in the fat body or the flight muscles of the mosquito; only *Dirofilaria* species develop in the Malpighian tubules. The 2 species formerly classified as *Dirofilaria*, *Pelecitis scapiceps* and *Loiana uniformis*, develop in the fat body, making it possible to distinguish them from *Dirofilaria* spp. However, larvae found developing in the Malpighian tubules could be any one of the 4 species mentioned earlier: *D. immitis*, *D. repens*, *D. striata* or *D. tenuis*. Once the L3 has left the site of development and migrated to the head, it may be difficult to identify the tissue in which it developed, especially in cases where the worm burden is low. Pathology characteristic of the development of *Dirofilaria* often can be observed in dissected Malpighian tubules, but again, this pathology will provide only genus level identification. Site of development cannot be used for identification when *en masse* dissection techniques are used because L3 are observed free of the mosquito and the individual from which they came cannot be identified.

The presence of infected vertebrate hosts also cannot be used as proof of the occurrence of a given parasite species. Animals such as raccoons and rabbits are ubiquitous, and these animals serve as hosts of *Dirofilaria* or closely related species indistinguishable from *D. immitis* (*D. tenuis* in raccoons, *L. uniformis* and *P. scapiceps* in rabbits). In most areas, it is unlikely that a site with mosquitoes and dogs will be completely lacking in other possible hosts or their mosquito-transmitted filariids.

Accurate identification of filariid L3 is necessary for the incrimination of vectors. None of the methods described is definitive, because there are many species of filariid nematodes transmitted by mosquitoes and several sympatric species are indistinguishable from *D. immitis*. Molecular techniques, like the PCR test (Scoles and Kambhampati 1995, Nicolas and Scoles 1997) can improve the accuracy of identifications and do not require expertise in worm morphology.

Older vector incrimination studies were based on one or more of the non-biochemical methods described above. Because of the inability of these methods to produce definitive identifications of parasites, many of the vector associations arising from these studies should be considered with caution.

Because the prevalence of *D. immitis* in mosquitoes in the field is often less than 0.2%, a large number of mosquitoes must be dissected to detect filariae. Complete dissection of individual mosquitoes would be the ideal method of collecting data since all of the mosquitoes with *D. immitis* in any developmental stage would be detected, but time and resources usually constrain this approach. Scoles et al. (1993) described an infectivity assay dissection technique that is a compromise. This technique allows rapid processing of large numbers of mosquitoes without significant loss of individual data. Although the technique may underestimate the overall parasite prevalence in the mosquito population (because the immature stages are not found), vector potential may be more accurately represented because only infective-stage

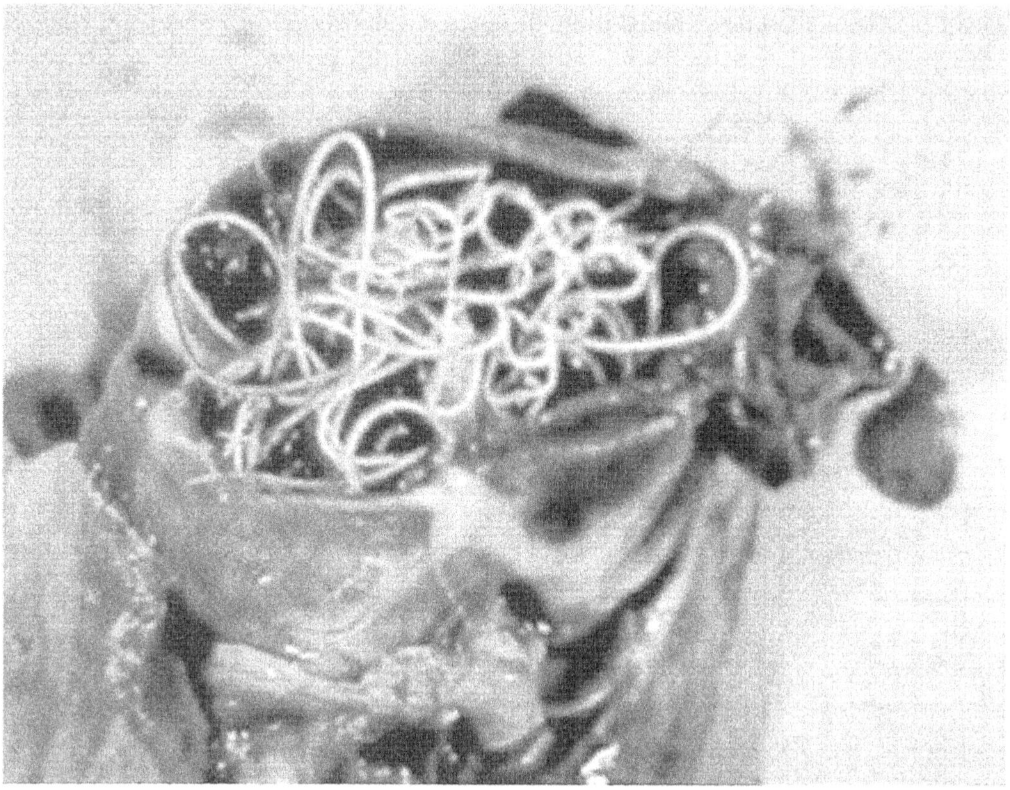

Figure 9.23. Adult *Dirofilaria immitis* in the heart of dog.

larvae are detected. Muller and Denham (1974) described an *en masse* dissection technique for processing of large numbers mosquitoes faster than the infectivity assay technique. However, *en masse* techniques do not furnish individual mosquito data, so only a minimum infection rate can be calculated. Further, the site of development in the mosquito cannot be determined with these techniques, so that parasite identifications cannot be aided by this information.

Many questions remain about the vector biology of canine heartworm in different regions and in different enzootic cycles. Newly identified enzootic foci of heartworm transmission occur in many areas of the country for which the vector species are unknown. Vectors of *Dirofilaria* to "emerging" hosts such as cats and ferrets await discovery; the same species that transmit worms to dogs may or may not be vectors.

Prevention and Control

Chemotherapy for infected dogs uses diethylcarbamazine and/or ivermectin as microfilariacides; however, treatments to reduce or eliminate adult worm burdens are complicated by adverse reactions. Therefore, prophylactic administrations of drugs are the preferred methods to prevent infection. They

Table 9.8. Species of mosquitoes found naturally infected with filariid nematodes presumed to be *Dirofilaria immitis*.

Species	State	Total reports	Reports with L3
Ae. vexans	Alabama, California, Connecticut, Indiana, Maryland, Michigan, Minnesota, New Jersey, New York, Oklahoma	14	8
An. punctipennis	Alabama, Iowa, Kentucky, Massachusetts, Maryland	7	3
Ae. trivittatus	Alabama, Iowa, Indiana, Oklahoma, Tennessee	6	6
An. quadrimaculatus	Massachusetts, Michigan, New York	4	3
Ae. sollicitans	Connecticut, North Carolina, New York	4	2
Ae. sticticus	Alabama, Massachusetts	3	3
Ae. taeniorhynchus	Florida, North Carolina	3	3
Ae. canadensis	Connecticut, Massachusetts, New Jersey	3	3
Cx. salinarius	Maryland, North Carolina, New Jersey	3	2
Cx. quinquefasciatus	Alabama, Florida, Louisiana	3	2
Ae. excrucians	Connecticut, Massachusetts	3	2
Ae. stimulans	Connecticut, Massachusetts	2	1
An. bradleyi	North Carolina	2	1
Ae. albopictus	Louisiana	2	1
Ps. ferox	Connecticut	1	1
Cx. nigripalpus	Florida	1	1
Ae. cantator	New Jersey	1	1
Ae. sierrensis	California	1	1
Ae. triseriatus	Indiana	1	1
An. freeborni	California	1	1
Cx. pipiens	Michigan	1	1

Table 9.9. Mosquitoes found with filariid nematodes presumed to be *Dirofilaria immitis*. Number infected shown as number with nematodes, number with L3 (in parentheses, if known), followed by total number dissected. (Table continued on next page).

State	Species	Number infected	Reference
Alabama	*Ae. sticticus*	10(1)/125	Johnson and Harrell (1986)
	Ae. trivittatus	1(1)/18	
	Ae. vexans	2(0)/28	
	An. crucians	1(0)/4	
	An. punctipennis	5(1)/64	
	Ae. vexans	6(6)/84	Tolbert and Johnson (1982)
	An. punctipennis	34(34)/326	
	Cx. quinquefasciatus	14(0)/520	
	Ae. sticticus	2(1)/55	Buxton and Mullen (1980)
	Ae. vexans	4(4)/1,599	
	An. punctipennis	9(2)/391	
California	*Ae. vexans*	8(3)/256	Walters and Lavoipierre (1982)
	Ae. sierrensis	28(2)/599	Walters (1996)
	Ae. vexans	37(7)/829	
	An. freeborni	5(1)/131	
Connecticut	*Ae. canadensis*	15(13)/483	Magnarelli (1978)
	Ae. stimulans	10(6)/541	
	Ae. excrucians	5(4)/214	
	Ae. sollicitans	11(0)/588	
	Ae. vexans	2(0)/223	
	Ps. ferox	3(2)/333	
Florida	*Cx. nigripalpus*	2(1)/403	Sauerman and Nayar (1983)
	Cx. quinquefasciatus	3(1)/200	
	Ae. taeniorhynchus	8(4)/264	
Indiana	*Ae. trivittatus*	22(19)/845	Pinger (1982)
	Ae. vexans	1(1)/904	
Iowa	*Ae. trivittatus*	25(7)/1,304	Christensen and Andrews (1976)
	An. punctipennis	2(0)/468	

should be administered before, during and after the mosquito season.

ANIMAL MODELS

The identification of appropriate laboratory animal models is a prerequisite for studies of pharmacology, immunity, biochemistry and molecular biology of medically important filariae. This requirement is challenging because of the strict specificity of species such as *O. vol-*

Table 9.9 (Continued). Mosquitoes found with filariid nematodes presumed to be *Dirofilaria immitis*. Number infected shown as number with nematodes, number with L3 (in parentheses, if known), followed by total number dissected.

State	Species	Number infected	Reference
Kentucky	*An. punctipennis*	2(0)/52	Courtney and Christensen (1983)
Louisiana	*Cx. quinquefasciatus*	33(24)/2,312	Villavaso and Steelman (1970)
	Ae. albopictus	2(0)/163	Comiskey and Wesson (1995)
Maryland	*Ae. vexans*	1(0)/20	Bickley et al. (1976)
	An. punctipennis	1(0)/54	
	Cx. salinarius	2(0)/142	
Massachusetts	*An. punctipennis*	4(0)/21	Phillips (1939)
	An. quadrimaculaltus	2(0)/10	
	Ae. excrucians	1(0)/1	
	Ae. canadensis	2(1)/258	Arnott and Edman (1978)
	Ae. stimulans group	3(0)/235	
	Ae. excrucians	5(4)/154	
	Ae. sticticus	1(1)/80	
Michigan	*Cx. pipiens*	1/1,025	Lewandowski et al. (1980)
	An. quadrimaculatus	3/1,350	
	Ae. vexans	2/32,500	
Minnesota	*Ae. vexans*	5(0)/4,747	Bemrick and Sandholm (1966)
New York	*Ae. vexans*	1/535	Todaro et al. (1977)
	An. quadrimaculatus	3/50	
North Carolina	*Ae. sollicitans*	10(4)/1,021	Parker (1986)
	Ae. taeniorhynchus	4(1)/139	
	An. bradleyi	3(0)/801	
	Cx. salinarius	2(1)/354	
	Ae. sollicitans	24(9)/2,533	Parker (1993)
	Ae. taeniorhynchus	21(5)/2,294	
	An. bradleyi	33(9)/2,641	
Oklahoma	*Ae. trivittatus*	8(1)/589	Afolabi et al. (1988)
	Ae. vexans	2(0)/674	
Tennessee	*Ae. trivittatus*	1(1)/530	Hribar and Gerhardt (1985)

vulus and *W. bancrofti* for primate hosts. The usefulness of a given animal model depends on the degree to which the parasite undergoes its natural course of development and on the spectrum of human disease and immune response resulting from the infection. The similarity of antifilarial drug metabolism to that occurring in humans also is a consideration. Rodents or other small laboratory animals provide the most practical subjects for laboratory study.

Table 9.10. Other studies of mosquitoes naturally infected with filariid nematodes presumed to be *Dirofilaria immitis*.

State	Species	Number infected	Reference*
Indiana and Michigan	*Ae. trivittatus*	11/3,613	Scoles (1997)
	Ae. vexans	1/4,133	
	Ae. triseriatus	1/637	
	An. quadrimaculatus	2/534	
Louisiana	*Ae. albopictus*	1/456	
New Jersey	*Ae. sollicitans*	Low VP, no L3	Crans and Feldlaufer (1974)
	Ae. vexans	Low VP, no L3	
	Ae. cantator	Moderate VP few with L3	
	Cx. salinarius	Moderate VP, few with L3	
	Ae. canadensis	High VP, many with L3	

*Study by Scoles utilized infectivity assay dissection technique that detects only infective larvae. Study by Crans and Feldlaufer defined a vectorial capacity (VP) level based on the presence of infective larvae (L3).

However, in some cases the best (or only) models available involve larger species, including captive primates.

Lymphatic Filariasis

To date, *B. malayi* and its zoonotic counterpart, *B. pahangi*, are the only medically important filariae to have been established in laboratory rodents. Ash and Riley (1970a, 1970b) demonstrated that both parasites develop to patency in the mongolian gerbil *Meriones unguiculatus*. Because they exhibit a stable microfilaremia, gerbils allow for long-term maintenance of *Brugia* spp. and for studies of vector-parasite interactions using laboratory-reared mosquitoes. Klei et al. (1982, 1990) used *Brugia*-infected gerbils to study immunopathological and other mechanisms underlying lymphadenopathy and other manifestations of lymphatic filariasis. Inbred strains of mice including congenitally athymic nude (Suswillo et al. 1980, Vincent et al. 1982) and SCID (Nelson et al. 1991) animals, have been used in conjunction with *B. pahangi* or *B. malayi* to examine the relative contributions of the cellular and humoral arms of the immune response to host resistance and disease.

Among the experimental models of lymphatic filariasis, the course of *B. pahangi* infection and the resulting lesions observed in domestic cats probably most closely mimic the theorized progression of infection and disease in humans. Denham and Fletcher (1987) used *Brugia*-infected cats to study the transitions from the symptomatic microfilaremic state to amicrofilaremic states characterized by varying degrees of lymphadenopathy. The success of all the laboratory models of *Brugia* spp. infection hinges upon the fact that a highly efficient and easily manipulated laboratory vector is available in the form of the "Liverpool selected" (f^m/f^m) strain of *Ae. aegypti* (Macdonald 1962a, 1962b).

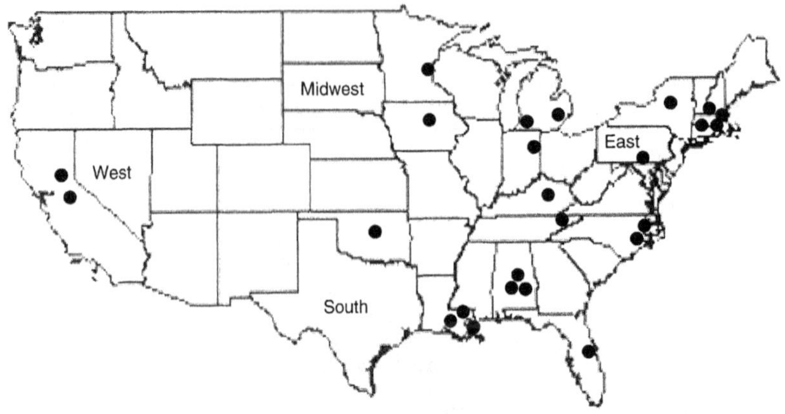

West	Midwest	South	East
Ae. sierrensis	*Ae. triseriatus*	*Ae. albopictus*	*Ae. canadensis*
Ae. vexans	*Ae. trivittatus*	*Ae. sollicitans*	*Ae. cantator*
An. freeborni	*Ae. vexans*	*Ae. sticticus*	*Ae. excrucians*
	An. punctipennis	*Ae. taeniorhynchus*	*Ae. sollicitans*
	An. quadrimaculatus	*Ae. trivittatus*	*Ae. sticticus*
	Cx. pipiens	*Ae. vexans*	*Ae. stimulans*
		An. bradleyi	*Ae. vexans*
		An. crucians	*Cx. salinarius*
		An. puncipennis	*Ps. ferox*
		Cx. nigripalpus	
		Cx. quinquefascitus	
		Cx. salinarius	

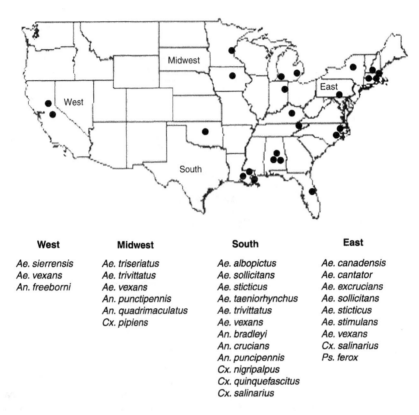

Figure 9.24. Regions of the USA where *Dirofilaria immitis* infects domestic dogs. For descriptions of species involved, see text and Tables 9.8 and 9.9.

Onchocerciasis

All attempts to date to induce patent *O. volvulus* infection in laboratory rodents have failed (Suswillo et al. 1977). However, Abraham et al. (1992, 1993) discovered that *O. volvulus* and a related parasite of domestic cattle, *O. lienalis*, will undergo partial development including the third- to fourth-stage molt and significant growth in inbred mice. The use of micropore chamber implants allows quantitative recovery of surviving larvae in experiments designed to evaluate vaccine candidates in these mice. Furthermore, by varying the pore size of the micropore chambers either to include or exclude cells, these investigators have evaluated the relative contributions of cellular and humoral factors in the innate and acquired responses of mice to early onchocercal infection. Other "partial infection" systems involving small laboratory animals have been used to investigate both protective and immunopathological responses to *Onchocerca* microfilariae. Carlow and Bianco (1987) and Carlow et al. (1988) exploited the ability of inbred mice to sustain living *O. lienalis* microfilariae in their skins for extended periods to study immune mediated clearance of these parasites. Donnelly et al. (1983, 1985) studied the local and systemic responses of mice and guinea pigs to subconjunctivally and intraocularly injected *O. lienalis* and *O. volvulus* microfilariae.

Nonhuman primates have found limited application as animal models of patent *O. volvulus* infection. Experimental infections have been induced in chimpanzees (Duke 1980, Prince et al. 1992) and mangabey monkeys (Eberhard et al. 1991); studies of the course of infection, including the elaboration of diagnostic antigens and trials of vaccine candidates, have been conducted in these hosts. However, the expense involved in breeding and maintaining primates and the ethical constraints against euthanasia make them impractical subjects for laboratory investigations, particularly vaccine or anthelmintic trials which require enumeration of adult parasites via postmortem examination.

More recently, *Onchocerca ochengi*, a parasite of cattle occurring sympatrically with *O. volvulus* in West Africa, has come to light as a potential model of *O. volvulus* infection (Trees et al. 1998). *Onchocerca ochengi* shares numerous biological attributes with *O. volvulus* including a simuliid vector and the propensity for adult worms to cluster in easily palpable subdermal nodules that occur primarily along the ventral midline. The possibility of isolating virtually all adult worms in the host by excision of these nodules constitutes a major advantage of this system over the various *O. volvulus*–primate models in which the majority of adult worms are found in bundles deep in intermuscular connective tissue. Significantly, McCall and Trees (1989) demonstrated that the temperate black fly *Simulium ornatum* can act as a laboratory vector of *O. ochengi*. This black fly is easily reared from larva to adult in the laboratory and inoculated with microfilariae via intrathoracic injection, thus making large numbers of L3 available for experimental infections in cattle. This abundance constitutes a major advantage over another nodule-forming bovine onchocercid, *Onchocerca gibsoni*, whose ceratopogonid vectors are not amenable to experimental infection on a large scale.

REFERENCES

Abraham, D., Eberhard, M.L., Lange, A.M., Yutanawiboonchai, W., Perler, F. and Lok, J.B. 1992. Identification of surrogate rodent hosts for larval *Onchocerca lienalis* and induction of protective immunity in a model system. J. Parasitol. **78**:447-453.

Abraham, D., Lange, A.M., Yutanawiboonchai, W., Trpis, M., Dickerson, J.W., Swenson, B. and Eberhard, M.L. 1993. Survival and development of larval *Onchocerca volvulus* in diffusion chambers implanted in primate and rodent hosts. J. Parasitol. **79**:571-582.

Afolabi, J.S., Ewing, S.A., Wright, R.E. and Wright, J.C. 1988. *Aedes trivittatus* and other potential vectors of canine dirofilariasis in Payne County, Oklahoma. Okla. Vet. **40**:75-79.

Agudelo-Silva, F. and Spielman, A. 1985. Penetration of mosquito midgut wall by sheathed microfilariae. J. Invertr. Pathol. **45**:117-119.

Akue, J.P., Egwang, T.G. and Devaney, E. 1994. High levels of parasite-specific IgG4 in the absence of microfilaremia in *Loa loa* infection. Trop. Med. Parasitol. **45**:246-248.

Anderson, R.C. 1992. Nematode parasites of vertebrates: their development and transmission. CAB International, Wallingford, UK. 578 pp.

Arnott, J.J. and Edman, J.D. 1978. Mosquito vectors of dog heartworm, *Dirofilaria immitis*, in Western Massachusetts. Mosq. News **38**:222-230.

Ash, L.R. and Riley, J.M. 1970a. Development of *Brugia pahangi* in the jird, *Meriones unquiculatuis*, with notes on infections in other rodents. J. Parasitol. **56**:962-968.

Ash, L.R. and Riley, J.M. 1970b. Development of subperiodic *Brugia malayi* in the jird, *Meriones unquiculatus*, with notes on infections in other rodents. J. Parasitol. **56**:969-973.

Bain, O. and Chabaud, A.G. 1986. Atlas des larves infestantes de filaries. Trop. Med. Parasitol. **37**:301-340.

Beerntsen, B.T., Severson, D.W., Klinkhammer, J.A., Kassner, V.A. and Christensen, B.M. 1995. *Aedes aegypti*: a quantitative trail locus (QTL) influencing filarial worm intensity is linked to a QTL for susceptibility to other mosquito-borne pathogens. Exp. Parasitol. **81**:355-362.

Bemrick, W.J. and Moorehouse, D.E. 1968. Potential vectors of *Dirofilaria immitis* in the Brisbane area of Queensland, Australia. J. Med. Entomol. **5**:269-272.

Bemrick, W.J. and Sandholm, H.A. 1966. *Aedes vexans* and other potential mosquito vectors of *Dirofilaria immitis* in Minnesota. J. Parasitol. **52:**762-767.

Bickley, W.E. 1976. Failure of *Culex salinarius* to transmit *Dirofilaria immitis* from dog to dog. Mosq. News **36:**366-367.

Bickley, W.E., Lawrence, G.S., Ward, G.M. and Shillinger, R.B. 1977. Dog-to-dog transmission of heartworm by *Aedes canadensis*. Mosq. News **37:**137-138.

Bickley, W.E., Mallack, J. and Seeley, D.C., Jr. 1976. Filaroid nematodes in field-collected mosquitoes in Maryland. Mosq. News **36:**92.

Blacklock, D.B. 1926. The development of *Onchocerca volvulus* in *Simulium damnosum*. Ann. Trop. Med. Parasitol. **29:**1-47.

Boatin, B., Molyneux, D.H., Hougard, J.M., Christensen, O.W., Alley, E.S., Yameogo, L., Seketeli, A. and Dadzie, K.Y. 1987. Patterns of epidemiology and control of onchocerciasis in west Africa. J. Helminthol. **71:**91-101.

Boreham, P.F.L. and Atwell, R.B. (eds). 1988. Dirofilariasis. CRC Press, Boca Raton, Florida. 249 pp.

Bryan, J.H. and Southgate, B.A. 1976. Some observations on filariasis in Western Samoa after mass administration of diethylcarbamazine. Trans. Royal Soc. Trop. Med. Hyg. **70:**39-48.

Bryant, J. 1935. Endemic retino-choroiditis in the Anglo-Egyptian Sudan and its possible relationship to *Onchocerca volvulus*. Trans. Royal Soc. Trop. Med. Hyg. **28:**523-532.

Buxton, B.A. and Mullen, G.R. 1980. Field isolations of *Dirofilaria* from mosquitoes in Alabama. J. Parasitol. **66:**140-144.

Calamari, C., Yamaeogo, L., Hougard, J.-M. and Leveque, C. 1998. Environmental assessment of larvicide use in the Onchocerciasis Control Programme. Parasitol. Today **14:**485-489.

Carlow, C.K. and Bianco, A.E. 1987. Transfer of immunity to the microfilariae of *Onchocerca lienalis* in mice. Trop. Med. Parasitol. **38:**283-286.

Carlow, C.K., Dobinson, A.R. and Bianco, A.E. 1988. Parasite-specific immune responses to *Onchocerca lienalis* microfilariae in normal and immunodeficient mice. Parasite Immunol. **10:**309-322.

Chan, M.S., Srividya, A., Norman, R.A., Pani, S.P., Ramaiah, K.D., Vanamail, P., Michael, E., Das, P.K. and Bundy, D.A.P. 1998. EPIFIL: a dynamic model of infection and disease in lymphatic filariasis. Am. J. Trop. Med. Hyg. **59:**606-614.

Chiang, G.L. 1993. Update on the bionomics of *Mansonia* vectors of brugian filariasis. S.E. Asian J. Trop. Med. Publ. Hlth. **24(2, Suppl.):**69-75.

Christensen, B.M. 1977. Laboratory studies on the development and transmission of *Dirofilaria immitis* by *Aedes trivittatus*. Mosq. News **37:**367-372.

Christensen, B.M. and Andrews, W.N. 1976. Natural infection of *Aedes trivittatus* (Coq.) with *Dirofilaria immitis* in central Iowa. J. Parasitol. **62:**276-280.

Cobbold, T.S. 1877. On *Filaria bancrofti*. Lancet **2:**495-496.

Collins, R.C. 1977. Development of *Onchocerca volvulus* in *Simulium ochraceum* and *Simulium metallicum*. Am. J. Trop. Med. Hyg. **28:**491-495.

Collins, R.C., Merino, M.E. and Cupp, E.W. 1981. Seasonal trends and diurnal patterns of man-biting activity of four species of Guatemalan black flies (Simuliidae). Am. J. Trop. Med. Hyg. **30:**728-733.

Colwell, E.J., Armstrong, D.R., Brown, J.D., Duxbury, R.E., Sadun, E.H. and Legters, L.J. 1970. Epidemiologic and serologic investigations of filariasis in indigenous populations and American soldiers in South Vietnam. Am. J. Trop. Med. Hyg. **19:**227-231.

Comiskey, N. and Wesson, D.M. 1995. *Dirofilaria* (Filarioidea: Onchocercidae) infection in *Aedes albopictus* (Diptera: Culicidae) collected in Louisiana. J. Med. Entomol. **32:**734-737.

Courtney, C.C. and Christensen, B.M. 1983. Field isolation of filarial worms presumed to be *Dirofilaria immitis* from mosquitoes in Kentucky. Mosq. News **43:**366-368.

Crans, W.J. and Feldlaufer, M.F. 1974. The likely mosquito vectors of dog heartworm in a coastal area of southern New Jersey. Proc. Calif. Mosq. Vect. Contr. Assoc. **42:**168.

Cupp, E.W., Bernardo, M.J., Kiszewski, A.E., Collins, R.C., Taylor, H.R., Aziz, M.A. and Greene, B.M. 1986. The effects of ivermectin on transmission of *Onchocerca volvulus*. Science **231:**740-742.

Dalmat, H.T. 1955. The black flies (Diptera, Simuliidae) of Guatemala and their role as vectors of onchocerciasis. Smithsonian Misc. Coll. Vol. 125. Smithsonian Institution, Washington, DC. 425 pp.

Davies, J.B. 1994. Sixty years of onchocerciasis vector control: a chronological summary with comments on eradication, reinvasion, and insecticide resistance. Annu. Rev. Entomol. **39:**23-45.

Demarquay, J.N. 1863. Note sur une tumeur des bourses contenant un lilquide laiteux (galactocèle de Vidal) et renferment des petits êtres vermifoirmes que l'on peut considérercomme des helminthes hematoides à l'état d'embryon. Gaz. Med. Paris **18:**665-667.

Denham, D.A. and Fletcher, C. 1987. The cat infected with *Brugia pahangi* as a model of human filariasis. Ciba Found. Symp. **127:**225-237.

Donnelly, J.J., Rockey, J., H., Bianco, A.E. and Soulsby, E.J. 1983. Aqueous humor and serum IgE antibody in experimental ocular *Onchocerca* infection of guinea pigs. Ophthal. Res. **15:**61-67.

Donnelly, J.J., Rockey, J.H., Taylor, H.R. and Soulsby, E.J.L. 1985. Onchocerciasis: experimental models of ocular disease. Rev. Infect. Dis. **7:**820-825.

Duke, B.O.L. 1980. Observations on *Onchocerca volvulus* in experimentally infected chimpanzees. Tropenmed. Parasitol. **31:**41-54.

Duke, B.O.L., Lewis, D.J. and Moore, P.J. 1966. *Onchocerca-Simulium* complexes. I. Transmission of forest and Sudan-savanna strains of *Onchocerca volvulus* from Cameroon by *Simulium damnosum* from various West African bioclimatic zones. Ann. Trop. Med. Parasitol. **60:**318-336.

Duke, B.O.L. and Wijers, D.J.B. 1958. Studies on loiasis in monkeys. I. The relationship between human and simian *Loa* in the rainforest zone of the British Cameroons. Ann. Trop. Med. Parasitol. **52:**158-173.

Duke, B.O.L., Zea-Flores, G. and Munoz, B. 1991. The embryogenesis of *Onchocerca volvulus* over the first year after a single dose of ivermectin. Trop. Med. Parasitol. **42:**175-180.

Eberhard, M.L., Dickerson, J.W., Boyer, A.E., Tsang, V.C., Zea-Flores, R., Walker, E.M., Richards, F.O., Zea-Flores, G. and Strobert, E. 1991. Experimental *Onchocerca volvulus* infections in mangabey monkeys (*Cercocebus atys*) compared to infections in humans and chimpanzees (*Pan troglodytes*). Am. J. Trop. Med. Hyg. **44:**151-160.

Eberhard, M.L. and Orhiel, T.C. 1981. Development and larval morphology of *Loa loa* in experimental primate hosts. J. Parasitol. **67:**556-564.

Eberhard, M.L. and Rabalais, F.C. 1976. *Dipetalonema viteae*: effects of hypo and hyperthermic stress on microfilaremia in the Mongolian jird, *Meriones unquiculatus*. Exp. Parasitol. **40:**5-12.

Furtado, A.F., Abath, F.G.C., Regis, L., Gomes, Y.M., Lucena, W.A., Furtado, P.B., Dhalia, R., Miranda, J.C. and Nicolas, L. 1997. Improvement and application of a polymerase chain reaction system for detection of *Wuchereria bancrofti* in *Culex quinquefasciatus* and human blood samples. Mem. Inst. Oswaldo Cruz **92:**85-86.

Garms, R., Cheke, R.A. and Sachs, R. 1991. A temporary focus of savanna species of the *Simulium damnosum* complex in the forest zone of Liberia. Trop. Med. Parasitol. **42:**181-187.

Gelband, H. 1994. Diethylcarbamazine salt in the control of lymphatic filariasis. Am. J. Trop. Med. Hyg. **50:**655-662.

Greene, B.M. 1990. Onchocerciasis. Pp. 429-439 *in* K.S. Warren and A. Mahmoud (eds.), Tropical and Geographical Medicine. McGraw Hill, New York.

Greene, B.M., Taylor, H.R., Cupp, E.W., Murphy, R.P., White, A.T., Aziz, M.A., Schulz-Key, H., D'Anna, S.A., Newland, H.S., Goldschmidt, L.P., Auer, C., Hanson, A.P., Freeman, S.V., Reber, E.W. and Williams, P.N. 1985. Comparison of ivermectin and diethylcarbamazine in the treatment of onchocerciasis. New England J. Med. **313:**133-138.

Grieve, R.B. and Lauria, S. 1983. Periodicity of *Dirofilaria immitis* microfilariae in canine and murine hosts. Acta Trop. **40:**121-127.

Grieve, R.B., Lok, J.B. and Glickman, L.T. 1983. Epidemiology of canine heartworm infection. Epidemiol. Rev. **5:**220-246.

Grove, D. 1990. A history of human helminthology. CAB International, Wallingford, Oxon, UK. 848 pp.

Grove, D.I., Valeza, F.S. and Cabrera, B.D. 1978. Bancroftian filariasis in a Philippine village: clinical, parasitological, immunological, and social aspects. Bull. Wld. Hlth. Organiz. **56:**975-984.

Gubler, D.J. and Battacharya, N.C. 1974. A quantitative approach to the study of bancroftian filariasis. Am. J. Trop. Med. Hyg. **23:**1027-1036.

Gyapong, J.O., Gyapong, M. and Adjei, S. 1996. The epidemiology of acute adenolymphangitis due to lymphatic filariasis in northern Ghana. Am. J. Trop. Med. Hyg. **54:**591-595.

Hairston, N.G. and de Meillon, B.A. 1968. On the inefficiency of transmission of *Wuchereria bancrofti* from mosquito to human host. Bull. Wld. Hlth. Organiz. **38:**935-941.

Hairston, N.G. and Jachowski, L.A. 1968. Analysis of the *Wuchereria bancrofti* population in the people of American Samoa. Bull. Wld. Hlth. Organiz. **38:**29-59.

Ham, P.J., Albuquerque, C., Baxter, A.J., Chalk, R. and Hagen, H.E. 1994. Approaches to vector control: new and trusted. 1. Humoral immune responses in blackfly and mosquito vectors of filariae. Trans. Royal Soc. Trop. Med. Hyg. **88:**132-135.

Ham, P.J. and Garms, R. 1988. The relationship between innate susceptibility to *Onchocerca*, and hemolymph attenuation of microfilarial motility *in vitro* using British and West African blackflies. Trop. Med. Parasitol. **39:**230-234.

Ham, P.J., Hagen, H.-E., Baxter, A.J. and Grunewald, J. 1995. Mechanisms of resistence to *Onchocerca* infection in blackflies. Parasitol. Today **11:**63-67.

Hawking, F. 1967. The 24-hour periodicity of microfilariae: biological mechanisms responsible for its production and control. Proc. Roy. Soc. Entomol. (B) **169:**59-76.

Hawking, F. 1976. The distribution of human filariasis throughout the world. Part II. Asia [Culicidae vectors]. Trop. Dis. Bull. **73:**967-1016.

Hawking, F. 1977. The distribution of human filariasis throughout the world. Part III. Africa. Trop. Dis. Bull. **74**:649-679.

Hawking, F. and Denham, D.A. 1976. The distribution of human filariasis throughout the world. Part I. the Pacific Region, including New Guinea. Trop. Dis. Bull. 73:347-373.

Hawking, F. and Gammage, K. 1968. The periodic migration of microfilariae of *Brugia malayi* and its response to various stimuli. Am. J. Trop. Med. Hyg. **17**:724-729.

Hendrix, C.M., Bemrick, W.J. and Schlotthauer, J.C. 1980. Natural transmission of *Dirofilaria immitis* by *Aedes vexans*. Am. J. Vet. Res. **41**:1253-1255.

Hendrix, C.M., Brunner, C.J. and Bellamy, L.K. 1986. Natural transmission of *Dirofilaria immitis* by *Aedes aegypti*. J. Am. Mosq. Contr. Assoc. **2**:48-51.

Hendrix, C.M., Wagner, M.J., Bemrick, W.J., Schlotthauer, J.C. and Stromberg, B.E. 1984. A scanning electron microscopic study of third-stage larvae of *Dirofilaria immitis*. J. Parasitol. **70**:149-151.

Hightower, A.J., Lammie, P.J. and Eberhard, M.L. 1993. Maternal filarial infection - a potential risk factor for microfilaremia in offspring? Parasitol. Today **9**:418-421.

Hissette, J. 1932. Memoire sur l'*Onchocerca volvulus* Leukart et ses manifestations oculaires au Congo Belge. Ann. Soc. Belg. Trop. Parasitol. Mycol. **12**:433-529.

Horsfall, W.R. 1972. Mosquitoes, their bionomics and relation to disease. Hafner, New York. 723 pp.

Hougard, J.-M., Yamégo, L., Sékétéli, A., Boatin, B. and Dadzie, K.Y. 1997. Twenty-two years of blackfly control in the Onchocerciasis Control Programme in West Africa. Parasitol. Today **13**:425-431.

Hribar, L.J. and Gerhardt, R.R. 1985. Wild-caught *Aedes trivittatus* naturally infected with filarial worms in Knox county, Tennessee. J. Am. Mosq. Contr. Assoc. **1**:250-251.

Hunter, J.M. 1992. Elephantiasis: a disease of development in north east Ghana. Soc. Sci. Med. **35**: 627-649.

Intermill, R.W. 1973. Development of *Dirofilaria immitis* in *Aedes triseriatus* Say. Mosq. News **33**:176-181.

Jingyuan, L., Zi, C., Xiaohang, H. and Zhaoping, T. 1992. Mass treatment of filariasis using DEC-medicated salt. J. Trop. Med. Hyg. **95**:132-135.

Johnson, W.E., Jr. and Harrell, L. 1986. Further study on the potential vectors of *Dirofilaria* in Macon County, Alabama. J. Parasitol. **72**:955-956.

Kartman, L. 1953. Factors influencing infection of the mosquito with *Dirofilaria immitis* (Leidy, 1856). Exp. Parasitol. **2**:27-78.

Katholi, C.R., Toe, L., Merriweather, A. and Unnasch, T.R. 1995. Determining the prevalence of *Onchocerca volvulus* infection in vector populations by polymerase chain reaction screening of pools of black flies. J. Infect. Dis. **172**:1414-1417.

Kazura, J.W., Bockarie, M., Alexander, N., Perry, R., Bockarie, F., Dagoro, H., Dimber, Z., Hyun, P. and Alpers, M.P. 1997. Transmission intensity and its relationship to infection and disease due to *Wuchereria bancrofti* in Papua New Guinea. J. Infect. Dis. **176**:242-246.

Kean, B.H., Mott, K.E. and Russell, A.J. 1978. Tropical medicine and parasitology. Classic investigations. Cornell University Press, Ithaca, New York. 673 pp.

Klei, T.R., Enright, F.M., Blanchard, D.P. and Uhl, S.A. 1982. Effects of presensitization on the development of lymphatic lesions in *Brugia pahangi*-infected jirds. J. Trop. Med. Hyg. **31**:280-291.

Klei, T.R., McVay, C.S., Dennis, V.A., Coleman, S.U., Enright, F.M. and Casey, H.W. 1990. *Brugia pahangi*: effects of duration of infection and parasite burden on lymphatic lesion severity, granulomatous hypersensitivity, and immune responses in jirds (*Meriones unquiculatus*). Exp. Parasitol. **71**:393-405.

Klion, A.D., Massougbodji, A., Sadeler, B.C., Ottesen, E.A. and Nutman, T.B. 1991. Loiasis in endemic and nonendemic populations: immunologically mediated differences in clinical presentation. J. Infect. Dis. **163**:1318-1325.

Kume, S. and Itagaki, S. 1955. On the life-cycle of *Dirofilaria immitis* in the dog as the final host. Brit. Vet. J. **111**:16-24.

Lehane, M.J. and Laurence, B.R. 1977. Flight muscle ultrastructure of susceptible and refractory mosquitoes parasitized by by larval *Brugia pahangi*. Parasitol. **74**:87-92.

Lehman, T., Cupp, M.S. and Cupp, E.W. 1995. Chemical guidance of *Onchocerca lienalis* microfilariae to the thorax of *Simulium vittatum*. Parasitol. **110**:329-337.

Leuckart, R. 1886. Die Parasiten des Menschen und die von ihnen herrührenden Ein Hand- und Lehrbuch für Naturforscher und Aertze (The parasites of man and the diseases which proceed from them. A textbook for students and practitioners, translated by W.E. Hoyle and J. Young). Vol. 1. Pentland, Edinburgh, Scotland. 1009 pp.

Levine, N.D. 1980. Nematode parasites of domestic animals and of man. Burgess, Minneapolis. 477 pp.

Lewandowski, H.B., Jr., Hooper, G.R. and Newson, H.D. 1980. Determination of some important natural potential vectors of dog heartworm in central Michigan. Mosq. News **40**:73-79.

Lewis, T.R. 1872. On a haematozoon inhabiting human blood, its relation to chyluria and other diseases. Pp. 241-266, Eighth Annual Report of the Sanitary Commissioner with the Government of India (1871). Appendix E.

Lewis, T.R. 1877. *Filaria sanguinis hominis* (mature form) found in a blood clot in a naevoid elephantiasis of the scrotum. Lancet **2:**453-455.

Li, J., Zhao, X. and Christensen, B.M. 1994. Dopachrome conversion activity in *Aedes aegypti*: significance during melanotic encapsulation of parasites and cuticular tanning. Insect Biochem. Mol. Biol. **24:**1043-1049.

Lok, J.B. 1988. *Dirofilaria* spp.: taxonomy and distribution *in* P.F.L. Boreham and R.B. Atwell (eds.), Dirofilariasis. CRC Press, Boca Raton, Florida.

Lok, J.B. and Abraham, D. 1992. Animal models for the study of immunity in human filariasis. Parasitol. Today **8:**168-171.

Low, G.C. 1900. A recent observation on *Filaria nocturna* in *Culex*: probable mode of infection in man. Brit. Med. J. **1:**1456-1457.

Ludlam, K.W., Jachowski, L.A. and Otto, G.F. 1970. Potential vectors of *Dirofilaria immitis*. J. Am. Vet. Med. Assoc. **157:**1354-1359.

Macdonald, W.W. 1962a. The genetic basis of susceptibility to infection with semi-periodic *Brugia malayi*. Ann. Trop. Med. Parasitol. **56:**373-382.

Macdonald, W.W. 1962b. The selection of a strain of Aedes aegypti susceptible to infection with semi-periodic Brugia malayi. Ann. Trop. Med. Parasitol. **56:**368-372.

Macdonald, W.W. and Ramachandran, C.P. 1965. The influence of the gene f^m (filarial susceptibility, *Brugia malayi*) on the susceptibility of *Aedes aegypti* to seven strains of *Brugia, Wuchereria* and *Dirofilaria*. Ann. Trop. Med. Parasitol. **59:**64-73.

Magnarelli, L.A. 1978. Presumed *Dirofilaria immitis* infections in natural mosquito populations of Connecticut. J. Med. Entomol. **15:**84-85.

Martin-Prevel, Y., Cosnefroy, J.Y., Tshipamba, P., Ngari, P., Chodakewitz, J.A. and Pinder, M. 1993. Tolerance and efficacy of single high-dose ivermectin for the treatment of loiasis. Ann. Trop. Med. Parasitol. **48:**186-192.

McCall, P.J. and Trees, A.J. 1989. The development of *Onchocerca ochengi* in surrogate temperate Simuliidae, with a note on the infective larva. Trop. Med. Parasitol. **40:**295-298.

McGreevy, P.B., McClelland, G.A.H. and Lavoipierre, M.M.J. 1974a. Inheritance of susceptibility to *Dirofilaria immitis* infection in *Aedes aegypti*. Ann. Trop. Med. Parasitol. **68:**97-109.

McGreevy, P.B., Theis, J.H. and Lavoipierre, M.M.J. 1974b. Studies on filariasis III. *Dirofilaria immitis*: emergence of infective larvae from the mouthparts of *Aedes aegypti*. J. Helminthol. **48:**221-228.

McReynolds, L.A., Poole, C., Hong, Y., Williams, S.A., Partono, F. and Bradley, J. 1993. Recent advances in the application of molecular biology in filariasis. S.E. Asian J. Trop. Med. Publ. Hlth. **24(2, Suppl.):**55-63.

Michael, E. 1993. Mathematical modelling of disease epidemiology. Parasitol. Today **9:**397-399.

Michael, E. and Bundy, D.A.P. 1997. Global mapping of lymphatic filariasis. Parasitol. Today **13:**472-476.

Michael, E., Bundy, D.A.P. and Grenfell, B.T. 1996. Reassessing the global prevalence and distribution of lymphatic filariasis. Parasitol. **112:**409-428.

Molyneux, D.H. and Davies, J.B. 1997. Onchocerciasis control: moving towards the millenium. Parasitol. Today **13:**418-425.

Muller, R.L. and Denham, D.A. 1974. A field technique for the recovery and preservation of infective filarial larvae from their vectors. Trans. Royal Soc. Trop. Med. Hyg. **68:**8-9.

Nayar, J.K., Knight, J.W. and Bradley, T.J. 1988. Further characterization of refractoriness in *Aedes aegypti* (L.) to infection by *Dirofilaria immitis* (Leidy). Exp. Parasitol. **66:**124-131.

Nayar, J.K., Knight, J.W. and Vickery, A.C. 1989. Intracellular melanization in the mosquito *Anopheles quadrimaculatus* (Diptera: Culicidae) against the filarial nematodes, *Brugia* spp. (Nematoda: Filarioidea). J. Med. Entomol. **26:**159-166.

Nayar, J.K. and Sauerman, D.M., Jr. 1975. Physiological basis of host susceptibility of Florida mosquitoes to *Dirofilaria immitis*. J. Insect Physiol. **21:**1965-1975.

Nelson, G.S. 1959. The identification of infective filarial larvae in mosquitoes: with a note on the species found in "wild" mosquitoes on the Kenya coast. J. Helminthol. **33:**233-256.

Nelson, G.S. 1960. The Identification of filarial larvae in their vectors. Indian J. Malariol. **14:**585-592.

Nelson, G.S. 1972. Onchocerciasis. Adv. Parasitol. **8:**173-224.

Nelson, F.K., Greiner, D.L.J., Schultz, L.D. and Rajan, T.V. 1991. The immunodeficient scid mouse as a model for human lymphatic filariasis. J. Exp. Med. **173:**659-663.

Newton, W.L. 1957. Experimental transmission of the dog heartworm, *Dirofilaria immitis* by *Anopheles quadrimaculatus*. J. Parasitol. **43:**589.

Nicolas, L. and Scoles, G.A. 1997. Multiplex polymerase chain reaction for detection of *Dirofilaria immitis* (Filariidea: Onchocercidae) and *Wuchereria bancrofti* (Filarioidea: Dipetalonematidea) in their common mosquito vector *Aedes polynesiensis* (Diptera: Culicidae). J. Med. Entomol. **34:**741-744.

Noble, E.R., Noble, G.A., Schad, G.A. and MacInnes, A.J. 1989. Parasitology: the biology of animal parasites, 6th ed. Lea and Febiger, Philadelphia. 574 pp.

Noyes, J.D. 1974. Changing geographic distribution of heartworm disease in the state of Illinois. Pp. 3-5 *in* H.C. Morgan, G. Otto, R.F. Jackson and W.F. Jackson (eds.), Proceedings of the 1974 Heartworm Symposium, Auburn, Alabama. Veterinary Medicine Publishing, Bonner Springs, Kansas.

Nutman, T.B. 1988. *Loa loa* infections in temporary residents of endemic regions: recognition of a hypersensitive syndrome with characteristic clinical manifestations. J. Infect. Dis. **154:**10-18.

Obiamiwe, B.A. 1977. Relationship between microfilarial density, the number of microfilariae ingested by mosquitoes and the proportion of mosquitoes with larvae. Ann. Trop. Med. Parasitol. **71:**491-500.

O'Conner, F.W. 1932. The aetiology of the disease syndrome in *Wuchereria bancrofti* infections. Trans. Royal Soc. Trop. Med. Hyg. **26:**13-33.

Omar, M.S. and Garms, R. 1975. The fate and migration of microfilariae of a Guatemalan strain of *Onchocerca volvulus* in *Simulium ochraceum* and *S. metalllicum* and the role of the buccopharyngeal armature in the destruction of microfilariae. Tropenmed. Parasitol. **26:**183-190.

Orihel, T.C. 1965. *Dirofilaria lutrae* sp. n. (Nematoda: Filarioidea) from otters in the southeast United States. J. Parasitol. **51:**409-413.

Orihel, T.C., Eberhard, M.L. and Lowrie, R.C., Jr. 1993. *Mansonella ozzardi*: the course of patency in experimentally-infected patas monkeys. Trop. Med. Parasitol. **44:**49-54.

Orihel, T.C. and Lowrie, R.C.J. 1975. *Loa loa*: development to the infective stage in an American deerfly, *Chrysops atlanticus*. Am. J. Trop. Med. Hyg. **24:**610-615.

Ottesen, E.A. 1990. The filariases and tropical eosinophilia. Pp. 407-428 *in* K.S. Warren and A. Mahmoud (eds.), Tropical and Geographical Medicine. McGraw Hill, New York.

Ottesen, E.A. 1992. Infection and disease in lymphatic filariasis: an immunological perspective. Parasitol. **104:**S71-S79.

Ottesen, E.A. 1993. Filarial infections. Infect. Dis. Clinics North Am. **7:**619-633.

Otto, G.F. 1975. Changing geographic distribution of heartworm disease in the United States. Pp. 1-2 *in* H.C. Morgan, G. Otto, R.F. Jackson and W.F. Jackson (eds.), Proceedings 1974 Heartworm Symposium, Auburn, Alabama. Veterinary Medicine Publishing, Bonner Springs, Kansas.

Otto, G.F. and Jachowski, L.A., Jr. 1981. Mosquitoes and canine heartworm disease. Pp. 17-32 *in* H.C. Morgan, G.F. Otto, R.F. Jackson, L.A. Jachowski, Jr. and C.H. Courtney (eds.), Proceedings of the 1980 Heartworm Symposium, Dallas, Texas. Veterinary Medicine Publishing, Edwardsville, KS.

Palmeri, J.R., Conner, D.H., Purnomo and Marwoto, H.A. 1983. Bancroftian filariasis. *Wuchereria bancrofti* infection in the silvered leaf monkey (*Presbytis cristatus*). Am. J. Pathol. **112:**383-386.

Parker, B.M. 1986. Presumed *Dirofilaria immitis* infections from field-collected mosquitoes in North Carolina. J. Am. Mosq. Contr. Assoc. **2:**231-233.

Parker, B.M. 1993. Variation in mosquito (Diptera: Culicidae) relative abundance and *Dirofilaria immitis* (Nematod: Filarioidea) vector potential in coastal North Carolina. J. Med. Entomol. **30:**436-442.

Parsons, A.C. 1909. *Filaria volvulus* Leuckhart, its distribution, structure and pathological effects. Parasitol. **1:**359-386.

Phillips, J.H. 1939. Studies on the transmission of *Dirofilaria immitis* in Massachusetts. Am. J. Hyg., D: Helminthol. **29:**121-128.

Phiri, J. and Ham, P.J. 1990. Enhanced migration of *Brugia pahangi* microfilariae through the mosquito midgut following N-acetyl-D-glucosamine ingestion. Trans. Royal Soc. Trop. Med. Hyg. **84:**462.

Pichon, C., Perrault, G. and Laigret, J. 1974. Rendement parasitaire chez les vecteurs de filarioses. Bull. Wld. Hlth. Organiz. **41:**517-524.

Pinger, R.R. 1982. Presumed *Dirofilaria immitis* infections in mosquitoes (Diptera: Culicidae) in Indiana, USA. J. Med. Entomol. **19:**553-555.

Prince, A.M., Brotman, B., Johnson, E.H., Jr., Smith, A., Pascual, D. and Lustigman, S. 1992. *Onchocerca volvulus*: immunization of chimpanzees with X-irradiated third-stage (L3) larvae. Exp. Parasitol. **74:**239-250.

Prout, W.T. 1901. A Filaria found in Sierra Leone. Brit. Med. J. **1:**209.

Rakita, R.M., White, A.C.J. and Kielhofner, M.A. 1993. *Loa loa* infection as a cause of migratory angioedema: report of three cases from the Texas Medical Center. Clin. Infect. Dis. **17:**691-694.

Ramachandran, C.P., Edeson, J.F.B. and Kershaw, W.E. 1960. *Aedes aegypti* as an experimental vector of *Brugia malayi*. Ann. Trop. Med. Parasitol. **54:**371-375.

Ramirez Perez, J. 1988. Human onchocerciasis foci and vectors in the American tropics and subtropics. PAHO Bull. **20:**381-402.

Rawlings, C. 1986. Heartworm disease in dogs and cats. W.B. Saunders, Philadelphia. 329 pp.

Ridley, H. 1945. Ocular onchocerciasis,including an investigation in the Gold Coast. Brit. J. Opthalmol. **10(Suppl.):**58.

Roberts, E., Milton, A. and Trpis, M. 1985. Laboratory transmission of dog heartworm (*Dirofilaria immitis*) by three species of mosquitoes (Diptera: Culicidae) from Eastern Maryland, USA. J. Med. Entomol. **22:**415-420.

Robles, R. 1917. Enfermidad nueva en Guatemala. Juven. Med. **17:**97-115.

Rothfels, K.H. 1979. Cytotaxonomy of black flies (Simuliidae). Annu. Rev. Entomol. **24:**507-539.

Roubaud, E. 1937. Nouvelles Recherches sur l'infection du moustique de la fièvre jaune par *Dirofilaria immitis* Leidy. Les races biologiques d'*Aedes aegypti* et l'infection filarienne. Bull. Soc. Pathol. Exot. **30:**511-519.

Roubaud, E., Colas-Belcour, J., Toumanoff, C. and Treillard, M. 1936. Recherche sur la transmission de *Dirofilaria immitis* Leidy. Bull. Soc. Pathol. Exot. **29:**1111-1120.

Samarawickrema, W.A., Spears, G.F.S., Sone, F., Ichimori, K. and Cummings, R.F. 1985. Filariasis transmission in Samoa 1. Relation between density of icrofilariae and larval density in laboratory-bred and wild-caught *Aedes (Stegomyia) polynesiensis* (Marks) and *Aedes (Finlaya) samoanus* (Gruenberg). Ann. Trop. Med. Parasitol. **79:**89-100.

Sasa, M. 1976. Human filariasis. A global survey of epidemiology and control. University Park Press, Baltimore, Maryland. 819 pp.

Sasa, M. and Tanaka, H. 1972. Studies on the methods for statistical analysis of the microfilarial periodicity survey data. S.E. Asian J. Trop. Med. Publ. Hlth. **4:**518-536.

Sauerman, D.A., Jr. and Nayar, J.K. 1983. A survey for natural potential vectors of *Dirofilaria immitis* in Vero Beach, Florida. Mosq. News **43:**222-225.

Schacher, J. 1962. Morphology of the microfilaria of *Brugia pahangi* and of the larval stages in the mosquito. J. Parasitol. **48:**679-692.

Schulz-Key, H., Albiez, E.J. and Buttner, D.W. 1977. Isolation of living adult *Onchocerca volvulus* from nodules. Tropenmed. Parasitol. **28:**428-430.

Scoles, G.A. 1997. Aspects of the vector biology of canine heartworm in the United States and the potential role of *Aedes albopictus*. Doctoral Dissertation thesis. University of Notre Dame, Notre Dame, Indiana. 178 pp.

Scoles, G.A., Dickson, S.L. and Blackmore, M.S. 1993. Assessment of *Aedes sierrensis* as a vector of canine heartworm in Utah using a new technique for determining the infectivity rate. J. Am. Mosq. Contr. Assoc. **9:**88-90.

Scoles, G.A. and Kambhampati, S. 1995. Polymerase chain reaction-based method for the detection of canine heartworm (Filarioidea: Onchocercidae) in mosquitoes (Diptera: Culicidae) and vertebrate hosts. J. Med. Entomol. **32:**864-869.

Severson, D.W., Mori, A., Zhang, Y. and Christensen, B.M. 1993. Linkage map for *Aedes aegypti* using restriction fragment length polymorphisms. J. Hered. **84:**241-247.

Severson, D.W., Mori, A., Zhang, Y. and Christensen, B.M. 1994. Chromosomal mapping of two loci affecting filarial worm susceptibility in *Aedes aegypti*. Insect Molec. Biol. **3:**67-72.

Shelly, A.J. 1988. Vector aspects of the epidemiology of onchocerciasis in Latin America. Annu. Rev. Entomol. **30:**337-366.

Siridewa, K., Karunanayake, E.H. and Chandrasekharan, N.V. 1996. Polymerase chain reaction-based technique for the detection of *Wuchereria bancrofti* in human blood samples, hydrocele fluid, and mosquito vectors. Am. J. Trop. Med. Hyg. **54:**72-76.

Slocombe, J.O.D. and McMillian, I. 1978. The geographic distribution of heartworm in Canada. Pp. 5-7 *in* G.F. Otto (ed.), Proceedings of the 1977 Heartworm Symposium, Atlanta, Georgia. Veterinary Medicine Publishing, Bonner Springs, Kansas.

Slocombe, J.O.D. and McMillian, I. 1981. Distribution of heartworm disease in Canada. Pp. 7-8 *in* G.F. Otto (ed.), Proceedings of the 1980 Heartworm Symposium, Dallas, Texas. Veterinary Medicine Publishing, Edwardsville, Kansas.

Southgate, B.A. 1992a. Intensity and efficiency of transmission and the development of microfilaremia and disease: their relationship in lymphatic filariasis. J. Trop. Med. Hyg. **95:**1-12.

Southgate, B.A. 1992b. The significance of low density microfilaraemia in the transmission of lymphatic filarial parasites. J. Trop. Med. Hyg. **95:**79-86.

Southgate, B.A. and Bryan, J.H. 1992. Factors affecting transmission of *Wuchereria bancroft* by anopheline mosquitoes. 4. Facilitation, limitation, proportionality and their epidemiological significance. Trans. Royal Soc. Trop. Med. Hyg. **86:**523-530.

Strong, R.P., Becquaert, J.C., Sandground, J.H. and Muñoz, O.M. 1934. Onchocerciasis with special reference to the Central American form of the disease. Harvard University Press, Cambridge, Massachusetts. 234 pp.

Suswillo, R.R., Nelson, G.S., Muller, R., McGreevy, P.B., Duke, B.O.L. and Denham, D.A. 1977. Attempts to infect jirds (*Meriones unquiculatus*) with *Wuchereria bancrofti, Onchocerca volvulus, Loa loa loa* and *Mansonella ozzardi*. J. Helminthol. **51:**132-134.

Suswillo, R.R., Owen, D.G. and Denham, D.A. 1980. Infections of *Brugia pahangi* in conventional and nude (athymic) mice. Acta Trop. **37:**327-335.

Sutherland, D.R., Christensen, B.M. and Lasee, B.A. 1986. Midgut barrier as a possible factor in filarial worm vector competency in *Aedes trivittatus*. J. Invertr. Pathol. **47:**1-7.

Takaoka, H. and Susuki, T. 1987. Epidemiology and control of Guatemalan onchocerciasis. Pp. 374-386 *in* K.C. Kim and R.W. Merritt (eds.), Black flies: Ecology, population management, and annotated world list. Pennsylvania State University Press, University Park.

Taylor, A.E.R. 1960. The development of *Dirofilaria immitis* in the mosquito *Aedes aegypti*. J. Helminthol. **34:**27-38.

Terwedow, H.A. and Craig, G.B., Jr. 1977. *Waltonella flexicauda*: development controlled by a genetic factor in *Aedes aegypti*. Exp. Parasitol. **41**:272-282.

Todaro, W.S., Morris, C.D. and Heacock, N.A. 1977. *Dirofilaria immitis* and its potential vectors in central New York State. Am. J. Vet. Res. **38**:1197-1200.

Toe, L., Merriweather, A. and Unnasch, T.R. 1994. DNA probe-based classification of *Simulium damnosum* s.l.-borne and human-derived filarial parasites in the onchocerciasis control program. Am. J. Trop. Med. Hyg. **51**:676-683.

Tolbert, R.H. and Johnson, W.E., Jr. 1982. Potential vectors of *D. immitis* in Macon County, Alabama. Am. J. Vet. Res. **43**:2054-2056.

Townson, H. and Chaithong, U. 1991. Mosquito host influences on development of filariae. Ann. Trop. Med. Parasitol. **85**:149-163.

Trees, A.J., Wood, V.L., Bronsvoort, M., Renz, A. and Tanya, V.N. 1998. Animal models — *Onchocerca ochengi* and the development of chemotherapeutic agents for onchocerciasis. Ann. Trop. Med. Parasitol. **92 (Suppl. 1)**:S175-S179.

Trpis, M., Childs, J.E., Fryauff, D.J., Greene, B.M., Williams, P.N., Munoz, B.E., Pacque, M.C. and Taylor, H.R. 1990. Effect of mass treatment of a human population on transmission of *Onchocerca volvulus* by *Simulium yahense* in Liberia, West Africa. Am. J. Trop. Med. Hyg. **42**:148-156.

Trpis, M., Duhrkopf, R.E. and Parker, K.L. 1981. Non-Mendelian inheritance of mosquito susceptibility to infection with *Brugia malayi* and *Brugia pahangi*. Science **211**:1435-1437.

Turell, M.J., Rossignol, P.A., Spielman, A., Rossi, C.A. and Bailey, C.L. 1984. Enhanced arboviral transmission by mosquitoes that concurrently ingested microfilariae. Science **225**:1039-1041.

Vajime, C.G. and Dunbar, R.W. 1975. Chromosomal identification of eight species of the subgenus *Edwardsellum* near and including *Simulium damnosum* Theobald (Diptera: Simuliidae). Tropenmed. Parasitol. **29**:473-482.

Valeza, F.S. and Grove, D.I. 1979. Bancroftian filariasis in a Philippine village: entomological findings. S.E. Asian J. Trop. Med. Publ. Hlth. **10**:51-61.

Vickery, A.C., Albertine, K.H., Nayar, J.K. and Kwa, B.H. 1991. Histopathology of *Brugia malayi*-infected nude mice after immune-reconstitution. Acta Trop. **49**:45-55.

Villavaso, E.J. and Steelman, C.D. 1970. Laboratory and field studies of the southern house mosquito, *Culex pipiens quinquefasciatus* Say, infected with the dog heartworm, *Dirofilaria immitis* (Leidy), in Louisiana. J. Med. Entomol. **7**:471-476.

Vincent, A.L., Vickery, A.C., Winters, A. and Sodeman, W.A., Jr. 1982. Life cycle of *Brugia pahangi* (Nematoda) in nude mice, CH3/HeN (nu/nu). J. Parasitol. **68**:553-560.

Wada, Y. 1982. Theoretical approach to the epidemiology of onchocerciasis in Guatemala. Japanese J. Med. Sci. Biol. **35**:183-196.

Wada, Y., Kimura, E., Takagi, M. and Tsuda, Y. 1995. Facilitation in *Anopheles* and spontaneous disappearance of filariasis: has the concept been verified with sufficient evidence? Trop. Med. Parasitol. **46**:27-30.

Walker, E.D., Torres, E.P. and Villanueva, R.T. 1998. Components of the vectorial capacity of *Aedes poicilius* for *Wuchereria bancroft* in Sorsogon province, Philippines. Ann. Trop. Med. Parasitol. **92**:603-614.

Walters, L.L. 1996. Risk factors for heartworm infection in northern California. Pp. 5-26 *in* M.D. Soll, D.H. Knight and W.C. Campbell (eds.), Proceedings of the 1995 Heartworm Symposium, Auburn, Alabama. American Heartworm Society, Batavia, Illinois.

Walters, L.L. and Lavoipierre, M.M.J. 1982. *Aedes vexans* and *Aedes sierrensis* (Diptera: Culicidae): potential vectors of *Dirofilaria immitis* in Tehama County, northern California, USA. J. Med. Entomol. **19**:15-23.

Wartman, W.B. 1947. Filariasis in American armed servicemen in World War II. Medicine (Baltimore) **12**:181-200.

Wattam, A. and Christensen, B.M. 1992. Further evidence that the genes controlling susceptibility of *Aedes aegypti* to filarial parasites function independently. J. Parasitol. **78**:1092-1095.

Webber, R.H. and Southgate, B.A. 1981. The maximum density of anopheline mosquitoes that can be permitted in the absence of continued transmission of filariasis. Trans. Royal Soc. Trop. Med. Hyg. **4**:499-506.

Weil, G., J., Jain, D.C. and Santhanam, S. 1987. A monoclonal antibody-based enzyme immunoassay for detecting parasite antigenemia in bancroftian filariasis. J. Infect. Dis. **156**:350-355.

World Health Organization. 1976. Epidemiology of onchocerciasis. WHO Tech. Rep. Ser. No. 579. World Health Organization, Geneva. 94 pp.

World Health Organization. 1987. Control of lymphatic filariasis, a manual for health personnel. World Health Organization, Geneva. 89 pp.

Wucherer, O.E. 1868. Noticia preliminar sobre vermes de una especie ainda nao descripta, encontrados na urina de doentes de hematuria intertropical no Brazil. Gaz. Med. Bahia **3**:97-99.

Yen, P.K.F. 1983. Taxonomy of Malaysian filarial parasites. Pp. 17-35 *in* J.W. Mak (ed.), Filariasis, Bulletin No. 19. Institute for Medical Research, Kuala Lumpur, Malaya.

Yen, P.K.F., Zaman, V. and Mak, J.W. 1982. Identification of some common infective filarial larvae in Malaysia. J. Helminthol. **56**:69-80.

Zhong, M., McCarthy, J., Bierwert, L., Lizotte-Waniewski, M., Chanteau, M., Nutman, T.B., Otteson, E.A. and Williams, S.A. 1996. A polymerase chain reaction assay for detection of the parasite *Wuchereria bancrofti* in human blood samples. Am. J. Trop. Med. Hyg. **54**:357-363.

Zielke, E. 1973. Untersuchungen zur Vererbund der Empfanglichkeit gegenuber der Hundefilarie *Dirofilaria immitis* bei *Culex pipiens fatigans* und *Aedes aegypti.* Tropenmed. Parasitol. **24**:36-44.

Zielke, E. 1977. On the escape of infective filarial larvae from the mosquito. Tropenmed. Parasitol. **28**:461-466.

Zielke, E. 1980. On the longevity and behaviour of microfilariae of *Wuchereria bancrofti, Brugia pahangi* and *Dirofilaria immitis* transfused to laboratory rodents. Trans. Royal Soc. Trop. Med. Hyg. **74**:456-458.

Zielke, E. and Kuhlow, F. 1977. On the inheritance of susceptibility for infection with *Wuchereria bancrofti* in *Culex pipiens fatigans.* Tropenmed. Parasitol. **28:** 68-70.

Zimmerman, P.A., Katholi, C.R., Wooten, M.C., Lang-Unnasch, N. and Unnasch, T.R. 1994. Recent evolutionary history of American *Onchocerca volvulus,* based on analysis of a tandemly repeated DNA sequence family. Molec. Biol. **11**:384-392.

Zimmerman, P.A., Toe, L. and Unnasch, T.R. 1993. Design of probes based upon analysis of a repeated sequence family. Molec. Biochem. Parasitol. **58**:259-267.

Chapter 10

Bacterial and Rickettsial Diseases

JOSEPH PIESMAN AND KENNETH L. GAGE
Centers for Disease Control and Prevention, Fort Collins, Colorado

INTRODUCTION

Diseases caused by bacterial and rickettsial pathogens transmitted by arthropods have caused incredible suffering throughout human history. The twin scourges, plague and typhus, may have killed more people than all wartime battlefield injuries combined (Zinsser 1934). Although improvements in sanitation and treatment of diseases have decreased these vetorborne agents around the world, recent events have shown how quickly the threat of these diseases can cause chaos in human society. Unlike diseases that prevail where poverty is high and living standards are low, bacterial and rickettsial agents transmitted by arthropods recently have emerged among the world's wealthiest residents. Lyme disease clearly is tied in the northeastern USA to the reversion of agricultural land to wealthy suburban communities, with accompanying regrowth of secondary forests and deer populations free from hunting. These changes have led to a reemergence of the blacklegged tick, *Ixodes scapularis*, and the 3 human diseases transmitted by these ticks: babesiosis, Lyme disease and human granulocytic ehrlichiosis. Dr. Harry Hoogstraal, the late eminent tick biologist, predicted that changing landscape conditions caused by human society would dramatically impact tickborne agents in the future. The emergence of 3 new agents associated with a single tick species late in the 20th Century emphasizes Dr. Hoogstraal's prediction. The plasticity of the vectorborne bacterial and rickettsial agents allows them to change and survive in parallel with human landscape changes.

Five major groups of arthropods transmit bacterial and rickettsial pathogens: ticks, mites, fleas, lice and sand flies. These groups were reviewed in Chapter 2. One cannot understand the interrelationship between a disease pathogen and its vector without knowing the biology of the arthropod group concerned. We are fortunate to have a comprehensive 2-volume set (Sonenshine 1991, 1993) that provides the basic knowledge of tick biology needed to understand the complex relationships between ticks, the pathogens they transmit and the hosts affected by these pathogens.

LYME DISEASE

European scientists first recognized a curious rash, termed *erythema chronicum migrans*,

B.F. Eldridge and J.D. Edman (eds.), Medical Entomology, Revised Edition, 377–413.

or *erythema migrans*, as early as 1909. Isolated observations in Europe associated this strange bull's eye-shaped spreading rash with (1) the bite of a tick, *Ixodes ricinus*, (2) a spirochete found in the skin near the site of the tick bite, and (3) with the fact that penicillin helped to alleviate the rash. The first report of *erythema migrans* in North America was in a Wisconsin resident in 1969. These isolated reports did not reach a wide scientific audience until, in 1975, a Connecticut housewife observed an epidemic of arthritis in her neighborhood and brought it to the attention of Dr. Allen Steere, a Yale University rheumatologist and epidemiologist. His intense and dramatic investigation quickly demonstrated that the epidemic around Lyme, Connecticut (Steere et al. 1986) was associated with *erythema migrans* and the bite of the blacklegged tick, *Ix. scapularis*. The agent causing this disease was unknown at the time, but the combined clinical picture of *erythema migrans*, arthritis, as well as neurological and cardiac sequelae, was named Lyme disease. Recalling European observations on *erythema migrans*, Dr. Steere discovered that Lyme disease responded to therapy with penicillin, thus suggesting a bacterial etiology. The search for the etiologic agent of Lyme disease continued until 1982, when Dr. Willy Burgdorfer and colleagues announced that they had isolated spirochetes from *Ix. scapularis* ticks from Shelter Island, New York (Burgdorfer et al. 1982). These spirochetes reacted with antibodies from the serum of Lyme disease patients and did not react with control sera. These previously unknown spirochetes were characterized further and named *Borrelia burgdorferi*, in honor of their discoverer. Although others may have seen this spirochete in ticks previously, Dr. Burgdorfer grasped the epidemiological significance of these spirochetes, in part because he had worked with relapsing fever spirochetes decades before while doing his thesis in Switzerland. This sequence of events proves that "chance favors the prepared mind."

After the discovery of the etiologic agent, understanding of the epidemic fell rapidly into place. Lyme disease, or Lyme borreliosis as European investigators prefer, is found in North America, Europe and Asia. The vector in eastern North America is *Ix. scapularis*, in the western USA *Ixodes pacificus*, in Western Europe *Ix. ricinus*, and in Asia *Ixodes persulcatus*. These ticks belong to the loosely defined taxonomic group called the *Ix. ricinus* complex. Countries with a particularly high incidence of Lyme disease include the USA, Germany, Switzerland, France, Sweden, Russia and Japan; countries with high socioeconomic status. To date, isolates of *B. burgdorferi* have been reported mainly from temperate regions of the Northern Hemisphere. One possible exception is an isolated cycle involving spirochetes closely related to *B. burgdorferi*, sea birds and *Ixodes uriae* (Olsen et al. 1995). This tick is distributed worldwide.

The classic areas where *Ix. ricinus* complex ticks can be found are dense deciduous forests with abundant leaf litter. These ticks require the leaf litter because they are extremely sensitive to desiccation. Heavily forested areas also have abundant deer populations. In the eastern USA, the white-tailed deer, *Odocoileus virginianus*, serves as the principal host for the adult stage of *Ix. scapularis*. Heavily forested areas also have abundant rodent populations, e.g., the white-footed mouse (*Peromyscus leucopus*) in the USA, and various *Apodemus* and *Clethrionomys* species in Europe. Early in our study of Lyme disease, Spielman et al. (1985) described a paradigm in highly enzootic regions of the northeastern USA which had an efficient system for maintaining *B. burgdorferi*. In this system, *P. leucopus* was the only important host for immature *Ix. scapularis*. Both larval and nymphal *Ix. scapularis* fed almost exclusively on this host and >80% of these mice were infected with *B. burgdorferi*. Mice acquired infection when exposed to infected nymphs in the spring and early summer months (April-July). The infected nymphs had picked up spirochetes from mice when

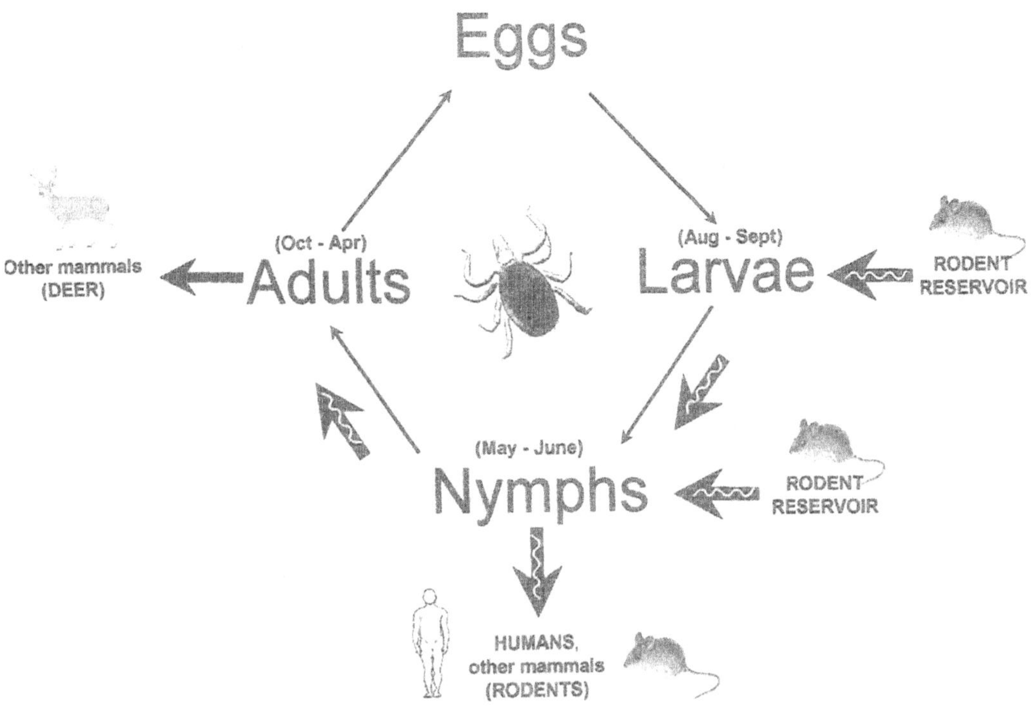

Figure 10.1. Seasonality of the Lyme disease parasite, *Borrelia burgdorferi*.

feeding as larvae the previous year (August-October). This inverted seasonality (Fig. 10.1), where nymphs transmitted the infection to näive mice early in the year, and these mice were maximally infectious to larvae several months later, infected a higher percentage of the tick population. In general, 25% of nymphal and 50% of adult *Ix. scapularis* are infected with *B. burgdorferi* in the northeastern USA. Where no such inverted seasonal feeding cycle exists, as in Europe or the southern USA, infection rates in *Ixodes* ticks generally are much lower. Adult *Ix. scapularis* feed on deer during the colder months of the year (October-April). In the northeastern USA, transmission by adults to deer is meaningless, since these hosts are considered to be refractory to infection by *B. burgdorferi*. Furthermore, although transovarial transmission of *B. burgdorferi* has been observed, the number of flat tick larvae transovarially infected is thought to be so low (0.7%, Piesman et al. 1986) that this route may not be important to the survival of the spirochete in North America. The principal driving force for survival of the spirochete is acquisition of spirochetes by larval *Ixodes* feeding on rodents, transstadial survival of the spirochete to the nymphal stage, and the subsequent transmission of the spirochete the next year by nymphs to näive rodents.

People are infected incidentally when a questing nymph attaches to them instead of a rodent.

Over the last 10 years, investigators have discovered numerous exceptions to this paradigm. Firstly, efficient enzootic cycles can take place in the absence of *Peromyscus*; research on Monhegan Island, off the coast of Maine, demonstrated that *Rattus norvegicus* served as an efficient reservoir host for *B. burgdorferi* in the total absence of *Peromyscus* (Smith et al. 1993). Moreover, in California (Brown and Lane 1992) and Colorado (Maupin et al. 1994), cycles involving *Ixodes* species and woodrats (*Neotoma* spp.) also have been highly efficient. Enzootic cycles involving rabbits and *Ixodes dentatus* may be important in maintaining spirochetes in nature. Finally, birds also may play an important supplemental role as reservoirs of *B. burgdorferi* and as phoretic hosts, serving to disperse spirochete-infected ticks (Stafford et al. 1995). Although these exceptions prove that the *Ix. scapularis-Peromyscus* cycle is not the only enzootic maintenance cycle for *B. burgdorferi* in nature, it is the most efficient one described to date, infecting approximately 25% of questing nymphs with spirochetes and consequently putting the public at extreme risk of acquiring Lyme disease.

The taxonomic classification of the principal vector of Lyme disease in the eastern USA has sparked controversy. When *Ixodes* ticks first were implicated in the transmission of Lyme disease, they were identified as *Ix. scapularis*. Subsequently, based on morphological and biological criteria, Spielman et al. (1979) described the ticks transmitting both Lyme disease and human babesiosis in the northeastern USA as a new species, *Ixodes dammini*. However, Oliver et al. (1993a) claimed that based on the ability of *Ix. scapularis* and *Ix. dammini* to mate and produce fertile offspring in the laboratory, the 2 forms were conspecific. Hutcheson et al. (1995) looked at the morphological differences between northern and southern populations of *Ix. scapularis* and concluded that although they varied from north to south, most of this variation was removed in a size-independent analysis and did not support the status of *Ix. dammini* as a distinct species. The debate now has moved on to molecular systematic studies. Sequence analysis of the 16S ribosomal RNA gene, encoded on the mitochondria, has found 2 distinct lineages of *Ix. scapularis;* one type is restricted to the southern USA and the other is distributed throughout the eastern USA (Rich et al. 1995). Although an argument can be made that this distinction supports the acceptance of *Ix. dammini* as a valid species, analysis of nuclear genes in addition to mitochondrial genes does not support recognition of *Ix. dammini* as a species because gene flow appears to be occurring throughout the range of *Ix. scapularis* (Norris et al. 1996). Although this debate has sparked interest in the systematics of ticks, it is probably of little relevance in evaluating the risk of Lyme disease spirochete transmission.

Risk of transmission of *B. burgdorferi* varies dramatically from region to region. In the northeastern USA, the majority of Lyme disease cases occur during the summer months (June-July), immediately following the peak of nymphal *Ix. scapularis* questing (May-June). Although the infection rate in adult *Ix. scapularis* is roughly twice that of nymphs, few cases of Lyme disease have their onset during the colder months when adults are feeding. This difference probably is due to the size differential. Adult *Ix. scapularis* are much larger than nymphs (Fig. 10.2) and consequently are detected quickly and removed by infested persons. Research with both the nymphal and adult stages of *Ix. scapularis* (Piesman et al. 1987b, 1991) has demonstrated that ticks rarely if ever transmit *B. burgdorferi* during the first 24 hr of attachment, inefficiently transmit disease pathogens during 24-48 hr of attachment, and transmit efficiently from 48 hr onwards. Therefore, transmission by adult ticks often is aborted by prompt tick inspection and removal. Thus, transmission in the northeastern USA hyperendemic regions is

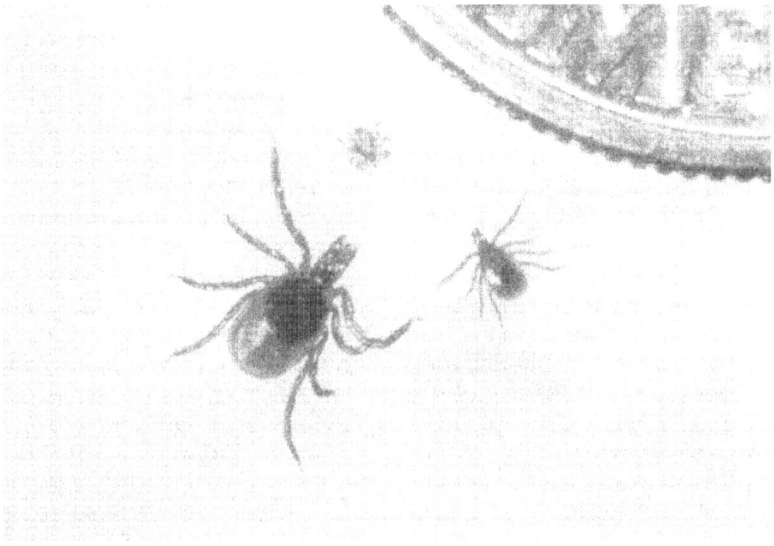

Figure 10.2. Larva, nymph and adult *Ixodes scapularis*, the blacklegged tick (courtesy of Russell Johnson).

largely linked to the number of questing infected nymphal *Ix. scapularis* (Piesman et al. 1987a). Abundance can be quantified as the number of infected nymphs collected per person-hour of flagging, or the number of infected nymphs collected per area flagged. A true quantification of risk is difficult, since human behavior is an important part of contact with ticks. Two individuals walking the same path through a forest may vary in their contact with leaf litter and hence vary in the number of ticks to which they are exposed; this difference is especially true when it comes to the nymphal stage, which tends to quest low down in the vegetation.

Estimates of risk in the northeastern USA often run high, with as many as 16.2 infected nymphs collected per person-hour on Nantucket Island, Massachusetts (Piesman et al. 1987a). Infection rates in California, where *Ix. pacificus* is the primary vector, generally are low (1-2% in both nymphs and adults); however, foci of higher infection rates in nymphal *Ix. pacificus* (13.6%) in Potter Valley, California have been reported (Clover and Lane 1995). Nymphal *Ix. pacificus* usually are encountered in recreational settings, whereas risk in the northeastern USA from *Ix. scapularis* is both recreational and residential (Maupin et al. 1991). In the southern USA, it is difficult to flag nymphal *Ix. scapularis*. A survey of St. Catherines and Sapelo Islands, Georgia found only 10 nymphal *Ix. scapularis* in 12.5 person-hour of flagging and none of these were infected with *B. burgdorferi*, although numerous adults were found infesting deer and *B. burgdorferi* was isolated from adult *Ix. scapularis* (Oliver et al. 1993b). Clearly, the transmission risk of Lyme disease in the southern USA is less than that in the northeast. Infection rates in nymphal *Ix. scapularis* in the Midwestern USA appear to be intermediate between those of the northeast and the south. More research is needed to define transmission risk. This analysis must be done at the local level. *Ixodes ricinus* in Europe also requires more research. Infection rates in both nymphal and adult *Ix. ricinus* vary among study sites; in general, nymphal infection rates (10-15%) are less in *Ix. ricinus* than they are in *Ix. scapularis* in the northeastern USA (Mejlon and Jaenson 1993, Kurtenbach et al. 1995). Interestingly, in areas of Russia where *Ix. persulcatus* is the primary vector, adults may be more important than nymphs in the transmission of Lyme disease to humans (Korenberg et al. 1996).

The availability and range of infected hosts influences the infection rate of *B. burgdorferi* in *Ixodes* ticks. In areas where rodents predominate and reptilian hosts (principally lizards) are rare or absent, infection rates tend to be high. By contrast, in areas where lizards predominate and rodents are less common, infection rates in ticks are extremely low. Thus, lizards may serve as a form of zooprophylaxis, diverting immature *Ixodes* ticks from highly competent rodent hosts (Manweiler et al. 1992). Lizards were thought to be totally refractory to *B. burgdorferi*, but recent evidence has shown that under certain conditions they may serve as reservoirs, albeit not as efficiently as rodents (Levin et al. 1996). The prevalence of lizards in the south and rodents in the north may partially explain the north-south gradient of spirochetal infection rates in *Ix. scapularis* along the Atlantic coast of the USA.

The spirochete populations found in ixodid ticks are quite diverse. The *B. burgdorferi* s.l. group of spirochetes includes at least 3 **genospecies** (species described on the basis of nucleotide sequences of genes): *B. burgdorferi* s.str., found in the USA and Europe, *Borrelia afzelii* (associated with the skin condition *acrodermatitis chronica atrophicans*), and *Borrelia garinii* in Eurasia (Baranton et al. 1992). These 3 species are found in the *Ix. ricinus* complex of ticks (i.e., *Ix. ricinus*, *Ix. persulcatus*, *Ix. scapularis* and *Ix. pacificus*) and show sufficient diversity based on restriction fragment length polymorphism (RFLP) analysis, as well as DNA hybridization and 16S rRNA gene sequence, to allow classification as separate genetic species. Recently, a new species of spirochete has been described from *Ixodes ovatus* in Japan: *Borrelia japonica* (Kawabata et al. 1993). Interestingly, all human isolates from Japan have been characterized as either *B. garinii* or *B. afzelii*, which are found in the human-biting *Ix. persulcatus*; to date *B. japonica*, found in the non-human-biting *Ix. ovatus*, has not been found to infect people. A similar situation may be occurring with the *Ix. dentatus*-rabbit cycle in North America. Apparently, some of these rabbit associated isolates (Marconi et al. 1995), as well as isolates from California (Kolbert et al. 1995), may represent an undescribed but distinct species of *Borrelia*. The number of genospecies recognized in the *B. burgdorferi* s.l. group presently is in a state of flux. Although diverse, the spirochetes isolated from ticks in the genus *Ixodes* are closely related. However, *Amblyomma americanum* ticks apparently contain a spirochete that differs from the other *Borrelia* from hard ticks. Based on 16S rRNA and flagellin gene analysis, the spirochetes in *A. americanum* tentatively have been described as a new species, *Borrelia lonestari* (Barbour et al. 1996). The answer to whether this newly described spirochete causes human disease awaits isolation and further characterization of the spirochete. In addition, isolation and characterization of spirochetes isolated from cattle and cattle-feeding ticks in Africa and North America, referred to as *Borrelia theileri*, is a worthwhile goal. Finally, there are several reports of spirochetes in *Haemaphysalis* ticks in Asia. When isolated, these too may differ from *B. burgdorferi*. We may find that *Borrelia* have evolved to be transmitted by individual species or closely related species of ticks. This system may provide an interesting way to study coevolution.

The prevention and control of Lyme disease has stimulated considerable research. The simplest and most effective way to kill Lyme disease vector ticks is with area-wide application of acaricides. Commonly used acaricides that have proven to be highly effective against *Ix. scapularis* include carbaryl, chlorpyrifos, diazinon, cyfluthrin, esfenvalerate and permethrin. When applied just before the peak of nymphal abundance, these acaricides can reduce dramatically the number of questing nymphs, especially in high use residential areas (Curran et al. 1993). Acaricides have been shown to reduce dramatically the risk of Lyme disease in residential, recreational and occupational settings. In addition, acaricides generally cost less than alternative methods. However, the public resists widespread use of acaricides. As with all pesticides, acaricides have some toxicity for humans and household pets. They also may affect nontarget species in the environment. When properly used, these acaricides pose minimal risk, but misuse and bad publicity have made many residents of Lyme disease endemic areas reluctant or totally opposed to them for control. Accordingly, the need for nonchemical means to control ticks has driven research toward creative alternatives.

One of the first alternative methods for Lyme disease control was using acaricides in a host-targeted, rather than an area-wide format. The habit of the principal reservoir host of *B. burgdorferi* in the northeastern USA, *P. leucopus*, of collecting cotton placed in nesting boxes, led to placing permethrin-treated cotton balls in cardboard tubes where the mice could have free access to them. For effective control of *Ix. scapularis*, tick infested areas are saturated with the tubes. On initial testing in Massachusetts, the number of ticks on mice and, more importantly, the number of questing nymphal *Ix. scapularis* capable of transmitting *B. burgdorferi* was reduced dramatically after application of the bait tubes (Mather et al. 1988). However, when this method was tested in Connecticut and New York, the number of infected questing nymphs did not decrease. Reasons suggested for this disparity include geographic differences in the density of hosts, differences in the role of *Peromyscus* as a primary host for *Ix. scapularis* immatures, the degree of saturation and placement of the tubes and the presence of alternative nesting material. Another way to treat rodents with pesticides involves the placement of bait-tubes containing liquid-dust permethrin. This preparation contains ingredients that help the permethrin bind to host fur and can be used to treat a variety of rodents for fleas and ticks (Gage et al. 1997). In addition to treating the rodent hosts of immature *Ix. scapularis*, the prospect of treating white-tailed deer, the principal host for adult ticks, is promising. Cattle can be dipped as they come to feed stations. These techniques have been modified to deliver topical pesticides to deer. The potential is great to dramatically reduce populations of *Ix. scapularis* at deer feeding stations. Systemic insecticides also hold great potential for reduction of tick populations. Ivermectin and related compounds kill ticks feeding on both deer and rodents exposed to treated foods. The best strategy for treating the greatest proportion of deer and rodent populations with systemic acaricides is an ongoing area of research. Host-targeted pesticide application is exciting to investigate and could be applied readily to the control of Lyme disease as well as other tickborne diseases.

Alternatives to conventional pesticides also may control Lyme disease vector ticks. Both soaps and desiccants have been effective in the laboratory and in the field against *Ix. scapularis* (Patrican and Allan 1995). The challenge will be to develop effective compounds with shorter activity periods in nature than those of most conventional chemical pesticides. Some natural products derived from trees and other forest products show promise as acaricides. In addition, pheromones may lure ticks to pesticide-treated collars or tags. Finally, biological control of ticks by parasitic wasps, fungi or nematodes is under evaluation.

Removing deer from target regions is another way to control Lyme disease. This separation can be achieved by active removal, as occurred on Great Island, Massachusetts (Wilson et al. 1988), or by fencing (Daniels and Fish 1995, Gray et al. 1992). Although effective, these methods take time to implement and there is public resistance both to deer reduction and fencing in residential areas. Control of rodents may be less controversial, but methods to achieve reduction of ubiquitous rodents such as *Peromyscus* are lacking.

Vegetation management also may impact tick populations. This can be achieved by burning or mowing, or even simple leaf litter removal. All these methods reduce populations of *Ix. scapularis,* but they are laborious and expensive and provide only temporary relief. Personal protection prevents Lyme disease: avoiding heavily infested wooded areas, wearing protective clothing and judiciously using of repellents. All of these components can be included in an educational campaign, and each has advantages and disadvantages. The future undoubtedly involves a more integrated approach. Integrated pest management has made progress in the 1990s. Lyme disease presents an excellent opportunity for application of these methods to control medically important arthropods.

Many challenges lie ahead for the public health community in the battle against Lyme disease. It needs improvements in treatment and diagnosis. Moreover, the availability of a vaccine against *B. burgdorferi,* based on the Outer Surface Protein A (Steere et al. 1998), is now a reality. If the vaccine proves to be safe and effective, medical entomologists will have to help determine when, where and how it should best be used to prevent Lyme disease, based on ecological risk factors.

RELAPSING FEVER

Sucking lice or ticks transmit relapsing fever spirochetes. Louseborne relapsing fever is transmitted by the body louse, *Pediculus humanus humanus,* and has only a single etiologic agent, *Borrelia recurrentis.* The disease appears in North and South America, Europe, Africa and Asia. During and immediately after World War II, 10 million people probably were infected by louseborne relapsing fever, with a 5% mortality rate (Burgdorfer 1976). Thus, approximately one-half million people may have died during this pandemic. Many of these victims also were co-infected with the etiological agent of louseborne typhus. During the 2nd half of the 20th Century, the incidence of louseborne relapsing fever has decreased dramatically around the world, with a single focus of disease in the highlands of Ethiopia and Sudan accounting for most of the cases reported to the World Health Organization. Dusting clothes with DDT controlled louseborne relapsing fever and typhus. In addition to proper bathing and washing of clothes, pyrethrins plus piperonyl butoxide, permethrin or lindane preparations now are recommended for louse control. Wars or natural disasters that force mass migrations and refugees to live in poor sanitary conditions always will pose a risk of body louse infestations and louseborne relapsing fever.

The manner in which *B. recurrentis* is transmitted by the body louse is fascinating. Lice ingest spirochetes circulating in an infected person's bloodstream. Spirochetes pass into the louse midgut and then some migrate through the gut epithelium into the hemolymph. Interestingly, these spirochetes are trapped in the hemolymph of the louse and do not invade the salivary glands or the ovaries of the louse. *Borrelia recurrentis* primarily reaches a susceptible human host when scratching crushes the louse. The spirochete-infected louse hemolymph thus is exposed to the bite site or scratch marks, offering entry for the spirochete into the host

Table 10.1. Diseases caused by vectorborne bacteria

Disease	Pathogen(s)	Vectors	Distribution
Bartonellosis (Carrión disease)	*Bartonella bacilliformis*	*Lutzomyia verrucarum*	South America (Andean nations)
Cat scratch disease	*Bartonella henselae*	*Ctenocephalides felis*	Worldwide
Lyme disease	*Borrelia burgdorferi*	*Ixodes scapularis, Ixodes pacificus, Ixodes ricinius, Ixodes persulcatus*	North America and Eurasia
	Borrelia garinii	*Ixodes ricinus, Ixodes persulcatus*	Eurasia
	Borrelia afzelii	*Ix. ricinus, Ix. persulcatus*	Eurasia
Plague	*Yersinia pestis*	*Xenopsylla cheopis* and other human-biting fleas	Worldwide
Relapsing fever (louse-borne)	*Borrelia recurrentis*	*Pediculus humanus humanus*	Worldwide
Relapsing fever (tick-borne)	*Borrelia duttonii*	*Ornithodoros moubata*	Africa
	Borrelia hispanica	*Ornithodoros erraticus*	Europe and North Africa
	Borrelia crocidurae	*Ornithodoros erraticus*	Europe and North Africa
	Borrelia persica	*Ornithodoros tholozani*	Africa and Asia
	Borrelia caucasica	*Ornithodoros verrucosus*	Europe and Middle East
	Borrelia hermsii	*Ornithodoros hermsi*	North America
	Borrelia turicatae	*Ornithodoros turicata*	North America
	Borrelia parkeri	*Ornithodoros parkeri*	North America
	Borrelia mazzottii	*Ornithodoros talaje*	Central and South America
	Borrelia venezuelensis	*Ornithdoros rudis*	Central and South America
Trench fever	*Bartonella quintana*	*Pediculus humanus humanus*	Worldwide
Tularemia	*Francisella tularensis*	*Dermacentor andersoni* and other ticks	Worldwide

bloodstream. A high degree of adaptation apparently has taken place at the spirochete-louse-human interface.

Tickborne relapsing fever is associated with soft ticks in the genus *Ornithodoros*. In contrast to louseborne relapsing fever, which is caused by a single agent transmitted around the world, tickborne relapsing fever is characterized by a high degree of focality, with different tick species transmitting their own individual species of *Borrelia* in clearly defined geographic foci. Excellent reviews of the diverse biology of tickborne relapsing fever include those by Felsenfeld (1979) and Burgdorfer (1976). These reviews list all the *Borrelia-Ornithodoros* combinations causing human disease (Table 10.1). Of all these diseases, relapsing fever in Africa caused by *Borrelia duttonii* and transmitted by *Ornithodoros moubata* is by far the best known. David Livingstone encountered relapsing fever during his explorations in Africa as early as 1854. It is thought that *O. moubata* originally was associated with warthog and porcupine dens, but it readily adapted to human dwellings made of mud and grass. Two subspecies are recognized: *Ornithodoros moubata moubata*, which is associated with humans, and *Ornithodoros moubata porcinus*, which is associated with pigs. Interestingly, pigs are not susceptible to *B. duttonii*, suggesting that these spirochetes are natural parasites of the tick, *O. moubata*. *Borrelia duttonii* probably was introduced to humans when their homes first became infested with warthog-related ticks. In Africa, residents of endemic areas quickly learned that if they traveled they would lose their immunity and become sick upon their return. To prevent reinfection, they often carried infected ticks with them when away, allowing these ticks to feed on them, thereby maintaining their immunity (Felsenfeld 1979).

The classic description of the developmental cycle of relapsing fever spirochetes in *Ornithodoros* ticks is the *B. duttonii-O. moubata* system (Burgdorfer 1951). Once the spirochetes are ingested along with the bloodmeal, they penetrate through the midgut to the hemocoel, then invade all tissues including salivary glands, ovaries and coxal glands. The coxal gland of soft ticks is a hemolymph-secreting organ that has its opening near the base of the legs (coxa) of the tick. This organ helps the tick concentrate bloodmeals by exporting hemolymph during rapid bloodfeeding. In contrast, hard ticks have several days to concentrate bloodmeals by the use of water-secreting cells in the salivary glands and thus, hard ticks lack coxal glands. When an infected *O. moubata* feeds on a host, small numbers of spirochetes probably pass into the host through saliva, but many more are placed onto the host skin in the copious coxal gland excretions (Fig. 10.3). These spirochetes in the coxal fluid enter the host, either through the bite site contaminated with coxal fluid, or directly through host skin. However, several North American species, e.g., *Ornithodoros hermsi*, produce much less coxal fluid than does *O. moubata*. Hence, salivary gland transmission of relapsing fever spirochetes probably is more important with North American species. Transovarial transmission works efficiently with relapsing fever spirochetes; filial infection rates of up to 98% are reported with *O. moubata* and 100% with *Ornithodoros turicata* (Burgdorfer 1976). Transovarial transmission of relapsing fever spirochetes probably is an important survival mechanism for these pathogens, since exposure to spirochetemic hosts may be rare.

The ecology of the 3 relapsing fever spirochete species found in North America is similar in that all 3 are associated with rodent reservoirs. By far, the agent most frequently causing human relapsing fever in North America is *Borrelia hermsii*, transmitted by *O. hermsi*. Yellow pine chipmunks (*Tamias amoenus*), pine squirrels (*Tamiasciuris hudsonicus richardsoni*) and meadow voles (*Microtus pennsylvanicus*) develop particularly high spirochetemias when infected with *B. hermsii*. This agent is found from British Columbia south to Arizona. Classic sce-

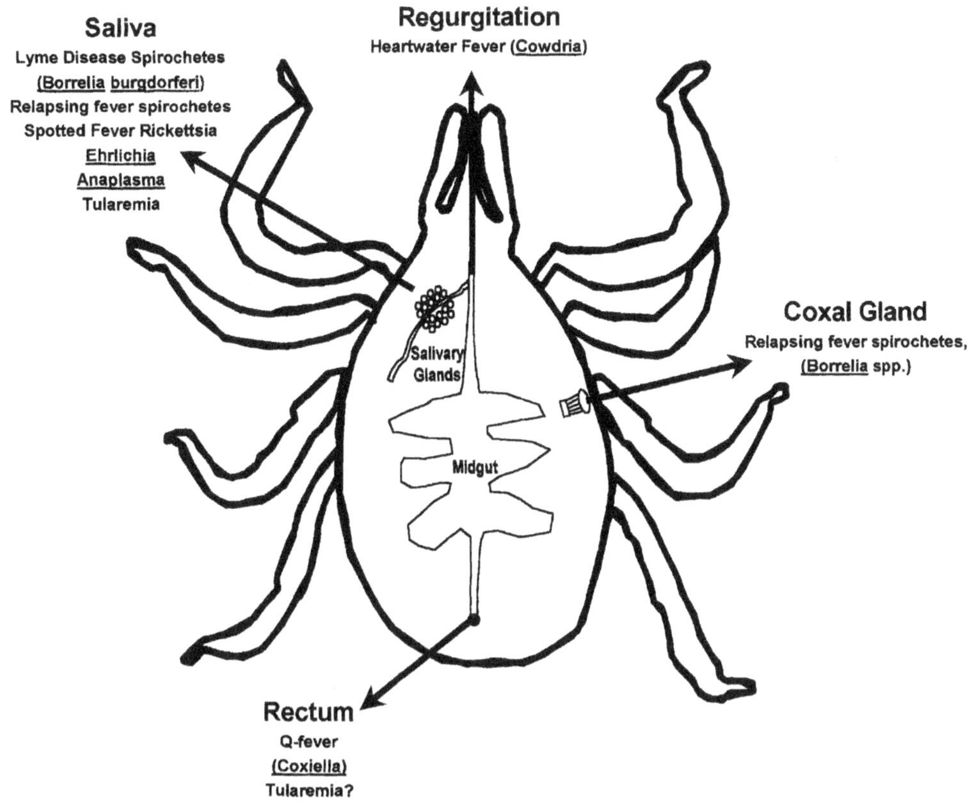

Figure 10.3. Routes of transmission of tickborne pathogens.

narios for relapsing fever occur in the Pacific northwestern USA, where summer cabins are closed for the winter and become refuges for rodents and *O. hermsi*. When cabins are opened and cleaned at the beginning of summer, rodent hosts disappear and are replaced by the humans who become victims of both the ticks and relapsing fever. An interesting variation of this theme occurred on the North Rim of the Grand Canyon in 1973. The attic space of rustic tourist cabins provided the perfect habitat for nests of chipmunks and pine squirrels. Populations of soft tick, *O. hermsi*, established themselves within these pine cone havens. When a plague epizootic occurred, decimating these rodent populations, the ticks in the attic lost their principal hosts. Consequently, they headed downstairs, fed on the tourists and caused an outbreak of tickborne relapsing fever with dozens of cases occurring (Boyer et al. 1977). A program of acaricidal spraying and rodent-proofing these cabins curtailed the epidemic. No significant relapsing fever occurred in these cabins for the subsequent two decades. When memory of the risk of relapsing fever faded and rodent-proofing standards were relaxed, another outbreak of relapsing fever resulted in these same cabins during the 1990s.

Borrelia parkeri generally causes no human disease. It is found in association with burrowing rodents along the Great Basin of the western USA, from California to Colorado. Isolated human cases of relapsing fever due to infection with *Borrelia turicatae* occur in the southern USA, particularly in people exploring caves. Such an outbreak recently occurred in Ozona, Texas. Interestingly, a spirochete isolated from a dog in Florida closely resembles *B. turicatae*. Finally, a new spirochete, *Borrelia coriaceae* was isolated from *Ornithodoros coriaceus* in northern California (Lane and Manweiler 1988). The reservoir of this agent is unknown. *Borrelia coriaceae* may be the agent of epizootic bovine abortion, but this suggestion has not been proven.

The relationship between each relapsing fever *Borrelia* species and its respective tick vector is highly specific. This linkage has been studied in detail using classical vector competence experiments with the 3 main North American relapsing fever spirochetes, *B. hermsii*, *B. parkeri* and *B. turicatae*. Xenodiagnosis was used to show that *B. hermsii* could infect only *O. hermsi* and not *Ornithodoros parkeri*, while *B. parkeri* could only infect *O. parkeri* (Davis 1956). Similarly, Brumpt (1933) showed the specificity of *B. turicatae* for *O. turicata*. The only exception to this specificity reported for *Borrelia-Ornithodoros* combinations was an aberrant *B. parkeri* strain that could be transmitted erratically by *O. turicata* (Davis and Burgdorfer 1955). Schwan et al. (1995) examined whether the naming of *Borrelia* species based on vector specificity is warranted. They confirmed that only *O. hermsi* could support the growth of and transmit *B. hermsii*. When *O. parkeri* and *O. turicata* ingested *B. hermsii*, the spirochetes migrated out of the midgut, but survived only in the central ganglion. Spirochetes in the "wrong" host did not persist in the salivary glands or coxal glands where transmission could be effected. Davis and Burgdorfer (1955) also reported that *B. parkeri* was capable of infecting *O. parkeri* and *O. turicata*, but usually disappeared from the salivary glands and coxal glands of *O. turicata* within 35 days. However, *B. parkeri* did persist in the central ganglia of *O. turicata* for at least 322 days post-infection. The basis for this tissue specificity based on vector species invites further study.

Unlike the Lyme disease spirochete, *B. burgdorferi*, which clearly is associated with various tissues in the vertebrate host, the relapsing fever spirochetes are localized in the bloodstream, constantly exposed to humoral immune responses. These spirochetes can stay one step ahead of the host's antibody defenses through antigenic variation, i.e., by shifts in surface proteins. In this respect, the relapsing fever spirochetes use a strategy similar to trypanosomes that cause African sleeping sickness. The variable antigenic types, called vmp (variable major proteins) in *Borrelia*, are analogous to the vsg proteins of *Trypanosoma*. The most intensive study of how relapsing fever spirochetes vary their surface proteins used *B. hermsii* as the model system. A total of 26 different serotypes have been identified among the progeny of a single *B. hermsii* cell (Barbour 1988). One change, from vmp serotype 7 to vmp serotype 21, has been well defined. Genes for the production of both vmp7 and vmp21 have been described. When vmp7 is in an active expression site, this protein is expressed. Meanwhile, vmp21 sits in a silent storage site. Variant cells arise which have, through a nonreciprocal recombinant event, moved the vmp21 gene into the active site. As the antibody response of the host eliminates vmp7 serotype spirochetes, vmp21-expressing spirochetes are selected quickly and begin to predominate. Thus, the relapsing nature of the human disease: fever occurs while the spirochete population is high, then antibodies catch up and eliminate the dominant vmp serotype, and the fever remisses. Subsequently, a new vmp type emerges and the fever relapses.

In addition to the *Borrelia* that cause human relapsing fever and Lyme disease, *Borrelia anserina* causes avian borreliosis. The principal vectors are ticks in the genus *Argas*. Vectors in Eurasia include *Argas persicus* and *Argas reflexus*, and in South America the principal vector is *Argas miniatus* (Felsenfeld 1979). Despite numerous reports of the red fowl mite, *Dermanyssus gallinae*, being involved in the transmission of *B. anserina*, reports of transmission of a *Borrelia* by vectors other than ticks require careful reexamination.

BARTONELLOSIS

Infection with *Bartonella bacilliformis* in the Andean nations of Peru, Colombia and Ecuador causes human bartonellosis (Carrión disease). Named after a scientist who experimentally infected himself and died in 1885, Carrión disease has 2 distinct clinical forms. **Verruga peruana** involves wart-like dermal eruptions of long duration, but rarely is fatal; **Oroyo fever** is characterized by a febrile anemia due to destruction of erythrocytes adhered to by the bacteria, with a case fatality rate of 10-40%. Fatalities often are linked to secondary septicemic infections with *Salmonella*. Verruga peruana may be marked by joint and muscle pain before the occurrence of dermal eruptions. Oroyo fever tends to be acute; Verruga is chronic. Sand flies in the genus *Lutzomyia* transmit the bacterium. These are the same sand flies that transmit leishmaniasis in South America (see Chapter 8). In the interandean valleys of Peru, including the Huayllacallan Valley, Dr. Humberto Guerra and colleagues showed that at elevations <2,000 m, *Lutzomyia verrucarum* is the dominant sand fly species, and residents suffer from Carrión disease. At elevations >2,000 m, *Lutzomyia peruensis* is the dominant sand fly, and residents suffer from leishmaniasis. Such a clear demarcation of endemic diseases at different elevations is a fascinating ecologic phenomenon. Humans may be the sole reservoir of *B. bacilliformis*. It is not known whether sand flies transmit the pathogen to humans via mechanical transmission or biological transmission.

TRENCH FEVER

Trench fever is caused by *Bartonella quintana* and transmitted by the body louse, *P. humanus humanus*. Transmission of the bacteria occurs when an infected louse is transferred from an infected to a susceptible human, and the louse or louse feces is crushed or rubbed into cuts on the skin or through mucous membranes (Gage et al. 1995). As with louseborne typhus and relapsing fever, trench fever results from the crowded conditions and poor personal hygiene that occurs during wars or the mass exodus of refugees. Over a million cases of trench fever occurred during World War I, and >80,000 cases during World War II (Vinson 1973). The disease is rare in modern times, but potentially risky during times of social upheaval. Both *B. quintana* and *B. bacilliformis* respond to antibiotic treatment with erythromycin, doxycycline or rifampin for an extended period of 4-6 weeks.

CAT SCRATCH DISEASE

The genus *Rochalimaea* was long included in the order Rickettsiales. However, molecular studies of pathogens causing cat scratch disease (CSD) have led to the conclusion that *Rochalimaea* should be removed from the order Rickettsiales and folded into the bacterial genus *Bartonella* (Brenner et al. 1993). The newly described agent of CSD, *Bartonella henselae*, causes a papule at the site of inoculation, followed by regional lymphadenopathy 7-50 days later. The disease is more serious in immunocompromised individuals, including those infected with HIV. The primary reservoir of *B. henselae* is the domestic cat, and it is mainly young kittens that infect people. Cat fleas may be involved in the transmission of *B. henselae* (Zangwill et al. 1993), and transmission of the etiologic agent of CSD by *Ctenocephalides felis* under carefully controlled conditions has been reported (Chomel et al. 1996).

PLAGUE

Plague is an acute and often fatal fleaborne zoonosis caused by infection with a gram-negative bacterium, *Yersinia pestis*. Rodents are the primary vertebrate hosts of *Y. pestis*, but other mammals, including humans, can be infected (Fig. 10.4). The plague bacterium probably originated in rodents and fleas that lived in the steppes of central Asia, then spread along international trade routes by infected commensal rats (primarily *Rattus rattus*) and fleas (primarily Oriental rat fleas, *Xenopsylla cheopis*). The 3 great plague pandemics (Justinian's Plague: 542–602 A.D.; the Black Death: 14th–16th centuries A.D.; and the so-called Modern Pandemic: 1894 to present), caused millions of human deaths (Pollitzer 1954). The first 2 pandemics also affected the history of Roman and Medieval civilizations. Yersin first identified the etiological agent of plague in 1894 at the beginning of the last pandemic. Four years later, Simond demonstrated transmission of this agent to healthy rats through the bites of infectious fleas (Gregg 1985).

The range of plague often expands during regional outbreaks. Most of these range extensions are temporary, but occasionally new foci of *Y. pestis* infections become established among previously unaffected populations of wild rodents and their fleas. During the current pandemic, new sylvatic rodent foci have appeared in southern Asia, Java, southern Africa and the Americas; many of these are still active today. Plague probably first entered the USA at the beginning of the 20th Century. Although a number of cities along the Gulf of Mexico and Pacific Coast experienced rat-associated epidemics of bubonic plague, only the San Francisco epidemic is believed to have contributed significantly to the establishment of sylvatic foci. Here, the disease spread quickly from commensal rats in the city to wild rodent populations in the surrounding countryside. Eventually, the disease spread to other rodent and flea populations living elsewhere in the western USA, thereby establishing wild rodent foci of infection that persist to the present day. Molecular studies support the contention that plague did not exist in the USA prior to the last pandemic (Guiyoule et al. 1994).

Plague typically causes severe and often fatal illness in untreated humans. Other mammals, including certain rodents, rabbits and hares (Leporidae), as well as members of the cat family (Felidae), also are highly susceptible and often die following infection. Non-mammalian vertebrates appear to be completely resistant to plague.

Mammals can become infected with *Y. pestis* after being bitten by infectious fleas, after handling or eating infected animals, or, rarely, after inhaling infectious respiratory droplets expelled from coughing infected mammals.

Plague occurs in 3 clinical forms: bubonic, septicemic and pneumonic. The most common of these is bubonic plague and the rarest is the pneumonic form. Persons with plague typically develop fever and other symptoms including headache, chills, muscle pains, malaise, prostration and diarrhea or other gastrointestinal complaints. Individuals with bubonic plague also develop extremely painful and swollen lymph nodes (buboes). These buboes characterize bubonic plague and occur in over 70% of the plague cases in the USA. Buboes appear most often in inguinal lymph nodes, less frequently in axillary nodes, and least often in cervical lymph nodes. Typically, a bubo develops in the node that is situated nearest to the site where *Y. pestis* first entered the body. Thus, a person who has been bitten on the leg by an infectious flea is likely to develop an inguinal bubo, while an individual who has handled an infected animal will more likely develop axillary buboes. Carnivores, particularly domestic cats and bobcats, that are susceptible to *Y. pestis* infection often develop submandibular buboes after ingesting infected prey.

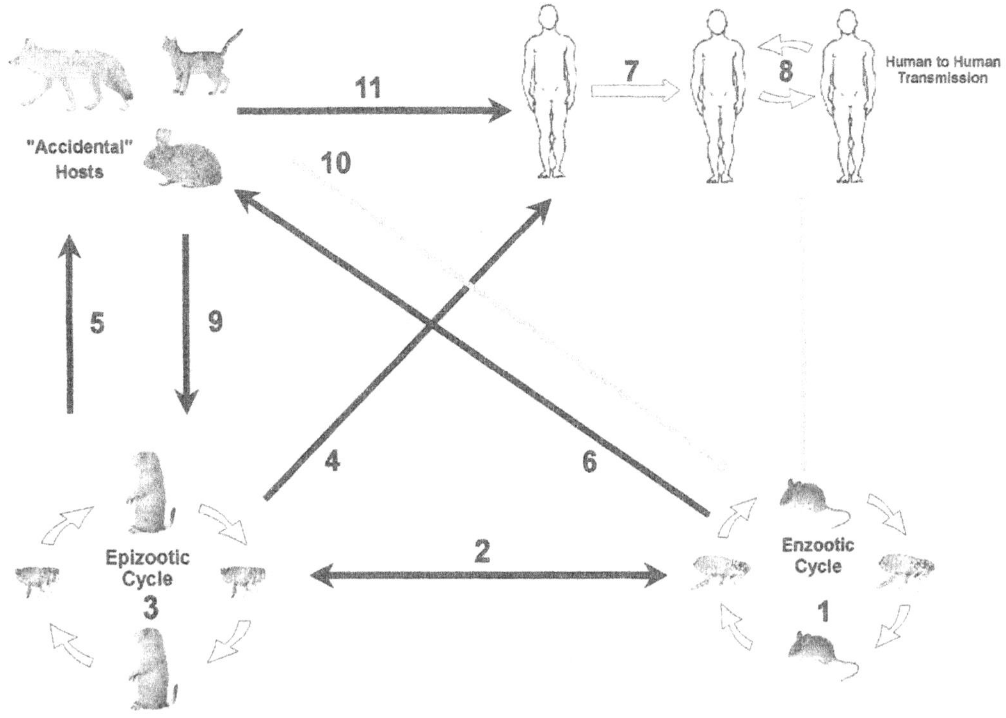

Figure 10.4. Natural cycle of *Yersinia pestis*. (1) Enzootic cycle involving moderately resistant enzootic or maintenance hosts; (2) transfer of *Y. pestis*-infected fleas from enzootic hosts to epizootic hosts; (3) epizootic cycle resulting in amplification and geographic spread of plague; involves certain wild rodents or occasionally *Rattus* spp.; (4) transmission to humans by rodent flea bites or through handling of infected epizootic hosts; (5) transmission to non-rodent mammals by rodent flea bites or consumption of infected epizootic hosts by carnivores, etc.; (6) transmission to non-rodent mammals by rodent fleas or consumption of infected enzootic hosts by carnivores; (7) transmission to humans by inhalation of infectious respiratory droplets expelled by patients with secondary pneumonic plague; (8) continued person to person transmission by inhalation of infectious respiratory droplets expelled from persons with primary pneumonic plague; (9, 10) transport of infected fleas by carnivores or other "transport hosts;" (11) transmission to humans via direct contact with infected animals or inhalation of infectious respiratory droplets expelled from cats with plague pneumonia; pets also can transport infectious fleas into human habitations; (12) transmission from enzootic hosts to humans.

Septicemic plague occurs when plague bacteria circulate freely and multiply in the bloodstream. In addition to the generalized signs and symptoms noted previously, persons with untreated septicemic plague are likely to die after developing overwhelming infections and various manifestations of systemic inflammatory response syndrome. Septicemic plague can occur with or without prior lymphadenopathy (buboes). Persons with pneumonic plague typically experience high fever, severe pneumonia, cough, and often have blood and plague bacteria in their sputum. The latter sign increases the likelihood that the disease will spread from person to person or animal to person if the infected host coughs and expels infectious material. Case fatality rates for untreated human bubonic cases range from 50–60%, while untreated pneumonic cases probably are always fatal (Poland and Barnes 1979). The disease can be treated effectively with certain antibiotics, such as streptomycin, gentamicin or tetracyclines, providing diagnosis and treatment occur early in the infection.

A wide variety of mammals have been infected with *Y. pestis*. In the USA alone, plague has been identified in 76 species of mammals (Barnes 1982). Although members of many mammalian orders have been infected, the disease is maintained primarily in complex enzootic and epizootic transmission cycles involving various wild rodent species and their fleas (Fig. 10.4).

The mammalian hosts of plague vary considerably from one region of the world to another. In Asia and the extreme southeast corner of Europe near the Caspian Sea, the hosts include various combinations of ground squirrels, marmots, gerbils, voles, wild mice and certain species of rats. Depending on the region examined, important hosts in African foci include indigenous rat or mouse species, gerbils and possibly springhares. In South America, naturally occurring plague infections have been reported in native rats and mice and various species of caviid rodents (relatives of the domestic guinea pig). The most important hosts in the USA are believed to be certain species of ground squirrels, prairie dogs, woodrats, mice and voles (Gage et al. 1995).

Carnivores that prey on rodents can play an important ecological role by transporting infected fleas from one area to another, but these animals rarely, if ever, serve as sources of infection for fleas. Some carnivores, such as coyotes and dogs, appear to be relatively resistant, while felids are very susceptible to fatal infections (Gage 1998). Rabbits and hares also commonly are reported to die during epizootics, but these animals probably are only "accidental" hosts of infection (von Reyn et al. 1976).

In sylvatic foci in the western USA, the disease often results in epizootics that cause high mortality (80–100%) among certain kinds of rodents, particularly various species of prairie dogs, ground squirrels, chipmunks and woodrats. Epizootics among these highly susceptible rodents (often termed epizootic or amplification hosts) and their fleas can lead to the rapid spread of *Y. pestis* from one host population to another of the same or different species. Occasionally, these epizootics can extend over hundreds to thousands of square kilometers. Risks of infection for certain "accidental" hosts, such as humans, carnivores and lagomorphs, are greatest during such periods of epizootic amplification. The factors that govern the occurrence of these epizootics are poorly documented, but probably include weather-related changes that influence the dynamics of rodent and flea populations.

In contrast to the above species, other rodents, such as kangaroo rats, reportedly suffer relatively little mortality during epizootics. Still other species, including deer mice (*Peromyscus maniculatus*), California voles (*Microtus californicus*) and grasshopper mice (*Onychomys leucogaster*), vary greatly in their susceptibility to plague-related mortality. These inter-populational variations in susceptibility may be due to the exposure of these populations to *Y. pestis* infection (Thomas et al. 1988).

In the USA, populations of partially resistant voles (*Microtus* spp.) or deer mice and their relatives (*Peromyscus* spp.) may act as enzootic hosts for plague, thus maintaining the infection between periods of epizootic amplification (Poland and Barnes 1979). Animals from resistant populations reportedly become bacteremic but rarely die when infected with *Y. pestis*. Infected mice and voles that circulate sufficient quantities of bacteria in their blood can infect feeding fleas. These animals also host a variety of flea species reported to be naturally infected with *Y. pestis*. Unlike the various epizootic hosts listed above, which usually reproduce only once or twice a year, the mice and vole species that may be enzootic hosts reproduce many times per year. High reproductive rates probably maintain plague in mice and voles because they introduce nonimmune individuals into local populations of these animals.

Although wild rodents and their fleas are essential for maintaining natural foci of plague, almost all large outbreaks of human plague are associated with epizootics among commensal rats (*R. rattus* and *R. norvegicus*) and their fleas, especially *X. cheopis*. Fortunately, rat-associated epidemics have not occurred in the USA since 1924. Instead, most human cases in this country occur singly or as clusters of 2–3 cases, and usually coincide with the occurrence of epizootics among susceptible wild rodent populations. During the last few decades, *Y. pestis* infections have been identified occasionally in individual commensal rats found in or near certain cities in the plague-enzootic western USA. Actual epizootics among *Rattus* spp., however, have not been reported. One possible explanation for the lack of reported epizootics among these suburban and rural rat populations is that these animals are not infested with significant numbers of *X. cheopis* or other competent flea vectors. Schwan et al. (1985) examined *R. rattus* in suburban environments in southern California and found that these animals harbored only small numbers of northern rat fleas (*Nosopsyllus fasciatus*) and mouse fleas (*Leptopsyllis segnis*), but no *X. cheopis*. The first two species generally are regarded as less efficient plague vectors than *X. cheopis*.

Complex factors determine whether or not a certain flea species will be an important vector of plague. More than 150 species of fleas, most of which infest wild rodents, have been reported to be naturally infected with *Y. pestis* (Pollitzer and Meyer 1961). Although each of these naturally infected flea species can be thought of as a potential vector of plague, most probably are relatively inefficient vectors of *Y. pestis* or infest hosts that play little or no role in the natural cycle of plague.

The importance of a given flea species as a vector of *Y. pestis* depends on:

1. The species of rodents infested.
2. The abundance of fleas during seasons of peak transmission.
3. The geographic distribution of the flea species.
4. The ability of the fleas to survive for extended periods in off-host environments (especially after their hosts have died during plague epizootics).
5. The likelihood that the flea will infest alternative hosts when their normal hosts are absent.
6. The capability of the flea to become blocked after ingesting *Y. pestis* while feeding on an infected host (Fig. 10.5).

The ability of a flea to become "blocked" is critical for efficient transmission of plague bacteria by these insects (Bacot and Martin 1914). Soon after being ingested by a competent flea vector, the plague bacilli begin to multiply in the flea's gut. Eventually, small brown colonies of *Y. pestis* appear in the proventriculus or anterior midgut of the infected flea (Pollitzer 1954). These colonies continue to increase in size until they coalesce and form sticky gelatinous masses that adhere to spines in the flea's proventriculus. If these masses become sufficiently large, they can occlude the gut tract and interrupt the normal intake of blood during feeding. At this point, the flea is said to be blocked and ingested blood cannot pass beyond the proventriculus into the midgut. Because ingested blood cannot reach the midgut for digestion, the flea begins to starve and will try repeatedly to feed on available hosts. During these attempts, the flea regurgitates viable *Y. pestis* and other components of the block into the feeding site in an unsuccessful effort to clear the proventricular

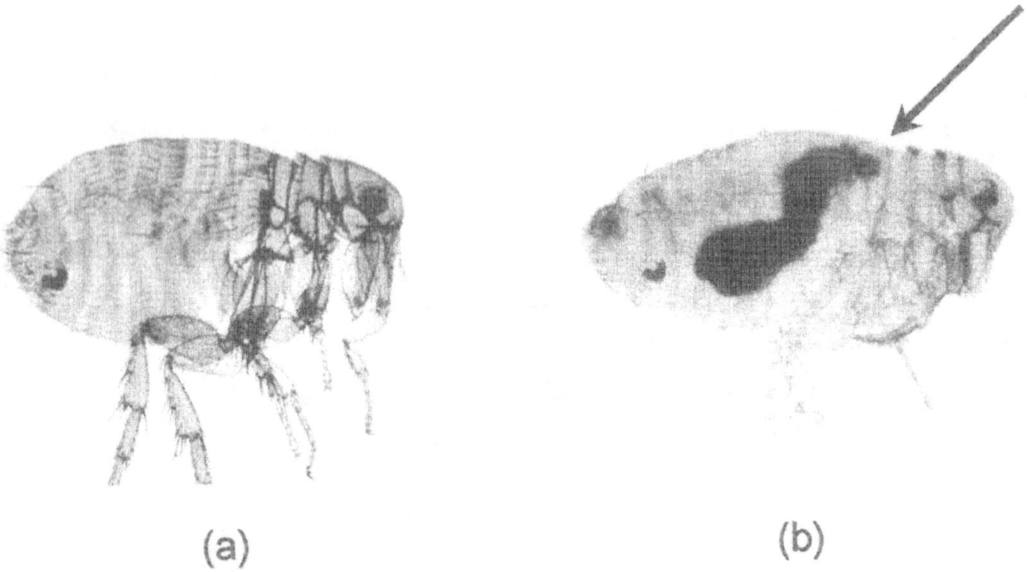

Figure 10.5. (a) Unfed, unblocked female *Xenopsylla cheopis*, the Oriental rat flea; (b) blocked female *X. cheopis*. Arrow indicates blocked proventriculus (photograph by Kenneth L. Gage).

obstruction. This regurgitation transmits the plague bacillus to susceptible mammalian hosts.

Formation of these proventricular blockages depends, at least in part, on the characteristics of the plague bacillus. Plague strains that bind Congo red in culture (pigment-positive strains) reportedly cause blockage formation in fleas, while strains that fail to bind this pigment (pigment-negative strains) lack such a capacity. Apparently, pigment-positive strains can bind iron-containing hemin to the surface of their cells, which makes them much more "sticky" than pigment-minus strains. This adhesiveness, of course, suggests that pigment-binding strains are more likely than pigment-negative ones to form colonies that will adhere to the spines in a flea's proventriculus and cause blockage formation (Bibikova 1977). Hinnebusch et al. (1996) demonstrated that a pigmented (*hms*$^+$) *Y. pestis* strain could develop blockages in fleas, while nonpigmented (*hms*$^-$) mutants derived from the same parent strain failed to do so. The ability of nonpigmented mutants to promote blockage formation could be restored by complementation of these mutants with a low copy number recombinant plasmid containing a combination of 3 genes comprising the hemin storage locus of *Y. pestis*.

Blocking also depends on temperature. Fleas such as *X. cheopis*, that are competent vectors, will become blocked when incubated at temperatures below approximately 28° C, but fail to do so at higher temperatures. Those held at higher temperatures can become cleared of infection (Cavanaugh 1971). The formation of blockages in *X. cheopis* reportedly takes 9–26 days at temperatures ranging from approximately 10°–27°C, but often takes longer (3 weeks to 4 months) in certain species of wild rodent fleas (Eskey and Haas 1940).

Cavanaugh (1971) hypothesized that temperature influenced block formation by affecting an enzyme that exhibits coagulase activity at temperatures ≤27.5°C (promotes block formation) or fibrinolysin activity at temperatures >27.5°C (degrades blockages). Such an enzyme (plasminogen activator) does exist on a 9.5 kb plasmid of *Y. pestis*, but recent studies indicate that plague strains lacking the 9.5 kb plasmid still can form proventricular blockages in *X. cheopis* (Hinnebusch et al. 1998). If the plasminogen activator enzyme was indeed critical for block formation, it would be expected that loss of the plasmid containing the gene that encodes for it would prevent blockage development.

Some fleas appear to be incapable of becoming blocked and, therefore, are unlikely to transmit plague efficiently. Nevertheless, transmission by apparently unblocked fleas has been reported. In most of these studies, transmission occurred only when very large numbers of infected fleas were placed on susceptible host animals. These results suggest that the unblocked fleas probably transmit *Y. pestis* to susceptible animals via contaminated mouthparts (mechanical transmission), an inefficient process compared to transmission of plague by blocked fleas.

Mechanical transmission is unlikely to be important unless the supposed mechanical vector abounds in host habitats and has frequent opportunities to move quickly from one susceptible host to another. For example, epidemiological studies have implicated *Pulex irritans*, the so-called human flea, as a vector in plague epidemics in Peru, Morocco and Kurdistan (Hopla 1980). This flea, which feeds readily on many species of mammals and domestic fowl, and apparently an inefficient experimental vector of plague, might act as a mechanical vector where there are numerous human dwellings or livestock rearing pens. Another potentially significant mechanical vector is *Malaraeus telchinum*, a flea common on certain vole and native mouse populations in the western USA (Kartman 1957). Reportedly, it does not become blocked when infected with *Y. pestis*, and transmits plague to experimental animals only when these hosts are infested with large numbers of fleas.

Risks of human plague can be reduced by applying appropriate insecticides to rodent burrows and runways, reducing the availability of food or shelter for rodents, avoiding sick or dead animals and rodent nests or burrows, treating dogs and cats with insecticidal flea powders to prevent transport of infected fleas into homes, and, under certain circumstances, using rodenticides to reduce the populations of plague-susceptible hosts in areas of high human risk. Applications of rodenticides always should be preceded by flea control to prevent hostless fleas from attacking nearby humans. A vaccine also is available, but its effectiveness has not been evaluated thoroughly, especially for protecting humans against airborne exposures. However, it might be useful for preventing fleaborne plague (Centers for Disease Control and Prevention 1996).

TULAREMIA

Tularemia occurs throughout much of the Northern Hemisphere but apparently is absent from the Southern Hemisphere. The etiologic agent, *Francisella tularensis*, is a gram-negative bacterium that infects a variety of arthropods and vertebrates. Two distinct variants (called biovars), A and B, exist. Type A strains usually are vectorborne and typically are more virulent than Type B strains. The latter often are waterborne. Both A and B variants occur in the USA, but type A strains are more common. Type B strains predominate in the Palearctic region. Symptoms of human tularemia typically include fever, chills, malaise, fatigue and lymphadenopathy. Respiratory symptoms also are common, but unlike those of plague, do not appear to be associated with risk of airborne human-to-human transmission. The most common clinical form of the disease (75–85% of cases) is ulceroglandular tularemia, where an ulcerative skin lesion develops at the site of the original infection and nearby lymph nodes become infected and swollen. Glandular, typhoidal, oculoglandular and oropharyngeal forms also occur. The skin ulcers of ulceroglandular tularemia usually appear at the site where *F. tularensis* initially enters the body. Humans become infected through direct contact with infected animals, bites of infected arthropods (especially ticks and tabanid flies), or contact with water contaminated with urine from bacteriuric rodents. The fatality rate for untreated cases is about 5–7%. Prompt treatment with appropriate antibiotics such as streptomycin generally is effective.

Although *F. tularensis* occasionally has been identified in birds, reptiles and even fish, the primary vertebrate hosts are mammals (Hopla 1974). To date, more than 100 species of mammals have been infected (Hopla and Hopla 1994). However, as with plague, certain mammal species are more important than others as hosts, especially microtine rodents (voles, muskrats and their allies) and lagomorphs.

Ticks play significant roles as both reservoirs and vectors in the enzootic cycles of *F. tularensis* transmission in both the Palearctic and Nearctic regions. Transmission cycles involving lagomorphs and their ticks are particularly important for maintaining *F. tularensis* in North America. The tick vectors present on rabbits and hares include various species within the genera *Amblyomma*, *Dermacentor*, *Haemaphysalis* and *Ixodes* (Hopla 1974). Many of these ticks also transmit *F. tularensis* to other mammals.

Francisella tularensis may multiply within its tick vectors, but the modes of transmission have been debated. Transstadial transmission of *F. tularensis* has been demonstrated in ticks, but most authorities doubt that transovarial transmission occurs. Some researchers have reported that *F. tularensis* can penetrate the gut barrier, pass through the hemocoel and eventually infect the salivary glands, thereby allowing this bacterium to be transmitted with the saliva to mammalian hosts during tick feeding. Others believe that transmission to mammals occurs because the feeding site has been contaminated with infectious tick feces.

Apparently, *F. tularensis* does not multiply significantly in deer flies (Tabanidae) and transmission by such flies is strictly mechanical. Jellison (1974) has stated that *Chrysops discalis* is the only important tabanid vector of tularemia in the USA. This species occurs throughout much of the western USA. Most human cases resulting from the bites of infectious tabanid flies occur in the Intermountain West.

Francisella tularensis also is commonly identified in various species of rodents. Among the most important rodent hosts in North America are various voles (*Microtus* spp. and *Clethrionomys* spp.), as well as muskrats (*Ondatra zibethicus*) and beavers (*Castor canadensis*). Although these rodents may be associated with vectorborne transmission cycles involving ticks, they also become chronically infected and can excrete viable *F. tularensis* in their urine.

According to Hopla and Hopla (1994), approximately 200 human cases occurred per year in the USA in the interval 1981 to 1991, but these authors believe that under-reporting is common. The incidence and modes of transmission of *F. tularensis* vary from one region to another. In the USA, most human cases (about 60%) occur in the south-central region. Although many cases there result from direct contact with infected animals, the most important source of exposure is via the bites of infected ticks, especially *A. americanum*. Important sources of exposure in the western USA include mechanical transmission by tabanid flies, biological transmission by ixodid ticks (particularly *Dermacentor andersoni*), and direct contact with infected animals, especially rabbits. Most cases in the midwestern USA occur in hunters and trappers that have directly contacted infected animals.

Waterborne tularemia infections in humans are relatively rare in North America, but occur frequently in the Palearctic region, often as a result of contact with water contaminated with urine from infected microtine rodents. Epidemics of tularemia among muskrat trappers in the northeastern USA have coincided with waterborne epizootics among the trappers' prey, but these human cases probably resulted from direct contact with infected animals rather than exposure to *F. tularensis*-contaminated water.

Risks of exposure to *F. tularensis* can be reduced by (1)treating clothing with approved pesticides, (2) treating clothing and skin with repellents, and (3) wearing gloves while handling potentially infected animals. Where risks of waterborne tularemia are high, water supplies must be treated appropriately or care must be taken to prevent contamination by urine from bacteriuric rodents. An experimental vaccine exists for laboratory personnel, but it is unlikely to be available widely in the near future.

RICKETTSIAE AND OTHER OBLIGATE INTRACELLULAR BACTERIA

A variety of obligate intracellular bacteria have infected many arthropod species. Among those that infect fleas, lice, mites and ticks are some that cause severe illness in humans or their domestic animals. Traditionally, these organisms are referred to collectively, albeit somewhat mistakenly, as rickettsiae, and have been grouped together within the Order Rickettsiales. Recent molecular studies, however, clearly demonstrate that this order is polyphyletic and should be redefined.

Rickettsioses

The genus *Rickettsia* contains species that cause severe and often fatal human diseases, including the etiologic agents of Rocky Mountain spotted fever (*Rickettsia rickettsii*) and louseborne typhus (*Rickettsia prowazekii*) (Table 10.2). Still other *Rickettsia* spp. appear to cause little pathogenesis in vertebrates and apparently are maintained in nature almost exclusively by transstadial and transovarial transmission in their tick hosts (Fig. 10.6). Regardless of their status as disease agents, these gram-negative obligate intracellular bacteria are all morphologically similar (typically coccoid to rod-shaped), grow free within the cytoplasm of host cells, and reproduce by simple binary fission. Almost all of the known species of *Rickettsia* can be grouped into one of two distinct serogroups: typhus (TG) and spotted fever (SFG). The primary vectors of *Rickettsia* spp. are lice, fleas, mites and ticks.

Symptoms of human rickettsial infection include high fever, severe headache, chills, malaise, myalgia and often a rash on the trunk (murine or louseborne typhus) or extremities (tickborne spotted fevers). A black crusty lesion, or eschar, also can appear at the feeding site of the vector. Such eschars are common in patients with Boutonneuse fever and the recently identified Japanese spotted fever, but are rare among cases of Rocky Mountain spotted fever or typhus. Cases of rickettsialpox, a relatively mild rickettsiosis, also exhibit a distinctive vesicular rash not observed in other forms of rickettsial illness.

Following infection of a susceptible host, rickettsiae invade and multiply in the endothelial cells lining the host's blood vessels. Multi-

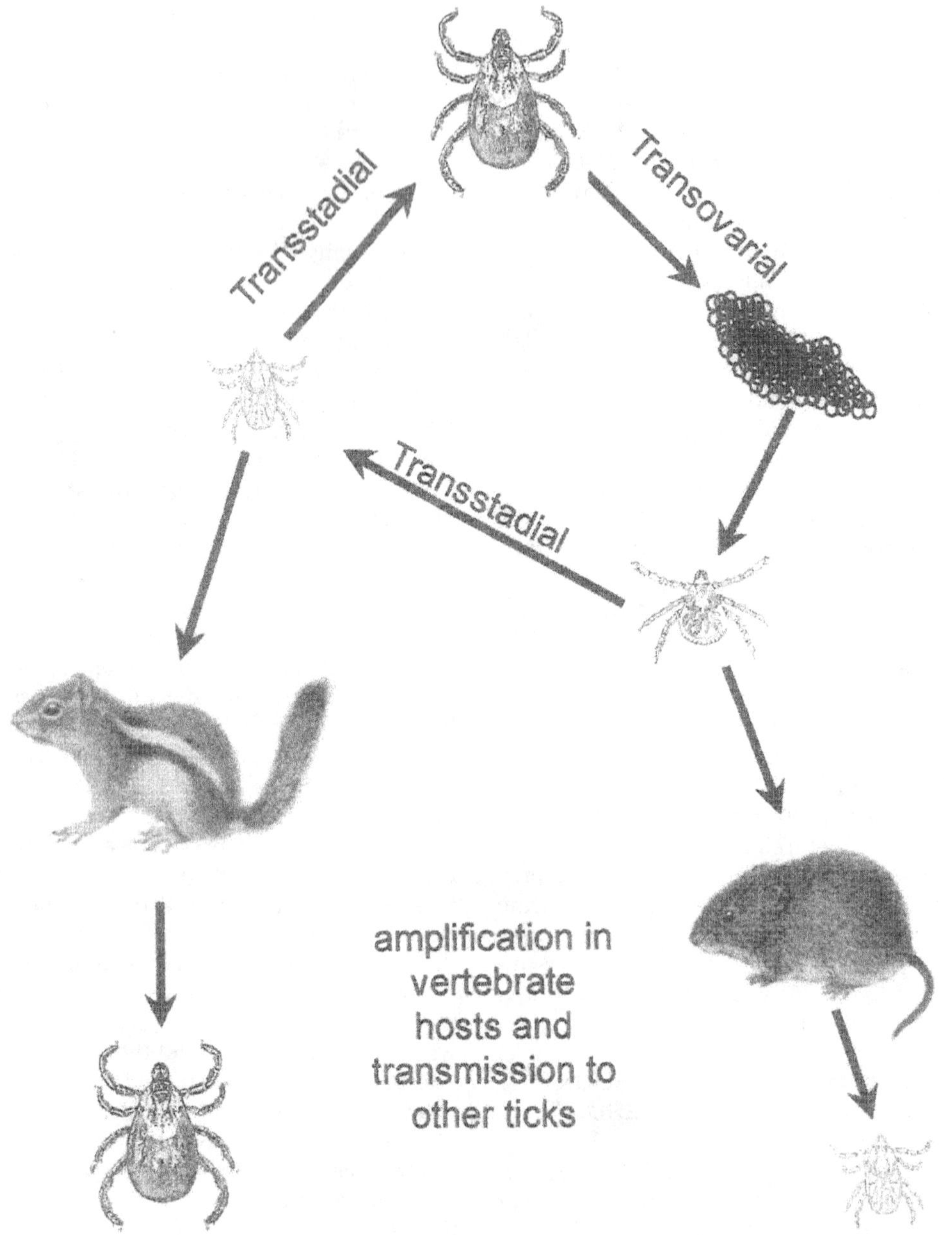

Figure 10.6. Natural cycle of *Rickettsia rickettsii* illustrating transovarian and transstadial transmission within ticks and horizontal transmission of spotted fever rickettsiae from tick to mammal to tick. The cycles for other spotted fever group rickettsiae are similar.

Table 10.2. Important human and livestock diseases caused by arthropodborne rickettsiae and other obligate intracellular bacteria.

Disease	Pathogen	Primary vectors	Primary vertebrate hosts	Distribution
Louseborne (epidemic) typhus	*Rickettsia prowazekii*	*Pediculus humanus humanus*	Humans, possibly other mammals	Africa, the Americas
Fleaborne (endemic or murine typhus)	*Rickettsia typhi* (*R. mooseri*)	*Xenopsylla cheopis*, other fleas	Rodents, opossums, cats, other small mammals	Tropical, subtropical, warm temperate regions
Rocky Mountain spotted fever	*Rickettsia rickettsii*	*Dermacentor andersoni, D. variabilis, Amblyomma cajennense*, probably *D. occidentalis*	Rodents, other small to medium-sized mammals	North, Central, South America
North Asian tick typhus	*Rickettsia sibirica*	*Dermacentor marginatus, Dermacentor* spp., *Haemaphysalis* spp.		Northern Asia
Boutonneuse fever	*Rickettsia conorii*	*Rhipicephalus sanguineus*, other ticks	Rodents, possibly other small to medium-sized mammals	Europe, Africa, Middle East, India
Queensland tick typhus	*Rickettsia australis*	*Ixodes holocyclus*	Small marsupials	Australia
Oriental spotted fever	*Rickettsia japonica*	Possibly *Dermacentor taiwanensis, Haemaphysalis flava*	Rodents, possibly other small to medium-sized mammals	Japan
Rickettsialpox	*Rickettsia akari*	*Liponyssoides sanguineus*	House mice (widespread), possibly field voles (Korea)	USA, former USSR, Korea, probably other areas
Scrub typhus	*Orientia tsutsugamushi*	*Leptotrombidium* spp.	Rodents (although mite hosts, not known to serve as reservoir or amplifying hosts)	Southeast Asia
Heartwater	*Cowdria ruminantium*	*Amblyomma hebraeum, A. variegatum*	Wild, domestic ruminants	Sub-Saharan Africa
Anaplasmosis	*Anaplasma marginale, A. ovis, A. centrale*	*Dermacentor* spp., *Amblyomma* spp., other ticks	Wild, domestic ruminants	Worldwide

plication of rickettsiae within these cells eventually destroys them, thereby causing the smaller blood vessels involved in microcirculation to lose their integrity and "leak" blood into the surrounding tissue spaces. Destruction of these vessels causes the characteristic rashes and more serious complications of rickettsial infections (Walker and Gage 1997). These infections can be treated with tetracyclines or chloramphenicol.

Tickborne rickettsiae

The most common arthropod hosts of *Rickettsia* spp. are various kinds of ticks, which serve as both vectors and reservoirs of infection. Ricketts first demonstrated the basic ecological role played by ticks in the transmission and maintenance of rickettsiae. In a series of elegant experiments, Burgdorfer (1980) showed that wood ticks (*D. andersoni*) fed on infected guinea pigs could transmit Rocky Mountain spotted fever to noninfected guinea pigs during subsequent feeding attempts. He also demonstrated that ticks could maintain the spotted fever agent within their own populations by transstadial and transovarial transmission. Most tickborne rickettsiae belong to the SFG, including *R. rickettsii*, the etiologic agent of Rocky Mountain spotted fever (RMSF) and other rickettsial species known to be pathogenic for humans (Table 10.2). Each of these pathogenic SFG rickettsiae is associated with a geographically distinct group of tick vectors and mammalian hosts (Burgdorfer 1980).

The only tickborne species of *Rickettsia* known to be pathogenic for humans in North America is *R. rickettsii*. The ecology and epidemiology of this rickettsia resembles other tickborne SFG rickettsiae and can be used as a general example. *Rickettsia rickettsii* is maintained in nature by transstadial and transovarial transmission within ticks, and horizontal transmission between certain ticks and their susceptible mammal hosts (Fig. 10.6). Venereal transmission of rickettsiae from infected male ticks to female ticks during mating does not appear to be an important means of transmission.

Transstadial and transovarial transmission are both efficient, with rates of nearly 100% for the former and 30–100% for the latter (Burgdorfer and Brinton 1975). If rickettsiae are to be transmitted transstadially or transovarially within ticks, they first must be ingested by the tick as it feeds. After entering the digestive tract of the tick, some rickettsiae will invade the cells of the midgut epithelium, multiply within these cells, and then escape into the hemocoel. Rickettsiae that penetrate the hemocoel can, in turn, infect hemocytes and be carried to other tissues within the tick, including those in the reproductive organs and salivary glands. Infection of the ovaries is essential for transovarial transmission. Likewise, transmission of rickettsiae from infected ticks to susceptible mammals (as described below) during tick feeding depends on the rickettsia's ability to establish itself successfully in the salivary glands.

Despite the obvious importance of transstadial and transovarial transmission, evidence exists that life expectancies for *R. rickettsii*-infected ticks are lower than those for noninfected ticks, and that infected females lay fewer viable eggs than noninfected females. These observations led McDade and Newhouse (1986) to conclude that long-term maintenance of *R. rickettsii* in nature requires not only transstadial and transovarial transmission but also horizontal transmission between ticks and certain mammal species. Horizontal transmission occurs when infectious ticks feed upon certain mammal species (primarily rodents), become infected with rickettsiae, and eventually circulate sufficient numbers of rickettsiae to serve as sources of infection for additional ticks that feed while the host is rickettsemic. Infection rates for *R. rickettsii* in *D. andersoni* range from 1–13.5% in western Montana and from less than 1% to slightly more than 10% in the eastern USA (Burgdorfer 1980).

In North America, the primary vectors of the fully virulent (R-like) strains of *R. rickettsii* are *D. andersoni* and *Dermacentor variabilis*. Immature stages of these ticks feed primarily on small- to medium-sized mammals. Among the mammalian hosts of immature *D. variabilis* and *D. andersoni* are certain rodent species that become rickettsemic when infected and act as amplifying hosts for transmitting rickettsiae to noninfected ticks. Although some of these amplifying hosts occasionally die following infection, many survive and are believed to be immune to reinfection with *R. rickettsii* and probably other related SFG rickettsiae (Gage and Jerrells 1992). Adults of *D. andersoni* and *D. variabilis* feed on medium- to large-sized hosts, including humans. Risk of exposure to *R. rickettsii* infection is greatest for humans during the spring months when adult *D. andersoni* or *D. variabilis* actively quest for hosts. Immature stages of these ticks are not important sources of transmission to humans.

Small numbers of human cases of a spotted fever-like illness also have been identified by serological means in California. The vector involved with these illnesses has not been identified conclusively, but might be *Dermacentor occidentalis*. A rickettsial serotype (364-D) closely related to R-like strains of *R. rickettsii* has been isolated from these ticks (Philip et al. 1981).

Amblyomma americanum, a common tick pest in the southeastern and south-central USA, is cited often as a vector of RMSF, but little evidence verifies its importance. Burgdorfer et al. (1974) failed to find evidence of *R. rickettsii* infections among 952 *A. americanum* ticks they examined from Tennessee. However, they did identify *R. rickettsii* in slightly more than 4% of the *D. variabilis* ticks sampled from the same area. Goddard and Norment (1988) also failed to find rickettsial infection in *A. americanum* ticks collected in the south-central USA, but they noted that *A. americanum* ticks often are infected with either *Rickettsia parkeri* or another SFG rickettsia referred to as WB-8-2. Neither of these rickettsiae is thought to be pathogenic for humans, although *R. parkeri* causes recognizable illness in guinea pigs.

In addition, strains of a second, less virulent serotype of *R. rickettsii* (Hlp-like serotype) occur in North America but are not known to cause human disease. These strains commonly are identified in the rabbit tick, *Haemaphysalis leporispalustris*, but also have been reported from *D. andersoni*.

Rickettsia rickettsii also has been identified in wild mammals and ticks in Central and South America. The primary vector in these regions appears to be *Amblyomma cajennense*. *H. leporispalustris* ticks in Central America, and *Amblyomma striatum* ticks in South America also have been infected (Fuentes et al. 1985). Some have claimed that *Rhipicephalus sanguineus* (European brown dog tick) transmits the disease in Mexico, but this hypothesis should be reevaluated.

As noted earlier for *A. americanum*, tick-rickettsia surveys often identify apparently nonpathogenic species of *Rickettsia* in various species of ticks. *Dermacentor andersoni* and *D. variabilis* alone are known to be hosts for at least 6 rickettsial serotypes or species, including *R. rickettsii* that are fully virulent for guinea pigs (R-like strains), the less virulent Hlp-like *R. rickettsii*, and serotypes such as *Rickettsia montana*, *Rickettsia rhipicephali*, *Rickettsia bellii* and *Rickettsia peacocki* (East Side Agent), which apparently are of low virulence or nonvirulent for guinea pigs (Gage et al. 1994, Niebyslki et al. 1997). Among these rickettsiae, only the R-like serotype of *R. rickettsii* is thought to be pathogenic for humans.

The biologies of the species or serotypes of presumably nonpathogenic SFG rickettsiae listed above are not well-understood, but they could be epidemiologically important because they might affect the distribution of virulent R-like strains of *R. rickettsii*. Specifically,

Burgdorfer et al. (1981) reported that *D. andersoni* ticks infected with East Side Agent could be superinfected with a strain of the virulent R-like serotype of *R. rickettsii*, but the latter rickettsia did not invade the ovaries of these ticks and, therefore, was not transmitted transovarially. This discovery led Burgdorfer et al. (1981) to propose that such "interference" between nonpathogenic and pathogenic rickettsiae could affect the geographic distribution of pathogenic species, including *R. rickettsii*.

In addition to the SFG rickettsiae discussed earlier, at least 2 additional species of non-SFG tickborne rickettsiae occur in North America. The first of these, *Rickettsia canada*, generally is considered to be a member of the TG. To date, it has been identified only in rabbit ticks (*H. leporispalustris*) collected in Ontario, Canada, and in California (Philip et al. 1982). This rickettsia is not believed to be pathogenic for humans, nor does it appear to cause illness in rabbits and snowshoe hares, which are the major hosts of *H. leporispalustris* ticks. Presumably, *R. canada* is maintained in nature primarily through transstadial and transovarial transmission in its tick host.

The other non-SFG rickettsia commonly reported from ticks in North America is *R. bellii*, which was first isolated from *D. variabilis* collected in Arkansas (Philip et al. 1983). Evidence indicates that this rickettsia does not belong to any of the established rickettsial serogroups. Limited tick surveys suggest that *R. bellii* is relatively common in both *D. andersoni* and *D. variabilis* populations in different geographic regions of the USA. This rickettsia is not known to cause disease in humans and is reported to be nonpathogenic for guinea pigs and voles.

Humans most likely contract tickborne rickettsial diseases when they intrude into the tick vectors' natural habitats and are exposed to questing and infected ticks. In the northern Rocky Mountain region (western Montana and nearby areas), most cases of RMSF occur among persons who have participated in outdoor recreational activities or have worked in mountain forests. Although most of the cases reported during the first few decades of the 20th Century occurred in the Rocky Mountains, this region now accounts for only a small percentage of the cases reported within the USA. At present, most cases occur in the southeastern and south-central USA. The increased incidence of the disease in these states over the past few decades is not thoroughly understood but may be related to increased suburbanization and other changes in land use, or to increased recognition and reporting of the disease.

Rickettsia conorii and a number of closely related serotypes are the etiologic agents of Boutonneuse fever and similar illnesses in certain regions of southern Europe, the Middle East, Africa and India. The primary vector of Boutonneuse fever is *R. sanguineus*. Other reported vectors include additional *Rhipicephalus* species and various *Amblyomma, Haemaphysalis, Hyalomma, Boophilus, Ixodes* and *Dermacentor* ticks. The wide range of vectors probably is due to the presence of previously unrecognized serotypes that have been confused with *R. conorii* (Kelly et al. 1994). Other tickborne SFG rickettsial pathogens include *Rickettsia sibirica, Rickettsia australis* and *Rickettsia japonica*. *Rickettsia sibirica*, the etiologic agent of North Asian tick typhus, is transmitted primarily by various species of *Dermacentor* and *Haemaphysalis*. *Rickettsia australis* occurs in northeast Australia and causes Queensland tick typhus. The primary vector of this rickettsia is *Ixodes holocyclus*, a tick that feeds mainly on small marsupials, but also will parasitize humans or other vertebrates. A previously unrecognized rickettsial disease caused by *R. japonica* (Oriental spotted fever) was discovered in 1986 (Walker and Fishbein 1991). Among the potential vectors are *Dermacentor taiwanensis* and *Haemaphysalis longicornis* (Takada et al. 1994, Uchida et al. 1995). Evidence indicates that ticks in regions of the world outside North America also are infected with a variety of presumably nonpathogenic rickettsiae.

Miteborne rickettsiae

Rickettsia akari is a miteborne member of the spotted fever group that causes rickettsialpox, a relatively mild rickettsiosis. Unlike the various tickborne spotted fevers, cases of rickettsialpox are encountered most commonly in urban settings, such as New York City, where human habitations are heavily infested with mice (Brettman et al. 1984, Lackman 1963). The vector of *R. akari* is *Liponyssoides sanguineus*, a mite commonly found infesting house mice (*Mus musculus*). Transstadial and transovarial transmission of *R. akari* occurs in the mite vector. Transmission from nymphal or adult *L. sanguineus* to susceptible mice also has been reported. Larval *L. sanguineus* do not feed on blood and, therefore, do not transmit *R. akari* to mice or humans and cannot become infected by feeding on rickettsemic mice. Although mice are infected readily with *R. akari*, they typically recover from these infections.

The global distribution of *R. akari* is poorly documented and little is known about possible cycles of infection involving species of mammals other than *M. musculus*. Isolates have been reported from the USA and the former Soviet Union, and the disease may occur wherever populations of *M. musculus* are infested with *L. sanguineus* mites. A single isolate of *R. akari* has been reported from a field vole (*Microtus fortis pelliceus*) in South Korea (Burgdorfer 1980), but the significance of this observation has yet to be determined.

Louseborne rickettsiae

Rickettsia prowazekii, the etiologic agent of louseborne or epidemic typhus, belongs to the TG of the genus *Rickettsia*. The signs and symptoms of the disease generally resemble those of other rickettsial infections, although the rash is more likely to appear on the trunk than on the extremities, as is common for illnesses caused by tickborne SFG rickettsiae. Louseborne typhus is treatable with the same antibiotics effective against SFG rickettsiae.

The body louse, *P. humanus humanus* transmits *R. prowazekii* from human to human and the disease usually is associated with persons living in conditions of poverty. Head lice (*Pediculus capitus*) are not considered important vectors of *R. prowazekii*, but they can become infected. Although louseborne typhus outbreaks now are restricted to a few impoverished areas in developing countries, the disease once was a feared killer in Europe and other regions, especially during times of war and famine (Zinsser 1934). Millions perished from louseborne typhus in eastern Europe during and after World War I.

Rickettsia prowazekii is pathogenic not only for its human hosts but also for its louse vector. Lice become infected after ingesting blood containing viable *R. prowazekii*. Once ingested, the rickettsiae invade the epithelial cells lining the louse's gut. This results in the destruction of these cells and the eventual death of the louse vector. Following the destruction of infected gut epithelial cells, large numbers of rickettsiae are released into the gut lumen where they can either invade other cells or be passed from the gut along with the louse's feces. Humans become infected when they rub this infectious fecal material into the feeding site or other breaks in the skin, or when they crush lice and contaminate their skin with *R. prowazekii*-infected material. Unlike *R. rickettsii* or other SFG rickettsiae in ticks, *R. prowazekii* typically does not cause a generalized infection in lice.

Lice are of little importance as reservoirs of infection because they live for less than two weeks after becoming infected with *R. prowazekii* and do not pass this rickettsia to their offspring via transovarial transmission. The actual reservoirs of the disease are humans, who apparently become chronically infected with *R.*

prowazekii unless adequately treated. When chronically infected persons become elderly or immunocompromised, they can suffer an attack of Brill-Zinsser disease, a form of recrudescent typhus that can occur many years to decades after the host's primary attack. Although these recrudescent cases generally are much milder than the initial illness, they can result in circulation of sufficient rickettsiae to serve as sources of infection for 1–5% of body lice feeding on these individuals (Walker and Gage 1997). Thus, people experiencing a case of Brill-Zinsser disease can serve as sources of new outbreaks of louseborne typhus in areas where the disease has been absent for many years. Such outbreaks are most likely to occur when many people are lousy and personal hygiene practices are interrupted because of war, famine, social unrest or other causes.

Although humans may be reservoirs of *R. prowazekii*, some researchers have attempted to identify extrahuman infection cycles. In general, they have failed or provided ambiguous results from serological tests (Burgdorfer et al. 1973). However, McDade et al. (1980) identified TG infections in eight individuals in the eastern USA who had typhus-like illness. None of these patients was believed to have been exposed to *P. humanus humanus*, but at least two had likely exposure to flying squirrels (*Glaucomys* spp.) or their ectoparasites. Bozeman et al. (1975, 1981) previously had succeeded in isolating a rickettsial agent from flying squirrels and their lice that appeared to be similar if not identical to *R. prowazekii* in serological assays. Additional analyses have suggested that this rickettsia is indeed *R. prowazekii* (McDade 1987).

Fleaborne rickettsiae

Rickettsia typhi (*Rickettsia mooseri*) is a fleaborne rickettsia that causes an illness similar to, but somewhat milder than, louseborne typhus. This illness, referred to as murine, fleaborne, or endemic typhus, appears throughout much of the world because of its association with *Rattus* spp. and their fleas (primarily *X. cheopis*, the Oriental rat flea). The disease rarely is fatal and often is misdiagnosed. Murine typhus appears to be common in many tropical and warm temperate regions but surveillance for it is almost nonexistent. Humans can become infected when *R. typhi*-infected fleas contaminate the feeding site or skin abrasions with infectious feces. In addition, Farhang-Azad et al. (1985) have presented evidence that transmission can occur by bite as well.

The primary maintenance cycle of *R. typhi* involves transmission between commensal rats and their fleas, particularly *X. cheopis*. There are reports of *R. typhi* remaining viable in brain tissues of rats for months after infection, but this could not be confirmed by others (Azad 1990). However, once adult fleas become infected, they probably remain so throughout the remainder of their lives.

Other rodents, as well as shrews, skunks, opossums and cats, also can become involved, along with their fleas, in the natural cycle of *R. typhi*. In addition to *X. cheopis*, certain fleas have been reported to be actual or potential vectors of *R. typhi*, including the cosmopolitan species *C. felis*, *Echidnophaga gallinacea*, *L. segnis*, *N. fasciatus*, and *P. irritans*. Other reported vectors include *Xenopsylla astia* in southern Asia and the eastern coast of Africa, *Xenopsylla brasiliensis* in Africa, South America and India, and *Monopsyllus anisus* in northern Asia (Azad 1990). In most instances, however, cycles involving mammals other than *Rattus* spp. or fleas other than *X. cheopis* probably are of secondary importance. Williams et al. (1992) reported a natural cycle in California involving opossums (*Didelphis virginianus*) and cat fleas (*C. felis*). Unlike *R. prowazekii* infections in humans and the human body louse, the murine typhus agent does not appear to be markedly pathogenic for either its rat hosts or its flea vectors.

After being ingested by a feeding flea, *R. typhi* enters the epithelial cells of the midgut where it multiplies and eventually escapes to the gut. These rickettsiae can, in turn, infect other epithelial cells or be released with the feces. Under favorable conditions, *R. typhi* can survive for as long as 3 days in flea feces, which increases the likelihood that viable rickettsiae eventually will be able to infect the host through the feeding site or an abrasion. In some instances, *R. typhi* apparently escapes to the hemocoel where it can become established in the ovaries of *X. cheopis* fleas, resulting in transovarial transmission (Farhang-Azad et al. 1985). The importance of such transovarial transmission is poorly understood and needs to be investigated further, but could represent an additional means of maintaining *R. typhi* in nature.

In most instances, humans acquire fleaborne typhus while working or living in areas infested with commensal rats and their fleas. In the USA, most cases occur in California or the southeastern and south-central states, particularly in ports along the Gulf Coast of Texas.

Recently, a previously unrecognized rickettsia distinct from *R. typhi* was identified in the ovaries and other tissues of infected *C. felis* fleas (Azad et al. 1992). Transovarial transmission of this rickettsia is reported to occur regularly within infected cat flea colonies. This agent also has been identified in opossums (*D. virginianus*) in Texas (Schriefer et al. 1994b) and was suggested to be the cause of a mild typhus-like illness in humans (Azad et al. 1997, Schriefer et al. 1994a). Recently, the name *Rickettsia felis* has been proposed for this organism (Higgins et al. 1996).

Scrub Typhus

Orientia tsutsugamushi, the etiologic agent of scrub typhus, is a miteborne rickettsia found in southeast Asia and nearby islands. Originally, this organism was classified within the genus *Rickettsia* because of its obligate intracellular nature, association with an arthropod vector, and the fact that it causes a rickettsia-like illness that develops a rash like that caused by most species of *Rickettsia*. However, these illnesses resemble each other because the primary target cells for both *O. tsutsugamushi* and *Rickettsia* spp. are endothelial cells lining the host's blood vessels. Scrub typhus organisms have been placed in their own genus (*Orientia*) because of significant differences in cell wall composition, including the lack of lipopolysaccharide, peptidoglycans and a slime layer (Walker and Gage 1997).

The larval stages of various species of trombiculid mites (chiggers) within the genus *Leptotrombidium* transmit *O. tsutsugamushi* to humans and other mammals. Nymphal and adult *Leptotrombidium* do not transmit the scrub typhus agent because, unlike the larvae, they are not parasitic on vertebrates, but instead feed on the eggs, quiescent stages or fresh carcasses of other small ground-dwelling arthropods. The principal hosts of the parasitic *Leptotrombidium* larvae are various species of field rats (*Rattus* spp.). In some regions, other small rodents (primarily voles and mice), insectivores and even birds can be important hosts for larvae.

The only known reservoirs of *O. tsutsugamushi* are the mite vectors, which maintain the rickettsiae via both transstadial and transovarial transmission. Attempts to infect larval *Leptotrombidium* by feeding them on infected mammalian hosts have caused transient infections in these larvae but have not established *O. tsutsugamushi* in the ovaries of infected females. This failure suggests that feeding of larvae on infected mammals plays little role in long-term maintenance of infection (Traub and Wisseman 1974, Takahashi et al. 1994). The possibility of venereal transmission of *O. tsutsugamushi* from males to females through transfer of infected spermatophores has yet to be investigated adequately.

Scrub typhus is a focal disease that persists only in areas that have certain well defined features. The basic elements that must be present for a focus to persist are *O. tsutsugamushi*, *Leptotrombidium deliense*-group mites, certain small mammal hosts (especially field rats of the genus *Rattus*), and "scrub" or transitional vegetation. The focality of scrub typhus depends upon the habitat requirements of the *Leptotrombidium* vectors. Optimal mite habitats have high humidity, moderate to high air temperatures, abundant hosts for the parasitic larval stages, and moist soil conditions that provides abundant prey for the free-living nymphal and adult stages. Such conditions are best met in those areas where the normal climax stage vegetation has been removed and replaced by transitional secondary (scrub) vegetation comprised of various mixtures of grasses, herbs and small woody species. Human activities likely to promote suitable scrub vegetation are clearing land for agriculture, logging operations and abandonment of fields.

EHRLICHIAE

Ehrlichia spp. are gram-negative intracellular bacteria that grow within membrane-bound vacuoles in circulating leukocytes or platelets (Walker and Gage 1997). Multiplication of ehrlichiae within host-cell vacuoles results in the formation of grape-like clusters, or morulae, that characterize the genus. The types of host cells infected vary between the different species of *Ehrlichia*, with some infecting primarily monocytes (*Ehrlichia canis*, *Ehrlichia chaffeensis*, *Ehrlichia sennetsu* and *Ehrlichia risticii*) and others occurring in granulocytes (*Ehrlichia equi*, *Ehrlichia phagocytophila* and *Ehrlichia ewingii*). One species, *Ehrlichia platys*, a pathogen of dogs, is found in platelets.

Ehrlichia spp. have long been known to cause disease in domestic animals such as dogs (*E. canis*, *E. ewingii*, *E. platys*), horses (*E. risticii*, *E. equi*), sheep and cattle (*E. phagocytophila*). One species, *E. sennetsu*, was reported to cause human disease in Japan as early as 1954 (Ristic et al. 1991). The latter species, however, is not believed to be transmitted by arthropods and recent molecular studies suggest that it is not closely related to other species of *Ehrlichia*.

Apparently, arthropodborne species of *Ehrlichia* were not identified as causative agents of human disease until 1986 when a case of ehrlichiosis occurred in a 51 year-old man with probable tick exposure in Arkansas (Maeda et al. 1987, Fishbein and Dawson 1991). Initially, these infections were thought to be caused by *E. canis*, the agent of a severe hemorrhagic disease of dogs that is transmitted by *R. sanguineus*. The etiologic agent of this and other ehrlichial infections contracted in the southern USA is believed to be *E. chaffeensis*, a previously unrecognized species that first was described from human blood samples by Anderson et al. (1991). The disease associated with *E. chaffeensis* infection now is referred to as human monocytic ehrlichiosis (HME). Common symptoms include fever, headache, anorexia, rigors, thrombocytopenia, nausea, myalgia and elevated production of certain liver enzymes (Walker and Fishbein 1991). The disease can be treated with certain antibiotics, including tetracycline. Fatality rates for HME have been reported to be 2–3% (Walker and Dumler 1996). Reducing exposure to ticks may prevent infection.

The ecology of *E. chaffeensis* is poorly understood. At present, the most likely vertebrate reservoirs appear to be white-tailed deer (*O. virginianus*). Deer living in the southeastern and south-central USA are reported to have antibodies reactive with *E. chaffeensis*. The vector of *E. chaffeensis* also has not been established, but an ehrlichial species presumed to be *E. chaffeensis* has been detected by polymerase chain reaction (PCR) in *A. americanum* ticks from Missouri, North Carolina, Kentucky and New Jersey. Additionally, most cases of HME have occurred within the range of *A. americanum*. A single PCR-positive *D. variabilis* also has been reported (Walker and Dumler 1996). Some HME cases have been reported from areas where this tick occurs but *A. americanum* does not. A few HME cases have been reported from areas outside the known ranges of both *A. americanum* and *D. variabilis*, yet it is not clear whether these cases have been transmitted by a different tick vector, misdiagnosed, or infected with another unrecognized ehrlichial agent.

Soon after *E. chaffeensis* was identified as the etiologic agent of HME, another ehrlichia closely related to or synonymous with *E. equi* was implicated as the cause of a different form of ehrlichiosis in humans, now referred to as human granulocytic ehrlichiosis (HGE) (Bakken et al. 1994, Goodman et al. 1996). As the name implies, the organisms are found primarily in granulocytes rather than monocytes. Symptoms of HGE usually include fever, chills, malaise, headache, myalgia and occasionally nausea, vomiting, cough and confusion.

Most cases of HGE occurred in the northeastern and upper midwestern USA, but another cluster of cases has been reported from northern California. The geographic occurrence of HGE resembles that observed for Lyme disease. Current evidence indicates that the two diseases are distributed similarly because both are transmitted by *Ix. scapularis* and probably *Ix. pacificus* ticks in the northeastern and upper midwestern states, and northern California, respectively. Pancholi et al. (1995) identified the HGE agent in a tick removed from a patient and from *Ix. scapularis* ticks collected in Wisconsin. Telford et al. (1996) demonstrated that white-footed mice (*P. leucopus*) can serve as a source of infection for laboratory-reared *Ix. scapularis*. In addition, they found that ticks infected by feeding on ehrlichemic mice can transmit the organism transstadially and also to mice during subsequent feedings. These authors also identified ehrlichiae in salivary glands of field-collected and laboratory-infected *Ix. scapularis*.

HEARTWATER

Cowdria ruminantium, a tickborne intracellular bacterium closely related to *Ehrlichia* spp., causes a serious disease of wild and domestic ruminants in sub-Saharan Africa. Heartwater also has been identified on certain islands in the Indian Ocean, Atlantic Ocean and Caribbean Sea where *Amblyomma variegatum* has been introduced (Yunker 1996). The introduction of *C. ruminantium* and suitable vector ticks to the Caribbean has raised concerns that the disease will spread in the tropical and subtropical regions of North and South America (Barre et al. 1987). Mortality among cattle, sheep and goats can range as high as 90%, but usually is less for indigenous stock than for recently imported animals. Wild ruminants, such as blesbuck, eland and wildebeest also suffer mortality from heartwater. The primary target for *C. ruminantium* is the endothelium of the heart, brain and spleen, but lymphocytes and phagocytes also have been found infected. Destruction of the endothelial target cells leads to hydropericardium or "heartwater," the characteristic pathology of the disease. Symptoms can include high fever, anorexia, lacrimation, depression, convulsions and other neurological signs. Very young animals of some species (<9 days old) appear to resist infection. *Cowdria ruminantium* can cause chronic infections in both domestic and wild ruminants, but the natural reservoir hosts are thought to be various species of wild African ruminants. The primary tick vectors are *Amblyomma hebraeum* and *A. variegatum*. Transstadial transmission of *C. ruminantium* is known to occur, and transovarial transmission has been reported (Bezuidenhout and Jacobsz 1986), but is thought by some to be rare or unimportant. Adult male *A. hebraeum* ticks reportedly can feed briefly on an infected host, detach, and then reattach and transmit the heartwater agent to another animal. The most common control technique is dipping cattle for ticks, but an intentional infection and follow-up antibiotic treatment method is used commonly in South Africa and efforts are underway to develop an effective vaccine (Bezuidenhout 1993).

ANAPLASMOSIS

Anaplasma spp. are intracellular bacteria that infect wild and domestic ungulates (Kuttler 1984). The most important species are *Anaplasma marginale*, *Anaplasma ovis*, and *Anaplasma centrale*. The primary vectors are ticks, but mechanical transmission via biting flies or contaminated fomites also has been reported. *Anaplasma* spp. can be transmitted both transstadially and transovarially in its tick vectors (Kocan 1986). Adult males also can feed briefly on infected hosts, acquire the agent, detach and then reattach to another host, thereby transmitting the disease. At least 20 different ticks have been reported to be infected, including species of *Dermacentor* and *Boophilus*. Mortality among infected ungulates can approach 50%, but subacute and chronic infections also have been reported, suggesting that both the vertebrate hosts and tick vectors are significant reservoirs of infection. Animals older than 18 months of age typically develop the most severe symptoms, which include primarily anorexia and anemia. Control and prevention methods include tick and biting fly control, prophylactic antibiotic therapy and vaccination.

Q FEVER

The etiologic agent of Q fever, *Coxiella burnetii*, is an obligate intracellular bacterium that commonly occurs in a wide variety of vertebrates, especially wild and domestic ungulates. Many species of ticks, fleas, lice and mites also have been reported to be infected. The most significant of these potential arthropod hosts probably are various ticks, including those within the ixodid genera *Dermacentor*, *Rhipicephalus*, *Haemaphysalis*, *Ixodes*, *Hyalomma* and *Amblyomma*. Argasid ticks within the genera *Ornithodoros*, *Argas*, *Boophilus* and *Otobius* also have been reported to be infected. Various authors differ on the ecological importance of *C. burnetii* infections in ticks or other ectoparasites. Some feel that these hematophagous arthropods simply ingest *C. burnetii* and briefly maintain it without subsequent transmission. Others believe that ticks are important reservoirs as indicated by reports of transstadial and transovarial transmission, transmission to susceptible mammals during tick feeding, and possible transmission via contaminated tick feces. *C. burnetii* reportedly can remain viable in tick feces for as long as a year (Stoenner 1980).

Although many human *C. burnetii* infections appear to be asymptomatic, acute Q fever in humans can cause fever, pneumonitis, hepatic and bone marrow granulomas, meningoencephalitis and endocarditis. In its chronic form, the disease can result in glomerulonephritis, osteomyelitis and life-threatening endocarditis.

Most humans become infected after contacting contaminated and infectious livestock products or wastes, including unpasteurized milk, urine, feces, hides, wool or placental material and other birth products. Persons also can be infected by inhaling infectious aerosols arising from these materials. Transmission to humans by the bites of infected arthropods is thought to be of little or no significance in the epidemiology of the disease.

REFERENCES

Anderson, B.E., Dawson, J.E., Jones, D.C. and Wilson, K.H. 1991. *Ehrlichia chaffeensis*, a new species associated with human ehrlichiosis. J. Clin. Microbiol. **29**:2838–2842.

Azad, A.F. 1990. Epidemiology of murine typhus. Annu. Rev. Entomol. **35**:553–569.

Azad, A.F., Radulovic, S., Higgins, J.A., Noden, B.H. and Troyer, J.M. 1997. Flea-borne rickettsioses: ecological considerations. Emerg. Infect. Dis. **3**:319–328.

Azad, A.F., Sacchi, J.B., Jr., Nelson, W.M., Dasch, G.A., Schmidtmann, E.T. and Carl, M. 1992. Genetic characterization and transovarial transmission of a typhus-like rickettsia found in cat fleas. Proc. Nat. Acad. Sci. **89**:43–46.

Bacot, A.W. and Martin, C.J. 1914. Observations on the mechanism of the transmission of plague by fleas. J. Hyg. (Cambridge) Plague Supplement **3**:423–439.

Bakken, J.S., Dumler, J.S., Chen, S.M., Eckman, M.R., Van Etta, L.L. and Walker, D.H. 1994. Human granulocytic ehrlichiosis in the upper midwest – a new species emerging? J. Am. Med. Assoc. **272**:212–218.

Baranton, G., Postic, D., Saint Girons, I., Boerlin, P., Piffaretti, J.C., Assous, M. and Grimond, P.A.D. 1992. Delineation of *Borrelia burgdorferi* sensu stricto, *Borrelia garinii* sp. nov., and group VS461 associated with Lyme borreliosis. Internat. J. System. Bacteriol. **42**:378–383.

Barbour, A.G. 1988. Antigenic variation of surface proteins of *Borrellia* species. Rev. Infect. Dis. **10**:S399–S402.

Barbour, A.G., Maupin, G.O., Teltow, G.J., Carter, C.J. and Piesman, J. 1996. Identification of an uncultivable *Borrelia* species in the hard tick *Amblyomma americanum*: possible agent of a Lyme disease-like illness. J. Infect. Dis. **173**:403-407.

Barnes, A.M. 1982. Surveillance and control of bubonic plague in the United States. Pp. 237–270 *in* M.A. Edwards and U. McDonnel (eds.), Animal Disease in Relation to Animal Conservation., Vol. 50. Symposium of the Zoological Society of London.

Barre, N.G., Uilenberg, G., Morel, P.C. and Camus, E. 1987. Danger of introducing heartwater onto the American mainland: potential role of indigenous and exotic *Amblyomma* ticks. Onderstepoort J. Vet. Res. **54**:405–417.

Bezuidenhout, J.D. 1993. The significance of recent highlights in heartwater research. Review Elev. Med. Vet. Pays. Trop. **46**:101–108.

Bezuidenhout, J.D. and Jacobsz, C.J. 1986. Proof of transovarial transmission of *Cowdria ruminantium* by *Amblyomma hebraeum*. Onderstepoort J. Vet. Res. **53**:31–41.

Bibikova, V.A. 1977. Contemporary views on the interrelationships between fleas and the pathogens of human and animal diseases. Annu. Rev. Entomol. **22**:23–32.

Boyer, K.M., Munford, R.S., Maupin, G.O., Pattison, C.P., Fox, M.D., Barnes, A.M., Jones, W.L. and Maynard, J.L. 1977. Tick-borne relapsing fever: an interstate outbreak originating at Grand Canyon National Park. Am. J. Epidemiol. **105**:469–479.

Bozeman, F.M., Masiello, S.A., Williams, M.S. and Elisberg, B.L. 1975. Epidemic typhus rickettsiae isolated from flying squirrels. Nature **255**:545–547.

Bozeman, F.M., Sonenshine, D.E., Williams, M.S., Chadwick, D.P., Lauer, D.M. and Elisberg, B.L. 1981. Experimental infection of ectoparasitic arthropods with *Rickettsia prowazekii* (GvF-16 strain) and transmission to flying squirrels. Am. J. Trop. Med. Hyg. **30**:253–263.

Brenner, D.J., O'Connor, S.P., Winkler, H.H. and Steigerwalt, A.G. 1993. Proposals to unify the genera *Bartonella* and *Rochalimaea*, with descriptions of *Bartonella quintana* comb. nov., *Bartonella vinsonii* comb. nov., *Bartonella henselae* com. nov., and *Bartonella elizabethae* comb. nov., and to remove the family *Bartonellaceae* from the order *Rickettsiales*. Internat. J. System. Bacteriol. **43**:777–786.

Brettman, L.R., Lewin, S., Holzman, R.S., Goldman, W.D., Marr, J.S., Kechijian, P. and Schinella, R. 1984. Rickettsialpox: report of an outbreak and a contemporary review. Medicine **60**:363–371.

Brown, R.N. and Lane, R.S. 1992. Lyme disease in California: a novel enzootic transmission cycle of *Borrelia burgdorferi*. Science **256**:1439–1442.

Brumpt, E. 1933. Etude du *Spirochaeta turicatae*, n. sp., agent de la fievre recurrente sporadique des Etats-Unis transmise par *Ornithodorus turicata*. C.R. Sean. Soc. Biol. **113**:1369–1372.

Burgdorfer, W. 1951. Analyse des infektionsverlaufes bei *Ornithodoros moubata* (Murray) und der naturlichen uebertragung von *Spirochaeta duttonii*. Acta Trop. **8**:193–262.

Burgdorfer, W. 1976. The epidemiology of relapsing fevers. Pp. 191–200 *in* R.C. Johnson (ed.), The biology of parasitic spirochetes. Academic Press, New York.

Burgdorfer, W. 1980. The spotted fever-group diseases. Pp. 279–301 *in* J.H. Steele (ed.), Zoonoses, Vol. 1. CRC Press, Boca Raton, Florida.

Burgdorfer, W., Barbour, A.G., Hayes, S.F., Benach, J.L., Grunwaldt, E. and Davis, J.P. 1982. Lyme disease—a tick-borne spirochetosis? Science **216**:1317–1319.

Burgdorfer, W. and Brinton, L.P. 1975. Mechanisms of transovarial infection of spotted fever rickettsiae in ticks. Ann. N.Y. Acad. Sci. **266**:61–72.

Burgdorfer, W., Cooney, J.C. and Thomas, L.H. 1974. Zoonotic potential (RMSF and tularemia) in the Tennessee Valley region. II. Prevalence of *Rickettsia rickettsii* and *Francisella tularensis* in mammals and ticks from the Land Between the Lakes. Am. J. Trop. Med. Hyg. **23**:109–117.

Burgdorfer, W., Hayes, S.F. and Mavros, A.J. 1981. Nonpathogenic rickettsiae in *Dermacentor andersoni*: a limiting factor for the distribution of *Rickettsia rickettsii*. Pp. 585–594 *in* W. Burgdorfer and R.L. Anacker (eds.), Rickettsiae and Rickettsial Diseases. Academic Press, New York.

Burgdorfer, W., Ormsbee, R.A., Schmidt, M.L. and Hoogstraal, H. 1973. A search for the epidemic typhus agent in Ethiopian ticks. Bull. Wld. Hlth. Organiz. **48:**563–569.

Cavanaugh, D.C. 1971. Specific effect of temperature upon transmission of the plague bacillus by the Oriental rat flea, *Xenopsylla cheopis*. Am. J. Trop. Med. Hyg. **20:**264–273.

Centers for Disease Control and Prevention. 1996. Prevention of plague. Recommendations of the Advisory Committee on Immunization Practices (ACIP). Morbidity and Mortality Weekly Report – Recommendations and Reports **45:**RR No.14.

Chomel, B.B., Kasten, R.W., Floyd-Hawkins, K., Chi, B., Yamamoto, K., Roberts-Wilson, J., Gurfield, A.N., Abbott, R.C., Pedersen, N.C. and Koehler, J.E. 1996. Experimental transmission of *Bartonella henselae* by the cat flea. J. Clin. Microbiol. **34:**1952–1956.

Clover, J.R. and Lane, R.S. 1995. Evidence implicating nymphal *Ixodes pacificus* (Acari: Ixodidae) in the epidemiology of Lyme disease in California. Am. J. Trop. Med. Hyg. **53:**237–240.

Curran, K.L., Fish, D. and Piesman, J. 1993. Reduction of nymphal *Ixodes dammini* (Acari: Ixodidae) in a residential suburban landscape by area application of insectides. J. Med. Entomol. **30:**107–113.

Daniels, T.J. and Fish, D. 1995. Effect of deer exclusion on the abundance of immature *Ixodes scapularis* (Acari: Ixodidae) parasitizing small and medium-sized mammals. J. Med. Entomol. **32:**5–11.

Davis, G.E. 1956. The identification of spirochetes from human cases of relapsing fever by xenodiagnosis with comments on local specificity of tick vectors. Exp. Parasitol. **5:**271–275.

Davis, G.E. and Burgdorfer, W. 1955. Relapsing fever spirochetes: an aberrant strain of *Borrelia parkeri* from Oregon. Exp. Parasitol. **4:**100–106.

Eskey, C.R. and Haas, V.H. 1940. Plague in the western part of the United States. Publ. Hlth. Bull. **254:**1–82.

Farhang-Azad, A. and Traub, R. 1985. Transmission of murine typhus rickettsiae by *Xenopsylla cheopis*, with notes on experimental infection and effects of temperature. Am. J. Trop. Med. Hyg. **34:**555–563.

Farhang-Azad, A., Traub, S. and Baqar, S. 1985. Transovarial transmission of murine typhus rickettsiae in *Xenopsylla cheopis* fleas. Science **227:**543–545.

Felsenfeld, O. 1979. Borreliosis. Pp. 79–96 *in* J.H. Steele (ed.), Bacterial, Rickettsial, and Mycotic Diseases, Vol. 1. CRC Press, Boca Raton, Florida.

Fishbein, D.B. and Dawson, J.E. 1991. Ehrlichiae. Pp. 1054–1058 *in* A. Balows, W.J. Hauser, Jr., K.L. Herrman, H.D. Isenberg and H.J. Shadomy (eds.), Manual of Clinical Microbiology., 5th ed. Am. Soc. Microbiol., Washington, DC.

Fuentes, L., Calderon, A. and Hun, L. 1985. Isolation and identification of *Rickettsia rickettsii* from the rabbit tick (*Haemaphysalis leporispalustris*) in the Atlantic zone of Costa Rica. Am. J. Trop. Med. Hyg. **34:**564–567.

Gage, K.L. 1998. Plague. Pp. 885–904 *in* L. Collier, A. Balows, M. Sussman and W.J. Hausler (eds.), Topley and Wilson's microbiology and microbial infections, 9th ed, Vol. 3. Edward Arnold, Ltd., London.

Gage, K.L. and Jerrells, T.R. 1992. Demonstration and characterization of antigens of *Rickettsia rhipicephali* that induce cross-reactive cellular and humoral immune responses to *Rickettsia rickettsii*. Infect. Immun. **60:**5099–5106.

Gage, K.L., Maupin, G.O., Montenieri, J., Piesman, J., Dolan, M. and Panella, N.A. 1997. Flea (Siphonaptera: Ceratophyllidae: Hystrichopsyllidae) and tick (Acarina: Ixodidae) control on wood rats using host-targeted liquid permethrin in bait tubes. J. Med. Entomol. **34:**46-51.

Gage, K.L., Ostfeld, R.S. and Olson, J.G. 1995. Nonviral vector-borne zoonoses associated with mammals in the United States. J. Mammol. **76:**695–715.

Gage, K.L., Schrumpf, M.E. and Karstens, R.H. 1994. DNA typing of rickettsiae in naturally infected ticks using a PCR-RFLP typing system. Am. J. Trop. Med. Hyg. **50:**247–260.

Goddard, J. and Norment, B.R. 1988. Spotted fever group rickettsiae in the lone star tick, *Amblyomma americanum* (Acari: Ixodidae). J. Med. Entomol. **23:**465–472.

Goodman, J.L., Nelson, C., Vitale, B., Madigan, J.E., Dumler, J.S., Kurtti, T.J. and Munderloh, U.G. 1996. Direct cultivation of the causative agent of human granulocytic ehrlichiosis. New England J. Med. **334:**209–215.

Gray, J.S., Kahl, O., Janetzki, C. and Stein, J. 1992. Studies on the ecology of Lyme disease in a deer forest in County Galway, Ireland. J. Med. Entomol. **29:**915–920.

Gregg, C.T. 1985. Plague – An ancient disease in the Twentieth Century. Revised Edition. University of New Mexico Press, Albuquerque. 169 pp.

Guiyoule, A., Grimont, F., Iteman, I., Grimont, P.A.D., Lefevre, M. and Carniel, E. 1994. Plague pandemics investigated by ribotyping of *Yersinia pestis* strains. J. Clin. Microbiol. **32:**634–641.

Higgins, J.A., Radulovic, S., Schriefer, M.E. and Azad, A.F. 1996. *Rickettsia felis*: a new species of pathogenic rickettsia isolated from cat fleas. J. Clin. Microbiol. **34:**671–674.

Hinnebusch, B.J., Fischer, E.R. and Schwan, T.G. 1998. Evaluation of the role of the *Yersinia pestis* plasminogen activator and other plasmid-encoded factors in temperature-dependent blockage of the flea. J. Infec. Dis. **178:**1406-1415.

Hinnebusch, B.J., Perry, R.D. and Schwan, T.G. 1996. Role of the *Yersinia pestis* hemin storage (*hms*) locus in the transmission of plague by fleas. Science **273:**367–370.

Hopla, C.E. 1974. The ecology of tularemia. Adv. Vet. Sci. Comp. Med. **18:**25–53.

Hopla, C.E. 1980. A study of the host associations and zoogeography of *Pulex*. Pp. 185–207 *in* R. Traub and H. Starcke (eds.), Fleas: proceedings of the International Conference on Fleas, Ashton Wold, Peterborough, UK, 21-25 June, 1977. A.A. Balkema, Rotterdam.

Hopla, C.E. and Hopla, A.K. 1994. Tularemia. Pp. 113–126 *in* G.W. Beran and J.H. Steele (eds.), Bacterial, rickettsial, chlamydial, and mycotic diseases, 2nd ed. CRC Press, Boca Raton, Florida.

Hutcheson, H.J., Oliver, J.H., Jr., Houck, M.A. and Strauss, R.E. 1995. Multivariate morphometric discrimination of nymphal or adult forms of the blacklegged tick (Acari: Ixodidae), a principal vector of the agent of Lyme disease in eastern North America. J. Med. Entomol. **32:**827–842.

Jellison, W.L. 1974. Tularemia in North America. University of Montana Foundation, Missoula. 276 pp.

Kartman, L. 1957. The concept of vector efficiency in experimental studies of plague. Exp. Parasitol. **6:**599–609.

Kawabata, H., Masuzawa, T. and Yanagihara, Y.Y. 1993. Genomic analysis of *Borrelia japonica* sp. nov. isolated from *Ixodes ovatus* in Japan. Microbiol. Immunol. **37:**843–848.

Kelly, P.J., Beati, L., Matthewman, L.A., Mason, P.R., Dasch, G.A. and Raoult, D. 1994. A new pathogenic spotted fever group rickettsia from Africa. J. Trop. Med. Hyg. **97:**129–137.

Kocan, K.M. 1986. Development of *Anaplasma marginale* Theiler in ixodid ticks: coordinated development of a rickettsial organism and its tick host. Pp. 472–505 *in* J.R. Sauer and J.A. Hair (eds.), Morphology, physiology, and behavioral biology of ticks. Ellis Horwood, Ltd., Chichester, UK.

Kolbert, C.P., Podzorski, D.S., Mathiesen, D.A., Wortman, A.T., Gazumyan, A., Schwartz, I. and Persing, D.H. 1995. Two geographically distinct isolates of *Borrelia burgdorferi* from the United States share a common unique ancestor. Res. Microbiol. **146:**415–424.

Korenberg, E.I., Vorobyeva, N.N., Moskvitina, H.G. and Gorban, L.Y. 1996. Prevention of Borreliosis in persons bitten by infected ticks. Infect. **24:**187–189.

Kurtenbach, K., Kampen, H., Dizij, A., Arndt, S., Seitz, H.M., Schaible, U.E. and Simon, M.M. 1995. Infestation of rodents with larval *Ixodes ricinus* (Acari: Ixodidae) is an important factor in the transmission cycle of *Borrelia burgdorferi* s.l. in German woodlands. J. Med. Entomol. **32:**807–817.

Kuttler, K.L. 1984. *Anaplasma* infections in wild and domestic ruminants: a review. J. Wildlife Dis. **20:**12–20.

Lackman, D.B. 1963. A review of information on rickettsialpox in the United States. Clin. Ped. **2:**296–301.

Lane, R.S. and Manweiler, S.A. 1988. *Borrelia coriaceae* in its tick vector *Ornithodoros coriaceus* (Acari: Argasidae), with emphasis on transstadial and transovarial infection. J. Med. Entomol. **25:**172–177.

Levin, M., Levin, J.F., Yang, S., Howard, P. and Apperson, C.S. 1996. Reservoir competence of the southeastern five-lined skink (*Eumeces inexpectatus*) and green anole (*Anolis carolinensis*) for *Borrelia burgdorferi*. Am. J. Trop. Med. Hyg. **54:**92–97.

Maeda, K., Morkowitz, N., Hawley, R.C., Ristic, M., Cox, D. and McDade, J.E. 1987. Human infection with *Ehrlichia canis*, a leukocytic rickettsia. New England J. Med. **316:**853–856.

Manweiler, S.A., Lane, R.S. and Tempelis, C.H. 1992. The western fence lizard *Sceloporus occidentalis*: evidence of field exposure to *Borrelia burgdorferi* in relation to infestation by *Ixodes pacificus* (Acari: Ixodidae). Am. J. Trop. Med. Hyg. **47:**328–336.

Marconi, R.T., Liveris, D. and Schwartz, I. 1995. Identification of novel insertion restriction fragment length polymorphism patterns, and discontinuous 23S rRNA in Lyme disease spirochetes: Phylogenetic analyses of rRNA genes and their intergenic spacers in *Borrelia japonica* sp. nov. and genomic group 21038 (*Borrelia andersonii* sp. nov.) isolates. J. Clin. Microbiol. **33:**2427–2434.

Mather, T.N., Ribeiro, J.M.C., Moore, S.I. and Spielman, A. 1988. Reducing transmission of Lyme disease spirochetes in a suburban setting. Ann. N.Y. Acad. Sci. **539:**402–403.

Maupin, G.O., Fish, D., Zultowsky, J., Campos, E.G. and Piesman, J. 1991. Landscape ecology of Lyme disease in a residential area of Westchester County, New York. Am. J. Epidemiol. **133:**1105–1113.

Maupin, G.O., Gage, K.L., Piesman, J., Montenieri, J., Sviat, S.L., VanderZanden, L., Dolan, M. and Johnson, B.J.B. 1994. Discovery of an enzootic cycle of *Borrelia burgdorferi* in *Neotoma mexicana* and *Ixodes spinipalpis* from northern Colorado, an area where Lyme disease is nonendemic. J. Infect. Dis. **170:**636–643.

McDade, J.E. 1987. Flying squirrels and their ectoparasites: disseminators of epidemic typhus. Parasitol. Today **3:**85–87.

McDade, J.E. and Newhouse, V.F. 1986. Natural history of *Rickettsia rickettsii*. Annu. Rev. Microbiol. **40:**287–309.

McDade, J.E., Shepard, C.C., Redus, M.A., Newhouse, V.F. and Smith, J.D. 1980. Evidence of *Rickettsia prowazekii* infections in the United States. Am. J. Trop. Med. Hyg. **29:**277–284.

Mejlon, H.A. and Jaenson, T.G.T. 1993. Seasonal prevalence of *Borrelia burgdorferi* in *Ixodes ricinus* in different vegetation types in Sweden. Scand. J. Infect. Dis. **25:**449–456.

Niebyslki, M.L., Schrumpf, M.E., Burgdorfer, W., Fischer, E.R., Gage, K.L. and Schwan, T.G. 1997. *Rickettsia peacocki sp. nov.*, a new species infecting wood ticks, *Dermacentor andersoni*, in western Montana. Internat. J. System. Bacteriol. **47:**446–452.

Norris, D.E., Klompen, J.S.H., Keirrans, J.E. and Black, W.C. IV. 1996. Population genetics of *Ixodes scapularis* (Acari: Ixodidae) based on mitochondrial 16S and 12S genes. J. Med. Entomol. **33:**78–89.

Oliver, J.H., Jr., Owsley, M.R., Hutcheson, H.J., James, A.M., Chen, C.S., Irby, W.S., Dotson, E.M. and McLain, D.K. 1993a. Conspecificity of the ticks *Ixodes scapularis* and *I. dammini* (Acari: Ixodidae). J. Med. Entomol. **30:**54–63.

Oliver, J.H., Jr., Chandler Jr., F.W., Luttrell, M.P., James, A.M., Stallknecht, D.E., McGuire, B.S., Hutcheson, H.J., Cummins, G.A. and Lane, R.S. 1993b. First isolation and transmission of the Lyme disease spirochete from southeastern United States. Proc. Nat. Acad. Sci. **90:**7371–7375.

Olsen, B., Duffy, D.C., Jaenson, T.G.T., Gylfe, A., Bonnedahl, J. and Bergstrom, S. 1995. Transhemispheric exchange of Lyme disease spirochetes by seabirds. J. Clin. Microbiol. **33:**3270–3274.

Pancholi, P., Kolbert, C.P., Mitchell, P.D., Reed, K.D., Dumler, J.S., Bakken, J.S., Telford, S.R., III and Persing, D.H. 1995. *Ixodes dammini* as a potential vector of human granulocytic ehrlichiosis. J. Infect. Dis. **172:**1007–1012.

Patrican, L.A. and Allan, S.A. 1995. Application of desiccant and insecticidal soap treatments to control *Ixodes scapularis* (Acari: Ixodidae) nymphs and adults in a hyperendemic woodland site. J. Med. Entomol. **32:**859–863.

Philip, R.N., Casper, E.A., Anacker, R.L., Cory, J., Hayes, S.F., Burgdorfer, W. and Yunker, C.E. 1983. *Rickettsia bellii* sp. nov.: a tick-borne rickettsia, widely distributed in the United States, that is distinct from the spotted fever and typhus biogroups. Internat. J. System. Bacteriol. **33:**94–106.

Philip, R.N., Casper, E.A., Anacker, R.L., Peacock, M.G., Hayes, S.F. and Lane, R.S. 1982. Identification of an isolate of *Rickettsia canada* from Canada. Am. J. Trop. Med. Hyg. **31:**1216–1221.

Philip, R.N., Lane, R.S. and Casper, E.A. 1981. Serotypes of tick-borne spotted fever group rickettsiae from western California. Am. J. Trop. Med. Hyg. **30:**722–727.

Piesman, J., Donahue, J.G., Mather, T.N. and Spielman, A. 1986. Transovarially acquired Lyme disease spirochetes (*Borrelia burgdorferi*) in field-collected larval *Ixodes dammini* (Acari: Ixodidae). J. Med. Entomol. **23:**219.

Piesman, J., Mather, T.N., Dammin, G.J., Telford, S.R., III, Lastavica, C.C. and Spielman, A. 1987a. Seasonal variation of transmission risk: Lyme disease and human babesiosis. Am. J. Epidemiol. **126:**1187–1189.

Piesman, J., Mather, T.N., Sinsky, R.J. and Spielman, A. 1987b. Duration of tick attachment and *Borrelia burgdorferi* transmission. J. Clin. Microbiol. **25:**557–558.

Piesman, J., Maupin, G.O., Campos, E.G. and Happ, C.M. 1991. Duration of adult female *Ixodes dammini* attachment and transmission of *Borrelia burgdorferi*, with description of a needle aspiration isolation method. J. Infect. Dis. **163:**895–897.

Poland, J. and Barnes, A.M. 1979. Plague. Pp. 515–559 *in* J.H. Steele (ed.), CRC Handbook Series in Zoonoses, Vol. 1. CRC Press, Boca Raton, Florida.

Pollitzer, R. 1954. Plague. World Health Organization, Geneva. 698 pp.

Pollitzer, R. and Meyer, K.F. 1961. The ecology of plague. Pp. 433–501 *in* J.F. May (ed.), Studies in Disease Ecology. Hafner, New York.

Rich, S.M., Caporale, D.A., Telford, S.R., III, Kocher, T.D., Hartl, D.L. and Spielman, A. 1995. Distribution of the *Ixodes ricinus*-like ticks of eastern North America. Proc. Nat. Acad. Sci. **92:**6284–6288.

Ristic, M., Holland, C.J. and Khondowe, M. 1991. An overview of research on ehrlichiosis. IVth International Symposium on Rickettsiae and Rickettsial Diseases. European J. Epidemiol. **7:**246–252.

Schriefer, M.E., Sacci, J.B., Jr., Dumler, J.S., Bullen, M.G. and Azad, A.F. 1994a. Identification of novel rickettsial agent infection in a patient diagnosed with murine typhus. J. Clin. Microbiol. **32:**949–954.

Schriefer, M.E., Sacci, J.B., Jr., Taylor, J.P., Higgins, J.A. and Azad, A.F. 1994b. Murine typhus: updated roles of multiple urban components and a second typhuslike rickettsia. J. Med. Entomol. **31:**681–685.

Schwan, T.G., Gage, K.L. and Hinnebusch, J. 1995. Analysis of relapsing fever spirochetes from the western United States. J. Spirochetal Tick-borne Dis. **2:**3–8.

Schwan, T.G., Thompson, D. and Nelson, B.C. 1985. Fleas on roof rats in six areas of Los Angeles County, California: their potential role in the transmission of plague and murine typhus to humans. Am. J. Trop. Med. Hyg. **34:**372–379.

Smith, R.P., Jr., Rand, P.W., Lacombe, E.H., Telford, S.R., III, Rich, S.M., Piesman, J. and Spielman, A. 1993. Norway rats as reservoir hosts for Lyme disease spirochetes on Monhegan Island, Maine. J. Infect. Dis. **167:**687–691.

Sonenshine, D.E. 1991. Biology of ticks. Vol. 1. Oxford University Press, New York. 447 pp.

Sonenshine, D.E. 1993. Biology of Ticks. Oxford University Press, New York. 465pp.

Spielman, A., Clifford, C.M., Piesman, J. and Corwin, M.D. 1979. Human babesiosis on Nantucket Island, U.S.A. Description of the vector *Ixodes (Ixodes) dammini*, n. sp. (Acarina: Ixodidae). J. Med. Entomol. **15**:218–234.

Spielman, A., Wilson, M.L., Levine, J.F. and Piesman, J. 1985. Ecology of *Ixodes dammini*-borne human babesiosis and Lyme disease. Annu. Rev. Entomol. **30**:439–460.

Stafford, K.C., III, Bladen, V.C. and Magnarelli, L.A. 1995. Ticks (Acari: Ixodidae) infesting wild birds (Aves) and white-footed mice in Lyme, Connecticut, U.S.A. J. Med. Entomol. **32**:453–466.

Steere, A.C., Snydman, D., Murray, P., Mensch, J., Main, A.J., Wallis, R.C., Shope, R.E. and Malawista, S.E. 1986. Historical perspective of Lyme disease. Zentrbl. Bakeriol. Mikrobiol. Hyg. **263**:3–6.

Steere, A.C., Sikand, V.K., Meurice, F., Parenti, D.L., Fikrig, E., Schoen, R.T., Nowakowski, J., Schmid, C.H., Laukamp, S., Buscarine, C. and Krause, D.S. 1998. Vaccination against Lyme disease with recombinant *Borrelia burgdorferi* outer surface lipoprotein A with adjuvant. New England J. Med. **339**:209-215.

Stoenner, H.G. 1980. Q Fever. Pp. 337–349 *in* J.H. Steele (ed.), CRC Handbook Series in Zoonoses., Vol. 2. CRC Press, Boca Raton, Florida.

Takada, N., Fujita, H., Yano, Y., Tsuboi, Y. and Mahara, F. 1994. First isolation of a rickettsia closely related to Japanese spotted fever pathogen from a tick in Japan. J. Med. Entomol. **31**:183–185.

Takahashi, M., Murata, M., Misumi, H., Hori, E., Kawamura, A., Jr. and Tanaka, H. 1994. Failed vertical transmission of *Rickettsia tsutsugamushi* (Rickettsiales: Rickettsiacaea) acquired form rickettsemic mice by *Leptotrombidium pallidum* (Acari: Trombiculidae). J. Med. Entomol. **31**:212–216.

Telford, S.R., III, Dawson, J.E., Katavalos, P., Warner, C.K., Kolbert, C.P. and Persing, D.H. 1996. Perpetuation of the agent of human granulocytic ehrlichiosis in a deer tick-rodent cycle. Proc. Nat. Acad. Sci. **93**:6209–6214.

Thomas, R.E., Barnes, A.M., Quan, T.J., Beard, M.L., Carter, L.G. and Hopla, C.E. 1988. Susceptibility to *Yersinia pestis* in the northern grasshopper mouse (*Onychomys leucogaster*). J. Wildlife Dis. **24**:327–333.

Traub, R. and Wisseman, C.L. 1974. The ecology of chigger-borne rickettsiosis (scrub typhus). J. Med. Entomol. **11**:237–303.

Uchida, T., Yan, Y. and Kitaoka, S. 1995. Detection of *Rickettsia japonica* in *Haemaphysalis longicornis* ticks by restriction fragment length polymorphism of PCR product. J. Clin. Microbiol. **33**:824–828.

Vinson, J.S. 1973. Louse-borne diseases worldwide: trench fever. Pp. 76–79, International Symposium on Lice and Louse-borne Diseases. Pan American Health Organization, Washington, DC.

von Reyn, C.F., Barnes, A.M., Weber, N.S. and Hodgin, U.G. 1976. Bubonic plague from exposure to a rabbit: A documented case, and a review of rabbit-associated plague cases in the United States. Am. J. Epidemiol. **104**:81–87.

Walker, D.H. and Dumler, J.S. 1996. Emergence of the ehrlichioses as human health problems. Emerg. Infect. Dis. **2**:18–29.

Walker, D.H. and Fishbein, D.B. 1991. Epidemiology of rickettsial diseases. European J. Epidemiol. **7**:237–245.

Walker, D.H. and Gage, K.L. 1997. *Rickettsia, Orientia, Ehrlichia*, and *Coxiella*. Pp. 371-380 in D. Greenwood, R.C.B. Slack and J.F. Pentherer (eds.), Medical microbiology, 15th Ed. Churchill Livingston, New York.

Williams, S.G., Sacci, J.B., Jr., Schreifer, M.E., Anderson, E.M., Fujioka, K.K., Sorvillo, F.J., Barr, A.R. and Azad, A.F. 1992. Typhus and typhus-like rickettsiae associated with opposums and their fleas in Los Angeles County, California. J. Clin. Microbiol. **30**:1758–1762.

Wilson, M.L., Telford, S.R., III, Piesman, J. and Spielman, A. 1988. Reduced abundance of immature *Ixodes dammini* (Acari: Ixodidae) following elimination of deer. J. Econ. Entomol. **25**:224–228.

Yunker, C.E. 1996. Heartwater in sheep and goats: a review. Onderstepoort J. Vet. Res. **63**:159–170.

Zangwill, K.M., Hamilton, D.H., Perkins, B.A., Regnery, R.L., Pilkaytis, B.D., Hadler, J.L., Cartter, M.L. and Wenger, J.D. 1993. Cat scratch disease in Connecticut — Epidemiology, risk factors, and evaluation of a new diagnostic test. New England J. Med. **329**:8–13.

Zinsser, H. 1934. Rats, lice and history. Little, Brown & Company, Boston. 301 pp.

Chapter 11

Arbovirus Diseases

BRUCE F. ELDRIDGE[1], THOMAS W. SCOTT[1], JONATHAN F. DAY[2] AND WALTER J. TABACHNICK[2]
[1]University of California, Davis, and [2]Florida Medical Entomology Laboratory, Vero Beach

THE ARBOVIRUSES

Arboviruses are a diverse group of microorganisms that share the common feature of being biologically transmitted to vertebrate hosts by arthropods. Arboviruses occur in nearly all parts of the world except the polar ice caps.

Currently there are more than 500 distinct viruses in this category. Nearly all of them are included in 5 families: Togaviridae, Flaviviridae, Bunyaviridae, Reoviridae and Rhabdoviridae (Karabatsos 1985, Monath 1988a). Some viruses are not known to produce diseases in humans or domestic animals, but about 100 cause human infections, and about 40 infect livestock (Brès 1988; Table 11.1).

History

Dmitri Ivanowski, a Russian scientist, reported in 1892 that the pathogenic agent causing tobacco mosaic disease would pass through a filter fine enough to hold back the smallest microorganisms then known. This finding was confirmed by Martinus Beijerinck, a Dutch microbiologist, who showed that the filterable agent could be cultured, and thus that the agent was a living organism (see Levine 1996). The first discovery of a virus of animals was made in 1898, when Loeffler and Frosch found that the cause of foot and mouth disease in cattle was a filterable virus. Walter Reed, James Carroll, A. Agramonte and Jesse Lazear in 1901 discovered that yellow fever could be produced in human volunteers by the injection of infected sera that had passed through a filter. Further, they proved that transmission was through the bite of infected mosquitoes (Reed et al. 1900). After these initial discoveries, filtration showed that these very small organisms caused many diseases of plants and animals. Montgomery found the first tickborne virus disease, Nairobi sheep disease, in 1917. In this same year, Daubney and Hudson reported the first arthropodborne zoonotic disease, which was caused by the Rift Valley fever virus.

Studies of the structure and composition of viruses began in the 1930s, using new techniques such as differential centrifugation and crystallography. In 1939, the then newly developed electron microscope first visualized a virus particle. In 1948, Sanford developed a method to culture single animal cells, thus accelerating discoveries in animal virology.

In the late 1930s, investigations on the immunologic relationships among arboviruses

B.F. Eldridge and J.D. Edman (eds.), Medical Entomology, Revised Edition, 415–460.

Table 11.1. Characteristics of virus families containing arboviruses.

Family	Nucleocapsid morphology	Genome	Vectors	Example
Togaviridae	Spherical, 70 nm, enveloped	Single stranded, linear, positive sense RNA, non-segmented	Mosquitoes, ticks, swallow bugs	VEE, EEE, WEE viruses
Flaviviridae	Spherical, 45-60 nm, enveloped	Single stranded, linear, positive sense RNA, nonsegmented	Mosquitoes, ticks	YF, SLE, JE viruses
Bunyaviridae	Spherical or pleomorphic, 80-120 nm, enveloped	Single stranded, circular, negative sense RNA, 3-segmented	Mosquitoes, biting midges, phlebotomine sand flies	LAC, TAH, RFV viruses
Reoviridae	Spherical, 60-80 nm, non-enveloped	Double stranded, linear, positive sense RNA, 10-27 segments	Ticks, biting midges	CTF, BLU, AHS viruses
Rhabdoviridae	Bullet-shaped, 130-380 nm x 70-85 nm, enveloped	Single stranded, linear, negative sense RNA, nonsegmented	Phlebotomine sand flies, black flies, tabanid flies	VS, BEF viruses

began, and Smithburn (1942) found that on the basis of the neutralization test, West Nile, Japanese encephalitis and St. Louis encephalitis viruses were related. This discovery led to classification of arboviruses based on serological reactions.

Characteristics

Viruses are relatively simple organisms. Mature, infectious virus particles, or virions, consist of a **genome** (a central core of nucleic acid containing genetic information) and a protein **capsid** surrounding the genome. The genome and capsid together are called the **nucleocapsid**. In some cases, the nucleocapsid is surrounded by an outer **envelope** consisting of one or more membranes formed at least in part from the plasma membrane of the infected cell (Fig. 11.1). The Reoviridae lack this envelope (Nibert et al. 1996). Virions of arboviruses vary in size and shape. Most are spherical, except for Rhabdoviridae, which are bullet-shaped. Spherical viruses have icosahedral symmetry. Virions range in size from about 45 nm in diameter (flaviviruses) to more than 380 nm in length (some rhabdoviruses).

The genome of almost all arboviruses contains RNA, and this also is true of most plant viruses that are transmitted by arthropods. A notable exception is African swine fever, a DNA virus (Beaty et al. 1988). The RNA may be single or double stranded, may be linear or circular, and may be positive or negative sense. Positive sense RNA can act as mRNA, and genomes with this type of RNA are termed infectious. On the other hand, genomes with negative sense RNA first must make positive sense RNA for transcription, and are termed noninfectious. Some viruses have genomes that are non-seg-

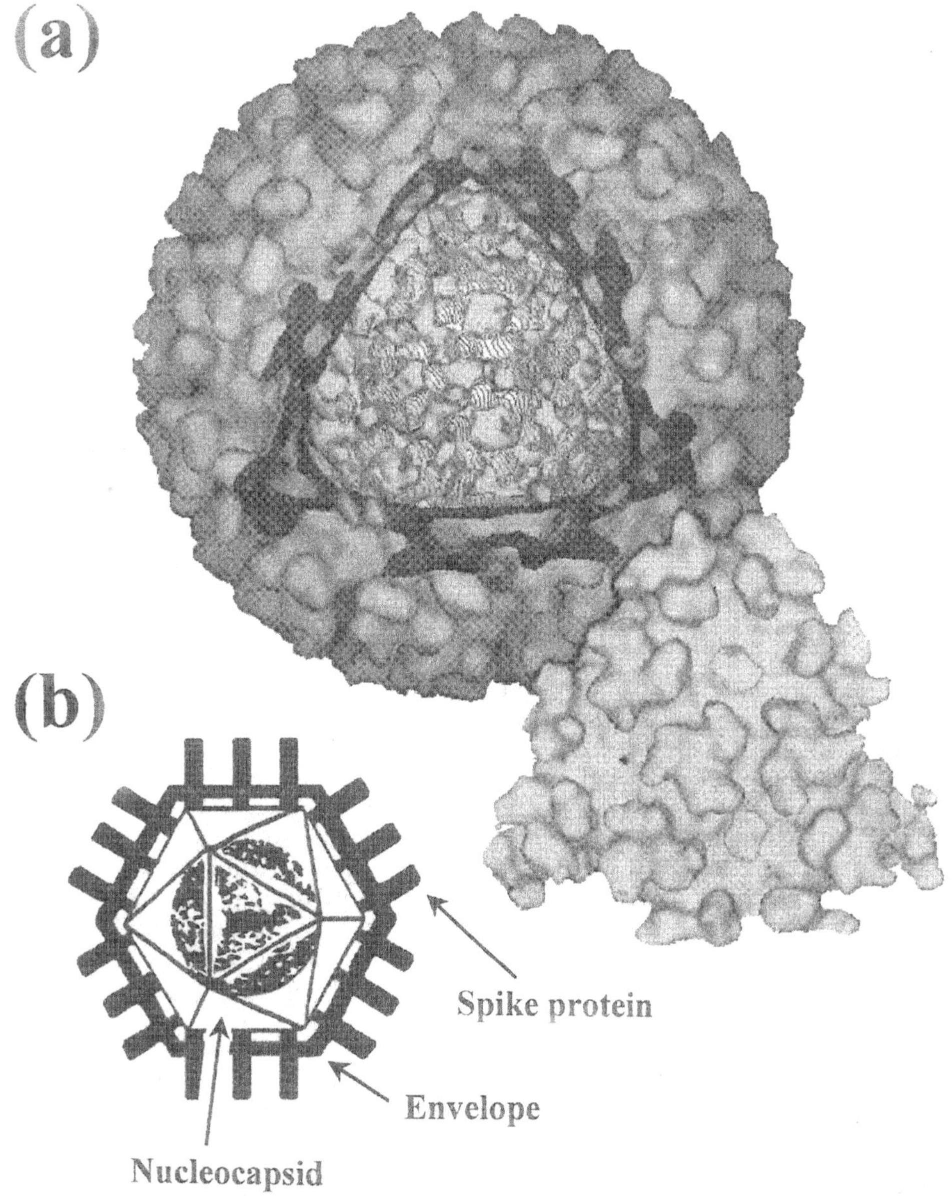

Figure 11.1. Sindbis virus: (a) Computer-generated depiction of virus with section of envelope removed to show nucleocapsid; (b) diagrammatic representation of viral structures (both courtesy of Stephen Fuller, European Molecular Biology Laboratory, Heidelberg, Germany).

mented (Alphaviridae, Flaviviridae and Rhabdoviridae), but others have genomes that are divided into distinct segments. Bunyaviruses have 3-segmented genomes; reoviruses have genomes with as many as 27 segments. Some authorities believe that viruses with segmented genomes are capable of rapid evolution because segment reassortment may occur during viral replication, leading to entirely new viruses (Beaty et al. 1988). Regardless of whether the genome is segmented or not, arboviruses with an RNA genome can change rapidly because of their high rates of mutation (Eigen 1991).

Ecology

By definition, arboviruses are biologically transmitted to vertebrate hosts by arthropods. Most are zoonoses and the natural vertebrate hosts are birds or mammals. The most important vectors, based on isolations of arboviruses from arthropods in nature, appear to be mosquitoes, followed by ticks, phlebotomine sand flies, biting midges, horse and deer flies, and black flies. Transmission of arboviruses is propagative by arthropod vectors, but can be augmented by mechanical means. Rift Valley fever virus can be transmitted mechanically by tsetse and mosquitoes (Hoch et al. 1985), and human infections can occur by direct contact with infected vertebrates. Other arboviruses also can be transmitted directly from vertebrate to vertebrate (e.g., Venezuelan equine encephalomyelitis virus, eastern equine encephalomyelitis virus), and there have been cases of **nosocomial** (related to hospitals) transmission in the case of Crimean-Congo hemorrhagic fever virus (Brès 1988).

Some arbovirus diseases, particularly those caused by alphaviruses and flaviviruses, are transmitted among birds by ornithophilic mosquitoes that do not feed on mammals. Mammalian infections may occur as a result of blood-feeding by other mosquito species that will feed on both birds and mammals. Such species are called **bridge vectors**.

Some viruses that are pathogenic for animals are transmitted mechanically by insects or by direct contact (e.g., equine infectious anemia virus). By definition, such viruses are not arboviruses, and are discussed in Chapter 12.

Classification

Arbovirus taxonomy has undergone gradual refinement since the discovery of these viruses. The first classifications were based on the type of disease produced (e.g., yellow fever, dengue). As immunologic relationships were studied, a classification of arboviruses emerged based on various serological tests, including the **hemagglutination inhibition** (HI) test (Casals 1961). Eventually, the HI test was used to separate viruses into major groupings (called serogroups), and **complement fixation** and **neutralization** tests were used to further differentiate viruses into subordinate taxonomic categories (Calisher and Karabatsos 1988).

The International Committee on Nomenclature of Viruses (ICNV) has adopted a universal system for virus taxonomy that uses the traditional categories of animal and plant taxonomy (order, family, subfamily, genus, species). This system is based upon many viral characteristics, including virion size, structure and antigenicity, and biological properties. A virus species is defined as "a polythetic class of viruses that constitutes a replicating lineage and occupies a particular ecological niche" (Murphy 1996). The concept of a virus species accommodates the inherent variability of viruses. We have used virus terminology consistent with the ICNV system, but have retained the traditional (but informal) superspecific taxa of serogroup and complex and the subspecific taxa of serotype and variety. Serotype is roughly equivalent to the subtype of Calisher and Karabatsos

(1988) and serogroup comes closest to the subfamily of Murphy (1996).

Arboviruses usually are named according to the disease they produce or the geographic location of their first isolation. Newly discovered arboviruses ordinarily are registered with the American Committee of Arboviruses (ACAV), a committee associated with the American Society of Tropical Medicine and Hygiene. At the time of registration, a short abbreviation of each virus is designated in accordance with rules established by ACAV. For example, the abbreviation for western equine encephalomyelitis virus (but not the disease), is WEE. ACAV maintains a catalog of arboviruses; the most recent edition was published in 1985 (Karabatsos 1985). It also includes a number of animal viruses that are not known to be arthropodborne (e.g., Marburg, Ebola).

Importance of Arboviruses

Infections in humans and livestock can result in febrile illnesses ranging from mild discomfort to severe influenza-like symptoms, and can cause encephalitis (inflammation of the brain), encephalomyelitis (inflammation of the brain and myelin sheath), and hemorrhagic fevers. Mortality rates can be relatively high, especially in infections resulting in **central nervous system** (CNS) involvement.

Dengue virus causes periodic epidemics involving thousands of cases. Until the 1950s, dengue was considered to cause only nonfatal febrile illnesses. However, a hemorrhagic form (called **dengue hemorrhagic fever**) has emerged as a disease of major public health importance. In Asia, fatality rates of 30–40% can occur in children if the disease is not recognized and treated (Gubler 1997). Aside from the death and suffering caused by dengue infections, monetary costs associated with prevention and treatment are substantial. Single epidemics of dengue may cost nearly $17 million for hospitalization and mosquito control (Brès 1988). In addition, loss of tourism may have further direct economic effects.

Yellow fever virus also continues to produce epidemics, especially in Africa. Some of these epidemics have occurred in urban areas and have had devastating effects in terms of suffering and death. An epidemic in eastern Nigeria in 1986 was estimated to have resulted in over 9,000 human cases and over 5,000 deaths. Although accurate cost analyses are not available, emergency immunization programs in Africa may have cost millions of dollars (Monath 1988b).

Some arbovirus diseases, such as eastern equine encephalomyelitis, can result in **sequelae** (residual symptoms) in individuals that recover from the initial infection. These damages can include mental retardation and Parkinsonism. In such situations, the effects of the disease can extend many years. Villari et al. (1995) estimate that a single case of eastern equine encephalomyelitis causing permanent disability may generate costs of $3 million over the lifetime of the patient.

Arboviruses that produce diseases primarily in domestic animals also can have important economic impacts. Costs associated with the 1971 Venezuelan equine encephalomyelitis epizootic from insecticide spraying, immunization, quarantine and surveillance have been placed at $20–$30 million (Brès 1988). Diseases such as bluetongue, vesicular stomatitis and African horse sickness not only produce substantial economic loss from death of animals, but also restrict movement of animals in international trade. This constraint can have enormous economic consequences.

TOGAVIRIDAE

The family Togaviridae includes 2 genera, *Alphavirus* and *Rubivirus* (rubella) (Murphy 1996). Within the family, only alphaviruses are

Table 11.2. Arthropodborne alphaviruses of public health and veterinary importance (family Togaviridae, genus *Alphavirus*).

Virus	Vectors	Affected hosts (other than wildlife)
Eastern equine encephalomyelitis (EEE)	*Culiseta melanura* (N. Amer.) *Culex (Melanoconion)* spp. (S. and Central Amer.) *Aedes* spp.	Horses, humans, certain birds
Western equine encephalomyelitis (WEE)	*Culex tarsalis* *Aedes melanimon* *Aedes dorsalis* *Cx.* (*Mel.*) spp.	Horses, humans, certain birds
Highlands J (HJ)	*Cs. melanura*	Domestic birds
Venezuelan equine encephalitis (VEE)	*Cx.* (*Mel.*) spp. *Psorophora* spp. *Ae.* spp.	Horses, humans
Chikungunya (CHIK)	*Aedes aegypti* *Aedes* (*Diceromyia*) spp. *Aedes* (*Stegomyia*) spp. *Mansonia* spp.	Humans
O'Nyong-nyong (ONN)	*Anopheles funestus* *Anopheles gambiae*	Humans
Ross River (RR)	*Cx.* spp. *Ae.* spp. *Anopheles* spp. *Ma.* spp.	Humans
Sindbis (SIND)	*Cx.* spp.	Humans
Barmah Forest (BF)	*Culex annulirostris* *Ae.* spp. *Coquillettidia* spp.	Humans

arthropodborne. The genus *Alphavirus* is composed of 7 complexes with 15 species and many serotypes and varieties (Calisher and Karabatsos 1988) (Table 11.2). Mosquitoes transmit all but one alphavirus. Fort Morgan virus, which is in the western equine encephalomyelitis complex, is transmitted by a cimicid, the swallow bug (*Oeciacus vicarius*), among cliff swallows and house sparrows (Reisen and Monath 1988).

It is not clear to what extent vertical transmission among arthropod and vertebrate hosts contributes to amplification of alphaviruses. Transovarial transmission among mosquito vectors has been reported for western equine encephalomyelitis (WEE), Ross River (RR) and Sindbis (SIN) viruses. During 1991 and 1992, in coastal California, WEE virus was recovered from adult *Aedes dorsalis* that were reared from field-collected larvae (Fulhorst et al. 1994). Subsequent attempts to duplicate that finding have not been successful (Reisen et al. 1996). Entomological studies in Australia indicate that several *Aedes* species may transmit alphaviruses vertically. Ross River and SIN viruses were recovered from adult *Aedes camptorhynchus* that were reared from field-collected larvae (Dhileepan et al. 1996). Ross River virus was recovered from male *Aedes vigilax* and *Aedes tremulus* collected from Western Australia (Lindsay et al. 1993).

Transplacental transmission in vertebrates has been reported for WEE and Venezuelan equine encephalomyelitis (VEE) viruses, but it is not believed to contribute to virus amplification (Scott 1988). Eastern equine encephalomyelitis (EEE) and Highlands J (HJ) viruses have been experimentally transmitted in infected semen from male to female turkeys, indicating that venereal transmission may occur among birds (Guy et al. 1995).

Alphavirus transmission cycles are most often enzootic, with birds and small mammals serving as vertebrate hosts. Humans usually, but not always (see Chikungunya and RR viruses), are dead-end alphavirus hosts because they do not circulate enough virus in their blood to infect vector mosquitoes. Transmission to humans and livestock is often by bridge vectors.

Alphaviruses have a broad geographic distribution, ranging through North and South America, Africa, the former USSR, Asia and Australia. Viruses in the *Alphavirus* genus can be divided into 2 categories: those associated with fever and polyarthritis (e.g., RR virus) and those associated with encephalitis (e.g., EEE virus).

Alphaviruses replicate in the cytoplasm of host cells. The genome is a single strand of positive sense RNA, approximately 11–12 kilobases in length. After replication, naked nucleocapsids push through the host cell membrane to form an envelope.

Eastern equine encephalomyelitis (EEE) virus

The initial isolate of EEE virus was recovered in 1933 from the brain of a dead horse in New Jersey (Giltner and Shahan 1933). An equine epizootic associated with that case included more than 1,000 horses along the eastern seaboard of the USA. Other North American epizootics have included more than 14,000 equines. Eastern equine encephalomyelitis virus probably was transmitted in North America long before it was recognized as a viral disease; there are clinical descriptions of cases reported as early as 1831. The first human cases of eastern equine encephalomyelitis were diagnosed in 1938. Since then, sporadic human infections, typically preceded by equine cases, have occurred each year throughout the North American distribution of EEE virus. Fundamental components of the epidemiology and transmission of EEE virus were defined during the 1940s and 1950s (Scott and Weaver 1989).

Eastern equine encephalomyelitis viral infection often causes a fatal illness in exotic gamebirds, equines and humans. Serologic surveys suggest that porcine infections by EEE virus probably are underdiagnosed and that EEE viral infections deserve more attention as a source of morbidity and mortality in swine. Although EEE viral infections can be viscerotropic, including myocarditis and pulmonary dysfunction in fatal cases, most are associated with acute necrotizing encephalitis. The mortality rate is 50–90% among infected humans and 80–90% for horses. Most humans or equines that recover suffer from neurologic sequelae. The onset of human and equine disease following infection is rapid and includes stupor, convulsions, coma and death. The **apparent to inapparent infection ratio** is high, i.e., a relatively high proportion of humans or horses that are infected become clinically ill.

EEE virus is the only species in the eastern equine encephalomyelitis complex. There are 2 varieties. The South American variety has been recovered from Central America to southern South America; it rarely has been detected in the Caribbean. The North American variety is regularly recovered in eastern North America

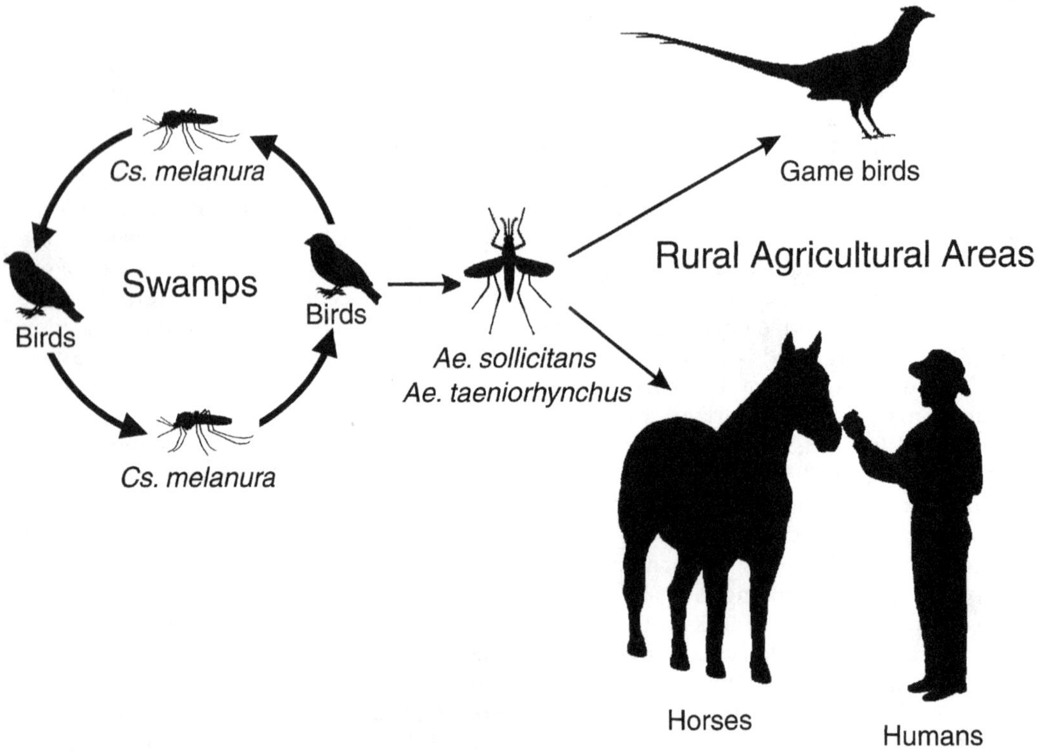

Figure 11.2. Life cycle of eastern equine encephalomyelitis virus.

and occasionally in the Caribbean (Scott et al. 1994).

In North America, the mosquito *Culiseta melanura* transmits virus in an enzootic cycle among songbirds in freshwater swamps (Fig. 11.2). Foci of transmission occur along coastal regions and in a few geographically distinct inland locations. A variety of other mosquito species are suspected to be bridge vectors that infect humans and horses with EEE virus (Morris 1988). Among captive gamebirds, transmission can occur from one bird to another by pecking and cannibalism. The mechanism by which EEE virus survives the winter when mosquitoes are not actively bloodfeeding may be through chronic relapsing avian host infections (Crans et al. 1994).

In Central and South America, mosquitoes in the subgenus *Melanoconion* of the genus *Culex* are considered EEE virus vectors. Although birds may be hosts in some locations, small mammals are considered the principal natural vertebrate hosts in tropical America. Transmission in the Caribbean is poorly understood.

Western equine encephalomyelitis (WEE) virus

The study of WEE virus began in the San Joaquin Valley of California during a 1930 encephalitis epizootic that included more than 6,000 horses. Mosquito transmission was demonstrated experimentally using *Aedes aegypti* in 1933. In 1938, WEE virus was isolated from the brain of child that had died from encephalitis. In 1941, naturally infected *Culex tarsalis* were collected. This sequence of events prompted more than 4 decades of detailed arbovirus research on WEE virus (Reeves 1990).

The severity of disease associated with WEE virus depends on the infected vertebrate host species. In humans, infections result in a range of responses from inapparent to headache to aseptic meningitis to encephalitis. Most clinically ill patients recover. The apparent to inapparent infection ratio is age dependent, with infants and children (0–4 yr of age) having the highest ratio and those >14 yr of age the lowest. Similarly, recovery with neurologic sequelae is most common for infants and young children. The mortality rate for western equine encephalomyelitis (3–15%) is low compared to that for eastern equine encephalomyelitis (50–90%). In horses, the clinical cases often are fatal (Fig. 11.3), but inapparent infections are common.

Western equine encephalomyelitis virus is comprised of 6 serotypes that appear to vary in their virulence characteristics. The geographic distribution of WEE virus ranges from western North America to Argentina.

Western equine encephalomyelitis virus transmission has been more thoroughly studied in North America than in South America. In the western half of North America, it is transmitted by *Cx. tarsalis* (Fig. 11.4) among songbirds in agroecosystems (Fig. 11.5). This species also serves as the epizootic and epidemic vector as well. The mechanism by which WEE virus survives the winter in these temperate habitats, when mosquitoes are not transmitting virus, has been elusive. In some North American locations, small mammals and *Aedes* spp. (*Aedes melanimon* and *Ae. dorsalis*) may be involved in WEE virus transmission. *Culex (Melanoconion) ocossa* and *Culex (Melanoconion) portesi*, respectively, may transmit WEE virus in enzootic cycles among small mammals and birds in Argentina and Brazil. Other mammal-feeding mosquitoes have been suggested as bridge vectors of WEE virus to humans. *Aedes albifasciatus* in Argentina has been incriminated as the vector transmitting WEE virus among European hares. This connection resembles the *Ae. melanimon*-WEE virus-Leporidae transmission system in North America.

Highlands J (HJ) virus

Highlands J virus is a member of the western equine encephalomyelitis complex. It is closely related to WEE virus and originally was considered to be an eastern serotype of WEE virus. Highlands J virus persists in the same ecological setting and transmission cycle as EEE virus. *Culiseta melanura* transmits HJ virus among songbirds in freshwater swamps in the eastern half of North America. HJ virus is characterized by low pathogenicity in mammals and has been associated rarely with equine or human disease (Karabatsos et al. 1988). Outbreaks of disease in penned domestic birds have been reported (Guy et al. 1994) but are mild compared to the devastating disease associated with EEE viral infections in domestic or exotic avian species.

Venezuelan equine encephalomyelitis (VEE) virus

Venezuelan equine encephalomyelitis virus is comprised of a number of enzootic and epizootic serotypes and varieties, all of which can cause illness in humans. Encephalitis, CNS involvement and death are rare, occurring most often in children. Most infected humans de-

Figure 11.3. Horse dying from western equine encephalomyelitis. Ground shows evidence of typical circular digging pattern of infected horses (courtesy of William C. Reeves).

velop undifferentiated influenza-like symptoms. Most infections are acquired from the bite of an infected mosquito, but laboratory infections can occur from aerosols while handling infected rodents or high concentrations of virus. Naturally acquired human infections typically are preceded by equine cases. Clinical signs in equines are characteristic of a generalized viral infection. Some infections may be lethal. In these cases, the animal may fall down, become comatose and dig a characteristic circular pattern in the soil and vegetation with its limbs (Fig. 11.3). Death occurs 5–14 days after exposure. A proportion of infected horses (10–15%) experience mild or inapparent infections and recover without neurologic sequelae.

Venezuelan equine encephalomyelitis viral diversity has been the most thoroughly studied and defined of the arboviruses causing encephalitis (Scott et al. 1994, Walton and Grayson 1988). Varieties IAB and IC are referred to as epizootic varieties because typically they are recovered only during equine epizootics. The other varieties (ID, IE, IF and II–IV) are maintained in sylvatic enzootic transmission cycles without being associated with human or equine epidemics. Venezuelan equine encephalomyelitis virus has been recovered from the southern USA, through Central and South America to northern Argentina.

Several mosquito species that feed on large mammals transmit epizootic VEE varieties (Walton and Grayson 1988). Equines are the am-

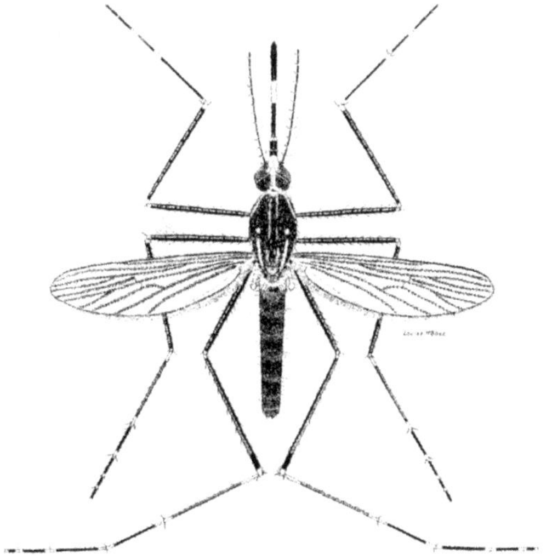

Figure 11.4. *Culex tarsalis*, the primary vector of western equine encephalomyeltis virus (drawing by Louise Horne).

plifying vertebrate host for epizootic virus transmission. Venezuelan equine encephalomyelitis virus moves swiftly through an area in wavelike patterns. After a large portion of the equine hosts have been infected and herd immunity is high, transmission ceases and epizootic virus can no longer be recovered. The 1995 epidemic in Colombia and Venezuela occurred almost 25 years after the previous major outbreak of Venezuelan equine encephalomyelitis in South America. The source of virus for these epizootics is unknown. Phylogenetic studies imply that epizootic viruses may be derived repeatedly from enzootic virus varieties (Rico-Hesse et al. 1995). Alternatively, enzootic viruses may persist in undetected foci of transmission and periodically expand their range of transmission that causes detectable disease.

Enzootic VEE varieties persist throughout the year in well characterized vector-host transmission cycles or in infected mosquitoes as they estivate during hot, dry times of the year. The types of mosquito species involved in transmitting enzootic viruses vary geographically, but vector-virus associations are more specific for enzootic than epizootic varieties. All documented enzootic vectors are in the genus *Culex*, subgenus *Melanoconion*. Small rodents are the principal vertebrate hosts; in some locations, birds may be involved in transmission cycles. Mosquito vectors of epizootic varieties feed most often on nonhuman hosts. This preference may explain why human disease associated with enzootic varieties is rare.

Chikungunya (CHIK) virus

Chikungunya was the word used by native people in Tanzania to describe the disease they experienced during a 1952–1953 epidemic (Jupp and McIntosh 1988). Chikungunya viral infections in humans produce an acute illness within 2 days of exposure that lasts for 3–5 days and is characterized by sudden onset of illness, fever, chills, headache, photophobia, arthralgia and rash. The symptoms of CHIK viral infections can be confused with dengue. Generalized myalgia with pain in the back and shoulders is common. Most patients recover in 5–7 days, but some severe cases can require months of recovery. Occasionally, the patient may never recover from articular dysfunction (Peters and Dalrymple 1990).

The geographic distribution of CHIK virus includes Africa and Asia. On both continents, *Ae. aegypti* maintain human to human transmission cycles. However, in Africa, nonhuman primates also are involved in transmission cycles with a variety of mosquito vector species. Troops of baboons and *Cercopithecus* monkeys are thought to serve as maintenance hosts from

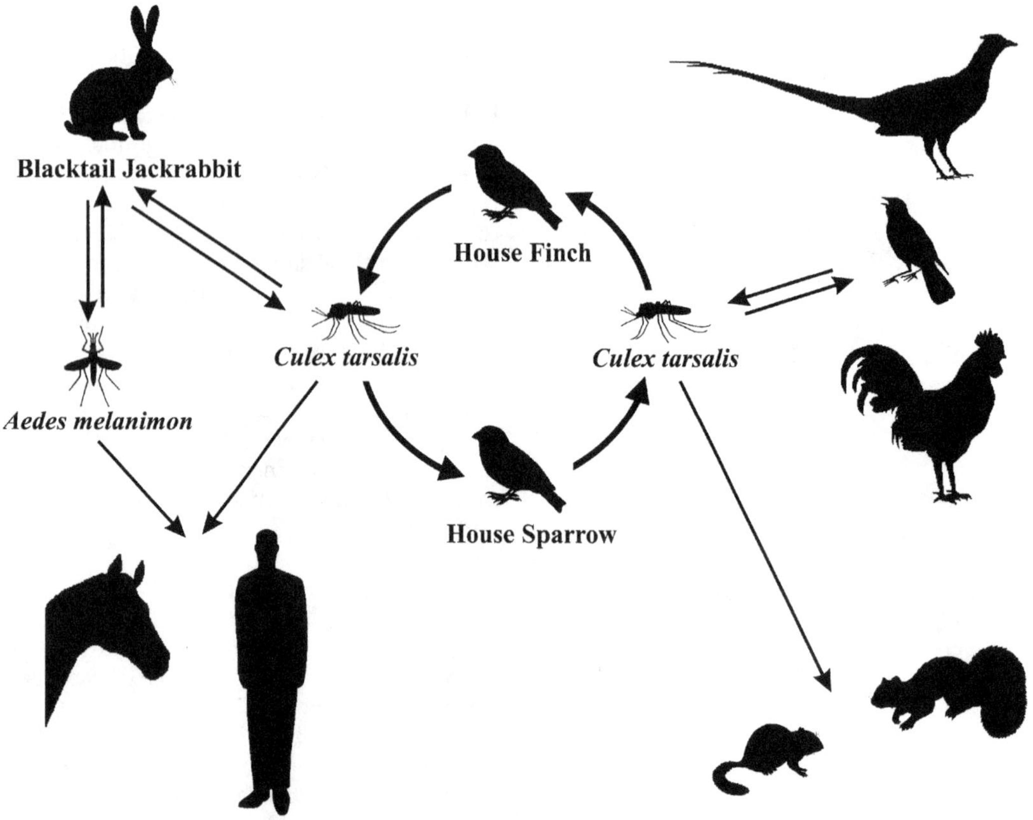

Figure 11.5. Life cycle of western equine encephalomyelitis virus in western North America (after a drawing by James L. Hardy).

which virus occasionally spreads to humans. Mosquitoes responsible for enzootic transmission live in the savannas and forests of tropical Africa and are classified in 2 subgenera of the genus *Aedes*: *Stegomyia* (*Aedes africanus*, *Aedes luteocephalus* and *Aedes opok*) and *Diceromyia* (*Aedes cordellieri*, *Aedes furcifer and Aedes taylori*). *Mansonia africana* may in some situations transfer virus from primate hosts to humans. Although primates in Asia do become viremic after experimental infection with CHIK virus, they are not thought to be important for maintaining virus transmission in that part of the world.

O'nyong-nyong (ONN) virus

O'nyong-nyong virus is classified as a serotype of CHIK virus and has been recovered from Kenya, Tanzania, Uganda, Malawi, Mozambique, Cameroon, the Central African Republic and Senegal. O'nyong-nyong virus occurs only in tropical Africa, whereas CHIK virus is more widely distributed in Africa, overlapping the distribution of ONN virus, and also occurring in Asia. O'nyong-nyong virus was first isolated in 1959 during an epidemic in Uganda. The word o'nyong-nyong is from a Ugandan dialect and refers to joints being weak. Detectable transmission ceased in the 1960s until it reappeared during a Ugandan epidemic during 1996. Cases during the 1996 epidemic were clustered around inland lakes in 3 districts of Uganda near Lake Victoria. Symptoms of sick people during 1996 were the same as those described for CHIK infection during the 1960s. In some villages, 80–90% of the residents were infected and approximately 75% of those infected became ill. ONN virus was isolated from several patients and one mosquito species, *Ma. africana.*

Fever, severe joint and back pain, lymphadenopathy and an irritating rash characterize human infections with ONN virus. Rash is apparent in 60–70% of the patients. All clinically ill patients recovered following symptomatic therapy. Serologic studies suggest that most infections result in overt symptoms.

Virus has been recovered most often from *Anopheles funestus* and *Anopheles gambiae.* Although mosquitoes of the genus *Anopheles* may be vectors of some bunyaviruses, ONN virus is the only arbovirus for which maintenance of transmission is dependent upon mosquitoes of this genus (Peters and Dalrymple 1990). *Anopheles funestus* was the most important vector in East Africa during the 1960s. It frequently bites people, has a distribution that correlates with that of the disease, and has relatively high infection rates. The observed clusters of o'nyong-nyong cases within houses or villages was attributed to the focal distribution and biting behavior of *An. funestus.* It is not known whether there are nonhuman vertebrate hosts for ONN viral transmission. Disease in vertebrates other than humans and laboratory mice has not been reported.

Ross River (RR) virus

Although never fatal, human infection with RR virus can be debilitating and protracted. Fever, rash and polyarthritis may accompany infection. Pain in the joints can last for weeks, sometimes even years, reducing productivity and harming the economy (Mackenzie and Smith 1996). Epidemics attributed to RR viral infections first were noted in rural Australia in 1928. Ross River virus is endemic, causing occasional epidemics in temperate and tropical Australia as well as in parts of Papua New Guinea. As many as 6% of all arthritis cases admitted to hospitals in Papua New Guinea were attributed to RR viral infection. Clusters of cases have been reported from coastal regions, river valleys and irrigated farmlands in Australia. Apparent to inapparent infection ratios are low. Most infected people probably experience mild illness or an inapparent infection.

Ross River virus originally was isolated from *Ae. vigilax* and 3 species of passerine birds. Mosquito vector studies have resulted in isolates from several species in at least 5 genera. Most isolates are from *Culex annulirostris, Ae. vigilax, Aedes normanensis, Anopheles amictus, Mansonia uniformis* and *Coquillettidia linealis* (Kay and Aaskov 1988). Laboratory studies indicated high vector competence for RR virus among a broad range of mosquito species. Nineteen of 22 species exposed to virus became infected. All of the 11 species infected and tested for their ability to transmit virus were competent vectors.

Mammals are considered the vertebrate hosts for enzootic transmission. Serologic studies of wild birds indicated that they seldom are infected. Kangaroos and wallabies are the most likely maintenance hosts. However, humans occasionally may transmit virus. The virus titer in some human sera has been high enough to infect vector mosquitoes. *Aedes aegypti* and *Aedes polynesiensis* may have been important vectors for a human-mosquito-human transmission cycle during the 1979–1980 epidemic in the South Pacific islands.

Sindbis (SIN) virus

Sindbis virus is the prototype alphavirus. It has been the subject of many laboratory-based studies on the biochemistry and replication of viruses. Sindbis virus infectious clones have been used to genetically engineer virus constructs that will interfere with arbovirus transmission by mosquitoes (Olson et al. 1996).

Sindbis virus first was isolated in 1952 from *Culex* mosquitoes collected in the Egyptian village of Sindbis, about 30 km north of Cairo. Its geographic range is one of the broadest of any arbovirus. It is present in Europe, Africa, Asia and Australia. Migrating, viremic avian hosts (Norder et al. 1996) may facilitate virus dispersal.

Recovery of SIN virus from febrile patients in Uganda during 1961 constituted the first known human illness associated with this virus. Most human infections probably are asymptomatic. The ratio of apparent to inapparent infections and rates of human exposure vary from one region of the world to another. For example, along the Nile Valley in Egypt, human infection is common, with seroconversion rates as high as 30%. Conversely, in India, where virus is isolated frequently, human antibody prevalence is only 3%. Variation in disease reported from different geographic locations has been attributed to variation in virus strains, differences in diagnosis and reporting of disease, and differences in human exposure to virus. Sindbis viral infections were considered to be a minor public health problem until epidemics occurred in South Africa during 1974 and northern Europe during 1981–1984. Australia has reported no epidemics.

Sindbis virus likely is maintained in a transmission cycle that includes ornithophilic mosquito vectors in the genus *Culex* and their avian hosts. Virus has been recovered from *Aedes* and *Culiseta* species that also may serve as vectors. The transmission cycle is complicated further by the suggestion that virus moves from the mosquito-avian cycle into humans by additional mosquito species that feed on both avian and human hosts, thus serving as bridge vectors. High natural antibody prevalence and successful laboratory infection studies support the conclusion that birds serve as the primary vertebrate hosts in nature. Humans are considered dead-end hosts, and thus do not contribute to virus amplification.

Semliki Forest (SF) virus

The first isolation of SF virus was from Uganda in 1944. Its geographic distribution is limited to sub-Saharan Africa. Its transmission cycle remains undefined, although it has been isolated from several mosquito species in the genus *Aedes* as well as from *An. funestus*. Serosurveys of humans suggest that people are exposed to SF virus, but no disease associated with natural infections has been reported. Accidental laboratory infections have varied from asymptomatic to a single case of encephalitis. High antibody prevalence in horses suggests that SF virus may deserve more detailed study from the perspective of veterinary health, although no clinical signs have been associated with infections (Peters and Dalrymple 1990).

Barmah Forest (BF) virus

The first BF viral isolate was from the Murray Valley of Australia in 1974 (Lindsay et al. 1995). The first human infection was reported in 1986 and the first associated human disease was reported in 1988. As many as 10% of all epidemic polyarthritis cases in northeastern Australia may be due to infection with BF virus (Mackenzie and Smith 1996).

Barmah Forest viral isolates have been recovered from *Cx. annulirostris* and several *Aedes* species (*Ae. normanensis*, *Aedes bancroftianus*, and *Ae. vigilax*) in Australia. *Culex annulirostris* was considered the principal vector. *Aedes camptorhynchus* was implicated as the principal vector during 1993, in association with an epidemic in southwestern Australia, with secondary involvement from *Cx. annulirostris* and *Coquillettidia* spp. Antibody prevalence studies suggest that birds serve as vertebrate hosts, although marsupials may be involved.

Human BF viral infections are characterized by fever and arthralgia similar to, but milder than, those associated with RR viral infections, which also occur in Australia. In the few clinical cases that have been described, polyarthralgia was present in only a minority of the patients. Arthralgias (acute nerve pain in the joints) are transient but can last for as long as 6 months.

FLAVIVIRIDAE

Arboviruses belonging to the family Flaviviridae are classified in a single genus: *Flavivirus*. Flaviviruses are spherical organisms measuring 45–60 nm in diameter and containing one segment of single-stranded positive-sense RNA.

Calisher and Karabatsos (1988) categorized the arthropodborne flaviviruses based on their antigenic relationships. The medically important flaviviruses are organized into 4 groups. The first 3 are primarily mosquitoborne, the last tickborne (Table 11.3):

1. Viruses not assigned to a complex. At least 19 flaviviruses are not assigned to a complex. Of these, rocio, Wesselsbron and yellow fever viruses are important pathogens of humans and/or domestic animals.
2. St. Louis encephalitis complex. This complex contains 4 disease-causing pathogens: St. Louis encephalitis virus, Japanese encephalitis virus, Murray Valley encephalitis virus and West Nile virus.
3. Dengue complex. The dengue complex contains only dengue virus, with 4 medically important serotypes: dengue-1, dengue-2, dengue-3 and dengue-4.
4. Russian spring-summer encephalitis complex. This complex contains 5 important tickborne pathogens of humans and/or domestic animals.

In addition to the vectorborne flaviviruses, the genus also includes viruses that are not transmitted by arthropods, such as Modoc and Rio Bravo viruses from the western USA and Cowbone Ridge virus from Florida.

Flaviviridae includes some of the most dangerous arboviruses in the world. Yellow fever virus has caused many urban epidemics with high mortality rates, and continues to break out in tropical Africa and South America. Dengue virus may be the largest cause of morbidity among all the arboviruses, and since the emergence of the dengue hemorrhagic shock syndrome in children, a serious cause of mortality as well.

Mosquitoborne Viruses not Assigned to an Antigenic Complex

Rocio (ROC) virus

Several epidemics of meningoencephalitis occurred in southern São Paulo State, Brazil beginning in 1975, caused by a previously undescribed flavivirus subsequently named rocio virus. The disease now is referred to as rocio encephalitis. The distribution of the virus is limited to southern São Paulo State, where epidemics continued to occur each summer until 1980. Adult human males were the most severely affected, suggesting that occupational risk associated with outdoor exposure to an arthropod vector may be an important factor in human disease occurrence. Initial symptoms include headache, fever, vomiting and weakness. Central nervous system symptoms may occur, with death or serious sequelae in the most serious infections. Field and laboratory studies have shown that *Psorophora ferox* and *Aedes*

Table 11.3. Arthropodborne flaviviruses of public health and veterinary importance (family Flaviviridae, genus Flavivirus).

Virus	Vectors	Affected hosts
Rocio (ROC)	*Aedes* spp. *Psorophora ferox*	Humans
Wesselsbron (WSL)	*Aedes* (*Ochlerotatus*) spp. *Aedes* (*Neomelaniconion*) spp. *Mansonia* spp. *Anopheles* spp.	Sheep
Yellow fever (YF)	*Aedes aegypti* *Aedes* (*Stegomyia*) spp. *Haemagogus* spp.	Humans
St. Louis encephalitis (SLE)	*Culex pipiens* complex *Culex nigripalpus* *Culex tarsalis*	Humans
Japanese encephalitis (JE)	*Culex tritaeniorhynchus* *Culex gelidus* *Culex vishnui* complex	Humans, swine
Murray Valley encephalitis (MVE)	*Culex annulirostris*	Humans, horses
West Nile (WN)	*Culex* spp. Ticks?	Humans
Dengue (DEN)	*Ae. aegypti* *Ae.* (*Stg.*) spp.	Humans
Russian spring-summer encephalitis (RSSE)	*Ixodes persulcatus*	Humans
Omsk hemorrhagic fever (OMSK)	*Dermacentor reticulatus* *Dermacentor marginatus* *Ixodes apromorphus*	Humans
Kyasanur Forest disease (KFD)	*Haemaphysalis spingera*	Humans
Louping ill (LI)	*Ixodes ricinus*	Sheep, goats, horses, cattle
Powassan (POW)	*Ixodes* spp. *Dermacentor* spp.	Horses, cattle, humans

scapularis mosquitoes are involved in the transmission of ROC virus.

Rocio virus infects a number of vertebrate animals. A 1975 serosurvey conducted in the epidemic area revealed antibodies to ROC virus in rodents, marsupials, bats, chickens, ducks and pigeons (Lopes et al. 1978). ROC virus has been isolated from wild birds and likely is maintained in a sylvan transmission cycle involving mosquitoes and wild birds. The means by which humans become infected are unknown. ROC virus is an example of a virus that emerged suddenly, produced several severe epidemics and disappeared just as mysteriously as it had appeared.

Wesselsbron (WSL) virus

Wesselsbron virus first was identified in 1955 in southern Africa. The virus is found throughout South Africa, as far north as the Central African Republic to Senegal in West Africa, and in East Africa as far north as Uganda. It has been recovered in Thailand. Floodwater breeding *Aedes* mosquitoes in the subgenera *Ochlerotatus* and *Neomelaniconion* serve as primary vectors of WSL virus. The virus also has been isolated from field-collected mosquitoes belonging to the genera *Mansonia, Culex* and *Anopheles*, as well as from the ixodid tick *Rhipicephalus muhsamae.*

Sheep, especially newborn lambs and pregnant ewes, suffer pathogenic infections of WSL virus producing fever, anorexia and listlessness. Mortality rates of 37% have been seen in experimental infections in lambs. Mild WSL virus infections are observed in nonpregnant cattle, horses, pigs, camels and humans, although infections of WSL virus are rare in humans. Most human cases have been accidental infections in laboratory workers. The incubation period in humans ranges from 2–7 days with onset marked by the sudden appearance of fever, headache, eye pain, body pains and neurologic dysfunction. Illness lasts from several days to several weeks and recovery usually is complete. WSL virus appears to circulate throughout the year in the warm and moist areas of southern Africa. In the inland plateau of South Africa winters are harsh, and transovarial transmission of virus from female mosquitoes to their drought-resistant eggs has been proposed as an overwintering mechanism.

Yellow fever (YF) virus

Historically, YF virus has been among the most important and the most dangerous of the mosquitoborne viruses. Yellow fever in humans was recognized first in the 17th Century, but mosquito transmission of YF virus was not demonstrated until the landmark studies undertaken by Major Walter Reed, Carlos Finlay and their colleagues in Cuba in 1900 (Sosa 1989). However, it was another 27 years before workers at the Rockefeller Foundation's West Africa Yellow Fever Commission isolated the etiologic agent and identified it unequivocally as a virus.

Numerous vertebrates serve as hosts of YF virus. Primates are especially susceptible to infection. Humans developing clinical disease may have headache, nausea, malaise and myalgia initially, followed by fever, vomiting, epigastric pain, prostration, dehydration and jaundice. The name yellow fever derives from this last disease sign, associated with liver malfunction. The most serious cases progress to hemorrhagic symptoms, liver and renal failure. Mortality is 20–50% in patients developing jaundice.

Yellow fever is endemic in the South American and African tropics, but absent in Asia. The reason is unknown. Monath (1988b) proposed 3 hypotheses to explain why YF virus is not found in Asia:

1. A combination of geographic, demographic and ecological factors prevent the movement of YF virus from Africa to Asia. Most major links between Africa and Asia are along the coast, where yellow fever is absent.
2. Immunologic cross-protection produced by exposure of humans to dengue virus, or perhaps to other related viruses, suppresses YF viremias, thus preventing transmission to mosquito vectors.
3. Asian strains of *Ae. aegypti* are less efficient vectors of YF virus than are strains from Africa and South America. There is some experimental evidence to support this.

There is considerable variability among populations of *Ae. aegypti*. A darkly colored sylvan form, *Aedes aegypti formosus*, predominates in Africa south of the Sahara. The larvae of this subspecies are found mainly in treeholes. Elsewhere, the predominant form is *Aedes aegypti aegypti*, which is lightly colored. The larvae of this form breed in a variety of artificial containers, including many present in and around human habitations (tires, flower pots, water jars). This is the domestic form of the species, and the one most frequently associated with yellow fever and dengue transmission (Tabachnick 1991).

Until the early part of the 20th Century, regions of Europe, North and Central America and the Caribbean were subject to periodic reinvasions of *Ae. aegypti* and reintroductions of YF virus that caused widespread yellow fever epidemics. Even today, portions of the world that sustain populations of *Ae. aegypti* mosquitoes risk periodic yellow fever epidemics. Between 1987 and 1991, 18,735 cases of yellow fever with 4,522 deaths were reported to the World Health Organization. This was the highest level of YF virus activity for any 5-year period since 1948 (Robertson et al. 1996). The difficulty of controlling urban populations of the epidemic vector of YF virus, *Ae. aegypti aegypti*, the increase of urban human populations living under substandard conditions, the continued presence of sylvan endemic foci, and problems of vaccine cost, distribution and use suggest that yellow fever will continue to be a serious world problem in the 21st Century.

In tropical America, there are 2 distinct YF virus transmission cycles (Fig. 11.6):

1. The jungle, or sylvan, cycle. YF virus in this cycle is transmitted by *Haemagogus* mosquitoes in the forest canopy to marmosets and monkeys, especially howler monkeys (*Alouatta* spp.), capuchin monkeys (*Cebus* spp.), night monkeys (*Aotus* spp.) and spider monkeys (*Ateles* spp.). The mechanism by which the virus survives dry periods in this cycle is unknown, but may involve (1) continued low-level transmission by the primary vector, (2) transovarial transmission, or (3) virus survival in long-lived secondary vectors, such as *Sabethes chloropterus* mosquitoes.

2. The urban cycle. This cycle begins when humans become infected in the jungle cycle by entering habitats that support enzootic and epizootic transmission. For example, infected *Haemagogus* mosquitoes that follow cut trees to the ground bite woodcutters harvesting tropical hardwood trees. People infected in this way return to their villages or cities thereby initiating urban transmission. The urban cycle in the New World involves a single mosquito subspecies, *Ae. aegypti aegypti*. This subspecies lives and breeds close to humans in rural villages as well as busy cities. Humans are the only vertebrate host in the urban cycle, thus making the urban yellow fever cycle a simple mosquito to human to mosquito transmission cycle.

In tropical Africa, there are 3 distinct transmission cycles:

1. The enzootic forest cycle. In this sylvan cycle, *Ae. africanus* is the principal mosquito vector, and forest dwelling monkeys (such as *Colobus abyssinicus* in Central Africa) are the principal vertebrate hosts. As in South America, humans entering forests to harvest wood or fruit sometimes are bitten by infected mosquitoes and, once infected, can transport virus between the forest and urban transmission cycles.

2. The moist savanna/gallery forest cycle. This cycle involves multiple mosquito vectors, including *Aedes bromeliae* in East and Central Africa and *Aedes vittatus* in West Africa. Many nonhuman primate species serve as intermediate amplification hosts in the savanna/gallery forest cycle. Viremic humans infected by mosquitoes in the forest and savanna zones introduce virus to humans in rural villages and towns.

3. The dry savanna/urban epidemic cycle. In arid sections of Africa, water is stored in large earthen jars. This practice produces large popu-

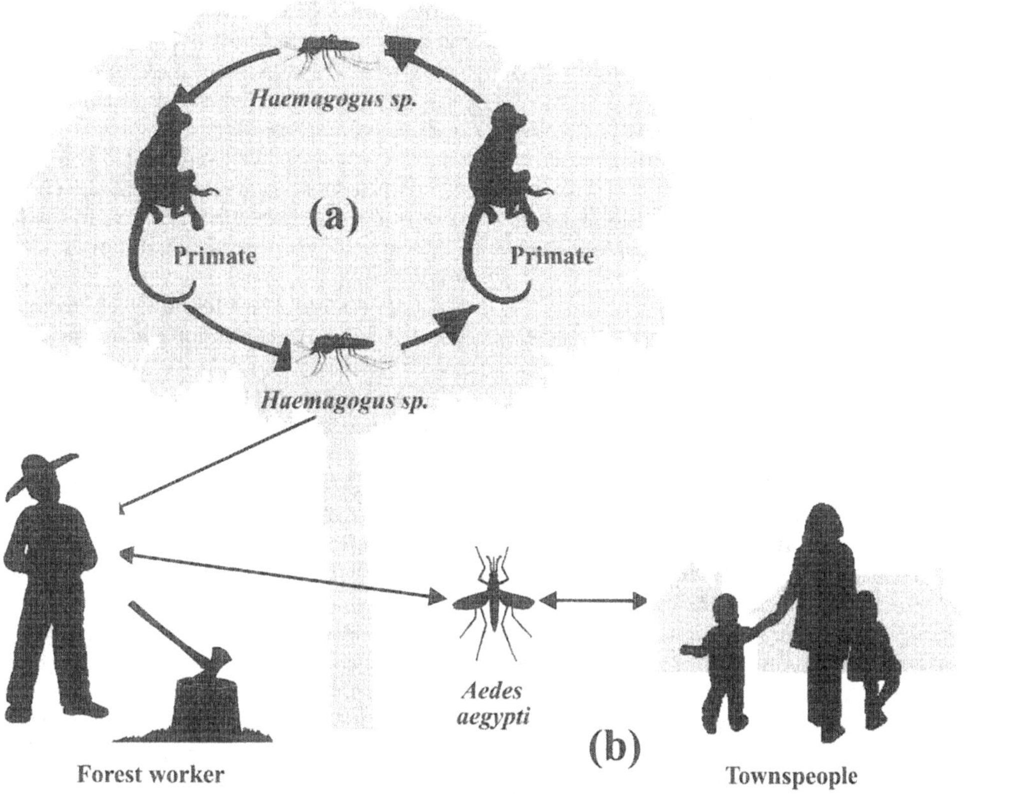

Figure 11.6. Yellow fever transmission cycles in the American tropics: (a) The jungle cycle; (b) the urban cycle. Transmission cycles in Africa are similar, but with different jungle vectors.

lations of domestic *Ae. aegypti aegypti* mosquitoes. People infected with YF virus outside of the village are bitten by mosquitoes when they return home. They thus can initiate an urban epidemic cycle in the semiarid habitats supporting large *Ae. aegypti aegypti* populations that breed in artificial containers.

The French developed a live yellow fever vaccine adapting wild-type virus to mouse brains in the 1930s. This vaccine was discontinued because of risk of postvaccination encephalitis. Theiler and Smith (1937) developed the currently used live vaccine, known as 17D, in the late 1930s by attenuating YF virus as a result of cultivation in cell culture. It is one of the most effective vaccines ever produced. The 17D vaccine confers long-lasting immunity with neutralizing antibodies persisting for up to 35 years. However, for purposes of International Travel Certificates, vaccinations are valid for only 10 years. The 17D vaccine is used in 2 ways to fight yellow fever. An emergency intervention method is used for rapid mass vaccination in anticipation of a yellow fever outbreak. An advantage of this technique is reduced cost because of fewer distribution problems. The vaccine also can be used for preventive immunization where entire at-risk populations are immunized. Preventive immunization has proved to be effective, but it is expensive and programs must be sustained to insure that children are vaccinated when they enter the population.

In spite of its effectiveness, the yellow fever vaccine has problems. The world supply is finite, with production limited to only 15 million doses annually. It is difficult to forecast demand in advance of epidemics, and a reliable cold-chain is required for shipment of the lyophilized vaccine (Monath 1988b).

Mosquitoborne Viruses Assigned to the St. Louis Encephalitis Antigenic Complex

St. Louis encephalitis (SLE) virus

St. Louis encephalitis virus is the most important mosquitoborne virus in the continental USA because of the frequency and size of epidemics and the number of reported human cases (see Monath 1980). St. Louis encephalitis virus first was recognized as a disease agent in 1933 when one of the single largest North American arthropodborne viral disease epidemics was recorded in St. Louis, Missouri. There were 1,095 reported cases and 201 deaths. Lumsden (1958) suspected that *Culex pipiens* complex mosquitoes were vectors, and this insight was supported by the isolation of SLE virus from *Cx. pipiens* females in Yakima, Washington in 1941 (Hammon et al. 1945).

The virus is found from southern Canada south to Argentina. However, the majority of human cases have occurred in the continental USA east of the Mississippi River basin.

St. Louis encephalitis virus infections in humans usually are asymptomatic, with only one of several hundred resulting in CNS symptoms. Patients with clinical disease may have fever, headache, aseptic meningitis and encephalitis. The risk of encephalitis increases with age. Mortality rates also are higher in older individuals.

St. Louis encephalitis infection occurs in many species of vertebrate animals, including most domestic pets and livestock. Signs of infection are rare in these animals, and only birds play any part in transmission cycles.

St. Louis encephalitis virus exists in 3 distinct transmission cycles in the USA:

1. An enzootic/epidemic *Cx. pipiens* complex mosquitoborne transmission cycle in the Midwest, especially along the Mississippi and Ohio River basins.

2. An enzootic/epidemic *Culex nigripalpus* mosquitoborne transmission cycle in south-central Florida (Fig. 11.7).

3. An enzootic/epidemic *Cx. tarsalis* mosquitoborne transmission cycle in California.

Enzootic transmission cycles are maintained between *Culex* mosquitoes and passeriform, columbiform and perhaps other birds. In Florida, SLE virus transmission by infected *Cx. nigripalpus* females has been correlated with rainfall that produces large spring broods of mosquitoes. When considerable numbers of vector mosquitoes coincide with large populations of non-immune wild birds (especially mourning doves and common grackles) and an abundance of SLE virus, a profusion of infected mosquitoes can result. Transmission to humans, who act as dead-end hosts for the virus, often results.

Human St. Louis encephalitis epidemics appear to be cyclic. Epidemics have been reported in Florida during 1958, 1959, 1961, 1977, 1980 and 1990. Similar observations have been made in other parts of the USA, especially in the upper mid-west.

Japanese encephalitis (JE) virus

Epidemics of Japanese encephalitis, known in Japan as "summer encephalitis," were recognized as early as 1870. A major epidemic struck Japan in 1924, involving 6,125 cases and 3,797 fatalities. Japanese encephalitis virus first

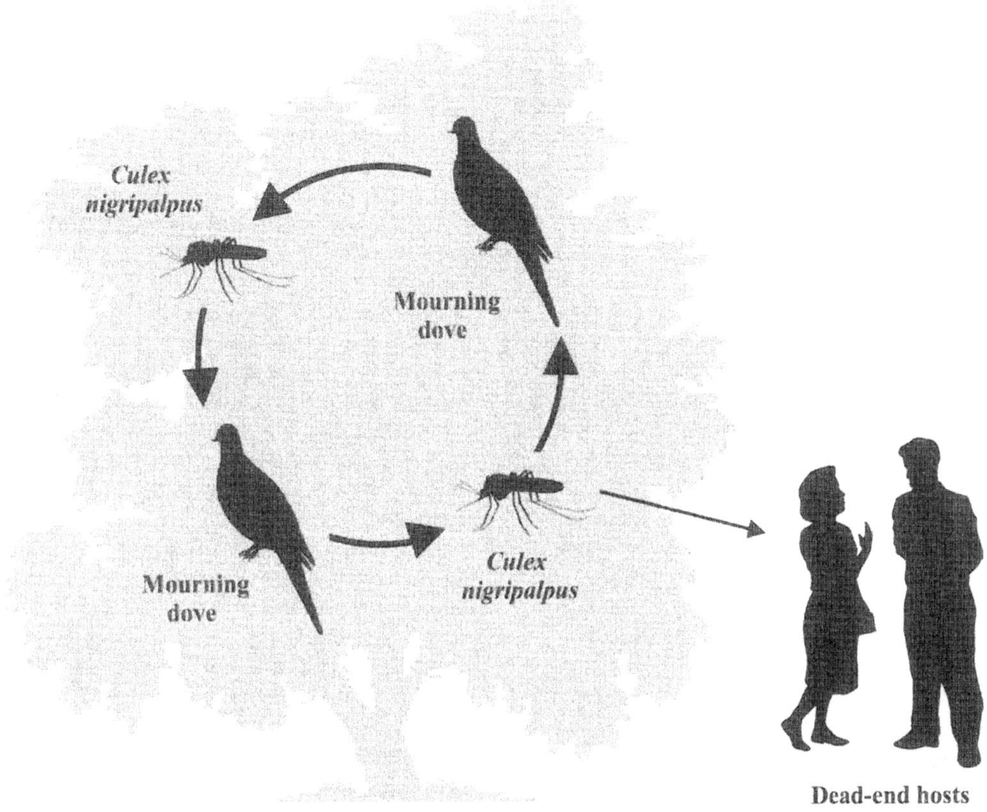

Figure 11.7. St. Louis encephalitis virus transmission cycle in Florida, USA.

was isolated in 1935 from the brain tissue of a fatal human encephalitis case in Japan. Beginning in the 1940s, human cases of encephalitis caused by Japanese encephalitis virus were documented in other Asian countries. By 1984, the known range of the virus had spread to all but the 2 western provinces in China and into Southeast Asia and west through Bangladesh into Nepal (Burke and Leake 1988).

Japanese encephalitis virus has been isolated from a large number of mosquito species. Laboratory and field studies indicate that *Culex tritaeniorhynchus* is the principal vector throughout most of the range of JE virus. Other mosquito species may be locally important JE virus vectors. For example, *Culex gelidus* is an important vector in Southeast Asia and members of the *Culex vishnui* group are important JE virus vectors in Taiwan and India.

Antibodies to JE have been detected from both domestic and wild animals. Domestic swine serve as the major amplification host during Japanese encephalitis epidemics. Virtually all infected swine develop viremias capable of infecting mosquitoes. Wild birds, especially ricefield birds such as water hens, bitterns, cattle egrets, pond herons, black-crowned night herons, plumed egrets and lesser egrets, also are amplification hosts for JE virus.

Japanese encephalitis is an important disease throughout its geographic range. It affects primarily school-age children, is highly lethal and produces permanent neurologic sequelae in survivors. The ratio of apparent to inapparent infections may be as high as 1:20, and in some epidemics one of every 4 infections may result in acute encephalitis (Burke and Leake 1988). The incidence of Japanese encephalitis in northern Thailand in 1998 exceeded that of poliomyelitis at its peak in the USA. Formalin-inactivated mouse-brain vaccines were tested in Japan in the 1930s and licensed in 1954. A protamine sulfate-precipitated vaccine (Biken vaccine) was developed in 1968 and tested in Thailand in the mid-1980s. This and other studies recommend a 3-dose schedule of immunization for protection against JE virus infection. Researchers continue to develop inexpensive, efficacious and safe JE virus vaccines.

Murray Valley encephalitis (MVE) virus

Murray Valley encephalitis virus first was isolated in 1951 from the postmortem brains of 3 patients who died of severe encephalitis in the Murray Valley of New South Wales and northern Victoria, Australia. MVE virus has been isolated in all states of Australia and in New Guinea.

Murray Valley encephalitis virus infection may cause severe symptoms in affected humans including fever, headache, vomiting, diarrhea, encephalitis and coma. Mortality rates in humans range from 20–40%, but as with most arboviral infections, the apparent to inapparent infection ratio is low. Domestic animals and livestock are susceptible to infection by MVE virus, but only horses suffer signs of disease.

Isolates of MVE virus have been made from 10 species of culicine mosquitoes including 6 species of *Culex*, 2 *Aedes* spp. and 2 *Anopheles* spp. However, the majority of field isolates are from *Cx. annulirostris*. Laboratory studies that show it to be a highly competent virus vector further strengthen the importance of this species as an MVE virus vector.

Field observations and laboratory experiments indicate that wild birds, particularly waterbirds in the orders Ciconiiformes (herons and allies), Pelecaniformes (pelicans and allies) and Anseriformes (waterfowl), are the most likely natural amplification and maintenance hosts for MVE virus. The Nankeen night heron (*Nycticorax caledonicus*) is an important amplification host. The size of populations of this heron display cyclic periods that appear to foreshadow Murray Valley encephalitis epidemics. Individual birds are sensitive to MVE virus infection and produce high titer viremias lasting 3–5 days. Other vertebrates that may have a peripheral involvement in MVE virus maintenance and amplification include feral pigs, domestic and feral rabbits, and gray kangaroos.

The epidemic transmission of MVE virus in Australia resembles that of SLE virus in North America because sporadic epidemics occur. Sporadic human cases are reported during non-epidemic periods. Fatality rates were high, especially during the early epidemics of Murray Valley encephalitis, and the distribution of human cases during recognized epidemics covers wide geographic areas (Boughton 1996).

West Nile (WN) virus

West Nile virus first was isolated from a human in the West Nile province of Uganda in 1937 (Hayes 1988). WN virus is distributed widely; human epidemics have been reported in Israel, France and South Africa. In addition to viral isolates made during epidemics, WN virus has been recovered from humans in Uganda, Egypt, India, Pakistan, Central African Republic, Nigeria, Russia, Congo and Ethiopia. In 1999, it was recovered from humans, birds and mosquitoes in the Eastern USA; this was the first evidence of this virus in the western hemisphere (Centers for Disease Control and Prevention 1999). It since has spread to nearly all the continental US states. Evidence of WN virus transmission also has been observed in Botswana, Madagascar, Cyprus, Czechoslovakia and Portugal.

West Nile virus has been isolated from mosquitoes and ticks, but the majority of isolates have been from mosquitoes, indicating the likelihood that they are the primary vectors of WN virus. In Egypt, Israel and South Africa, *Culex univittatus* has been identified as the major WN virus vector. In India and Pakistan, mosquitoes from the *Cx. vishnui* complex likely are involved with WN virus transmission. Several field isolates of WN virus have been made from ticks. Additionally, ticks have been shown to transmit the virus in the laboratory. Their involvement in natural transmission cycles has not been well defined, but it is unlikely that ticks provide an important link in the natural epidemic transmission cycle of WN virus. However, ticks may play a role in virus dissemination or overwintering.

West Nile virus has been isolated from a number of vertebrate hosts including horses, camels, dogs, rodents and bats. Wild birds appear to serve as the primary amplification hosts for WN virus. Hooded crows and house sparrows appear to be particularly important amplification hosts.

West Nile virus causes a febrile dengue-like illness in humans. Clinical disease usually is mild and fatalities have not been reported, but cases in young children occasionally are severe. Human cases tend to be sporadic, with an apparent to inapparent infection ratio of 1:325 reported in Romania. Epidemics are rare, but when they do occur, hundreds (Israel, 1950) to thousands (South Africa, 1974) of clinical cases and hospitalizations are reported.

Mosquitoborne Viruses Assigned to the Dengue Antigenic Complex

Dengue virus (DEN-1, DEN-2, DEN-3, DEN-4)

One of 4 closely related serotypes of dengue virus causes **dengue** and **dengue hemorrhagic fever**. For a thorough treatment of these diseases, the reader should consult Gubler and Kuno (1997). The first clinical description of dengue, or breakbone fever, was made following an outbreak of "bilious remitting fever" in Philadelphia during the summer and autumn of 1780. However, the etiology of the disease was not discovered until 1944. The DEN-1 serotype first was isolated from soldiers who became ill in Hawaii, India and New Guinea. Three strains of the DEN-2 serotype first were isolated in New Guinea. In 1956, 2 additional serotypes (DEN-3 and DEN-4) were isolated from patients with hemorrhagic disease in the Philippines.

Beginning in 1779 and 1780, pandemics of dengue-like illness occurred at intervals ranging from 10–30 yr in Asia and the Americas. The advent of modern diagnostic virology in the 1930s more accurately identified the geographic and epidemiologic range of the dengue serotypes. All 4 dengue serotypes circulate in the large urban centers of Asia. With the spread of *Ae. aegypti* following World War II and the arrival of jet airplane travel in the 1960s, all 4 dengue serotypes have spread throughout the tropical world.

Epidemics caused by multiple dengue serotypes have occurred with increasing frequency and severity. Epidemics have occurred in Southeast Asia, the Pacific region, Africa and tropical America. Epidemic hemorrhagic dengue has become a leading cause of hospitalization and death among children in several affected countries. Cases currently are so widespread that they present a global pandemic, and so dengue is classified as an emerging disease (Gubler 1997).

Dengue virus may be maintained in 3 separate transmission cycles:

1. An urban cycle involving humans and *Ae. aegypti aegypti* in the Old and New World tropics.

2. A rural/suburban cycle involving humans and *Ae. aegypti aegypti* or peridomestic *Aedes albopictus* in Asia, and *Ae. polynesiensis* in the South Pacific.

3. A forest cycle involving nonhuman primates and forest *Aedes* spp. Forest transmission cycles may exist in Asia, Africa and tropical America.

Isolation of dengue virus from pools of male mosquitoes in Africa, *Ae. aegypti* larvae and males in Burma, and adult mosquitoes collected as eggs in Trinidad indicates that transovarial transmission of dengue virus may be involved in all these transmission cycles, but probably at very low rates (Rodhain and Rosen 1997).

Infection with dengue virus causes a spectrum of illness ranging from inapparent infection to severe and sometimes fatal hemorrhagic disease, known as dengue hemorrhagic fever (DHF) or **dengue shock syndrome** (DSS). Classical dengue fever generally is reported in adults and is characterized by sudden onset of fever, headache, nausea, rash and muscle and joint pain. DHF/DSS typically are diseases of children, but adults may be affected as well, especially in the Western Hemisphere where immunity is less likely to be gained in early life. Early symptoms resemble those reported for classical dengue fever, but are followed by hemorrhagic manifestations (e.g., bleeding from the nose and gums), signs of circulatory failure and profound shock with undetectable pulse and blood pressure (DSS) (Fig. 11.8). Appropriate administration of intravenous fluids significantly reduces mortality. Presently, 2 hypotheses explain the occurrence of DHF/DSS:

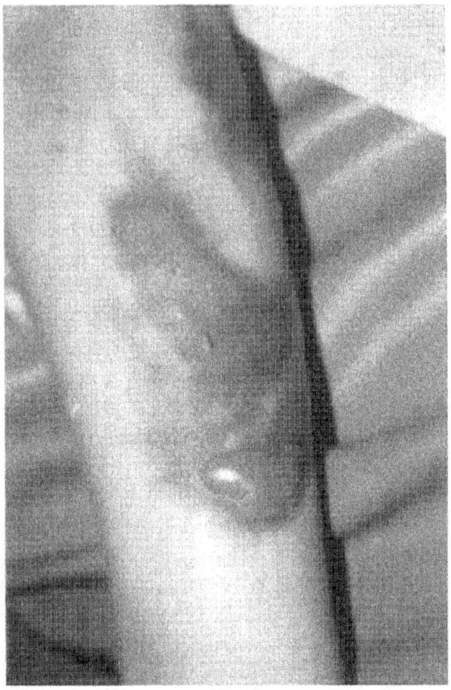

Figure 11.8. A case of dengue hemorrhagic fever from Thailand, showing swollen arm and severe hemorraging (Courtesy of World Health Organization)

1. Immune Enhancement. There is no lasting cross-protective immunity between the dengue serotypes, so an individual may have as many as 4 dengue infections during his or her lifetime. The immune enhancement hypothesis states that people experience antibody-enhanced viral infections. (For contrasting points of view on immune enhancement, see Halstead 1970 and Rosen 1977).

2. Virus Virulence. Occasionally, DSS is observed in primary infections where there is no preexisting antibody and no chance of immune enhancement. Laboratory and field observations indicate that there is strain variation, in terms of disease severity, among the dengue virus serotypes. Some dengue strains are asso-

ciated more with explosive human epidemics, high viremias and DHF/DSS manifestations than others (Gubler and Trent 1994).

Because there is no available human dengue vaccine, nor any specific treatment, vector control is the primary means to prevent and control dengue. Contemporary programs encourage (1) less reliance on organized chemical control programs, and (2) increased community participation to eliminate *Ae. aegypti* breeding sites (Gubler 1989). Efforts to develop a **polyvalent** (against all 4 serotypes) vaccine face formidable challenges.

Tickborne Viruses Assigned to the Russian Spring-summer Encephalitis Antigenic Complex

Russian spring-summer encephalitis (RSSE) virus

Russian spring-summer encephalitis virus was isolated in 1937 from the blood of patients suffering severe encephalitis and from ticks collected in the far eastern regions of Russia. This isolation followed an outbreak of severe encephalitis of unknown origin in the same region in 1932. The distribution of RSSE virus correlates with that of the tick vector, *Ixodes persulcatus*, which is restricted to eastern Russia where the highest incidence of disease in humans is reported.

Ixodes persulcatus ticks are active from late April to early June (Fig. 11.9). The population density of these ticks is linked to the population levels of their vertebrate hosts. These are 3-host ticks, with separate hosts for larvae, nymphs and adults. Forest rodents serve as the principal hosts for larval ticks and for RSSE virus. They have a large and rapid population turnover and develop viremias high enough to infect vector ticks. Birds and medium-sized mammals, including hedgehogs, shrews and moles, serve as hosts for nymphal ticks. Large vertebrates, such as wild deer, domestic animals and humans, serve as blood sources for adult ticks. Large mammals do not contribute significantly to direct virus transmission because their viremias are low, they are long-lived, and once infected with RSSE virus they have lifetime immunity and are removed effectively from the transmission cycle. However, large mammals can contribute indirectly to virus transmission by serving as abundant sources of blood for successful reproduction in adult ticks.

Russian spring-summer encephalitis virus infection in humans generally is severe, with a case fatality rate of 8–54%. Symptoms occur after an incubation period of 10–14 days. Initial symptoms include fever, headache, nausea and neurological involvement. Later, patients may develop aseptic meningitis without signs of brain dysfunction, while others progress to encephalitis with disrupted consciousness.

Omsk hemorrhagic fever (OMSK) virus

A disease of unknown origin occurred in the Omsk district of West Siberia, Russia during the springs of 1945 and 1946. An agent identified as OMSK virus was isolated in 1947 from an acute-phase blood specimen of a patient with disease symptoms consistent with those observed in patients from the same area in 1945–46. The geographic distribution of OMSK virus is restricted to the northern and southern forest-steppe landscape zone of western Siberia (Lvov 1988).

Dermacentor reticulatus ticks are the primary vectors of OMSK virus in the northern forest-steppe region of the Omsk district. *Ixodes apronophorus* ticks are vectors of OMSK virus in the grassy marshes of the western Siberian lowland regions. All developmental stages of this tick species feed on small vertebrates, especially rodents, including muskrats. *Dermacentor marginatus* ticks serve as the primary vector of OMSK virus in the southern forest-steppe region of western Siberia. The virus was isolated from pools of *Coquillettidia richiardii, Aedes*

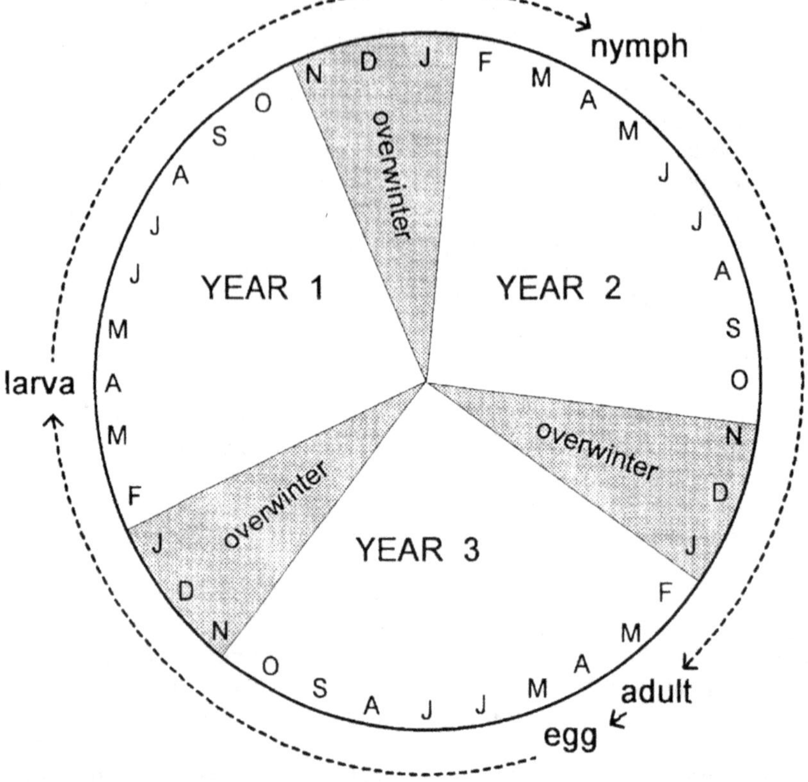

Figure 11.9. Russian spring-summer encephalitis virus transmission cycle.

flavescens and *Aedes excrucians*, indicating that mosquitoes may be involved in OMSK virus transmission.

Muskrats spread rapidly after their introduction to western Siberia in 1928. These rodent populations fluctuate in approximately 10-year cycles. Intense Omsk hemorrhagic fever epizootics often are associated with peak rodent populations. The 1946–1950 rodent epizootic coincided with the first known Omsk hemorrhagic fever epidemic in humans in 1946. Between 1945 and 1949, human morbidity and mortality was reported in endemic regions of West Siberia. Following 1949, Omsk hemorrhagic fever cases declined, then virtually disappeared. Since the early 1950s, only sporadic cases of Omsk hemorrhagic fever have been reported, mostly among muskrat hunters and members of the hunter's families.

Kyasanur Forest disease (KFD) virus

A human epidemic of a disease similar to typhoid fever was reported in the Shimoga district of Karnataka State of India in January of 1957 and 1958. Many sick and dying monkeys were observed in forested areas prior to the start of the epidemic, leading to the hypothesis that the monkey mortality was caused by sylvan yellow fever, a disease unknown on the Indian subcontinent. However, villagers observed that only humans who contacted monkeys became sick. A virus, later identified as KFD virus, was isolated from postmortem specimens of blood and visceral organs collected from monkeys and from the blood of sick humans in nearby villages.

The onset of Kyasanur Forest disease in humans is sudden, with chills and severe frontal headache. Fever follows these symptoms. Diarrhea and vomiting may occur and rarely, hemorrhaging. The case mortality rate is 8–10% (Banerjee 1988).

The geographic distribution of clinical cases in humans and viral isolates have been restricted to the Karnataka State in southwestern India. An average of 450 human cases of Kyasanur Forest disease is reported each year in Karnataka State. Fewer than 40 cases were reported in 1961 but more than 1,000 were reported in 1976, 1977 and 1983.

Kyasanur forest disease virus has been isolated from 10 species of *Haemaphysalis* ticks collected at sites where the virus is known to cycle. Most viral isolates have come from *Haemaphysalis spinigera*, a tick found most often on the forest floor. Infection rates in this species are as high as 1:10 during periods when human and monkey cases are reported. *Haemaphysalis spinigera* is a 3-host tick. Adults of both sexes feed on large domestic animals, such as cattle, goats and sheep, and on large wild mammals, such as spotted deer and Indian bison. Larvae feed on small mammals, such as rats and porcupines and ground-living birds. Porcupines may be important amplification hosts for KFD virus. Larval ticks molt to nymphs that feed on humans, monkeys, other medium-sized mammals and medium-sized ground-living birds. *Haemaphysalis spinigera* nymphs are the most important vectors of KFD virus to humans.

Louping ill (LI) Virus

A disease called louping ill was described in sheep in the Scottish highlands more than 200 years ago (Reid 1988). The etiological agent was identified as a filterable virus in 1931. LI virus occurs primarily in the upland sheep grazing areas of the western coast of the UK. The virus also is reported throughout much of rural Ireland. Encephalitis in sheep caused by LI virus, or a closely related virus, also has been reported in Bulgaria, Turkey, Norway and Spain. *Ixodes ricinus* ticks transmit LI virus.

Louping ill is primarily a disease of sheep. Most cases occur in lambs over one year of age and ewes retained for breeding. LI virus affects the CNS; the most prevalent result is encephalitis. Case mortality may be 60% when susceptible sheep are moved into enzootic areas.

The vertebrate amplification portion of the LI virus transmission cycle is poorly understood. Virus has been isolated from small mammals, including wood mice, field voles and common shrews. In laboratory studies, roe deer had viremias sufficient to infect ticks. Small mammals groom themselves, ingesting large numbers of ticks and perhaps infecting themselves via the oral route. Red grouse, ptarmigan and willow grouse were infected with LI virus in the laboratory. All 3 species developed high and sustained viremias for up to 8 days, but also suffered high mortality. Their importance as natural amplification hosts is questionable. Of the domestic animals tested, only sheep developed virus titers capable of infecting ticks and thus probably serve as amplification hosts in areas where the virus is endemic. Human infections with LI virus are rare. A serologic survey of slaughterhouse workers indicated that 8% had experienced LI virus infection.

Louping ill is a significant problem in sheep. Various methods of control have been employed, including chemical dips for tick control and the use of vaccines.

Powassan (POW) virus

Powassan virus first was isolated in 1958 near the town of Powassan, Northern Ontario, Canada, from brain tissue of a child who suffered a fatal case of encephalitis. POW virus is distributed throughout southern Canada and the northern USA as far south as West Virginia. Serologic evidence from animals from Sonora suggests that POW virus is present and circulating in Mexico. Also, it has been reported from the Pirmor'ye region of southeastern Russia, the People's Republic of China and Southeast Asia (Artsob 1988).

Powassan virus has been isolated from 4 North American ticks: *Ixodes cookei, Ixodes marxi, Ixodes spinipalpus* and *Dermacentor andersoni.* In Russia, the virus has been isolated from ticks (*Haemaphysalis neumanni, Ix. persulcatus* and *Dermacentor silvarum*) and mosquitoes (*Aedes togoi* and *Anopheles hyrcanus*). Virus and antibody have been detected from at least 38 species of North American mammals. Small rodents appear to be the primary amplification hosts of POW virus. Three North American enzootic transmission cycles have been proposed:

1. Arboreal squirrels and *Ix. marxi* ticks in eastern North America.
2. Medium-sized rodents, carnivores and *Ix. cookei* ticks in eastern and midwestern North America.
3. Small- and medium-sized mammals and *Ix. spinipalpus* ticks in northwestern North America.

Human infections of POW virus in North America and Russia are uncommon. Most human cases are diagnosed as encephalitis. Symptomatic infections most commonly are reported in individuals younger than 15 years.

BUNYAVIRIDAE

The family Bunyaviridae encompasses a large group of arthropodborne viruses that share morphological, morphogenic and antigenic properties (Schmaljohn 1996). Categorically, these viruses are referred to as bunyaviruses. There are 5 genera currently recognized in the family. One of them, the genus *Topsovirus*, includes viruses that infect more than 400 species of plants. The other 4, *Bunyavirus, Hantavirus, Nairovirus* and *Phlebovirus*, infect animals. Phleboviruses generally are associated with phlebotomine sand flies, although mosquitoes transmit Rift Valley fever virus. Viruses in the genus *Hantavirus* are transmitted horizontally between vertebrate hosts either directly, or through infected urine or feces. Apparently, hantaviruses do not have arthropod vectors. The remaining 3 genera contain arboviruses. Some of the most important bunyaviruses are characterized in Table 11.4.

The genomes of all bunyaviruses have 3 single-stranded RNA segments, designated large (L), medium (M), and small (S).

The Genus *Bunyavirus*, Bunyamwera Serogroup

Bunyamwera (BUN) virus

Bunyamwera virus originally was isolated from a pool of *Aedes* mosquitoes from Uganda. It, along with a group of related viruses (Ilesha, Germiston, Bwamba and Tataguine), infects humans in Africa and causes febrile illness. Symptoms frequently include headache and anthralgia. The disease tends to be more serious in children.

Mosquitoes transmit BUN virus and related viruses; all are most prevalent in East, West and Central Africa, but occur as far south as Natal. No BUN virus epidemics are known. BUN virus has been isolated most frequently from *Aedes circumluteolus* and *Aedes pembaensis*. It also has been isolated from *Ma. africana, Ma. uniformis* and *Culex* spp. (Karabatsos 1985). BUN virus has infected a wide variety of ani-

Table 11.4. Arthropodborne bunyaviruses of public health and veterinary importance.

Virus	Genus	Vectors	Affected hosts
Bunyamwera (BUN)	*Bunyavirus*	*Aedes* spp. *Mansonia* spp. *Culex* spp.	Humans
Cache Valley (CV)	*Bunyavirus*	*Ae.* spp. *Psorophora* spp. *Anopheles* spp.	Humans, cattle
Northway (NOR)	*Bunyavirus*	*Ae.* (*Ochlerotatus*) spp. *An.* spp.	Humans?
California encephalitis (CE)	*Bunyavirus*	*Aedes melanimon* *Aedes dorsalis*	Humans
Jamestown Canyon (JC)	*Bunyavirus*	*Ae.* (*Och.*) spp. *Culiseta inornata*	Humans
LaCrosse (LAC)	*Bunyavirus*	*Aedes triseriatus* *Aedes hendersoni*	Humans
Snowshoe hare (SSH)	*Bunyavirus*	*Ae.* (*Och.*) spp.	Humans
Tahyna (TAH)	*Bunyavirus*	*Aedes vexans* *Cx.* spp.	Humans
Trivittatus (TVT)	*Bunyavirus*	*Aedes trivittatus* *Ae. triseriatus*	Humans
Oropouche (ORO)	*Bunyavirus*	*Culicoides paranensis*	Humans
Crimean-Congo hemorrhagic fever (CCHF)	*Nairovirus*	*Hyalomma* spp.	Humans
Nairobi sheep disease (NSD)	*Nairovirus*	*Rhipicephalis appendiculatus* *Amblyomma variegatum*	Sheep, goats
Rift Valley fever (RVF)	*Phlebovirus*	*Ae.* (*Neomelaniconion*) spp. *Culex pipiens* Simuliidae *Culicoides* Ticks?	Sheep, cattle, humans
Sand fly fever (SFF)	*Phlebovirus*	*Phlebotomus* spp. *Lutzomyia* spp.	Humans

mals, including goats and sheep. Wild rats may be involved in enzootic cycles.

Cache Valley (CV) virus

The first isolation of CV virus was from a pool of *Culiseta inornata* collected in Cache Valley, Utah (Karabatsos 1985). Since then, it has been isolated from many other species of mosquitoes in the genera *Anopheles*, *Aedes* and *Psorophora*. Widespread in North America, it is more common east of the Rocky Mountains. Circumstantial evidence suggests that CV and related viruses may be responsible for defects in human and other animal fetuses (Calisher and Sever 1995). Cache Valley virus has been implicated as the cause of a human case of hemorrhagic disease in the Eastern USA.

Northway (NOR) virus

Northway virus originally was isolated from a pool of unidentified *Aedes* collected in Northway, Alaska, a small village close to the Canadian border, and about 320 km southeast of Fairbanks (Calisher et al. 1974). It has been isolated subsequently from numerous species of mosquitoes in Canada and California, including *Aedes communis*, *Aedes sierrensis*, *Cs. inornata* and *Anopheles freeborni*.

The public health and veterinary significance of NOR virus is unknown. Antibodies to NOR virus have been detected in large ungulates and in humans in Alaska and Canada, but no human or animal disease has been attributed to this virus.

The Genus *Bunyavirus*, California Serogroup

The California Serogroup of arboviruses includes a number of viruses that infect animals, including humans, and in some cases cause serious CNS disorders. Symptoms in humans with California serogroup infections range from mild, nearly asymptomatic infections to those with respiratory system involvement, to severe CNS diseases such as encephalitis. The 2 most important viruses from a public health standpoint are Lacrosse virus in North America, and Tahyna virus in Europe. Both result in significant numbers of reported illnesses in humans annually, and as is the case with most arboviruses, many more undiagnosed cases. It is impossible to establish the numbers of human disease cases resulting from individual California serogroup viruses in the USA. Most laboratories do not identify individual viral species involved with infections, and cases are reported only as California serogroup infections. Most infections are assumed to be caused by LaCrosse virus because of the geographic location of cases and the presence of its primary vector, *Aedes triseriatus*.

All California serogroup viruses are believed to be transmitted transovarially in mosquitoes. Based on serologic relationships, Calisher and Karabatsos (1988) recognized 4 viruses within the California serogroup: California encephalitis, Melao, trivittatus and Guaroa, with several subtypes and varieties within each group. For convenience, most of these subtypes are treated here as viral species.

California encephalitis (CE) virus

California encephalitis virus was the first virus of this serogroup discovered, and is thus the prototype for the group. A viral isolate was recovered from a pool of *Ae. melanimon* mosquitoes from the Central Valley of California in 1943. Subsequently, 3 cases of encephalitis in humans in California were caused by a virus serologically related to the original isolate. The virus was named California encephalitis. No additional human cases of encephalitis caused by CE virus were detected until 1998, when a case from northern California was diagnosed retrospectively. California encephalitis virus has occurred throughout the western USA, and the virus has been isolated from several mosquito species. *Aedes melanimon* has yielded most of the isolates in California (Reeves 1990) and *Ae. dorsalis* most of the isolates from Utah (Smart et al. 1972). However, many isolates made outside the state of California were not serotyped, and their designation as CE virus must be considered equivocal. Further, there may be questions about specific identifications of mosquitoes in the *Ae. dorsalis* group. Antibodies to CE virus have been detected in a wide variety of mammals, including opossums, gray foxes, raccoons, skunks, porcupines and dusky-footed woodrats.

Jamestown Canyon (JC) virus

Jamestown Canyon virus occurs throughout most of North America, and has been implicated in numerous cases of human disease, including encephalitis. Snowpool or other early spring mosquitoes of the genus *Aedes* are the primary vectors, with the species involved varying from region to region. In the western USA, *Aedes tahoensis* is the primary vector, in the upper Midwest, *Aedes provocans*. In New York State, still other *Aedes* species appear to be involved, including *Aedes stimulans* and members of the *Aedes punctor* group. Jamestown Canyon virus originally was isolated from a pool of *Cs. inornata*, but this species does not seem to be involved as a vector in most areas. The public health significance of JC virus is unresolved in spite of its association with human disease in the midwestern USA (Grimstad et al. 1986). Antibodies to JC virus have been detected in humans, but confirmed cases of disease caused by JC virus are rare. In addition to humans, many other large mammals are infected in nature with JC virus, including deer and other cervids, and domestic livestock.

La Crosse encephalitis (LAC) virus

La Crosse virus is probably the second most important mosquitoborne arbovirus in North America (behind SLE virus). An average of 25–35 human cases is reported annually, mostly from the upper midwestern and mid-Atlantic USA. LAC virus infections cause febrile illnesses in humans accompanied by a variety of symptoms including fever, headache, nausea and vomiting, nuchal rigidity, lethargy, seizures and coma. Antibodies to LAC virus have been detected in various domestic animals, but these animals have not shown any clinical signs as a result of the infections. La Crosse virus occurs over most of the eastern USA; viral activity is most intense in forested areas where the primary vector, *Ae. triseriatus*, is found. *Aedes hendersoni*, a sibling species, also may be a vector. In the eastern USA, a variant of LAC virus probably is transmitted by *Aedes canadensis* (Grimstad 1988). LAC virus is perpetuated in a natural cycle involving mosquitoes and small mammals, especially gray squirrels and chipmunks (Fig. 11.10).

La Crosse virus was the first arbovirus demonstrated conclusively to be transovarially transmitted by mosquitoes (Watts et al. 1973). It also was the first arbovirus shown to be venereally transmitted in the mosquito vector (Thompson and Beaty 1977). Many other California serogroup viruses since then have been shown to be transmitted transovarially in mosquito vectors.

Larvae of *Ae. triseriatus* develop in rot holes of hardwood trees throughout much of North America east of the Rocky Mountains. *Aedes hendersoni* has a similar geographic distribution, but there are spatial and temporal differences in oviposition sites. *Aedes triseriatus* oviposits at ground level, *Ae. hendersoni* in holes several meters higher. Because these species occur in widely dispersed tree holes, control strategies are limited and there are few effective control programs. Most human cases of LaCrosse en-

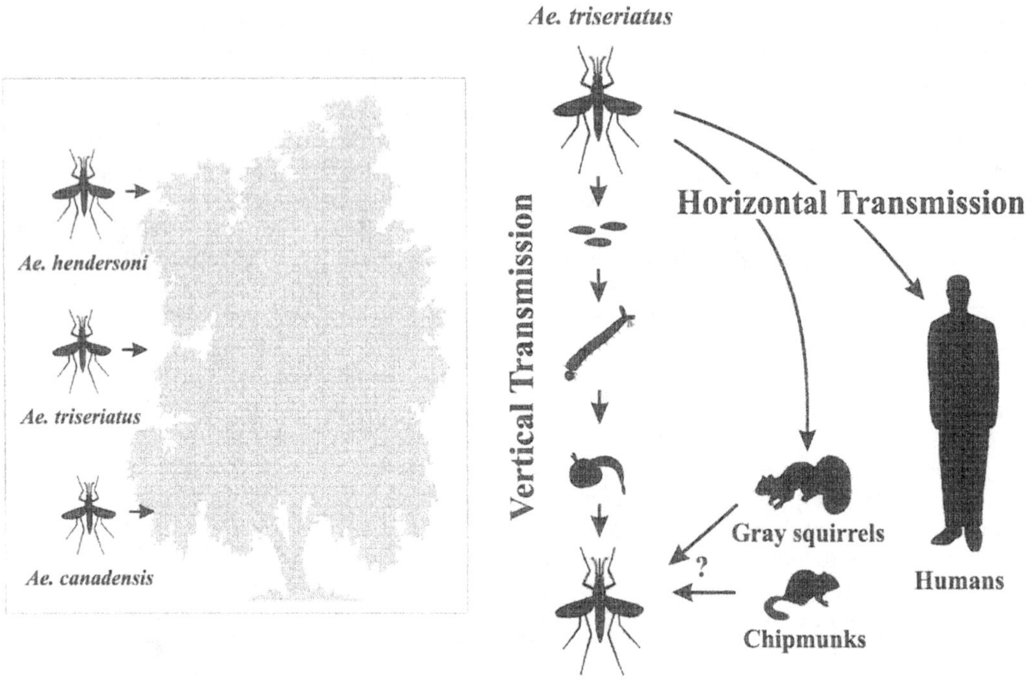

Figure 11.10. La Crosse encephalitis virus transmission cycle.

cephalitis are associated with sources of *Ae. triseriatus* breeding near houses, suggesting that control should be intensified in these situations.

Snowshoe hare (SSH) virus

Snowshoe hare virus is closely related to LAC virus, but occurs in the northwestern USA and Canada. As the name suggests, small mammals such as rabbits serve as vertebrate hosts. SSH virus has been isolated from at least 12 species of *Aedes*, plus species of *Culex*, *Culiseta* and *Anopheles*. Most isolations have been from snowpool *Aedes* in the subgenus *Ochlerotatus*, and these likely are the primary vectors. Snowshoe hare virus probably causes illness in people, and confirmed human cases of encephalitis have been reported (Fauvel et al. 1980).

Tahyna (TAH) Virus

Tahyna virus is common throughout central Europe and Asia and causes considerable human illness. The virus also has been detected in Africa. Infections in humans may result in a febrile influenza-like disease. The most seriously affected patients may suffer from bronchopneumonia, myalgia and meningoencephalitis. There have been numerous outbreaks of disease in humans. Suspected vectors of TAH virus include *Aedes vexans*, *Aedes caspius*, *Aedes cantans* and *Aedes cinereus*. Transovarial transmission has been demonstrated in most of these species. In addition, virus has been isolated from 7 other *Aedes* species, and there have been a few isolations from mosquitoes in other genera (*Culex*, *Anopheles*, *Culiseta* and *Mansonia*). Virus also has been isolated from larvae of *Culiseta annulata* and from sentinel hares in central Europe. Antibodies have been detected in many mammals, including ground squirrels, rabbits, foxes, wild boars, deer, cattle and swine. No clinical signs have been detected in animals other than humans. Humans are accidental hosts of TAH virus, with wild mammals serving as reservoirs.

Trivittatus (TVT) virus

Trivittatus virus is a human pathogen, and can cause disease with CNS symptoms. It occurs throughout the USA, but most isolations from mosquitoes are from the upper Midwest. Isolations have been made most frequently from *Aedes trivittatus* and *Aedes infirmatus*, and infrequently from other *Aedes*, *Culex* and *Anopheles* spp. Eastern cottontail rabbits, fox squirrels, opossums and raccoons show evidence of infection on the basis of antibodies against TVT virus detected in their systems (Pinger et al. 1975).

The Genus *Bunyavirus*, Simbu Serogroup

Oropouche (ORO) virus

Oropouche virus occurs in the Amazon basin of South America. It causes a disease in humans called Oropouche fever. In Brazil it also is called Febre de Mojuí, named for the village Mojuí dos Campos where the first epidemic of the disease occurred in 1975. Oropouche fever is a benign febrile illness in humans, often with chills, myalgia, anthralgia, photophobia and dizziness; nausea, vomiting and diarrhea also may occur.

Oropouche virus infections have resulted in a number of epidemics, some resulting in thousands of cases. The virus is not known to cause disease in any other animal, although several nonhuman primates can be infected. Circumstantial evidence suggests that ORO virus is maintained in nature in a cycle involving monkeys, sloths and birds.

The vector relationships have not been fully established, but most virus isolations have been from *Culicoides paraensis* and *Culex quinquefasciatus*. *Culicoides paraensis* has been incriminated as a vector based on experimental transmission of ORO virus by wild-caught midges that had fed on viremic humans (Pinheiro et al. 1982). Larvae of vector midges develop in water held in banana stalks and discarded coconut husks. Disruption of these habitats may control this disease.

The Genus *Nairovirus*

Crimean-Congo hemorrhagic fever (CCHF) virus

Crimean-Congo hemmorrhagic fever virus has caused sporadic outbreaks of human disease in Asia, Europe and Africa. Most outbreaks have been small, but there have been epidemics in the former USSR and Bulgaria involving 200–700 human cases in a single year. The disease is prevalent in desert and semiarid areas of Eurasia and North Africa. Symptoms include headache, dizziness, fever, stiffness and chills; in severe cases, rashes and hemorrhagic lesions occur. Overall, the case fatality rate varies from 2–5%. Crimean-Congo hemorrhagic fever is unusual among arboviruses because a high proportion of individuals infected with the virus develop clinical symptoms. One out of 5 persons infected may develop symptoms of disease (Watts et al. 1988). Domestic animals are known to be susceptible to infection by CCHF virus, yet there is no evidence that any disease signs develop. The disease may be considered to be of veterinary importance because movement of livestock may be restricted when there is evidence of CCHF virus infection.

Vectors of CCHF virus are ticks in the genus *Hyalomma*, but virus isolations have been made from many other tick species, mostly ixodid 2-host or 3-host ticks. The virus can be transmitted transstadially and transovarially in ticks. Peaks in occurrence of the disease usually coincide with peaks in tick density during warmer parts of the year. Humans also may become infected by contact with infected patients.

Nairobi sheep disease (NSD) virus

Nairobi sheep disease occurs in East Africa and Somalia, and can result in serious losses in non-immune sheep populations. Epizootics can kill one-half to two-thirds of flocks (Davies 1988). Sheep, and to a lesser extent goats, are the principle vertebrate hosts, and most other domestic animals are not susceptible to infection. The virus is transmitted to sheep by ticks, primarily *Rhipicephalus appendiculatus*, and the virus is transmitted transstadially and transovarially in this tick. The tick *Amblyomma variegatum* also can transmit the virus, but the geographic distribution of Nairobi sheep disease is more closely correlated with *R. appendiculatus*. Control of the disease is almost entirely through control of the vector tick using acaricides.

The Genus *Phlebovirus*

Rift Valley fever (RVF) virus

Rift Valley fever is a zoonotic disease responsible for fatal infections in domestic animals and severe infections in humans. Serious epizootics have occurred in Africa. Rift Valley fever is most severe in sheep and cattle, and is considered a dangerous disease of livestock. Accordingly, it is an International Office of Epizootics (OIE) List A disease. The diseases on this list cause non-tariff trade barriers with significant economic losses because they restrict trade in domestic livestock. Infections result in abortion in a high percentage of pregnant animals, and high case fatality rates in lambs and calves. Horses, goats, pigs and camels also are susceptible, but infections are less severe, or are subclinical. Symptoms in humans include fever, myalgia, headache, retro-orbital pain and anorexia. The human case fatality rate is unknown, but human fatalities have been in the hundreds during epizootics in Africa.

A variety of arthropods have been implicated as vectors of RFV virus. Most virus isolations have been made from mosquitoes in several genera, but simuliids and *Culicoides* also have been found infected, and there has been an isolation from a *Rhipicephalus* tick. RVF virus is one of the few phleoboviruses not transmitted by phlebotomine sand flies.

The ecology of RVF virus is complex, consisting of epizootic and interepizootic phases. The virus probably is maintained in the interepizootic phase by floodwater *Aedes (Neomelaniconion)* mosquitoes in which the virus is transmitted transovarially. Species such as *Aedes (Neomelaniconion) macintoshi* are competent vectors, presumably responsible for enzootic transmission of RVF virus. Heavy seasonal rainfall apparently leads to flooding of *Aedes* habitats (called dambos), resulting in abnormally high densities of mosquitoes and causing the virus to spill over into other vertebrate and invertebrate hosts. Horizontal transmission to domestic animals, and to humans, is probably by various species of mosquitoes, including those in the *Cx. pipiens* complex. Transmission also can occur directly between vertebrate hosts. Humans that contact infected animals (veterinarians, abattoir workers, etc.) are at risk (Meegan et al. 1980).

Sand fly fever and other phleboviruses

There is a group of phlebotomine-transmitted viruses of which the sand fly fever viruses Naples (SFN) and Sicilian (SFS) are the best known. Collectively, they are known as Phlebotomus fever viruses. Rift Valley fever virus also is a member of this group of viruses, but it differs from the other phleboviruses in several respects. It is not sand flyborne and it produces different signs of disease in vertebrate animals.

Sand fly fever Naples and SFS viruses both occur in the Mediterranean region, extending south into Egypt and east as far as China. The primary vector of SFS and SFN viruses is *Phlebotomus papatasi*. Therefore, the disease caused by these viruses has been called papatasi fever. It also has been known as 3-day fever based on duration of symptoms.

Sand fly fever is most important when nonimmune individuals enter endemic areas. Several thousand cases occurred in troops stationed in the Mediterranean region during World War II. Transmission tends to be highly focal because of the short flight range of the vectors. Several vector species, including *P. papatasi*, *Phlebotomus perniciosus* and *Phlebotomus perfiliewi*, are peridomestic and highly anthropophilic. However, the New World vectors in the genus *Lutzomyia* tend to be more sylvan, and human infections are less common.

Sand fly fever Naples and SFS viruses can be transovarially transmitted from parent to offspring. Barnett and Suyemoto (1961) were the first to discover this when they isolated SFS virus from male sand flies collected from several Asian countries. Although SFN, SFS and related viruses have been isolated from various animals, there is no evidence of disease in animals other than humans, nor of a maintenance roll for any animal other than the sand fly vector.

All of the phlebotomus fever viruses known to cause human disease produce nearly identical influenza-like symptoms. Symptoms are acute, last for 2–5 days, and are marked by headache, fever, back pain, myalgia, retro-orbital pain and malaise. No recorded deaths have occurred from infections (Tesh 1988).

Phlebotomus fevers occur throughout the world wherever phlebotomine sand flies abound. They are especially common in Europe, Africa, central Asia and the New World tropics.

REOVIRIDAE

Members of the Reoviridae have virions that are from 60–80 nm in size, nonenveloped and possess double protein capsid shells. Genomes contain double stranded, positive sense, linear RNA divided into 10–27 segments.

The family Reoviridae originally was named because the members of the group usually were isolated from respiratory and gastrointestinal tracts, hence the prefix "reo" (respiratory-enteric-orphan viruses). The family now consists of 9 genera including *Orthoreovirus*, *Orbivirus*, *Rotavirus*, *Coltivirus*, *Aquareovirus*, *Cypovirus*, *Fijivirus*, *Phytoreovirus* and *Orzavirus* (Murphy et al. 1995). Only the orbiviruses and coltiviruses contain arboviruses that infect animals (Table 11.5).

The *Orbivirus* genus contains 13 serogroups in addition to several unclassified viruses. Several groups contain viruses that are transmitted by species of biting midges in the genus *Culicoides*. The *Culicoides*-transmitted orbiviruses of veterinary importance include African horse sickness, bluetongue, epizootic hemorrhagic disease, ibaraki, palyam and equine encephalosis viruses. Changuinola, lebombo, orungo and kemerovo viruses are orbiviruses that have been implicated with human disease; each has been associated with at least one human case, generally a self-limiting

Table 11.5. Arthropodborne reoviruses of public health and veterinary importance.

Virus	Genus	Vectors	Affected hosts
Bluetongue (BLU)	*Orbivirus*	*Culicoides* spp.	Sheep, goats, cattle
African horse sickness (AHS)	*Orbivirus*	*Culicoides* spp.	Horses
Epizootic hemorrhagic disease (EHD)	*Orbivirus*	*Culicoides* spp.	Deer
Equine encephalosis (EE)	*Orbivirus*	*Culicoides* spp.	Horses
Palyam (PAL)	*Orbivirus*	*Culicoides* spp.	Cattle
Colorado tick fever	*Coltivirus*	*Dermacentor andersoni*	Humans

febrile illness (Monath and Guirakhoo 1996). Chanquinola is transmitted by phlebotomines. Lebombo has been isolated from *Ma. africana* and from *Ae. circumluteolus* mosquitoes. Orungo virus is distributed widely in Africa, but the severity of disease symptoms is not known. It has been isolated from *An. funestus*, *An. gambiae*, *Aedes dentatus* and *Culiseta perfuscus*. Kemerovo viruses have been isolated from humans with febrile illness in eastern Europe, and are transmitted by ticks (e.g., *Ix. persulcatus* and *Hyalomma anatolicum*).

The *Coltivirus* genus contains 3 serotypes, but only Colorado tick fever virus is a recognized arbovirus that infects humans and causes a fatal human disease.

Bluetongue (BLU) virus

This is the type virus for the genus *Orbivirus*. BLU virus consists of 24 different serotypes (BLU1–24). Bluetongue virus is distributed throughout most of the tropical and subtropical world and infects ruminants wherever suitable *Culicoides* vectors are found (Tabachnick 1996). The major vectors of BLU virus are: *Culicoides imicola* in Africa; a member of the *Culicoides variipennis* complex, *Culicoides sonorensis* in North America and *Culicoides brevitarsis* in Australia.

Bluetongue virus causes disease in sheep characterized by high temperature, hyperemia and swelling of the buccal and nasal mucosa, profuse salivation, swollen tongue, hemorrhage in the mucosal membranes of the mouth, oral erosions and hemorrhages in the coronary bands of the hoof that results in lameness (Fig. 11.11). Vomiting may occur that can lead to pneumonia and frequently death. Sheep mortality may range from 5–50%. Cattle develop a prolonged viremia lasting several weeks but only rarely (<5% of infected animals) show clinical signs of bluetongue. However, due to their prolonged viremia, cattle serve as reservoirs for BLU virus. Viremic bulls are capable of passing the virus to noninfected cows venereally. As a result, in an effort to protect sheep populations, countries that are BLU virus-free place non-tariff trade restrictions on cattle, cattle semen and sheep from bluetongue enzootic regions. Bluetongue is an OIE List A disease along with Rift Valley fever.

Bluetongue virus is well characterized (Roy 1996). The 10 RNA segments of the genome have been sequenced and the 7 structural and

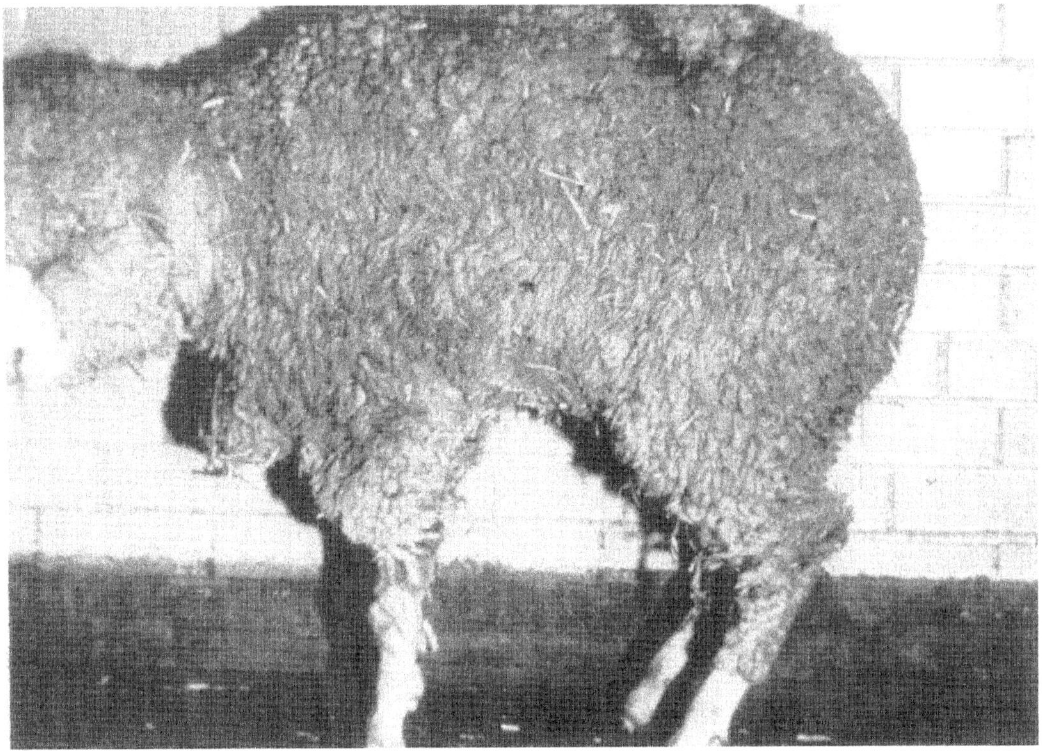

Figure 11.11. Sheep suffering from bluetongue (courtesy of Bennie I. Osburn).

3 nonstructural proteins encoded by the segments have been identified. The genetic diversity among the BLU serotypes has shown higher genetic similarity within geographic regions, supporting the concept of geographic topotypes and the evolution of the group along geographic lines (Gould et al. 1992).

Bluetongue disease proves the need for information to reduce the economic impact of arboviral diseases of cattle. The risk of BLU virus spreading to new regions depends on the presence of suitable *Culicoides* vectors. Hence, understanding *Culicoides* vector competence for BLU viruses is essential to predict risk. Studies on the epidemiology of bluetongue and the distribution and vector competence of *Culicoides* in the USA have lead to proposals that the northeastern USA should be considered a BLU virus-free region. The North American vector *C. sonorensis* is absent from this region and there is little serologic evidence of active BLU virus transmission among cattle. Establishment of bluetongue-free regions within countries would provide greater economic opportunity for trade in livestock and livestock germplasm. Several attenuated bluetongue vaccines are available in different countries for specific serotypes. Recombinant vaccines against specific BLU virus epitopes are being developed.

African horse sickness (AHS) virus

The virus that causes African horse sickness consists of at least 9 serotypes (AHS 1–9; Hess 1988). AHS virus serotypes generally have been confined to sub-Saharan Africa; however, on occasion the virus has caused epizootics elsewhere, e.g., the Middle East, Pakistan, Afghanistan, North Africa and the Iberian Peninsula.

African horse sickeness virus causes severe disease in equids and is an OIE List A disease. Clinical signs of the **peracute** (very acute) form of African horse sickness include fever, pulmonary edema, pleural and pericardial effusions. Death occurs within a few hours of onset in over 90% of the animals with this form of the disease. A less virulent form called the cardiac or subacute form is characterized by a fever, edema of the head and neck and progressive signs of cardiac failure and tamponade (compression or fluid in the heart). Death occurs in about 50% of the animals within 4–8 days. Horses are more susceptible than mules and donkeys. Dogs, zebras and camels can be infected naturally with the virus. No natural human infections of AHS virus have been reported. However, 4 laboratory workers exposed to lyophilized neuroadapted AHS virus in a vaccine factory in South Africa became ill with encephalitis and chorioretinitis (Van Der Meyden et al. 1992).

The primary vector of AHS virus is *C. imicola.* This species is found in sub-Saharan Africa and has been implicated as the vector in almost all recent African horse sickness epizootics.

The Republic of South Africa distributes attenuated AHS vaccines against several of the AHS serotypes. Recombinant vaccines against specific AHS epitopes are being developed. Although the North American BLU virus vector, *C. sonorensis,* has been shown to be a capable AHS virus vector in the laboratory, AHS virus has never appeared in the Western Hemisphere.

Epizootic hemorrhagic disease (EHD) virus

Epizootic hemorrhagic disease virus consists of 10 serotypes (EHD 1–10). EHD virus is closely related to BLU virus and is transmitted by members of the *C. variipennis* complex in North America. The same vectors of BLU virus also are likely vectors of EHD virus. EHD virus has been isolated in Australia, Japan and Africa.

Epizootic hemorrhagic disease virus causes a severe disease in deer. Other ruminants (e.g., elk, sheep and cattle) may be infected but rarely show clinical signs of illness. Clinical signs of illness in deer include fever, mucosal congestion, drooling, recumbency and terminal convulsions. Many deer may hemorrhage from the mouth, nares and rectum. Mortality in deer has been reported as high as 90%, but this is probably an overestimate, as serologic surveys have revealed high prevalence rates in normal deer (Nettles et al. 1992). EHD virus is an important infectious disease agent in wild deer populations in the USA.

Equine encephalosis (EE) virus

There are 7 known serotypes of EE virus. These serotypes are confined to South Africa where they cause a disease in horses similar to the fatal peracute form of African horse sickness. The virus is transmitted by various species of *Culicoides.*

Palyam (PAL) virus

Palyam virus consists of 2 serotypes: Kasba and Chuzan. They have been found in South Africa and Japan, respectively, and both are believed to be transmitted by species of *Culicoides,* probably *Culicoides oxystoma.* The Chuzan serotype has been implicated as the cause of congenital abnormalities in cattle in Japan. Clinical signs of this disease include hydroencephaly and cerebellar hyperplasia. The specifics of the

Table 11.6. Arthropodborne rhabdoviruses of public health and veterinary importance.

Virus	Genus	Vectors	Affected hosts
Vesicular stomatitis (VSV)	*Vesiculovirus*	*Lutzomyia* spp. *Culicoides* spp. Simuliidae Muscoid flies Chloropidae	Horses, cattle, swine, humans
Bovine ephemeral fever (BEF)	*Ephemerovirus*	*Culicoides* spp. *Culex* spp. *Aedes* spp.	Humans

disease transmission cycle, the viruses and their vectors, remain unknown. An inactivated whole virus vaccine against the Chuzan serotype is marketed in Japan.

Colorado tick fever (CTF) virus

Colorado tick fever virus is a member of the genus *Coltivirus*. It contains 12 double stranded RNA segments and can cause human disease. CTF virus occurs principally in forest habitats between 1,000 and 3,000 m in elevation in the Rocky Mountains of the USA and Canada. This region is the range of the principal tick vector, *D. andersoni*. CTF virus has been isolated from several other tick species, but the role of these species in transmission of the virus is uncertain. Humans are a dead-end host for the virus and do not serve as a source of infection for ticks. A variety of mammalian species serve as hosts for infecting ticks, e.g., porcupines, coyotes, ground squirrels, woodrats and deer mice.

Colorado tick fever in humans has an average incubation period of 4 days, but onset can range 1–19 days after the bite of an infected tick. Symptoms include fever, chills, headache, myalgia, malaise and retro-orbital pain. Gastrointestinal problems may occur. The acute phase of the illness may last 5–10 days. The disease may necessitate hospitalization in about 20% of cases. There were 1,432 documented human cases in the USA between 1980 and 1988. However, since the disease is easily confused with other diseases, it probably has been underreported. There have been 3 fatalities attributed to Colorado tick fever, all children (Monath and Guirakhoo 1996).

RHABDOVIRIDAE

Members of the family Rhabdoviridae are distributed widely in nature and have infected vertebrates, invertebrates and many plant species. There are 5 genera, 4 of which contain viruses that are transmitted by arthropods. Viruses in one genus, *Lyssavirus,* containing the rabies virus, do not infect arthropods and are transmitted only among vertebrates. The important arbovirus diseases of humans and other animals are found in the genus *Vesiculovirus,* containing vesicular stomatitis virus, and in the genus *Ephemerovirus,* containing bovine ephemeral fever virus (Table 11.6).

The Rhadoviruses are large (130–380 nm long, 70–85 nm in diameter), bullet-shaped viruses that are enveloped and contain single stranded RNA consisting of 5–6 genes (Wagner and Rose 1996). The genome generally is compact. For example, vesicular stomatitis virus has only 2 nucleotides between each of the genes encoding 5 mRNAs.

Vesicular stomatitis (VS) virus

Vesicular stomatitis virus is the type virus in the genus *Vesiculovirus*. The 3 primary serotypes of VS virus are: vesicular stomatitis New Jersey (VSNJ), Indiana (VSI) and Alagoas (VSA). VSNJ and VSI serotypes are responsible for most animal disease; the VSA serotype was isolated during an epizootic in Brazil. The nucleotide sequence has been determined for both VSI and VSNJ serotypes (Nichol et al. 1993). VS virus is found in the Western Hemisphere and is enzootic in the eastern USA, Mexico, Central America, Colombia, Ecuador and Peru.

Vesicular stomatitis is a disease of horses, cattle and swine and is considered one of the oldest known diseases of livestock. VS virus infections of humans have been recorded, although reports of actual human disease are infrequent. In humans, the disease is characterized as an acute influenza-like illness that is self-limiting. Due to the short duration of the illness and its nonspecific nature, most human cases caused by VS virus probably go unrecognized.

Vesicular stomatitis of livestock is characterized by a short febrile period and the appearance of papules and vesicles in the mouth, on the udder, in interdigital areas and on the coronary band. Profuse salivation is often a prominent sign. Common complications are secondary bacterial infections, myiasis and mastitis. Case fatality is low, but the disease can cause appreciable economic losses, primarily when it affects dairy cattle and swine.

Differential diagnosis must be made rapidly because of its resemblance to foot and mouth disease. The OIE characterizes vesicular stomatitis as a List A disease that could spread rapidly across international borders causing serious loss to agriculture. Therefore, the presence of VS virus justifies restricting the export of livestock from an affected country to VS virus-free areas. During a 1995 vesicular stomatitis epizootic in the southwestern USA, horses and livestock from affected states were restricted in their transport to unaffected states and to several countries, e.g., Canada and members of the European Union. The economic costs of lost trade, closed sales, closed rodeos and livestock shows were ~$50 million during this epizootic.

The modes of transmission of VS virus are complex. The virus spreads both by contact transmission between animals and by arthropod transmission (Webb and Holbrook 1989). Arthropods play a significant role in the transmission of the VSI serotype in tropical areas. The sand flies *Lutzomyia longipalpis* and *Lutzomyia shannoni* transmit the VSNJ serotype in Georgia in the USA. The VSNJ serotype in the southwestern USA has been isolated from a variety of field-collected arthropods, e.g., *Culicoides*, Simuliidae, Chloropidae, Anthomyiidae and *Musca*. *Simulium vittatum* has been shown to be a capable vector in the laboratory (Cupp et al. 1992), and VS virus-infected *C. sonorensis* transmitted the virus to vertebrates under laboratory conditions. The relative roles of arthropod transmission and animal contact transmission in the spread of VS virus remain unknown. Effective control that involves vaccination, animal quarantines and arthropod vector control requires more information on the primary modes of VS virus transmission.

Bovine ephemeral fever (BEF) virus

Bovine ephemeral fever virus is a member of the genus *Ephemerovirus*. The virus can be found in Africa, the Middle East, Asia and Australia. Bovine ephemeral fever is a disease of cattle but it also infects a range of ruminants subclinically.

Mortality in cattle usually is low during epizootics, but morbidity levels can be high. The disease is of short duration, with fever lasting

Table 11.7. Arthropodborne African swine fever-like viruses of public health and veterinary importance.

Virus	Genus	Vectors	Affected hosts
African swine fever	Unnamed	*Ornithodoros* spp.	Swine

2–3 days accompanied by stiffness, lameness and a major drop in milk production. The disease eventually may destroy the animal, particularly those having permanent lameness. Several countries have tested vaccines in an effort to control the disease. BEF virus testing is sometimes required for cattle shipment to BEF virus-free regions.

The virus has been isolated from a variety of hematophagous insects including species of *Culicoides, Culex* and *Aedes*. The specific species involved in the transmission of BEF virus and their role in the epidemiology of the disease remain unknown.

African Swine Fever-like Viruses

African swine fever (ASF) virus

African swine fever virus is in an unnamed family and unnamed genus for African swine fever-like viruses (Table 11.7). Only one serotype has been identified. ASF virus is the only known arbovirus with a DNA genome.

African swine fever is a highly contagious disease of swine. ASF virus infects domestic and wild pigs, warthogs, bushpigs and giant forest hogs. The virus is endemic in most of southern Africa and on the Iberian Peninsula. African swine fever outbreaks have occurred in Europe and in the Caribbean.

Strains of ASF virus differ in virulence from nonpathogenic to those causing near 100% mortality. The virus may incubate for 5–15 days and clinical signs may range from a peracute form with death occurring rapidly, to a mild chronic form with transient, recurring fever, possible pneumonia, lameness, skin lesions and secondary infections. Pigs may be chronically infected and excrete virus for 6 weeks after infection.

African swine fever virus transmission may occur via contact, infected meat, fomites and biting flies. Biological transmission occurs via soft ticks in the genus *Ornithodoros*. The reservoir for the virus in Africa is warthogs and their associated ticks, *Ornithodoros moubata porcinus*. The virus spreads to domestic swine when these contact infected ticks. Contact transmission has been an important mode for ASF virus spread during outbreaks in Europe. During the acute phase of the disease (high fever, skin blotching, recumbency, diarrhea and abortion, with mortality near 100% within 7 days), the virus is shed in high concentrations in excretions and secretions from infected pigs. African swine fever is an OIE List A disease.

EMERGING VIRUS DISEASES

Emerging diseases are defined as those that are spreading rapidly or those that are newly recognized (Morse 1993). Some of the most feared human and animal viral diseases seem to emerge suddenly from either obscurity or complete absence. Hantavirus infections in the USA and AIDS worldwide are 2 examples of virus diseases that are newly recognized and spreading. They have emerged with little warning. KFD virus was discovered only in the 1950s. La Crosse encephalitis was unknown before the 1960s. The dengue viruses have emerged as a great public health problem in Asia, the Caribbean basin and countries in South America. In

1999, an epidemic of arboviral encephalitis occurred in New York City. The virus responsible eventually was identified as a West Nile-like virus, an agent previously known only from Africa, Asia and eastern Europe (Centers for Disease Control and Prevention 1999).

Many reasons have explained why emerging virus diseases seem to show up with increasing frequency. Increased air travel, increased animal trade and movement, better diagnostic capabilities, increased awareness, and increases in contacts between humans and their domestic animals with wild animal disease cycles have been suggested. As a result, there has been new emphasis on world surveillance for emerging diseases and for basic research on vectorborne disease ecology. This emphasis has included greater interest in the evolution of virus-host relationships. The RNA viruses are believed to have high mutation rates, and consequently, great potential for evolving new virus types. Certain alphaviruses are believed to have arisen by recombination of the genomes of other alphaviruses. Western equine encephalomyelitis virus is believed to have descended from a recombination event between EEE-like and SIN-like viruses (Hahn et al. 1988). Similarly, BF virus may have descended from a recombinant ancestor (Lee et al. 1997). Recombination could occur when 2 different viruses co-infect either a mosquito or vertebrate host (Kuhn et al. 1996). Arboviruses with segmented genomes may have even greater opportunities for diversification through reassortment of the segments during dual infections with more than one virus. A number of arboviruses, including BLU virus and LAC virus, apparently reassort in nature (Beaty et al. 1988). An essential feature of arthropodborne pathogens is the role of the vector in influencing the biology of the pathogens and the effects of the vector on the host. The arthropod vector is more than a flying syringe and influences viral evolution, pathogenesis and host immunity (Tabachnick 1998). To predict and control emerging arboviral pathogens will require detailed understanding of the complex interactions between host, virus and vector.

Climatological models have projected a gradual warming worldwide over the next 100 years, with a rise in mean temperature of 1.0–3.5°C and an increase in precipitation of 7–15% (Feddema and Mather 1992). There is considerable interest in the possible effect such an increase might have on the transmission of arboviruses. Jetten and Focks (1997) used predictive models to evaluate possible changes in dengue transmission worldwide. Reeves et al. (1994) discussed possible changes in WEE and SLE virus transmission in California based on current studies of temperature effects on a variety of virologic and entomologic factors.

Only about 100 years have passed since the discovery of the first arbovirus. The enormous changes in research methods and our understanding of the diversity in all aspects of arbovirus disease cycles suggest that we have only begun to comprehend the factors involved in the appearance and reappearance of these organisms.

CONTROL OF ARBOVIRUS DISEASES

Because there are no specific treatments available for arbovirus diseases, patients receive supportive care only. For control of these diseases in humans and domestic animals the emphasis is on preventive measures, either vector control or vaccination. There is a highly effective vaccine against YF virus for general use in humans, and vaccines against selected other arboviruses for laboratory workers and military personnel. In the USA, horses are vaccinated routinely against WEE and EEE viruses. An effective vaccine to protect humans against JE virus infection has been available for many years. There have been attempts to develop a vaccine against DEN virus. Recent efforts have been promising, but a licensed vaccine currently is not available. Other control measures resemble those used for all arthropodborne diseases: surveillance, avoidance of exposure to arthropods (personnel protective measures) and suppression of vector populations. Reeves and Milby (1990) reviewed the concepts and strategies for control of arbovirus vectors.

Control of arbovirus diseases in livestock includes the quarantine of animals from infected herds and the restriction of international trade in both animals and animal germ plasm.

REFERENCES

Artsob, H. 1988. Powassan encephalitis. Pp. 29–49 *in* T.P. Monath (ed.), The arboviruses: epidemiology and ecology, Vol. 4. CRC Press, Boca Raton, Florida.

Banerjee, K. 1988. Kyasanur Forest disease. Pp. 93–116 *in* T.P. Monath (ed.), The arboviruses: epidemiology and ecology, Vol. 3. CRC Press, Boca Raton, Florida.

Barnett, H.C. and Suyemoto, W. 1961. Field studies on sandfly fever and Kala-azar in Pakistan, in Iran, and in Baltistan (Little Tibet) Kashmir. Trans. N.Y. Acad. Sci. **23:**609–617.

Beaty, B.J., Trent, D.W. and Roehrig, J.T. 1988. Virus variation and evolution: mechanisms and epidemiological significance. Pp. 59–85 *in* T.P. Monath (ed.), The arboviruses: epidemiology and ecology, Vol. 1. CRC Press, Boca Raton, Florida.

Boughton, C.R. 1996. Australian arboviruses of medical importance: a handbook for general practitioners and other clinicians. Royal Australian College of General Practitioners, Melbourne, Australia. 67 pp.

Brès, P. 1988. Impact of arboviruses on human and animal health. Pp. 1–18 *in* T.P. Monath (ed.), The arboviruses: epidemiology and ecology, Vol. 1. CRC Press, Boca Raton, Florida.

Burke, D.S. and Leake, C.J. 1988. Japanese encephalitis. Pp. 63–92 *in* T.P. Monath (ed.), The arboviruses: epidemiology and ecology, Vol. 3. CRC Press, Boca Raton, Florida.

Calisher, C.H. and Karabatsos, N. 1988. Arbovirus serogroups: definition and geographic distribution. Pp. 19–57 *in* T.P. Monath (ed.), The arboviruses: epidemiology and ecology, Vol. 1. CRC Press, Boca Raton, Florida.

Calisher, C.H., Lindsey, H.S., Ritter, D.G. and Sommerman, K.M. 1974. Northway virus: a new Bunyamwera group arbovirus from Alaska. Canad. J. Microbiol. **20:**219–223.

Calisher, C.H. and Sever, J.L. 1995. Are North American Bunyamwera serogroup viruses etiologic agents of human congenital defects of the central nervous system? Emerg. Infect. Dis. **1:**147–151.

Casals, J. 1961. Procedures for identification of arthropod-borne viruses. Bull. Wld. Hlth. Organiz. **24:**723–734.

Centers for Disease Control and Prevention. 1999. Outbreak of West Nile-like viral encephalitis – New York, 1999. MMWR **48:** 845-849.

Crans, W.J., Caccamise, D.F. and McNelly, J.R. 1994. Eastern equine encephalomyelitis in relation to the avian community in a coastal cedar swamp. J. Med. Entomol. **31:**711–728.

Cupp, E.W., Mare, C.J., Cupp, M.S. and Ramberg, F.B. 1992. Biological transmission of vesicular stomatitis virus (New Jersey) by *Simulium vittatum* (Diptera: Simuliidae). J. Med. Entomol. **29:**137–140.

Davies, F.G. 1988. Nairobi sheep disease. Pp. 191–203 *in* T.P. Monath (ed.), The arboviruses: epidemiology and ecology, Vol. 3. CRC Press, Boca Raton, Florida.

Dhileepan, K., Azuolas, J.K. and Gibson, C.A. 1996. Evidence of vertical transmission of Ross River and Sindbis viruses (Togaviridae: *Alphavirus*) by mosquitoes (Diptera: Culicidae) in Southeastern Australia. J. Med. Entomol. **33:**180–182.

Eigen, M. 1991. Viral quasispecies. Sci. Am. **269:**42–49.

Fauvel, M., Artsob, H., Calisher, C.H., Davignon, L., Chagnon, A., Skvorc, R.R. and Belloncik, S. 1980. California group virus encephalitis in three children from Quebec: clinical and serologic findings. Canad. Med. Assoc. J. **122:**60–64.

Feddema, J.J. and Mather, R.J. 1992. Hydrological impacts of global warming over the United States. Pp. 50–62 *in* S. K. Majumdar, L. S. Kalkstein, B. M. Yarnal, E. W. Miller and L. M. Rosenfeld (eds.), Global climate change: implications, challenges and mitigation measures. Pennsylvania Academy of Sciences, Easton.

Fulhorst, C.F., Hardy, J.L., Eldridge, B.F., Presser, S.B. and Reeves, W.C. 1994. Natural vertical transmission of western equine encephalomyelitis virus in mosquitoes. Science **263:**676–678.

Giltner, L.T. and Shahan, M.S. 1933. The 1933 outbreak of infectious equine encephalomyelitis in the eastern states. N. Am. Vet. **14:**25.

Gould, A.R., Pritchard, K.A. and Pritchard, L.I. 1992. Phylogenetic relationships between bluetongue and other orbiviruses. Pp. 452–460 *in* T.E. Walton and B.I. Osburn (eds.), Bluetongue, African horse sickness and related orbiviruses. CRC Press, Boca Raton, Florida.

Grimstad, P.R. 1988. California group viruses. Pp. 99–136 *in* T.P. Monath (ed.), The arboviruses: epidemiology and ecology, Vol. 2. CRC Press, Boca Raton, Florida.

Grimstad, P.R., Calisher, C.H., Harroff, R.N. and Wentworth, B.B. 1986. Jamestown Canyon virus (California serogroup) is the etiologic agent of widespread infection in Michigan humans. Am. J. Trop. Med. Hyg. **35**:376–386.

Gubler, D.J. 1989. *Aedes aegypti* and *Aedes aegypti*-borne disease control in the 1990s: top down or bottom up. Am. J. Trop. Med. Hyg. **40**:571–578.

Gubler, D.J. 1997. Dengue and dengue hemorrhagic fever: its history and resurgence as a public health problem. Pp. 1–22 *in* D.J. Gubler and G. Kuno (eds.), Dengue and dengue hemorrhagic fever. CAB International, Wallingford, Oxon, UK.

Gubler, D.J. and Kuno, G. (eds). 1997. Dengue and dengue hemorrhagic fever. CAB International, Wallingford, Oxon, UK. 478 pp.

Gubler, D.J. and Trent, D.W. 1994. Emergence of epidemic dengue/dengue hemorrhagic fever as a public health problem in the Americas. Infec. Agents Dis. **2**:383–393.

Guy, J.S., Barnes, H.J., Ficken, M.D., Smith, L.G., Emory, W.H. and Wages, D.P. 1994. Decreased egg production in turkeys experimentally infected with eastern equine encephalitis virus and Highlands J virus. Avian Dis. **38**:563–571.

Guy, J.S., Siopes, T.D., Barnes, H.J., Smith, L.G. and Emory, W.H. 1995. Experimental transmission of eastern equine encephalitis virus and Highlands J virus via semen of infected tom turkeys. Avian Dis. **39**:337–342.

Hahn, C.S., Lustig, S., Strauss, E.G. and Strauss, J.H. 1988. Western equine encephalomyelitis virus is a recombinant virus. Proc. Nat. Acad. Sci. **85**:5997–6001.

Halstead, S. 1970. Observations related to pathogenesis of dengue hemorrhagic fever. VI. Hypotheses and discussion. Yale J. Biol. Med. **42**:350–362.

Hammon, W.M., Reeves, W.C., Benner, S.R. and Brookman, B. 1945. Human encephalitis in the Yakima Valley, Washington, 1942, with 49 isolations (western equine and St. Louis types) from mosquitoes. J. Am. Med. Assoc. **128**:1133–1139.

Hayes, C.G. 1988. West Nile fever. Pp. 59–88 *in* T.P. Monath (ed.), The arboviruses: epidemiology and ecology, Vol. 5. CRC Press, Boca Raton, Florida.

Hess, W.R. 1988. African horse sickness. Pp. 1–18 *in* T.P. Monath (ed.), The arboviruses: epidemiology and ecology, Vol. 2. CRC Press, Boca Raton, Florida.

Hoch, A.L., Gargan, T.P., II and Bailey, C.L. 1985. Mechanical transmission of Rift Valley fever virus by hematophagous Diptera. Am. J. Trop. Med. Hyg. **34**:188–193.

Jetten, T.H. and Focks, D.A. 1997. Potential changes in the distribution of dengue transmission under climate warming. Am. J. Trop. Med. Hyg. **57**:285–297.

Jupp, P.G. and McIntosh, B.M. 1988. Chikungunya virus disease. Pp. 137–157 *in* T.P. Monath (ed.), The arboviruses: epidemiology and ecology, Vol. 2. CRC Press, Boca Raton, Florida.

Karabatsos, N. (ed.) 1985. International catalogue of arboviruses including certain other viruses of vertebrates, 3rd ed. American Society of Tropical Medicine and Hygiene, San Antonio, Texas. 1147 pp.

Karabatsos, N., Lewis, A.L., Calisher, C.H., Hunt, A.R. and Roehrig, J.T. 1988. Identification of Highlands J virus from a Florida horse. Am. J. Trop. Med. Hyg. **19**:603–606.

Kay, B.H. and Aaskov, J.G. 1988. Ross River virus disease (epidemic polyarthritis). Pp. 93–112 *in* T.P. Monath (ed.), The arboviruses: epidemiology and ecology, Vol. 4. CRC Press, Boca Raton, Florida.

Kuhn, R.J., Griffin, D.E., Owen, K.E., Niesters, H.G.M. and Strauss, J.H. 1996. Chimeric Sindbis-Ross River viruses to study interactions between alphavirus nonstructural and structural regions. J. Virol. **70**:7900–7909.

Lee, E., Stocks, C., Lobis, P., Hislop, A., Straub, J., Marshall, I., Weir, R. and Dalgarno, L. 1997. Nucleotide sequence of the Barmah Forest virus genome. J. Virol. **227**:509–514.

Levine, A.J. 1996. The origins of virology. Pp. 1–14 *in* B.N. Fields, D.M. Knipe and P.M. Howley (eds.), Fundamental virology, 3rd ed. Lippincott-Raven, Philadelphia.

Lindsay, M.D.A., Broom, A.K., Wright, A.E., Johansen, C.A. and Mackenzie, J.S. 1993. Ross River virus isolations from mosquitoes in arid regions of western Australia: Implications of vertical transmission as a means of persistence of the virus. Am. J. Trop. Med. Hyg. **49**:686–696.

Lindsay, M.D.A., Johansen, C.A., Smith, D.A., Wallace, M.J. and Mackenzie, J.S. 1995. An outbreak of Barmah Forest virus disease in the south-west of Western Australia. Med. J. Austral. **162**:291–294.

Lopes, O.S., Sacchetta, L.A., Coimbra, T.L.M., Pinto, G.H. and Glasser, C.M. 1978. Emergence of a new arboviral disease in Brazil. II. Epidemiologic studies on 1975 epidemic. Am. J. Epidemiol. **108**:394–401.

Lumsden, L.L. 1958. St. Louis encephalitis in 1933. Publ. Hlth. Rep. **73**:340–353.

Lvov, D.K. 1988. Omsk hemorrhagic fever. Pp. 205–216 *in* T.P. Monath (ed.), The arboviruses: epidemiology and ecology, Vol. 3. CRC Press, Boca Raton, Florida.

Mackenzie, J.S. and Smith, D.W. 1996. Mosquito-borne viruses and epidemic polyarthritis. Med. J. Austral. **164**:90–92.

Meegan, J.M., Khalil, G.M., Hoogstraal, H. and Adham, F.K. 1980. Experimental transmission and field isolation studies implicating *Culex pipiens* as a vector of Rift Valley fever virus in Egypt. Am. J. Trop. Med. Hyg. **29:**1405–1410.

Monath, T.P. (ed.) 1980. St. Louis encephalitis. American Public Health Association, Washington, DC. 680 pp.

Monath, T.P. (ed.) 1988a (1988-1989). The arboviruses: epidemiology and ecology. 5 vol. CRC Press, Boca Raton, Florida. 1,319 pp.

Monath, T.P. 1988b. Yellow fever. Pp. 139–231 *in* T.P. Monath (ed.), The arboviruses: epidemiology and ecology, Vol. 5. CRC Press, Boca Raton, Florida.

Monath, T.P. and Guirakhoo, F. 1996. Orbiviruses and Coltiviruses. Pp. 1735–1766 *in* B.N. Fields, D.M. Knipe and P.M. Howley (eds.), Fields virology, Vol. 2. Lippincott-Raven, Philadelphia.

Morris, C.D. 1988. Eastern equine encephalomyelitis. Pp. 1–20 *in* T.P. Monath (ed.), The arboviruses: epidemiology and ecology, Vol. 3. CRC Press, Boca Raton, Florida.

Morse, S.S. 1993. Emerging viruses. Oxford University Press, New York. 317 pp.

Murphy, F.A. 1996. Virus taxonomy. Pp. 15–57 *in* B.N. Fields, D.M. Knipe and P.M. Howley (eds.), Fields virology, 3rd ed, Vol. 1. Lippincott-Raven, Philadelphia.

Murphy, F.A., Fauquet, C.M., Bishop, D.H.L., Ghabrial, S.A., Jarvis, A.W., Martelli, G.P., Mayo, M.A. and Summers, M.D. 1995. The classification and nomenclature of viruses: The sixth report of the International Committee on Taxonomy of Viruses. Springer-Verlag, Vienna and New York. 586 pp.

Nettles, V.F., Hylton, S.A., Stallknecht, D.E. and Davidson, W.R. 1992. Epidemiology of epizootic hemorrhagic disease viruses in wildlife in the USA. Pp. 238–248 *in* T.E. Walton and B.I. Osburn (eds.), Bluetongue, African horse sickness and related orbiviruses. CRC Press, Boca Raton, Florida.

Nibert, M.L., Schiff, L.A. and Fields, B.N. 1996. Reoviruses and their replication. Pp. 691–730 *in* B.N. Fields, D.M. Knipe and P.M. Howley (eds.), Fundamentals of virology, 3rd ed. Lippincott-Raven, Philadelphia.

Nichol, S.T., Rowe, J.E. and Fitch, W.M. 1993. Punctuated equilibrium and positive Darwinian evolution in vesicular stomatitis virus. Proc. Nat. Acad. Sci. **90:**10424–10428.

Norder, H., Lundstrom, J.O., Kosuch, O. and Magnius, L.O. 1996. Genetic relatedness of Sindbis virus strains from Europe, Middle East, and Africa. Virol. **222:**440–445.

Olson, K.E., Higgs, S., Gaines, P.J., Powers, A.M., Davis, B.S., Kamrud, K.I., Carlson, J.O., Blair, C.D. and Beaty, B.J. 1996. Genetically engineered resistance to dengue-2 virus transmission in mosquitoes. Science **272:**884–886.

Peters, C.J. and Dalrymple, J.M. 1990. Alphaviruses. Pp. 713–761 *in* B.N. Fields (ed.), Fields virology, 2nd ed, Vol. 2. Raven Press, New York.

Pinger, R.R., Rowley, W.A., Wong, Y.W. and Dorsey, D.C. 1975. Trivittatus virus infections in wild mammals and sentinel rabbits in central Iowa. Am. J. Trop. Med. Hyg. **24:**1006–1009.

Pinheiro, F.P., Travassos da Rosa, A.P.A., Gomes, M.L.C., LeDuc, J.W. and Hoch, A.L. 1982. Transmission of Oropouche virus from man to hamster by the midge *Culicoides paraensis*. Science **215:**1251–1253.

Reed, W., Carroll, J., Agramonte, A. and Lazear, J.W. 1900. The etiology of yellow fever. A preliminary note. Philad. Med. J. **6:**790–796.

Reeves, W.C. 1990. Epidemiology and control of mosquito-borne arboviruses in California, 1943–1987. Calif. Mosq. Vect. Contr. Assoc., Sacramento. 508 pp.

Reeves, W.C., Hardy, J.L., Reisen, W.K. and Milby, M.M. 1994. Potential effect of global warming on mosquito-borne arboviruses. J. Med. Entomol. **31:**324–332.

Reeves, W.C. and Milby, M.M. 1990. Strategies and concepts for vector control. Pp. 383–430 *in* W.C. Reeves (ed.), Epidemiology and control of mosquito-borne arboviruses in California, 1943–1987. Calif. Mosq. Vect. Contr. Assoc., Sacramento.

Reid, H.W. 1988. Louping Ill. Pp. 117–136 *in* T.P. Monath (ed.), The arboviruses: epidemiology and ecology, Vol. 3. CRC Press, Boca Raton, Florida.

Reisen, W.K., Hardy, J.L., Chiles, R.E., Kramer, L.D., Martinez, V.M. and Presser, S.B. 1996. Ecology of mosquitoes and lack of arbovirus activity at Morro Bay, San Luis Obispo County, California. J. Am. Mosq. Contr. Assoc. **12:**679–687.

Reisen, W.K. and Monath, T.P. 1988. Western equine encephalomyelitis. Pp. 89–137 *in* T.P. Monath (ed.), The arboviruses: epidemiology and ecology, Vol. 5. CRC Press, Boca Raton, Florida.

Rico-Hesse, R., Weaver, S.C., De Siger, J., Medina, G. and Salas, R.A. 1995. Emergence of a new epidemic-epizootic Venezuelan equine encephalitis virus in South America. Proc. Nat. Acad. Sci. **92:**5278–5281.

Robertson, S.E., Hull, B.P., Tomori, O., Bele, O., LeDuc, J.W. and Esteves, K. 1996. Yellow fever, a decade of reemergence. J. Am. Med. Assoc. **276:**1157–1162.

Rodhain, F. and Rosen, L. 1997. Mosquito vectors and virus-vector relationships. Pp. 45–60 *in* D.J. Gubler and G. Kuno (eds.), Dengue and dengue hemorrhagic fever. CAB International, Wallingford, Oxon, UK.

Rosen, L. 1977. *The Emperor's New Clothes* revisited, or reflections on the pathogenesis of dengue hemorrhagic fever. Am. J. Trop. Med. Hyg. **26:**337–343.

Roy, P. 1996. Orbiviruses and their replication. Pp. 1709–1734 *in* B.N. Fields, D.M. Knipe and P.M. Howley (eds.), Fields virology, 3rd ed, Vol. 2. Lippincott-Raven, Philadelphia.

Schmaljohn, C.S. 1996. Bunyaviridae: The viruses and their replication. Pp. 649–673 *in* B.N. Fields, D.M. Knipe and P.M. Howley (eds.), Fundamental virology, 3rd ed. Lippincott-Raven, Philadelphia.

Scott, T.W. 1988. Vertebrate host ecology. Pp. 257–280 *in* T.P. Monath (ed.), The arboviruses: epidemiology and ecology, Vol. 1. CRC Press, Boca Raton, Florida.

Scott, T.W. and Weaver, S.C. 1989. Eastern equine encephalomyelitis virus: Epidemiology and evolution of mosquito transmission. Adv. Virus Res. **37:**277–328.

Scott, T.W., Weaver, S.C. and Mallampalli, V.L. 1994. Evolution of mosquito-borne viruses. Pp. 293–324 *in* S.S. Morse (ed.), The evolutionary biology of viruses. Raven Press, New York.

Smart, K.L., Elbel, R.E., Woo, R.F.N., Kern, E.R., Crane, G.T., Bales, G.L. and Hill, B.W. 1972. California and western encephalitis viruses from Bonneville Basin, Utah in 1965. Mosq. News **32:**282–289.

Smithburn, K.C. 1942. Differentiation of the West Nile virus from the viruses of St. Louis and Japanese B encephalitis. J. Immunol. **44:**25–31.

Sosa, O. 1989. Carlos J. Finlay and yellow fever: A discovery. Bull. Entomol. Soc. Am. **35**(2):23–25.

Tabachnick, W.J. 1991. Evolutionary genetics and arthropod-borne disease: the yellow fever mosquito. Am. Entomol. **37:**14–24.

Tabachnick, W.J. 1996. *Culicoides variipennis* and bluetongue-virus epidemiology. Annu. Rev. Entomol. **41:**23–43.

Tabachnick, W.J. 1998. Arthropod-borne pathogens: issues for understanding emerging infectious diseases. Pp. 411–429 *in* R.M. Krause (ed.), Emerging infectious diseases. Academic Press, New York.

Tesh, R.B. 1988. Phlebotomus fevers. Pp. 15–27 *in* T.P. Monath (ed.), The arboviruses: epidemiology and ecology, Vol. 4. CRC Press, Boca Raton, Florida.

Theiler, M. and Smith, H.H. 1937. Use of yellow fever virus modified by *in vitro* cultivation for human immunization. J. Exp. Med. **65:**787–800.

Thompson, W.H. and Beaty, B.J. 1977. Venereal transmission of LaCrosse (California encephalitis) arbovirus in *Aedes triseriatus* mosquitoes. Science **196:**530–531.

Van Der Meyden, C.H., Erasmus, B.J., Swanepoel, R. and Prozesky, O.W. 1992. Encephalitis and chorioretinitis associated with neurotropic African horse sickness in laboratory workers. South African Med. J. **81:**451–454.

Villari, P., Spielman, A., Komar, N., McDowell, M. and Timperi, R.J. 1995. The economic burden imposed by a residual case of eastern encephalitis. Am. J. Trop. Med. Hyg. **52:**8–13.

Wagner, R.R. and Rose, J.K. 1996. Rhabdoviridae : the viruses and their replication. Pp. 561–573 *in* B.N. Fields, D.M. Knipe and P.M. Howley (eds.), Fundamental virology, 3rd ed. Lippincott-Raven, Philadelphia.

Walton, T.E. and Grayson, M. 1988. Venezuelan equine encephalomyelitis. Pp. 203–231 *in* T.P. Monath (ed.), The arboviruses: epidemiology and ecology, Vol. 4. CRC Press, Boca Raton, Florida.

Watts, D.M., Ksiazek, T.G., Linthicum, K.J. and Hoogstraal, H. 1988. Crimean-Congo hemorrhagic fever. Pp. 177–222 *in* T.P. Monath (ed.), The arboviruses: epidemiology and ecology, Vol. 2. CRC Press, Boca Raton, Florida.

Watts, D.M., Pantuwatana, S., DeFoliart, G.R., Yuill, T.M. and Thompson, W.H. 1973. Transovarial transmission of LaCrosse virus (California encephalitis group) in the mosquito, *Aedes triseriatus*. Science **182:**1140–1141.

Webb, P.A. and Holbrook, F.R. 1989. Vesicular stomatitis. Pp. 1–29 *in* T.P. Monath (ed.), The arboviruses: epidemiology and ecology, Vol. 5. CRC Press, Boca Raton, Florida.

Chapter 12

Mechanical Transmission of Disease Agents by Arthropods

LANE D. FOIL[1] AND J. RICHARD GORHAM[2]
[1]*Louisiana State University, Baton Rouge and* [2]*Uniformed Services University of the Health Sciences, Bethesda, Maryland*

INTRODUCTION

Mechanical transmission means the transfer of pathogens from an infected host or a contaminated substrate to a susceptible host, where a biological association between the pathogen and the vector is not necessary. The vectors in this case are not restricted to arthropods. Birds, rats, mice, other animals and even humans can serve as mechanical vectors; thus, vector ecologists must know about non-arthropod taxa and their roles in transmitting disease agents.

This chapter separates mechanical transmission by arthropods into two types: (1) when agents are transferred directly between two hosts (**direct mechanical transmission**) and (2) when arthropods transmit pathogens picked up from substrates contaminated by secretory and/or excretory products of infected hosts (**indirect, or contaminative, mechanical transmission**). Table 12.1 lists examples of the various types of mechanical transmission. **Hematophagous** (bloodfeeding), as well as non-biting insects, that are attracted to open sores or lacrimal secretions of infected vertebrates are involved in direct transmission. Ants and roaches are examples of arthropods incriminated in indirect transmission. House flies and related Diptera may be involved in both types.

Because of the worldwide importance of biologically transmitted diseases such as malaria and yellow fever, there is a tendency to overlook the importance of the mechanical transmission of disease agents by arthropods. Biological transmission of pathogens generally has attracted more attention from medical entomologists than has mechanical transmission. Nevertheless, there is ample evidence that arthropods play important roles as mechanical vectors of pathogens causing important diseases of humans and domestic animals. The literature on this subject is extensive, but scattered, and rarely has been assembled into a single source.

FACTORS AFFECTING MECHANICAL TRANSMISSION BY ARTHROPODS

With mechanical transmission, there is no development or multiplication of an agent within a vector. Unlike most pathogens biologically transmitted by arthropods, there usually are multiple routes of infection for mechanically transmitted agents. For agents in blood, there

B.F. Eldridge and J.D. Edman (eds.), Medical Entomology, Revised Edition, 461–514.

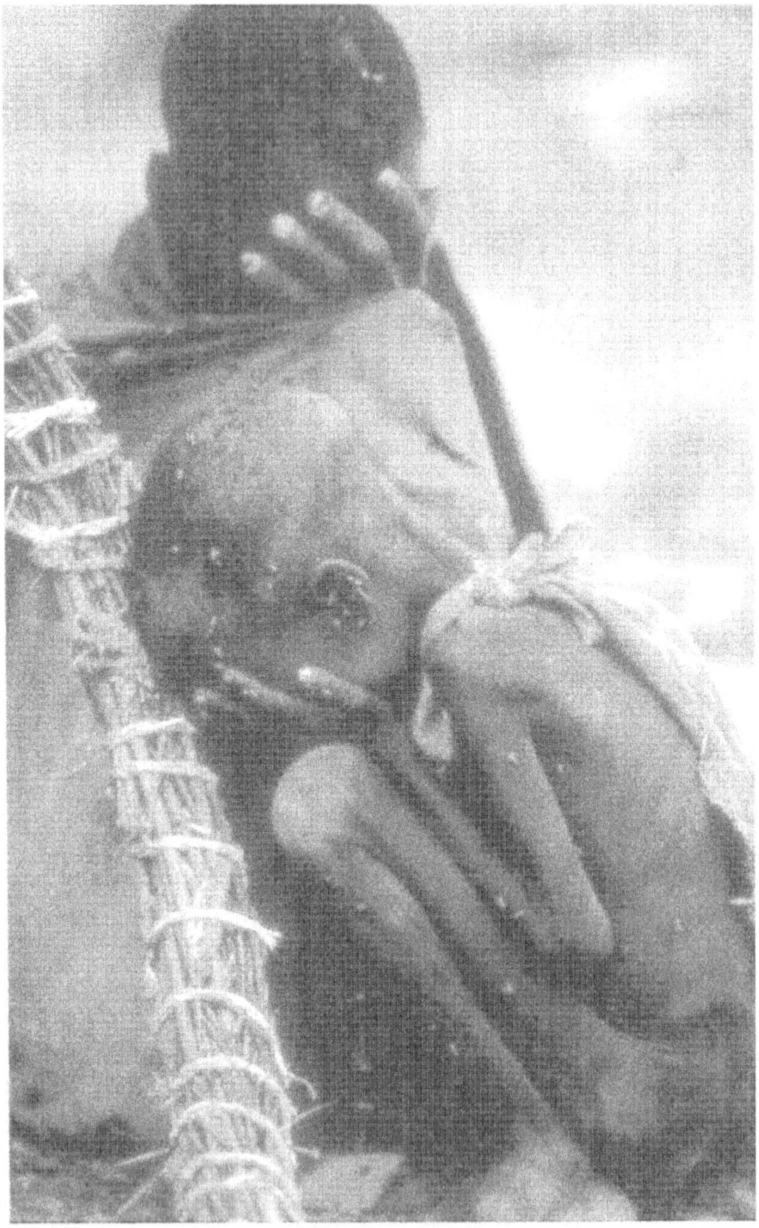

Figure 12.1. Despair overwhelms a southern Sudanese man and boy. Flies contribute to the transmission of pathogenic organisms during wars and natural disasters (photograph by Robert Caputo/Aurora).

Table 12.1. Examples of indirect (contaminative) and direct mechanical transmission of vertebrate pathogens by arthropods.

Agents of	Source	Vector(s)	Host exposure
	Indirect (contaminative) mechanical transmission		
Botulism (toxin)	Carcasses	Maggots	Ingestion of maggots
Enteric disease	Offal	Flies feeding on carcasses	Ingestion of contaminated meat
Enteric disease	Offal	Flies feeding on animal food	Ingestion of contaminated food
Enteric disease	Feces	Flies feeding on human food	Ingestion of contaminated food
Enteric disease	Feces	Flies feeding on animal food	Ingestion of contaminated food
Marek's disease	Carcasses	Beetles	Ingestion of insects in food
Enteric disease	Feces	Flies feeding on oral mucosa	Ingestion of pathogen or contaminated vector
Nosocomial infections	Host secretions, fomites	Ants, roaches, flies	Ingestion of contaminated food or other substances
	Direct mechanical transmission beween live hosts		
Yaws	Host secretions	Eye gnats	Insects feeding on wounds
Summer mastitis	Host secretions	Head flies	Agent introduced at site of feeding by another arthropod
Infectious bovine keratosis	Eye secretions	Face flies	Enters eye at lesion caused by arthropod feeding or mechanical trauma
Myxomatosis	Skin	Mosquitoes and fleas	Insect bite
Equine infectious anemia	Blood	Horse flies	Insect bite
Hepatitis B and cat scratch disease (proposed)	Blood	Arthropods	Insect defecates on, is smashed on or is ingested by second host
Anaplasmosis	Blood secondary to another feeding site	Eye gnats	Insects feeding on eyes or secondary bloodmeal

Table 12.2. Estimates of the quantity of blood retained or transferred between hosts by various agents.

Agent and circumstances	Amount (nanoliters) retained (R) or transferred (T)	Reference
Neck of a 20 ml syringe following blood collection and expulsion	1.37×10^5 to 1.42×10^5 (R)	Gunn (1993)
22-gauge needle after blood collection	1×10^3 (T)	Napoli and McGowan (1987)
Horse fly after interrupted bloodmeal	10 (R)	Foil et al. (1987)
22-gauge needle after intramuscular inoculation	9 (R)	Knaus et al. (1993)
Deer fly after interrupted bloodmeal	6 (T)	Knaus et al. (1993)
Horse fly after interrupted bloodmeal	5 (T)	Foil et al. (1987)
Stable fly after bloodmeal (based on microscopic examination of mouthparts)	0.03 (R)	Weber et al. (1988)
Mosquito after bloodmeal (based on transmission trials)	1×10^{-3} to 1×10^{-4} (T)	Miike (1987)
Bed bug after bloodmeal (based on experiments with artificial membrane feeders)	1×10^{-3} to 1×10^{-4} (T)	Ogston (1981)

may be needles, surgical procedures, transfusions and tattooing. For livestock, procedures like rectal palpation, shearing and ear tagging may be means of transmission. Vampire bats may serve as vectors. For agents in body secretions, a compromised barrier of the recipient (e.g., an open wound or arthropod bite) may be required to serve as a portal of entry of the agent. For many agents that may be mechanically transmitted by arthropods, there also may be the direct contamination of food and water by excretions, as well as transfer of agents on the hands of humans. Some agents may be biologically transmitted by arthropods under most circumstances, but transmitted mechanically under others.

Mechanically transmitted pathogens may multiply not only within the original source host, but also in various environmental media such as water, soil or manure. More detailed reviews on agents of environmental contamination of arthropods may be found in Greenberg (1971, 1973), Gorham (1994) and Olson (1998).

Morphology of Arthropods

Some arthropods are involved in the fecal-oral route of contaminative transmission due to their morphological features. A number of arthropods can transfer pathogens between substrates under experimental conditions. The integument of domestic ants and flies has multiple fine setae, and bacteria and parasite eggs from contaminated substrates may readily adhere to the setae. Domestic roaches also can pick up and disperse pathogenic bacteria (Gazivoda and Fish 1985). Hematophagous arthropods that are pool-feeders (**telmophagous**) and have blade-like mouthparts (e.g., tabanid flies) are more likely to be involved in mechanical transmission than vessel-feeding (**solenophagous**) arthropods with needle-like mouthparts (e.g., mosquitoes). This is because a larger surface area of the mouthparts is in contact with the bloodmeal resulting in the potential for a larger quantity of blood retained (Table 12.2).

Behavior

Feeding and other behavioral patterns of arthropods also determine the ability to transmit pathogens mechanically. Typically, arthropods closely associated with people and domestic animals (such as ants, flies and roaches) have been implicated in mechanical transmission. They feed frequently on various contaminated organic materials, either liquid or solid, then move to food surfaces.

Given sufficient time, individuals from a colony of pharaoh ants (*Monomorium pharaonis*) will explore every potential feeding site in a large building (Fowler et al. 1993). Greenberg and Bornstein (1964) determined that flies dispersed up to 5 km from a slaughterhouse where they had become contaminated with *Salmonella* bacteria.

Regurgitation is an important process in digestion for a number of flies (e.g., *Musca* spp.). Non-biting flies cannot consume solid food without first breaking it down using enzyme-laden saliva. For the house fly, food must be reduced to particles <0.05 mm in diameter (Hewitt 1914, Bolton and Hansens 1970) before uptake through spongiform mouthparts. The fact that flies may feed sequentially on human or animal feces and human or animal food, regurgitating a portion of the previously ingested feces onto the food, is well known.

Akkerman (1933) observed regurgitation in adult American roaches (*Periplaneta americana*) and confirmed the report by Barber (1914) that *Vibrio cholerae* may persist in the regurgitate. However, regurgitation is not essential to ingestion for any of the common domestic roach species; their chewing mouthparts break up the food into particles small enough to be swallowed. Regurgitation is important in the feeding behavior of many muscoid flies, and the frequent regurgitation by *Musca autumnalis* is an important factor in transmitting *Moraxella bovis*, a bacterium associated with conjunctivitis in vertebrates. Bloodsucking muscoids such as the stable fly regurgitate under artificial conditions, but no convincing studies support this as a natural means of transmission.

Defecation patterns of arthropods also are significant. Pathogens may be ingested, pass through the alimentary canal and then be deposited on food or other surfaces. If the pathogens do not multiply during passage through the alimentary canal, this is, by definition, mechanical transmission. Filth flies and cockroaches often defecate on substrates while feeding. Pathogens may be found in roach feces, but the isolation rate for any given pathogen tends to be low compared to pathogens found in the feces of filth flies (Ostrolenk and Welch 1942, Rueger and Olson 1969).

The mechanical transmission of bloodborne agents of humans is rare, although tabanids transmit the agents of tularemia, Lyme disease and anthrax. In contrast, mechanical transmission is important where hematophagous insects are abundant on livestock (Figs. 12.6, 12.7). For bloodborne agents, the most frequent mechanism of mechanical transmission is via contaminated mouthparts of the arthropod following interrupted bloodmeals. Foil (1989) and Davies (1990) reviewed factors that contribute to multiple and interrupted bloodfeeding by insects. The most efficient mechanical vectors are telmophagous. The mouthparts of tabanids have approximately 10 nl of blood residue following an interrupted bloodmeal. As agents approach 10^6 infectious particles (IP) per ml of blood (e.g., equine infectious anemia virus), transmission by individual tabanids may occur (Table 12.2). Similarly, at least one in 10 tabanid flies can transmit infectious agents such as bovine leukemia virus when titers exceed 10^5 IP per ml of blood. This amount would be a capacity of transmission equal to 0.1 according to the terminology of Leclercq (1969). Stable fly mouthparts have been estimated to retain 0.03 nl of blood following an interrupted feed. As agents such as murine leukemia virus approach 10^8 IP per ml, single stable flies can transfer infections. When pathogens such as Venezuelan equine encephalomyelitis virus and Rift Valley fever virus exceed 10^8 IP per ml of blood, solenophagous arthropods such as bed bugs and mosquitoes, as well as small arthropods such as biting midges and mites, can transfer viruses mechanically. Similarly, the density of infectious particles within the skin of hosts (e.g., myxoma virus) can reach levels where individual insects can transmit the pathogens.

Other ways hematophagous insects may introduce infectious agents mechanically include being ingested by, smashed on or defecating on the host. This type of transmission may be particularly important in the case of environmentally resistant pathogens such as *Francisella tularensis* and hepatitis B virus. The ingestion by vertebrate hosts of arthropods that normally would be considered biological vectors is probably rare, but experimental studies show that rodents can be infected by eating mosquitoes infected with La Crosse encephalitis or with malarial parasites. Thus, this method of transmission should not be ignored (Edman and Spielman 1988).

Secondary hematophagy is bloodfeeding by arthropods from wounds created by another arthropod or from other trauma such as veterinary procedures. Sometimes, this behavior can contribute to mechanical transmission of diseases such as yaws and dermatophilosis. For transmission of bloodborne agents, secondary bloodfeeders would feed first on a lesion on an infected animal and subsequently introduce the agent at a portal of entry of a susceptible host. Many omnivorous muscoid flies, as well as certain other hematophagous muscoid flies (e.g., horn flies) feed on such wounds.

Survival of Pathogens Within Arthropods

Many factors affect the passage of pathogens through arthropods, such as the quantity and number of species of ingested bacteria, the type of food source, the previously established gut flora (Greenberg 1968, Greenberg et al. 1970) and the presence of antibacterial substances in the insect gut (Wollman 1921, Duncan 1926). Akkerman (1933) observed that *V. cholerae* died out in the foregut of *P. americana*. Other cockroaches have antibacterial substances in the gut (Zhang Ran et al. 1990).

Cockroach nymphs are just as capable of bacterial passage as adults. In flies, pathogenic bacteria acquired during the larval stage may not always be found in the adult (Greenberg 1965). Maggots have been associated with epizootics of botulism (limberneck) in birds (Greenberg 1973). Ingesting larvae of *Cochliomyia* and *Lucilia* spp. that have fed on carcasses or contaminated food containing *Clostridium botulinum* toxins can produce botulism in pheasants and poultry.

Passage of helminth eggs, protozoan cysts and bacteria through arthropod guts may favor survival because it protects them from desiccation (Rivault et al. 1993). Numerous experiments (e.g., Barber 1914) suggest that virulence is retained while the pathogen remains within the alimentary canal of the transport host and for at least a short period after being expelled in vomitus or feces.

Other Factors

The titer of infectious agents, the persistence of agents and the infectiousness of agents at portals of entry to the host are major factors in the probability of mechanical transmission. Similarly, the number and type of insects coming into contact with hosts are important. The probability of mechanical transmission is influenced by a number of other factors, including the immune status of hosts, the number of pathogens or parasites in the infectious source, and the ability of pathogens to survive outside of arthropods. The role of host immunity is outside the scope of this book, as is a discussion of the relationship between infection and size of pathogen inoculum. However, the number of infectious particles present on or within individual arthropods should be considered when evaluating their role as mechanical vectors. House flies typically carry large numbers of bacteria (2.5–30 million) on their external surfaces (Ostrolenk and Welch 1942); in contrast, relatively few helminth eggs or protozoan cysts are carried externally by flies or cockroaches.

Pathogens with high environmental stability lend themselves to mechanical transmission. Examples of highly stable microorganisms are *Pseudomonas aeruginosa* (Hurst and Sutter 1966) and rotaviruses (Ansari et al. 1988). Experiments by Klowden and Greenberg (1977) demonstrated that bacteria (*Salmonella* spp.), when placed on the bodies of cockroaches that subsequently were killed, survived on the roach carcasses for as long as 60 days.

The survival of pathogens outside a host also is influenced by the medium in which the pathogens are located. Most bacteria survive longer in moist environments. The survival of even the most labile viruses may be prolonged if protected by blood or serum. Feces, whether vertebrate or arthropod, protect some bacterial pathogens (Blaser et al. 1980).

PUBLIC HEALTH AND VETERINARY SIGNIFICANCE

Discussions of mechanical transmission may leave the impression that such routes of infection are the dominant means of transmission. Often, this is not the case. Certainly, the discussions that follow are not meant to exaggerate mechanical transmission when it is not important.

Because pathogens mechanically transmitted by arthropods also are nearly always transmitted by other means, it is difficult to assess the public health and veterinary significance of mechanical transmission. Many laboratory studies addressed this question, as well as numerous reports of recovery of pathogenic microorganisms from arthropods in nature. However, there have been few epidemiological studies of mechanical transmission by arthropods. An exception is the classic study by Watt and Lindsay (1948) in which they demonstrated that diarrheal disease in people was significantly less in towns with intensive chemical control of house flies, compared to towns without fly control.

Throughout history, crowded conditions, accompanied by inadequate facilities for sanitation, have encouraged massive fly populations and epidemics of enteric diseases. Most of the disease cases probably resulted from direct fecal contamination of food or drinking water. Even when fly populations are dense, flies play a secondary role to direct fecal contamination of water and food in epidemics of enteric diseases (Weil 1943). Yet, history, especially military history, is filled with examples of how good sanitary practices (with relatively fewer flies) compare favorably with poor sanitary practices (with concurrent dense fly populations) relative to the incidence of enteric diseases. Baker-Jones (1941) ruled out food and water as likely sources of infection during an epidemic of typhoid fever in what was then Southern Rhodesia, and concluded that filth flies were the primary carriers of the bacteria. Concurrent monitoring of disease incidence and vector population density has in some instances demonstrated a lowering of disease incidence following fly population suppression (Gaud and Faure 1951, Cohen et al. 1991). Dense fly populations often accompany bacterial dysentery, especially in tropical regions (Hornick 1977, Levine and Levine 1991).

Military Camps and Operations

Dysentery and other enteric diseases traditionally afflict military forces. Moving a military group into an area endemic for enteric diseases at a time when weather conditions favor the development of filth flies enhances the risk of infections.

Typhoid fever and dysentery attacked US Army camps in the southern USA during the Spanish-American War in 1898. The Army commissioned a special team, consisting of Walter Reed, V.C. Vaughan and E.O. Shakespeare, to investigate, with special attention to typhoid fever. The team directly connected flies and typhoid and set the standards for camp sanitation to prevent future epidemics (Vaughan 1899). In World War I, a US Army entomologist noted a lack of sanitation in a camp in southwest France. He reported an explosion in the house fly population, followed by an epidemic of dysentery (Simmons 1923).

By the time of World War II, most armies had established regulations calling for fly control to reduce enteric disease. The British enforced such regulations with great success in the Middle East (Fairley and Boyd 1943).

Refugee Camps

When warfare or civil strife crowds large numbers of people into camps lacking sanitary facilities, the refugees have no alternative but to defecate on the ground. Surface waters, often the only source of water in the camps, quickly become contaminated. Cases of bacillary dysentery (shigellosis), cholera and viral diseases (e.g., rotaviral enteritis and hepatitis A), and conjunctivitis may soon appear, as among African refugee camps in recent years (Anonymous 1996). In these refugee camps, area-wide chemical fly control often is not effective. The ready availability of exposed feces meets the food requirements of fly larvae, so populations of synanthropic flies explode.

Hospital Infections

Infections acquired by patients as a direct result of their confinement in hospitals are called **nosocomial** infections. In hospitals and similar institutions (including veterinary facilities), ants, flies and cockroaches may mechanically transmit pathogens to patients. These insects may be attracted to secretions and excretions from patients, including vomitus, blood, pus, sweat, tears, urine, mucosal secretions, serum seepage, sputum and feces (Šrámová et al. 1992). Other infectious sources include contaminated organic matter such as medical wastes and garbage. The quantity of infectious material transferred by insects under such conditions often may be minuscule when compared to humans and surgical instruments as transport vectors.

Studies of nosocomial infections often have demonstrated the isolation of pathogenic agents from arthropods collected in hospitals. Fotedar and Banerjee (1992) recovered opportunistic fungi of 5 genera from cockroaches collected in hospital wards in India. Daniel et al. (1992) isolated 25 species (116 strains) of bacteria from arthropods collected inside hospital buildings in Prague. In Great Britain, Beatson (1972) isolated *Pseudomonas, Salmonella* and *Staphylococcus* bacteria from pharaoh ants in hospitals and *Bordetella bronchiseptica* from these ants in a veterinary isolation unit. LeGuyader et al. (1989) isolated 29 bacterial species from cockroaches from hospitals in France, including *Staphylococcus aureus* and *Escherichia coli*. Fotedar et al.

(1991b) isolated drug-resistant *Klebsiella* from cockroaches in hospitals in India, and later found experimentally that *P. aeruginosa* fed to cockroaches was excreted for up to 114 days (Fotedar et al. 1993).

Confined Animals

Mechanical transmission of infectious agents by arthropods to confined animals is important because it affects animal production and because animals may play a role in zoonotic disease transmission. Problems with confined cattle, swine and poultry arise frequently because of mechanically transmitted diseases. Mastitis of cattle under confinement is one of the most common diseases associated with mechanical transmission by flies. Companion animals also can be the source of pathogens and parasites that may be picked up by mechanical vectors (Wilton 1963).

For poultry and swine, the most important mechanism of dissemination of pathogens is transport of infected animals or movement of contaminated fomites or personnel. However, arthropods may play a significant part in mechanical transmission of agents of a number of diseases.

The mechanical transmission of pathogens of poultry diseases can serve as a model for other diseases of confined animals. Many arthropod species are associated with poultry manure and litter, and transmission of agents of poultry diseases has been associated with these pests. The mechanical transmission of bacteria by flies and beetles is a major problem. Poultry products are sources of zoonotic infections caused by serotypes of *Salmonella* and *Campylobacter jejuni*. The organisms can be isolated in most production systems from food, litter, live birds, carcasses and insects, particularly flies. *Campylobacter jejuni* has been isolated from the gut of cockroaches (Umunnabuike and Irokanulo 1986). Shane et al. (1985) demonstrated the transfer of *C. jejuni* from chickens known to be fecal excreters to noninfected chickens via house flies that were confined with the infected chickens for 5 days.

The lesser mealworm, *Alphitobius diaperinus*, has been incriminated as a mechanical vector of *Salmonella typhimurium* (McAllister et al. 1994). Beetles fed chicken feed inoculated with *S. typhimurium* were infectious when fed to chicks; the bacteria survived for at least 28 days within the beetles. Jones et al. (1991) isolated 5 serotypes of *Salmonella* from litter beetles and cockroaches collected from broiler houses and processing plants. Kopanic et al. (1994) collected cockroaches naturally infected with *Salmonella* spp. from a poultry hatchery and feed mill in North Carolina. They demonstrated experimentally that cockroaches can become infected with *S. typhimurium* by feeding on a contaminated food source, can infect each other as well as food and water, and can transport the bacteria to hatchery eggs in an incubator.

Studies on the involvement of the lesser mealworm in the transmission of the gamma herpes virus, the causal agent of Marek's disease, are discussed later in this chapter. The lesser mealworm also is a possible reservoir for infectious bursal disease virus and *E. coli* in poultry production systems (McAllister et al. 1995, 1996).

ARTHROPODS ASSOCIATED WITH INDIRECT (CONTAMINATIVE) MECHANICAL TRANSMISSION

There are reports of isolations from arthropods for virtually every pathogen that occurs in feces or other decomposing organic matter. Although pathogens have been recovered from many other arthropods, including mites (Mlodecki and Burzynska 1956), ants, flies and cockroaches are the most frequent source of such isolations.

Figure 12.2. The house fly on trial (from Greenberg 1973). By the turn of the 20th Century, the threat of the house fly to public health had been recognized.

Beetles, Moths and Other Food Pests

Food pests as a group frequently are involved in the mechanical transmission of pathogens, although not all food pests are of any sanitary significance. Even when arthropods contaminate food during processing before sterilization, arthropod fragments may still cause allergic responses (Bernton and Brown 1969). Pantry pests develop mainly in processed food, and thus are considered to be **obligatory** pests. Included in this category are the warehouse beetle (*Trogoderma variabile*), granary weevil (*Sitophilus granarius*), confused flour beetle (*Tribolium confusum*) and Mediterranean flour moth (*Anagasta kuehniella*). Pantry pests probably transmit few pathogens because their activities usually are confined to infested food. However, they can transfer microorganisms from infested to noninfested food products (Cravedi and Quaroni 1983, Schuster et al. 1972). Arthropod pests associated with food that are **opportunistic** pests (e.g., German cockroach, pharaoh ant, common silverfish, house fly and Australian spider beetle) are more important in mechanical transmission. The presence of even one such arthropod in a federally inspected food facility is a technical violation of US Food and Drug Administration regulations (Gorham 1994). Field pests (those that infest food crops in the field before harvest) are not important agents in mechanical transmission of disease.

House Flies and Related Diptera

Greenberg (1971) tabulated the literature describing the associations between flies and pathogenic and nonpathogenic organisms. These associations included 29 families containing 346 species of flies; his systematic list of organisms associated with flies contained 363 types, strains or species of viruses, bacteria, fungi, protozoa and helminths. Greenberg (1973) reviewed all pertinent studies on dermatophytes, so this group will not be discussed further in this chapter.

Interest in the role of **synanthropic** (associated with humans) flies as potential mechanical vectors arose early in the history of medical entomology, especially with regard to enteric infections (Greenberg 1965). The house fly (Fig. 12.2) was an early target as a mechanical carrier of typhoid fever in studies in the USA (Howard 1911) and in England (Graham-Smith 1914).

West (1951) summarized information about flies and disease in his book, *The Housefly, Its Natural History, Medical Importance, and Control.* West and Peters (1973) assembled the world literature on the house fly.

Musca domestica has received most of the attention in connection with transmission of enteric disease agents (Levine and Levine 1991, Olsen 1998), but other fly species have been implicated in mechanical transmission in epidemic situations as well (Sturtevant 1918, Linhares 1981, Janisiewicz et al. 1999). Expanding geographic ranges of some species of flies may be a factor in changes in the epidemiology of some diseases. *Chrysomya megacephala* apparently was unknown on Guam before the Japanese invasion during World War II; that species quickly took on a major role as a mechanical vector of intestinal parasites on the island (Harris and Down 1946). In addition to their role as indirect vectors of enteric pathogens, house flies and other muscoids that are attracted to eyes, open lesions and other sources of pathogens on infected vertebrates may directly transmit these disease agents to noninfected hosts. Yaws, trachoma and dermatophilosis are diseases whose pathogens may be transmitted in this manner.

Arthropods frequently associated with the mechanical transmission of infectious agents from eyes or lesions of a host are flies in the families Chloropidae and Muscidae. Infectious agents of lesions or eyes generally cannot penetrate intact skin or cornea and some form of compromise (trauma or arthropod bite) at the portal of entry is required for infection. Members of the family Chloropidae have been referred to as eye gnats, ulcer flies and yaws flies; these names reflect the persistent feeding on body secretions from common sources. The involvement of eye gnats in the transmission of agents of human and animal eye diseases as well as yaws, mastitis and anaplasmosis has been documented. Sometimes eye gnats may accelerate the spread of agents to hosts. For example, Bassett (1970) investigated an epidemic of acute bacterial glomerulonephritis in Trinidad and concluded that *Hippelates* flies feeding on skin lesions contributed to the rapid spread of *Streptococcus pyogenes* throughout the island.

Cockroaches

Like flies, cockroaches have been associated with a vast number of pathogens. Literature on the medical importance of cockroaches was collected in two publications. Roth and Willis (1957) covered the associations of 18 species of cockroaches with pathogens. Roth and Willis (1960) covered biotic associations of cockroaches with pathogenic and nonpathogenic organisms and compiled a list of pathogens that included 33 kinds of bacteria, 2 species of fungi, 7 species of helminths, 2 kinds of protozoans and one virus.

Cockroaches are not as mobile and usually not as obvious as house flies. Nevertheless, they often move widely within and between structures and can contact human food (Fig. 12.3). Cockroaches carry an array of bacterial species that are part of their normal gut flora. However, the isolation of vertebrate pathogens that are not part of the normal cockroach gut flora suggests that cockroaches are capable of acquiring and disseminating these pathogens in their environment (LeGuyader et al. 1989, Fotedar et al. 1991a).

Ants

The volume of literature on the role of ants as mechanical vectors is much smaller than for either flies or cockroaches. Barber (1914) noted that *V. cholerae* could be recovered from ants and cockroaches. Hughes et al. (1989) captured ants in a veterinary clinic laboratory and, using blood agar plates, isolated species of *Serratia, Citrobacter, Klebsiella, Enterobacter, Proteus, Staphylococcus* and *Yersinia*, including *Yersinia pestis*.

HEMATOPHAGOUS ARTHROPODS ASSOCIATED WITH DIRECT MECHANICAL TRANSMISSION

The following discussion of potential vectors of agents of systemic diseases (skin and blood) stresses hematophagous arthropods. Important characteristics of arthropods are morphology and behavior. Characteristics of pathogens that influence mechanical transmission by arthropods include the pathogen titer in blood or other tissue fluids and the environmental stability of the pathogen.

Morphological characteristics of arthropods that can influence mechanical transmission include body size and the form of mouthparts associated with bloodfeeding. Relevant aspects of feeding behavior include the amount of pain associated with the bite, the persistence in bloodfeeding, the habit of interrupted feeding and the propensity of the particular arthropod species to attack certain sites on the body of the host (Figs. 12.4, 12.5, 12.6, 12.7). A feeding site on the back can be exempt from tail swipes by livestock or hand slaps by primates. The habit of taking interrupted or multiple bloodmeals by arthropods is controlled by a complex of factors, including mobility and density of the vectors. There also is considerable interspecific variation for this trait. Age, reproductive status and nutritional state of the arthropod can affect the persistence of individuals, and thus the likelihood of interrupted feeding (Edman and Spielman 1988, Walker and Edman 1985).

Interrupted bloodfeeding contributes to **multiple bloodmeals** in arthropods (more than a single bloodmeal within a single gonotrophic cycle or mixed feedings). The interval between an interrupted feed and a second feed is critical for environmentally labile agents such as retroviruses, where tabanids and stable flies can move quickly between hosts, but not for agents like pox viruses, where the organism can persist on the mouthparts of mosquitoes and fleas through normal feeding cycles.

Tabanid flies may be the prototype of mechanical vectors of bloodborne agents. There are several reviews of the importance of tabanids as mechanical vectors, e.g., Krinsky (1976) and Foil (1989). Because of tabanid flies' method of bloodfeeding and the morphology of their mouthparts, they retain relatively large amounts of blood after feeding (Table 12.2). Tabanids are highly mobile insects, and females of some species are not persistent bloodfeeders. Their bites are painful, and host defensive behavior contributes to a high incidence of interrupted feeding.

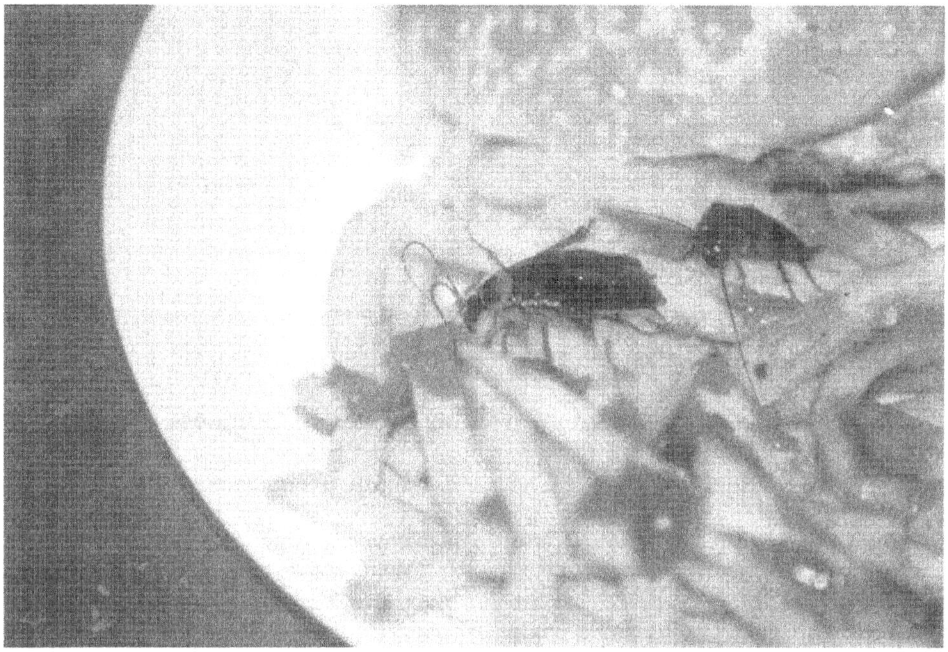

Figure 12.3. Several American roaches get a quick snack on an unattended burger and fries. One wonders about the sanitary status of their previous meal (photograph by Glen O'Remus and Andrew Mackay).

Arthropod factors intertwine with pathogen factors in mechanical transmission. When the titer of infectious agents is high (Fig. 12.5), even mites and ceratopogonids have been implicated in transmission of Venezuelan equine encephalitis and Rift Valley fever viruses, respectively (Hoch et al. 1985, Durden et al. 1992). When the agent is stable on or within the arthropod, rapid mobility of vectors may not be important, as with ticks and bed bugs, potential vectors of myxoma virus (Chapple and Lewis 1965). However, when agents are environmentally labile, the incidence of mixed feeding is a crucial variable. Multiple feeding by vectors also is an important factor in mechanical transmission of agents between hosts. The study of mixed feeding has evaluated vectorial capacity of vectors of agents such as malaria that can be transmitted by probing of mosquitoes. **Patent mixed feeds** (that can be identified by standard serological techniques) are defined as those obtained from at least two different host species within one gonotrophic cycle, while cryptic mixed feeds are from the same host species. The incidence of mixed feeding by mosquitoes on humans varies from 7–19% for *Anopheles* spp. (Burkot et al. 1988) up to 50% for *Aedes aegypti* (Scott et al. 1993). Multiple bloodmeals may be the norm for some species (Briegel and Horler 1993).

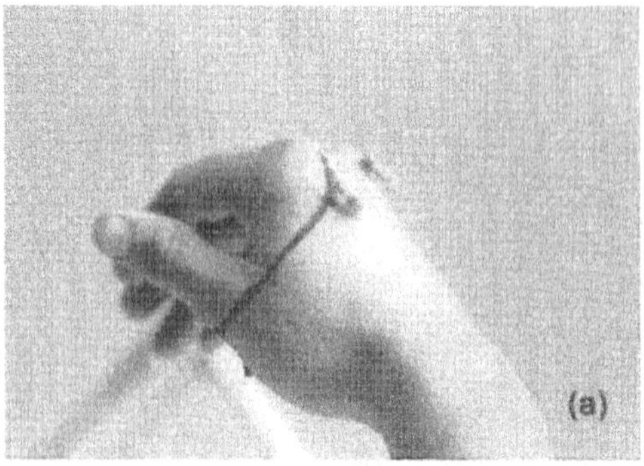

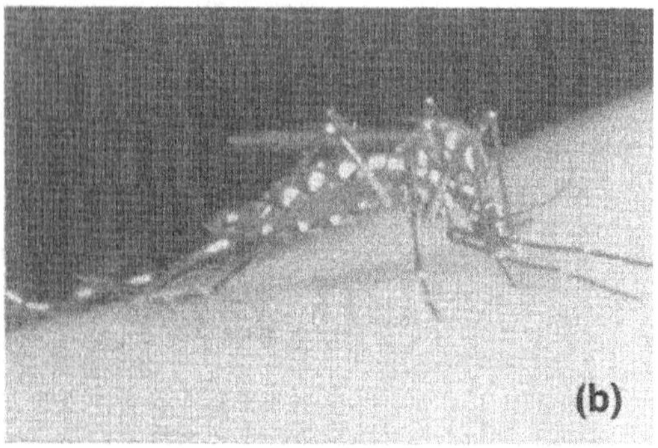

Figure 12.4. Bloodfeeding by Diptera: (a) Horse fly feeding on human hand (photograph by Lane D. Foil); (b) mosquito feeding on human hand (photograph by Leonard E. Munstermann).

PATHOGENIC ORGANISMS MECHANICALLY TRANSMITTED BY ARTHROPODS

The agents described in the following sections are those for which there is experimental evidence of mechanical transmission by arthropods, or for which there is reasonable epidemiological or epizootiological evidence of mechanical transmission. Table 12.3 provides examples of pathogens transferred mechanically to susceptible hosts, but for which biological or contagious transmission is the accepted route. Although references to mechanical transmission by arthropods for the agents in this table may be found, we do not consider this an important factor in the epidemiology of these agents.

Before the recognition of the importance of biological transmission of pathogens by arthropods, there were many attempts to prove mechanical transmission of disease agents, some of which were successful. Two such examples: the mechanical transmission of yellow fever virus and *Leishmania* parasites. However, after biological transmission mechanisms were demonstrated for these and other biologically transmitted pathogens, most early studies were deemphasized.

Viruses

Contaminative transmission

Poliomyelitis virus

During the first half of the 20th Century, poliomyelitis was one of the most serious viral diseases of humans. Because the polio season often coincided with the fly season, speculation connected arthropods and polio (Rogers 1989). Isolations of poliovirus were made from wild cockroaches (Dow 1954) and flies (Melnick and Dow 1948). Experimentation showed that cockroaches, filth flies and even the biting fly *Stomoxys calcitrans* could transmit poliovirus to experimental animals (Ward et al. 1945). Most investigations centered on the house fly, but Francis et al. (1948), experimenting in Michigan and Tennessee, found fly species such as *Phaenicia sericata* more likely to visit human feces than the house fly. Generally, poliovirus was recovered from all species of flies that fed on infected human feces, but no evidence proved that arthropods were important in the epidemiology of polio (Greenberg 1973).

After the beginning of the polio oral vaccination program, the possibility that insects might be vehicles of immunization (Barnett 1960) resurfaced with evidence that flies could

Table 12.3. Disease agents for which there is some evidence of mechanical transmission by arthropods, but with biological transmission by hematophagous arthropods or non-arthropod routes being the primary means of transmission. (Table continued on next page).

		Means of Transmission		
Disease agent	**Primary vertebrate host**	**Primary means**	**Mechanical by arthropods**	**Reference**
Erysipelothrix rhusiopathiae (bacterium)	Swine	Direct contact	Mosquitoes, stable flies, horse flies	Wellman (1950)
Mycobacterium farcinogens	Cattle	Contact with soil	Ticks	Al-Janabi et al. (1975)
Streptococcus suis	Swine	Respiratory route	House flies	Enright et al. (1987)
Leishmania spp.	Humans	Sand flies	Stable flies, tsetse	Lightner and Roberts (1984)
Pseudorabies virus	Livestock	Respiratory and oral route	House flies	Medveczky et al. (1988)
Swinepox virus	Swine	Direct contact, transplacental	Sucking lice	Schwarte and Biester (1941)
African swine fever virus	Swine	Direct contact	Stable flies	Mellor et al. (1987)
Foot and mouth virus	Livestock	Direct contact	Ticks, various insects	Hyslop (1970)
Rinderpest virus	Cattle	Direct contact	Tsetse, horse flies	Hornby (1926), Krinsky (1976)
Rift Valley fever virus	Livestock	Mosquitoes	Mosquitoes, tsetse, other insects	Hoch et al. (1985)
Bovine viral diarrhea virus	Livestock	Direct contact, iatrogenic	Horse flies, stable flies	Tarry et al. (1991)
California encephalitis virus	Livestock	Mosquitoes	Tabanid flies	Main et al. (1979)
Pasteurella multocida	Livestock and poultry	Oral and respiratory route	House flies, biting flies	Nieschulz and Kraneveld (1929)

Table 12.3 (Continued). Disease agents for which there is some evidence of mechanical transmission by arthropods, but with biological transmission by hematophagous arthropods or non-arthropod routes being the primary means of transmission.

Disease agent	Primary vertebrate host	Means of Transmission: Primary means	Means of Transmission: Mechanical by arthropods	Reference
Papillomavirus	Rabbits	Contact through abraded skin	Mosquitoes, assassin bugs	Dalmat (1957)
Ross River virus	Livestock, wildlife	Mosquitoes	Mosquitoes	Kay (1982)
Western equine encephalomyelitis virus	Birds	Mosquitoes	Mosquitoes	Barnett (1956)
Yellow fever virus	Primates	Mosquitoes	Triatomid bugs, stable flies	Hoskins (1934)
Vesicular stomatitis virus	Livestock	Black flies, biting midges	Horse flies, stable flies, mosquitoes	Ferris et al. (1955)
Babesia ovata	Cattle	Ticks	Ticks	Takahashi et al. (1984)
Newcastle disease virus	Birds	Direct contact	*Fannia canicularis*	Rogoff et al. (1975)

transmit the vaccine strain of the poliovirus (Riordan et al. 1961, Sheremetiev 1964).

Even though flies and cockroaches can acquire polioviruses under both natural and experimental conditions and can transmit them to susceptible nonhuman primates (Melnick and Penner 1947), their role in human polio was found to be minimal at best. Paffenbarger and Watt (1953) reached this conclusion after comparing two communities, one with fly control, the other without, during the course of an epidemic of poliomyelitis. Numerous other enteroviruses were isolated from arthropods during the active period of search for links between insects and poliomyelitis epidemics, particularly the coxsackieviruses (Melnick and Dow 1948).

Other viruses

Coxsackieviruses (Dalldorf et al. 1949, Dalldorf 1950) and Echoviruses (Paul et al. 1962) were discovered in connection with investigations of poliomyelitis. Both viruses have been isolated from wild populations of flies (Melnick et al. 1954). Cockroaches have served as experimental mechanical vectors of polioviruses and coxsackieviruses (Roth and Willis 1957) but there is little evidence that cockroaches play any significant role in their dissemination in nature. Similarly, arthropods are not important in the epidemiology of hepatitis A or C viruses, and no role has been established for arthropods in the dissemination of emerging viruses such as Ebola.

Direct mechanical transmission

Hepatitis B virus

The hepatitis B virus (HBV) has a worldwide distribution and is transmitted among humans

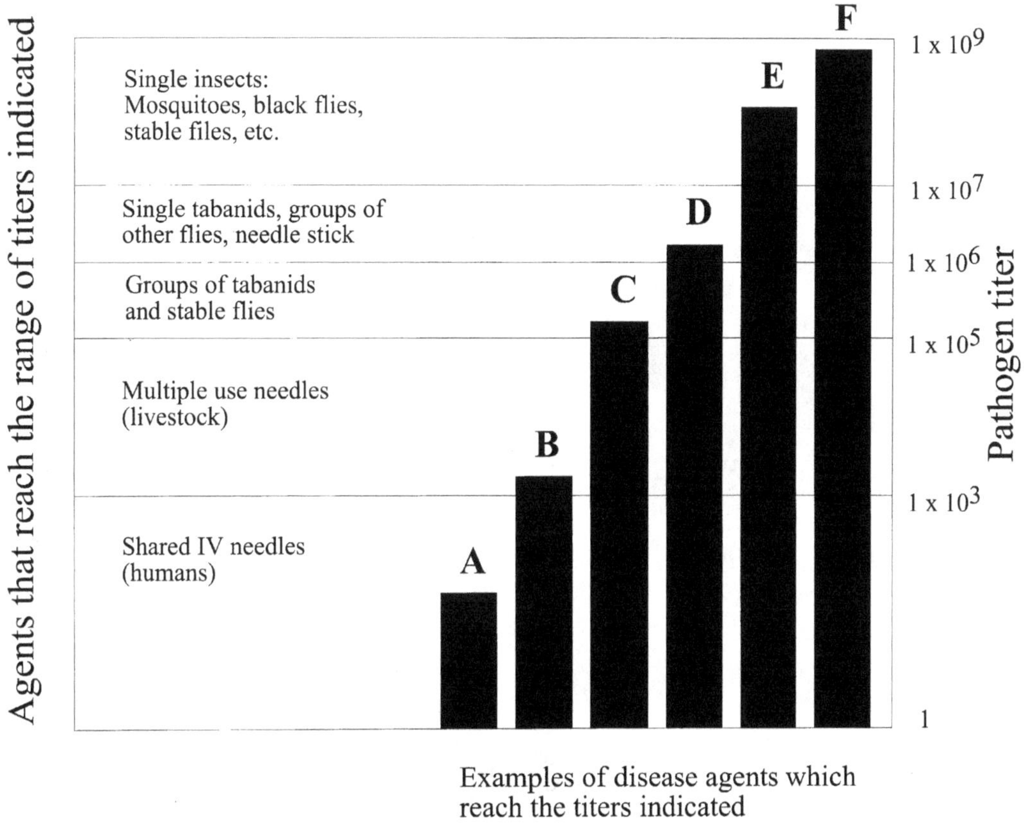

A HIV

B *Anaplasma marginale*, EIA virus

C BLU, BVD virus

D EIA virus, *Anaplasma marginale* (titer during acute disease)

E Myxoma virus, *F. tularensis*

F VEE, RVF, Fowlpox viruses

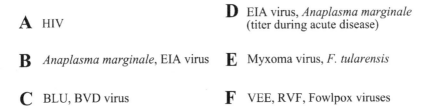

Figure 12.5. Relationship between method of feeding, pathogen titer, and disease transmission capability.

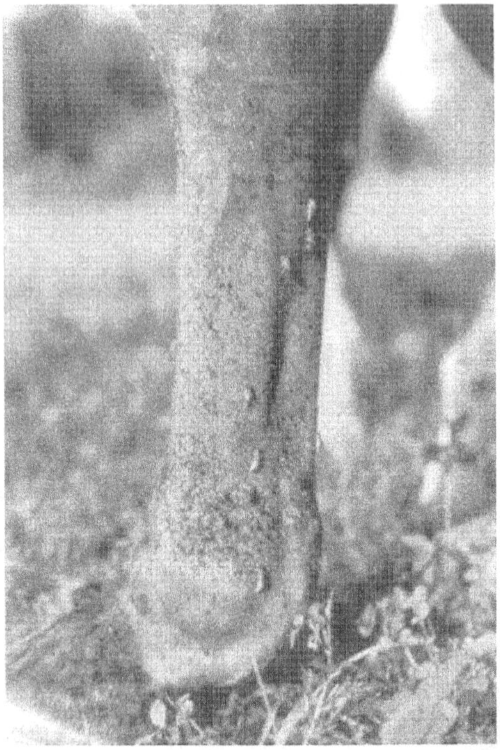

Figure 12.6. Horse flies around the leg of a horse. Each of the lesions on the leg represents a previous tabanid feeding site. On the day this photograph was taken, a landing rate of over 1,000 flies per hour was recorded for this horse. (photograph by Lane D. Foil).

by the transfer of blood, blood products and other body fluids. Hepatitis B virus is the most common cause of chronic liver disease. Humans are the only significant reservoir of infection. There are over 1 million HBV carriers in the USA and over 100 million carriers worldwide. Baruch Blumberg's Nobel Prize acceptance lecture (Blumberg 1977) described the method of discovery of Australia antigen (Au) and ideas on the potential involvement of insects in the transmission of HBV.

The concentration of HBV in the serum of humans can range from undetectable to 10^5–10^9 virions per ml. Humans and chimpanzees become infected following injection with serum dilutions from 10^{-7}–10^{-8} from infected patients. The virus remains viable in dried serum stored at 25°C for over a week. Epidemics usually are linked to the transfusions of infectious blood or blood products, but sexual promiscuity, illicit injectable drug use, tattooing, sharing of razors and acupuncture are high-risk activities. However, the transmission factors discussed above cannot account for the prevalence of HBV in some areas. Jupp et al. (1991) proposed that in Africa, mechanical transmission of HBV by bed bugs (*Cimex lectularius*) and tampan ticks (*Ornithodorus moubata*) may account for epidemiological patterns of HBV infection. Mosquitoes, bed bugs and tampan ticks collected in Africa, where Au frequencies in humans were high, found a high incidence of Au but no evidence of biological transmission (Blumberg et al. 1973, Jupp and Lyons 1987). Transstadial transmission of HBV has been demonstrated in tampan ticks and bed bugs. The hepatitis B surface antigen (HBsAG) has persisted in the tampan tick for up to 779 days and up to 52 days in bed bugs, while another hepatitis B antigen (HBeAG) remained in tampan ticks for up to 16 days. HBV can be excreted in bed bug feces; bed bugs (groups of 40–80) feeding on rabbits and guinea pigs transmit HBV up to 52 days after feeding on infectious human blood (Jupp and McElligott 1979). Ogston (1981) reported that the bed bug, *Cimex hemipterus*, transferred between 0.15 and 0.18 picoliters (1 pl = 10^{-12} liters) in an interrupted feeding between artificial feeders. Groups of 80 bugs could transfer 14.4 pl that would contain HBV virions at titers of 10^8–10^9 per ml. Transmission trials for bed bugs, tampan ticks and mosquitoes, using chimpanzees as recipients, have been unsuccessful, but experiments were restricted because of low

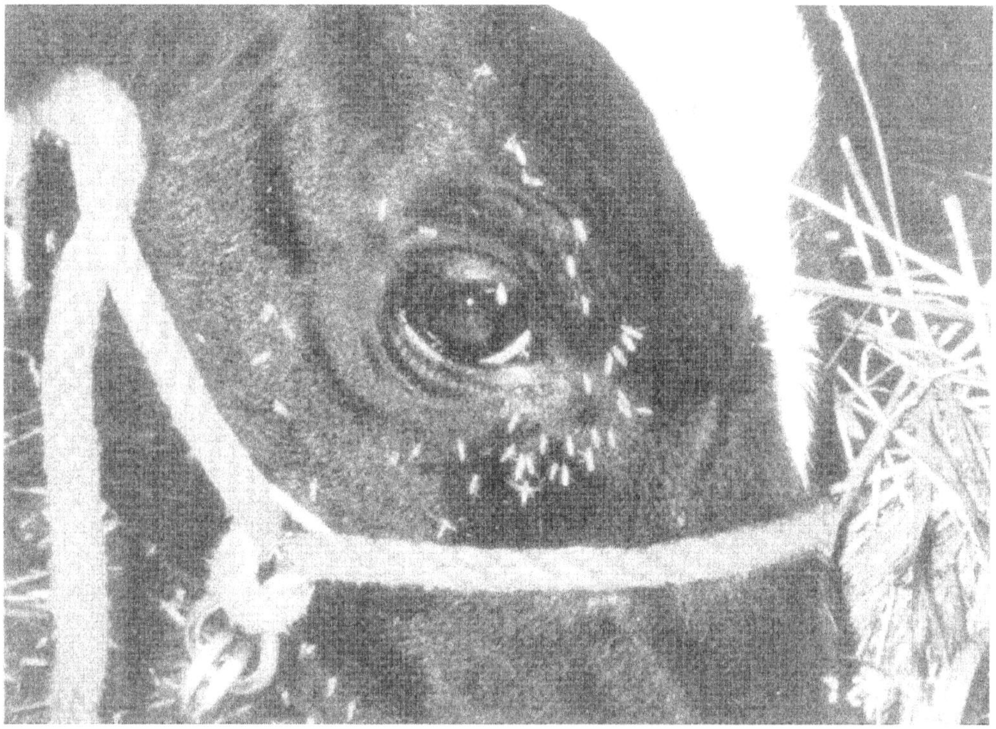

Figure 12.7. Isolated epizootics of anaplasmosis and several other bloodborne agents of livestock have been attributed to mechanical transmission by mosquitoes when populations are extremely high (photograph by Reid R. Gerhardt).

numbers of experimental animals. Since interrupted feeding between humans by bed bugs or ticks is infrequent, the crushing of an arthropod or exposure to feces or coxal fluid from arthropods that had fed previously and molted are more likely routes of infection. No studies suggest that mechanical transmission by arthropods plays a major role in the epidemiology of HBV generally, but the studies discussed above suggest that mechanical transmission of HBV by arthropods may be important under special circumstances. There is no evidence for replication or dissemination of hepatitis C virus in insects (Silverman et al. 1996).

Retroviruses

The retroviruses are a diverse group of viruses that generally are species specific. The identification of retroviruses as agents of livestock diseases dates back to the early 1900s. The important retroviruses of humans and domestic animals are classified in two subfamilies, the oncoviruses and the lentiviruses. The oncoviruses are associated with chronic lymphomas and leukemia. The lentiviruses induce immunodeficiency, inflammatory and hemolytic diseases. Certain retroviruses are transmitted vertically either **epigenetically** (mother to progeny transplacentally or via milk), e.g., visna and Human T-lymphotropic virus, or **genetically** (from both parents), e.g., avian and murine leukosis viruses. However, most retroviruses are transmitted horizontally; sexual transmission as a major route is confined to human and murine retroviruses. There currently is no evidence of propagation of vertebrate-associated retroviruses in arthropods. Studies of insect transmission of bovine and equine retroviruses have served as models to evaluate the possible transmission of other retroviruses as well as other bloodborne agents of humans and animals. Foil and Issel (1991) summarized details of many of the important studies on retrovirus transmission.

Equine infectious anemia. Commonly referred to as swamp fever, this disease has been recognized for approximately 100 yr. The equine infectious anemia (EIA) virus is found worldwide and is confined to the family Equidae. EIA virus is a particularly good model since the epizootiology is not clouded by varying biological transmission cycles or vertebrate hosts other than equids. Within 1–4 weeks of exposure, either a fatal, acute disease or a chronic infection in horses can occur. Signs of equine infectious anemia are weight loss, progressive weakness, anorexia, intermittent fever, hemorrhage, anemia and dependent edema of the legs, brisket and abdomen. The chronic infection is associated with recurrent fevers that may occur in intervals from 1–12 months. Periodic febrile episodes are associated with increases in plasma-free virus that results from release of new antigenic variants; maximum viremia is found during the febrile periods. The majority of EIA virus-infected horses are inapparent carriers; these animals exhibit no overt signs of disease. Infected horses harbor the virus for life.

Before a diagnostic test to detect EIA virus infection was available, epizootics of equine infectious anemia occurred at gathering points for horses such as race tracks. Most outbreaks resulted from **iatrogenic** (caused by the activity of a physician or a veterinarian) transmission, such as the use of contaminated needles, syringes or surgical instruments. In an epizootic at the Rockingham Park racetrack in 1947, over 10% of the horses died from the disease. Coggins and Norcross (1970) developed the agar-gel immunodiffusion (AGID) test for detecting EIA infections; this test remains the standard. Equine infectious anemia was designated as a reportable disease in 1973; federal and some state guidelines were established to reduce the transport of infected horses. From 1974 to 1985, the national percentage of reported positive AGID tests dropped from 2.5% to 0.45% (Issel et al. 1988).

Transmission from acutely infected horses can occur by bites of tabanids, but transmission from inapparent carriers usually is iatrogenic. Since EIA virus cannot survive on the mouthparts of insects for more than an hour, transport of horses usually spreads the virus. Controlling the spread of the disease during epizootics requires consideration of possible insect vectors, and involves community cooperation and education. The primary goal is to prevent iatrogenic transmission. Test and slaughter is one management strategy for livestock diseases. Separating infected animals from susceptible animals by at least 200 m is recommended; this distance was based first on common sense and later confirmed as an appropriate strategy based upon tabanid behavior (Foil et al. 1983). Since the transfer of the agent is via blood, any blood-feeding arthropod can be a mechanical vector. Managing certain potential vectors, e.g., stable flies, is recommended. The value of the animals determines implementation of other measures. Housing animals can be effective since most tabanids will not enter barns. Repellents can protect individual animals from vectors; but insecticides are not included in preventive programs.

Reviews on mechanical transmission of EIA virus by insects are available (Issel and Foil 1984, Foil et al. 1988). Concentration of EIA virus can reach at least 10^6 infectious particles per ml of blood in acute, febrile infections. The incubation period of the acute form of the disease is short enough to allow frequent observations of the association with insects or iatrogenic modes of transmission; transmission by a single contaminated needle or a single fly is possible. Early observations were made on vectors for the **parenteral** (routes other than the digestive tract) transfer of blood among horses. The common name for equine infectious anemia, swamp fever, and the geographic and temporal aspects of EIA virus transmission suggest insect transmission. In the USA, the EIA antibody is highest in the Gulf Coast states, presumably associated with high vector populations. Epizootics occur consistently during periods of insect activity; observations on tabanid activity frequently accompany reports on epizootics, but other Diptera (e.g., stable flies) obviously are important alternate vectors. Studies of epizootics resulting in large numbers of inapparent infections in horses have been associated with peaks of tabanid activity [landing rates up to 1,000 per hr (Fig. 12.6; Foil et al. 1984, 1988)]. However, since the titer of EIA virus in afebrile carriers rarely exceeds 10^3 HID (horse infectious doses)/ml, epizootics stemming from horses with inapparent infections are unusual.

Mosquitoes are not considered potential vectors of EIA virus. Transmission studies using various species, including *Psorophora columbiae, Aedes taeniorhynchus* and *Aedes sollicitans*, in groups ranging from 40–300+, have been negative. It is estimated that between 10^{-9}–10^{-10} ml of blood are transferred to a second host by a mosquito following an interrupted feeding. Since transmission of EIA is considered to require considerably higher amounts of blood, mosquito involvement is unlikely (Miike 1987). Furthermore, Williams et al. (1981) found that EIA virus can survive on needles for up to 96 hr, but on mosquito mouthparts for only about 1 hr.

Several studies (Foil and Issel 1991) have demonstrated transmission of EIA virus by groups of 50–200 stable flies. Regurgitation by stable flies during a second feeding also has been described using *in vitro* techniques (Butler et al. 1977), but results of EIA virus transmission trials do not indicate a high rate of regurgitation. Weber et al. (1988) estimated by measurements made from electron micrographs that the internal mouthparts of stable flies would retain 0.03 nanoliters (1 nl = 10^{-9} liters) of blood, and this estimate is consistent with the transmission trials for EIA virus (200 flies could transfer 6 nl of blood, or a possible 6 HID).

Tabanids are the most important vector of EIA virus. Hawkins et al. (1976) demonstrated EIA virus transmission by the bite of a single horse fly (*Tabanus fuscicostatus*) in one of seven trials; Foil et al. (1983) reported transmission with six deer flies *(Chrysops flavidus)*. Hawkins et al. (1976) also obtained virus transmission with groups of 25 *T. fuscicostatus* that were held for 3–30 min between the initial interrupted meal and the second feeding. However, groups held for 4 and 24 hr did not transmit. Foil et al. (1987) reported that a bloodmeal residue of ca 5 nl (theoretically with 5 HID when fed upon a host with 10^6 infectious units per ml) would be transferred to a second host following an interrupted feeding by *T. fuscicostatus*. Knaus et al. (1993) showed that the amount of blood adhering to 22-gauge needles following a percutaneous needle-stick was approximately 9 nl; the amount of blood on the mouthparts of small to medium-sized tabanids ranged from 6–12.5 nl. Thus, the vector potential of horse flies for EIA virus would be approximately 200 times greater than stable flies and 6,000–60,000 times greater than mosquitoes.

Human retroviruses. The transfer of blood or other body fluids containing human immunodeficiency virus (HIV) between humans is the major route of HIV transmission. There have been no indications that insect transmission could be a major factor in the spread of this virus (Friedland and Klein 1987). However, since the mid-1980s the possibility that insects transmit AIDS has been a recurrent topic.

In 1985, there were unsubstantiated claims that a number of AIDS cases from Belle Glade, Florida, were related to insect transmission. Seroepidemiologic studies demonstrated convincingly that the cases were associated with sexual activity and intravenous drug use (Castro et al. 1988). Studies by a French research group also created interest; HIV-like nucleic acid sequences were found in insects collected from two areas with endemic HIV in central Africa, but not from insects from Paris (Becker et al. 1986). Cockroaches, tsetse and antlions from Zaire, and mosquitoes, ticks and bed bugs from the Central African Republic were all positive by an assay for HIV proviral DNA. The presence of the nucleic acid sequences in non-blood-feeding insects suggested that the results were laboratory artifacts. Subsequent efforts could not confirm the French studies; other studies found no evidence to incriminate insects as biological vectors of HIV. Furthermore, the virus does not survive for long periods in arthropods (Jupp and Lyons 1987, Srinivasan et al. 1987, Webb et al. 1989).

The potential for mechanical transmission of HIV between hosts by insects during interrupted feeding has been considered. Jupp and Lyons (1987) failed to demonstrate transmission after feeding four groups of 100 bed bugs on a blood-virus mixture (4.4 X 10^5 tissue culture infectious doses (TCID)/ml) and then transferring them to a noninfected membrane feeder to complete the meal. Webb et al. (1989) reported similar results. Both studies used concentrations of HIV far higher than would ever occur in human blood. This test further demonstrated how unlikely the bed bug would be as a mechanical vector of HIV.

There have been estimates that the amount of HIV in 1 ml blood from an asymptomatic HIV-infected human would average 0.06 TCID compared to 7 TCID from a symptomatic patient (Ho et al. 1989). The quantity of blood transferred between hosts has been estimated to be less than a picoliter for individual vessel-feeding insects, bed bugs or mosquitoes (Miike 1987), which means that the rate of transmission via interrupted feeding would occur at an incidence of one per 1 million to 10 million events. This estimate does not include an estimate of the frequency of mixed feeding by the insects (Burkot et al. 1988). In contrast, the quantity of blood transferred by intravenous use of a 22-gauge needle is approximately 1 microliter (1 µl = 10^{-6} liters) (Napoli and McGowan 1987). The amount of blood remaining in the neck of a syringe following expulsion of blood can range from 135–150 µl (Gunn 1993).

HTLV-1 has been associated with adult T-cell leukemia, tropical spastic paraparesis and myelopathy (Quinn et al. 1989). Miller et al. (1986) conducted a seroepidemiologic survey in an urban community of Trinidad and reported significant relationships between seropositivity for HTLV-1 antibody and poor quality housing and proximal water courses. He concluded that insects may be important in HTLV-1 transmission under certain circumstances. Murphy et al. (1989) responded to the Miller study by reporting a survey conducted in Jamaica. Serum samples from 200 volunteers (90 HTLV-1 positive and 110 HTLV-1 negative) were screened for 5 arboviruses reported to occur in Jamaica; the authors found no evidence of correlation between prevalence of HTLV-1 and arboviruses. The major routes of transmission for HTLV-1 appear to be similar to HIV with the addition of breast-feeding (Quinn et al. 1989).

Bovine retroviruses. Enzootic bovine leukosis is associated with bovine leukemia (BL) virus; there are 2 distinct forms of the disease: persistent lymphocytosis (an increase in lymphocyte counts that persists for at least 3 months) and lymphosarcoma. Antibody responses following exposure to BL virus are seen within 4 months, but the disease may not appear for years. Bovine lymphosarcoma or lymphocytic leukemia usually is found in cattle >3 yr old. Proliferation of neoplastic lymphocytes results in tumors that are disseminated widely in the lymph nodes, spinal cord, abomasum, heart, uterus, retrobulbular lymphatics, kidney and spleen. Clinical signs usually are weight loss, anemia, lameness, exopthalmosis and decreased milk production. Genetic predisposition to developing lymphosarcoma is well described. Persistent lymphocytosis (PL) is considered a separate disease. Approximately two-thirds of cattle that develop lymphosarcoma have PL for 2–3 yr. The percentage of lymphocytes that have infectious BL virus varies among animals, but PL usually is associated with a high infectious titer of blood. Diagnosis of enzootic bovine leukosis is based upon hematology and serologic tests such as the AGID. Policies on control of enzootic bovine leukosis vary geographically. Test and removal of antibody positive animals control the disease. Eradication programs have been successful in some countries in Europe. Enzootic bovine leukosis is a reportable disease in countries of the European Economic Community.

Horizontal transmission is the most important route under natural conditions. The majority of cattle become infected after the age of 18 months (Ferrer 1980, Evermann et al. 1987). Transmission of BL virus via milk and *in utero* transmission occur. Venereal transmission has been reported but is not considered to be a major route of transmission (Van der Maaten et al. 1981). Iatrogenic transmission of BLV has been described in association with tattooing, dehorning, palpating and vaccinating (Henry et al. 1987).

For bovine leukemia, the most sensitive method of infection is intradermal or intracutaneous inoculation. The most infectious bovine tissue, secretion or excretion is blood. With PL, the volume of blood required to transfer infection can be as low as 50–100 µl (Foil et al. 1989). Epizootics have been described during winter housing periods (Wilesmith et al. 1980), but there is evidence for involvement of insects in BL virus transmission (Bech-Nielsen et al. 1978). During a 2-year study of 3,328 cattle in 3 areas of France, Manet et al. (1989) found a significant correlation between the density and seasonal activity of tabanids and the rate of seroconversion to BL virus.

Foil et al. (1989) demonstrated BL virus transmission by horse flies (*T. fuscicostatus*), initially fed on cattle, to goats and dairy calves. Using mouthpart residue estimates (10 nl) for *T. fuscicostatus*, groups of 10–20 flies would transfer from 50–200 nl of blood to recipients; 100 nl of the donor blood was shown to be infectious for calves and goats. These studies suggested that tabanids may be involved in horizontal transmission of BL virus among cattle; studies on mechanical transmission of BL virus by insects other than tabanids have met with limited success (Weber et al. 1988). The importance of insects in the epizootiology of BL virus probably varies by geographic area. Presently, iatrogenic transmission is considered to be the primary mechanism of spread for the bovine immunodeficiency (BI) virus.

Other retroviruses. Reticuloendothelial virus (REV) is a retrovirus of poultry in the avian leukosis-sarcoma complex of avian neoplastic diseases such as leukosis, osteopetrosis and erythroblastosis. Both horizontal and vertical transmission are important for viruses of this complex. The RE virus is associated with myeloblastosis, a leukemia that can result in over 10^6 myeloblasts per ml of blood. After REV was introduced into Australian commercial chicken flocks in 1979 by contaminated vaccines, the virus was transmitted the following summer and fall. Mechanical transmission of REV by mosquitoes (approximately 40 *Culex annulirostris*) that previously had been fed on infectious chickens was demonstrated (Motha et al. 1984). Insect-proof pens protected sentinel chickens from infection, while unprotected sentinels seroconverted and REV was recovered from engorged and nonengorged *Culex* mosquitoes. Studies by Fischer et al. (1973) on the arthropod transmission of friend (murine) leukemia virus are useful in the effort to evaluate the potential role of insects in mechanical transmission. Mosquitoes and stable flies were fed on donors and then transferred to recipient suckling mice in groups of 1–20. Transmission by single stable flies occurred in 12% of the trials when the virus titer was $10^{7.5}$ virus particles per ml. Single mosquitoes transferred the virus when the titer approached 10^9.

There is little evidence to indicate a role for insects in the transmission of feline retroviruses. Feline immunodeficiency virus and feline leukemia virus both are transmitted primarily by cat bite.

Arboviruses

Arboviruses are, by definition, viruses that are transmitted biologically by arthropods (Chapter 11). They are classified in 5 families: Togaviridae, Flaviviridae, Bunyaviridae, Reoviridae and Rhabdoviridae. Mechanical transmission may supplement biological transmission of some arboviruses where vertebrate host viremias are very high. Studies demonstrating this fact include Semliki Forest virus (Woodall and Bertram 1959) and Nodamura virus (Scherer and Hurlbut 1967). However, the impact of mechanical transmission in the epidemiology of these agents rarely is addressed. This section will cover arboviruses and related viruses that may be mechanically transmitted by arthropods.

Family Flaviviridae, genus Pestivirus. Most flaviviruses associated with arthropods are classified in the genus *Flavivirus* and are transmitted biologically. Viruses in the genus *Pestivirus* are not transmitted biologically by arthropods, but mechanical transmission may be involved. The two major diseases of livestock caused by pestiviruses are hog cholera and bovine virus diarrhea. Hog cholera (HC) virus infections are confined to swine, and bovine virus diarrhea (BVD) virus infections occur in cattle, swine, sheep and deer. BVD virus causes border disease in sheep (Tarry et al. 1991).

The economic importance of hog cholera, an acute, highly contagious disease of swine, has led to successful eradication programs in the USA, Canada and parts of Europe. Signs of HC virus infection are anorexia and fever followed by depression, diarrhea, vomiting and discharge from the eyes. Encephalomyelitis occurs in the majority of fatal infections. Degeneration in the walls of the smaller vessels of the internal organs can result in petechial hemorrhages. Chronically infected animals may succumb to secondary infections. Congenital infections and abortions occur. Since there appear to be no reservoirs of HC virus outside of swine, the success of eradication programs is based upon elimination of infected pigs. Antiserum therapy can save herds when early diagnoses are made; attenuated vaccines also are available. The US Hog Cholera Eradication program from 1962–1978 eradicated HC virus from the USA.

Hog cholera virus is transmitted primarily between animals by contact; the virus is stable and can remain infective for up to 3 days at 50°C. Persistent infections (several months) have been described, indicating that pigs can serve as a reservoir for the virus. Undercooked pork, ultimately fed to pigs, also has caused infections. Iatrogenic transmission likewise is associated with the spread of the disease.

Tidwell et al. (1972) presented epidemiological evidence supporting mechanical transmission of HC virus by various Diptera; high populations of flies were associated with infected swine herds. The authors demonstrated mechanical transmission of HC virus from infected pigs to susceptible pigs by transfers by as few as 5 horse flies *(Tabanus lineola)* following interrupted feeding. Morgan and Miller (1976) showed that HC virus could be transmitted by house flies, stable flies and face flies up to 48 hr after contact with infected donors. Stewart et al. (1975) demonstrated mechanical transmission by groups of 50 *Ae. aegypti* in 2 of 8 groups fed on donor pigs (viremia measured at 10^6 PFU – plaque forming units – per ml of blood) and transferred to susceptible pigs.

Bovine viral diarrhea is distributed worldwide, frequently associated with abortion and fetal abnormalities. The virus is transmitted vertically in cattle. The principal reservoirs of BVD virus appear to be congenitally infected cattle.

Winter is the peak season of BVD virus transmission, suggesting that under normal circumstances arthropods play a minor role in transmission. Iatrogenic transmission by needles and nose tongs has been demonstrated (Gunn 1993). However, Tarry et al. (1991) demonstrated mechanical transmission from a donor bullock with approximately $10^{4.5}$ $TCID_{50}$ per ml of serum by the interrupted feeding of both stable flies and tabanids (*Haematopota pluvialis*) in groups of 50 transferred to calves or sheep. The bullock was a persistently infected, asymptomatic animal. These experiments suggest that mechanical transmission of BVD virus may occur during warmer parts of the year. Mechanical transmission may be a factor in the spread of BVD virus among farms by insects during epizootics.

Family Togaviridae, genus Alphavirus. Chikungunya (CHIK) virus is one of several alphaviruses known to produce arthritis in humans. Primates are considered to be the primary reservoir of CHIK virus epidemics in Asia and Africa, and *Aedes* mosquitoes are the proven biological vectors. Rao et al. (1968) showed that individual *Ae. aegypti* mosquitoes could mechanically transmit CHIK virus between viremic ($>10^9$ LD_{50}/ml blood) and noninfected mice for up to 8 hr after an interrupted feeding.

Numerous field studies have implicated black flies, biting midges and other Diptera as mechanical vectors of eastern equine encephalitis (EEE) virus (Chamberlain and Sudia 1961). These authors also demonstrated mechanical transmission of EEE virus; using insect pins, stable flies and mosquitoes, they showed that the virus could be transmitted mechanically for up to 70 hr at 26.6°C.

Venezuelan equine encephalomyelitis (VEE) virus studies repeatedly have demonstrated the role of mechanical transmission by arthropods in nature. Since the virus reaches titers of 10^8 mouse LD_{50} per ml in equids, mechanical transmission is highly probable for many bloodsucking insects and acarines. Black flies have been incriminated as mechanical vectors (Eddy et al. 1972). Durden et al. (1992) demonstrated mechanical transmission by groups of hematophagous mites (*Dermanyssus gallinae*) between suckling mice when donors had blood titers of $10^{7.4}$–$10^{7.8}$ PFU per ml. Transmission occurred up to 16 hr after feeding on the donor. Similar studies had been reported previously for EEE and western equine encephalomyelitis viruses.

Family Bunyaviridae, genus Bunyavirus. Rift Valley fever (RVF) virus causes an acute febrile disease of humans, sheep and cattle in Africa and the Middle East. Fatality rates range from 10% in cattle to 100% in lambs. Humans are exposed by vectors or contact with infected animals or aborted fetuses. Mortality in zoonotics can be >90%. The disease in humans can have dengue-like symptoms. Biological transmission by many different species of mosquitoes is well documented. Isolations have been made from *Simulium* spp. and *Culicoides* spp. Dissemination of the virus over long distances may be via windborne vectors (Sellers et al. 1982). The RVF viremia can reach 10^9–10^{10} virus particles per ml in cattle and sheep. Mechanical transmission of the virus by mosquitoes, tsetse, stable flies and *Culicoides* spp. has been demonstrated (Hoch et al. 1985).

Poxviruses

Viremias of the poxviruses, largest of the animal viruses, often are associated with pathological changes of the skin. Poxviruses are classified in the family Poxviridae. Poxviruses infect a variety of animals, including humans. The importance of arthropods in the epidemiology of these viruses ranges from nil (smallpox) to being the major route of transmission (myxomatosis). Infection by poxviruses normally occurs by aerosol or cutaneous exposure, depending upon the viral species. Nine poxviruses can cause diseases in humans. In spite of the dermotropic nature of some of these viruses, insect transmission as a possible route of human exposure currently is considered important only for the rare zoonosis caused by tanapox virus (genus *Yatapoxvirus*). Insects have been implicated in the epizootiology of many of the poxvirus diseases of livestock, including myxomatosis, lumpy skin diseases, swinepox and fowlpox (Fig. 12.8). The involvement of insects in the transmission of certain other animal poxviruses is assumed, but not yet demonstrated (e.g., Shope fibroma virus).

The poxviruses are enveloped viruses with double-stranded DNA, but they lack icosahedral symmetry. After the primary infection, there normally are primary and secondary viremias that ultimately result in skin disease. Pox lesions in the skin progress from macules to papules to vesicles to pustules (the pock) with a depressed center and a raised, erythematous perimeter. The pustules rupture and ultimately heal, sometimes leaving a scar. The quantity of infectious poxvirus particles that occurs in the skin (available for the mechanical transmission by arthropods) is both species and strain specific and also influenced by vertebrate host immunity.

Myxoma virus. Myxoma virus is a poxvirus in the genus *Myxomavirus*. The natural hosts for myxoma virus are rabbits of the genus *Sylvilagus* in the Americas. The virus is maintained within *Sylvilagus* rabbit populations by mechanical transmission by insects that feed on localized skin tumors (where virus titers are highest). Myxomatosis was recognized as a fatal disease of domestic European rabbits (*Oryctolagus cuniculus*) in South America in the 1800s. Mosquitoes from the native rabbit enzootic cycle introduced myxoma virus into domestic rabbit populations. The European rabbit was released into Australia in 1859; for lack of predators, it became a major agricultural pest. Using myxoma virus as a biological control agent for rabbits arose from the knowledge of the epizootics in domestic rabbits in the Americas.

Myxomatosis is a highly contagious and fatal disease in rabbitries; most domestic rabbit breeds are susceptible. Initial signs of the disease include swelling of the eyelids and purulent discharge. Subsequently, tumorous masses appear over most of the body and death occurs within 1–2 weeks.

Shope fibroma virus is related closely to myxoma virus. The virus is not spread between rabbits by contact. Biting insects may be involved in the natural transmission. This virus has been used to vaccinate breeding stocks in Europe against myxomatosis, but attempts to introduce the virus and allow insect transmission to establish immunity in wild rabbit populations have not been successful. Myxoma and fibroma viruses produce only localized tumors in *Sylvilagus* rabbits, but myxoma virus produces viremia in domestic rabbits.

Transmission mechanisms for fibroma and myxomaviruses are similar. Myxoma virus is extremely stable environmentally under certain conditions. This stability probably is associated with a dense protein matrix within the cytoplasmic inclusion (Fig. 12.8). For example, Chapple and Lewis (1965) fed rabbit fleas, *Spilopsyllus cuniculi*, on infected rabbits and were able to infect rabbits with fleas held for as long as 105 days. This longevity explains why fleas can serve as reservoirs in deserted burrows. The virus also has been viable for up to 220 days on the mouthparts of mosquitoes. Fenner and Ratcliffe (1965) reviewed research on potential arthropod vectors of myxoma virus and concluded that every arthropod induced to bite an infected rabbit was capable of mechanical transmission. Their list included mosquitoes, biting flies, fleas, lice, ticks and mites.

Myxoma virus was introduced into Australia in 1950 to control the European rabbit populations that plagued agriculture. The virus was assumed to be highly contagious and would spread. This assumption was adopted in spite of findings by Bull et al. (1954) indicating that the agent would not spread in the absence of appropriate vectors. There were several unsuccessful releases at various sites in the spring and winter of 1950; successful releases were achieved only at sites in the Murray River Valley where mosquitoes were available as vectors. Within 2 weeks of the establishment of an epizootic, the disease spread by jumps up to 960 km along water courses. Within 2 yr, the rabbit population was reduced from 600 million to 100 million. In Australia, a highly virulent strain could be maintained in nature only during the summer when mosquito populations were high. Rabbits infected with highly virulent strains have short survival times. There is only a short period during which the virus can be found at high enough titers in the skin for arthropod transmission. Rabbits infected with moderately attenuated strains can remain infectious for up to 30 days longer than those infected with highly virulent strains. It was predicted that releases of the highly virulent strain would be required yearly, since it could not overwinter. However, within the first year, mutants of the virus appeared that were moderately virulent and thus would survive the winter because of the increased survival time of the infected rabbits. Over time, the original strain was replaced by strains of lower virulence. These strains predominate in areas where arthropod transmission is the primary means of dissemination.

Myxoma virus was introduced accidentally into Europe in 1952; within 2 yr an epizootic had spread throughout continental Europe and into Great Britain. When the myxomatosis epizootic occurred in the United Kingdom, the European rabbit flea, *S. cuniculi*, was the most important vector. Considering the efficiency of the rabbit flea as a vector of myxoma virus, *S. cuniculi* was introduced into Australia in 1966 to extend transmission periods and the range of the virus (Rosamond et al. 1977). Releases were made in the Malee region of Victoria, and flea populations were established over a 3-yr period. The flea was introduced later into southwestern Australia. Attempts to introduce a Spanish rabbit flea, *Xenopsylla cuniculi*, into arid zones of Australia (Rosamond et al. 1977) further the work to introduce vectors to expand the range in which myxoma virus survives year around.

In addition to changes in the virus, there have been changes in the resistance of rabbits. Inherited resistance to myxoma infection in wild rabbits was described by 1958 in Australia and by 1977 in Great Britain (Ross and Sanders 1984).

Fowlpox viruses. The genus *Avipoxvirus* includes the bird poxviruses that cause canarypox, pigeonpox, turkeypox and fowlpox.

Fowlpox is characterized by papules that progress to dry scabs. These papules are located on the head of chickens, especially on the comb and wattles. Young mature birds that have not been exposed or vaccinated are at greatest risk of infection. The disease causes loss of appetite, and hens may stop laying. It may be fatal, but birds that recover from infection have solid immunity. Presence of the lesions and microscopic demonstration of intracytoplasmic inclusion bodies (Fig. 12.8) confirm the diagnosis of fowlpox; vaccines are available. Fowlpox virus may be maintained in domestic or wild birds and also is environmentally stable on insects, remaining infectious for up to 210 days on mosquitoes.

Fowlpox virus is transmitted primarily by contact or mechanical transmission by insects. Transmission requires a skin barrier compromised by abrasions or insect bites. Direct transmission between birds due to pecking may be an important route of transmission, but many poultry facilities maintain birds in small groups to limit this type of transmission. Transmission among flocks by insects appears to be a major route of dissemination. Vector transmission is observed most frequently in the fall, which may be related to flock susceptibility, vectors or migratory bird reservoirs (Lee et al. 1958).

Kligler (1929) demonstrated mechanical transmission of fowlpox virus by individual mosquitoes (*Culex pipiens* and *Ae. aegypti)* feeding on the heads of infected chickens and then transferring within 2 hr to susceptible birds. Transmission by a *Cx. pipiens* mosquito fed 14 days previously also was reported. Considering the environmental stability of the virus on the mouthparts of insects, frequency of feeding or interrupted feeding is not as important as a close association between potential vectors and poultry. Lee et al. (1958) reported that 8 of 9 virus isolations from mosquitoes in Australia were from *Culex quinquefasciatus*, a species that feeds almost exclusively on poultry in that region. Fukuda et al. (1979) demonstrated the transfer of fowlpox virus via *Culicoides arakawae* fed on an infected host 7 days previously, demonstrating that interrupted feeding is not required for the mechanical transmission of fowlpox virus. They suggested that vector density as well as body size of potential mechanical vectors should be considered. As illustrated in Fig. 12.8, the fasicle of mosquitoes and biting midges contact dense fields of intracytoplasmic inclusion bodies during the normal feeding process.

Capripox viruses. Diseases caused by viruses of the genus *Capripoxvirus* include goatpox, sheeppox and lumpy skin disease of cattle and wildlife. Strains of capripoxviruses of goats and sheep can be host specific or can cause malignant pox diseases affecting both sheep and goats. Capripoxvirus infections of goats and sheep are found worldwide. Sheeppox, the most damaging of the pox diseases of livestock, occurs in Africa, Europe, Asia and the Middle East. Signs of goatpox and sheeppox appear within 1–2 weeks of exposure; fever, conjunctivitis and nasal discharge are followed within 1–2 days by generalized pox lesions of the skin. Mortality rates are as high as 50%, and there is reduced meat and wool production. These poxviruses can remain viable for up to 6 months in dried crust. They can be transmitted by direct contact among animals.

Stable flies are efficient vectors of capripoxviruses. The viruses remain infectious for up to 4 days in flies, and the agents have been transmitted by flies feeding upon an infectious meal 24 hr previously (Mellor et al. 1987). A capripoxvirus known as the Neethling virus causes lumpy skin disease of cattle and wildlife. The disease has been described in Africa and the Middle East. Signs of the disease are pyrexia, general lymphadenitis, nodular cutaneous lesions and lameness. Hide damage and reduction in meat and milk production cause economic losses. Insect transmission of lumpy skin disease virus is considered important. Transmission of the virus has been associated with seasonal peaks of arthropods. The virus is not transmitted by direct contact between cattle (Woods 1988). Vaccines are available for the capripoxviruses of livestock.

Poxvirus diseases in humans. Tanapoxvirus is the only poxvirus known to be mechanically transmitted to humans by insects. Human tanapox has been described in Kenya and Zaire, where it is suggested that the virus exists in wildlife hosts. Monkeys injected with the virus produce lesions. Serological evidence of infection has been shown in wild monkeys. In human skin, the typical pock-like lesions are found singularly or in pairs. Residents of endemic areas and animal keepers in contact with infected monkeys are at risk. Mechanical transmission by arthropods, particularly mosquitoes, is con-

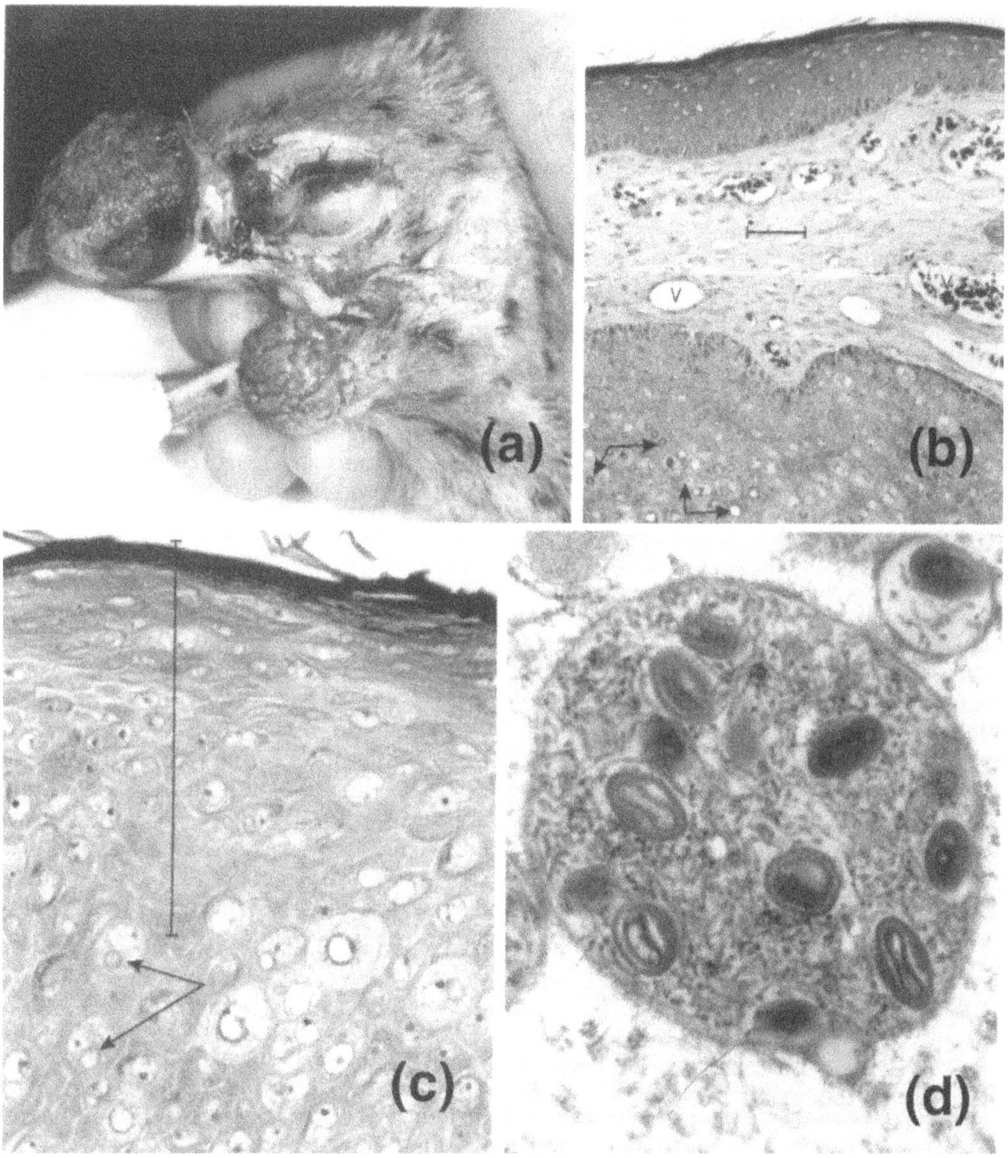

Figure 12.8. Poxviruses: (a) Head of a juvenile ostrich with multiple proliferative and ulcerative poxvirus lesions; (b) hematoxylin- and eosin-stained proliferative poxvirus skin lesion in a canary (16X). Arrows indicate pox inclusion bodies, letter v refers to dermal venule (feeding site for mosquitoes) and bar indicates 0.05 mm, which is the estimated internal diameter of the fascile of a mosquito; (c) hematoxylin- and eosin-stained proliferative poxvirus skin lesion in a canary (40X). Arrows indicate pox inclusion bodies, the bar indicates the estimated length of the fascile of *Culicoides pulicaris* (Jobling 1927); (d) electron micrograph (53,330X) of a poxvirus inclusion containing individual virus particles (photographs of ostrich and electron micrograph by Thomas Tully and Alvin Camus. Photographs of histological lesions by Fred Enright. Preparation of composite figure by William Henk).

sidered a probable route of human exposure. Manson-Bahr and Downie (1973) found a correlation between antibody titers against West Nile virus (a flavivirus) and tanapoxvirus in residents of the Tana River valley. Other poxvirus zoonoses include orf, paravaccinia, cowpox and monkeypox. Paravaccinia virus (milker's nodules), orf virus and cowpox virus are all transmitted to humans from livestock by direct contact; however, transmission of poxviruses among animals by arthropods may be important in viral maintenance in nature.

Smallpox was one of the most important diseases in history. Jenner initiated vaccination in humans with cowpox. This strategy ultimately eradicated the smallpox virus. Smallpox infection was spread primarily by aerosol; infection occurred by inhalation of virus released from the oropharynx of humans. Unlike myxoma, low titers of the smallpox virus were found in the skin lesions with large amounts of interferon. Greenberg (1973) discussed the minor historical associations of smallpox and flies, which occurred mostly in hospitals during smallpox epidemics.

Herpesviruses

Over 80 species of herpesviruses are classified in the family Herpesviridae. Latent infections within the natural host are a general property among the herpesviruses. Most herpesviruses have affinities for epithelial cells. Herpesvirus infections of cattle and poultry have been associated with arthropods. Bovine herpesvirus-2 (BHV-2), a dermatrophic herpesvirus with worldwide distribution, causes bovine herpes mammillitis and pseudo-lumpy skin disease. Marek's disease viruses (=gallid herpesviruses 2 and 3) cause an oncogenic disease of young poultry, affecting the nervous system as well as other organs.

Bovine herpesvirus-2. Bovine mammillitis, inflammation of the teat skin, is caused by local infections with poxviruses and BHV-2. Lesions of the teat also are signs of vesicular stomatitis, foot and mouth disease and rinderpest. Bovine ulcerative mammillitis can occur within 3–7 days after infection with BHV-2; skin damage is required for virus penetration. There is swelling of the teat wall; vesicles rapidly appear and the epithelium sloughs, resulting in decreased milk production, failure to nurse calves and secondary bacterial mastitis (inflammation of the mammary gland). Strong immunity exists upon recovery. In herds that have been exposed previously, the disease is confined to heifers. Transmission patterns for BHV-2 can be consistent with arthropod transmission, i.e., peak incidence in summer and fall, but some epizootics do not fit this pattern. In dairies, isolation and hygienic procedures have little impact on the transmission of the virus. The disease can occur in beef herds (Gibbs et al. 1973). Although the epizootiology suggests the involvement of the mechanical transmission of BHV-2 by arthropods, transmission trials are lacking (Gibbs et al. 1973). Fly control is recommended to minimize the risk of epizootics.

Pseudo-lumpy skin disease also is caused by BHV-2. After infection and a mild febrile reaction, disseminated, superficial skin nodules appear. The firm nodules are round and flat with a slightly depressed center. There is an absence of lymphadenopathy. Pseudo-lumpy skin disease has been associated with breeding habitats and seasonal abundance of hematophagous insects in Africa. The BHV-2 of pseudo-lumpy skin disease is not transmitted by direct contact. The virus has been isolated from muscoid flies feeding on infected cattle. Connor and Mukangi (1986) reported a concurrent outbreak of acute *Trypanosoma vivax* infection and pseudo-lumpy skin disease in cattle in Tanzania, and noted the presence of a large number of *Stomoxys* and some tabanids. Bovine herpesvirus-1 (BHV-1) can cause different diseases in cattle including respiratory, genital, meningoencephalitis and abortions. BHV-1 is transmitted readily by contact from infected cattle.

Face flies supposedly play a role in the transmission of BHV-1 induced keratoconjunctivitis and dermal disease. However, controlled studies with face flies do not support this assertion (Johnson et al. 1991).

Marek's disease. Although Marek's disease, or avian leukosis, has been associated with a complex of viruses, the primary etiologic agent of the disease is a gamma herpesvirus, one of the DNA tumor viruses. It has been recognized as a major threat to poultry throughout the world. Young chickens >3 weeks old are the primary hosts at risk, but other birds are susceptible. The primary symptoms are paralysis of legs, wings and neck. Commercial vaccines are available. It is possible to manage the disease by establishing isolated disease-free flocks.

Respiratory spread of Marek's disease virus is the primary route of transmission, but insect transmission may be important in certain production systems. Eidson et al. (1966) suggested that an environmental factor affected the incidence of the disease. They noted that chickens that had died of leukosis had larvae and adults of the lesser mealworm *Alphitobius diaperinus* (a darkling beetle) feeding in the subcutis. The only source of extracellular infectious virus particles is feather follicles. When homogenates of beetles collected from poultry house litter were fed to one-day-old birds, up to 85% of treated birds developed leukosis by 4 weeks of age. Chicks placed on litter containing beetles also developed tumors. Surface sterilization did not decrease the infectiousness of the beetles, but autoclaving did.

Rickettsiae

The order Rickettsiales contains Gram-negative bacteria that are obligate, intracellular organisms of the reticuloendothelial vascular cells or erythrocytes of vertebrates. All instances of mechanical transmission of rickettsiae by arthropods are by direct mechanical transmission.

Anaplasmosis

Anaplasmosis is an infectious, noncontagious disease characterized by anemia, weakness, high fever, constipation, icterus, loss of appetite, depression, dehydration and labored respiration (see Chapter 10). The causal agent is *Anaplasma marginale.* The disease is most prevalent in cattle 1–2 years of age and older with higher mortality in the older animals (Jones and Brock 1966).

Anaplasma centrale causes a milder form of disease. Exposure to *A. centrale* may protect against severe disease on a subsequent exposure to *A. marginale.*

Ticks are known biological vectors of both *A. centrale* and *A. marginale.* Over 20 species of ticks have been shown capable of transstadial transmission. Transovarial transmission also has been described for at least 7 species of ticks. The epizootiology of anaplasmosis and bovine babesiosis is similar in regions where *Boophilus* ticks are the biological vectors of both agents. In these areas, anaplasmosis and babesiosis often are found to be enzootically stable. Young animals (which are not as susceptible to disease as older animals) are infected, favoring continued transmission (Payne and Osorio 1990). However, in many large geographic areas throughout the world, anaplasmosis is transmitted mechanically both by insects and humans. Interestingly, some strains of *A. marginale* do not infect for ticks; in fact, only one of 3 standard isolates tested in the USA infected ticks. A taillike appendage of the rickettsial organism may be a marker for tick transmissibility. It is present in a strain of *A. marginale* that infects ticks, but lacking in a strain that does not infect ticks (Smith 1987). *Trypanosoma vivax* also is transmitted by both biological and mechanical modes. Changes in the kinetoplast of trypanosomes has been correlated with the selection of clones that are only mechanically transmitted. The importance of mechanical transmission of *A. marginale* was recognized soon after the fact that ticks were biological vectors became known. Even with successful eradication of *Boophilus* ticks and babesiosis, anaplasmosis persisted in the USA in the first part of the 20th Century (Ewing 1981). Iatrogenic transmission associated with typical veterinary procedures such as vaccination, castration and dehorning is important (Dikmans 1950). Reeves and Swift (1977) reported transmission by a single intramuscular injection with an automatic syringe previously used to inject a carrier cow.

Cyclic rickettsemia during persistent infections ranges from approximately 10^4–10^6 infected cells/ml in normal carrier cattle. At peak rickettsemias, horse flies would be candidate vectors, but when the rickettsemia is $<10^4$ infected cells/ml, iatrogenic transmission would be more probable than mechanical transmission by arthropods. Wiesenhutter (1975) suggested iatrogenic transmission from carrier animals can produce acute infections that fuel epizootics associated with insect transmission. Kieser et al (1990) also found rickettsemias $>10^8$ in acute infections, which places most hematophagous Diptera as potential mechanical transmission vectors.

Mosquitoes (2 genera), tabanids (9 species) and stable flies have been incriminated as vectors of *A. marginale* by immediate and delayed transfer feeding from acutely infected donors to susceptible animals (Ewing 1981). In prospective studies, mosquito and tabanid population peaks have been associated with periods of anaplasmosis transmission (Steelman et al. 1968, Wilson 1968). Epizootics of anaplasmosis have been associated with explosive populations of tabanids, and these flies generally are accepted as important mechanical vectors of *A. marginale.* The importance of other insects is unknown.

Secondary hematophagy may be significant in the mechanical transmission of anaplasmosis. Roberts and Love (1973) concluded from transmission trials and observations that eye gnats (*Hippelates pusio* and others) were vectors of *A. marginale* by secondary bloodfeeding and subsequent feeding on the eyes of susceptible animals. Davis et al. (1970) confirmed that *A. marginale* could be transmitted by the ocular route. Infection through other mucous membranes also is possible, although this route is more likely for agents from infectious sources such as feces or secretions.

The epizootiology of anaplasmosis demonstrates clearly that agents persistent in the vertebrate host can be maintained in areas following control of biological vectors or in regions naturally devoid of biological vectors or reservoir hosts. Several rickettsial agents cause diseases in vertebrates. For some, the life cycle within arthropods either has not been discovered or does not exist. The only zoonotic rickettsial agent for which isolations from arthropods has not been described is the environmentally stable *Coxiella burnetii*, the causative agent of Q Fever (Krinsky 1976).

Ehrlichioses

Potomac horse fever (equine monocytic ehrlichiosis), caused by *Ehrlichia risticii*, first was recognized as an acute diarrheal disease of horses in the Potomac River valley in the late 1970s. Anorexia, dehydration, fever, colic and laminitis are clinical signs. The agent is not transmitted by direct contact; most disease cases occur June–September. Transmission trials with ticks and fleas as potential biological vectors were unsuccessful (Schmidtmann et al. 1988). Because the transmission period corresponds to the seasonal activity of dipteran pests of equids, there have been attempts to establish a role of tabanids as mechanical vectors of the agent (Levine and Griffin 1992). However, *E. risticii* has been found to occur in snails and their trematode cercariae, which would account for the association between *E. risticii* infections and aquatic habitats (Reubel et al. 1998). Diseases caused by similar agents for which biological vectors have not been described and for which mechanical transmission has been suggested are equine ehrlichiosis *(Ehrlichia equi)*, bovine petechial fever *(Cytoecetes ondiri)*, contagious ophthalmia *(Colesiota conjunctivae)*, sennetsu rickettsiosis (*Ehrlichia sennetsu*) and Jembrana.

Other rickettsiae

Mechanical transmission of *Eperythrozoon ovis* between sheep has been demonstrated by the interrupted feeding of *Cx. annulirostris* mosquitoes. Mosquitoes fed on a carrier lamb, then fed on a recipient lamb within 10 min, transmitted the organism (Daddow 1980). Mason and Statham (1991) demonstrated *E. ovis* transmission using up to 10^{-8} ml of blood from chronically infected sheep. *Haemobartonella muris* reportedly is transmitted both mechanically and biologically by the rat louse, *Polyplax spinulosa*. MacWilliams (1987) suggested that ectoparasites as well as cat bites were important in the transmission of *Haemobartonella felis*.

Bacteria

Contaminative transmission

Enteric bacteria

The importance of arthropods in the spread of fecalborne viral diseases is minimal when compared to fecal-oral and contaminated water transmission. On the other hand, bacteria that cause enteric diseases such as strains of *Escherichia* (Iwasa et al. 1999), *Salmonella* and *Shigella* have been associated with arthropods, particularly flies, since enteric bacteria first were described. An inoculum of only 10–100 *Shigella* organisms can cause shigellosis in susceptible hosts (Levine and Levine 1991). Most of the enteric diseases are distributed globally, but appear only intermittently in epidemic form. The more usual situation is that carriers of inapparent infections join with small numbers of acute cases to serve as sources of infection (Watt and Hardy 1945). Under these endemic conditions, etiologic agents are available almost universally; all that is required for an outbreak is an efficient means to transfer the pathogen from infected to noninfected hosts. Infected food handlers with poor personal hygiene serve well, as does untreated or inadequately treated drinking water.

The importance of arthropods in the transmission of fecalborne disease pathogens varies seasonally and geographically. It is especially influenced by the amount of exposed feces serving as infectious sources and by the larval habitats of potential vectors. There are many reports of a correlation between the incidence of enteric bacterial disease and environmental conditions or personal habits of residents. Higher incidences have been associated with communities with large populations of filth flies, large quantities of exposed human feces and the frequent exposure of food to flies. For example, Oo et al. (1989) cultured *Escherichia coli*, *V. cholerae*, *Salmonella* spp. and *Shigella* spp. from 5–77% of house fly pools prepared from randomly collected house flies in Myanmar. Yao et al. (1929) showed a striking coincidence between the numbers of flies captured in traps and the numbers of cases of enteric diseases in Beijing, China. A similar correspondence between fly abundance and human diarrheal disease rate was observed in Thailand (Echeverria et al. 1983).

Not all studies have shown a correlation between fly control and incidence of enteric diseases. In some cases, the peak of abundance for flies did not correspond with the highest incidence of diarrhea. In other cases, fly populations remained relatively constant, but the incidence of enteric diseases waxed and waned throughout the year (Greenberg 1973). A study of risk factors involved in cases of diarrhea among Malaysian children by Knight et al. (1992) found that protection of food from flies was not a risk factor. These contrasting sets of observations emphasize the complexity of interactions among people, pests and pathogens.

Direct mechanical transmission

Anthrax

Bacillus anthracis causes anthrax, a disease of many kinds of mammals. The endospores of *B. anthracis* are endemic in many areas of the world. Germination of the spores is associated with flooding in warm geographic areas with alkaline soils. Susceptibility and symptomatology vary among host species and among routes of exposure. Humans and horses are moderately susceptible to infection, cattle and sheep less so. Carnivores and pigs have low susceptibility. The cutaneous disease in humans is characterized by an initial pruritic vesicle that may be surrounded by a small ring of erythema. This eruption is followed by the development of a central necrosis and black eschar, which dries and sloughs. Septicemia associated with untreated cutaneous infections can result in a 5–20% fatality rate.

Anthrax is an occupational disease of butchers, farmers, hide handlers and veterinarians. *Bacillus anthracis* is an aerobic, Gram-positive, spore-forming rod. Infection and pathogenesis of *B. anthracis* are related to capsule formation that protects against phagocytosis and the production of a multicomponent exotoxin. Spores can remain viable in the soil for over 50 yr and for at least several days on insects.

For humans, the normal routes of exposure to *B. anthracis* are respiratory, oral and cutaneous. Gastrointestinal anthrax, caused by ingestion of contaminated meat, rarely is reported. Respiratory exposure and associated pulmonary disease (woolsorter's disease) normally results from occupational exposure to hides, hair or wool contaminated with spores. It is severe, with up to 85% fatality. Cutaneous lesions are associated with direct contact with infected animals or contaminated animal products and spores penetrating a compromised skin barrier. Ingestion of soilborne spores normally infects herbivores. When germination occurs, the vegetative stages ultimately invade the bloodstream; death occurs as quickly as 12 hr after first signs of infection. Since effective vaccines are available, vaccination of livestock is recommended in endemic areas.

Mechanical transmission of *B. anthracis* by interrupted feeding of insects has been demonstrated repeatedly. Cases of cutaneous anthrax in humans frequently have been associated with the feeding sites of insects, in particular tabanids (Krinsky 1976). Mechanical transmission of *B. anthracis* was demonstrated many years ago for tabanids, stable flies, horn flies and mosquitoes. Turell and Knudson (1987) showed single stable flies or mosquitoes could transmit *B. anthracis* when flies were fed first on guinea pigs with bacteremias of $10^{8.4}$–$10^{8.7}$ colony forming units per ml.

Spirochetosis

Insect transmission may be a factor for 2 species of *Treponema* that infect humans. Both cause disabling and disfiguring infections of children and adolescents in tropical and subtropical areas. *Treponema pertenue* is the causative agent of yaws, a disease occurring throughout tropical areas of most of the world. *Treponema carateum*, the causative agent of pinta, is confined to the New World tropics.

The treponemes are coiled, Gram-negative spirochetes that can be visualized by silver staining or dark-field microscopy. Members of the genus *Treponema* have been studied extensively because of the importance of *Treponema pallidum*, the causative agent of syphilis. The spirochetes reproduce by binary fission and do not produce exotoxins. The treponemes are inactivated rapidly by slight changes in environmental conditions.

Yaws and pinta are considered to be diseases of childhood but are not congenital. The primary infection for yaws begins as a lesion at the portal of entry (mother yaw) within 2–8 weeks from exposure (Fig. 12.9). Within a year, secondary lesions can appear, often as widely disseminated granulomatous papules. A tertiary form of the disease, which can appear 30–40 yr after infection, includes destructive lesions of the periosteum, bones and skin. Pinta normally affects impoverished natives <20 yr old; the lesions are confined to the skin, producing dyschromic (hyperpigmentation and depigmentation) and hyperkeratotic lesions. Control of yaws and pinta is based upon management of the primary and secondary lesions with antibiotics.

Contact is the primary route of transmission for yaws and pinta. The organisms cannot penetrate unbroken skin; thus, abrasions and other lesions are necessary portals of entry. Peaks of transmission of yaws and pinta coincide with rainy seasons. When insects are not implicated, impoverished conditions, including crowding and poor hygiene, contribute to transmission. For centuries, flies have been implicated in the transmission of yaws and pinta by observations that flies, having fed on lesions and then on cuts or abrasions of others, could transmit the agents (Greenberg 1973). Eye gnats (including *Hippelates flavipes*, the "yaws fly") and other flies have been shown to be important in transmission, but the diseases also exist outside the distribution of eye gnats. Transmission of *T. carateum* has been demonstrated from pinta lesions to humans and monkeys with abraded skin by interrupted feeding of *Hippelates* spp. and house flies. Transmission of *T. pertenue* between humans by flies has been demonstrated; the organism has been isolated repeatedly from chloropids and muscids (Greenberg 1973). Since treponemes survive in the digestive system of flies for <8 hr, anterior transmission caused by contaminated mouthparts and regurgitation is considered to be the most important route of fly transmission.

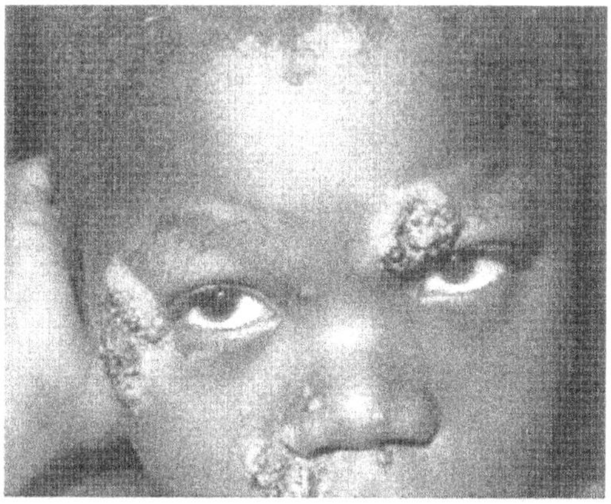

Figure 12.9. An early disseminated yaws lesion on a young boy (photograph by Leo Lanoie, courtesy of Wayne M. Meyers and Armed Forces Institute of Pathology).

Borreliosis

Lyme disease, caused by *Borrelia burgdorferi*, is biologically transmitted by ticks (see Chapter 10). However, numerous isolates of *B. burgdorferi* have been made from tabanids and mosquitoes (Magnarelli et al. 1986). *Erythema chronicum migrans* has been associated with transmission of *B. burgdorferi* by tabanids (Luger 1990). Mechanical transmission of other *Borrelia* spp. that usually are transmitted by ticks also has been reported, e.g., *Bartonella anserina* by fowl mites and mosquitoes (Harwood and James 1979).

Tularemia

Tularemia, or rabbit fever, caused by *F. tularensis*, is a facultative intracellular bacterium (see Chapter 10). The bacteria were isolated from ground squirrels in Tulare County, California, and were differentiated from *Y. pestis* in the early 1900s. *Francisella tularensis* subsequently was shown to be the causative agent of "deer fly fever" in Utah. Mechanical transmission trials with *Chrysops discalis* fed on rabbits up to 4 days after feeding on infected rabbits were successful (Francis and Mayne 1921). The epidemiology of this zoonosis varies geographically relative to biovars (geographic subspecies of the bacterium), local reservoirs, potential vectors and even local hunting and trapping practices, such as skinning lagomorphs and rodents (Hopla and Hopla 1994). In the USA, >60% of the human cases have been attributed to arthropod transmission in some areas (Taylor et al. 1991), whereas in other areas, >80% of the cases have been shown to be caused by animal contact exposure (Hopla and Hopla 1994). Depending upon the biovar, *F. tularensis* can be of veterinary importance for sheep and rabbits. The parenteral exposure considered the minimal infectious dose for humans is 10 bacteria. Levels in the blood and internal organs of rabbits can reach 10^7 bacteria per ml. In most geographic areas, arthropod exposure during hunting seasons is limited, so seasonal aspects of clinical cases of tularemia can aid in discovering the local mechanisms of human exposure. Biological transmission of *F. tularensis* by ticks is important for enzootic maintenance and human exposure in certain areas of the USA. However, in the western USA and parts of Russia, tabanids (particularly *Chrysops*) are important mechanical vectors; mosquitoes are thought to be mechanical vectors in parts of Russia and Sweden. Mechanical transmission is important in maintenance cycles in wildlife from which infections of humans and domestic animals originate. Olssufiev (1941) demonstrated mechanical transmission from water voles (*Arvicola amphibius*) to guinea pigs with mosquitoes (*Aedes vexans*). Hopla (1980) demonstrated that fleas are important mechanical vectors of *F. tularensis* and that fleas could be important in a maintenance cycle in microtine rodents. Hopla and Hopla (1994) reviewed studies that suggest a possible mechanical transmission role for mites, flies and sucking lice.

Brucellosis

Bacteria of the genus *Brucella* cause diseases mainly in livestock. One of the clinical signs of human brucellosis, a zoonosis, is an influenza-type syndrome with fever that may persist for 2–4 weeks if untreated; relapses occur, as reflected by the common name undulent fever. *Brucella melitensis* first was described as the agent of Malta fever during an epidemic in the Mediterranean island of Malta in 1900 that claimed the lives of 2,000 British soldiers. The boiling of goat milk was found to reduce greatly the incidence of the disease in troops. However, scientists of the British Malta Fever Commission were able to transmit the bacterium experimentally to a monkey by mosquitoes that had fed on brucellosis patients. Russian studies of possible vectors of *Brucella* spp. were summarized by Rementsova (1987).

Brucella spp. remain viable in milk, water and soil for up to 4 months. They also persist for long periods in arthropods. Homogenates of bloodfed arthropods are infectious for laboratory animals. However, *Brucella* bacteremias rarely exceed 10^3 bacteria per ml, suggesting that mechanical transmission on insect mouthparts is unimportant. For *Brucella abortus*, the aborted fetus and associated fluids are highly infectious, reaching 10^{10}–10^{13} bacteria per ml of fluid. Contact with aborted fetuses is the primary route of exposure for livestock and wildlife. Since *Brucella* spp. are highly infectious via oral or ocular exposure, flies landing on aborted fetuses and the feces of animals or humans may be a source of subsequent infections (Cheville et al. 1989).

Bartonellosis

The agent of Carrión disease, or *verruga peruana* (*Bartonella bacilliformis*), is transmitted by phlebotomine sand flies, but propagation has not been demonstrated in the vectors.

Bartonella henselae, considered to be the causative agent of human cat scratch disease (CSD), is associated with bacillary angiomatosis and bacillary peliosis in patients with defects in cellular immunity (Tompkins 1997). Higgins et al. (1996) demonstrated that cat fleas can maintain *B. henselae* and excrete viable organisms in their feces for up to 9 days after ingesting an infected bloodmeal, which may indicate that infection with *B. henselae* can occur by means of flea feces inoculation at a cutaneous site (Tompkins 1997). The agent of trench fever, *Bartonella quintana*, multiplies in the lumen of the digestive tract of lice. Chomel et al. (1996) demonstrated that fleas transferred from *B. henselae* bacteremic cats to SPF kittens were capable of mediating an infection. Foil et al. (1998) showed that cats injected with feces from fleas fed on cats infected with *B. henselae* for 4 days became positive by culture for *B. henselae* at 1–2 weeks post-exposure. The fleas that fed on positive cats for 4 days (likely before propagation in the fleas) were caged and fed on susceptible cats for 6 days, but no infection was detected by this route of exposure.

Considering the importance of bacillary angiomatosis associated with *B. henselae* in AIDS patients, the role of arthropods in maintenance cycles should be investigated thoroughly.

Infectious primary lesions

Corynebacteria. Corynebacterium pseudotuberculosis produces a chronic, purulent skin disease of horses, sheep and goats (but rarely cattle and humans) in North and South America, Europe and Africa. The disease is referred to as caseous lymphadenitis in sheep and goats and ulcerative lymphangitis or pigeon fever in horses. *Corynebacterium pseudotuberculosis* is the predominant bacterium isolated from lesions, but mixed infections similar to those of summer mastitis are common.

The bacteria are facultative anaerobes. The organism produces an exotoxin, phospholipase D, which is important in virulence. *Corynebacterium pseudotuberculosis* does not form spores, but the bacteria can survive in the soil for extended periods. However, this is not considered to be an important factor in transmission.

The primary signs of disease are suppuration and necrosis of lymph nodes. The local infection spreads to regional lymph nodes that enlarge with greenish pus and then ulcerate. The organism can be recovered from the lesions.

Normal spread of *C. pseudotuberculosis* is directly from open lesions of animals to abraded skin of a susceptible host. Distribution of lesions often corresponds to areas fed on by arthropods. Seasonal patterns of transmission further suggest involvement of arthropods. Addo (1983) isolated *C. pseudotuberculosis* from house flies collected from infected horses. Humphries and Gibbons (1942) isolated the organism from engorged *Dermacentor albipictus* females and their progeny. Although most of the evidence of arthropod transmission is circumstantial, undoubtedly mechanical transmission is important in the spread of this organism (Scott 1988). Sanitation and fly control are recommended as preventive measures.

Dermatophilus congolensis. This bacterium causes dermatophilosis, an exudative skin disease of livestock, wildlife and occasionally humans. Dermatophilosis occurs worldwide; it can cause weight loss and death as well as reduce the value of hides and wool. The organism once was thought to be a fungus because of its morphology, but now it is known to be an actinomycete (fungus-like bacterium). The organism is branched and appears as bundles of cocci often referred to as "railroad tracks." *D. congolensis* appears to be confined to the skin of described hosts. Outbreaks occur when climatic conditions such as rainy periods occur; moisture releases infectious, motile zoospores.

Dermatophilosis is a crusting, exudative dermatosis with variable distribution on the body relative to host and site of exposure. In horses, crusted lesions usually are found dorsally (rain scald), but lesions also are found on the face and legs. The lesions in humans are self-limiting, pustular lesions that appear within 2–7 days of exposure to an infected wildlife or domestic host. Hosts that are immunocompromised (e.g., by drugs or parasitic infections) are more susceptible. Tick control effectively reduces the disease in cattle.

The essential factors associated with transmission of *D. congolensis* are skin damage and moisture. Moisture activates the infectious stage; exposure to long periods of rain can partially reduce the protective barriers of the skin. Skin damage can be induced by biting arthropods, abrasive vegetation or mechanical trauma such as shearing. Arthropods are not important in the transmission of *D. congolensis* in many situations, but in certain areas, arthropods clearly are associated with infections. Arthropods may be vectors, or they may assist transmission by providing a portal of entry while bloodfeeding. Studies indicate that hypersensitivity to feeding ticks may affect local immunocompetence of the skin as a barrier. Macadam (1962) demonstrated transmission of *D. congolensis* from cattle to rabbits via tick bites. Richard and Pier (1966) demonstrated that *D. congolensis* could be transmitted mechanically by stable flies and house flies; stable flies transmitted the organism for up to 24 hr after feeding on infective donors. The incidence of bovine dermatophilosis has been associated with the distribution of *Amblyomma variegatum* ticks; tick control measures with acaricides reduce the incidence in cattle (Plowright 1956). Although prevalence of dermatophilosis lesions is correlated to the abundance of *A. variegatum* ticks, the distribution of lesions is not correlated to tick feeding sites. This fact has encouraged speculation about fly transmission.

Eye diseases

Trachoma. Chlamydia trachomatis causes trachoma, an infection of the conjunctival and corneal epithelial cells, as well as congenital eye diseases. Insects are implicated only in the spread of *C. trachomatis* infections that result in conjunctivitis. Trachoma affects over 400 million people worldwide. Blindness occurs in 6 million cases.

Chlamydia trachomatis is a Gram-negative, obligate intracellular parasite. Infectious elementary bodies are released from infected cells and enter susceptible cells by endocytosis. The elementary bodies can remain viable in the environment for several days.

Chlamydia trachomatis can be spread among people via direct contact. In some endemic areas of trachoma, eye-frequenting insects, including *Musca sorbens, M. domestica, Siphunculina* spp. and *Hippelates* spp., are important mechanical vectors; in other areas, flies are less important. Taylor (1988) indicated that flies were most important in areas with episodes of bacterial conjunctivitis; a survey in Tanzania showed a significant correlation between the household density of synanthropic flies (particularly *M. sorbens*) and the incidence of active inflammatory trachoma in children. Taylor (1988) suggested that local control measures could effectively reduce fly numbers and associated eye diseases.

Haemophilus influenzae. This agent has been clearly associated with eye infections transmitted by insects. *Haemophilus* spp. are small, Gram-negative coccobacilli that require blood-derived factors for growth. Different biotypes of *H. influenzae* are associated with different clinical diseases. Biotype III, formerly known as *Haemophilus aegyptius,* is the biotype associated most frequently with conjunctivitis. In the USA, *H. influenzae* biotype III has been recognized as the cause of acute seasonal bacterial conjunctivitis (gnat sore eyes) since the 1920s. Recurrent epidemics have been associated consistently with the presence of eye gnats, particularly *H. pusio.* Buehler et al. (1983) described an epidemic in southeastern Georgia involving over 2,000 reported cases of conjunctivitis. Increased populations of eye gnats were reported, but the authors failed to isolate bacteria from eye gnats collected from a first grade classroom with a high incidence of acute conjunctivitis. Recently, (Harrison et al. 1989) described *H. influenzae* biotype III as the agent of Brazilian purpuric fever (BPF). Brazilian purpuric fever first was recognized in epidemics in children in

Brazil in which *H. influenzae* biotype III caused a disease characterized by high fever, abdominal pain, vomiting, hemorrhagic skin lesions, vascular collapse and death. There was an interval of 1–60 days between the occurrence of conjunctivitis and acute illness in children.

Chloropid flies have been studied for a possible association with outbreaks of acute conjunctivitis that were precursors to BPF in Brazil (Paganelli and Sabrosky 1993). Several different chloropid species in the genera *Hippelates, Liohippelates* and *Siphunculina* were collected from children's eyes. *Haemophilus influenzae* biotype III has been isolated from homogenates of *Liohippelates peruanus* and *Hippelates neoproboscidens* collected from the eyes of children.

The association of eye-frequenting insects with conjunctivitis caused by mixed or singular infections of bacteria of the genera *Streptococcus, Staphylococcus, Moraxella, Haemophilus, Corynebacterium, Neisseria* and *Pneumococcus* has been observed repeatedly (Greenberg 1971). The insects mentioned as potential vectors for trachoma and agents of acute conjunctivitis are potential contributors to mixed ocular infections of humans and animals.

Infectious bovine keratitis. Moraxella bovis is considered to be the primary causative agent of pinkeye or infectious bovine keratitis (IBK), a highly contagious disease of cattle worldwide. Concurrent bacterial, rickettsial, mycoplasmal and viral infections can be linked to this and other eye diseases of livestock. Several species of *Moraxella* infect humans including *Moraxella lacunata*, which is associated with conjunctivitis of humans living under poor hygienic conditions.

There are hemolytic and nonhemolytic strains of *M. bovis*. Only the hemolytic form produces conjunctivitis. Because nonhemolytic strains can produce both hemolytic and nonhemolytic forms on blood agar, Pugh et al. (1966) suggested that ultraviolet light may induce mutations in nonhemolytic strains to produce hemolytic forms in animal hosts. Maximum UV exposure occurs when vector populations are highest.

Economic losses in cattle production from IBK infections are from reduced weight gain, reduced milk production, as well as eye disfigurement and blindness. Cattle are the only reservoirs of *M. bovis*. Prevention of IBK is aided by the control of flies that feed on the eyes of cattle, in particular face flies. For treatment, antibiotic therapy administered topically or parenterally are recommended. Providing shade for animals may help prevent infections.

An increase in cattle eye disorders, in particular IBK, was observed after the introduction of the face fly into North America around 1952 (Fig. 12.10). Correlations between the number of face flies per animal and the percentage of cattle in herds contracting IBK were reported, and control of face flies reduced the incidence of pinkeye (Gerhardt et al. 1976). *Moraxella bovis* cannot penetrate the intact conjunctiva. Mechanical damage caused by environmental factors such as dust and insect feeding can contribute to infections (Shugart et al. 1979). Glass and Gerhardt (1984) demonstrated the ability of face flies to transmit *M. bovis*. In other regions, different insects (particularly eye gnats) that feed on the lachrymal and nasal secretions of bovids have been incriminated in the mechanical transmission of *M. bovis* and other ocular infections of livestock.

Mastitis

Mastitis is the most common and economically important disease of dairy cattle, but it can be an important disease of domestic animals worldwide. Bovine mastitis will be used as the example of this complex disease. The focus will be on bacterial infections rather than predisposing factors such as viral infections. Over 130 different microorganisms have been isolated from the milk of cattle with mastitis; all reported associations of these organisms with flies have

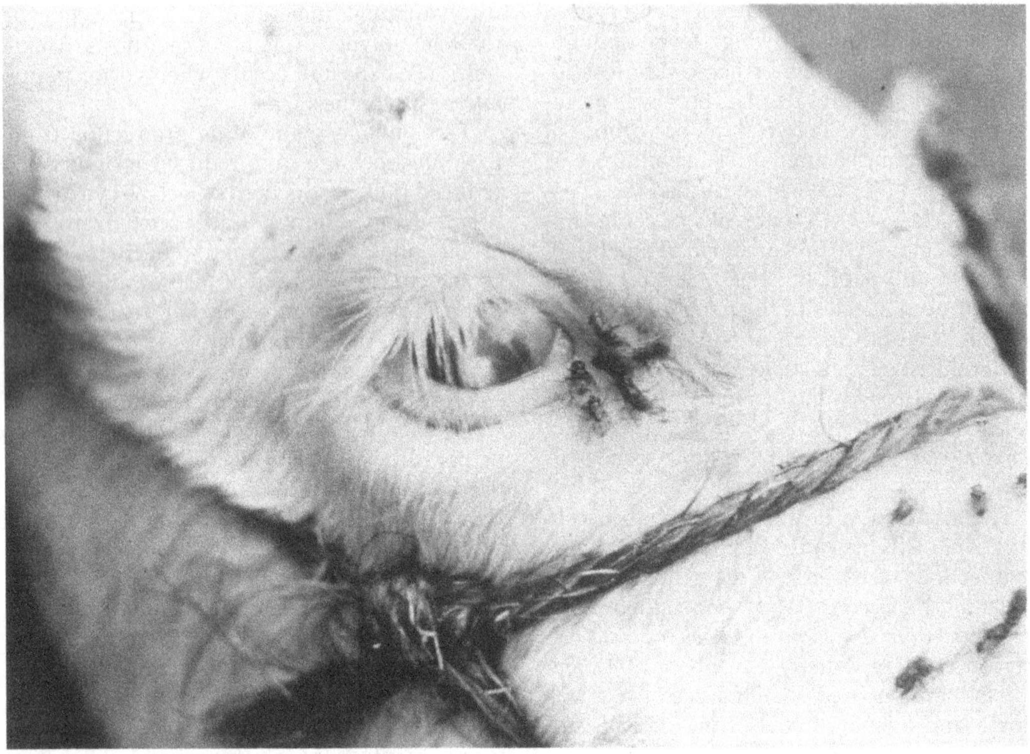

Figure 12.10. The face fly, *Musca autumnalis*, on and around the eye of a cow with infectious bovine keratitis (photograph by Reid R. Gerhardt).

been cataloged by Greenberg (1971). Flies that frequent the secretions of udders and other secretory organs have been associated with the transmission of mastitis. For this chapter, the discussion will be divided into mastitis of lactating dairy cattle and summer mastitis, an example of mastitis of mixed bacterial origin.

Mastitis of lactating dairy cattle. The causal agent of mastitis of lactating dairy cattle is transmitted mechanically between cattle by milking or by other types of direct contact (Fig. 12.11). Theoretically, mechanical transmission by flies can supplement any of these routes of exposure. The infection of the mammary gland usually is established within 2 hr of milking when the teat canal is least resistive to invasion by microorganisms. The contagious microorganisms most frequently causing mastitis are *S. aureus*, *Streptococcus agalactiae* and *Mycoplasma* spp. The genera of bacteria frequently associated with infections from the environment are *Streptococcus*, *Escherichia*, *Klebsiella*, and *Enterobacter*. Herd management practices that reduce the incidence of acute mastitis in dairies include sanitation, elimination of chronic carrier animals and strategically timed antibiotic therapy. Fly transmission of organisms that cause mastitis in lactating dairy animals probably is less frequent in modern dairies. Where there is evidence of involvement by flies in infections, eye gnats, *M. domestica* and other species of *Musca, Muscina, Morellia* and *Hydrotaea* should be considered potential vectors. Biting flies such as tabanids, stable flies and horn flies may be associated with transmission in a range of pasture production systems. Owens et al. (1998) clearly established that horn flies (*Haematobia irritans*) can be involved in the mechanical transmission of *S. aureus*-induced mastitis.

Summer mastitis. Pasture or summer mastitis is an acute mastitis of cattle caused by a mixed bacterial infection that produces purulent abscesses of the skin. As the name suggests, the peak incidence of the syndrome occurs during summer. The disease occurs in non-lactating beef and dairy animals, but primarily it affects the dairy industry, with 20–60% of the dairy herds developing signs of disease in enzootic areas. Summer mastitis is enzootic in Great Britain and many other parts of Europe, but also has been reported in the USA, Australia, Brazil, Zimbabwe and Japan.

Actinomyces pyogenes (formerly *Corynebacterium pyogenes*) and various obligate anaerobic bacteria such as *Peptostreptococcus indolicus, Streptococcus disgalactiae, Fusobacterium necrophorum, Bacteroides melaninogenicus* and other bacteria are associated with the disease (Madsen et al. 1990). These bacteria do not grow at the pH of milk (6.6). Synergism among bacteria is important in most polybacterial infections. Many of the organisms that contribute to summer mastitis commonly are found in suppurative infections such as foot diseases, abscesses, pneumonia and metritis. These bacteria also are found on the mucosal membranes of the respiratory and reproductive tracts as well as on the skin of healthy animals.

Clinical signs of summer mastitis include fever, anorexia and stiff gait. The infected area in the forequarters of the udder is painful, swollen and hard. There is an associated yellow-green secretion that is thick, creamy, foul-smelling and highly attractive to flies, particularly *Hydrotaea irritans* (Fig. 12.11). Antibiotic therapy effectively returns animals to health, but there often is residual reduction in milk production and weight loss. Bacteria can continue to be shed for 2–3 yr after resolution of the mastitis. Antibiotic use and fly control appear to provide equivalent benefits. Cattle maintained indoors do not develop summer mastitis, perhaps because of reduced contact with flies. Poor hygienic conditions increase the probability of transmission outside of the summer months.

There is a consistent agreement between the peak incidence of summer mastitis with the peak abundance of *H. irritans*, in spite of geographical differences in seasonal distributions of *H. irritans* (Thomas et al. 1987). The bacteria survive from 7–17 days within the guts of the flies. Flies contaminate surfaces by regurgitation of the gut contents during subsequent feedings. At the beginning of the fly season, few organisms can be found, but during the peak of the season, bacteria can be isolated from 10–20% of *H. irritans* sampled from udders.

Protozoa

Contaminative transmission

Enteric parasites

Reports of isolations of pathogenic protozoans from insects are not common, and reports of experimental transmissions even less so (Greenberg 1973). *Entamoeba histolytica* has been the subject of many surveys and experiments with flies and cockroaches. Worldwide, there are about 500 million new cases of amebiasis annually, with about 100,000 fatalities. Among parasitic diseases, amebiasis ranks second to malaria as a cause of death worldwide.

The basic experimental work with fly transmission of intestinal protozoans was done by Root (1921). He demonstrated that the cysts of *E. histolytica* (as well as *Entamoeba coli* and *Enta-*

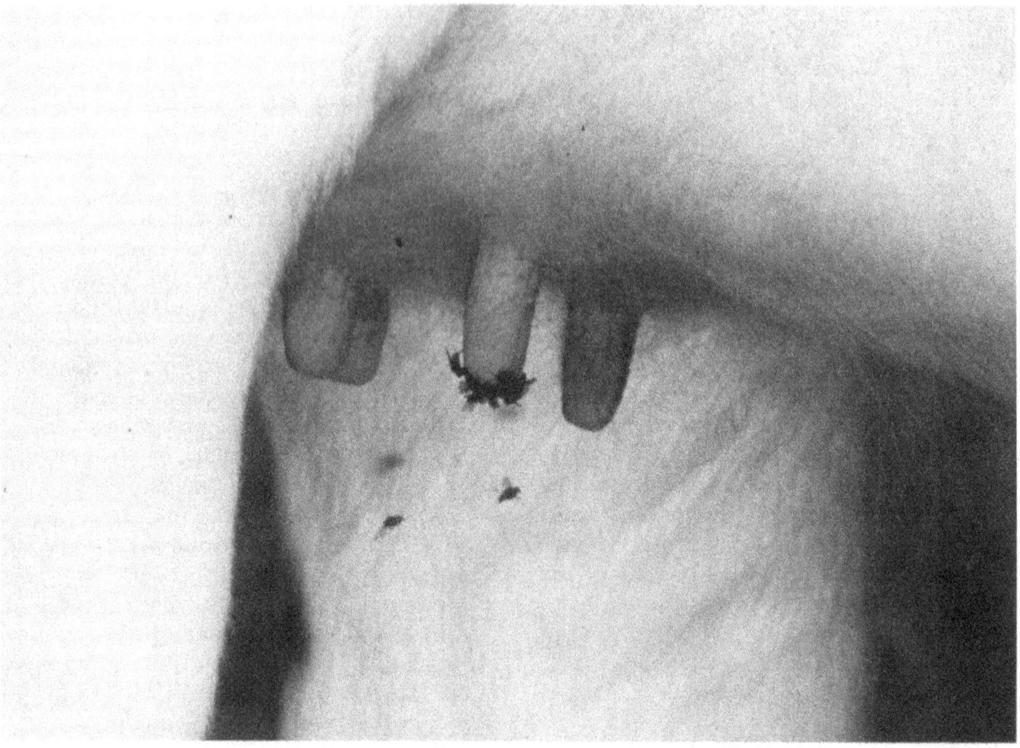

Figure 12.11. Flies (*Hydrotea irritans*) feeding on the teat of a cow with summer mastitis (photograph by Gerthin Thomas).

moeba nana) survive passage through the intestinal tract of house flies. He also showed that the numbers of viable cysts declined rapidly, half being dead 15 hr after ingestion. No viable cysts survived after 49 hr. Harris and Down (1946) showed that flies deposited cysts on a kitchen table. *Entamoeba histolytica* also has been recovered from cockroaches, but the total number of isolations reported is small (Roth and Willis 1957).

Sarcocystis and other coccidia have been isolated from wild-caught flies collected from dog feces (Markus 1980). House flies have contaminated milk with viable *Toxoplasma gondii* oocysts for approximately 24 hr following contact with infected cat feces (Chinchilla and Ruiz 1976).

Direct mechanical transmission

Trypanosomes

Biologically transmitted hemoflagellates are covered in Chapter 8. There is conclusive evidence for mechanical transmission by arthropods of 2 species of *Trypanosoma* that cause animal diseases: *Trypanosoma evansi* and *T. vivax*. *Trypanosoma evansi* is transmitted mechanically worldwide, and *T. vivax* is transmitted biologically by tsetse in sub-Saharan Africa and mechanically transmitted outside of tsetse ranges in Africa and in the New World.

Trypanosoma evansi (Fig. 12.12) was the first trypanosome shown to cause disease in domestic livestock. Infections also have been reported in dogs and wildlife in Central and South America, Africa and Asia. *T. evansi* may have evolved from *Trypanosoma brucei* and subsequently been carried out of tsetse range via transport camels.

Trypanosoma evansi has a wide host range. There may be strain differences in pathogenicity. When näive animal populations are exposed, either by transportation of the animals into an endemic area or transportation of a carrier animal into a nonendemic area, explosive epizootics with high morbidity and mortality can occur. Clinical signs include anemia, fever, edema, anorexia and ataxia. For livestock other than equids, the disease, called sura, usually is chronic in endemic areas. Sura decreases livestock production and reduces work capability of draft animals.

Transplacental transmission of *T. evansi* and *T. vivax* has been reported, but the extent of this type of transmission has not been determined. Small animal reservoir hosts for *T. evansi* exist. Vampire bats can be both host and vector of *T. evansi* (Hoare 1972).

Conclusive evidence incriminates tabanids as vectors. Nieschulz and Kraneveld (1929) in Indonesia were able to transmit *T. evansi* by single tabanids, but the probability of transmission dropped precipitously when flies were not fed until 24 hr after the initial feeding. Moutia (1928) demonstrated mechanical transmission of *T. evansi* with *Stomoxys nigra* from infected guinea pigs to noninfected guinea pigs and dogs.

Trypanosoma vivax is the only species of tsetse-transmitted trypanosome that has become permanently established outside of Africa. In contrast with *T. evansi*, mechanical transmission by tabanids was demonstrated only recently by Raymond (1990). He successfully infected a calf with *T. vivax* by the interrupted feeding of *Cryptotylus unicolor*. Otte and Abuabara (1991) transferred *T. vivax* infections with 3 species of *Tabanus* to noninfected cattle after feeding flies on steers with a parasitemia of 5×10^5 parasites per ml. Mihok et al. (1995) were unable to incriminate various species of African Stomoxyinae based on transmission experiments with *T. vivax* and mice.

Even where the primary route of transmission is biological, mechanical transmission may be important and may have an adverse effect on the success of vaccination programs based upon metacyclic or metacyclic-derived vaccines.

Other protozoa

Besnoitiosis is a protozoan disease of cattle, horses and other livestock caused by protozoans of the genus *Isospora* (formerly *Besnoitia*) (Tadros and Laarman 1982). Besnoitiosis occurs in Africa, southern Europe, Asia and South America. *Isospora besnoiti*, the causative agent of bovine besnoitiosis, is the only sporozoan for which the mechanical transmission by arthropods has been shown to be important.

After an incubation period of 6–10 days, there is a febrile disease with generalized lymphadenopathy. The skin disease presents itself as thickened skin, lost hair and oozing fissures. Heavy-walled pseudocysts are found within the subcutis and connective tissue of livestock. Chronically infected livestock are the primary reservoir.

The life cycle of *Isospora* spp. is unusual among the Sporozoa. The organisms have a demonstrated heteroxenous life cycle with cyclic development occurring in cats, and intermediate tissue cysts occurring as intermediate stages in livestock. Mechanical transmission by insects may be the primary method of transfer. The dermis is heavily parasitized with millions of cysts, favoring mechanical transmission by hematophagous flies. Insect transmission is consistent with the seasonal occurrence of the disease in summer and autumn. Bigalke (1968)

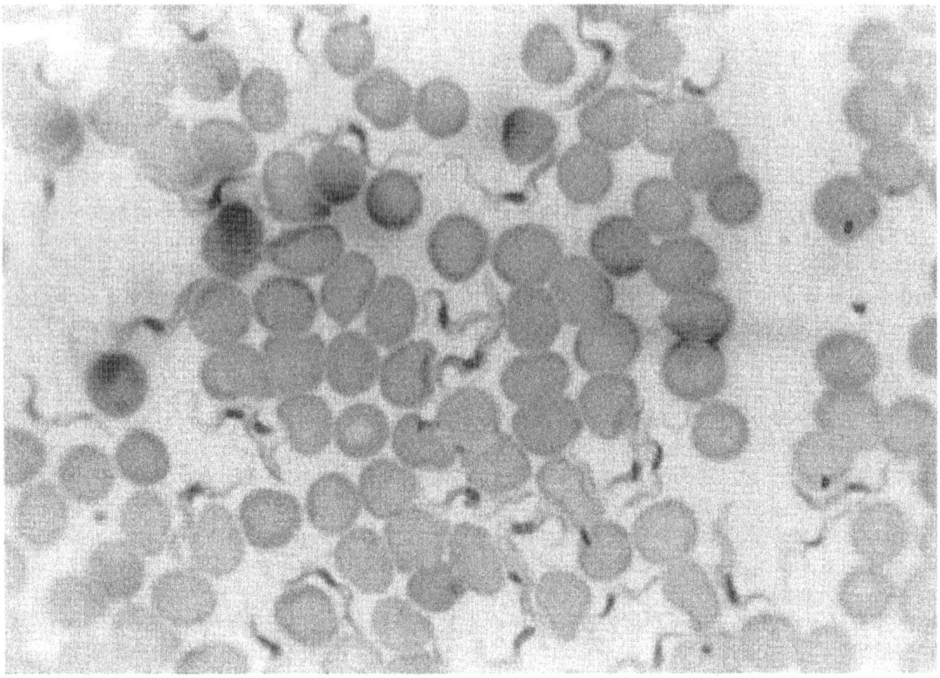

Figure 12.12. The horse strain of *Trypanosoma evansi* in the blood of a rat (photograph by Roberto A.M.S. Silva).

studied the transmission of *Isospora besnoiti*. Tsetse fed on chronically infected cattle and then transferred to susceptible cattle transmitted the agent readily if large groups of flies (30–100) were used. Similarly, bovid-to-bovid transmission was demonstrated with large groups of tabanids (*Tabanus denticornis* and *Atylotus nigromaculatus*) and stable flies.

Helminths

Chapter 1 mentioned the role of arthropods as intermediate hosts of helminths. However, there is considerable evidence that arthropods also may serve as contaminative mechanical vectors. Helminth eggs pass through cockroaches with little change in viability (Young and Babero 1975), and eggs of 7 species of helminths have been isolated from cockroaches contaminated under natural conditions (Roth and Willis 1957).

The eggs of at least 17 species of parasitic helminths have been recovered from filth flies under natural conditions (Greenberg 1973). Tod et al. (1971) collected infectious third-stage larvae of parasitic nematodes of the genera *Oesopagostoma* and *Ostertagia* clinging to the abdomens of psychodid flies. The authors proposed that the nematodes mounted the flies during the period following adult fly emergence and suggested that the flies may play a limited role in the dispersal of worm larvae. Filth fly transmission of helminth eggs also has been demonstrated experimentally (Goddeeris 1980,

Harris and Down 1946). Filth flies can transport helminth eggs either on the body surface or within the gut. The primary limiting factor in the ability of flies to ingest eggs is the diameter of openings in their spongiform mouthparts. Some helminth eggs apparently are too large to pass (Goddeeris 1980). Helminth eggs do not survive well on exterior surfaces of flies because the eggs are susceptible to desiccation, and grooming flies remove many of the them (Heinz and Brauns 1955).

Transmission dynamics of helminths in flies vary. Lawson and Gemmell (1990) demonstrated that the eggs of *Taenia hydatigena* and *Taenia pisiformis* will infect susceptible animal hosts after the hosts ingested calliphorid flies (mostly *Calliphora stygia*). Flies must carry the eggs for at least 24 hr, the time required for the eggs to become infective (Lawson and Gemmell 1990). The authors also showed how pigs fed on food contaminated by blow flies that had fed on dog feces containing proglottids of *T. hydatigena* became infected. Eggs of *Ascaris lumbricoides* regurgitated by *Musca vicina* or *M. domestica* within 2 hr after ingestion continued to develop normally. The larvae of both *A. lumbricoides* and hookworms were viable when regurgitated up to 8 hr post-ingestion (Dipeolu 1982).

Although helminth eggs can be found consistently on the external surfaces and in the guts of synanthropic flies, epidemiological studies have not provided evidence of an important role for flies in transmission of helminths in nature. In a community where 16 species of parasitic helminths and protozoans were recovered from human feces, only 6 of the species were isolated from house flies (Gupta et al. 1972). Here, fly transmission appeared minimal. Umeche and Mandah (1989) examined 5,000 house flies for presence of helminth eggs in Calabar, Nigeria, where parasites were abundant in both humans and dogs. Larvae or ova of *A. lumbricoides, Strongyloides stercoralis, Toxocara canis* and *Ancylostoma caninum* were found in the guts of 0.8%, 1.8%, 2.1% and 6.2% of the flies, respectively. A high correlation between rates of parasitism in human populations and the presence of parasites in or on house flies rarely has been demonstrated (Umeche and Mandah 1989). However, Sulaiman et al. (1989) reported that 79.5% of *Chrysomya megacephala* collected in Malaysia had eggs or larvae of *A. lumbricoides, Trichuris trichura, Necator americanus* and/or *Ancylostoma duodenale* on the external body surfaces or internally.

When Monzon et al. (1991) compared the rates of helminth ova attachment to the external surfaces of *M. domestica* and *C. megacephala* collected in Manila, Philippines, they found that 41.9% of the *C. megacephala*, but only 9.4% of the *M. domestica*, were positive. Clearly, adult foraging behavior has a major impact on whether different fly species could be important in the dispersal of helminths. An appreciation for the interactions of flies and vertebrates can be obtained from Greenberg (1971, 1973).

In spite of the lack of evidence for fly transmission from epidemiological studies, experimental studies suggest that fly transmission should not be undervalued. Ample evidence shows that when fly control is practiced in combination with general sanitation and hygiene, rates of parasitism invariably decline (Levine and Levine 1991).

Fungi

Direct contact of some kind transmit most fungal infections of animals. It is unlikely that mechanical transmission by arthropods plays a significant role for any fungal species.

Gip and Svensson (1968) and Kamyszek (1969) demonstrated mechanical transmission of *Trychophyton meniagrophytes* by house flies; Fotedar et al. (1991a) claimed natural transmission of *Aspergillus niger* by roaches in hospitals.

Zygomycosis is a word used for diseases caused by fungi classified in the Phylum Zygomycota. Certain *Conidiobolus* and

Basidiobolus species that infect humans and other animals also have been isolated from arthropods, but oral or respiratory contact with infected feed debris or bedding is considered the normal route of exposure (Rippon 1988, Zahari et al. 1990). If the fungi that infect vertebrates are obligate, insect-infecting organisms, then this would be a form of biological, not mechanical transmission.

REFERENCES

Addo, P.B. 1983. Role of the common house fly (*Musca domestica*) in the spread of ulcerative lymphangitis. Vet. Rec. **113:**496–497.

Akkerman, K. 1933. Researches on the behaviour of some pathogenic organisms in the intestinal canal of *Periplaneta americana* with reference to the possible epidemiological importance of this insect. Acta Leidensia **8:**80–120.

Al-Janabi, B.M., Branagan, D. and Danskin, D. 1975. The trans-stadial transmission of the bovine farcy organisms, *Nocardia farcinica*, by the ixodid *Amblyomma variegatum* (Fabricius, 1794). Trop. Anim. Hlth. Prod. **7:**205–209.

Anonymous. 1996. Morbidity and mortality surveillance in Rwandan refugees—Burundi and Zaire, 1994. Morbid. Mortal. Weekly Rpt. **45:**104–107.

Ansari, S.A., Sattar, S.A., Springthorpe, V.S., Wells, G.A. and Tostowaryk, W. 1988. Rotavirus survival on human hands and transfer of infectious virus to animate and nonporous inanimate surfaces. J. Clin. Microbiol. **26:**1513–1518.

Baker-Jones, E. 1941. A fly-borne epidemic of enteric fever. The Medical Officer **65:**65-67.

Barber, M.A. 1914. Cockroaches and ants as carriers of the vibrios of Asiatic cholera. Philipp. J. Sci. **9A:**1–4.

Barnett, H.C. 1960. The incrimination of arthropods as vectors of disease. Proc. 11th Intern. Congr. Entomol **2:**341-345.

Bassett, D.C.J. 1970. *Hippelates* flies and streptococcal skin infection in Trinidad. Trans. Royal Soc. Trop. Med. Hyg. **64:**138–147.

Beatson, S.H. 1972. Pharaoh's ants as pathogen vectors in hospitals. Lancet **I:**425–427.

Bech-Nielsen, S., Piper, C.E. and Ferrer, J.F. 1978. Natural mode of transmission of the bovine leukemia virus: role of bloodsucking insects. Am. J. Vet. Res. **39:**1089–1092.

Becker, J.-L., Hazan, U., Nugeyre, M.-T., Rey, F., Spire, B., Barré-Sinoussi, F., Georges, A., Teuliéres, L. and Chermann, J.-C. 1986. Infection de cellules d'insectes en culture par le virus HIV, agent du SIDA, et mise en évidence d'insectes d'origine africaine contaminés par ce virus. C. R. Acad. Sci. Paris **303:**303–306.

Bernton, H.S. and Brown, H. 1969. Insect allergy: the allergenic potentials of the cockroach. Southern Med. J. **62:**1207–1210.

Bigalke, R.D. 1968. New concepts on the epidemiological features of bovine besnoitiosis as determined by laboratory and field investigations. Onderstepoort J. Vet. Res. **35:**3–138.

Blaser, M.J., Hardesty, H.L., Powers, B. and Wang, W.L. 1980. Survival of *Campylobacter fetus* subsp. *jejuni* in biological milieus. J. Clin. Microbiol. **11:**309–313.

Blumberg, B.S. 1977. Australia antigen and the biology of hepatitis B. Science **198:**17–25.

Blumberg, B.S., Wills, W., London, W.T. and Millman, I. 1973. Australia antigen in mosquitoes. Feeding experiments and field studies. Res. Comm. Chem. Pathol. Pharmacol. **6:**719–732.

Bolton, H.T. and Hansens, E.J. 1970. Ability of the house fly, *Musca domestica*, to ingest and transmit viable spores of selected fungi. Ann. Entomol. Soc. Am. **63:**98–100.

Briegel, H. and Horler, E. 1993. Multiple blood meals as a reproductive strategy in *Anopheles* (Diptera: Culicidae). J. Med. Entomol. **30:**975–985.

Buehler, J.W., Holloway, J.T., Goodman, R.A. and Sikes, R.K. 1983. Gnat sore eyes: seasonal, acute conjunctivitis in a southern state. Southern Med. J. **76:**587–589.

Bull, L.B., Ratcliffe, F.N. and Edgar, G. 1954. Myxomatosis: its use in the control of rabbit populations in Australia. Vet. Rec. **66:**61–62.

Burkot, T.R., Graves, P.M., Paru, R. and Lagog, M. 1988. Mixed blood feeding by the malaria vectors in the *Anopheles punctulatus* complex (Diptera: Culicidae). J. Med. Entomol. **25:**205–213.

Butler, J.F., Kloft, W.J., DuBose, L.A. and Kloft, E.S. 1977. Recontamination of food after feeding a ^{32}P food source to biting Muscidae. J. Med. Entomol. **13:**567–571.

Castro, K.G., Lieb, S., Jaffe, H.W., Narkunas, J.P., Calisher, C.H., Bush, T.J. and Witte, J.J. 1988. Transmission of HIV in Belle Glade, Florida: lessons for other communities in the United States. Science **239:**193–197.

Chamberlain, R.W. and Sudia, W.D. 1961. Mechanism of transmission of viruses by mosquitoes. Annu. Rev. Entomol. **6:**371–390.

Chapple, P.J. and Lewis, N.D. 1965. Myxomatosis and the rabbit flea. Nature **207:**388–389.

Cheville, N.F., Rogers, D.G., Deyoe, W.L., Krafsur, E.S. and Cheville, J.C. 1989. Uptake and excretion of *Brucella abortus* in tissues of the face fly (*Musca autumnalis*). Am. J. Vet. Res. **50:**1302–1306.

Chinchilla, M. and Ruiz, A. 1976. Cockroaches as possible transport hosts of *Toxoplasma gondii* in Costa Rica. J. Parasitol. **62:**140–142.

Chomel, B.B., Kasten, R.W., Floyd-Hawkins, K., Chi, B., Yamamoto, K., Roberts-Wilson, J., Gurfield, A.N., Abbott, R.C., Pedersen, N.C. and Koehler, J.E. 1996. Experimental transmission of *Bartonella henselae* by the cat flea. J. Clin. Microbiol. **34:**1952–1956.

Coggins, L. and Norcross, N.L. 1970. Immunodiffusion reaction in equine infectious anemia. Cornell Vet. **60:**330–335.

Cohen, D., Green, M., Block, C., Slepon, R., Ambar, R., Wasserman, S.S. and Levine, M.M. 1991. Reduction of transmission of shigellosis by control of houseflies (*Musca domestica*). Lancet **337:**993–997.

Conner, R.J. and Mukangi, D.J.A. 1986. Concurrent outbreak of pseudo-lumpy skin disease and acute *Trypanosoma vivax* infection in cattle. Trop. Anim. Hlth. Prod. **18:**127–132.

Cravedi, P. and Quaroni, S. 1983. Modi e mezzi di diffusione dei microrganismi negli alimenti ad opera degli insetti. Proc. Dif. Antiparassit. Industr. Aliment. Protez. **3:**167–172.

Daddow, K.N. 1980. *Culex annulirostris* as a vector of *Eperythrozoon ovis* infection in sheep. Vet. Parasitol. **7:**313–317.

Dalldorf, G. 1950. The coxsackie viruses. Bull. N.Y. Acad. Med. **26:**329–335.

Dalldorf, G., Sickles, G.M., Plager, H. and Gifford, R. 1949. A virus recovered from the feces of "poliomyelitis" patients pathogenic for suckling mice. J. Exp. Med. **89:**567–582.

Dalmat, H.T. 1957. Arthropod transmission of rabbit papillomatosis. J. Exp. Med. **108:**9–20.

Daniel, M., Srámová, H., Absolonová, V., Dedicová, D., Lhotová, H., Masková, L. and Petrás, P. 1992. Arthropods in a hospital and their potential significance in the epidemiology of hospital infections. Folia Parasitol. **39:**159–170.

Davies, C.R. 1990. Interrupted feeding of blood-sucking insects: causes and effects. Parasitol. Today **6:**19–22.

Davis, H.E., Dimopoullos, G.T. and Roby, T.O. 1970. Anaplasmosis transmission: inoculation by the ocular route. Res. Vet. Sci. **11:**594–595.

Dikmans, G. 1950. The transmission of anaplasmosis. Am. J. Vet. Res. **11:**5–16.

Dipeolu, O.O. 1982. Laboratory investigations into the role of *Musca vicina* and *Musca domestica* in the transmission of parasitic helminth eggs and larvae. Internat. J. Zoonoses **9:**57–61.

Dow, R.P. 1954. A note on domestic cockroaches in South Texas. J. Econ. Entomol. **48:**106–107.

Duncan, J.T. 1926. On a bactericidal principle present in the alimentary canal of insects and arachnids. Parasitol. **18:**238–252.

Durden, L.A., Linthicum, K.J. and Turell, M.J. 1992. Mechanical transmission of Venezuelan equine encephalomyelitis virus by hematophagous mites (Acari). J. Med. Entomol. **29:**118–121.

Echeverria, P., Harrison, B.A., Tirapat, C. and McFarland, A. 1983. Flies as a source of enteric pathogens in a rural village in Thailand. Appl. Environ. Microbiol. **46:**32–36.

Eddy, G.A., Marin, D.H. and Johnson, K.M. 1972. Epidemiology of the Venezuelan equine encephalomyelitis virus complex. Proc. 3rd Intern. Conf. Equine Infect. Dis. (Paris), pp. 26–145.

Edman, J.D. and Spielman, A. 1988. Blood-feeding by vectors: physiology, behavior, and vertebrate defense. Pp. 153–189 *in* T.P. Monath (ed.), The arboviruses: epidemiology and ecology, Vol. 1. CRC Press, Boca Raton, Florida.

Eidson, C.S., Schmittle, S.C., Goode, R.B. and Lal, J.B. 1966. Induction of leukosis tumors with the beetle *Alphitobius diaperinus*. Am. J. Vet. Res. **27:**1053–1057.

Enright, M.R., Alexander, T.J.L. and Clifton-Hadley, F.A. 1987. Role of houseflies (*Musca domestica*) in the epidemiology of *Streptococcus suis* type 2. Vet. Rec. **121:**132–133.

Evermann, J.F., DiGiacomo, R.F. and Hopkins, S.G. 1987. Bovine leukosis virus: understanding viral transmission and the methods of control. Vet. Med. **82:**1051–1058.

Ewing, S.A. 1981. Transmission of *Anaplasma marginale* by arthropods. Proc. 7th Nat. Anaplasmosis Conf., pp. 395–423.

Fairley, N.H. and Boyd, J.S.K. 1943. Dysentery in the Middle East with special reference to sulphaguanidine treatment. Trans. Royal Soc. Trop. Med. Hyg. **36:**253–286.

Fenner, F. and Ratcliffe, F.N. 1965. Myxomatosis. Cambridge University Press, New York. 379 pp.

Ferrer, J.F. 1980. Bovine lymphosarcoma. Adv. Vet. Sci. Comp. Med. **24:**2–68.

Ferris, D.H., Hanson, R.P., Dicke, R.J. and Roberts, R.H. 1955. Experimental transmission of vesicular stomatitis virus by Diptera. J. Infect. Dis. **96:**184–192.

Fischer, R.G., Luecke, D.H. and Rehacek, J. 1973. Friend leukemia virus (FLV) activity in certain arthropods. III. Transmission studies. Neoplasma **20:**255–260.

Foil, L.D. 1989. Tabanids as vectors of disease agents. Parasitol. Today **5:**88–96.

Foil, L.D., Adams, W.V., Jr., Issel, C.J. and Pierce, R. 1984. Tabanid (Diptera) populations associated with an equine infectious anemia outbreak in an inapparently infected herd of horses. J. Med. Entomol. **21**:28–30.

Foil, L.D., Adams, W.V., McManus, J.M. and Issel, C.J. 1987. Bloodmeal residues on mouthparts of *Tabanus fuscicostatus* (Diptera: Tabanidae) and the potential for mechanical transmission of pathogens. J. Med. Entomol. **24**:613–616.

Foil, L.D., Adams, W.V., Jr., McManus, J.M. and Issel, C.J. 1988. Quantifying the role of horse flies as vectors of equine infectious anemia. Pp. 189–195 *in* D.G. Powell (ed.), Equine infectious diseases. Proc. Fifth Intern. Conf. University Press, Lexington, Kentucky.

Foil, L., Andress, E., Freeland, R.L., Roy, A.F., Rutledge, R., Triche, P.C. and O'Reilly, K.I. 1998. Experimental infection of domestic cats with *Bartonella henselae* by inoculations of *Ctenocephalides felis* (Siphonaptera: Pulicidae) feces. J. Med. Entomol. **35**: 625-628.

Foil, L.D., French, D.D., Hoyt, P.G., Issel, C.J., Leprince, D.J., McManus, J.M. and Seger, C.L. 1989. Transmission of bovine leukemia virus by *Tabanus fuscicostatus*. Am. J. Vet. Res. **50**:1771–1773.

Foil, L.D. and Issel, C.J. 1991. Transmission of retroviruses by arthropods. Annu. Rev. Entomol. **36**:355–381.

Foil, L.D., Issel, C.J., Adams, W.V. and Meek, C.L. 1983. Mechanical transmission of equine infectious anemia by deer flies (*Chrysops flavidus*) and stable flies (*Stomoxys calcitrans*). Am. J. Vet. Res. **44**:155–156.

Fotedar, R. and Banerjee, U. 1992. Nosocomial fungal infections—study of the possible role of cockroaches (*Blattella germanica*) as vectors. Acta Trop. **50**:339–343.

Fotedar, R., Banerjee, U. and Shriniwas. 1993. Vector potential of the German cockroach in dissemination of *Pseudomonas aeruginosa*. J. Hospit. Infect. **23**:55–59.

Fotedar, R., Banerjee, U., Shriniwas and Verma, A. 1991a. Cockroaches (*Blattella germanica*) as carriers of microorganisms of medical importance in hospitals. Epidemiol. Infect. **107**:181–187.

Fotedar, R., Shriniwas, Banerjee, U., Samantray, J.C., Nayar, E. and Verma, A. 1991b. Nosocomial infections: cockroaches as possible vectors of drug-resistant *Klebsiella*. J. Hospit. Infect. **18**:155–159.

Fowler, H.G., Bueno, O.C., Sadatsune, T. and Montelli, A.C. 1993. Ants as potential vectors of pathogens in hospitals in the state of São Paulo, Brazil. Insect Sci. Appl. **14**:367–370.

Francis, E. and Mayne, B. 1921. Experimental transmission of tularemia by flies of the species *Chrysops discalis*. Publ. Hlth. Rep. **36**:1738–1746.

Francis Jr., T., Brown, G.C. and Penner, L.R. 1948. Search for extrahuman sources of poliomyelitis virus. J. Am. Med. Assoc. **136**:1088–1093.

Friedland, G.H. and Klein, R.S. 1987. Transmission of the human immunodeficiency virus. New England J. Med. **317**:1125–1134.

Fukuda, T., Goto, T., Kitaoka, S., Fujisaki, K. and Takamatsu, H. 1979. Experimental transmission of fowl pox by *Culicoides arakawae*. Nat. Inst. Anim. Hlth. Quart. **19**:104–105.

Gaud, J. and Faure, P. 1951. Effet de la lutte antimouches sur l'incidence des maladies oculaires dans le Sud marocain. Bull. Soc. Pathol. Exot. **44**:446–448.

Gazivoda, P. and Fish, D. 1985. Scanning electron microscopic demonstration of bacteria on tarsi of *Blattella germanica*. J. New York Entomol. Soc. **93**:1064–1067.

Gerhardt, R.R., Parrish, G., Snyder, R.Q. and Freeland, R.D. 1976. Incidence of pinkeye in relation to face fly control. Tennessee Farm & Home Sci. **97**:14–15.

Gibbs, E.P.J., Johnson, R.H. and Osborne, A.D. 1973. Experimental studies of the epidemiology of bovine herpes mammillitis. Res. Vet. Sci. **14**:139–144.

Gip, L. and Svensson, S.A. 1968. Can flies cause the spread of dermatophytosis? Acta Dermat.-Venereol. **48**:26–29.

Glass, H.W., Jr. and Gerhardt, R.R. 1984. Transmission of *Moraxella bovis* by regurgitation from the crop of the face fly (Diptera: Muscidae). J. Econ. Entomol. **77**:399–401.

Goddeeris, B. 1980. The role of insects in dispersing eggs of tapeworms, in particular *Taeniarhynchus saginatum*. I. Review of the literature. Ann. Soc. Belge Méd. Trop. **60**:195–201.

Gorham, J.R. 1994. Food, filth and disease: a review. Pp. 627–638 *in* Y.H. Hui, J.R. Gorham, K.D. Murrell and D.O. Cliver (eds.), Foodborne disease handbook, Vol. 3. Marcel Dekker, New York.

Graham-Smith, G.S. 1914. Flies in relation to disease: nonbloodsucking flies. Cambridge University Press, Cambridge, England. 389 pp.

Greenberg, B. 1965. Flies and disease. Sci. Am. **213**:92–99.

Greenberg, B. 1968. Model for destruction of bacteria in the midgut of blow fly maggots. J. Med. Entomol. **5**:31–38.

Greenberg, B. 1971. Flies and disease. Vol. 1, Ecology, classification and biotic associations. Princeton University Press, Princeton, New Jersey. 856 pp.

Greenberg, B. 1973. Flies and disease. Vol. 2, Biology and disease transmission. Princeton University Press, Princeton, New Jersey. 447 pp.

Greenberg, B. and Bornstein, A.A. 1964. Fly dispersion from a rural Mexican slaughterhouse. Am. J. Trop. Med. Hyg. **13**:881–886.

Greenberg, B., Kowalski, J.A. and Klowden, M.J. 1970. Factors affecting the transmission of *Salmonella* by flies: natural resistance to colonization and bacterial interference. Infect. Immun. **2**:800–809.

Gunn, H.M. 1993. Role of fomites and flies in the transmission of bovine viral diarrhoea virus. Vet. Rec. **132:**584–585.

Gupta, S.R., Rao, C.K., Biswas, H., Krishnaswami, A.K., Wattal, B.L. and Raghavan, G.S. 1972. Role of the housefly in the transmission of intestinal parasitic cysts/ova. Indian J. Med. Res. **60:**1120–1125.

Harris, A.H. and Down, H.A. 1946. Studies of the dissemination of cysts and ova of human intestinal parasites by flies in various localities on Guam. Am. J. Trop. Med. Hyg. **26:**789–800.

Harrison, L.H., Silva, G.A., Pittman, M., Fleming, D.W., Vranjac, A. and Broome, C.V. 1989. Epidemiology and clinical spectrum of Brazilian purpuric fever. J. Clin. Microbiol. **27:**599–604.

Harwood, R.F. and James, M.T. 1979. Entomology in human and animal health, 7th ed. Macmillan, New York. 548 pp.

Hawkins, J.A., Adams, W.V., Wilson, B.H., Issel, C.J. and Roth, E.E. 1976. Transmission of equine infectious anemia virus by *Tabanus fuscicostatus*. J. Am. Vet. Med. Assoc. **168:**63–64.

Heinz, H.J. and Brauns, W. 1955. The ability of flies to transmit ova of *Echinococcus granulosus* to human foods. South African J. Med. Sci. **20:**131–132.

Henry, E.T., Levine, J.F. and Coggins, L. 1987. Rectal transmission of bovine leukemia virus in cattle and sheep. Am. J. Vet. Res. **48:**634–636.

Hewitt, C.G. 1914. The housefly, *Musca domestica* Linn.; its structure, habits, development, relation to disease, and control. Cambridge University Press, Cambridge, England. 382 pp.

Higgins, J.A., Radulovic, S., Jaworski, D.C. and Azad, A.F. 1996. Acquisition of the cat scratch disease agent *Bartonella henselae* by cat fleas (Siphonaptera: Pulicidae). J. Med. Entomol. **33:**490–495.

Ho, D.D., Moudgil, T. and Alam, M. 1989. Quantitation of human immunodeficiency virus type 1 in the blood of infected persons. New England J. Med. **321:**1621–1625.

Hoare, C.A. 1972. The trypanosomes of mammals: a zoological monograph. Blackwell Scientific Publications, Oxford, England. 749 pp.

Hoch, A.L., Gargan, T.P., II and Bailey, C.L. 1985. Mechanical transmission of Rift Valley fever virus by hematophagous Diptera. Am. J. Trop. Med. Hyg. **34:**188–193.

Hopla, C.E. 1980. A study of the host associations and zoogeography of *Pulex*. Pp. 185–207 *in* R. Traub and H. Starcke (eds.), Fleas: Proceedings of the International Conference on Fleas, Ashton Wold, Peterborough, UK, 21-25 June, 1977. A.A. Balkema, Rotterdam.

Hopla, C.E. and Hopla, A.K. 1994. Tularemia. Pp. 113–126 *in* G.W. Beran and J.H. Steele (eds.), Bacterial, rickettsial, chlamydial, and mycotic diseases, 2nd ed. CRC Press, Boca Raton, Florida.

Hornby, H.E. 1926. Studies in rinderpest immunity. 2. Methods of infection. Vet. J. **82:**348–355.

Hornick, R.B. 1977. Bacillary dysentery. Pp. 562–569 *in* P.D. Hoeprich (ed.), Infectious diseases, 2nd ed. Harper & Row, Hagerstown, Maryland.

Hoskins, M. 1934. An attempt to transmit yellow fever virus by dog fleas. J. Parasitol. **20:**299–303.

Howard, L.O. 1911. House flies. Farmers' Bulletin No. 459. US Department of Agriculture, Washington, DC. 16 pp.

Hughes, D.E., Kassim, O.O., Gregory, J., Stupart, M., Austin, L. and Duffield, R. 1989. Spectrum of bacterial pathogens transmitted by Pharaoh's ants. Lab. Anim. Sci. **39:**167–168.

Humphries, F.A. and Gibbons, R.J. 1942. Some observations on corynebacterial infections. Canad. J. Compar. Med. Vet. Sci. **6:**35–45.

Hurst, V. and Sutter, V.L. 1966. Survival of *Pseudomonas aeruginosa* in the hospital environment. J. Infect. Dis. **116:**151–154.

Hyslop, N.S.G. 1970. The epizootiology and epidemiology of foot and mouth disease. Adv. Vet. Sci. Comp. Med. **14:**261–307.

Issel, C.J. and Foil, L.D. 1984. Studies on equine infectious anemia virus transmission by insects. J. Am. Vet. Med. Assoc. **184:**293–297.

Issel, C.J., Rushlow, K., Foil, L.D. and Montelaro, R.C. 1988. A perspective on equine infectious anemia with an emphasis on vector transmission and genetic analysis. Vet. Microbiol. **17:**251–286.

Iwasa, M., Makino, S.-I., Asakura, H., Kobori, H. and Morimoto, Y. 1999. Detection of *Escherichia coli* 0157:H7 from *Musca domestica* (Diptera: Muscidae) at a cattle farm in Japan. J. Med. Entomol. **36:**108-112.

Janisiewicz, W.J., Conway, W.S., Brown, M.W., Sapers, G.M., Fratamico, P. and Buchanan, R.L. 1999. Fate of *Escherichia coli* 0157:H7 on fresh-cut apple tissue and its potential for transmission by fruit flies. Appl. Environ. Microbiol. **65:**1-5.

Jobling, B. 1927. The structure of the head and mouthparts in *Culicoides pulicaris* L. (Diptera Nematocera). Bull. Entomol. Res. **18:**211-236.

Johnson, G.D., Campbell, J.B., Minocha, H.C. and Broce, A.B. 1991. Ability of *Musca autumnalis* (Diptera: Muscidae) to acquire and transmit bovine herpesvirus-1. J. Med. Entomol. **28:**841–846.

Jones, E.W. and Brock, W.E. 1966. Bovine anaplasmosis: its diagnosis, treatment, and control. J. Am. Vet. Med. Assoc. **149:**1624–1633.

Jones, F.T., Axtell, R.C., Rives, D.V., Schneider, S.E., Tarver, S.E., Walker, R.L. and Wineland, M.J. 1991. A survey of salmonella contamination in modern broiler production. J. Food Protec. **54:**502–507.

Jupp, P.G. and Lyons, S.F. 1987. Experimental assessment of bedbugs (*Cimex lectularius* and *Cimex hemipterus*) and mosquitoes (*Aedes aegypti formosus*) as vectors of human immunodeficiency virus. AIDS **1:**171–174.

Jupp, P.G. and McElligott, S.E. 1979. Transmission experiments with hepatitis B surface antigen and the common bedbug (*Cimex lectularius* L.). South African Med. J. **56:**54–57.

Jupp, P.G., Purcell, R.H., Phillips, J.M., Shapiro, M. and Gerin, J.L. 1991. Attempts to transmit hepatitis B virus to chimpanzees by arthropods. South African Med. J. **79:**320–322.

Kamyszek, F. 1969. The viability of spores of *Trychophyton meniagrophytes* in the alimentary tract of the housefly (*Musca domestica* L.). Med. Weteryn. **25:**163–165.

Kay, B.H. 1982. Three modes of transmission of Ross River virus by *Aedes vigilax* (Skuse). Aust. J. Exp. Biol. Med. Sci. **60:**339–344.

Kieser, S.T., Eriks, I.S. and Palmer, G.H. 1990. Cyclic rickettsemia during persistent Anaplasma marginale infection of cattle. Infect. Immunol. **58:** 1117-1119.

Kligler, I.J. 1929. Transmission of fowl pox by mosquitoes. J. Exp. Med. **49:**649–660.

Klowden, M.J. and Greenberg, B. 1977. Effects of antibiotics on the survival of *Salmonella* in the American cockroach. J. Hyg. (Cambridge) **79:**339–345.

Knaus, R.M., Foil, L.D., Issel, C.J. and Leprince, D.J. 1993. Insect blood meal studies using radiosodium ^{24}Na and ^{22}Na. J. Am. Mosq. Contr. Assoc. **9:**264–268.

Knight, S.M., Toodayan, W., Caique, W.C., Kyi, W., Barnes, A. and Desmarchelier, P. 1992. Risk factors for the transmission of diarrhoea in children: a case-control study in rural Malaysia. Intern. J. Epidemiol. **21:**812–818.

Kopanic Jr., R.J., Sheldon, B.W. and Wright, C.G. 1994. Cockroaches as vectors of *Salmonella*: Laboratory and field trials. J. Food Protec. **57:**125–132.

Krinsky, W.L. 1976. Animal disease agents transmitted by horse flies and deer flies (Diptera: Tabanidae). J. Med. Entomol. **13:**225–275.

Lawson, J.R. and Gemmell, M.A. 1990. Transmission of taeniid tapeworm eggs via blowflies to intermediate hosts. Parasitol. **100:**143–146.

Leclercq, M. 1969. Entomological parasitology; the relations between entomology and medical sciences, 1st ed. Pergamon Press, Oxford, England. 158 pp.

Lee, D.J., Fenner, F. and Lawrence, J.J. 1958. Mosquitoes and fowl pox in the Sydney area. Aust. Vet. **34:**230–237.

LeGuyader, A., Rivault, C. and Chaperon, J. 1989. Microbial organisms carried by brown-banded cockroaches in relation to their spatial distribution in a hospital. Epidemiol. Infect. **102:**485–492.

Levine, B. and Griffin, D.E. 1992. Persistence of viral RNA in mouse brains after recovery from acute alphaviral encephalitis. J. Virol. **66:**6429–6435.

Levine, O.S. and Levine, M.M. 1991. Houseflies (*Musca domestica*) as mechanical vectors of shigellosis. Rev. Infect. Dis. **13:**688–696.

Lightner, L. and Roberts, L.W. 1984. Mechanical transmission of *Leishmania major* by *Glossina morsitans* (Diptera: Glossinidae). J. Med. Entomol. **21:**243.

Linhares, A.X. 1981. Synanthropy of Muscidae, Fanniidae and Anthomyiidae (Diptera) in the city of Campinas, São Paulo, Brazil. Rev. Brasil. Entomol. **25:**231–243.

Luger, S.W. 1990. Lyme disease transmitted by a biting fly. New England J. Med. **322:**1752.

Macadam, I. 1962. Bovine streptothricosis: production of lesions by the bites of the tick *Amblyomma variegatum*. Vet. Rec. **74:**643–645.

MacWilliams, P.S. 1987. Erythrocytic rickettsia and protozoa of the dog and cat. Vet. Clin. N. Am. Small Anim. Prac. **17:**1443–1461.

Madsen, M., Sorensen, G.H. and Aalbaek, B. 1990. Summer mastitis in heifers: a bacteriological examination of secretions from clinical cases of summer mastitis in Denmark. Vet. Microbiol. **22:**319–328.

Magnarelli, L.A., Anderson, J.F. and Barbour, A.G. 1986. The etiologic agent of Lyme disease in deer flies, horse flies, and mosquitoes. J. Infect. Dis. **154:**355–358.

Main, A.J., Brown, S.E., Wallis, R.C. and Elston, J. 1979. Arbovirus surveillance in Connecticut. II. California serogroup. Mosq. News **39:**552–559.

Manet, G., Guilbert, X., Roux, A., Vuillaume, A. and Parodi, A.L. 1989. Natural mode of horizontal transmission of bovine leukemia virus (BLV): the potential role of tabanids (*Tabanus* spp.). Vet. Immunol. Immunopathol. **22:**255–263.

Manson-Bahr, P.E.C. and Downie, A.W. 1973. Persistence of tanapox in Tana River Valley. Brit. Med. J. **2:**151–153.

Markus, M.B. 1980. Flies as natural transport hosts of *Sarcocystis* and other Coccidia. J. Parasitol. **66:**361–362.

Mason, R.W. and Statham, P. 1991. The determination of the level of *Eperythrozoon ovis* parasitemia in chronically infected sheep and its significance to the spread of infection. Aust. Vet. J. **68:**115–116.

McAllister, J.C., Steelman, C.D., Newberry, L.A. and Skeeles, J.K. 1995. Isolation of bursal disease virus from the lesser mealworm, *Alphitobius diaperinus* (Panzer). Poult. Sci. **74:**45–49.

McAllister, J.C., Steelman, C.D., Skeeles, J.K., Newberry, L.A. and Gbur, E.E. 1996. Reservoir competence of *Alphitobius diaperinus* (Coleoptera: Tenebrionidae) for *Escherichia coli* (Eubacteriales: Enterobacteriaceae). J. Med. Entomol. **33**:983–987.

McAllister, J.C., Steelman, C.D. and Skeeles, J.K. 1994. Reservoir competence of lesser mealworm (Coleoptera: Tenebrionidae) for *Salmonella typhimurium* (Eubacteriales: Enterobacteriaceae). J. Med. Entomol. **31**:369–372.

Medveczky, I., Kovács, L., Kovács Sz., F. and Papp, L. 1988. The role of the housefly, *Musca domestica*, in the spread of Aujeszky's disease (pseudorabies). Med. Vet. Entomol. **2**:81–86.

Mellor, P.S., Kitching, R.P. and Wilkinson, P.J. 1987. Mechanical transmission of capripox virus and African swine fever virus by *Stomoxys calcitrans*. Res. Vet. Sci. **43**:109–112.

Melnick, J.L. and Dow, R.P. 1948. Poliomyelitis in Hidalgo County, Texas, 1948. Poliomyelitis and coxsackie viruses from flies. Am. J. Hyg. **58**:288–309.

Melnick, J.L., Emmons, J., Coffey, J.H. and Schoof, H. 1954. Seasonal distribution of coxsackie viruses in urban sewage and flies. Am. J. Hyg. **59**:164–184.

Melnick, J.L. and Penner, L.R. 1947. Experimental infection of flies with human poliomyelitis virus. Proc. Soc. Exp. Biol. Med. **65**:342–346.

Mihok, S., Maramba, O., Munyoki, E. and Kagoiya, J. 1995. Mechanical transmission of *Trypanosoma* spp. by African Stomoxyinae (Diptera: Muscidae). Trop. Med. Parasitol. **46**:103–105.

Miike, L. 1987. Do insects transmit AIDS? Staff Paper 1. Office of Technology Assessment, US Congress, Washington, DC. 43 pp.

Miller, G.J., Pegram, S.M., Kirkwood, B.R., Beckles, G.L.A., Byam, N.T.A., Clayden, S.A., Kinlen, L.J., Chan, L.C., Carson, D.C. and Greaves, M.F. 1986. Ethnic composition, age and sex, together with location and standard of housing as determinants of HLTV-I infection in an urban Trinidadian community. Intern. J. Cancer **38**:801–808.

Mlodecki, H. and Burzynska, H. 1956. Materials for the hygenic evaluation of foodstuffs contaminated by storage mites. I. Bacteriologic investigations of foodstuffs contaminated by storage mites. Ann. Polish Inst. Hyg. **7**:419–423.

Monzon, R.B., Sanchez, A.R., Tadiaman, B.M., Najos, O.A., Valencia, E.G., de Rueda, R.R. and Ventura, J.V.M. 1991. A comparison of the role of *Musca domestica* (Linnaeus) and *Chrysomya megacephalala* (Fabricius) as mechanical vectors of helminthic parasites in a typical slum area of metropolitan Manila. S.E. Asian J. Trop. Med. Publ. Hlth. **22**:222–228.

Morgan, N.O. and Miller, L.D. 1976. Muscidae (Diptera): experimental vectors of hog cholera virus. J. Med. Entomol. **36**:657-660.

Motha, M.X.J., Egerton, J.R. and Sweeney, A.W. 1984. Some evidence of mechanical transmission of reticuloendotheliosis virus by mosquitoes. Avian Dis. **28**:858–867.

Moutia, A. 1928. Surra in Mauritius and its principal vector *Stomoxys nigra*. Bull. Entomol. Res. **19**:211–216.

Murphy, E.L., Calisher, C.H., Figueroa, J.P., Gibbs, W.N. and Blattner, W.A. 1989. HTLV-I infection and arthropod vectors. New England J. Med. **321**:1146.

Napoli, V.M. and McGowan, J.E., Jr. 1987. How much blood is in a needlestick? J. Infect. Dis. **155**:828.

Nieschulz, O. and Kraneveld, F.C. 1929. Experimentelle Untersuchungen über die Übertragung der Buffelseuche durch Insekten. Zentrbl. Bakteriol. Parasitenk. **113**:403–417.

Ogston, C.W. 1981. Transfer of radioactive tracer by the bedbug *Cimex hemipterus* (Hemiptera: Cimicidae): a model for mechanical transmission of hepatitis B virus. J. Med. Entomol. **18**:107–111.

Olsen, R.A. 1998. Regulatory action criteria for filth and other extraneous materials. III. Review of flies and foodborne enteric disease. Regulatory Toxicol. Pharmacol. **28**: 199-211.

Olssufiev, N.G. 1941. The role of mosquitoes in transmitting the tularemia infection to wild rodents, birds and domestic animals (In Russian with English summary). Tr. Mil. Med. Acad. Red Army S.M. Kyrova **25**:190–197.

Oo, K.N., Sebastian, A.A. and Aye, T. 1989. Carriage of enteric bacterial pathogens by house flies in Yangon, Myanmar. J. Diarrhoeal Dis. Res. **7**:81–84.

Ostrolenk, M. and Welch, H. 1942. The common house fly (*Musca domestica*) as a source of pollution in food establishments. Food Research **7**:192–200.

Otte, M.J. and Abuabara, J.Y. 1991. Transmission of South American *Trypanosoma vivax* by the neotropical horsefly *Tabanus nebulosus*. Acta Trop. **49**:73–76.

Owens, W.E., Oliver, S.P., Gillespie, B.E., Ray, C.H. and Nickerson, C. 1998. Role of horn flies (*Haematobia irritans*) in *Staphylococcus aureus*-induced mastitis in dairy heifers. Am. J. Vet. Res. **59**: 1122-1124.

Paffenbarger, Jr., R.S. and Watt, J. 1953. Poliomyelitis in Hidalgo County, Texas, 1948. Epidemiologic observations. J. Hyg. (Cambridge) **58**:269–287.

Paganelli, C.H. and Sabrosky, C.W. 1993. *Hippelates* flies (Diptera: Chloropidae) possibly associated with Brazilian purpuric fever. Proc. Entomol. Soc. Wash. **95**:165–174.

Paul, J.R., Horstmann, D.M., Riordan, J.T., Opton, E.M., Neiderman, J.C., Isacson, P. and Green, R.A. 1962. An oral poliovirus vaccine trial in Costa Rica. Bull. Wld. Hlth. Organiz. **26**:311–329.

Payne, R.C. and Osorio, O. 1990. Tick-borne diseases of cattle in Paraguay. I. Seroepidemiological studies on anaplasmosis and babesiosis. Trop. Anim. Hlth. Prod. **22:**53–60.

Plowright, W. 1956. Cutaneous stretothricosis of cattle: introduction and epizootiological features in Nigeria. Vet. Rec. **68:**350–355.

Pugh, G.W., Hughes, D.E. and Mcdonald, T.J. 1966. The isolation and characterization of *Moraxella bovis*. Am. J. Vet. Res. **27:**957–962.

Quinn, T.C., Zacarias, F.R.K. and St. John, R.K. 1989. HIV and HTLV-I infectious in the Americas: a regional perspective. Medicine **68:**189–209.

Rao, T.R., Paul, S.D. and Singh, K.R.P. 1968. Experimental studies on the mechanical transmission of chikungunya virus by *Aedes aegypti*. Mosq. News **28:**406–408.

Raymond, H.L. 1990. *Tabanus importunus*, vecteur mecanique experimental de *Trypanosoma vivax* en Guyane Francaise. Ann Parasitiol. Hum. and Comp. **65:**44–46.

Reeves, J.D., III and Swift, B.L. 1977. Iatrogenic transmission of *Anaplasma marginale* in beef cattle. Vet. Med. Small Anim. Clin. **72:**5, 911–912, 914.

Rementsova, M.M. 1987. Brucellosis in wild animals. Amerind Publishing Company, New Delhi, India. 323 pp.

Richard, J.L. and Pier, A.C. 1966. Transmission of *Dermatophilus congolensis* by *Stomoxys calcitrans* and *Musca domestica*. Am. J. Vet. Res. **27:**419–423.

Riordan, J.T., Paul, A., Yoshioka, I. and Horstmann, D.M. 1961. The detection of poliovirus and other enteric viruses in flies. Am. J. Hyg. **74:**119–136.

Rippon, J.W. 1988. Medical mycology, 3rd ed. W.B. Saunders Company, Philadelphia. 797 pp.

Rivault, C., Cloarec, A. and Le Guyader, A. 1993. Bacterial contamination of food by cockroaches. J. Environ. Hlth. **55:**21–22.

Roberts, R.H. and Love, J.N. 1973. The potential of *Hippelates pusio* Loew (Diptera: Chloropidae) as a vector of anaplasmosis. Proc. 6th Nat. Anaplasmosis Conf., pp. 21–122.

Rogers, N. 1989. Dirt, flies, and immigrants: explaining the epidemiology of poliomyelitis, 1900-1916. J. Hist. Med. Allied Sci. **44:**486–505.

Rogoff, W.M., Carbrey, E.C., Bram, R.A., Clark, T.B. and Gretz, G.H. 1975. Transmission of Newcastle disease virus by insects: detection in wild *Fannia* spp. (Diptera: Muscidae). J. Med. Entomol. **12:**225–227.

Root, F.M. 1921. Experiments on the carriage of intestinal Protozoa of man by flies. Am. J. Hyg. **1:**131–153.

Rosamond, C.H., Shepherd, C.H. and Edmonds, J.W. 1977. Myxomatosis: the transmission of highly virulent strain of myxoma virus by the European rabbit flea *Spilopsyllus cuniculi* (Dale) in the Mallee region of Victoria. J. Hyg. (Cambridge) **79:**405–409.

Ross, J. and Sanders, M.F. 1984. The development of genetic resistance to myxomatosis in wild rabbits in Britain. J. Hyg. (Cambridge) **92:**255–261.

Roth, L.M. and Willis, E.R. 1957. The medical and veterinary importance of cockroaches. Smithsonian Misc. Coll. **134:**1–147.

Roth, L.M. and Willis, E.R. 1960. The biotic associations of cockroaches. Smithsonian Misc. Coll. **141:**1-470.

Reubel, G.H., Barough, J.E. and Madigan, J.E. 1998. Production and characterization of *Ehrlichia risticii*, the agent of Potomac horse fever, from snails (Pleuroceridae: *Juga* spp.) in aquarium culture and genetic comparison to equine strains. J. Clin. Microbiol. **36:**1501-1511.

Rueger, M.E. and Olson, T.A. 1969. Cockroaches (Blattaria) as vectors of food poisoning and food infection organisms. J. Med. Entomol. **6:**185–189.

Scherer, W.F. and Hurlbut, H.S. 1967. Nodamura virus from Japan: a new and unusual arbovirus resistant to diethyl ether and chloroform. Am. J. Epidemiol. **86:**271–285.

Schmidtmann, E.T., Robl, M.G. and Carroll, J.F. 1988. Attempted transmission of *Ehrlichia risticii* by field-captured *Dermacentor variabilis* (Acari: Ixodidae). Am. J. Vet. Res. **47:**2393–2395.

Schuster, D.J., Mills, R.B. and Crumrine, M.H. 1972. Dissemination of *Salmonella montevideo* through wheat by the rice weevil. Environ. Entomol. **1:**111–115.

Schwarte, L.H. and Biester, H.E. 1941. Pox in swine. Am. J. Vet. Res. **2:**136–140.

Scott, D.W. 1988. Large animal dermatology. W.B. Saunders Co., Philadelphia. 487 pp.

Scott, T.W., Clark, G.G., Lorenz, L.H., Amerasinghe, P.H., Reiter, P. and Edman, J.D. 1993. Detection of multiple feeding in *Aedes aegypti* (Diptera: Culicidae) during a single gonotrophic cycle using a histologic technique. J. Med. Entomol. **30:**94–99.

Sellers, R.F., Pedgley, D.E. and Tucker, M.R. 1982. Rift Valley fever, Egypt 1977: disease spread by windborne insect vectors? Vet. Rec. **110:**73–77.

Shane, S.M., Montrose, M.S. and Harrington, K.S. 1985. Transmission of *Campylobacter jejuni* by the housefly (*Musca domestica*). Avian Dis. **29:**384–391.

Sheremetiev, N.N. 1964. Dynamics of isolation of poliomyelitis vaccinal strains from flies after vaccination. J. Microbiol. Epidemiol. Immunol. **41:**102–106.

Shugart, J.I., Campbell, J.B., Hudson, D.B., Hibbs, C.M., White, R.G. and Clanton, D.C. 1979. Ability of the face fly to cause damage to eyes of cattle. J. Econ. Entomol. **72:**633–635.

Silverman, A.L., McCray, D.G., Gordon, S.C., Morgan, W.T. and Walker, E.D. 1996. Experimental evidence against replication or dissemination of hepatitis C virus in mosquitoes (Diptera: Culicidae) using detection by reverse transcriptase polymerase chain reaction. J. Med. Entomol. **33:**398–401.

Simmons, P. 1923. A house fly plague in the American expeditionary force. J. Econ. Entomol. **16:**357–363.

Smith, C.E.G. 1987. Factors influencing the transmission of western equine encephalomyelitis virus between its vertebrate maintenance hosts and from them to humans. Am. J. Trop. Med. Hyg. **37:**33S–39S.

Srámová, H., Daniel, M., Absolonová, V., Dedicová, D., Jedlicková, Z., Lhotová, H., Petrás, P. and Subertová, V. 1992. Epidemiological role of arthropods detectable in health facilities. J. Hospit. Infect. **20:**281–292.

Srinivasan, A., York, D. and Bohan, C. 1987. Lack of HIV replication in arthropod cells. Lancet **I:**1094–1095.

Steelman, C.D., Foote, L.E., Roby, T.O. and Hollon, B.F. 1968. Seasonal occurrence of moquitoes feeding on dairy cattle and the incidence of anaplasmosis in southern Louisiana. Proc. 5th Nat. Anaplasmosis Conf., pp. 178–182.

Stewart, W.C., Carbrey, E.A., Jenny, E.W., Kresse, J.I., Snyder, M.L. and Wessman, S.J. 1975. Transmission of hog cholera virus by mosquitoes. Am. J. Vet. Res. **36:**611–614.

Sturtevant, A.H. 1918. Flies of the genus *Drosophila* as possible disease carriers. J. Parasitol. **5:**84–85.

Sulaiman, S., Sohadi, A.R. and Jeffrey, J. 1989. Human helminth parasite burdens on cyclorrhaphan flies (Diptera) trapped at an aboriginal settlement in Malaysia. Bull. Entomol. Res. **79:**625–629.

Tadros, W. and Laarman, J.J. 1982. Current concepts on the biology, evolution and taxonomy of tissue cyst-forming eimeriid coccidia. Adv. Parasitol. **20:**293–468.

Takahashi, K., Wada, A., Kawai, S., Yokota, H., Kurosawa, T. and Sonoda, M. 1984. Investigations of isolation, transmission and virulence of bovine *Babesia* sp. in Hokkaido (Abstract). Rev. Appl. Entomol. **72:**179.

Tarry, D.W., Bernal, L. and Edwards, S. 1991. Transmission of bovine virus diarrhoea virus by blood feeding flies. Vet. Rec. **128:**82–84.

Taylor, H.R. 1988. A simple method for assessment of association between synanthropic flies and trachoma. Am. J. Trop. Med. Hyg. **38:**623–637.

Taylor, J.P., Istre, G.R., McChesney, T.C., Satalowich, F.T., Parker, R.L. and McFarland, L.M. 1991. Epidemiologic characteristics of human tularemia in the Southwest-central states, 1981-1987. Am. J. Epidemiol. **133:**1032–1038.

Thomas, G., Prijs, H.J. and Trapman, J.J. 1987. Factors contributing to differential risk between heifers in contracting summer mastitis. Pp. 30–35 *in* G.T. Over, H.J. Over, U. Vecht and P. Nansen (eds.), Summer mastitis. Martinus Nijhoff, Dordrecht, Germany.

Tidwell, M.A., Dean, W.D., Tidwell, M., Combs, G.P., Anderson, D.W., Cowart, W.O. and Axtell, R.C. 1972. Transmission of hog cholera virus by horseflies (Tabanidae: Diptera). Vet. Res. **33:**615–622.

Tod, M.E., Jacobs, D.E. and Dunn, A.M. 1971. Mechanisms for the dispersal of parasitic nematode larvae. 1. Psychodid flies as transport hosts. J. Helminthol. **45:**133–137.

Tompkins, L.S. 1997. Of cats, humans, and *Bartonella*. New England J. Med. **337:**1916–1917.

Turell, M.J. and Knudson, G.B. 1987. Mechanical transmission of Bacillus anthracis by stable flies (*Stomoxy calcitrans*) and mosquitoes (*Aedes aegypti* and *Aedes taeniorhynchus*). Infection and Immunity **55:** 1859-1861.

Umeche, N. and Mandah, L.E. 1989. *Musca domestica* as a carrier of intestinal helminths in Calabar, Nigeria. East African Med. J. **66:**349–352.

Umunnabuike, A.C. and Irokanulo, E.A. 1986. Isolation of *Campylobacter* subsp. *jejuni* from oriental and American cockroaches caught in kitchens and poultry houses in Vom, Nigeria. Internat. J. Zoonoses **13:**180–186.

Van der Maaten, M.J., Miller, J.M. and Schmerr, M.J.F. 1981. *In utero* transmission of bovine leukemia virus. Am. J. Vet. Res. **42:**1052–1054.

Vaughan, V.C. 1899. Some remarks on typhoid fever among our soldiers during the late war with Spain. Am. J. Med. Sci. **118:**10-24.

Walker, E.D. and Edman, J.D. 1985. The influence of host defensive behavior on mosquito (Diptera: Culicidae) biting persistence. J. Med. Entomol. **22:**370–372.

Ward, R., Melnick, J.L. and Horstmann, D. 1945. Poliomyelitis virus in fly-contaminated food collected at epidemic. Science **101:**491–493.

Watt, J. and Hardy, A.V. 1945. Studies of the acute diarrheal diseases. XIII. Cultural surveys of normal populations. Publ. Hlth. Rep. **60:**261–273.

Watt, J. and Lindsay, J.R. 1948. Diarrheal disease control studies. I. Effect of fly control in a high morbidity area. Publ. Hlth. Rep. **63:**1319–1334.

Webb, P.A., Happ, C.M., Maupin, G.O., Johnson, B.J.B., Ou, C.Y. and Monath, T.P. 1989. Potential for insect transmission of HIV: experimental exposure of *Cimex hemipterus* and *Toxorhynchites amboinensis* to human immunodeficiency virus. J. Infect. Dis. **160:**970–977.

Weber, A.F., Moon, R.D., Sorensen, D.K., Bates, D.W., Meiske, J.C., Brown, C.A., Rohland, N.L., Hooker, E.C. and Strand, W.O. 1988. Evaluation of the stable fly (*Stomoxys calcitrans*) as a vector of enzootic bovine leukosis. Am. J. Vet. Res. **49:**1543–1549.

Weil, A.J. 1943. Progress in the study of bacillary dysentery. J. Immunol. **46:**13–46.

Wellman, G. 1950. Rotlaufübertragung durch Verschiedene blutsaugende Insektenarten auf Tauben. Zentrbl. Bakteriol. Parasitenk. **155:**109–115.

West, L.S. 1951. The housefly, its natural history, medical importance, and control. Comstock, Ithaca, New York. 584 pp.

West, L.S. and Peters, O.B. 1973. An annotated bibliography of *Musca domestica* Linnaeus. Dawsons, Folkestone, UK 743 pp.

Wiesenhutter, E. 1975. Research into the relative importance of Tabanidae (Diptera) in mechanical disease transmission. I. The seasonal occurrence and relative abundance of Tabanidae in a Dar es Salaam dairy farm. J. Nat. Hist. **9:**377–384.

Wilesmith, J.W., Straub, O.C. and Lorenz, R.J. 1980. Some observations on the epidemiology of bovine leukosis virus infection in a large dairy herd. Res. Vet. Sci. **28:** 10–16.

Williams, D.L., Issel, C.J., Steelman, C.D., Adams, W.V., Jr. and Benton, C.V. 1981. Studies with equine infectious anemia virus: transmission attempts by mosquitoes and survival of virus on vector mouthparts and hypodermic needles, and in mosquito tissue culture. Am. J. Vet. Res. **42:**1469–1473.

Wilson, B.H. 1968. Observations on horse fly abundance and the incidence of anaplasmosis in a herd of dairy cattle in southern Louisiana. Proc. 5th Nat. Anaplasmosis Conf., p. 173.

Wilton, D.P. 1963. Dog excrement as a factor in community fly problems. Proc. Hawaiian Entomol. Soc. **18:**311–317.

Wollman, E. 1921. Le rôle des mouches dans le transport des germes pathogènes étudié par la méthode des élevages aseptiques. Ann. Inst. Pasteur **35:**431–449.

Woodall, J.P. and Bertram, D.S. 1959. The transmission of Semliki Forest virus by *Aedes aegypti* L. Trans. Royal Soc. Trop. Med. Hyg. **53:**440–444.

Woods, J.A. 1988. Lumpy skin disease—a review. Trop. Anim. Hlth. Prod. **20:**11-17.

Yao, H.Y., Yuan, I.C. and Huie, D. 1929. The relation of flies, beverages and well water to gastro-intestinal diseases in Peiping. Nat. Med. J. China **15:**410–418.

Young, P.L. and Babero, B.B. 1975. Studies on the transmission of helminth ova by cockroaches. Proc. Okla. Acad. Sci. **55:**169–174.

Zahari, P., Hirst, R.G., Shipton, W.A. and Campbell, R.S.F. 1990. The origin and pathogenicity of *Basidiobolus* species in northern Australia. J. Med. Vet. Mycol. **28:**461–468.

Zhang Ran, Chien Lien-sheng and Chang, J.T. 1990. Induction and isolation of an antibacterial peptide in *Periplaneta americana*. Acta Entomol. Sinica **33:**7–13.

Chapter 13

Surveillance for Arthropodborne Diseases

BRUCE F. ELDRIDGE
University of California, Davis

INTRODUCTION

Surveillance is the act of watching over something. A survey is a general study or inspection. Applied to infectious diseases, surveillance and surveys both involve the systematic collection of data relating to the occurrence or likely occurrence of disease cases. Surveillance programs continue indefinitely, whereas surveys are of limited duration. Surveillance may involve detecting cases within human and animal populations, monitoring infectious agents in natural vectors or reservoir animals, seroconversions in sentinel animals, estimating vectors or reservoir host population abundance, and assessing weather patterns.

Arthropodborne disease surveillance provides the basis for predicting disease outbreaks in humans or other animals, and thus permits timely intervention to avoid or abort such outbreaks. Most human and animal health professionals consider disease surveillance to be an important component of preventive medicine. In 1992, a committee of the Institute of Medicine (chartered by the US National Academy of Science) stated, "The key to recognizing new or emerging infectious diseases, and to tracking the prevalence of more established ones, is surveillance." This committee went on to say, "Surveillance is important to any disease control effort. It is absolutely essential if that effort's goal is eradication" (Lederberg et al. 1992). Smallpox, for example, was eradicated only after the development of a comprehensive surveillance program. In 1971, when the smallpox eradication program began, only about 1% of the cases was being reported. By 1979, when the disease was declared eradicated by the World Health Organization (WHO), a global reporting network had been developed which searched house-by-house for new cases.

The key element of human and animal disease surveillance is case detection and reporting. Surveillance can be **passive** (the reporting of cases by public health or veterinary workers) or **active** (the seeking out of disease cases by epidemiologists). Case detection is often the only component of many surveillance programs, and some writers use the terms surveillance and case detection nearly synonymously. However, medical entomologists assert that surveillance encompasses a broader range of activities. This definition is because in vectorborne diseases, monitoring other components of pathogen life cycles (e.g., enzootic activity, vec-

B.F. Eldridge and J.D. Edman (eds.), Medical Entomology, Revised Edition, 515–538.

tor abundance and vector infection rates) gives advanced warning of outbreaks.

Few worldwide programs for the surveillance of vectorborne diseases go beyond case detection and reporting. Surveillance systems are expensive to maintain, and during periods when vectorborne diseases are at low levels, there is little public enthusiasm to maintain them. Even case reporting varies from one geographic region to another, according to the status of a given disease.

PRINCIPLES OF SURVEILLANCE

Surveillance programs for vectorborne diseases are based on several concepts. One is that various factors in nature, such as high vector population densities, precede human and animal disease cases, so that estimates of these phenomena can provide an early warning of outbreaks. Further, environmental factors such as weather patterns, which may influence factors such as vector density, also may be used as surveillance indicators.

Secondly, there is an inverse relationship between the amount of advance warning provided by environmental factors and their predictive value (Table 13.1). For instance, depth of snowpack in mountainous regions is related distantly to the likelihood of the occurrence of mosquitoborne human viral diseases in certain parts of the world (Fig. 13.1). This is because after the snowpack melts, spring flooding may result in abnormally high densities of vector mosquitoes. Depth of snowpack may be estimated months in advance of the mosquito season, thus giving ample advance warning of a potential outbreak. However, factors unrelated to snowpack, such as presence of virus, temperature patterns, flood control measures and herd immunity in populations of vertebrate hosts, also influence the possibility of outbreaks. In contrast, the seroconversion by sentinel animals indicates that infective vectors are present in an area, and theoretically, highly predictive of potential human cases.

To be useful in predicting disease outbreaks, surveillance systems must be based on a thorough knowledge of the epidemiology of the disease of interest, as Chapter 6 explains. In fact, many of the methods employed in surveillance are simply applications of the research methods used to determine epidemiological relationships.

Surveillance programs depend upon sound sampling methodology, and on reliable methods of identifying vectors and pathogens. Modern approaches using molecular tools such as PCR and DNA hybridization should improve identification capabilities, but a lack of specialists trained in field ecology and arthropod taxonomy may make the task difficult.

WORLDWIDE DISEASE SURVEILLANCE SYSTEMS

Surveillance schemes vary widely throughout the world, ranging from well-organized international efforts to small community programs. Global surveillance of infectious diseases involves health care providers, medical researchers, country ministries of health, WHO, the Pan American Health Organization and the International Office of Epizootics (OIE).

In the USA, a number of agencies conduct surveillance for vectorborne diseases. The Centers for Disease Control and Prevention (CDC) work with individual states in monitoring the incidence of specific diseases. The Departments of Agriculture and Defense also maintain surveillance both in the USA and overseas. Cases of certain human infectious diseases are reported to the National Notifiable Diseases Surveillance System (NNDSS). The CDC notifiable diseases associated with arthropods are listed in Table 13.2 (Wharton et al. 1990).

Table 13.1. Conceptual framework for a mosquitoborne arbovirus surveillance program (from Eldridge 1987).

Surveillance Method	Surveillance Indicator	Predictive Value	Advance Warning
Measurement of snow depth	Snow depth	Moderately low	Several months
Analysis of flood forecasts	Standing water	Moderate	Several weeks
Trapping of adult mosquitoes	Mosquito vector density	Moderately high	Several weeks
Testing mosquito pools for virus	Infected vector density	Moderately high	1-2 weeks
Testing sera from sentinel chickens	Infective vector density	High	A week or less
Analysis of indvidual cases	Number of human and animal disease cases	High	Days
Analysis of case summaries	Temporal or geographic trends	Highest	Days

Many other developed countries have comprehensive surveillance systems for communicable diseases. For example, the Australian Communicable Diseases Network (Curran et al. 1997) includes vectorborne diseases such as malaria, dengue and encephalitis.

On a worldwide basis, responsibility for maintaining surveillance systems rests generally with local jurisdictions. A few explosive diseases, including yellow fever and plague, are reportable to WHO on the basis of international health regulations. Other diseases, including malaria and louseborne typhus, though not required to be reported, are under WHO surveillance. The American Public Health Association classifies human communicable diseases based on their surveillance priorities (Benenson 1995). Table 13.3 lists the important arthropodborne diseases in this classification. Priorities are based not on the severity or importance of the disease, but rather on the practical benefits of surveillance in preventing cases. Therefore, malaria is placed in Class 1 only in non-endemic areas (e.g., the USA). Where malaria cases occur in endemic areas, they are placed in Class 3. It is noteworthy that 5 of the 8 diseases in the highest category (Class 1) are arthropodborne.

The international reporting agency for diseases of livestock is the OIE. The OIE is an intergovernmental organization created in 1924 by an international agreement signed by 144 countries. Periodically, OIE publishes the International Animal Health Code containing lists of important infectious diseases of animals. The OIE List A contains diseases having the potential for rapid spread, irrespective of national borders, with serious socioeconomic or public health consequences. These diseases are of major importance in the international trade of animals and animal products. The arthropodborne diseases on List A are vesicular stomatitis, Rift Valley fever, bluetongue, African horse sickness and African swine fever. These List A diseases are reportable because their presence in a nonendemic area may change animal export regu-

Table 13.2. Diseases currently (1996) reportable to the US Centers for Disease Control and Prevention.

Arthropodborne	Non-arthropodborne
Encephalitis, primary[1]	Anthrax
Lyme disease	Meningococcal disease
Malaria	Leprosy
Tularemia[1]	
Yellow fever	
Aseptic meningitis[1]	
Plague	
Rocky Mountain spotted fever	

[1] Includes non-arthropod transmitted disease cases.

lations from the region. The economic consequences of reducing exports can be severe.

The OIE List B diseases are those considered important within countries, but less important in international trade of animals and animal products. List B arthropodborne diseases include screwworm, Nairobi sheep disease, Japanese encephalitis and Venezuelan equine encephalomyelitis.

Surveillance for List A pathogens may be essential for the establishment of regions for disease-free status. The US Animal and Plant Health Inspection Service of the US Department of Agriculture conducts yearly bluetongue serologic surveys of cattle in the northeastern USA (Pearson et al. 1992). The absence of antibody in this survey has supported the claim that this is a bluetongue-free region (Walton et al. 1992). This judgment has reduced restrictions on US cattle exports from this region to bluetongue-free countries such as Canada.

Surveillance of vectorborne diseases usually targets diseases that pose the most serious risk for a particular area (i.e., high mortality and morbidity) and situations where cases are sporadic. The components of the surveillance system also vary depending upon the diseases being monitored. As mentioned previously, surveillance systems often consist only of the routine detection and reporting of human or animal disease cases. More comprehensive programs combine approaches such as the trapping and testing of vectors for the presence of pathogens. Few programs incorporate all of the components listed in Table 13.3. Even well organized surveillance efforts focus on only a few critical diseases, using only one or two surveillance approaches for each.

COMPONENTS OF SURVEILLANCE

Weather Data

Weather data may predict the probability of outbreaks of certain arthropodborne diseases. Temperature and precipitation are the two most widely used weather features for surveillance. Weather may affect vectors, pathogens and reservoir hosts.

Temperature patterns frequently are associated with outbreaks of vectorborne disease, especially in temperate areas of the world. Because arthropods are poikilothermic animals, their physiological processes vary in rate with ambient temperature. This linkage is particularly true for reproduction and development, and consequently, vector populations tend to grow faster when the weather is warm. Temperature patterns also influence the rate of snowmelt, and thus the extent of runoff and flooding. For vectors that have aquatic immature stages, such as mosquitoes, a combination of warm temperatures and resulting large amounts of standing water from runoff and flooding may favor larval development.

Table 13.3. Surveillance categories for arthropodborne diseases (American Public Health Association classification)[1].

Class	Criteria	Diseases
1	Case report universally required by international health regulations, or as a disease under surveillance by WHO	Plague, yellow fever, louseborne typhus, relapsing fever, malaria[2]
2	Case report regularly required wherever disease occurs	Rocky Mountain spotted fever, fleaborne typhus, encephalitis[3]
3	Selectively reportable in recognized endemic areas	Most mosquitoborne encephalitides, vesicular stomatitis, babesiosis, ehrlichiosis, visceral leishmaniasis, bancroftian filariasis, Lyme disease, trypanosomiases, tularemia, scrub typhus
4	Obligatory report of epidemic, no case report required	Dengue
5	Official report ordinarily not justifiable	Cutaneous and mucosal leishmaniasis, loiasis, onchocerciasis, pediculosis

[1] Agricultural interests require reporting of OIE List A diseases from non-endemic areas, i.e., African horse sickness, African swine fever, bluetongue, Rift Valley fever and vesicular stomatitis.
[2] Class 1 for non-endemic areas only. For endemic areas, Class 3.
[3] Should be reported under specific disease (e.g., St. Louis encephalitis); otherwise report as encephalitis (other forms) or aseptic meningitis.

Precipitation is another important weather factor in surveillance. The significance of snowpack in surveillance systems already has been mentioned. The snowpack in the Sierra Nevada of California in 1951-52 was one of the highest on record, and the summer of 1952 saw a serious outbreak of western equine encephalomyelitis among people living in the Central Valley. Precipitation that falls as rain and leads to flooding can effect mosquitoborne disease outbreaks. Outbreaks of Rift Valley fever, a tropical African mosquitoborne virus disease, accompany heavy rainfall that floods ground depressions (called dambos), resulting in heavy populations of the *Aedes* vectors. Precipitation data not only predict flooding, but also provide advance information on the availability of water to farmers for irrigation, and thus standing water for the breeding of mosquitoes and other vectors with aquatic immature stages (Reeves and Milby 1980). Ironically, lack of precipitation in irrigated farming areas also may lead to increased mosquito populations because of increased irrigation.

Vectorborne disease surveillance systems seldom incorporate weather data, partly because of the lack of a thorough understanding of the relationship between weather patterns and disease outbreaks. However, there have been attempts to correlate weather patterns with arthropodborne diseases. Hess and Hayes (1967) felt that a combination of higher than average precipitation combined with cool spring temperatures predisposed outbreaks of WEE virus in horses and people. Moore et al. (1978) correlated rainfall patterns and dengue virus transmission in Puerto Rico.

Fig. 13.1. Snow survey crew (courtesy of California State Department of Water Resources).

Weather data are available in a variety of forms. The US National Weather Service maintains detailed weather data for North America, and most state agencies provide flood forecasts, snowpack, temperature, precipitation, runoff and reservoir storage for multiple stations. Often, these data are available as computer files that can be incorporated into predictive models and geographic information systems (GIS). Weather and climate forecasts are available from the Climate Prediction Center of the US National Oceanic and Atmospheric Administration. Weather forecasts state the probability conditions for the following 1–5 day and provide climate forecasts for the following 1–3 months.

For model development, several sources, in some cases going back many years, provide historical data. These data are critical in understanding differences between outbreak and non-outbreak years. For special studies, small, automated weather stations can be purchased that also output data as electronic files.

There is a need for reliable predictive models that include weather data as inputs to predict disease outbreaks. Unfortunately, few such models have been developed even for the best-studied arthropodborne diseases. Recently Mount et al. (1997) developed a model (called LYMESIM) incorporating 20 years of historical weather data to simulate the transmission of Lyme disease by blacklegged ticks. This program permits the manipulation of over 180 epidemiological and environmental variables to simulate vector-host-pathogen relationships.

Arthropod Surveillance

Arthropod sampling is the most important activity associated with the surveillance of vectorborne diseases. Routine sampling estimates population levels of both infected and noninfected arthropods. For the former purpose, arthropods are collected alive and assayed for pathogen infection in laboratories equipped with adequate biosafety containment. Methods of handling vary according to the vectors and pathogens involved. For surveillance programs, individual arthropods usually are trapped, identified, counted by sex, assembled into pools and chilled or frozen for later assay. After suitable preparation, pathogens may be detected by direct microscopic observation, or by inoculation of the test material into microbiological media, vertebrate tissue cultures or live animals. A variety of methods, commonly some type of immunoassay, may identify pathogens from positive samples.

Figure 13.2. The New Jersey light trap (courtesy of Robert K. Washino).

Figure 13.3. The CDC miniature light trap, with dry ice (courtesy of William K. Reisen).

Estimating vector density

One of most common uses of arthropod sampling data in surveillance is the estimation of vector density. This technique is based on the concept (though not always true) that high vector densities are associated with outbreaks of vectorborne diseases (Reeves and Milby 1980). The correlation between disease cases in human or animal hosts and estimates of vector density often is difficult to demonstrate, and will vary according to a number of factors (Eldridge 1987). In the case of malaria and dengue, transmission of pathogens to humans appears to take place in endemic areas with low mosquito densities. Furthermore, malaria transmission may occur even when few females survive long enough to complete an extrinsic incubation period (Mattingly 1969). On the other hand, associations between vector density and viral activity have been demonstrated for several arboviruses (Moore et al. 1978, Olson et al. 1979), onchocerciasis (Collins 1979) and Chagas disease (Gurtler et al. 1987).

Many sampling tools are available for arthropods. The choice of which tool to use will depend on the arthropod sought and the specific objectives of the surveillance operation. If the isolation of pathogens from arthropod specimens is desired, methods must be used to pre-

Figure 13.4. Truck-mounted trap for collection of flying insects (courtesy of Robert K. Washino).

vent deterioration of the specimens before they can be tested for pathogens in the laboratory. If specimens are identified and pooled in the field, they must be maintained chilled or frozen until they can be prepared and tested in the laboratory.

Surveillance programs have been established for many different arthropods, including mosquitoes, hard ticks, fleas, triatomid bugs, tsetse, black flies, phlebotomine sand flies, screwworm flies, house flies and tabanid flies.

Light traps may be used for various species of flying insects, including mosquitoes, biting gnats, sand flies and some horse flies (Service 1993). The number of individuals collected may be increased greatly if dry ice or some other form of carbon dioxide is used instead of, or in addition to, a light source. A traditional mosquito sampling device is the New Jersey light trap (Fig. 13.2). This trap was invented many years ago, and it has been supplanted for many applications by smaller battery-powered traps. The New Jersey trap only estimates relative population density and specimens typically are killed in this type of trap. If specimens are to be tested later for pathogens such as viruses, a large container without poison must be used as the collection receptacle.

Many commercial traps can be used with electric lamps, CO_2, or both. Smaller than the New Jersey trap, these traps are battery-powered. Most descend from the CDC miniature light trap (Fig. 13.3).

Some traps not based on principles of attraction are located to intercept flying insects. These traps may be stationary (e.g., the Malaise trap) or moving (e.g., the truck trap, Fig. 13.4). These traps collect small numbers of a wide variety of insect species.

Sampling devices usually are biased toward certain vector species and certain components of vector populations. For example, mosquito traps that use CO_2 will selectively trap females seeking a bloodmeal but trap few gravid or bloodfed females. Male mosquitoes rarely enter CO_2 traps. The biases of any type of collection device must be understood when analyzing data, and factored into conclusions based on the sampling. Probably the most important principle of systematic sampling is to use the same kind of trap throughout a study in order to minimize the effects of the biases inherent in different kinds of traps. Even using a single type of trap yields data that are nearly always relative rather than absolute, and estimates based on numbers collected over time at the same site may or may not be correlated with changes in density.

Mosquitoes

Mosquito abundance is monitored by sampling both adults and larvae. Adults usually are collected in either New Jersey light traps or a form of the CDC miniature trap supplemented with dry ice. Abundance is expressed commonly as the number of adults collected per trap per night (usually designated per trap-night). The universal sampling device for mosquito larvae and pupae is the long-handled dipper (Fig. 13.5). Population densities are estimated

Figure 13.5. Sampling mosquito larvae with standard one-pint dipper (photograph by Bruce F. Eldridge).

based on larvae or pupae obtained per dip. Mosquitoes whose immature stages develop in tree holes and artificial containers are often sampled using various kinds of oviposition traps (Fig. 13.6). These traps commonly are used to sample the egg stage of the yellow fever mosquito, *Aedes aegypti*, and the Asian tiger mosquito, *Aedes albopictus*. Treehole mosquito larvae also can be sampled using suction devices such as a household turkey baster.

Surveillance programs for dengue often estimate mosquito density based on larval frequency. A common expression for this type of estimate is the Breteau index, or the number of houses with water containers positive for *Ae. aegypti* per 100 houses sampled (Moore et al. 1978).

Ticks

Ticks and other flightless ectoparasites require different types of sampling devices. Because hard ticks in search of a bloodmeal often "quest" on grass and other vegetation, they may be collected by dragging a large square of cloth over the ground (called a "tick drag"; Fig. 13.7). Where dense vegetation prevents this approach, the cloth can be fashioned into a flag that can be waved across vegetation. The size of the cloth should be standardized for estimates of density, as should the time and distance involved in dragging. Another approach to surveillance of ticks is the "tick walk," where a person wearing white clothing walks along a standardized path and then all ticks present on the clothing are removed and identified. CO_2 also attracts hard ticks, and tick traps based on this prin-

Figure 13.6. An ovitrap used for sampling eggs of *Aedes* mosquitoes (courtesy of Paul Reiter).

Figure 13.7. Sampling ticks by dragging (courtesy of Durland Fish).

ciple also have been developed (Hair et al. 1972). Ticks also can be collected directly from animal hosts, but this method selectively samples stages and species which spend long periods of time on the hosts, and is further complicated by movement of the hosts (Varma 1993).

Fleas

Some fleas are monitored by livetrapping mammal hosts, then after anesthetizing the mammal with a volatile chemical such as ether, removing the fleas. The fleas also are anesthetized, and either drop off or are removed by combing or brushing. Other ectoparasites are collected in the same way. Flea abundance estimates are essential components of plague surveillance programs (Barnes 1982).

Triatomid bugs

Surveys to determine the presence and abundance of triatomid bugs are challenging. Few traps effectively attract triatomids, and accurate estimates of bug abundance may require the dismantling of mammal nests or houses. This destructive procedure is rarely practical in surveillance programs, and most estimates are based on manual catches during house searches for a set time period. Pinchin et al. (1981) found that manual catches were more productive if supplemented by a flushing agent such as pyrethrum (the FO method). Gurtler et al. (1993) developed an approach in which bugs that are knocked down by a controlled insecticidal application are counted (the KD method). The FO method is more sensitive, but the KD method is faster and better suited for standardization.

Tsetse

The trapping of tsetse has received considerable attention, both for estimation of fly density and for the control of flies based on the "trapping out" concept. Numerous trap designs have been tested, many of them consisting of colored panels covered with sticky substances. Others utilize natural (e.g., cattle urine) and synthetic (e.g., octenol) chemical attractants. The Vavoua trap consists of colored panels hung below a cone of white netting, and is highly effective in capturing riverine species (Laveissière and Grébaut 1990). This trap also will attract other biting muscoid flies such as stable flies (Mihok et al. 1995).

Black flies

Some kinds of light traps attract black flies. Emergence traps have been tried with limited success in stream habitats. Shipp (1985) compared silhouette traps (panels of dark-colored cloth on rigid frames baited with dry ice), sticky traps and suction traps for their effectiveness in attracting adult black flies. He found that the silhouette traps attracted Canadian black fly species. Schmidtmann (1987) developed a simulated-ear-insect trap (SEIT) that consists of a tripod-mounted Styrofoam container with dry ice to which a mock sticky ear is attached (Fig. 13.8). He found the SEIT would attract simuliid fly pests of livestock that fed preferentially on the ears.

Biting midges

Biting midges (Ceratopogonidae) have been sampled using a variety of light traps that can be targeted to specific segments of the adult population (Holbrook and Bobian 1989). An important feature for biting midge sampling is to use very fine mesh bags so these tiny flies cannot escape.

Sand flies

Some phlebotomine sand fly species are not attracted to light traps, so other methods of collection must be used. Flies can be aspirated from diurnal resting areas such as the deep buttresses of certain tropical trees. Young and Duncan (1994) reviewed the various kinds of traps that are effective for sand flies, including the Shannon trap and the Disney trap.

House flies and related Diptera

Densities of muscoid Diptera can be estimated in various ways. A device known as the Scudder fly grill has been used for this purpose for many years (Scudder 1947). A density index is established based on the number of flies present on a quadrant of a wooden grill after a set period of time. Various kinds of baited traps work well for many types of flies. Traps baited with meat scraps or similar materials can collect screwworm flies and other saprophagous Diptera. Screwworm flies can be attracted with odor-baited traps (Green et al. 1993).

Horse flies have been collected in a variety of traps, many based on visual cues. Some of these are referred to as canopy traps, the Malaise trap being a well-known example. It may be used with or without bait, and will trap various types of flying Diptera. This trap consists of large screen baffles suspended beneath a fabric pyramid. A collection container is located at the top of the pyramid. The Manning trap is similar, but uses a spherical dark-colored target suspended from the apex of the trap to attract flying insects. Catts (1970) designed a simple portable canopy trap for sampling of tabanids.

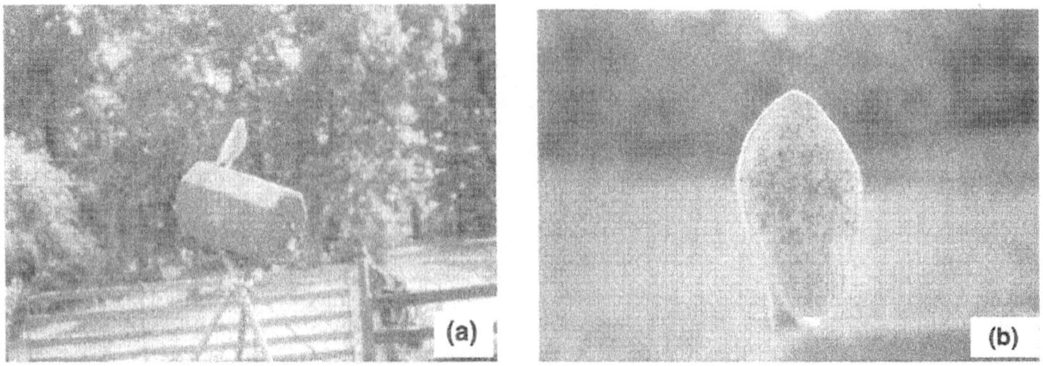

Figure 13.8. The simulated-ear-insect trap (SEIT): (a) General view of trap; (b) black flies captured on sticky surface of mock ear (photograph by Edward T. Schmidtmann).

Detection and identification of pathogens from arthropods

Surveillance for some diseases involves the systematic detection and identification of pathogens from samples of vectors. This approach ordinarily is most useful with zoonotic diseases in which pathogens may be detected in vectors before the appearance of human or animal disease cases. It typically has been taken with mosquitoborne viral diseases (Reeves et al. 1990), and also for some tickborne diseases such as Lyme disease.

For pathogen isolation, arthropods must be collected alive and then processed for testing. Accurate arthropod identification of species is extremely important, especially if the vectors for particular diseases are not well established, with retrospective vector incrimination a goal (see Chapter 6). After collecting, sorting, identifying and counting by sex, individual arthropods are placed in vials and transported singly or in pools to a central laboratory where they can be assayed for the presence of pathogens (Fig. 13.9). Specimens either are maintained alive until testing, or are frozen. Ordinarily, pools are tested for only one or a few kinds of suspected pathogens.

To detect pathogens, internal organs of individual arthropods are dissected and examined for pathogens (e.g., malarial parasites in salivary glands) or pools of arthropods are inoculated, after suitable preparation, into either susceptible laboratory vertebrate animals such as mice, guinea pigs or embryonated chicken eggs, or into a susceptible cell culture (Fig. 13.10). Arboviruses also have been isolated by inoculation of arthropod tissue into susceptible mosquitoes of the genus *Toxorhynchites* (Rosen 1981). The current trend is to avoid using vertebrate animals for this purpose because of their cost and because social pressure discourages using experimental animals where satisfactory alternatives exist.

After a suitable incubation period, evidence of infection is provided by illness or death (in animals) or by cytopathic effect (in cells). Infected tissue or cells then are transferred to other susceptible media for identification.

Traditionally, microscopic examination detected larger parasites such as plasmodia, trypanosomes or nematodes. However, in many cases specific identification cannot be obtained from microscopic preparations, and some type of immunoassay may need to be used instead.

Detailed procedures for immunoassays can be found in various publications, such as

Matthews and Burnie (1991), Calisher and Monath (1988) and Calisher and Shope (1988). Enzyme immunoassays (EIAs) are special types of immunoassays that utilize antigens or antibodies conjugated to an enzyme for detection of a positive reaction (Fig. 13.11a). Horseradish peroxidase is a commonly used enzyme for this purpose. EIAs and other immunoassays can detect unknown antigens (pathogens) or antibodies, depending upon how they are designed. For detection, unknown samples containing antigen are added to special plastic plates, then overlaid with known antibodies that are conjugated to some kind of detection system. If the antibody-detector conjugate binds to the unknown antigen, it will remain in the test plates after rinsing, and will produce a color reaction when a substrate is added, thereby identifying the pathogen. This color reaction indicates a positive test for the particular known antibody. In a variation of this design, called a "sandwich" assay, a known "capture" antibody first is added to the plates, overlaid with unknown antigen, then further overlaid with a different known antibody-detector complex.

Older immunologic assays still are used: the **neutralization test**, the **hemagglutination-inhibition test**, the **complement-fixation test**, the **agar gel immunodiffusion test** and the **fluorescent antibody test** (Matthews and Burnie 1991).

The significance of pathogens isolated and identified from arthropods varies considerably according to the particular disease being monitored, and the ecological circumstances. Human pathogens may be isolated from arthropods that are not primary vectors (do not come into direct contact with human or animal hosts of interest), and thus may not have a direct relationship to disease outbreaks in these hosts. Because of that association, some surveillance systems may restrict testing of arthropods to known primary vectors.

There have been attempts to correlate pathogen isolation rates with human and animal disease outbreaks. Longshore et al. (1960) and Meyers et al. (1960) were unable to demonstrate a direct correlation between virus isolation rates in the primary vector, *Culex tarsalis*, and human cases of western equine encephalomyelitis for the period 1953–1957. However, Reeves (1990) points out that in each year studied, the first isolations of WEE were detected in mosquitoes 1–2 months before the first human case.

Estimates of natural malarial infections in anopheline vectors may be based on the percentage of females that have sporozoites in salivary glands (called the sporozoite rate), or malarial oocysts on the stomach walls (called the oocyst rate) (Gilles 1993). This process often was tedious, requiring dissections of mosquitoes, until the development of several sensitive antibody- and nucleic acid-based tests to detect and identify sporozoites (Wirtz and Burkot 1991). Malarial infection rates in mosquito populations tend to be low even in areas where transmission is active (although not as low as viral infection rates in arbovirus vectors). Other entomological components of a malarial survey are estimates of mosquito density, human biting rate and longevity (Gilles 1993). In contrast to surveys, surveillance for malaria ordinarily is done in non-endemic areas, and depends almost entirely on passive case detection. In endemic areas, surveys involving active case detection are more common (Reisen 1994).

It is difficult to isolate and identify pathogens from ticks; specialized laboratories usually perform such tasks. **Direct and indirect immunofluorescence assays** (DFA and IFA) usually detect spirochetes causing Lyme disease in individual ticks. These tests resemble EIAs, except that instead of using an enzyme-antibody complex which produces a color change for detection, a specific antibody is labeled with a fluorescent material such as fluorescein isothiocyanate so that positive specimens will give off a greenish glow under special lighting for microscopic examination. Although the IFA test requires one more step than does the DFA,

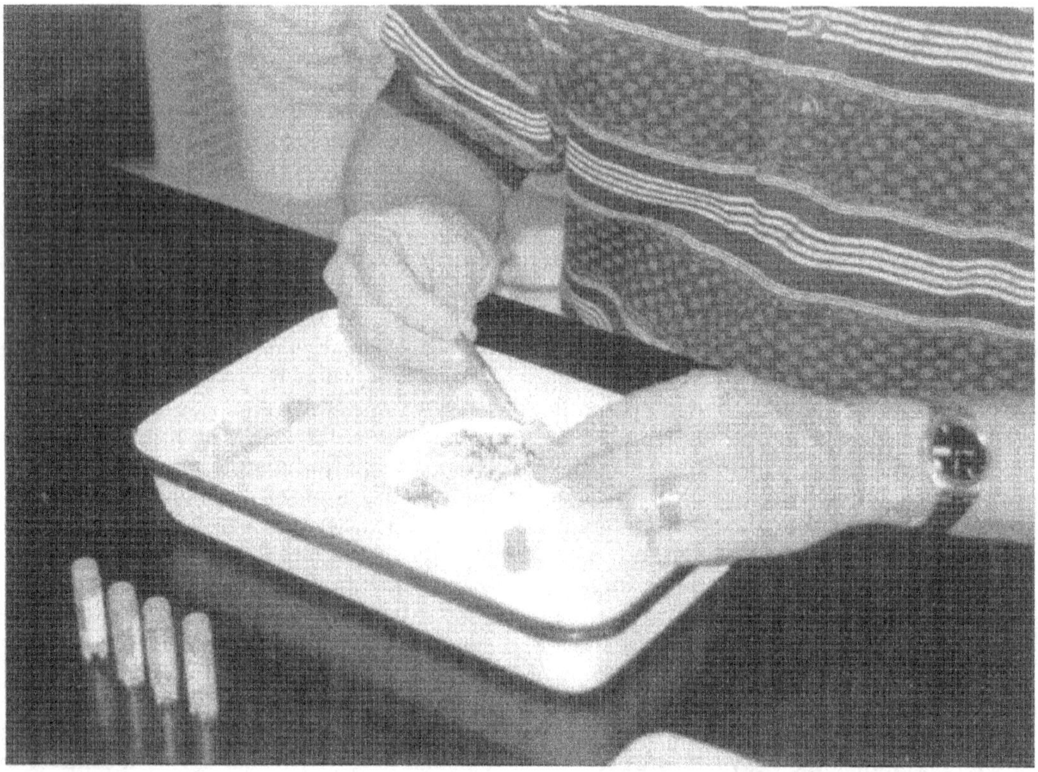

Figure 13.9. Pooling mosquitoes for isolation of arboviruses (photograph by Bruce F. Eldridge).

it is preferred because it produces a stronger reaction in positive samples, and because the reagents are readily available commercially (Kramer 1994).

A classic means of detecting pathogens in vertebrate host blood is to feed noninfected (clean) susceptible vectors on hosts, then after a suitable incubation period examine the vectors for pathogens. This technique is called **xenodiagnosis**, and before the advent of modern diagnostic tools was used frequently to detect trypanosomes in suspected Chagas disease patients. It also has been used to diagnose cases of tickborne relapsing fever (Davis 1956).

Vertebrate Host Surveillance

For diseases that involve ectoparasite vectors such as plague, it often is more efficient to test the vertebrate hosts (usually rodents) for infection rather than the fleas found on the vertebrates. Carnivorous mammals can become infected with plague bacilli after eating infected rodents, so carnivores also are often sampled for plague antibodies (Barnes 1982). Domestic and feral cats recently have been recognized as important hosts of plague bacilli, and serosurveys of these animals may provide information about plague activity (Chomel et al. 1994).

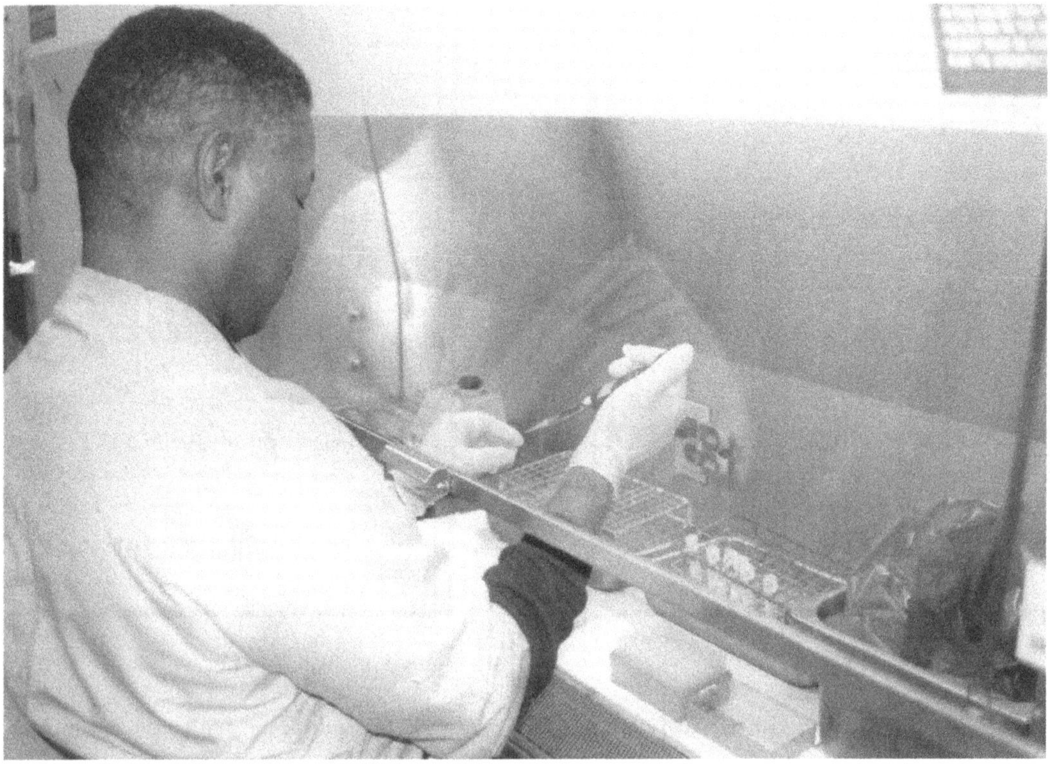

Figure 13.10. Isolation of arboviruses using cell cultures (photograph by Bruce F. Eldridge).

Naturally occurring hosts

Sampling natural reservoir hosts most frequently is done in connection with plague surveillance studies. Surveillance of plague transmission from sylvatic foci in the western USA is based on routine searches for dead and dying rodents (especially ground squirrels, prairie dogs, marmots, wood rats and chipmunks), and culturing plague bacilli from dead or dying animals (Fig. 13.12). In active surveillance programs, healthy animals are trapped and bled for the same purpose. In most instances, fleas are removed from animals, counted and identified. Most programs establish a network among forest rangers and other outdoor workers to report large die-offs of wild rodents.

For certain zoonotic arbovirus diseases with avian hosts, wild birds may be trapped and bled to detect evidence of infection.

Vertebrate animals as sentinels

Vertebrate animals frequently are employed as sentinels for arboviruses. Their movement is limited so there is less labor in using sentinel animals than trapping and bleeding wild animals; also, more specific information is provided on time and place of infection (Reisen et al. 1994). The usual sentinel animal for arboviruses with ornithophilic vectors is the domes-

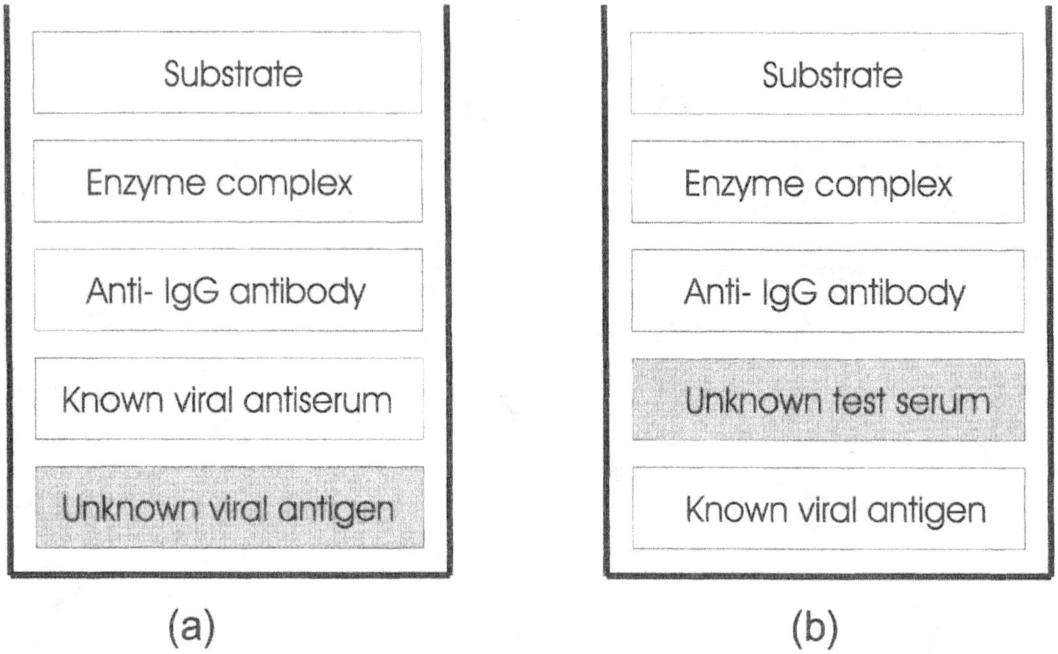

Figure 13.11. Generalized diagram for an indirect enzyme immunoassay (EIA) for detection of (a) antigen, and (b) antibodies.

tic chicken. Chickens tolerate a wide range of weather conditions and adult birds rarely show signs of infection. However, they are highly sensitive to some arbovirus infections and readily produce antibodies. Because infections rarely result in high viremias, it is unlikely that sentinels will infect mosquitoes. Flocks of 5–10 birds are placed in special hutches, and small blood samples are taken periodically from the chickens' combs and placed on filter papers for later immunological testing (Fig. 13.13). Samples usually are taken weekly or biweekly during the mosquito season. Using sentinel chickens has advantages over testing mosquito pools for virus. The former costs less and provides evidence of infective mosquitoes, whereas the latter provides evidence only of infected mosquitoes. Theoretically, then, the occurrence of antibodies in chicken flocks should be the most reliable predictors of human disease outbreaks short of actual cases. However, this premise assumes that the vector species infecting sentinels frequently contacts susceptible human hosts. Another advantage to using sentinel chickens is that tests for viral antibodies can be completed in a few days, whereas isolation and identification of viruses from mosquito pools can take a week or more. This time may be reduced in the future with tests such as PCR, but such tests have yet to be evaluated for surveillance programs. The value of chickens as arbovirus sentinels is controversial. Most studies have shown that seroconversions in sentinels precedes the appearance of human cases by

Figure 13.12. Collection of small rodents for plague testing (courtesy of Charles R. Smith, California Department of Health Services).

several weeks to a month or more. However, Crans (1986) reported that sentinel chickens did not provide advance warning of eastern equine encephalomyelitis cases in New Jersey in 1984.

Sentinel herds of livestock have helped monitor transmission of arthropodborne animal pathogens. In contrast to serologic surveys that provide only historical information of past infections, sentinels detect pathogens and recent transmission. A cattle sentinel herd program in the Caribbean Basin documented transmission and over 300 bluetongue virus isolates (Gibbs et al. 1992).

Serologic surveys of dogs have detected leishmaniasis in various parts of the world. However, their use for making quantitative estimates of leishmanial activity has been limited because of the relatively high proportion of asymptomatic canine cases that escaped detection. Sensitive detection methods such as PCR and immunoblotting now are available to determine more accurately the degree of infection in dogs (Berrahal et al. 1996).

Methods of antibody detection

Vertebrate animals infected with many arthropodborne pathogens develop antibodies that persist for several months (Calisher et al. 1986). The test used for detection depends upon the class of antibody involved. With modifications, the same immunological tests described above for the identification of pathogens (antigens) also may be used for identification of an-

Figure 13.13. A hutch for maintenance of a flock of sentinel chickens (photograph by William K. Reisen).

tibodies against those antigens. For hemagglutinating antibodies, there is the hemagglutination-inhibition (HI) test; for neutralizing antibodies, the neutralization test. Currently, many programs detect antibodies using some form of enzyme immunoassay (Hildreth and Beaty 1983). There are several factors to be considered in deciding which test to employ. The HI test takes a relatively long time to perform, and HI antibodies can cross-react with antigenically-related pathogens (Calisher et al. 1986). Although enzyme immunoassays require specific reagents, many are available from commercial sources for the most common arthropodborne pathogens, particularly arboviruses. Calisher et al. (1986) found these tests are highly specific and can differentiate between IgM antibody (less persistent, and thus indicative of recent or chronic infections) and IgG antibody (more persistent, and thus indicative of historic past infections).

An example of an enzyme immunoassay for the detection of immunoglobulins in sentinel chicken sera is that used in the California Vectorborne Disease Surveillance System (Fig. 13.11b). For a detailed description of this test, the reader should refer to Reisen et al. (1994). In summary, 96-well plates are coated with known viral antigen prepared from Vero cell cultures. Plates are blocked with 2% casein to minimize nonspecific binding and false positives. Later, diluted sera to be tested are added to the wells. After rinsing, biotinylated goat anti-chicken IgG conjugate is added. After further washing, a peroxidase-labeled avidin-biotin complex is added to the wells. Finally, after suitable incubation, a substrate for the peroxidase enzyme is added and the plates are examined using spectrophotometry. If the chicken sera contain viral antibody, it will become fixed to the antigen. If all unfixed antibody is removed after washing, the test is negative. If the test is positive, the antibody-peroxidase complex will have bonded to the antibody-antigen complex, so when the substrate is added a visible color reaction will result.

Case Detection

The careful diagnosis and recording of cases of vectorborne diseases in humans and livestock should be the focal point of any surveillance system. Although little advance warning of disease outbreaks may result from the detection of cases, information on the geographic location of index cases, and the geographic spread of cases over time, may be valuable. Further, cases should alert public health and agricultural officials to the near certainty of a public health or agricultural emergency and the need to implement emergency response plans. For many diseases, such as malaria in non-endemic areas, or vesicular stomatitis epizootics in US livestock, case detection may be the sole component of a surveillance system.

Case definitions and reporting criteria

Human and animal diseases, like all biological organisms, are named in accordance with rules of zoological, botanical and microbiological nomenclature. The names of infectious human diseases follow the International Classification of Diseases, and the pathogens causing the diseases follow nomenclature in Bergey's Manual of Determinative Bacteriology (Holt 1994). There are specific case definitions for every infectious human disease. These definitions may include both clinical findings and laboratory test results. Case definitions change from time to time, with the consequent change in the apparent incidence of the disease involved. Lyme disease in the USA is a recent example of this. Case definitions for infectious diseases under surveillance have been published (Centers for Disease Control and Prevention 1997).

Methods of diagnosis, confirmation and reporting

Although the diagnosis of infectious diseases had advanced in recent years, accurate diagnoses for some vectorborne diseases still are difficult, time-consuming and expensive. In many instances, local hospitals and clinics will not have the necessary equipment and reagents, and tissue and blood samples will have to be submitted to a centralized testing facility for diagnosis. In some cases, these facilities may be located long distances from the point where samples were collected. A key factor in the submission of patient samples for diagnosis is the awareness and interest of treating physicians. In the USA, reportable communicable diseases are published in the Morbidity and Mortality Weekly Report, produced by CDC, but printed, sold and distributed by the Massachusetts Medical Society. The Weekly Epidemiological Record, published by WHO, reports cases on a worldwide basis. Both of these publications also contain summaries and interpretive articles.

Problems with case detection

All reportable diseases should be diagnosed promptly and accurately, and all confirmed cases should be duly reported to public health and veterinary agencies responsible for assembling and maintaining statistics on such diseases. Unfortunately, surveillance and reporting systems are never perfect; consequently, most vectorborne diseases are underreported (Fig. 13.14). Diagnosis and confirmation of most diseases is a 2-step process. For example, arboviral encephalitis may be diagnosed first as aseptic meningitis on the basis of local testing that shows spinal fluid from a patient does not contain bacteria or fungi. A more specific

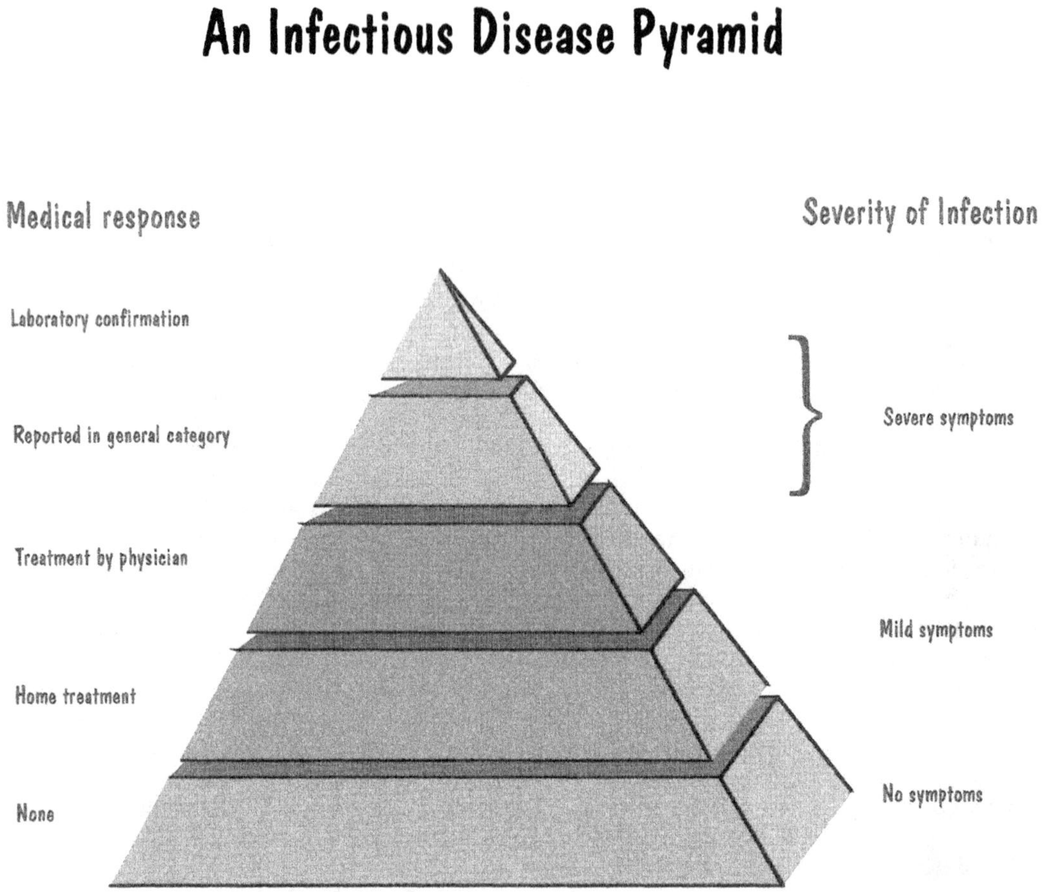

Figure 13.14. A hypothetical infectious disease pyramid to illustrate the range of host reactions to infection (left) and medical response (right).

diagnosis would require submission of blood or spinal fluid by the medical facility treating the patient to a centralized laboratory. Because at this point detailed information of the etiology of the disease often is not likely to change the course of treatment, physicians will not have a strong inducement to submit samples for specific diagnoses. However, the future development of specific antiviral agents may increase the demand for specific diagnoses. Previous studies have estimated that 5–10% of the cases diagnosed as aseptic meningitis actually are caused by an arthropodborne virus (Benenson 1995).

Misdiagnosis of animal pathogens could be catastrophic. Failure to differentially diagnose vesicular stomatitis from foot and mouth disease could delay disease control and spread the latter, highly contagious disease even farther.

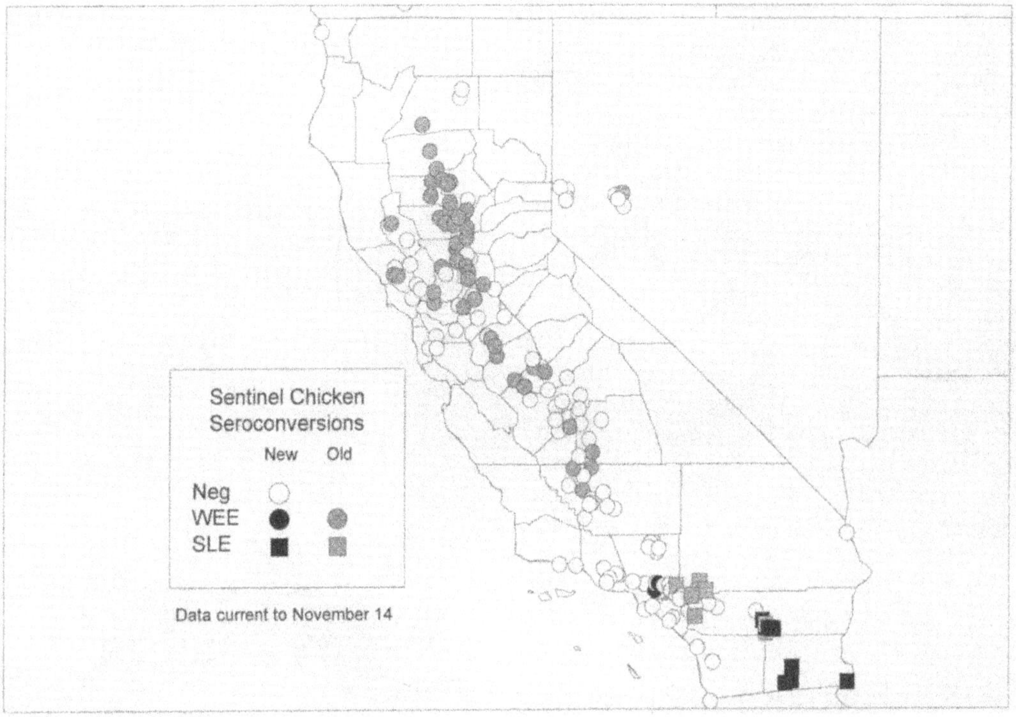

Figure 13.15. Map showing serocoversion of sentinel chicken flocks to arboviruses in California.

One of the problems with case detection of emerging diseases is the lack of accepted test standards. Craven et al. (1996) reported a low concordance of test results for Lyme disease even between CDC laboratories and consulting academic reference centers. Considerable coordination and cooperation is required to solve these problems.

THE FUTURE OF SURVEILLANCE

Experts agree that national and worldwide surveillance of infectious diseases, including those that are vectorborne, is essential to the control of emerging infections. As the world population grows and becomes increasingly mobile, there will be an increasing likelihood of vectorborne disease outbreaks. Trade will increase animal movement with concomitant increased risk of entry of exotic pathogens. This most likely will lead to the greater use of vectorborne disease surveillance systems. These systems will benefit from scientific and technical improvements in diagnosis, knowledge of vectorborne disease systems, and reporting. As development and improvements in computerized **geographic information systems** (systems that display data graphically in multilayer maps) take place, reporting and graphic representations of epidemiological data will become more accurate, rapid and useful (Clarke et al. 1996). In the future, it should be possible to follow seasonal and year-to-year trends in disease incidence, and by overlaying geographic data on climate, vegetation and other factors, make valuable predictions on the potential for future outbreaks of disease. Information on vector-pathogen-host interactions will add to this greater predictability of risk.

Electronic communication has improved surveillance activities considerably, especially international efforts, through greater and more rapid exchange of epidemiologic information. Vacalis et al. (1995) reported that since 1985, data on notifiable diseases from the USA has been transmitted to CDC as part of the National Electronic Telecommunications System for Surveillance (NETSS). Several electronic mail lists exchange information on vectorborne diseases. The development of the World Wide Web will provide further opportunities to report data from surveillance systems. In the USA, data generated by the California Vectorborne Disease Surveillance System are posted weekly to a World Wide Web site featuring maps of seroconversions of sentinel chickens to arboviruses and isolations of viruses from mosquitoes (Fig. 13.15).

Surveillance for arthropodborne diseases demands money and labor. Programs lacking the support of government agencies, medical and veterinary care givers and the public will be of limited use in preventing disease outbreaks in humans and domestic animals. However, surveillance activities encompassing modern procedures for testing and reporting are an essential component of programs to reduce the number of disease cases. Development and operation of such programs will continue to challenge medical entomologists and other public health and agricultural workers worldwide.

REFERENCES

Barnes, A.M. 1982. Surveillance and control of bubonic plague in the United States. Pp. 237–270 *in* M.A. Edwards and U. McDonnel (eds.), Animal disease in relation to animal conservation., Vol. 50. Symposium of the Zoological Society of London.

Benenson, A.S. 1995. Control of communicable diseases manual, 16th ed. American Public Health Association, Washington, DC. 577 pp.

Berrahal, F., Mary, C., Roze, M., Berenger, A., Escoffer, K., Lamouroux, D. and Duncan, S. 1996. Canine leishmaniasis: identification of asymptomatic carriers by polymerase chain reaction and immunoblotting. Am. J. Trop. Med. Hyg. **55:**273–277.

Calisher, C.H., Freemount, H.N., Vesely, W.L., El-Kafrawi, A.O. and Al-Deen Mahmud, M.I. 1986. Relevance of detection of immunoglobulin M antibody response in birds used for arbovirus surveillance. J. Clin. Microbiol. **24:**770–774.

Calisher, C.H. and Monath, T.P. 1988. Togaviridae and Flaviviridae: the alphaviruses and flaviviruses. Pp. 414–434 *in* E.H. Lennette, P. Halonen and F.A. Murphy (eds.), Laboratory diagnosis of infectious diseases, principles and practices. Springer-Verlag, New York.

Calisher, C.H. and Shope, R.E. 1988. Bunyaviridae: the bunyaviruses. Pp. 626–646 *in* E.H. Lennette, P. Halonen and F.A. Murphy (eds.), Laboratory diagnosis of infectious diseases, Principles and practices. Springer-Verlag, New York.

Catts, E.P. 1970. A canopy trap for collecting Tabanidae. Mosq. News **30:**472–474.

Centers for Disease Control and Prevention. 1997. Case definitions for infectious conditions under public health surveillance. Morbid. Mortal. Weekly Rpt. **46:**1–55.

Chomel, B., Jay, M.T., Smith, C.R., Kass, P.H., Ryan, C.P. and Barrett, L.R. 1994. Serological surveillance for plague in dogs and cats, California, 1970-1991. Comp. Immunol. Microbiol. Infect. Dis. **17:**111–123.

Clarke, K.C., McLafferty, S.L. and Tempalski, B.J. 1996. On epidemiology and geographic information systems: a review and discussion of future directions. Emerg. Infect. Dis. **2:**85–92.

Collins, R.C. 1979. Onchocerciasis transmission potential of four species of Guatemalan Simuliidae. Am. J. Trop. Med. Hyg. **28:**72–75.

Crans, W.J. 1986. Failure of chickens to act as sentinels during an epizootic of eastern equine encephalitis in southern New Jersey. J. Med. Entomol. **23:**626–629.

Craven, R.B., Quan, T.J., Bailey, R.E., Dattwyler, R., Ryan, R.W., Sigal, L.H., Steere, A.C., Sullivan, B., Johnson, B.J.B., Dennis, D.T. and Gubler, D.J. 1996. Improved serodiagnostic testing for Lyme disease: results of a multicenter serologic evaluation. Emerg. Infect. Dis. **2:**136–140.

Curran, M., Harvey, B., Crerar, S., Oliver, G., D'Souza, R., Myint, H., Rann, C. and Andrews, R. 1997. Australia's notifiable diseases status, 1996. Communicable Diseases Intelligence **21:**281–316.

Davis, G.E. 1956. The identification of spirochetes from human cases of relapsing fever by xenodiagnosis with comments on local specificity of tick vectors. Exp. Parasitol. **5**:271–275.

Eldridge, B.F. 1987. Strategies for surveillance, prevention, and control of arbovirus diseases in western North America. Am. J. Trop. Med. Hyg. **37**:77S–86S.

Gibbs, E.P.J., Homan, E.J. and Team, I.B. 1992. The sentinel herd as a tool for epidemiological studies for bluetongue virus infections in the Caribbean and Central America. Pp. 90–98 *in* T.E. Walton and B.I. Osburn (eds.), Bluetongue, African horse sickness and related orbiviruses. CRC Press, Boca Raton, Florida.

Gilles, H.M. 1993. Epidemiology of malaria. Pp. 124–163 *in* H.M. Gilles and D.A. Warrell (eds.), Bruce-Chwatt's essential malariology, 3rd ed. Edward Arnold, London.

Green, C.H., Hall, M.J.R., Fergianai, M., Chirico, J. and Husni, M. 1993. Attracting adult New World screwworm, *Cochliomyia hominivorax*, to odour-baited targets in the field. Med. Vet. Entomol. **7**:59–65.

Gurtler, R.E., Schweigmann, N.J., Cecere, M.C., Chuit, R. and Wisnevesky-Colli, C. 1993. Comparison of two sampling methods for domestic populations of *Triatoma infestans* in north-west Argentina. Med. Vet. Entomol. **7**:238–242.

Gurtler, R.E., Wisnivesky-Colli, C., Solarz, C., Lauricella, M.A. and Bujas, M.A. 1987. Dynamics of transmission of *Trypanosoma cruzi* in a rural area of Argentina. II. Household infection patterns in children and dogs relative to the density of infected *Triatoma infestans*. Bull. Panam. Hlth. Organiz. **21**:280–292.

Hair, J.A., Hoch, A.L., Barker, R.W. and Sempter, P.J. 1972. A method of collecting nymphal and adult lone star ticks, *Amblyomma americanum* (L.) (Acarina: Ixodidae) from woodlots. J. Med. Entomol. **9**:153–154.

Hess, A.D. and Hayes, R.O. 1967. Seasonal dynamics of western encephalitis virus. Am. J. Med. Sci. **253**:333–348.

Hildreth, S.W. and Beaty, B.J. 1983. Application of enzyme immunoassays (EIA) for the detection of La Crosse viral antigens in mosquitoes. Pp. 303–312 *in* C.H. Calisher and W.H. Thompson (eds.), California serogroup viruses. Alan R. Liss, Inc., New York.

Holbrook, F.R. and Bobian, R.J. 1989. Comparison of light traps for monitoring adult *Culicoides variipennis* (Diptera: Ceratopogonidae). J. Am. Mosq. Contr. Assoc. **5**:558–562.

Holt, J.G. 1994. Bergey's manual of determiniative bacteriology, 9th ed. Williams & Wilkins, Baltimore, Maryland. 787 pp.

Kramer, V.L. 1994. Lyme disease in California: an information handbook and interagency guidelines. California Mosquito and Vector Control Association, Elk Grove. 31 pp.

Laveissière, C. and Grébaut, P. 1990. Recherche sur les pièges à glossines (Diptera: Glossinidae). Mise au point d'un modèle économique: le piège "Vavoua." Trop. Med. Parasitol. **41**:185–192.

Lederberg, J., Shope, R.E. and Oaks, S.C., Jr. 1992. Emerging Infections. National Academy Press, Washington, DC. 294 pp.

Longshore, W.A., Jr., Lennette, E.H., Peters, R.H., Loomis, E.C. and Meyers, E.G. 1960. California encephalitis surveillance program: relationship of human morbidity and virus isolation from mosquitoes. Am. J. Hyg. **71**:389–400.

Matthews, R.C. and Burnie, J.P. 1991. Handbook of serodiagnosis in infectious diseases. Butterworth-Heinemann, Oxford, England. 220 pp.

Mattingly, P.F. 1969. The biology of mosquito-borne disease. American Elsevier, New York. 184 pp.

Meyers, E.G., Loomis, E.C., Fujimoto, F.Y., Ota, M.I. and Lennette, E.H. 1960. California encephalitis surveillance program: mosquito virus relationships. Am. J. Hyg. **71**:368–377.

Mihok, S., Kang'ethe, E.K. and Kamau, G.K. 1995. Trials of traps and attractants for *Stomoxys* spp. (Diptera: Muscidae). J. Med. Entomol. **32**:283–289.

Moore, C.G., Cline, B.L., Ruiz-Tiben, E., Lee, D., Romney-Joseph, H. and Rivera-Correa, E. 1978. *Aedes aegypti* in Puerto Rico: environmental determinants of larval abundance and relationship to dengue virus transmission. Am. J. Trop. Med. Hyg. **27**:1225–1231.

Mount, G.A., Haile, D.G. and Daniels, E. 1997. Simulation of blacklegged tick (Acari: Ixodidae) population dynamics and transmission of *Borrelia burgdorferi*. J. Med. Entomol. **34**:461–484.

Olson, J.G., Reeves, W.C., Emmond, R.W. and Milby, M.M. 1979. Correlation of *Culex tarsalis* population indices with the incidence of St. Louis encephalitis and western equine encephalomyelitis in California. Am. J. Trop. Med. Hyg. **28**:335–343.

Pearson, J.E., Gustafson, G.A., Shafer, A.L. and Alstad, A.D. 1992. Distribution of bluetongue in the United States. Pp. 128–139 *in* T.E. Walton and B.I. Osburn (eds.), Bluetongue, African horse sickness and related orbiviruses. CDC Press, Boca Raton, Florida.

Pinchin, R., Fanara, D.M., Castleton, C.W. and Oliveira Filho, A.M. 1981. Comparison of techniques for detection of domestic infestations with *Triatoma infestans* in Brazil. Trans. Royal Soc. Trop. Med. Hyg. **75**:691–694.

Reeves, W.C. 1990. Epidemiology and control of mosquito-borne arboviruses in California, 1943–1987. California Mosquito and Vector Control Association, Sacramento. 508 pp.

Reeves, W.C. and Milby, M.M. 1980. Assessment of the relative value of alternative approaches for surveillance and prediction of arboviral activity. Proc. Calif. Mosq. Vect. Contr. Assoc. **48:**3–7.

Reeves, W.C., Milby, M.M. and Reisen, W.K. 1990. Development of a statewide arbovirus surveillance program and models of vector populations and virus transmission. Pp. 431–459 *in* W.C. Reeves (ed.), Epidemiology and control of mosquito-borne arboviruses in California, 1943–1987. California Mosquito and Vector Control Association, Sacramento.

Reisen, W.K. 1994. Guidelines for malaria surveillance and control in California. California Mosquito and Vector Control Association, Sacramento. 14 pp.

Reisen, W.K., Presser, S.B., Lin, J., Enge, B., Hardy, J.L. and Emmons, R.W. 1994. Viremia and serological responses in adult chickens infected with western equine encephalomyelitis and St. Louis encephalitis viruses. J. Am. Mosq. Contr. Assoc. **10:**549–555.

Rosen, L. 1981. The use of *Toxorhynchites* mosquitoes to detect and propogate dengue and other arboviruses. Am. J. Trop. Med. Hyg. **30:**177–183.

Schmidtmann, E.T. 1987. A trap/sampling unit for ear-feeding black flies. Canad. Entomol. **119:**747–750.

Scudder, H.I. 1947. A new technique for sampling the density of housefly populations. Publ. Hlth. Rep. **62:**681–686.

Service, M.W. 1993. Mosquito ecology. Field sampling methods, 2nd ed. Chapman & Hall, London. 988 pp.

Shipp, J.L. 1985. Comparison of silhouette, sticky, and suction traps with and without dry-ice bait for sampling black flies (Diptera: Simuliidae) in Central Alberta. Canad. Entomol. **117:**113–117.

Vacalis, T.D., Bartlett, C.L.R. and Shapiro, C.G. 1995. Electronic communication and the future of international public health surveillance. Emerg. Infect. Dis. **1:**34–35.

Varma, M.R.G. 1993. Ticks and mites (Acari). Pp. 597–658 *in* R.P. Lane and R.W. Crosskey (eds.), Medical insects and arachnids. Chapman & Hall, New York.

Walton, T.E., Tabachnick, W.J., Thompson, L.H. and Hall, F.R. 1992. An entomologic and epidemiologic perspective for bluetongue regulatory changes for livestock movement from the USA and observations on bluetongue in the Caribbean Basin. Pp. 952–960 *in* T.E. Walton and B.I. Osburn (eds.), Bluetongue, African horse sickness and related orbiviruses. CRC Press, Boca Raton, Florida.

Wharton, M., Chorba, R.L., Vogt, V.L., Morse, D.L. and Buehler, J.W. 1990. Case definitions for public health surveillance. Morbid. Mortal. Weekly Rpt. **39:**1–43.

Wirtz, R.A. and Burkot, T.R. 1991. Detection of malarial parasites in mosquitoes. Pp. 77–106 *in* K.F. Harris (ed.), Advances in Disease Vector Research, Vol. 8. Springer-Verlag, New York.

Young, D.G. and Duncan, M.A. 1994. Guide to the identification and geographic distribution of *Lutzomyia* sand flies in Mexico, the West Indies, Central and South America (Diptera: Psychodidae). Associated Publishers, American Entomological Institute, Gainesville, Florida. 881 pp.

Chapter 14

Management of Arthropodborne Diseases by Vector Control

DONALD E. WEIDHAAS AND DANA A. FOCKS
USDA Center for Medical, Agricultural and Veterinary Entomology, Gainesville, Florida

THE DEVELOPMENT OF VECTOR CONTROL METHODS

The discovery in the late 1800s that arthropods transmit a variety of pathogenic microorganisms to vertebrate animals, including humans, soon led to the development of means to reduce disease transmission by controlling arthropod vectors. Some of these methods already had been developed for the management of plant pests, and were modified to control vectors of animal diseases. By the early 1900s, several of the approaches to vector management had been developed. Such approaches included the reduction of contact between arthropod vectors and vertebrate hosts (e.g., window and door screens, arthropod repellents, bednets), the application of chemical pesticides, management of the environment to reduce or eliminate arthropod vectors and the use of biological control agents. Some programs combined several approaches. This combination is a fundamental principle of modern programs of Integrated Pest Management (IPM) or Integrated Vector Management (IVM). This approach for mosquito control has been covered by Laird and Miles (1983, 1985).

At the beginning of the 20th Century, vector management techniques were crude by today's standards, and lacked many of the materials and approaches now available. New discoveries in arthropod physiology and microbiology led to more efficient and safer materials. Greater knowledge of the ecology of vector-host-pathogen systems led to improved approaches to vector management. Table 14.1 summarizes in chronological order the most significant developments in vector control.

After a century or more of research, a wealth of information is available on the life cycles of both disease organisms and their vectors, as well as the many variables that influence disease prevalence and incidence. This information, combined with new approaches for vector management (such as attractants, repellents, traps, biorational pesticides and environmental manipulations) has led to the development of comprehensive IVM programs to control arthropodborne diseases. Modern powerful and inexpensive computers, combined with new computer programs and mathematical techniques, have allowed models for the simulation of populations and disease transmission. Such models promote a better understanding of the transmission and prevalence of diseases. They

B.F. Eldridge and J.D. Edman (eds.), Medical Entomology, Revised Edition, 539–563.

Table 14.1. Historical developments in vector control.

Year(s)	Development	Remarks
1870–1910	Establishment of the relationship between arthropods and infectious disease organisms	Filariasis, malaria, yellow fever, dengue, Texas cattle fever, etc.
1870–1910	Development of vector control methods and protection from exposure to vectors	Bednets, window screens, repellents, natural pesticides, environmental manipulations, etc.
1910–present	Establishment of local mosquito and vector abatement agencies	
1910–present	Establishment of federal and state vector-borne disease control programs in USA and elsewhere	Example: malaria control in USA
1910–1960s	Insecticidal and acaricidal dips for control of ectoparasites on cattle	Used to control several tickborne diseases of cattle, including anaplasmosis and babesiosis
1930s	First synthetic pesticides	DDT
1940–1960	Development of new classes of synthetic pesticides	Organophosphates, carbamates, synthetic pyrethroids
1950s	Toxic baits for vector control	House flies, eye gnats
1960s	Development of traps for vector control	Tsetse, stable flies, horse flies
1960s	Development of sterile male release programs	Screwworm eradication programs
1960s	Research on genetic control of vectors	
1960s	Development of insect growth regulators	Methoprene, other JH-mimics
1960–present	Development of integrated vector management programs	
1980s	Development of microbial pesticides	Bti and *B. sphaericus*
1980s	Simulation models to aid vector control programs	
1990s	Emphasis on community-based vector control programs	
1990s	Research on genetic manipulation of vectors using molecular biology techniques	

also can assist in predicting outcomes of various management strategies against both the vector and the disease.

In recent years, there has been increased emphasis on public education and the organization of small community-based programs to control vectorborne diseases, especially in developing countries. This emphasis stems from the realization that large government-sponsored, vertically-organized programs cannot provide all of the resources necessary for sustained protection and control.

METHODS OF VECTOR CONTROL AND MANAGEMENT

Management of the Environment

Completely unaltered and unmanaged environments can be the source of arthropod vectors and the pathogens they transmit, but such environments are exceptional and generally exist only in remote areas of the world where there is little contact between arthropods and people. However, disturbed, changing and poorly managed environments often are significant sources of vectorborne diseases. The regrowth of old fields in the northeastern USA may be an important factor, along with the explosion of deer populations, in the emergence of Lyme disease and other tickborne diseases of humans as public health problems. Dams, reservoirs, agricultural irrigation systems, water retention ponds, waste treatment facilities and urban centers without water supply or waste removal systems frequently generate vector and disease problems. Historically, there have been many attempts to prevent or ameliorate vectorborne disease problems by environmental management. However, a general lack of planning, insufficient resources and the magnitude and complexity of the problems have limited the effectiveness of this approach. Construction of dams that result in the production of black flies in rapids downstream of the structures or mosquitoes in the shallow margins of the reservoir above illustrate how artificial structures create vector problems as does poor planning for the prevention of such easily anticipated problems. For a thorough examination of the relationship between human activities and the occurrence of vectorborne disease problems, the reader should consult Service (1989).

Prior to World War II, environmental management techniques were used extensively. Ault (1994) reviewed many of these methods commonly used to control vectors of yellow fever and malaria, which included cleaning of streets and yards, improved garbage collection, thorough screening of doors and windows of hospitals and construction of water supply and waste removal systems. The spectacular success of DDT and other synthetic chemical pesticides after World War II resulted in a de-emphasis of environmental management techniques. However, the combined problems of vector resistance to insecticides and concerns about the safety and environmental impact of chemical pesticides renewed interest in the integration of all available control methods into comprehensive vector and disease management schemes.

Modern vector management programs rely heavily on environmental manipulation, also referred to as **source reduction** or **physical control**. Some of the pre-World War II methods, such as the draining of wetlands for mosquito control, have become obsolete, but many others continue to form the basis of vector control programs for a variety of arthropodborne diseases. There are restrictions placed on environmental manipulations in most industrialized countries, and projects nearly always require careful planning and approval. Environmental manipulations for vector management must be cost-effective, and must consider other environmental values, such as habitat improvement for wildlife.

Aquatic environments

Many arthropods of medical importance have aquatic immature stages. The most important of these are mosquitoes, followed by black flies and *Culicoides* midges. Because the larval biology of these major groups differs, the types of environmental management also differ. For mosquitoes, the management of water depth and emergent vegetation remains an important principle for limiting the area suitable for mosquito larvae to develop. Many of the present techniques for mosquito control based on these principles were developed for the control of malaria vectors in reservoirs constructed by the Tennessee Valley Authority (TVA) in the 1920s and 1930s. Manipulation of these 2 factors is very difficult in wetlands created as waterfowl habitat because some of the same factors that result in suitable habitat for waterfowl also are conducive to mosquito breeding (gradually sloping margins, emergent vegetation, extensive areas of shallow water). Flooded pastures and rice fields present similar problems. Since these habitats are artificially maintained, proper timing of flooding may be a suitable approach to minimize problems with vector arthropods.

Water and vegetation management is accomplished more easily in irrigation systems consisting of ditches and drains. Modern methods of maintaining grades in irrigated fields have reduced mosquito habitat in irrigation tail waters. On the other hand, interplanting and no-till agricultural methods may have the opposite effect because they create standing irrigation water with emergent vegetation.

There may be little need for management of permanently flooded habitats if predators of mosquito larvae are plentiful. Walton et al. (1990) determined that pesticide applications were necessary for the control of *Culex tarsalis* larvae in experimental ponds in southern California only early in the season.

Completely different strategies are needed for mosquitoes such as *Aedes aegypti* that breed in artificial containers like water jars, tin cans and discarded tires. Mosquito control may be accomplished by eliminating such aquatic habitats by removing containers or tipping them over to prevent water from collecting in them. In the case of water containers used to supply household needs, adequate covers may prevent access by ovipositing female mosquitoes. Many community participation programs reduce or eliminate small water-holding containers.

Disposal of used tires is a worldwide problem, and thousands have accumulated in dump sites where they serve as breeding places for mosquitoes such as *Aedes albopictus*. Since such piles are nearly impossible to treat adequately with insecticides, the only sustainable approach to mosquito control is to eliminate the piles.

Coastal saltmarshes and mangrove swamps present special problems for mosquito control. Older practices involving grid ditches, impoundments, or the installation of tide gates to restrict inflow of seawater are now considered obsolete and destructive of estuarine habitats. Newer approaches are favored that take advantage of the flushing action of tides to minimize mosquito breeding (Resh and Balling 1983, Fig. 14.1).

Terrestrial environments

Environmental management is essential to control strictly terrestrial vectors. Effective control of house flies and related Diptera whose larvae develop in manure and other organic media requires a combination of cultural, biological and chemical methods. Routine applications of chemical pesticides are ineffective, and chemicals should be regarded only as supplemental (Axtell and Arends 1990).

Environmental management has controlled ticks under some circumstances (Ginsberg 1993). Bloemer et al. (1990) described a management plan for lone star ticks (*Amblyomma americanum*) in recreational areas that combined acaricides, vegetation management and exclu-

Figure 14.1. Using hand labor for ditch maintenance for management of saltmarsh mosquitoes (photograph by John R. Rusmisel, courtesy of Alameda County Mosquito Abatement District).

sion of white-tailed deer. Burning has been tested to reduce Lyme disease. Mather et al. (1993) discovered that a single burning of forest understory reduced the overall abundance of tick nymphs, but not of infected nymphs. Rotation of cattle among pastures may reduce infestations of single-host ticks.

Insecticides and Acaricides

The use of insecticides and acaricides for vector management goes back many years. The earliest materials were naturally occurring compounds such as sulfur and arsenic. Later, plant-derived pesticides such as rotenone and pyrethrum were formulated. Development and use of highly effective synthetic chemical pesticides started during World War II when special emphasis was placed on protecting military personnel from arthropods and arthropodborne diseases.

Pesticides can be classified in a variety of ways: their chemical structure, their mode of action, their type of formulation and their method of use. Common types of pesticides identified by their chemical structure include chlorinated hydrocarbons, organophosphates and carbamates. In terms of action, pesticides that kill on contact with vectors are considered contact pesticides. Those that must be ingested by vectors are known as stomach poisons, and those that enter the respiratory system of vectors as gases are called fumigants. Categories of formulations include dusts, granules and emulsifiable concentrates. Pesticides for vector control are used in different ways, including aerial and ground sprays and topical skin repellents on humans.

Chemical pesticides may be the method of choice against serious problems of disease transmission by arthropods. Many states and communities have emergency response plans for such situations. Properly developed plans will include insecticide susceptibility levels for potential arthropod vectors and the availability of pesticides and pesticide application equipment. Gratz and Jany (1994) review the role of chemical pesticides in vector control programs. However, IVM programs in which pesticides are combined with other approaches offers the best solution to sustained vector control and prevention of vectorborne diseases.

Synthetic pesticides

DDT, a highly effective and long-lasting residual insecticide against a variety of insects, fought many important vectors. At the time of its introduction, DDT's effectiveness as a vector control agent and its long residual activity were spectacular. The success of DDT led to the development of other chlorinated hydrocarbons, e.g., chlordane, lindane and dieldrin. However, within a few years physiological resistance to insecticides by arthropods became recognized. Species that had been readily controlled by these compounds could no longer be controlled even when dosages were increased to high levels. Laboratory tests comparing the susceptibility of strains of insects from nontreated areas to the susceptibility of strains from insecticide-treated areas confirmed the existence of genetically-controlled physiological resistance. Arthropods collected in nature and colonized in laboratories proved that strains susceptible or resistant to various chemical pesticides could be maintained indefinitely, thus providing material for toxicological research. Further, susceptible strains could be selected artificially for resistance by subjecting them to sublethal doses of insecticides, and resistant strains could be produced in just a few generations. The problem of pesticide resistance led to increased research on the mode of action of these insecticides. The first approach to solving the problem of resistance was the development of new classes of pesticides. Starting in the 1950s, compounds belonging to classes known as organophosphates and carbamates were developed to replace chlorinated hydrocarbons that no longer were effective. Other chemicals were synthesized based on the structure of the naturally occurring botanical insecticides called pyrethrins (extracted from *Chrysanthemum* flowers). These materials, called synthetic pyrethroids, are highly effective against many types of vectors, but have very low toxicity against vertebrate animals, except fish.

Another complicating factor in pesticide resistance was discovered in connection with research on the mode of action of pesticides. Resistance to one compound within a given class of pesticides often conferred resistance to other compounds within that class. This phenomenon, known as **cross-resistance**, explained why some populations of vectors resisted certain pesticides even though there had been no known previous exposure to them.

The 1960s saw a growing awareness of the potential damage to the environment from the agricultural overuse or misuse from some of these highly toxic and slowly degrading compounds. This awareness was heightened by the publication of the controversial book *Silent Spring* (Carson 1962). At about this time, the authority in the USA for the registration of pesticides was transferred from the Department of Agriculture to the newly created Environmental Protection Agency (EPA). Shortly thereafter, registration of DDT and several other chlorinated hydrocarbons was suspended, although their use continues in other parts of the world. Over the years, EPA restricted or suspended the use of many pesticides for vector control, including many organophosphate and carbamate compounds.

The World Health Organization (WHO) has a policy to eliminate DDT worldwide. This is a controversial issue because reduction of DDT use has helped the worldwide increase in malaria incidence (Roberts et al. 1997).

Other synthetic pesticides used for vector control include formamidines (amitraz), macrocyclic lactones (ivermectin and related compounds) and phenylpyrazoles (fipronil). Ivermectin is used widely to control several insectborne parasitic infections, including dirofilariasis (dog heartworm).

Biorational pesticides

In recent years, entirely new types of pesticides have been developed. Some of these interfere with developmental or reproductive processes of arthropods; others are based on highly specific microbial toxins. Because these materials have negligible vertebrate toxicity and are much less environmentally disruptive, they are called "biorational" or "third generation" pesticides.

Insect growth regulators

Insect growth regulators (IGRs) are compounds that retard or inhibit growth of an insect, ultimately causing death. IGRs have controlled mosquito larvae, cockroaches, fleas, flies and other vectors. In addition to their relatively high degree of environmental safety, these materials are less likely to induce physiological resistance in vectors, perhaps because IGRs mimic the function of natural substances. IGRs grew out of research on the factors regulating growth and development of insects. Chemicals within the insect itself regulated growth. The chemical structure of the first insect juvenile hormone, JH I, was reported in 1967, followed by identification of 2 additional structures, JH II and JH III. Bowers et al. (1965) first synthesized the compound later named JH III in 1965 and recognized its biological properties. The elucidation of these structures led to synthesis of related compounds and their evaluation as potential control materials. Since many of the new chemicals were not closely related to the JH compounds, but affected physiological mechanisms of growth and development, the term IGR came into use. The first IGR registered for use as an insecticide was methoprene, in 1975. Since then many products and formulations have been developed. Special formulations to provide extended and controlled release of the active ingredient have overcome some of the IGR's limitations, e.g., availability of the active ingredient at specific developmental times. IGRs have been used in the environment with a high margin of safety to fish, birds, other wildlife and aquatic organisms. Staal (1975) provided basic information on the development of these compounds, plus references to their characteristics and properties. Mulla (1995) reviewed IGRs in vector control including formulation, biological activity, field efficacy and safety.

Microbial pesticides

Several naturally occurring microbial organisms produce protein toxins that kill mosquitoes and black flies. The most widely used of these is *Bacillus thuringiensis* var. *israelensis* (Bti). Because the mode of action results from toxins produced by these organisms, and because they are applied with conventional pesticide application equipment, some classify these microbes as chemical pesticides. But these materials derive from naturally occurring pathogens, and extensive testing has shown them largely harmless to vertebrate animals as well as most invertebrates, so others consider them to be biological control agents.

Larvicidal oils

Mosquito larvae are vulnerable to disruption of the air-water interface because they must come to the surface of the water to breath through air tubes, or siphons. Petroleum products such as kerosene were used nearly 100 years ago to kill mosquito larvae. Diesel oil, light distillate domestic fuel oils and weed oils also have been used for mosquito control, but nearly all such products are no longer used because of environmental concerns. Oils can control mosquito larvae at relatively low doses, but tend to be nonspecific in their actions. Recent research has shown that some hydrocarbons may be safe and efficient larvicides, and in addition to killing mosquito larvae, may also inhibit oviposition by several *Culex* species (Beehler and Mulla 1996).

Plant-derived insecticides

Searches for compounds derived from plants are based on the assumption that naturally occurring compounds disrupt the environment less. This is a common public perception. The most effective of the plant-derived insecticides for mosquitoes and other Diptera are the pyrethrins, extracted from *Chrysanthemum* flowers. A new botanical pesticide is derived from the neem tree (*Azadirachta indica*). The active ingredient of extracts from the tree is known as azadirachtin (AZ). Azadirachtin affects various behaviors in a wide variety of agricultural pests, as well as some pests of medical importance, including mosquitoes, fleas and biting lice. Dhar et al. (1996) found that AZ, reetha, the Indian soapnut tree (*Sapindus mukorossi*) and garlic (*Allium sativum*) impaired reproduction in anopheline mosquitoes. Azadirachtin also may repel mosquitoes (Sharma et al. 1993) and phlebotomine sand flies (Sharma and Dhiman 1993).

Application of pesticides for vector control

The methods and equipment used for the application of pesticides are extensive. The many types of sprayers and dusters range from small hand-held units to large vehicle-mounted ground and aerial units with pumps powered by internal combustion engines and electric motors. Modern pesticide application equipment controls dosage rate and spray patterns better than earlier devices. Commercial equipment monitors vehicle speed, tank pressure and flow rate; some provide automatic operation based on feedback loops from sensors. Sprayers can be mounted on both helicopters and conventional fixed-wing aircraft. Choice of application equipment is based largely on choice of pesticide formulation. Generally, each formulation (see below) requires specialized application equipment.

The numerous types of formulations of pesticides used in such applications include powders and dusts, emulsifiable concentrates, oil solutions, concentrated ultra-low volume (ULV) formulations, granules and briquettes. Briquettes are pieces of charcoal, plaster of Paris, or similar substances impregnated with a pesticide such as methoprene. They release toxic levels of pesticides over a period of weeks or months. Slow release materials are used against many arthropods, including pesticide-impregnated cotton for control of tick vectors of Lyme disease (Deblinger and Rimmer 1991). Neck bands and tail tags that slowly release insecticides have controlled ectoparasites on livestock (Miller and George 1994).

In addition to the various types of equipment and formulations for pesticide application, there are various application strategies. ULV technology releases minute amounts of highly concentrated pesticides. Special equipment (Fig. 14.2) atomizes the pesticide into very small droplets. These droplets are so small and light that they can be used effectively only in very still air when atmospheric conditions favor the droplets falling to the ground. ULV insecticides primarily control flying insects. Mount et al. (1996) reviewed the use of ULV for mosquito control.

The worldwide malaria eradication effort of the late 20th Century was based on spraying the interior walls of houses with DDT. This approach is called a residual treatment, because vectors are killed after coming in contact with surfaces treated with insecticides. The residual effect of DDT lasted for many months. For some vectors and parasites, systemic insecticides are

Figure 14.2. An ultra-low volume insecticide sprayer (courtesy of Clark Equipment Company).

used. An example would be an insecticide fed to cattle to control cattle grubs.

A relatively recent innovation has been the use of synthetic pyrethroids to treat bednets, clothing and wide mesh window and doorway nets for personal protection against flying insect vectors. Chapter 15 describes this approach in detail.

Toxic baits

Attractive toxic baits that lure arthropod vectors to them make insecticides more effective and reduces the amount and area of coverage required for effective control. Baits or attractants with toxicants for house flies, ants, stable flies and tsetse exemplify successful technology.

Biological Control

The classical definition of biological control is the regulation by natural enemies of an organism's population density at a lower level than otherwise would occur (DeBach 1974). Biological control usually implies some kind of management strategy. A population that regulates itself without human intervention usually is called natural control. Death is the most important population control mechanism and so studies stress natural mortality factors. These include weather in addition to mortality by natural enemies. Natural mortality factors maintain stable populations of arthropod vectors. In a species where a single female can produce hundreds of eggs or progeny in her lifetime, stable populations, i.e., ones in which density is neither increasing or decreasing, imply high levels of natural mortality. For example, if one female in a stable population produces 200 eggs (100 females and 100 males), an average of 99 females (or 99%) will have died before they produce eggs. A slight reduction in natural mortality from 99% to 90% would allow a tenfold increase in density in the next generation. This is a sizable increase, particularly if sustained over several generations. Consequently, except where eradication is the goal, vector control programs should avoid disrupting habitats and the complex interactions of the biological organisms that serve as regulating factors.

Millions of years of adaptation have resulted in stable interactions between arthropod vectors and their natural mortality factors. In undisturbed environments, natural mortality may maintain vector populations at levels low enough to prevent serious disease transmission. However, in environments disrupted by human activities, natural mortality frequently has been lost or is insufficient to prevent large increases in vector density and disease transmission. In such cases, biological control must modify the balance between arthropod vectors and their natural enemies.

Classical biological control typically introduces a single exotic arthropod parasite or predator species (DeBach 1974). The basis of this approach is that the newly-established exotic species will suppress population densities of the arthropod vector below the threshold needed for transmission of pathogens. Considerable research is required for successful introductions of biological control agents. First, there must be a search for predators, parasites or disease pathogens of the target arthropod species (or a closely related species) in its original homeland (or area other than the problem area). This quest must be followed by collection and identification of potential biological control species and testing of agents for adverse affects on the environment. Finally, the agents are released into the problem area by what are known as inoculative releases. There have been many highly successful inoculative releases, particularly against agricultural crop pests. There also have been unsuccessful inoculative releases. To circumvent the limitations of inoculative releases, a different type of release was developed known as an inundative or augmentative release in which natural enemies are mass-collected or mass-produced and released. Inundative releases generally are required to achieve a significant level of control.

Fish have been used as biological control agents against mosquitoes for many years. Lacey and Orr (1994) noted that *Gambusia affinis* (Fig. 14.3), the most commonly used fish for biological control of mosquitoes, is the most widely distributed fish in the world. Fishes in the genera *Tilapia, Poecilia, Fundulus, Gasterosteus* and *Lucania* also have been used for mosquito control, while *Aphanius* fish in the *Panchax* group are used in Africa and Asia. Mosquito control agencies in the USA commonly collect or maintain colonies of fish for distribution in control programs. As with other control schemes, the effects on the environment must be considered before releasing fish. For a review of the use of larvivorous fish for mosquito control, the reader

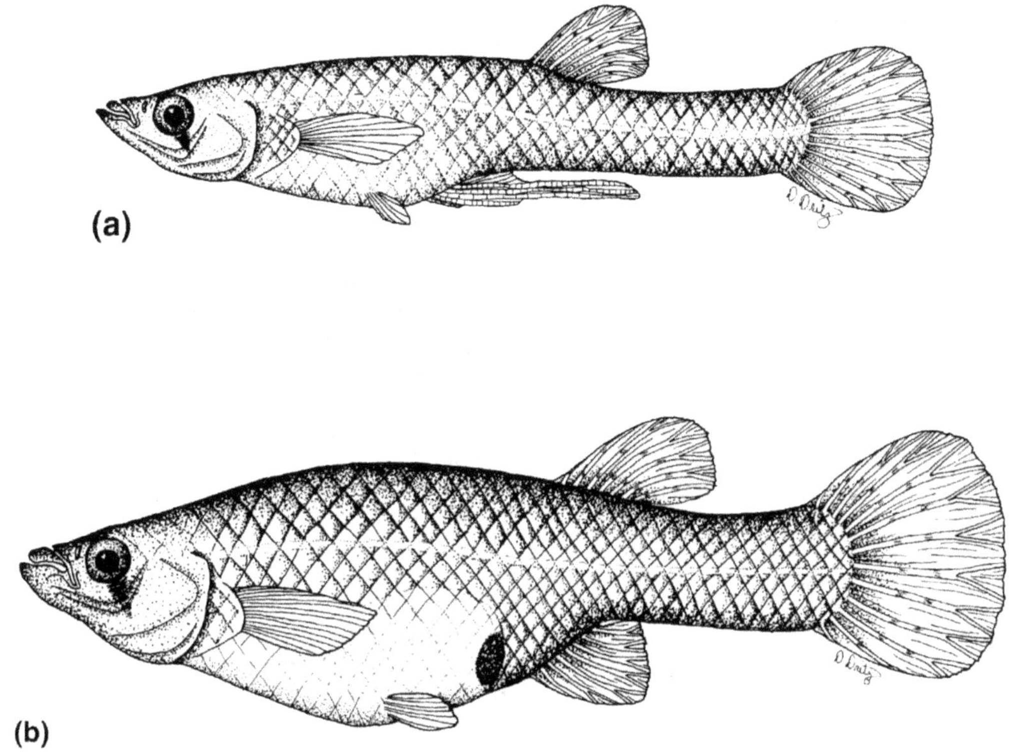

Figure 14.3. Mosquitofish, *Gambusia affinis*: (a) Male; (b) female (drawings by Deborah A. Dritz).

should consult Gerberich and Laird (1985). Swanson et al. (1996) reviewed advantages and disadvantages of using fish as biological control agents against mosquito larvae.

The use of the predaceous mosquito, *Toxorhynchites* spp., as a biological control agent has been reviewed by Steffan and Evenhuis (1981) and Focks (1985, 1991). Adults are large, colorful and non-bloodfeeding insects. The larvae also are large and feed on smaller mosquito larvae breeding in natural and artificial containers. Large-scale programs have been developed to mass-rear this mosquito and release it for the control of *Aedes* breeding in artificial and natural containers. Female *Toxorhynchites* can search out natural and artificial containers for egg deposition. The larvae consume large numbers of mosquito larvae and survive long periods in the absence of prey.

Copepods, fungi of the genus *Coelomomyces* and *Lagenidium* and mermethid nematodes have been tested as biological control agents for mosquitoes (Chapman 1974). *Lagenidium giganteum* appears to be especially promising as a biological control agent in non-polluted freshwater habitats. However, even though none of these organisms have proven effective for inoculative releases, they have been effective in inundative releases. Problems in mass-production and distribution need to be solved for their further development.

The most successful and widely used biological control agents against vector species, particularly mosquitoes and black flies, are 2 bacteria: Bti and *Bacillus sphaericus* (Bs). Both species are available commercially. There is a wide variety of Bti formulations available for various mosquito and black fly larval habitats. Formulations of Bs are less readily available, but are effective in highly organic habitats such as sewer lagoons and in tidal marshes where Bti is less effective.

Parasites and predators of house flies are important mortality factors, and management of moisture levels in manure piles on poultry and livestock ranches can enhance effects of naturally occurring organisms. Augmentative releases of the hymenopterous parasite *Spalangia* have lowered population densities of house flies in poultry facilities and other confined livestock operations (Axtell and Arends 1990).

Traps

Traps are used as survey tools to estimate vector density and to judge the need for vector control operations, as discussed in Chapter 13. The use of traps for control of vectors has long been a dream of vector control specialists. Although traps used for surveys do remove vector arthropods from the environment, most traps do not remove a sufficiently large number to reduce significantly the biting level or reproductive capacity of the population. This is true even when attractants such as CO_2, lactic acid, light, octenol or color are used to increase the attractiveness of the traps to vectors.

The use of attractive, toxic baits for house flies and eye gnats (Mulla et al. 1990) represented one of the first effective trapping systems for vector abatement. The technique of using traps in this way is known as **trapping out**. Population control of stable flies, *Stomoxys calcitrans*, with the Williams trap (Williams 1973) greatly reduced population density (Meifert et al. 1978, Rugg 1982). This trap used light reflective panels that attract adult stable flies. Two panels were mounted vertically and at right angles on a ground stake. Cloth netting treated with a quick-acting synthetic pyrethroid (Permethrin) covered the panels. Recent research on attractants responsible for bringing tsetse to their animal hosts has successfully controlled this insect. Traps have dark colored cloth panels sprayed with butanone or acetone to attract tsetse and deltamethrin to kill them. These traps reduced tsetse density only 3%, but reduced trypanosomiasis cases 99% after one year (Knols et al. 1993).

Traps employing a sticky surface (called sticky traps) and baited with CO_2 reduced horse fly populations in Louisiana (Wilson 1968). Traps containing baffles to prevent the escape of insects that enter them (called box traps) have been used effectively in the Cape Cod area of Massachusetts to reduce populations of horse flies (Wall and Doane 1980).

Field studies and mathematical calculations based on life history theory can estimate the potential of trapping out methods to reduce densities of arthropod vectors. The mark-release-recapture method can estimate the percentage of the total wild population captured in traps. Estimates of trapping efficiency are based on the assumption that if some percentage (say 1%) of insects marked and released are recaptured in the traps, then 1% of the wild population would be trapped by the same traps. This calculation is based on the assumption (which may not be valid) that the marked insects disperse and behave exactly as do the wild insects. Capturing 1% or less (consistent with results usually reported for mark-release-recapture studies) would have little noticeable effect on the number of biting arthropods in an area.

Weidhaas and Haile (1978) described a life history model for determining the degree of trapping required for vector management (Fig.

Table 14.2. Increase in density of hypothetical uncontrolled population of screwworm flies resulting from 5-fold increase over 6 generations.

Generation	Number of insects
1	1 million
2	5 million
3	25 million
4	125 million
5	625 million
6	3.125 billion

Table 14.3. Decrease in density of hypothetical population of screwworm flies with 90% kill of female flies per generation.

Generation	Number of Insects	Fertile individuals after 90% mortality
1	1 million	100,000
2	½ million	50,000
3	¼ million	25,000
4	125,000	12,500
5	62,500	6,250
6	31,250	3,125

14.4). The average daily mortality required to maintain or reduce populations by given amounts per generation can be estimated by this model. First, R_t is set to some value between 1 (no reduction from one generation to the next) and 0 (complete reduction from one generation to the next) and then the life history parameters (S_i, m and S_a) are estimated. Average daily mortality required is calculated as $(1 - S_t)$. Results from this model can evaluate the results of mark-release-recapture studies to determine the amount, if any, of population reduction that would be expected from a given trapping system.

Autocidal Methods

Dr. E.F. Knipling of the US Department of Agriculture conceived the concept of using a pest organism against itself. He developed the idea while working on the biology and control of the screwworm (*Cochliomyia hominivorax*) in Texas in the late 1930s. The successful development of this concept through the sterilization of the screwworm by gamma radiation (Knipling 1979) led to genetic mechanisms to reduce the population in other arthropod vectors.

Autocidal control of screwworms combined its high biotic potential with its relatively low density of flies in nature (in the order of hundreds per square mile). While searching for a method that would have advantages over the use of insecticides, Knipling derived relatively simple models of the effect of insecticides, not in terms of the numbers of insects killed, but on the ability of an arthropod population to increase its numbers from generation to generation. The models demonstrated how the average density of a population would change from generation to generation with and without control. Further, the models could compare the efficacy of insecticide treatments and sterilizing factors in terms of population regulation. In the absence of data on absolute densities and growth rates of natural screwworm populations, Knipling's early models assumed that a 5-fold rate of increase per generation would be the maximum the populations could attain. Table 14.2 illustrates the potential growth of an insect population over 6 generations at a rate of

Table 14.4. Decrease in density of hypothetical population of screwworm flies with 90% sterilization of flies per generation.

Generation	Number of Insects	Sterile:Fertile Insects	Percent Sterility	Fertile Females Remaining
1	1 million	9:1	90.0	100,000
2	½ million	18:1	94.7	26,316
3	131,579	68:1	98.6	1,896
4	9,480	940:1	99.9	10
5	10	180,747:1	100	0
6	0			

5-fold increase per generation when uncontrolled.

In 6 generations, such a population would increase to 3,125 times its original density. The theoretical effect of insecticide treatments of populations growing at a rate of 5-fold per generation is shown in Table 14.3. A treatment that kills 90% of the females reduces density 50% each generation and 97% in 6 generations.

Table 14.4 shows the theoretical advantage of releasing sterilized insects into the environment over insecticide treatments. Producing a 90% level of sterility in wild females by the release of sterile insects would require an initial release rate of 9 sterile insects for each fertile wild insect (9/9 + 1), in this example, 9 million. Once density reduction begins, ratios increase to levels that lead to extremely high levels of sterility and ultimately, elimination in 6 generations.

The advantages for sterility illustrated above depend on the mass-production and release of sterilized insects that are of high quality, able to compete in the field against wild insects and not seriously reduced in mating ability by rearing or the sterilizing or releasing methods. Knipling showed that the autocidal approach has an even greater advantage over conventional pesticide application when sterilization is by direct application to a wild population. This discovery led to an extensive search for chemical sterilizing agents. Certain compounds causing sterility in both males and females were too hazardous in natural environments. Eventually, effective and safe insect growth regulators were formulated that prevent development as well as cause sterility.

The mass-production and release of insects sterilized by gamma irradiation for control of the screwworm first was demonstrated on the island of Curaçao. The screwworm fly was eliminated from the southeastern USA by a release campaign in 1956–1957, the first successful large-scale application of the sterile male technique. Since that time the screwworm fly has been eliminated from the southwestern USA, Mexico and most Central American countries, and in countries such as Libya where it already had been introduced.

Sterile insect control programs have many requirements: (1) well-organized and budgeted support from industry and government; (2) well-equipped factories and equipment to produce, sterilize and package millions of insects per week; (3) suitable ground and aerial systems for distributing and releasing the insects; (4) means of maintaining quality and fitness of

$$R_0 = \frac{S_i \times m \times (S_a^{\,d})}{1 - (S_a^{\,c})}$$

$$R_t = \frac{S_i \times m \times (S_a^{\,d} \times S_t^{\,d})}{1 - (S_a^{\,c} \times S_t^{\,c})}$$

R_0 = Population growth rate per generation with only natural mortality
R_t = Population growth rate per generation after both natural mortality and trapping
S_i = Probablility of survival from egg to adult
S_a = Average daily survival of adults from natural mortality
S_t = Average daily survival from trapping (1-average daily mortality from trapping)
m = Average number of eggs per live female oviposition
d = Preoviposition time in days
c = Egg laying cycle in days

Figure 14.4. Weidhaas and Haile (1978) model for determining the degree of trapping required for vector management.

released insects; and (5) field surveys to ensure the effectiveness of the program.

Modern release approaches combine with other approaches in IVM programs. In the screwworm control programs, cattle are surveyed for active cases of screwworm and treated with insecticidal smears or dips. Agreement with both governments and private groups and individuals are made for quarantines to prevent entry into areas where the fly no longer exists. Surveys detect any introductions.

Sterile release technology is feasible for anopheline and culicine mosquitoes, although no operational programs have been developed. This lack may be due to the complexity, size, costs and organization needed for such large scale programs against non-isolated populations of vectors. The technology applies also to tsetse control. A combination of sterile fly releases and attractant bait-trapping systems for tsetse presents interesting possibilities.

Genetic engineering can eliminate the need to rear females for release. In the past, it has been impractical to avoid the inclusion of females from releases. If they are included, they must be completely sterilized and yet play no significant role in the effectiveness of the technology. Eliminating the need to rear females for release increases the number of males that can be produced for release and eliminates the competition beween released and wild females for mates. Such a system based on insecticide resistance was demonstrated in field tests with *Anopheles albimanus* (Kaiser et al. 1978).

Genetic Manipulations of Disease Vectors

The successful screwworm program increased interest in developing genetic mechanisms that could either sterilize wild populations or permit population replacement with nonvector strains of insects. Reared and mass-produced insects made sterile by genetic techniques (e.g., cytoplasmic incompatability, chromosomal translocations or compound chromosomes) did not improve the sterile insect release technique. Gwadz (1994) showed that development of malaria parasites in mosquitoes could be interrupted. This research led to the idea of genetically-engineered mosquito strains that are refractory to the development and transmission of malaria parasites, and of replacing wild vector populations with these non-transmitting strains. The possibilities of such mechanisms, although difficult and complex, are tantalizing. Gwadz (1994) suggested: "The growing awareness of the positive potential of genetic engineering and its rapidly expanding role in the production of plant and animal food products, pharmaceuticals and vaccines and an extraordinary range of new applications will, in the next decade, make the use of genetically engineered products an unremarkable and often necessary part of daily life."

Genetic Manipulation of Endosymbionts of Disease Vectors

Many of the schemes for genetic manipulation of vector arthropods have faced a major hurdle in developing methods for the stable insertion of foreign genes into natural populations. One solution to this problem is to express foreign anti-parasite or antiviral gene products in naturally occurring symbiotic bacteria harbored by vector arthropod species in a way that interferes with transmission (Beard et al. 1993). An approach being tested experimentally in the field involves the transformation of *Rhodococcus rhodnii*, a bacterial endosymbiont of *Rhodnius prolixus*, to express cercropin A, a peptide lethal to *Trypanosoma cruzi*, the causative agent of Chagas disease (Durvasula et al. 1997).

Anti-arthropod Vaccines

The discovery that vertebrate hosts may develop specific antibodies against certain ectoparasites, especially ticks, has led to the successful development of vaccines that prevent successful attachment and bloodfeeding. Chapter 4 covered this topic in detail in the discussion of immunological reactions to arthropods.

Integrated Vector Management

Integrated Vector Management (IVM) uses all available methods to reduce vector abundance and disease transmission. Review of such methods shows the large number of possibilities for improving vectorborne disease programs. Development of modern pesticides largely has been blamed for the lack of greater use of IVM. Certainly, IVM was practiced before the widespread use of DDT and other synthetic pesticides. Early control programs for yellow fever and malaria involved various methods in combination: source reduction, ditching, draining, water management, repellents, screening, biological control and chemical insecticides. Each vector problem sufficiently differed to require different combinations of technologies. The development of modern pesticides temporarily reduced dependence on IVM. They were initially so cost effective and successful in preventing or eliminating disease it would have been unwise not to use them extensively. The gradual movement away from total reliance on pesticides for reasons discussed earlier has led to increased attention to IVM. Present-day control programs by mosquito and vector control agencies rely heavily on survey, environmental management, source reduction, biorational pesticides and biological control. When surveys indicate unacceptable densities of vectors in terms of annoyance or potential disease transmission, pesticides may be used. Protection against Lyme disease depends largely on the use of repellents and treated clothing. Only in approved and highly used areas of known tick infestations are insecticidal treatments considered. An outbreak of mosquito-borne encephalitis may initiate aerial applications of adulticides in large areas to protect against potential disease transmission. After the threat has passed the treatments can be discontinued and normal IVM continued.

Community-based Programs

There is a trend to emphasize community-based programs in many vector control situations. In industrialized countries, vector control services are usually government supported. Most such programs operate at the county or city level with some additional support and regulation provided by the state and federal governments using funding obtained largely by local taxation. Often, special tax districts are established for this purpose. The need for and the objectives of such programs may be publicized through educational programs at schools and aimed at the local population through television and the news media.

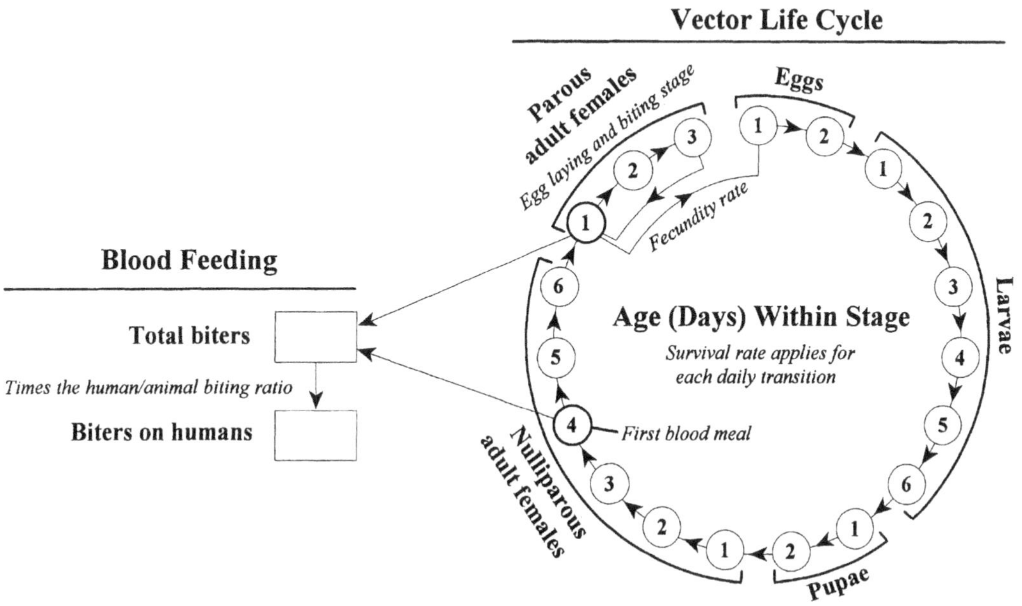

Figure 14.5. Graphic representation of a life table model for anopheline mosquitoes (adapted from Weidhaas and Haile 1977).

Historically, in Third World countries, vector control has been accomplished through large, vertically-organized programs sponsored and financially supported by governments and donor countries. As funding for such programs has decreased and the complexity of the problem of vector and disease control throughout the world has become better appreciated, community participation has been stressed. Gubler and Clark (1994) summarized community-based integrated control of *Ae. aegypti*. Service (1993) discussed self-help methods in community-based programs and reviewed successes and failures of this approach. Community-based programs also are becoming popular with abatement districts to reduce numbers of *Ae. albopictus* in the southeastern USA.

SIMULATION MODELS OF VECTOR AND DISEASE MANAGEMENT

Basic Principles

Controlling the transmission of arthropod-borne disease pathogens does not necessarily require quantification of all of the variables and their interactions that make transmission possible. However, quantification helps to understand which factors are of major importance to determine when, where and to what degree transmission and disease will occur and helps to design optimal and cost effective control strategies. Early attempts to quantify disease transmission led to mathematical expressions that defined vectorial capacity and other epidemiological concepts (Chapter 6). More recently, powerful and inexpensive personal computers have been used to create simulation models that can mimic the density of vectors and the prevalence and incidence of a disease with and without the application of various control technologies.

For many years ecologists have been developing the life table approach to define the growth of insect populations and the major factors that affect that growth. This approach is ideal for modeling of density and growth of vectors (controlled and uncontrolled) over time and the prevalence and incidence of diseases caused by their associated disease pathogens. Many of these models are based on the concept of generation time, i.e., the time required for an arthropod to complete development.

Examples of Models for Arthropod Vectors

Mosquitoes and house flies

Weidhaas (1986) reviewed computer simulation models used to estimate the effect of control strategies on malaria transmission and house fly density. Fig. 14.5 is a life table diagram used by Haile and Weidhaas (1977) to simulate densities of field populations of *An. albimanus* and study the efficacy of mosquito control strategies. Each box represents the number of individuals in a particular stage and age class. Knowledge of the values of life history parameters (average daily survival of the stages, daily growth rate, development times for eggs, larvae and pupae, preoviposition time, egg-laying cycle, number of eggs per oviposition and sex ratio) allows estimates of numbers for all age classes and stages and the daily advancement of the numbers in the table through time. Using a similar approach, it is possible to model the development of malaria parasites in anopheline vectors. Other models can simulate the development of a malaria parasite in a human population. These models in combination can be used to calculate daily values for the entire malaria life cycle, including the number of vectors becoming infected from biting infective humans and the number of humans becoming infected from infective vector bites. Various control technologies can be evaluated not only on the basis of vector density but also on disease prevalence and incidence.

The effect of various control strategies on house fly density also has been modeled. One such model incorporates the effects of a parasite, a predator and the release of sterile males on house fly density. With estimates of the life history parameters for each organism (house fly, predator and parasite), it is possible to study the interaction of the host fly with both the parasite and the predator. Simulations using this approach were compared to field studies on the release of the hymenopteran parasite, *Spalangia endius*, against the house fly, *Musca domestica*. General trends of the degree of parasitism generated by the simulation model agreed with those reported for the field trial (Weidhaas 1986).

Floodwater mosquito model

A dynamic life table model of *Psorophora columbiae* in a southern Louisiana rice agroecosystem simulated the population dynamics of this mosquito (Focks and McLaughlin 1988). In the model, immature development depends on temperature. Since this mosquito is a floodwater species, i.e., eggs remain dormant on damp soil until flooded, a hydrologic model was incorporated to estimate when sufficient flooding of egg deposition sites would occur for the hatching of eggs. This model permitted evaluation of a variety of control strategies including: (1) insecticidal treatment of adults; (2) insecticidal treatment of larvae in irrigation water; (3) elimination of diapausing eggs; (4) insecticidal treatment of large animal hosts and (5) amount of land devoted to the first crop of rice (Focks et al. 1988). The results were highly credible and agreed with available field data. The model showed cattle in the vicinity

Figure 14.6. Modern marsh managment for mosquito control using Thiokel® ditching equipment (photograph by John R. Rusmisel, courtesy of Alameda County Mosquito Abatement District).

regulated the density of this mosquito and that relocation of cattle may be an effective control measure.

Aedes aegypti—dengue transmission model

Focks et al. (1995) developed a pair of stochastic simulation models to describe the daily dynamics of dengue transmission in urban environments. A weather-driven dynamic life-table model of the population dynamics of container-breeding breeding mosquitoes provided vector input to a virus transmission model. The transmission model followed the growth of a human population according to country- and age-specific birth and death rates. Daily estimates of adult mosquito survival, gonotrophic development, weight and emergence from the vector model were input to describe the biting mosquito population in the transmission model. Survival and emergence values affected vector density; gonotrophic development and female weight affected biting frequency. Viremia level in humans and ambient temperature influenced the probability of transfer of virus from human to mosquito. It was possible to simulate concurrent epidemics involving different dengue serotypes. The vector and transmission models were built on site-specific data from an extensive literature of factors influencing the growth and development of the vector and development of the virus.

Black fly—onchocerciasis model

Davies et al. (1988) described a computer simulation model of the black fly, *Simulium damnosum*, that included the development of filarial worms causing onchocerciasis in flies and humans. The modeling of the life cycle of the black fly resembled that shown in Fig. 14.5 for anopheline mosquitoes. Inputs included development times of the immature stages, preoviposition times, egg-laying cycles, the number of eggs per oviposition, the sex ratio and average daily survival for the various stages. The beginning numbers for the various stages and age classes and the daily updating of these numbers can be developed from field-derived data on human biting rates and on life table analysis. Field data from the Onchocerciasis Control Program (OCP) in West Africa made it possible to reconstruct density curves typical of known field populations. Data on the development of *Onchocerca volvulus* in human patients was obtained by clinical research. In spite of the necessity of estimating certain variables needed to make the simulation run, the trends of the values for biting rates and prevalence obtained in the computer simulation approximated those observed in field studies.

Ticks

Computer simulation models have been developed for several species of ticks (the American dog tick, *Dermacentor variabilis*, the lone star tick, *A. americanum*, the black-legged tick, *Ixodes scapularis* and *Boophilus* ticks) based on life table analysis and temperature driven development of tick stages. They have simulated tick population dynamics and the transmission of several disease pathogens, e.g., Rocky Mountain spotted fever and babesiosis (Cooksey et al. 1990, Haile and Mount 1987, Haile et al. 1992, Mount and Haile 1987, 1989).

EXAMPLES OF INTEGRATED VECTOR MANAGEMENT PROGRAMS

The management of arthropodborne disease problems by IVM requires a thorough knowledge of vector ecology and epidemiology of particular diseases. IVM programs assume that population levels of vectors must be reduced to levels sufficient to disrupt transmission of pathogens or to levels where severe annoyance and/or economic loss is eliminated.

Mosquitoes

Environmental management is fundamental to IVM programs for mosquitoes. Such methods may include source reduction by the removal of artificial and natural water-holding containers, vegetation management, ditching (Fig. 14.6), draining, disking, burning, water management and fluctuation of water levels. Many mosquito problems can be considered man-made in that discarded and even useful containers, structures such as dams and modifications of the environment for human purposes all create standing water in which a variety of mosquitoes can develop. Any effort to control mosquitoes requires thorough surveys and documentation of the areas producing mosquitoes and the species of mosquito causing the problem. Only in this way can appropriate and effective environmental management techniques be selected. When mosquitoes breed in natural environments, special concerns about the environment and wildlife must be evaluated, and permits generally are required by regulatory agencies to ensure compatibility between mosquito control methods and habitat quality.

Under some conditions, mosquitofish (*Gambusia*) are included in IVM programs for mosquitoes (Fig. 14.3). Some climates are too cold for the fish to survive the winter, and special rearing facilities are needed to produce sufficient quantities of fish for mosquito control. Mosquitofish are generalist predators and may disrupt some environments, especially disturbed environments providing little vegetative cover for other fishes (Swanson et al. 1996). Other biological control methods are under development. The best environmental and biological controls sometimes can be insufficient to prevent serious mosquito problems. In such cases, surveys of an area can identify aquatic habitats that may be treated with larvicides to reduce the prevalence of mosquitoes. Most programs use biorational insecticides for this purpose, especially IGRs and microbial pesticides. Such materials can be obtained in a variety of formulations including various types of sprays, granules and briquettes. After adult mosquitoes emerge and begin seeking bloodmeals it is necessary to use conventional insecticides applied from ground or aerial application equipment. The most appropriate means for managing mosquitoes over large areas is through locally organized control districts, adequately funded and staffed. Such agencies can use modern technology and computer programs to survey mosquito densities, levels of disease pathogens and records of treatments.

Black Flies

Sustainable black fly control has proven to be a significant challenge. Vegetation removal from streams was a component of some programs before the advent of synthetic pesticides (Davies 1994). Because the larvae develop in fast running streams, obtaining prolonged contact by chemical pesticides has been difficult. Slow release formulations have been tried with some success using both conventional and microbial pesticides. Some newer formulations of the microbial pesticide Bti disperse easily in streams, although their application is not as easy as with organophosphate insecticides. Current recommendations are to use Bti during periods of low stream flow and organophosphates during other times, with the choice of material depending upon the occurrence of pesticide resistance in black flies (Guillet et al. 1990). The introduction of ivermectin as a microfilaricide has improved disease control in humans. Further, a greater understanding of variation in vector competence among sibling species of *Simulium* should permit a more focused approach to control (Shelly 1988).

The OCP is an outstanding example of a successful IVM program. The program was established in the late 20th Century to eliminate onchocerciasis in 7 west African countries: Benin, Burkino Faso, Ivory Coast, Ghana, Mali, Niger and Togo. Later expanded to Guinea, Guinea-Bissau, Senegal and Sierra Leone, the program originally was based on weekly applications of temephos to streams. When *S. damnosum* cytospecies developed resistance to temephos, Bti was substituted. In recent years, larviciding has been supplemented by treatment of human infections with ivermectin. Molyneux (1995) and Hougard et al. (1997) reviewed the results of this highly successful program.

Cattle Grubs

Scholl (1993) discussed the status of IVM programs for cattle grubs. Trial programs using sterile male release technology combined with area-wide insecticidal treatments have shown promise. The use of ivermectin as a systemic treatment against grubs has been highly effective, but there are some concerns about the possibility of physiological resistance in grubs resulting from the use of ivermectin in this way. Future emphasis will be placed on the development of anti-grub vaccines in addition to current methods.

Tsetse

Removal trapping, residual sprays of insecticides applied to vegetation used by tsetse as resting sites, clearing of trees, and animal management have controlled tsetse (World Health Organization 1986). Extensive behavioral studies of tsetse have led to the design of efficient traps based on both olfactory and visual cues. Because of the relatively low population growth rate of tsetse, sterile male release technology appears feasible on theoretical grounds. Community-based programs utilizing pyramid traps have been highly effective in some African communities (Colvin and Gibson 1992).

House Flies and Relatives

There is perhaps no category of medically important arthropods more difficult to control by application of pesticides than house flies and related insects. Consequently, IVM programs offer the only real chance of success. Successful programs combine management of solid wastes, use of biological control agents, insecticidal baits and traps (Axtell and Arends 1990). In poultry houses and similar operations, adequate air circulation to promote drying of manure is critical. Staged manure removal also is needed to preserve the parasite and predator complex of flies. Straight pesticide applications are of limited value because of high levels of resistance in most fly populations, and the negative effects of pesticides on parasites and predators.

Triatomid Bugs

Schofield (1994) summarized the problems associated with the control of vectors of Chagas disease. Many approaches (insecticides, housing modifications, juvenile hormone mimics and other insect growth regulators, traps, chemosterilants, genetic manipulations and various biological control agents) have been tried. Insecticides and housing modifications are the most effective methods. Chlorinated hydrocarbon insecticides were used for many years for house spraying, but recently, synthetic pyrethroid compounds (permethrin and deltamethrin) have supplanted the older materials. House improvements (such as improved foundations, floors, walls, roofs as well as removal of domestic animals from houses) offer the most sustained control, but funds and cultural customs are limiting factors.

Ticks

The severe impact of ticks on humans and domestic animals has emphasized IVM programs worldwide. Most programs include insecticidal treatments, both for area-wide tick control and for direct treatment of livestock with sprays or dips. Repellents have protected humans from tick infestations. Many non-chemical methods have been used for tick control, including pasture rotation to protect livestock and vegetation removal for protection of both humans and domestic animals. However, for control of certain tickborne diseases (e.g., Lyme disease), vegetation removal may be counterproductive, and in local situations, reduction of deer populations or systemic treatment of deer with acaricides may be more effective (Spielman et al. 1985).

REFERENCES

Ault, S.K. 1994. Environmental management: a re-emerging vector control strategy. Am. J. Trop. Med. Hyg. **50(Suppl.):** 35-49.

Axtell, R.C. and Arends, J.J. 1990. Ecology and management of arthropod pests of poultry. Annu. Rev. Entomol. **35:**101–126.

Beard, C.B., O'Neill, S.L., Tesh, R.B., Richards, F.F. and Aksoy, S. 1993. Modification of arthropod vector competence via symbiotic bacteria. Parasitol. Today **9:**179–183.

Beehler, J.W. and Mulla, M.S. 1996. Larvicidal oils modify the oviposition behavior of *Culex* mosquitoes. J. Vect. Ecol. **21:**60–65.

Bloemer, S.R., Mount, G.A., Morriss, T.A., Zimmerman, R.H., Barnard, D.R. and Snoddy, E.L. 1990. Management of lone star ticks (Acari: Ixodidae) in recreation areas with acaracide applications, vegetative management, and exclusion of white-tailed deer. J. Med. Entomol. **27:**543–550.

Bowers, W.S., Thompson, M.J. and Ubel, E.C. 1965. Juvenile and gonadotropic activity of 10,11-epoxyfarnesenic acid methyl ester. Life Sci. **4:**2323–2331.

Carson, R. 1962. Silent spring. Houghton Mifflin Co., Boston. 368 pp.

Chapman, H.C. 1974. Biological control of mosquito larvae. Annu. Rev. Entomol. **19:**33–59.

Colvin, J. and Gibson, G. 1992. Host-seeking behavior and management of tsetse. Annu. Rev. Entomol. **37**:21–40.

Cooksey, L.M., Haile, D.G. and Mount, G.A. 1990. Computer simulation of Rocky Mountain spotted fever transmission by the American dog tick (Acari: Ixodidae). J. Med. Entomol. **27**:671–680.

Davies, J.B. 1994. Sixty years of onchocerciasis vector control: a chronological summary with comments on eradication, reinvasion, and insecticide resistance. Annu. Rev. Entomol. **39**:23–45.

Davies, J.B., Weidhaas, D.E. and Haile, D.G. 1988. Models as aids to understanding onchocerciasis. Pp. 396–407 *in* K.C. Kim and R.W. Merritt (eds.), Black flies: ecology, population management, and abbreviated world list. Pennsylvania State University Press, University Park.

DeBach, P. 1974. Biological control by natural enemies. Cambridge University Press, London. 323 pp.

Deblinger, R.D. and Rimmer, D.W. 1991. Efficacy of a permethrin-based acaracide to reduce the abundance of *Ixodes dammini* (Acari: Ixodidae). J. Med. Entomol. **28**:708–711.

Dhar, R., Dawar, H., Garg, S., Basir, S.F. and Talwar, G.P. 1996. Effect of volatiles from neem and other natural products on gonotrophic cycle and oviposition of *Anopheles stephensi* and *An. culicifacies* (Diptera: Culicidae). J. Med. Entomol. **33**:195–201.

Durvasula, R.V., Gumbs, A., Panackal, A., Kruglov, O., Aksoy, S., Merrifield, R.B., Richards, F.F. and Beard, C.B. 1997. Prevention of insect-borne disease: an approach using transgenic symbiotic bacteria. Proc. Nat. Acad. Sci. **94**:3274–3278.

Focks, D.A. 1985. *Toxorhynchites*. Pp. 42–45 *in* H.C. Chapman (ed.), Biological control of mosquitoes. American Mosquito Control Association, Fresno, California.

Focks, D.A. 1991. Data sheet on *Toxorhynchites*. Wld. Hlth. Organiz. **1**:1–10.

Focks, D.A., Daniels, E., Haile, D.G. and Keesling, J.E. 1995. A simulation model of the epidemiology of urban dengue fever: literature analysis, model development, preliminary validation, and samples of simulation results. Am. J. Trop. Med. Hyg. **53**:489–506.

Focks, D.A. and McLaughlin, R.E. 1988. Computer simulation of management strategies for *Psorophora columbiae* in the rice agroecosystem. J. Am. Mosq. Control Assoc. **4**:399–413.

Focks, D.A., McLaughlin, R.E. and Smith, B.M. 1988. A dynamic life table model of *Psorphora columbiae* in the southern Louisiana rice agroecosystem with supporting hydrologic submodel. Part I. Analysis of literature and model development. J. Am. Mosq. Control Assoc. **4**:266–281.

Gerberich, J.B. and Laird, M. 1985. Larvivorous fish in the biocontrol of mosquitoes, with a selected bibliography of recent literature. Pp. 47–76 *in* M. Laird and J.W. Miles (eds.), Integrated mosquito control methodologies, Vol. 2. Academic Press, London.

Ginsberg, H.S. (ed.) 1993. Ecology and environmental management of Lyme disease. Rutgers University Press, New Brunswick, New Jersey. 224 pp.

Gratz, N.G. and Jany, W.C. 1994. What role for insecticides in vector control programs? Am. J. Trop. Med. Hyg. **50(6, Suppl.)**:11–20.

Gubler, D.J. and Clark, G.A. 1994. Community-based integrated control of *Aedes aegypti*: a brief overview of current programs. Am. J. Trop. Med. Hyg. **50(6, Suppl.)**:50–60.

Guillet, P., Kurtak, D.C., Philippon, B. and Meyer, R. 1990. Use of *Bacillus thuringiensis israelensis* for onchocerciasis control in West Africa. Pp. 187–201 *in* H. de Barjac and D.J. Sutherland (eds.), Bacterial control of mosquitoes and black flies. Rutgers University Press, New Brunswick, New Jersey.

Gwadz, R.W. 1994. Genetic approaches to malaria control: how long the road? Am. J. Trop. Med. Hyg. **50(6, Suppl.)**:116–125.

Haile, D.G. and Mount, G.A. 1987. Computer simulation of population dynamics of the lone star tick, *Amblyomma americanum* (Acari: Ixodidae). J. Med. Entomol. **24**:356–359.

Haile, D.G., Mount, G.A. and Cooksey, L.M. 1992. Computer simulation of management strategies for American dog tick (Acari: Ixodidae) and Rocky Mountain spotted fever. J. Med. Entomol. **27**:686–696.

Haile, D.G. and Weidhaas, D.E. 1977. Computer simulation of mosquito populations (*Anopheles albimanus*) for comparing the effectiveness of control technologies. J. Med. Entomol. **13**:553–567.

Hougard, J.-M., Yamégo, L., Sékétéli, A., Boatin, B. and Dadzie, K.Y. 1997. Twenty-two years of blackfly control in the Onchocerciasis Control Programme in West Africa. Parasitol. Today **13**:425–431.

Kaiser, P.E., Seawright, J.A., Dame, D.A. and Josyln, D.A. 1978. Development of a genetic sexing system for *Anopheles albimanus*. J. Econ. Entomol. **71**:766–771.

Knipling, E.F. 1979. The basic principles of insect population suppression and management. Agriculture Handbook No. 512. US Department of Agriculture, Washington, DC. 659 pp.

Knols, B.G.J., Willemse, L., Flint, S. and Mate, A. 1993. A trial to control the tsetse fly, *Glossina morsitans centralis*, with low densities of odour-baited targets in West Zambia. Med. Vet. Entomol. **7**:161–169.

Lacey, L.A. and Orr, B.K. 1994. The role of biological control of mosquitoes in integrated vector control. Am. J. Trop. Med. Hyg. **50(6, Suppl.)**:97–115.

Laird, M. and Miles, J.W. (eds). 1983. Integrated mosquito control methodologies, Vol 1. Academic Press, London and New York. 369 pp.

Laird, M. and Miles, J.W. (eds). 1985. Integrated mosquito control methodologies, Vol 2. Academic Press, London and New York. 444 pp.

Mather, T.N., Duffey, D.C. and Campbell, S.R. 1993. An unexpected result from burning vegetation to reduce Lyme disease transmission risks. J. Med. Entomol. **30**:642–645.

Meifert, D.W., Patterson, R.S., Whitfield, T., LaBrecque, G.C. and Weidhaas, D.E. 1978. Unique attractant-toxicant system to control stable fly populations. J. Econ. Entomol. **71**:290–292.

Miller, J.A. and George, J.E. 1994. Efficacy of a combination neckband and tailtag containing amitraz or cyhalothrin K against lone star ticks on pastured cattle. J. Agric. Entomol. (S.C. Entomol. Soc.) **11**:165–176.

Molyneux, D.H. 1995. Onchocerciasis control in West Africa: current status and future of the Onchocerciasis Control Programme. Parasitol. Today **11**:399–402.

Mount, G.A., Bierry, T.L. and Haile, D.G. 1996. A review of ultralow-volume aerial sprays of insecticides for mosquito control. J. Am. Mosq. Control Assoc. **12**:601–618.

Mount, G.A. and Haile, D.G. 1987. Computer simulation of area-wide management strategies for the lone star tick, *Amblyomma americanum* (Acari: Ixodidae). J. Econ. Entomol. **24**:523–531.

Mount, G.A. and Haile, D.G. 1989. Computer simulation of population dynamics of the American dog tick (Acari: Ixodidae). J. Econ. Entomol. **26**:60–76.

Mulla, M.S. 1995. The future of insect growth regulators in vector control. J. Am. Mosq. Control Assoc. **11(2, Part 2)**:269–273.

Mulla, M.S., Axelrod, H. and Wargo, M.J. 1990. Chemical attractant formulations against the eye gnat *Hippelates collusor* (Diptera: Chloropidae). Bull. Soc. Vect. Ecol. **15**:156–165.

Resh, V.N. and Balling, S.S. 1983. Tidal circulation alteration for salt marsh mosquito control. Environ. Manag. **7**:79–84.

Roberts, D.R., Laughlin, L.L., Hsheih, P. and Legters, L.J. 1997. DDT, global strategies, and a malaria control crisis in South America. Emerg. Infect. Dis. **3**:295–302.

Rugg, D. 1982. Effectiveness of Williams traps in reducing the number of stable flies (Diptera: Muscidae). J. Econ. Entomol. **75**:857–859.

Schofield, G.J. 1994. Triatominae: biology and control. Eurocommunica Publ., West Sussex, UK. 80 pp.

Scholl, P.J. 1993. Biology and control of cattle grubs. Annu. Rev. Entomol. **38**:53–70.

Service, M.W. (ed.) 1989. Demography and vector-borne diseases. CRC Press, Boca Raton, Florida. 402 pp.

Service, M.W. 1993. Community participation in vector-borne disease control. Ann. Trop. Med. Parasitol. **87**:223–234.

Sharma, V.P., Ansari, M.A. and Razdan, R.K. 1993. Mosquito repellent action of neem (*Azadirachta indica*) oil. J. Am. Mosq. Control Assoc. **9**:359–360.

Sharma, V.P. and Dhiman, R.C. 1993. Neem oil as a sand fly (Diptera: Psychodidae) repellent. J. Am. Mosq. Control Assoc. **9**:505–507.

Shelly, A.J. 1988. Vector aspects of the epidemiology of onchocerciasis in Latin America. Annu. Rev. Entomol. **30**:337–366.

Spielman, A., Wilson, M.L., Levine, J.F. and Piesman, J. 1985. Ecology of *Ixodes dammini*-borne human babesiosis and Lyme disease. Annu. Rev. Entomol. **30**:439–460.

Staal, G.B. 1975. Insect growth regulators with juvenile hormone activity. Annu. Rev. Entomol. **20**:417–460.

Steffan, W.A. and Evenhuis, N.L. 1981. Biology of *Toxorhynchites*. Annu. Rev. Entomol. **26**:159–181.

Swanson, C., Cech, J.J., Jr. and Piedrahita, R.H. 1996. Mosquitofish: biology, culture, and use in mosquito control. Mosq. Vect. Control Assoc. California, Elk Grove and Univ. California, Davis. 88 pp.

Wall, W.J. and Doane, O.W. 1980. Large scale use of box traps to study and control saltmarsh greenhead flies (Diptera: Tabanidae) on Cape Cod, Massachusetts. Environ. Entomol. **9**:371–375.

Walton, W.E., Tietze, N.S. and Mulla, M.S. 1990. Ecology of *Culex tarsalis* (Diptera: Culicidae): factors influencing larval abundance in mesocosms in Southern California. J. Med. Entomol. **27**:57–67.

Weidhaas, D.E. 1986. Models and simulations for biological control of flies. Pp. 57–68 *in* R.S. Patterson and D.A. Rutz (eds.), Biological control of muscoid flies, Vol. 61. Misc. Publ. Entomol. Soc. Am., College Park, Maryland.

Weidhaas, D.E. and Haile, D.G. 1978. A theoretical model to determine the degree of trapping required for insect population control. Bull. Entomol. Soc. Am. **24**:18–20.

World Health Organization. 1986. Epidemiology and control of African trypanosomiasis. Report of a WHO expert committee. WHO Tech. Rep. Ser. No. 739. Geneva, Switzerland. 127 pp.

Williams, D.F. 1973. Sticky traps for sampling populations of *Stomoxys calcitrans*. J. Econ. Entomol. **66:** 1279–1280.

Wilson, B.H. 1968. Reduction of tabanid populations on cattle with sticky traps baited with dry ice. J. Econ. Entomol. **61:**827–829.

Chapter 15

Prevention and Control of Arthropodborne Diseases

JAMES F. SUTCLIFFE
Trent University, Peterborough, Ontario, Canada

The prevention and control of arthropodborne diseases combines tactics aimed at disease vectors, pathogens and hosts. This chapter emphasizes preventive measures. These are measures, approaches or practices that place a barrier (physical, chemical, immunologic, behavioral or ecological) between disease pathogens and susceptible hosts. In general, preventive measures apply to individual or to small groups of hosts, although some larger scale applications of preventive measures (such as vaccination) also are discussed. Preventive measures include repellents and other forms of individual protection, chemoprophylaxis and chemotherapy, and vaccines. Vector control strategies used for prevention of specific diseases are mentioned briefly. Chapter 14 discussed general approaches to vector control.

PERSONAL PROTECTION

Vector Avoidance

Throughout the ages, humans have sought ways to protect themselves and their livestock from arthropod attack. In some cases, these tactics recognized the connection between arthropods and diseases but in others, vector avoidance integrated into cultures without an appreciation of linkage. Cattle-raising cultures, such as the Masai and many tribes of southern Africa, apparently knew that tsetse bites could give their livestock nagana and carefully avoided fly belts (areas where tsetse occur). If forced to drive cattle through tsetse-infested areas, they would do so at night when tsetse are inactive (McKelvey 1973). In West Africa, where the distribution of tsetse was less patchy, the relationship of tsetse and nagana was not so readily discernible. Nonetheless, travelling traders from arid areas knew from bitter experience that their pack animals quickly sickened and died when taken into forested zones. In what is today central Mali, where the forested southern zone gives way to the arid Sahel, the so-called "silent trade" is thought to have evolved as a way to carry on commerce while preventing contact and possible disease transmission between southern buyers and northern sellers. This was an honor-based bartering system where one group would lay out their offer (goods or payment) in the marketplace and withdraw allowing the other group to approach to peruse, counteroffer and consummate transactions. Despite the obvious temptation to cheat in this system,

B.F. Eldridge and J.D. Edman (eds.), Medical Entomology, Revised Edition, 565–620.

it apparently worked well, although the participants did not know exactly how, and protected the traders' livestock from nagana (McKelvey 1973).

Examples of cultural avoidance abound. Many highly fertile river valleys of West Africa were abandoned by agrarian peoples apparently because high rates of onchocerciasis-caused impaired vision and other conditions beset communities in those locales (Remme and Zongo 1989). Like the northern traders, these people did not link the simuliids coming out of the rivers with disease; nonetheless, abandoning the valleys alleviated the problem. Unfortunately, leaving the valleys also increased population pressures on the higher, less fertile inter-riverine plateaus and severely stretched local resources. The result was unhealthy communities because of malnutrition and crowding due to a little nematode carried by a little fly.

The word "malaria" harkens to a time when this disease was thought by ancient Romans to be due to the foul air of the Pontine Marshes near the city of Rome. After the fall of the Roman Empire, the marshes, partly drained as the city grew, reestablished themselves and the fertile *campagna* around the marshes became mosquito-ridden and unhealthy again. For centuries, the *campagna* was abandoned between June and September. Only drainage of the marshes and mosquito control in more recent times alleviated the problem. Malaria from the Pontine Marshes is thought to account for the fact that Rome, a city of over a million people in the time of Augustus Caesar (27 BC–14 AD), did not achieve that population again until the 1930s (Stage 1944).

Although the many connections made during the 19th Century between arthropod attack and disease transmission clarify the importance of vector avoidance, most people avoid biting arthropods not because of the risk of illness but because attacks annoy. Where the discomfort factor is great enough and an acceptable and even attractive alternative is to spend time indoors when insect pests are most active, disease prevention may be a by-product. Major decreases in the number of human disease cases caused by western equine encephalomyelitis (WEE) and St. Louis encephalitis (SLE) viruses in California in recent decades were attributed in part to television and air conditioning that took people from their porches and evening-biting *Culex* mosquitoes into the protection of their climate-controlled houses (Gahlinger et al. 1986). In this case, culture and technology accidently prevented disease.

All cultures have customs, beliefs, living habits and priorities that affect the occurrence of arthropodborne (and other) diseases. For the medical entomologist and public health practitioner who aim to prevent arthropodborne disease, the needed adjustments to customs often are obvious and basic but frustratingly difficult to achieve. For instance, transmission of onchocerciasis in West Africa could be greatly curtailed if the people at risk wore western-style trousers and shirts since these would prevent many black fly bites. This is a simple but often unrealistic approach because people do not change centuries of customs for what are, to them, abstract notions. It could be argued that such intransigence comes from ignorance and that if people fully understood the risks they would change their habits instantly. If this were true, no one in the technologically advanced, well-educated nations of the world would smoke or get AIDS.

In reality, a myriad of motivations govern people's lives so that the behavior that results often is illogical to those with specific goals (such as prevention of arthropodborne diseases). In fact, people rarely modify their behavior because of the risk of disease except in the case of major epidemics. More immediate concerns usually prevail. For instance, Reuben (1989) summarized a study documenting that men and older boys in a village in India preferred to spend evening hours sitting on their porches socializing even though they were being bitten the entire time by culicine mosquitoes, some of which undoubtedly were carrying Japanese encephalitis (JE) virus. Moreover, the men and boys also slept outside claiming it was cooler and that the annoyance from bed bugs (not disease vectors) indoors was too great. Custom and comfort easily overrode knowledge of the abstract encephalitis risk. In this same study, women and girls were found to be much less at risk for infection by JE virus because they spent the evening hours indoors engaged in food preparation and child care.

All arthropodborne disease transmission is embedded in similar social complexity and viable preventive measures must recognize and accommodate it or they are destined to fail.

Protective Dress

When humans cannot avoid biting arthropods, the single best piece of advice is to dress appropriately and wear clothing that provides a physical barrier to bites. Proper clothing depends on the arthropod. Mosquitoes can slip their slender mouthparts through many materials. Therefore, loose fitting clothing made of tightly woven material best prevents their bites. Black flies and ticks, on the other hand, must contact the skin directly to bite and will crawl up pant legs, sleeves and waistbands to reach skin. Thus, all openings that these arthropods might crawl through should be eliminated by tucking trouser legs into socks, taping shirt sleeves at the wrists, tucking shirts into trousers and buttoning collars. Wearing a hat also is recommended since black flies, ticks, tabanids and other insects readily crawl into the hair to bite the scalp. When Lyme disease transmission by ticks is a possibility, people should examine themselves or each other no matter how carefully they dress. Pets also should be checked thoroughly each day for adherent nymphal ticks.

Color and texture of material in clothes also are important considerations in dressing for protection against biting arthropods. Black flies, for instance, land most readily on materials that have a rough texture and dark, non-reflective color (Davies 1972). Unfortunately, such materials are found in blue denim jeans, flannel shirts and other popular outdoor wear.

Several specialized items provide extra protection from biting fly attack. Headnets of sufficiently fine mesh (some headnets have mesh coarse enough to allow small black flies to crawl through) are useful if properly worn. Unfortunately, headnets are most effective under conditions of light or intermediate activity only. Fly jackets that incorporate a built-in headnet also are available. Protective clothing impregnated with repellent materials also is potentially useful and is discussed in the "Repellents" section of this chapter.

Protective dress is practical in many conditions but hot weather and heavy work, especially when combined with moderate or low arthropod biting intensities, often make people less conscientious about protective practices. Sharp et al. (1995) studied the risk factors associated with contracting dengue in American servicemen involved in Operation Restore Hope in Somalia in 1992–93 such as failure to use supplied repellents, keep sleeves rolled down and treat uniforms with permethrin. They found that overall compliance of soldiers using personal protective measures was poor despite a thorough prior education process. For instance, 70% of those sampled used repellents less than once a day and only 25% kept their sleeves rolled down at all times. If soldiers specially trained and equipped to avoid vectorborne diseases in the field do not observe basic precautions, it is hardly surprising that the general public, despite all the best public health information, allows vector contact.

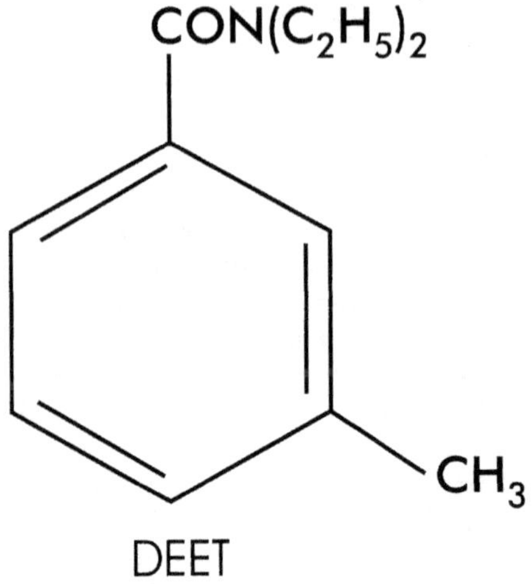

Figure 15.1. Structural formula for n,n-diethyl-m-toluamide (DEET).

Repellents

When avoidance of biting arthropods is not feasible, the application of one of a group of substances referred to generically as **repellents** may protect from bites. Repellents are chemically diverse materials and it is not known whether any of them prevent bloodsucking by actually repelling (inducing negative chemotaxis) host seeking arthropods. For example, diethylmethylbenzamide (n,n-diethyl-m-toluamide, or **DEET**) does not cause host-seeking arthropods to fly away from prospective hosts, but it does appear to prevent them from landing once they have come close. The sensory and behavioral mechanisms for this effect are not known. McIver (1981) speculated on these mechanisms for the repellent effect and suggested that the nonpolar DEET molecule (Fig. 15.1) rends large holes in the lipid membranes of the sensory dendrites of the mosquito's olfactory sensilla causing a flood of sensory input that the insect's central nervous system cannot integrate. The result would be a disruption of host seeking responses. Whatever DEET's effect on sensory membranes, it appears to be a general one since DEET's odor affects many different olfactory sensilla on the mosquito, including the lactic acid-sensitive pegs on the antennae and oviposition attractant-sensitive sensilla on the genitalia of *Aedes aegypti* (Kuthiala et al. 1992). Apparently, DEET renders the mosquito anosmic to lactic acid. Geier et al. (1996) showed that *Ae. aegypti* will not respond in the host seeking mode if lactic acid is not present in the mix of host volatiles. This insight into how DEET works at the sensory level in mosquitoes is the exception in our understanding of repellent mechanisms in general. Davis (1985) reviewed the literature relating to concepts of the modes of action of insect repellent substances and found no chemical commonality among them. He suggested that evidence supports a number of modes of action (depending on the repellent in question) including interference with the reception of host stimuli, stimulation of sensory inputs for competing behaviors and switching of the sensory message to one of repulsion or avoidance. How repellents work should be a priority area for research since understanding the sensory and behavioral mechanisms underlying their effects could help design better repellents and show us how to use existing repellents more effectively.

Although repellent use seems to have the potential to prevent arthropodborne disease, Gupta and Rutledge (1994) pointed out that few studies have tested this notion. In their review, they suggested that the clearest evidence for repellents affecting disease transmission came from a World War II study linking dibutyl phthalate use by Australian troops with a reduction in miteborne scrub typhus. The applicability of repellents to arthropodborne disease prevention should be investigated more widely.

Repellents are classifiable in terms of the way they are used and their source. Most commercial repellents are applied to the skin as **topical** preparations in the form of creams, lotions or sprays while **systemic** repellents are ingested. In practice, no systemic repellents are in use or known for humans or domestic animals. Claims that taking vitamin B_1 (thiamine) results in protection from mosquitoes have not been substantiated (Khan et al. 1969, cited in Curtis et al. 1990). Similarly, though some claim that eating bananas or garlic provides protection from mosquitoes and other biting insects, scientific evidence of these effects is lacking. **Natural product** repellents derive from plants and plant extracts and many, although not all, **synthetic** repellents (repellents synthesized in pure form) are modeled on naturally occurring molecules. Virtually every traditional culture on the globe has within its pharmacopoeia plants and plant products to cure or prevent arthropod infestations and irritation. That plants produce a wealth of **secondary compounds** that affect insect behavior is no coincidence. Most insect species are herbivorous and plants have evolved a potent arsenal of chemicals to protect themselves at least partly from insect attack. Hundreds of plant materials have insecticidal or repellent effects though few have been investigated.

Testing and evaluation of prospective repellents

New repellents are likely to come from plant substances, but assessment of these substances means more than simply crushing a bit of the plant or making a crude extract of it and observing the effects on arthropod biting behavior. Plant substances that repell insects typically are complex mixtures of compounds, only some (perhaps one) of which are active. For instance, oil of citronella (discussed below) consists of various acyclic monoterpenes including borneol, geraniol, citronellol and citronellal. Before the active ingredient (citronellal) could be identified, the complex oil had to be extracted, purified, analyzed and the constituents tested individually for their activity. Such isolation-purification is necessary because it allows those developing the product for the market to assure quality. This control is accomplished either by analyzing lots of purified plant extracts for their active ingredients (since various factors such as plant age and growing conditions affect levels of many secondary compounds in the tissues) or by synthesizing a pure form of the active ingredient.

Methods of testing prospective repellents vary but usually take the form of laboratory screenings in which test animals, artificial substrates or human arms (Fig. 15.2) are treated with the material and **complete protection time**, the time between skin/substrate treatment and the first confirmed bite (a second bite within 30 min of a first bite), is measured (Schreck 1977). This criterion has been criticized: the most tolerant biters in the test (least affected by the repellent) as opposed to the biters of average tolerance will determine whether the substance is evaluated positively or not. Thus, repellent testing commonly also measures ED_{50}s or ED_{90}s at a range of repellent dosages.

The "white rat" of candidate repellent screening traditionally has been the yellow fever mosquito, *Ae. aegypti*, since it is readily colonized and epidemiologically important. However, it has long been recognized that this species' responses to repellents do not necessarily represent those of all mosquitoes, let alone the responses of other biting arthropods. Rutledge et al. (1983) compared the sensitivities of *Ae. aegypti* and 4 other mosquito species to 31 commercial and experimental repellents and found that the sensitivity shown by *Ae. aegypti* bore little relationship to patterns shown by other species. *Aedes aegypti* proved to be considerably less sensitive to several substances than some of the other species. Repellent assessments should rely on the responses of several species of insect lest promising candidates be eliminated erroneously.

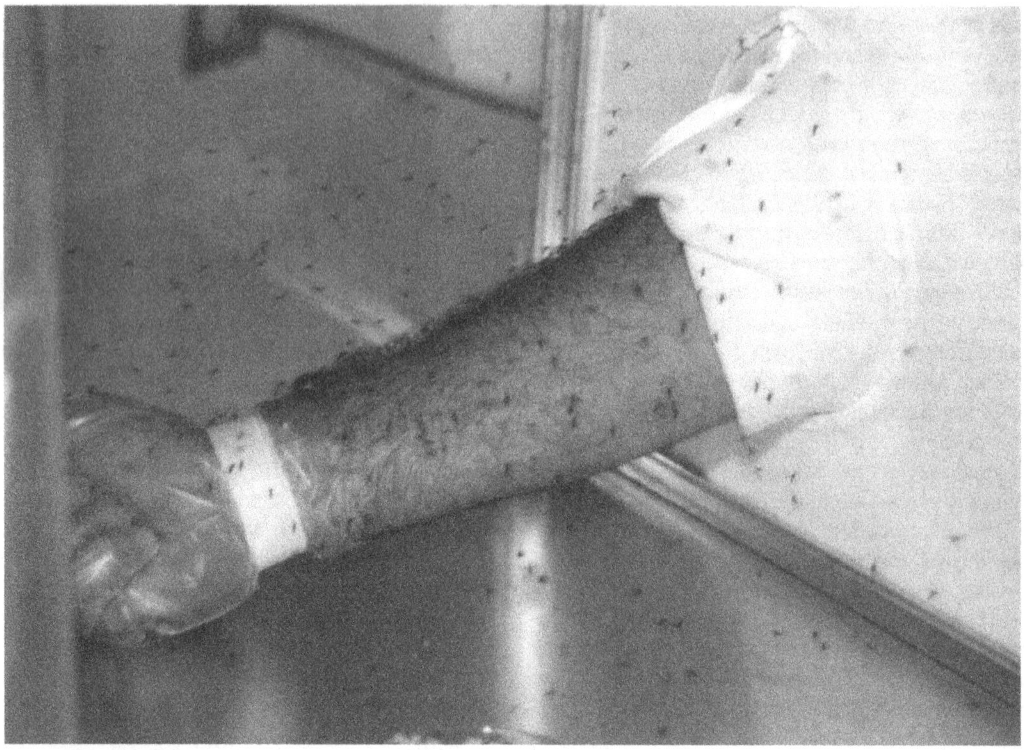

Figure 15.2. Testing of potential insect repellents on a human arm (courtesy of USDA Center for Medical, Agricultural and Veterinary Entomology).

The value of an animal model to replace humans in laboratory repellent screenings has been recognized, since using humans in this capacity raises ethical issues and since the toxic properties of many materials being tested have not been fully evaluated. Rutledge et al. (1994) compared repellency of DEET and several other repellents for *Ae. aegypti* when applied to mice and a human arm. In general, the test materials protected the arm better, but it was possible to calculate correction factors that made the mouse data a reliable predictor of the arm model. While animal models for repellent testing seem promising and desirable, it is important to recognize that different biting arthropod species react to animal hosts according to evolved host preferences. Thus, correction factors for every biting arthropod-animal model combination will have to be determined. Some combinations will not work.

Natural repellents

Curtis et al. (1990) discussed plant-derived repellents, focussing on a number of essential oils. The best known of these is oil of citronella, broadly effective against most biting arthropods, but used primarily as a mosquito repellent since the late 19th Century. Oil of citronella originally was extracted from nardus grass (*Cymbopogon nardus)* and is composed of several monoterpene aldehydes including citronel-

Table 15.1. Number of female mosquitoes collected from forearms of human volunteers in evening tests in areas with (Plant) and without (Control) "mosquito plants" (from Cilek and Schreiber 1994).

Trial	*Aedes albopictus*		*Culex quinquefasciatus*	
	Plant	Control	Plant	Control
1	27	42	66	18
2	18	22	9	8
3	77	58	116	139
4	19	21	187	101
5	11	15	232	105
6	110	104	157	44
7	10	13	186	426
8	2	0	294	123
Totals	274	275	1,247	964

lal, the active fraction. Although its effectiveness is generally more short-lived than DEET's, oil of citronella enjoys considerable popularity in citronella candles and as a topical repellent preparation because of its pleasant odor and natural source. These advantages notwithstanding, the efficacy of citronella candles is debatable. Lindsay et al. (1996) found the smoke, heat and light from citronella candles account for the small repellent effect for *Aedes* mosquitoes. Recent questions about the safety of DEET (discussed below) and a general unease with "non-natural" substances have further revived the popularity of oil of citronella.

Certain plants producing a lemony smell are claimed to contain oil of citronella and are currently being sold with the claim that they will rid yards of mosquitoes. Cilek and Schreiber (1994) tested the so-called "citrosa plant," or "mosquito plant," a cross between the African lemon geranium and the English finger bowl geranium. The number of *Aedes albopictus* and *Culex quinquefasciatus* females landing on the arms of 2 volunteers were counted in areas with and without geraniums. Although the geraniums were planted at 10 times the density recommended for effective repellency by the supplier, experimental and control site landing rates for either species exhibited no important differences (Table 15.1).

When applied topically, the active component of citronella is concentrated oil extracted from citronella plants or the synthetic equivalent. Concentrated and free of the tissues of the plant, repellent vapor pressures are said to be greater than the intact plant can generate, and thus far more effective. Claims like those made for mosquito plants should be treated with skepticism.

Pyrethrum, one of a family of compounds called pyrethrins, is a natural insecticide that has been extracted from the flowers of the pyrethrum daisy *(Chrysanthemum cinerariaefolium)* for over 100 yr. The natural pyrethrins are highly photo-labile but have low mammalian toxicity and are, therefore, best suited to applications as indoor fumigants, mosquito coils and short-lived flit sprays used to clear mosquitoes and other flying insects from houses. Pyrethrins, in addition to their insect toxicity, also are excito-repellents that cause insects to become hyperactive. This increased activity often causes shorter landing times and greater undirected flight that, in turn, translates into fewer bites. The natural pyrethrins have served as a model for a number of synthetic analogues, the pyrethroids, which have even greater insecticidal and excito-repellent properties. Some of these are photostable and persistent. The pyrethroids act against a broad spectrum of biting arthropods (e.g., mites and ticks, lice, reduviids, fleas and various biting flies) and find important medical applications as treatments for clothing and bednets.

See Curtis et al. (1990) for further information on natural repellents, including some of the promising substances that have been found in the extensive surveys undertaken in China over the past several decades.

Synthetic repellents

Early repellents

The first insect repellents to be synthesized were dimethyl phthalate (DMP) and dibutyl phthalate (DBP), patented by Moore and Buc in 1929. Prior to that, the only non-plant substance commonly used as a topical repellent was sulfur dusted onto skin and clothes or taken in a mixture with molasses. Other synthetic repellents were developed by the end of World War II, notably indalone and ethyl hexanediol (Rutgers 612), that together with DMP and DBP were the mainstays of protection from mosquitoes, mites and ticks for troops in the Pacific during World War II. Indalone, DMP and DBP still are commercially available, but DEET and the synthetic pyrethrins such as permethrin largely have replaced them.

DEET

DEET was first synthesized in 1954 and because of its superior and broad effectiveness, quickly became the main topical repellent, a status it still holds today. DEET has been tested against virtually all types of biting arthropods. Variations in effectiveness occur, but this substance is useful against many mosquito species (including several important vectors) the tsetse *Glossina morsitans*, black fly species in North America and Africa, and various phlebotomines, ceratopogonids, bed bugs and ticks. The fact that DEET also is a repellent for certain nonvectors (sweat bees and leeches) suggests that it affects these organisms through some common sensory mechanism.

As a topical repellent, DEET comes in various formulations including pastes, creams, sprays and alcohol solutions. Strengths of topical preparations range from ~15% up to 100%. Protection times in laboratory and field studies where the subjects are inactive readily exceed 10 hr (Gupta and Rutledge 1991; Fig. 15.3). Under normal conditions where people are active, protection times are much shorter (usually <5 hr). DEET is lost through absorption into the skin, evaporation and physical abrasion.

The US Department of Defense has sponsored extensive research to develop a personal protection system for armed service personnel that will work anywhere in the world. Part of the effort has been to develop DEET formulations that are longer lasting and more pleasant to use (less smelly and oily, with less damaging plasticizing effects) than the standard US military 75% DEET in ethanol formulation currently in use. Controlled release formulations developed by 3M Company and Biotek Corporation employ 35% and 45% DEET, respectively, complexed with slow-release systems (e.g., acrylate microcapsules) in a polymer that forms a film on skin. In tests of these formulations against the military standard, all formulations provided several hours of complete protection and even longer periods of substantial protec-

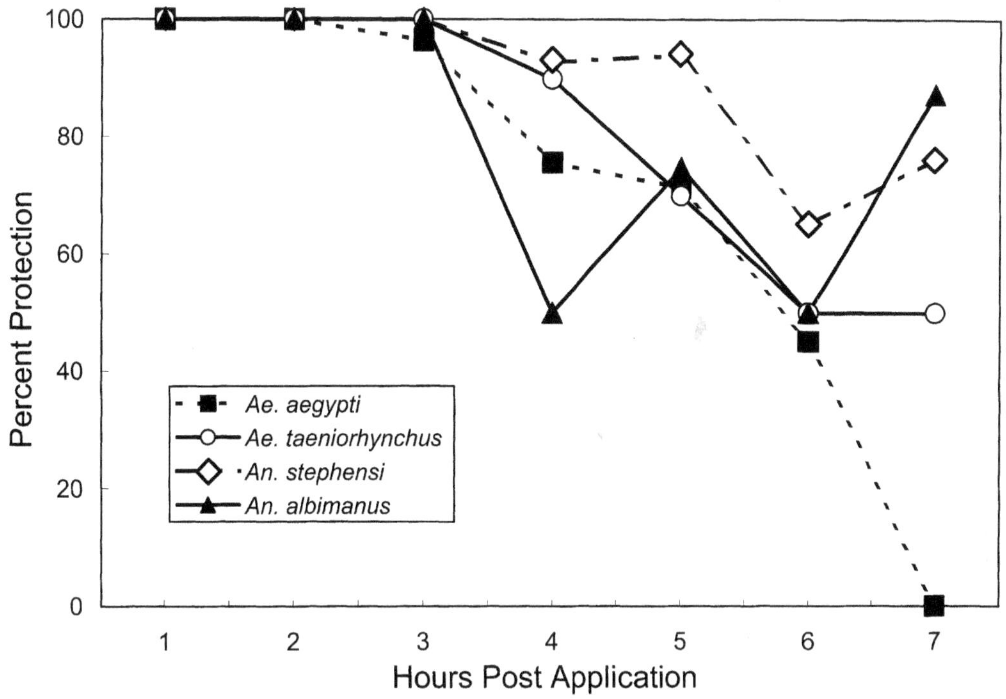

Figure 15.3. Percent protection from bites on volunteers using US Army standard repellent (75% DEET) in simulated tropical climate (from Gupta and Rutledge 1991).

tion from *Aedes taeniorhynchus* in the Florida Everglades and from caged *Anopheles quadrimaculatus*, but the military standard was judged marginally superior (Schreck and Kline 1989). In contrast, environmental chamber tests of the same 3 formulations against *Ae. aegypti, Ae. taeniorhynchus, Anopheles stephensi* and *Anopheles albimanus*, judged overall protection with the 3M formulation superior (Gupta and Rutledge 1991). While these tests are inconclusive, the 3M and Biotek formulations contain half or less as much DEET as the military standard and test subjects judged them to be easier and more pleasant to apply. If these factors promote greater use in the field and fewer side effects than from the high level DEET exposure in the military standard, the controlled-release formulations would be the formulations of choice.

Soap formulations of DEET may have promise. "Mosbar" soap contains 20% DEET and 0.5% permethrin in bar soap which is applied by lathering normally and letting the soap dry on the skin. Mosbar has been tested by volunteers in southern India against various *Culex* and *Aedes* species (Mani et al. 1991) and in coastal Peru and Ecuador (Kroeger et al. 1997). In India, 89–100% protection from bites was achieved in nightly tests. In South America, the degree of protection depended on activity levels of the volunteers. Protective efficacy was 81% for inactive volunteers and 52% in physically active volunteers 6 hr after application. While this approach may seem attractive, an annual cost of $4.50 per person (estimated by Kroeger et al. 1997) might be excessive and it is not known how much DEET exposure would result from the regular use of this soap and whether this level of exposure would be a healthy one, especially for children.

The safety of DEET has been questioned since its introduction in the mid-1950s although usually there have been only relatively minor side effects such as skin rashes and skin hypersensitivity. In recent years, with new fears of Lyme disease added to old worries about various mosquitoborne encephalitis viruses, the sale and use of DEET-containing products in the USA has increased greatly. It is estimated that up to 100 million people per year in the USA currently use DEET. Several cases of neurotoxicity and convulsions now have been linked circumstantially to liberal application of DEET to children. A significant proportion of the general public and some legislators have accepted the concept that strong formulations should not be applied to children. New York State has passed legislation banning the sale of any product containing more than 30% DEET. The chemical industry disputes the evidence that DEET can be connected to these cases (Osimitz and Grothaus 1995).

Concerns about DEET's effects on human physiology arise in part from the fact that when applied in ethanol solution to the skin, a proportion of the DEET is absorbed. While estimates of the amount absorbed vary, new slow release formulations (e.g., the Biotek and 3M formulations discussed above) make less DEET available per unit time and may reduce amounts of absorption while achieving the repellency levels of more concentrated preparations in ethanol. Domb et al. (1995) reported the efficacy (against *Ae. aegypti* and *An. stephensi*) and dermal uptake of 10% and 20% DEET formulations in a liposphere system. The 10% preparation was effective against both species for a minimum of 3.5 hr and the 20% preparation for 6.3 hr. Transdermal absorption of DEET from the 10% liposphere preparation was a third that of the 10% ethanol preparation. Further work on new formulations of this mainstay of personal protection is needed to reassure and protect the public.

Pyrethroids

DEET-impregnation of clothing reduces biting of various arthropods including mosquitoes (Grothaus et al. 1976), ceratopogonids (Harlan et al. 1983), black flies (Renz and Enyong 1983) and ticks (Schreck et al. 1980) usually for a few days or weeks. Persistent, photostable synthetic analogues of pyrethrum such as permethrin and deltamethrin have both repellent and insecticidal properties. They have largely replaced DEET in clothing impregnation since their low vapor pressure results in months of repellency and their toxicity reduces the biting population. Pyrethroids also are used extensively to treat bednets and curtains.

Pyrethroid-impregnated clothing reduced mosquito biting substantially (see, for example Lillie et al. 1988, Schreck et al. 1978, 1980), so the US Department of Defense incorporated treated uniforms into its personal protection system along with DEET treatment of exposed skin areas. This combined system was tested for its effectiveness against *Ae. taeniorhynchus* in the Florida Everglades (Schreck and Kline 1989). Permethrin-treated battle dress uniforms provided 99.9% protection from bites through the material. Uncovered arms of volunteers were completely protected by DEET repellents for between 11.3–12.5 hr, and while volunteers' heads were bitten significantly more, complete protection times ranged from 6.5–10.1 hr. These results illustrate the importance of using a combination of skin repellent and impregnated clothing. Hundreds of bites would have resulted through untreated clothing whether or not DEET was used on exposed skin. Many of the mosquitoes deterred from biting through impregnated uniforms probably would have diverted to the arms and head.

Repellents used on livestock

Many factors dictate how repellents are used to protect livestock from arthropod attack and their effectiveness. Livestock and other domestic animals represent a diverse group that fill different roles (as pets, food animals, draft animals, for wool, etc.) and occur in a wide range of climatic and physical situations (the home, in pastures and feedlots, ranges, etc.). Add to these factors the many arthropods that may attack domestic animals and one glimpses the complexity of the topic. Only a few aspects of this complexity will be discussed here. For more detailed reviews of this topic organized by arthropod group the reader is referred to the articles published by the Office of International Epizootics (Uilenberg 1994).

Repellents used on livestock resemble those used on humans: DEET, certain long residual activity pyrethroids and a few other compounds. Pyrethroids are both excito-repellent and highly toxic to arthropods so they provide the most popular alternatives where frequent re-treatment of livestock would be difficult and costly.

A popular way to use pyrethroids and other insecticides on cattle and horses is in the form of treated ear tags and ribbons. These have successfully effected long term relief from various muscoids including stable flies, horn flies and filth-associated flies. Parashar et al. (1989) reported that 10% permethrin-impregnated ear tags protected horses from stable flies (*Stomoxys calcitrans*) and from the tabanid *Haematopota dissimilis* for up to 2 months, but that hippoboscids (*Hippobosca maculata*) on horses were completely unaffected. The authors suggested that since hippoboscids spend almost all of their time on the host, the repellency of permethrin, which affected host finding in stable flies and tabanids, had no opportunity to affect ked behavior. In a follow-up study, Parashar et al. (1991) completely controlled *H. maculata* on horses for up to 3 months by applying deltamethrin preparations directly. In this instance, the treatment worked not because of its excito-repellent activity but because of its direct toxic effects on the keds.

New formulations for permethrins and other pyrethroids are under study to extend their effective periods and lower their costs on livestock. Meyer and Hunter (1991) demonstrated that a microencapsulated formulation of permethrin retains its effectiveness against stable flies on treated hair from dairy cows for up to 7 days, while the standard emulsifiable preparation was no longer detectable after 3 days. An effective period extended like this would more than halve the costs and workload associated with fly control in dairies.

Other compounds also protect livestock and domestic animals from arthropods though not all compounds work against all pests. For instance, Folz et al. (1986) reported that Amitraz (a triazapentadiene compound) applied topically to dogs is a highly effective tick repellent/detachment agent for several weeks but has little effect on fleas. Some substances marketed for other purposes appear to have insect repellent properties. Avon's Skin-So-Soft®, a proprietary bath oil, was shown by Fehrer and Halliwell (1987) to reduce flea infestations on dogs.

Perhaps application is the single largest problem in using repellents for livestock. Many livestock animals are pastured or even free range for much of the year. In these instances, methods of application must be found that are practical and result in long term protection. Aside from the impregnated ear tags discussed above, one of the simplest methods of repellent and insecticide application to animals is the pour or dip. In the former, a solution of the agent is poured on the back of the animal and allowed to spread to other parts of the body passively or through the animal's grooming activities. In some cases, self-applicators, devices that cattle can be trained to walk through, are used to apply repellents and insecticides. Self applicators are used in horn fly control because they are well-suited to the application of chemicals to the head, neck and back of the animal. Some self-applicators use strips of fabric suspended from above to brush flies off, achieving a measure of physical control. Suspended burlap bags or rubber or cloth applicators can be filled or impregnated with repellent/ insecticidal compounds, achieving chemical control.

Self applicators are less useful for the control of other types of flies such as stable flies, black flies and tabanids because these insects tend to bite on the extremities or on the undersides of the abdomen where such devices are not effective at depositing chemicals. Therefore, chemical control of these flies on cattle, horses and sheep depends on direct spraying of repellents/insecticides to the extremities. Depending on the type of operation (e.g., a few animals in a barn or stable vs. many on a free range), direct application may be relatively simple or a major and expensive undertaking. Thus, it usually is advisable to monitor pest population levels to determine when such measures are needed. See Foil and Hogsette (1994) for a comprehensive discussion of measures to control tabanids, stable flies and horn flies.

Bednets

Bednets have long been used for relief from the annoyance of night-biting arthropods. In the tropics, important night-biters include *Anopheles* mosquitoes. Thus, bednets have protected against malaria (and other vectorborne diseases) since long before the definitive connection was made at the beginning of the 20th Century between the malaria parasite and mosquitoes.

The first bednets, made of silk, originated in ancient China. Their basic design has changed greatly over the past 1,000 yr. Today, bednet size and material reflect the circumstances and local customs under which they are used. The reader should see Sexton (1994) for a discussion of some of these. The classic bednet is a conical shroud of white cotton mesh, though polyester and nylon are modern replacement materials (Fig. 15.4). However, many other bednet shapes are used. The best bednets for western-style beds have rectangular ceiling frames that permit the mesh to drop straight-sided to the edges of the mattress. This drape prevents the problem common with conical bednets in which the material sags inwards on the sleeper, making bites through the mesh more likely. Bednets also come in models for hammocks, sleeping mats and various-sized arrangements including family beds.

Figure 15.4. Bednets for sale in an open air market (courtesy of the Canadian International Development Research Corporation).

Figure 15.5. Family group in front of large bednet (courtesy Canadian International Development Research Corporation).

Under ideal conditions, a well-maintained bednet used nightly and properly can provide almost complete protection from night-biting arthropods. Proper use in this context means rolling the net tightly during the day to prevent entry of mosquitoes as they look for daytime resting sites, and lowering it at bedtime, tucking the edge under the mattress or mat to prevent mosquitoes coming in under or around the sides. Ideally, the sleeper should wear long sleeves and trousers or sleep completely under the covers so that bare arms and legs do not inadvertently touch the net where they could be bitten through the mesh. The net should be free of holes through which hungry mosquitoes could enter. A sleeper under a torn bednet is virtually a captive meal; a net in this condition may be worse than useless.

Historically, bednets were used by European colonials much more than by the colonized. In modern times, bednets have been used by visitors (tourists and the military) to the tropics and by certain ethnic communities in southern China, parts of Latin America and Africa and Papua New Guinea. Increasing malaria parasite drug resistance, combined with vector resistance to insecticides spread in the environment, now makes it clear that new tools and approaches are needed in malaria control. Until recently, bednets have been used for personal protection on an individual basis. Bednets have not been perceived as well-suited to widespread use by indigenous populations because they are difficult to maintain and they lack acceptance by the general public. Furthermore, though regular bednets protect the individual from malaria vectors, their use in no way affects the vector population itself, neither in terms of its longevity nor its vector potential. Vector mosquitoes that are denied access to one host by a bednet presumably deflect to those without bednets. Unless virtually everyone in the household and a very large proportion of the community uses bednets regularly, little advantage is gained in a public health sense (Fig. 15.5).

The recent development of bednets impregnated with quick knockdown pyrethroid insecticides of long stability and low mammalian toxicity (e.g., permethrin, deltamethrin) has made bednets much more potent tools in malaria prevention. Insecticide-impregnated bednets (IBNs) have all the advantages of untreated bednets (UBNs) plus several others due to the insecticidal and excito-repellent properties of the pyrethroids used on them (Fig. 15.6). In the absence of an effective vaccine, IBNs have been embraced by many international development agencies and the World Health Organization (WHO) as having an important role to play in limiting future malaria spread. The bednet now is seen as a tool in the arsenal of community health in the tropics as well as a device for personal protection.

Attempts to measure the effects of IBNs on mosquitoes have produced variable results, although trends have emerged (Tables 15.2, 15.3). In general, mosquitoes contacting IBNs experience high mortality rates and low feeding success compared to mosquitoes exposed to UBNs. These effects are attributable directly to the in-

Figure 15.6. Woman treating bednet by washing in permethrin solution (courtesy of Canadian International Development Research Corporation).

secticidal properties of the pyrethroids used. Just as importantly, pyrethroids are excito-repellents that discourage entry of mosquitoes into houses where they are used to hasten the departure of mosquitoes already in the house. A study by Arredondo-Jimenez et al. (1997) on *An. albimanus* in Mexico provides specific examples of several effects. Mosquitoes exposed to lambdacyhalothrin-treated (30mg/m^2) polyester bednets for as little as 3 min exhibited mortality rates of 40–55%. Exposure times of 15 min produced 90–100% mortality. Mean resting time of mosquitoes on IBNs was 6–9.5 min (depending on material), providing sufficient contact time for a large percentage of mosquitoes to acquire a lethal pyrethroid dose. Using unfed mosquitoes introduced into experimental houses over a period of 17 weeks, these authors found that bloodfeeding success rates were 8–15% in houses with polyester IBNs compared to 23–24% in houses with UBNs. Mosquitoes were much more likely to leave houses with IBNs (65–78% of sample within 3 hr) than they were to leave houses with UBNs (53–59% of sample within 3 hr), and those leaving IBN-houses showed mortality rates of up to 50% compared to less than 10% for mosquitoes leaving UBN houses. Other studies (e.g., Hossain and Curtis 1989) have shown that IBNs are effective even when torn since mosquitoes are less likely to remain in contact with the mesh long enough to find the holes and are more likely to die even if they enter and feed successfully.

Results of investigations vary from species to species and from study to study but the overall message is that IBNs provide better protection from night-biting mosquitoes than UBNs (or nothing at all) and they often do so in ways that profoundly affect the vector status of the mosquito population. Carnevale et al. (1988, cited in Curtis et al. 1990) described a test of IBNs in villages in a malaria hyperendemic part of Burkina Faso. Even though human collectors at collecting stations in both control (no bednets) and IBN villages received approximately the same number of bites in evening collections in the second year of the study, the malaria risks in each village were vastly different. Mosquitoes caught in the IBN village were on average younger (more nulliparous females), presumably because contact with IBNs had killed many mosquitoes before they could feed or while they were feeding. As a result, the population had much lower **sporozoite rates** (percentage of vector population carrying infective sporozoites) meaning that for people in the IBN village, the chance of a bite being infective (the **entomological inoculation rate** or **EIR**) was reduced by 90%. Even those not possessing or using IBNs in the IBN village received benefits because of the effects on the EIR. This fact trans-

Table 15.2. Effects of impregnated bednet use on selected entomologic factors (from Sexton 1994).

Entomologic factor	Location of study where entomologic factor was:	
	Improved	Unaffected
House entry	Burkina Faso, Kenya, Tanzania	—
Mosquito infection (sporozoite) rate	Burkina Faso, Kenya, Papua New Guinea, Tanzania	Papua New Guinea, The Gambia
Indoor bloodfeeding	Burkina Faso, China, Kenya, Papua New Guinea, Solomons, Surinam, Tanzania, The Gambia	Papua New Guinea, The Gambia
Outdoor bloodfeeding	Burkina Faso, China	Indonesia
House exiting	Burkina Faso, Kenya, Papua New Guinea, Surinam, Tanzania, The Gambia	—
Mosquito mortality	Burkina Faso, Kenya, Solomons, Surinam, Tanzania, Papua New Guinea	Papua New Guinea, The Gambia
Parity rate	Burkina Faso, China, Kenya, Papua New Guinea, Tanzania	Indonesia, Papua New Guinea

Note: Listing of the same country in both "improved" and "unaffected" columns reflects two or more studies with differing results.

forms the IBN from being a device useful only for personal protection to a public health tool. The authors estimated that the IBN village realized an overall 59% annual reduction in the chance of having a malaria attack.

Not every attempt to find a significant effect of IBNs on mosquito population features or vectorial capacity has been successful. Thomson et al. (1995) found no effect on sporozoite rates in populations of *Anopheles gambiae* s.l. in several villages included under the The Gambia's successful National Impregnated Bednet Program. They concluded that there is no evidence for an epidemiologically significant "mass killing effect" brought about by the use of IBNs in these villages, and further that the protective effect of the program is due to a high degree of personal protection provided by the IBNs, not by effects of IBNs at the community level. Similarly, Quinones et al. (1997) could not explain how IBNs protected children from malaria in a series of Gambian villages because they found no differences in mosquito density, population age structure, sporozoite rates or reproductive rates between IBN and UBN villages. Clearly, the interaction between vector mosquitoes and IBNs is complex. Fully effective use of IBNs probably is not possible without a much better understanding of these interactions.

IBNs have been tested in numerous field trials in Central and South America, Africa, China, southeast Asia, India, Papua New Guinea, Indonesia and other locales. Curtis et al. (1990) provided detailed summaries of a number of such trials. Most studies concluded that IBNs do indeed reduce the prevalence of malaria. To

Table 15.3. Effects of impregnated bednet use on selected malariometric factors (from Sexton 1994).

Malariometric factor	Location of study where malariometric factor was:	
	Improved	Unaffected
Fever	Burkina Faso, China, Kenya, Tanzania	Burkina Faso
Malaria mortality	The Gambia	—
Malaria incidence	China, Kenya, Papua New Guinea, Tanzania, Thailand	Burkina Faso, Papua New Guinea
Malaria prevalence	Burkina Faso, China, Kenya, Malaysia, Papua New Guinea, The Gambia, Vietnam	Burkina Faso, Mali, The Gambia
Spleen size	Burkina Faso, Mali, Papua New Guinea, The Gambia	The Gambia
Packed red cell volume	The Gambia	The Gambia
Parasitemia	Burkina Faso, Kenya, Tanzania, The Gambia	Kenya, Tanzania

Note: Listing of the same country in both "improved" and "unaffected" columns reflects two or more studies with differing results.

test this overall impression, Choi et al. (1995) performed a meta-analysis on 10 field trials that measured the affect of IBNs on the number of episodes of malaria. This analysis confirmed that IBNs reduce malaria episodes by 24% over UBNs and by 50% over no bednets at all.

In order to judge the success of bednet trials, it might seem that the most obvious measurement would be the reduction in malaria cases. However, Sexton (1994) lists several other malariometric factors, including packed red cell volume (as a measure of anemia), parasitemia level (as a measure of disease severity) and spleen size that may have greater relevance than the absolute number of malaria episodes. For instance, in addition to finding fewer childhood cases of malaria as a result of IBN use in The Gambia, Snow et al. (1988) also observed a significantly lower number of high parasitemias (and therefore, fewer cases of severe illness) in children sleeping under IBNs. In hyperendemic areas where the annual EIR is anywhere from 200–2,000 or more, IBN bite protection efficiency would have to be close to 100% to provide reasonable assurance that a person would not contract malaria. Furthermore, when the law of averages finally catches up, illness severity will be much greater than usual since human immunity to malaria, an incomplete immunity that requires constant reinforcement through malaria challenge, would have lapsed. It is more important in hyperendemic areas that IBNs reduce parasitemia, and therefore disease severity, in young children than reduce cases of malaria *per se*. Thus, children still would be exposed to malaria and gradually acquire partial immunity while being spared debilitating illness or death.

The role of IBNs in areas where malaria transmission is unstable, seasonal or hypoendemic differs from their role in the hyperendemic locales discussed above. Where transmission is unstable, the general population has little acquired immunity because the malaria challenge is not sustained. Malaria infections, when they do happen in areas of unstable transmission, are more likely to have severe consequences and are not as likely to be confined to children. Ironically, per case morbidity and mortality in areas of low malaria incidence is often much greater than in areas of hyperendemicity. This ratio is illustrated by Trape et al. (1993) who found that despite low prevalence in the urban setting in Dakar, malaria was responsible for a disproportionate number of school absences and fevers requiring medical attention. Given that the annual EIR in Dakar is between 0.25 and 1.0, an IBN program with the aim of decreasing the number of malaria attacks by eliminating infectious bites is both realistic and desirable.

To be effective at the community level, IBNs must be both cost-effective for the people or governments buying them and a large proportion of the community must use them. Costs for bednets vary with size, material and other factors, but single or 2-person bednets made of polyester cost anywhere between $3–$6.50. Recommendations usually are to re-treat the IBN with insecticide every 6 months at a cost of $0.50 (Richards et al. 1993). In households with 2 or more adults and several children, this expense can drain a subsistence income in a developing country. If the cost of the nets is out of reach for the family, governments or foreign-aid agencies must pay for them. In their cost-benefit analysis of IBN programs in Malawi and Cameroon, Brinkmann and Brinkmann (1995) found that the cost of IBNs compared favorably with the health care costs represented by the cases of malaria these programs would prevent. IBN programs also compared favorably to residual DDT spraying methods, both in terms of the costs of material and labor required to protect each household and the level of training and centralization needed to run each type of effort. DDT programs are centralized and require specialized personnel, whereas successful IBN initiatives can be community-based. Add to these benefits the improved economic performance that can be expected from healthy communities and there seems little doubt that subsidized IBN programs eventually will more than pay for themselves.

The success of IBN programs rests ultimately on the willingness of the members of the community to use their bednets and maintain them properly. Even the most highly-educated and motivated western tourist may at times have difficulty complying with all of the best practices of bednet use. Bednets can be inconvenient, confining and hot. Sharp et al. (1995) found that the occurrence of dengue in American soldiers in Somalia as part of Operation Restore Hope was significantly higher among those that did not use bednets regularly. Soldiers sleeping during the day after standing overnight watches often would not use their bednets because of the heat. Because of the obvious disincentives to using IBNs, work must be done at the community level to convince people of the advantages of bednets. The logical approach would be to educate the community about the advantages of IBNs in preventing malaria. However, this concept may be too abstract to serve as a strong motivation. Richards et al. (1993) asked 296 participants in an IBN study in Guatemala for their top reasons for using bednets. Only 11% indicated "to improve health" and only 7% specified "to prevent malaria." The most common responses were "to avoid mosquito bites" (56%) and "to improve sleep" (21%). Perhaps these and similar practical advantages of bednets (privacy, protection from bed bugs, reduviids, snakes, etc.) as opposed to the abstract epidemiological ones, should be used as the selling points for IBN programs.

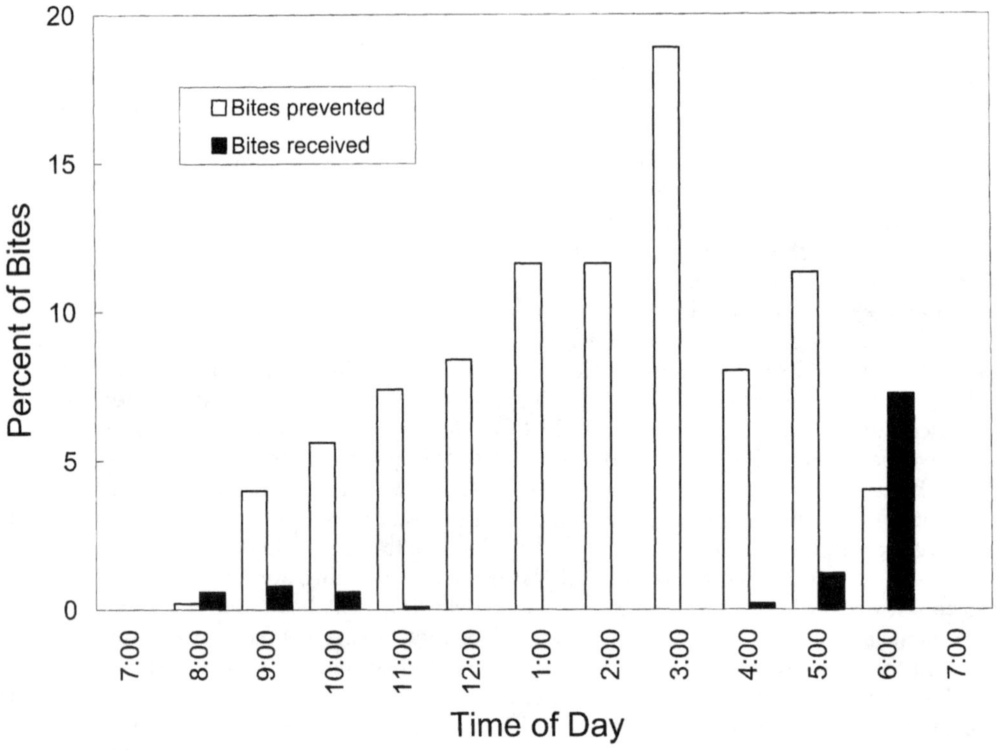

Figure 15.7. *Anopheles gambiae* biting intensity. "Bites prevented" indicates protection afforded by documented bednet use in The Gambia. "Bites received" indicates biting exposure due to nonuse of, or absence of, bednets through the night (from Snow et al. 1988).

Insecticide-Treated Curtains

Insecticide-treated curtains (ITCs) for windows, doors, eaves and walls also may find applications in malaria control since some situations are not well-suited to IBN use and since it is not practical to be in bed for all of the hours that *Anopheles* mosquitoes are active. Snow et al. (1988) plotted *An. gambiae* s.l. biting activity for each hour of the night against the proportion of each hour people in a Gambian village spent outside of IBNs. They found that assuming everyone has an IBN and uses it while in bed, about 10% of person-hours were spent out of IBNs and therefore exposed to potentially infective bites (Fig. 15.7). Because compliance with IBN use is never 100%, this estimate probably is optimistic. Curtis et al. (1990) also pointed out that even brief periods spent out of IBNs at night might be disproportionately risky because mosquitoes frustrated from biting by bednets may pounce when opportunities to feed present themselves.

Beach et al. (1993) compared IBNs to ITCs (both made of 50% nylon-polyester and impregnated with $0.5 g/m^2$ of permethrin) hung in windows and doorways for their affects on clinical, parasitologic and entomologic indicators of malaria prevalence in several villages in western Kenya. In comparison to levels in unprotected control villages, both IBNs and ITCs reduced *Plasmodium falciparum* infections in children <6 yr old, and reduced the prevalence of high parasitemias and EIRs. In another approach, Crook and Baptista (1995) examined the affects of permethrin-impregnated wall curtains on malaria indicators in a suburb of Maputo, Mozambique. Indoor biting of both vector species (*An. gambiae* s.l. and *Anopheles funestus*) was reduced in curtained houses, and the proportion of children exhibiting malaria parasitemias in curtained houses and in neighboring houses dropped over the course of the study.

Insecticide-treated curtains, while outperformed by IBNs in controlled parallel studies, appear to have potential in malaria prevention and should be examined more carefully. According to Beach et al. (1993), the cost of providing the average house with ITCs is 30% less than providing the occupants with IBNs. Furthermore, the per person cost of curtains drops as the number of occupants increases and everyone in the house enjoys the full effect of curtains, including those who will not use a bednet. ITCs installed in windows, doors and eaves also may provide protection from other diseases such as yellow fever and dengue that are transmitted by day-biting mosquitoes.

CHEMOPROPHYLAXIS AND CHEMOTHERAPY

The use of drugs in connection with arthropodborne diseases takes two forms of **chemoprophylaxis** (the use of drugs to prevent disease) and **chemotherapy** (the use of drugs to treat disease). Chemotherapy also may be a *de facto* preventative if reducing the size of the reservoir of infected individuals decreases the transmitting proportion of the vector arthropod population. For this reason, and because some chemotherapeutic agents also are used in prophylaxis, the roles of drugs in prevention and treatment of arthropodborne diseases are both considered here. Table 15.4 summarizes preventive and therapeutic agents for arthropodborne diseases other than malaria.

Arboviral Diseases

Finding compounds to prevent and treat viral infections has proven more difficult than finding antibacterial agents because viruses, being much simpler than bacteria, present few targets for such agents to work against. In addition, most of the vulnerable processes available as targets are normal cellular functions that the virus takes over. Extensive host cell damage could result from the application of antiviral agents. Although a few antivirals are in use, these generally are not applied to arboviral diseases. Treatment for viral (including arboviral) diseases ordinarily consists of general supportive care while the patient's immune system deals with the viral challenge.

Rickettsial and Bacterial Diseases

Antibiotics can readily treat arthropodborne rickettsial and bacterial diseases, although for the most part, drugs are not used to prevent them.

Most rickettsial diseases (epidemic typhus, murine typhus, Rocky Mountain spotted fever, Q fever) are treated with tetracyclines such as doxycycline as a first choice and if individual sensitivity, pregnancy or other contra-indications dictate, with chloramphenicol. These antibiotics rapidly effect **clinical cures** (elimination of symptoms), decrease the number and severity of relapses, and if given for a long enough period, also bring about **radical cures** (elimination of the pathogen from the body). A radical cure is desirable since individuals harboring subclinical levels of the organism in the

Table 15.4. Chemoprophylactic and chemotherapeutic agents and vaccine status for selected human arthropodborne diseases. For information on malaria, see Tables 15.5 and 15.7. (Table continued on next page).

Disease (pathogen)	Chemoprophylactic agent	Chemotherapeutic agent	Vaccine Status
Bacterial diseases			
Plague (*Yersinia pestis*)	Tetracycline	Tetracycline, streptomycin, chloramphenicol	Killed bacterial vaccine; not in general use
Tularemia (*Franciscella tularensis*)	None	As with plague	None
Typhus diseases (*Rickettsia* spp.)	None	Tetracycline, chloramphenicol	No longer available
Lyme disease (*Borrelia burgdorferi*)	None	Tetracycline or penicillin in early stages	OspA, B and C vaccines under development
Recurring fevers (*Borrelia* spp.)	None	Tetracycline, penicillin	None
Protozoan diseases (other than malaria)			
Chagas disease (*Trypanosoma cruzi*)	None	Nifurtimox or benznidazole for acute phases, none for chronic phases	None
Gambian sleeping sickness (*Trypanosoma brucei gambiense*)	Pentamidine and suramin effective, but not generally used	Pentamidine and suramin in early stages, melarsoprol if CNS involvement	None
Rhodesian sleeping sickness (*Trypanosoma brucei rhodesiense*)	None	Suramin in early stages, melarsoprol if CNS involvement	None
Kala azar (*Leishmania* spp.)	None	Pentavalent antimonials (e.g., sodium stibogluconate) as first choice, pentamidine in resistant cases	None
Cutaneous leishmaniasis (*Leishmania* spp.)`	None	Pentavalent antimonials as first choice, ketoconazole or amphotericine B in resistant cases	None, although some cultural practices result in immunization against Oriental sore

Table 15.4 (Continued).

Disease (pathogen)	Chemoprophylactic agent	Chemotherapeutic agent	Vaccine status
		Filarial diseases	
Onchocerciasis (*Onchocerca volvulus*)	None	Diethylcarbamazine (DEC) microfilariacidal, suramin adulticidal, but is neurotoxic, ivermectin microfilariacidal and drug of choice	None
Bancroftian filariasis (*Wuchereria bancrofti*)	None	DEC treatment of choice, ivermectin clearance of microfilariae less sustained than by DEC	None
Brugian filariasis (*Brugia malayi*)	None	DEC treatment of choice, ivermectin clears microfilariae slowly	None
Loiasis (*Loa loa*)	Weekly DEC treatment, not commonly used	As with brugian filariasis	None

body can be sources of future infection for others and since repeated sub-radical treatments may lead to development of drug resistance. Application of PCR methods to assay the presence of *Orientia tsutsugamushi* (causative agent of scrub typhus) in the blood of infected individuals (Murai et al. 1995) showed that even though symptoms no longer were present after 2 days of antibiotic therapy, up to 6 days more treatment were needed before all rickettsiae were eliminated. This time lag illustrates the importance of continuing antibiotic treatment after an apparent cure to avoid recurrence of symptoms and development of drug resistant strains.

The bacterial diseases plague and tularemia are treatable readily with streptomycin, A-tetracycline and chloramphenicol, in order of preference. These antibiotics work dramatically, especially when used early in the course of the illness.

Preferred treatments for *Borrelia* infections are tetracycline as a first choice and penicillin as a second choice. In general, antibiotic treatment of these infections is straightforward and effective. However, the spirochete causing Lyme disease (*Borrelia burgdorferi*) appears to be able to enter cells of some tissues, thus escaping the antibiotic when administered over the usual 30-day course of treatment. The result is a chronic infection characterized by arthritic and neurologic symptoms. There is much controversy about whether the chronic form of Lyme disease really exists and if so, how best to treat it (see Marshall 1995). Advocates of aggressive treatment claim that prolonged courses of antibiotic treatment are needed to achieve sufficiently high drug titers within cells, joints and connective tissues where the spirochetes are thought to hide.

Protozoan Diseases

Leishmaniasis

Sodium stibogluconate (Pentostam) or meglumine antimonate (Glucantine) treat visceral leishmaniasis (kala azar) caused by *Leishmania* spp. These drugs contain pentavalent antimony, a less toxic alternative to the trivalent antimonials (e.g., stibophen, antimony sodium tartrate) that preceded them. Most cases respond well to pentamonials although cost and the fact that they must be administered by injection limit their use. Inconsistencies in the effectiveness of pentavalent antimonials, particularly in India, have been interpreted as drug resistance in some strains (Callahan and Beverley 1991); however, Franco et al. (1995) found that some lots of Glucantine consisted of up to 16% of the trivalent form of antimony and suggested that product instability also may contribute to variations in treatment efficacy.

Glucantine injections also treat cutaneous leishmaniasis, but WHO recommendations do not call for this expensive treatment unless facial lesions fail to heal. Recent tests of topically-applied antibiotic ointment containing paromomycin for this condition are promising although a double-blind study failed to show differences in parasite levels in lesions of control and experimental individuals (Ben Saleh et al. 1995).

Trypanosomiasis

Treatment of American trypanosomiasis (Chagas disease) is highly problematical since an effective, affordable drug sufficiently free of adverse side effects is not available. Nifurtimox, a nitrofuran compound, and benznidazole, are the only drugs that have found some use in treatment of Chagas disease. Nifurtimox seems to attack the intracellular stages of the parasite. Although effective in treating acute phases of the disease, achieving up to 90% cure rates in clinical tests, neurologic side effects (including convulsions) in up to 70% of cases makes both drugs impractical alternatives for mass treatment.

Treatment options for African trypanosomiases offer greater potential. Another arsenical, Melarosoprol, largely has replaced the arsenical drug Tryparsamide, used for many years to treat Gambian trypanosomiasis. This newer compound is especially useful in treating cerebrospinal infections, but must be given under clinical supervision. The nonmetallic drugs suramin (germanin) and pentamidine also are well-established as treatments for African trypanosomiasis. Suramin is used against the early stages of both Gambian and Rhodesian trypanosomiasis. Since neither drug readily enters the cerebrospinal fluid, they are of no value in treating cases of cerebrospinal trypanosomiasis. Both suramin and pentamidine maintain high levels in the tissues following treatment making them potentially effective prophylactic agents.

Filariasis

Historically, treatment of microfilarial infection with *Wuchereria bancrofti, Brugia malayi* and *Loa loa* has been with **diethylcarbamazine** (DEC), a piperazine derivative developed in response to the thousands of cases of filariasis contracted by American soldiers in the Pacific during World War II. This drug kills microfilariae and adults of these species and can readily effect cures. DEC is largely without direct side effects, although severe reactions are common after treatments because of the great numbers of microfilariae that die suddenly, producing skin and systemic reactions. While DEC is an effective treatment for individuals with these filariases, it also has effects at the community level. For instance, Chadee et al. (1995) followed up on a 1980 mass DEC treatment of almost 600 people in a rural community in Trinidad. Even though no vector control had been undertaken in the intervening 12 years, the proportion of those tested that were positive for *W. bancrofti* had dropped from 14.5% in 1980 to 0% in 1992 because the reproducing adult worm population had been eliminated.

The antibiotic **ivermectin** recently has been shown to be safe for humans and to have significant microfilarial clearing properties. Reactions to large doses of ivermectin are less severe than to DEC, so ivermectin has been assessed for its ability to clear *W. bancrofti* infections in single large doses. Addiss et al. (1993) working in Haiti, and Kazura et al. (1993) working in Papua New Guinea, both found that ivermectin rapidly decreased microfilarial load after single doses. Kazura and co-workers found that DEC produced slower but more sustained decreases in microfilaremia. Although these conclusions differ somewhat, it is clear that ivermectin, either alone or in combination with DEC, has an important role to play in treatment and community level control of *W. bancrofti* infections. Ottesen and Campbell (1994) reviewed the status of ivermectin in human medicine.

Onchocerca volvulus microfilarial loads also are suppressed by DEC treatment, but in cases where *O. volvulus* microfilaremias are high, severe skin and systemic reactions from the thousands of dead and dying microfilariae often result within 16 hr of DEC treatment and last for several days. DEC has no effect on *O. volvulus* adults. A second agent, suramin, has been employed in combination with DEC as a radical cure for onchocerciasis.

Several studies also have shown that ivermectin effectively reduces *O. volvulus* microfilarial loads and that it does so both through direct toxic action on microfilariae (Duke et al. 1991a) and through longer term effects on the adult worms (Duke et al. 1991b, Duke et al. 1992). The pharmaceutical company Merck, Sharp and Dohme now donates ivermectin in the form of Mectizan™ to the Onchocerciasis Control Program (OCP) in West Africa and to countries in Central and South America which are endemic for onchocerciasis. Several community-based trials in Africa and Central America have confirmed ivermectin's ability to reduce microfilarial loads while producing fewer adverse effects than DEC. Benefits of ivermectin use in some trials have been greater than simple suppression of microfilarial loads and accompanying symptoms. Guatemalan studies where ivermectin was supplied to the people of test villages once every 6 months for 3 yr reduced new cases more than 80%, because microfilarial loads were reduced below infective thresholds for the vector black flies (Cupp et al. 1992). Similar findings in African studies (e.g., Boussinesq et al. 1995, Somo et al. 1993, Trpis et al. 1990) allay fears that reduced microfilarial loads might not lower transmission by more efficient vector species such as those in West Africa. Ivermectin's effects, though dramatic, are not sufficient in themselves to control onchocerciasis, especially where participation rates within the community are low and where movement of infected individuals brings new sources of infection from outside (e.g., Rodriguez-Perez et al. 1995). Ideally, ivermectin therapy should be combined with long-term vector control as is being done in the OCP and in Guatemala. This approach should interrupt the transmission cycle of the parasite, which with no animal reservoirs, eventually will clear from the human population. Cupp et al. (1992) briefly reviewed this topic.

Malaria

Prevention and treatment with drugs

The complexity of the malaria parasite's life cycle provides many potential targets for chemotherapy and chemoprophylaxis. True **causal prophylaxis** of malaria means elimination of the sporozoites in the brief few minutes they circulate before invading liver cells. Unfortunately, no drug that can do this is available. However, targeting other stages of the parasite can achieve effective prophylaxis. Certain agents (e.g., pyrimethamine) kill pre-erythrocytic schizonts, but

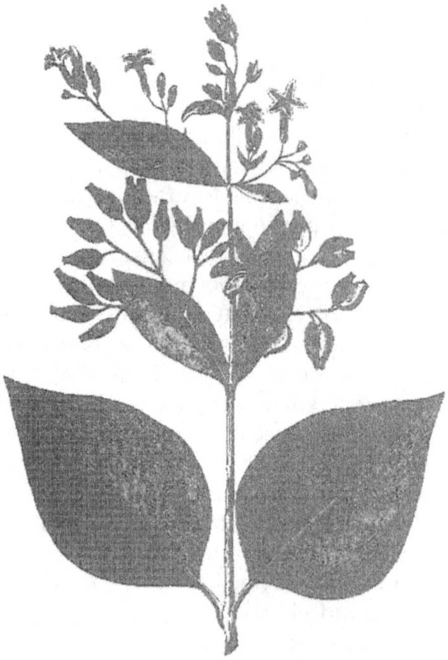

Figure 15.8. The cinchona tree (courtesy of Pan American Health Organization/World Health Organization).

most malaria prevention is actually by **suppressive treatment** of the erythocytic stages (trophozoites) by agents such as chloroquine, quinine and mefloquine. Suppressive treatment must be maintained for as long as the hepatic stages exist. This is a short period for *P. falciparum* malaria, but *Plasmodium vivax* may remain in the liver for months or years in the form of the hypnozoite. Elimination of the hypnozoite constitutes a radical cure. See Table 15.5 for a summary of drugs used in the prevention and treatment of malaria.

History of antimalarial drugs

Jesuit missionaries with the Spanish in South America learned very early, perhaps from aboriginal people, of the malaria-preventing properties of the bark of the cinchona tree (Fig. 15.8).

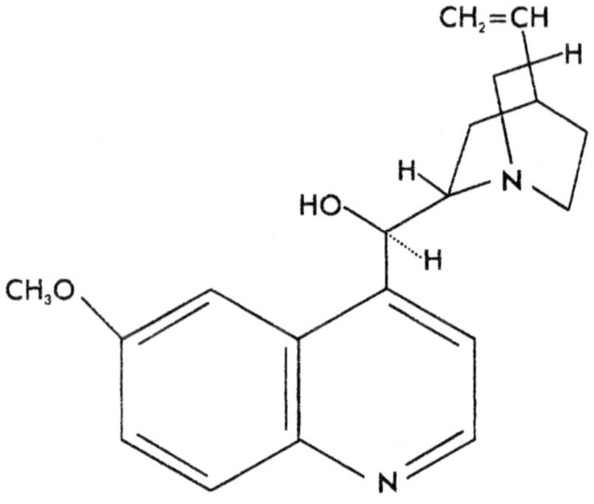

Figure 15.9. Structural formula for quinine.

The Jesuits introduced the bark (which became known as Jesuit's bark or Cardinal's bark) to Europe by 1638. Although the effectiveness of different batches of the bark and the preparations made from it varied greatly, and the medicine often was pedaled by charlatans, it found extensive use (mainly as a therapeutic) in a still-malarious Europe and in the tropics where malaria was seriously inhibiting European colonial activities. Significant advances in the preparation and supply of high quality bark extract were made in the early 19th Century. In 1820, Pelletier and Caventou isolated and purified the alkaloid **quinine** (Fig. 15.9), the active ingredient in cinchona bark. Charles Ledger determined which of the several species of cinchona tree (eventually named *Cinchona ledgeriana*) produces the most quinine.

In the first half of the 20th Century, quinine was considered by some to be the "magic bullet" that would rid the world of malaria as a disease. Quinine was, and still is, an effective clinical cure, but there are no magic bullets. Quinine was used widely prior to the introduction of chloroquine but now is reserved generally as a last-ditch treatment for multi-drug resistant malaria. The potential for using quinine as a preventive (suppressive) treatment never was great because its long term administration commonly produces unacceptable side effects. **Cinchonism** is the term applied to the cluster of symptoms (varying degrees of visual and hearing disturbances, gastrointestinal upset, breathing difficulties, renal damage and disorientation) that can occur after a single dose of quinine. These symptoms more often are associated with taking quinine over a long period of time for prevention or with large doses taken for curative purposes.

Quinine's toxicity eventually led to a search for other, less harmful antimalarials. The I.G. Farben Company of Germany tested a large number of 4-aminoquinoline compounds in the 1930s looking for new antimalarial agents. Among these was **chloroquine** (called Resochin by I.G. Farben; Fig. 15.10). Ironically, chloroquine erroneously was thought to be toxic, and as a result, was not widely used by the Germans during the World War II. The Americans reappraised the 4-aminoquinolines based on German-French field trial data captured in 1941. The US government synthesized chloroquine and, finding no toxicity problems, used it to replace atebrine, the antimalarial then in use by Allied troops in World War II.

Chloroquine, in addition to being relatively nontoxic, is inexpensive, easily synthesized and highly effective in prevention and cure for all human malarias. Used extensively by military and civilian travelers to the tropics since the 1940s, chloroquine was the mainstay of personal protection from malaria for several decades. Clinical cure is achieved with larger and more frequent dosages.

Chloroquine never has been used extensively to prevent malaria in native populations of endemic areas. This reluctance apparently is because the massive infrastructure that would be required to distribute the drug is expensive and partly because of fear of the side effects of long-term chloroquine treatment. Chloroquine is less toxic than quinine but has been associated with retinopathy (Lange et al. 1994) and other side effects. Further, Desowitz (1991) claimed that fear of massive overpopulation in developing tropical countries if malaria-associated childhood mortality were eliminated also prevented the international community from embracing population-wide use of chloroquine. Another "miracle chemical," the insecticide DDT, also influenced how chloroquine was used in indigenous populations. DDT used as a residual insecticide in houses to kill alighting mosquitoes, was the main weapon of the Global Malaria Eradication Program (1955–1970) spearheaded by the WHO. By the time it became clear that the eradication goal of the program was unattainable (see Desowitz 1991 for a discussion of reasons for this failure), some strains of *P. falciparum* had become chloroquine-resistant. The subsequent history of development of antimalarials is largely a response to the appearance of chloroquine resistance.

Evolution of resistance to antimalarial drugs

As with any trait of living organisms, the sensitivity of malaria parasites to antimalarials varies from population to population. For instance, some types of falciparum malaria around the Mediterranean were long recognized to be up to 8 times less sensitive to quinine than strains elsewhere. This is **natural resistance** and should not be confused with **acquired resistance**, which arises as a result of selective pressure exerted by a drug. "Resistance" as used here refers to evolved or acquired resistance unless otherwise stated.

Moore and Lanier (1961) published the first report of chloroquine-resistant malaria in *P. falciparum* infections in Colombia. Resistance appeared virtually simultaneously in this species in Southeast Asia and has since spread to, or appeared *de novo* in, almost all falciparum malaria endemic areas of the world (Fig. 15.11). The evolution of resistance often has been associated with the mass use of chloroquine or

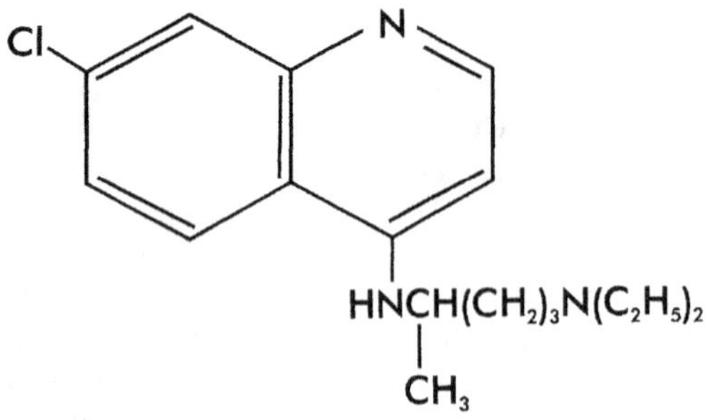

Figure 15.10. The structural formula for chloroquine.

other antimalarials in prevention programs. Once established, chloroquine resistance is stable even in the absence of drug selection pressure. The relatively low doses used in the preventive mode apparently favor resistant parasite mutants that arise spontaneously or occur naturally in the population. Curative doses, which are much higher than prophylactic doses, are less likely to select for resistance.

Resistance develops to some classes of antimalarials more rapidly than to others. Although chloroquine resistance now is widespread in *P. falciparum*, several decades of intensive chloroquine use preceded the first reports of resistance. In general, the 4- and 8-aminoquinolines (e.g., chloroquine, quinine, primaquine) less readily promote resistant strains than other classes of antimalarial compounds such as the pyrimethamines and biguanides (e.g., chloroguanil). This difference may be related to the modes of action in these groups of compounds. Various theories attempt to account for the antimalarial activities of the 4- and 8-aminoquinilones (see Slater 1993 for a review of several theories).

Pyrimethamines, biguanides and sulfonamide antibiotics interfere with folic acid synthesis in parasites by interfering with the enzyme dihydrofolate reductase or with the incorporation of para-aminobenzoic acid into folic acid. Not surprisingly, cross resistance is related to mode of action. Cross resistance commonly occurs between pyrimethamines, chloroguanides and sulfonamides as a group and between chloroquine and other 4- and 8-aminoquinolines. Strains exhibiting such cross resistance are known as **multi-drug resistant**.

Recent studies of malaria population genetics provide insights into circumstances that may favor the appearance and spread of unwanted traits in the parasite. Paul et al. (1995) suggested that lower malaria transmission rates provide greater opportunities for inbreeding within malaria strains since fewer strains are likely to be circulating in the host at a given time. Inbreeding, in turn, fixes traits such as drug resistance. They suggested that extremely high transmission rates, as found in many parts of sub-Saharan Africa, work against the appearance and spread of drug resistant malarias. Indeed, chloroquine resistance in Tanzania, where individuals receive several hundred infective bites per year, has spread much less quickly than in Papua New Guinea, where annual biting intensity is between 40 and 200 bites. It is ironic that efforts to reduce malaria transmission through the use of vaccines, antimalarial drugs and vector control may actually promote the evolution of strains of the parasite resistant to control measures.

The WHO currently recommends chloroquine for falciparum malaria prophylaxis only in Central America, the southern half of South America and the Middle East (including Egypt and Turkey). However, these recommendations are conservative, and chloroquine still may be suitable for use in many parts of Africa. A 10-yr study (1982–1991) by Guiguemende et al. (1994) of malaria resistance in Burkina Faso indicated the first resistance to therapeutic levels of chloroquine in *P. falciparum* infections in 1983. Incidence of resistant cases rose to 16% in 1990 but dropped to 3–6% in most areas by the end of the study. These authors still recommend chloroquine for clinical cure of malaria in most

Table 15.5. Chemoprophylactic and chemotherapeutic agents for human malaria. (Table continued on next page).

Drug	Mode of action	Prophylactic application	Therapeutic application	Remarks
Chloroquine	Interferes with hemoglobin breakdown	All malaria types; widespread *P. falciparum* resistance (CR)	Used for all but CR malaria	Inexpensive, safe for children and pregnant women; has side effects (insomnia, nausea, dizziness); CR limits use
Mefloquine	Dihydrofolate reductase inhibitor	Replaces chloroquine in most CR areas	Emergency self-treatment if not used prophylactically	Expensive, side effects preclude use by pilots; not safe for children <7 kg or during early pregnancy; resistance in Southeast Asia
Proguanil	Dihydrofolate reductase inhibitor	Combined with chloroquin for CR malaria	Not used	Safe for pregnant women; side effects (stomach upset, hair loss, mouth ulcerations); replaces mefloquine and doxycyline if these are not available
Pyramethamine	Dihydrofolate reductase inhibitor	Combined with sulfadoxine as Fansidar®; combined with dapsone as Malprim	Fansidar® used as emergency self-treatment when mefloquine prophylaxis fails	Fansidar® no longer used prophylactically because of some fatal reactions; Maloprime replaces mefloquine and doxycycline if these are not available
Doxycyline	Antibiotic effect	First choice to replace mefloquine if resistance present or otherwise unusable	Combined with quinine for multidrug falciparum resistance	Inexpensive; daily dosage required for prophylaxis; not safe for children <8 yr and pregnant women; side effects (phototoxic skin reactions)
Halofantrine	Dihydrofolate reductase inhibitor	Not used	Standby self-treatment in place of Fansidar®; used for multidrug resistant malaria	High or prolonged doses may cause fatal heart arrhythmias

Table 15.5 (Continued). Chemoprophylactic and chemotherapeutic agents for human malaria.

Drug	Mode of action	Prophylactic application	Therapeutic application	Remarks
Quinine	Dihydrofolate reductase inhibitor	Not used	Combined with tetracycline or doxycycline for high grade multidrug resistant falciparum malaria	Therapeutic dosages often cause cinchonism
Primaquine	May be oxidizing agent in cells	Side effects preclude sustained prophylactic use	Used as "terminal prophylaxis" for radical cure of vivax and ovale malaria	Used after cessation of prophylaxis; not recommended during pregnancy; some *P. vivax* resistance
Qinghaosu derivatives	Free radical damage to membranes	Not used	Rapidly effective against severe and uncomplicated vivax and falciparum malaria	Little evidence of adverse reactions; neurotoxicity in dogs after large doses; not widely available or acceptable

parts of the study area because of the low levels of resistance to this affordable drug.

Other antimalarial drugs

Despite the effectiveness and early availability of chloroquine, atebrine (quinacrine) was the first substance other than quinine used widely in malaria prevention and therapy. Atebrine was an effective antimalarial in extensive tests undertaken by the US military during World War II to replace quinine since sources of cinchona bark were under enemy control. Atebrine, an 8-aminoquinoline synthesized in Germany in 1930, is more toxic than quinine, often causing yellow skin, darkened urine, stomach upsets and occasional psychotic episodes. Atebrine was abandoned quickly as an antimalarial with the rediscovery of chloroquine in the mid-1940s.

Mefloquine (Fig. 15.12) (Lariam® in the USA and Canada) is closely related structurally to quinine. It is currently the WHO-recommended malaria preventative for most *P. falciparum* endemic areas although its cost per dose is many times that of chloroquine. Mefloquine taken weekly may produce adverse reactions (usually stomach upset) in a significant proportion of users and is not considered appropriate for airline pilots or heavy equipment operators because it reputedly affects fine motor control. Lower doses (300mg/week vs. 600mg/week) decrease the number and severity of adverse reactions, but a small percentage of people cannot use mefloquine at all. In such cases, doxycycline, a tetracycline antibiotic, is recommended, although it comes with its own set of adverse reactions including photosensitivity and increased risk of vaginal yeast infection.

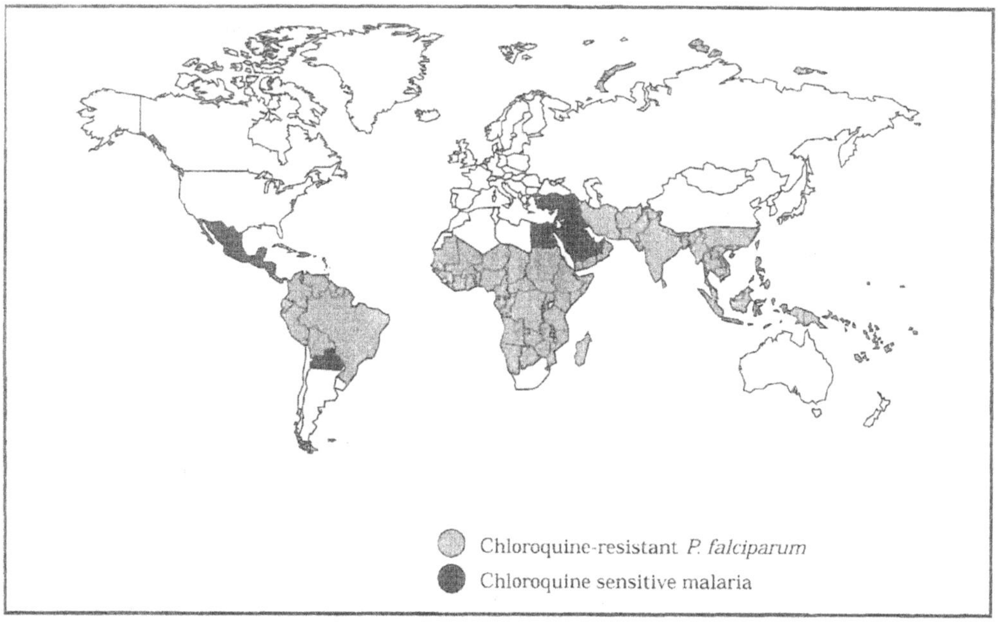

Figure 15.11. Global chloroquine resistance in *Plasmodium falciparum* (courtesy US Centers for Disease Control and Prevention).

Doxycycline is not recommended for pregnant women. Mefloquine resistance (in almost 60% of falciparum malaria cases) now has surfaced along the Thai-Cambodian border (Fontanet et al. 1993). Several studies (e.g., Brasseur et al. 1992, Cowman et al. 1994) indicate that development of quinine resistance is co-selected with mefloquine resistance. This coincidence raises the disturbing possibility of widespread resistance to quinine, currently the last resort for multi-drug resistant malaria treatment.

Several combinations of drugs also have been used in malaria chemoprophylaxis. Fansidar®, a combination of the sulfonamide sulfadoxine and pyrimethamine, originally advanced as a preventative for falciparum malaria areas of chloroquine resistance, no longer is recommended because it causes severe, sometimes fatal, skin reactions when used as a treatment after having been used prophylactically. Fansidar® now is used as a standby curative for individuals experiencing presumptive malaria attacks despite having taken a prophylactic such as chloroquine or mefloquine. Fansidar® should be taken within the first 12 hr of the onset of symptoms but only if more than 24 hr away from qualified medical assistance and only if the patient is not sulfa-drug sensitive. Proguanil (Paludrine) and chloroquine have been used widely as preventatives in areas resistant to straight chloroquine. This combination has the disadvantage of producing high rates of nausea and other side effects.

See Wyler (1993) for a summary of drug alternatives in malaria chemoprophylaxis and for an analysis of risk factors that help determine whether chemoprophylaxis is needed and justified.

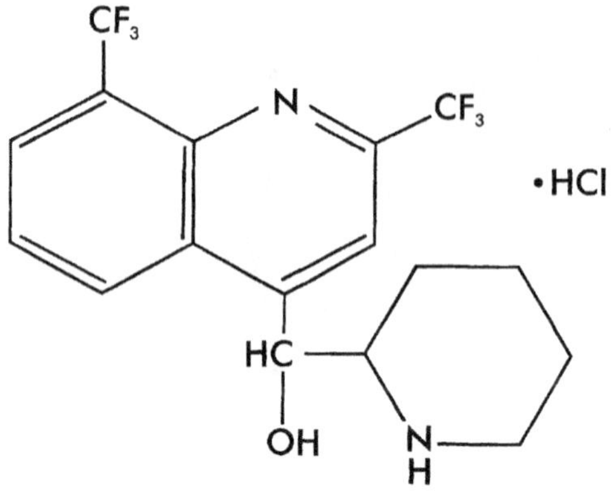

Figure 15.12. Structural formula for mefloquine.

Prospects for new antimalarial drugs

The occurrence of multi-drug resistant falciparum malaria has spurred efforts to develop new families of drugs, or new ways of using old drugs. Extracts of sweet wormwood (*Artemisia annua*) have been used in Chinese folk medicine for centuries to decrease fever. The active ingredient, artemisinin (*qinghaosu*), is a 16-carbon lactone endoperoxide, with impressive antimalarial properties. Orally administered *qinghaosu* was tested as a treatment for falciparum and vivax malarias in a study by Nguyen et al. (1993) in Vietnam where chloroquine and mefloquine resistance occurs. Parasitemias rapidly lowered over the course of treatments although recrudescence was a problem after treatment stopped. The authors concluded that because of the rapidity of its effects, *qinghaosu* may work best in combination with longer acting agents such as mefloquine, chloroquine or antibiotics.

Qinghaosu has been derivatized to a number of more expensive, but more effective, compounds. The most important of these are **arteether**, **artemether** and **artensuate**. The latter two currently are in clinical trials, but already are used widely in China. The WHO has selected arteether for development as an emergency treatment for cerebral malaria. Its high lipophilicity permits better accumulation in brain tissues than is possible with artemether and there is evidence that arteether metabolites are less toxic than those of artemether.

While arteether and other derivatives of *qinghaosu* are potent, they may have as yet unrecognized side effects. For instance, Brewer et al. (1994) suspected that a toxic metabolite of arteether given in high doses produced brain stem lesions, neurological symptoms and death in dogs and rats. Although dosages used in tests were higher than human clinical doses, this finding illustrates the need for more extensive investigations before drugs are made widely available. See Hien and White (1993) for a review of aspects of *qinghaosu* and its derivatives.

Other classes of chemicals with an array of purported modes of action also are under investigation for possible use in malaria prevention and treatment. For instance, new inhibitors of folic acid synthesis may hold promise. Canfield et al. (1993) reported that PS-15, a precursor of WR99210, a known inhibitor of dihydrofolate reductase, more effectively cured *Aotus* monkeys infected with multi-drug resistant *P. falciparum* than proguanil or WR99210 itself. They propose that PS-15 represents a potential new family of antimalarials.

A number of substances that modulate resistance to chloroquine also have been identified. These include verapamil, desipramine, chlorpromazine, cyproheptadine and others that reverse chloroquine resistance in mouse models. The precise way these compounds work is not known. In general, they appear to slow the rate of chloroquine removal from the food vacuole of the erythrocytic stage of the parasite, perhaps by blocking ion channels or by competitively binding chloroquine transport molecules (Slater 1993). De et al. (1993) reported that WR268954, a pyrrolidino-alkaneamine, decreases by 90-fold the 50% *in vitro* inhibitory concentration of chloroquine for resistant *P. falciparum*. They speculated that the molecule competitively inhibits the chloroquine transport molecule, thus preventing chloroquine's rapid clearance. Kyle et al. (1993) compared the abilities of several modulators to affect chloroquine resistant falciparum malaria in *Aotus* monkeys and found that chlorpromazine most effectively cleared parasitemia while cyproheptadine and verapamil were not effective (although they worked well *in vitro*) and were associated with increased levels of toxicity.

New drugs can, in theory, be developed around any aspect of the parasite's metabolism. Thus, iron chelators, substances that make iron unavailable to the parasite's enzyme systems, also are under investigation as chemotherapeutic agents. Gordeuk et al. (1993) administered desferrioxamine B, the only clinically-available iron chelator, to a group of Zambians infected with asymptomatic *P. falciparum*. The result was a rapid and long-term reduction in the number of ring forms in the subjects' blood. The authors proposed that iron chelators deserve further evaluation as antimalarials.

To date, most chemoprophylactic and therapeutic agents for malaria have been directed at the blood stages of the parasite because this stage is accessible, relatively long lasting and is responsible for most of the malaria pathology. However, the parasite's complex life cycle offers many targets for chemicals and other agents. Primaquine, for instance, attacks the latent stages (hypnotozoites) of *P. vivax*, making it the only agent available for radical cure of relapsing-type malaria. Schizonticides normally employed in suppressive treatment also affect the gametocyte stages and thus have potential to block transmission to the mosquito. Some of these include pyrimethamine, cycloguanil and proguanil. Little is known about these properties of antimalarial drugs, but given the high degree of schizonticidal resistance, it seems prudent to target other life cycle stages of this parasite.

VACCINES

Parasitic organisms (bacteria, viruses, protozoan and metazoan parasites) possess proteins (**antigens**) on their external surfaces and elsewhere that may evoke immune responses in the host that allow the host to avoid overt disease or help to shorten the course and severity of disease. Antigens that evoke responses of the immune system are referred to as **immunogens** and are said to be **immunogenic**. Immune responses often confer long-term resistance (**acquired immunity**) after initial exposure to the microorganism. Artificially prepared and administered immunogenic antigens are referred to as **vaccines** (Fig. 15.13) and when properly designed evoke long-lasting **artificial immunity** with few or no symptoms of the disease. Not many vaccines are in active use for bacterial, protozoan or filarial diseases (Table 15.4). Several vaccines have been developed for arboviral diseases (Table 15.6) and a wide range of vaccine candidates for malaria are under investigation (Table 5.7).

The first widely-used vaccine took advantage of the fact that the cowpox virus (*Vaccinia*), which produces a mild illness in humans, induces cross-immunity to the *Variola* virus, which causes a serious human disease, smallpox. Recognizing that milkmaids did not generally contract smallpox, Edward Jenner, an English physician, suspected earlier cowpox exposure protected them. In 1796, Jenner immunized others against smallpox by scratching small amounts of the fluids (containing *Vaccinia* virus) from cowpox blisters into the skin of well individuals.

Strong cross-reactions such as the one between *Vaccinia* and *Variola* viruses are rare, so it has been necessary to develop other methods of producing non-disease causing parasite antigens and introducing them into the host to induce immunity. Basch (1994) details the types of vaccines and their sources. Summarized briefly, vaccines are classified as: **conventional**,

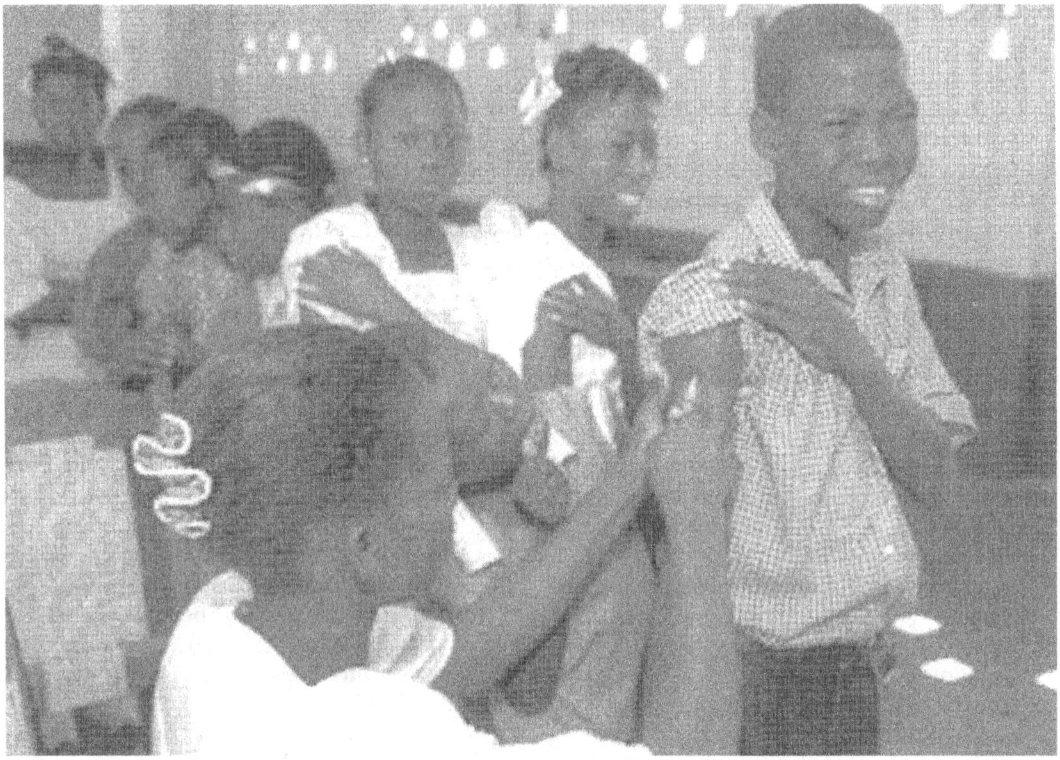

Figure 15.13. School children being vaccinated in Haiti (photograph by Armando Waak, courtesy of Pan American Health Organization/World Health Organization).

deriving from all or part of living organisms; **synthetic**, deriving from specifically synthesized immunogenic molecules; or **recombinant**, deriving from recombinant DNA procedures. Most currently-used conventional vaccines are made from organisms weakened by serial passage through non-host tissues; or by other methods (**attenuated vaccines**), from whole organisms killed by heat, radiation or chemical treatment (**killed vaccines**); from bacterial products such as toxins (**bacterial product vaccines**); or from parts of pathogenic organisms (**constituent** or **subunit vaccines**). An exciting new approach introduces DNA that codes for the antigen directly into the host genome (**DNA vaccines**), thus eliminating the need to extract and purify antigen made by recombinant means. In the future, vaccines made by synthetic or recombinant means will replace many conventional vaccines. This replacement is desirable for immunogen purity and production efficiency. For instance, a recombinant version of the yellow fever vaccine is imminent. See Basch (1994) for summaries of immune responses elicited by vaccines.

Bacterial and viral culture methods have aided vaccine research immensely by providing a supply of non-contaminated immunogenic material for vaccine production. On the other hand, progress towards vaccines for eukaryotic parasites has been slowed in part because these organisms often are impossible to

Table 15.6. Properties and status for vaccines for selected arbovirus diseases.

Disease or pathogen	Vaccine status
Bluetongue	Attenuated virus vaccines available for most strains; given as polyvalent vaccine; used mainly in South Africa and Israel
California serogroup viral diseases	None available
Chikungunya virus	Several inactivated virus vaccines available, but none in general use
Colorado tick fever	Inactivated virus vaccine developed but not licensed
Dengue	Active development for vaccine against all 4 strains; DEN-2 attenuated vaccine confers low rate of protection in humans; evidence of reversion to wild type in some candidate vaccines; risk of predisposition to dengue hemorrhagic fever if immunization not against all strains
Eastern equine encephalomyelitis	PE-6 inactivated vaccine used only for laboratory workers at risk; effective vaccine for horses available
Japanese encephalitis	Inactivated virus vaccine used only for humans with long-term risk; available vaccine for pigs and horses uneconomical
O'Nyong-nyong virus	None available
Oropouche fever	None available
Phlebotomus fevers	None available; multiple, poorly cross-reactive viral strains would require polyvalent vaccine
Rift Valley fever	Attenuated vaccine for sheep; inactive vaccine for horses; inactivated vaccine for laboratory workers only
St. Louis encephalitis	Synthetic polypeptide vaccine under development
Tickborne encephalitis	Inactivated vaccine in use since early 1980s
Venezuelan equine encephalomyelitis	Attenuated live and inactivated vaccine used for horses; human use restricted to laboratory workers at risk; immunization of horses
West Nile fever	None available
Western equine encephalomyelitis	Most horses immunized annually with inactivated virus vaccine; none available for general use
Yellow fever	Live attenuated vaccine (17D) confers long-term immunity in humans

culture or are culturable only through parts of their life cycles (e.g., only the blood stages of *P. falciparum* are currently in culture). Early work on malaria vaccines involved isolation of parasite material from hosts for experimental vaccination of others. This practice on humans was acceptable at one time and provided important insights, but it is not acceptable practice today, especially against the backdrop of HIV, HTLV, viral hepatitis and other bloodborne pathogens.

Vaccines for human use must go through an exhaustive series of tests *in vitro* and, where possible, in model hosts before being tested on human volunteers. Strict guidelines govern testing candidate vaccines in humans and assuring continued quality control (see Parkman and Hardegree 1994). Phase I trials involve short term experiments with small numbers of subjects to determine safety and effective dosage levels, immunization timetables and delivery systems. Adjuvants that stimulate the immune system deliver some vaccines. These combinations of factors must be evaluated thoroughly for efficacy and safety. Phase II trials are larger and longer-term and assess safety as well as effective dosage levels and timetables. Phase III trials involve large numbers of subjects in double-blind experiments that test candidate vaccines against placebos. Phase III trials must be conducted in a wide range of situations to test the vaccine's effectiveness against various parasite strains and interactions with climate, population demographics and ethnicity, nutrition and health of the population. This information must be known before the vaccine is made available for general use. Phase III trials also help to point out practical problems with vaccine distribution and maintenance of a cold chain from the point of shipping to the point of use.

The lengthiness and expense of human vaccine development and testing often prevents vaccines from being approved for general use. The size of the markets for many arthropodborne diseases would not repay the costs of vaccine development. Even where potential markets are large (e.g., malaria), they are mostly in developing countries that cannot afford to pay free market costs. Furthermore, the costs of vaccine storage and the need to turn vaccine stores over regularly to avoid loss of activity or spoilage often makes retention of large stores of vaccines as a safeguard against disease outbreaks non-cost-effective. Even where effective, economical vaccines are known, at any given time supplies may be small or difficult to access. For instance, when bubonic and pneumonic plague broke out in India in 1994, only 40,000 doses of vaccine were on hand in Indian government laboratory freezers. Before the vaccine could be released for use it had to be tested for potency and safety, causing a delay of many days (Bagla 1994). See "Bumps on the Vaccine Road" (Cohen 1994) for a further discussion of some issues that govern vaccine development and marketing.

Arboviral Diseases

Monath (1988, 1989) contains a complete discussion of vaccines used against arboviruses. Jong and McMullen (1995) provide detailed information on preventive measures and treatments available for a wide range of arthropodborne and other diseases.

Testing and licensing of vaccines for use in livestock and pets is not as strictly controlled. Several effective vaccines against zoonotic disease organisms available for animal use are not available for general human use, e.g., WEE virus.

Yellow fever

The yellow fever vaccine was the first vaccine developed against an arboviral disease. Of the 2 original yellow fever vaccines developed in the 1930s, only the 17D vaccine still is in use. This vaccine and its counterpart, the French neurotropic vaccine (FNV), are both live vaccines containing viruses attenuated by serial passage through mouse and chicken embryo tissue (17D) or mouse brain tissue (FNV). Although highly effective and largely responsible for the disappearance of yellow fever from West Africa, FNV was retired in the mid-1960s primarily because it produced higher than acceptable rates of encephalitis through reversion to the wild type and could not be used in children <10 yr of age. The 17D vaccine is safe for children as young as 6 months and produces long-lasting immunity in more than 99% of recipients after a single dose. Re-immunization for yellow fever is required after 10 yr for travel to many countries, although circulating antibodies have been detected in immunized individuals up to 30 yr following the single dose.

Hundreds of millions of doses of 17D vaccine have been administered with few serious side effects. The most serious side effect is encephalitis in very young children although recovery usually is complete. Several million doses are given in Africa and other endemic areas annually. The military and travelers to endemic countries (several countries require travelers to have valid yellow fever immunization certificates if travelling from endemic countries) also use vaccine. Recently, The Gambia, Nigeria and Ghana included yellow fever vaccine in their Extended Program of Immunization (EPI) packages. The EPI is a WHO-sponsored program originally designed to provide children in many African countries with vaccinations against poliomyelitis, tetanus and diphtheria. The WHO currently is studying inclusion of yellow fever vaccine in other endemic countries. The EPI provides an effective cold-chain (17D vaccine must be stored between +5° and –30°C to maintain its potency), allowing distribution to all parts of these countries.

Monath and Nasidi (1993) modeled the cost-effectiveness of including 17D vaccine into the EPI of Nigeria. They estimated a cost of $0.65 per immunized person versus $7.84 per person when emergency immunization is used at the time of an outbreak. They also predicted that wide scale use of yellow fever vaccine in the EPI for Nigeria would decrease transmission of the disease since it would gradually clear infection from the populace.

Japanese encephalitis

Millions of doses of the vaccine to protect against infection by Japanese encephalitis (JE) virus have been given in Japan, China and parts of Southeast Asia where it is partly responsible for bringing this disease under control. Based on inactivated virus preparations, the vaccine confers protection on up to 95% of those receiving it. Although effective, this vaccine usually is not recommended unless one is travelling into endemic areas for long periods. While yellow fever vaccine is useful preventively in the case of outbreaks, Japanese encephalitis vaccine is not as useful since it must be given as a series of 2 or 3 inoculations over 2–6 weeks. The effectiveness of 2 vs. 3 doses in eliciting immunity are examined by Poland et al. (1990), who recommend 3 doses to North American travelers who are at significant risk.

The current Japanese encephalitis vaccine is an inactivated virus usually raised in baby hamster kidney cell culture or mouse brain tissue. Progress is being made towards a recombinant vaccine made from highly conserved regions of the JE virus envelope glycoprotein, the E protein (Brandt 1990).

Another approach to Japanese encephalitis control is the immunization of pigs, the main domestic reservoir for the virus. While pig immunization reduced the number of pigs carrying virus, the cost effectiveness of this approach is doubtful since pig turnover necessitates constant immunization of new stock (Wada 1988, cited in Mogi and Sota 1991). Another limitation of this approach: if vaccinations are done during the transmission season, pigs still can become infected because the time required to raise the host's immune response extends into weeks. A live, attenuated vaccine used in China protects horses from JE virus infection.

Dengue

The vaccines for dengue (DEN) virus illustrate some of the complexity inherent in vaccine development and use. Although effective vaccines do not yet exist for DEN virus (see Monath 1988, 1989; Brandt 1990), the eventual successful candidates will have to be effective against all 4 DEN serotypes to be useful. Anything less than a quadrivalent vaccine may result in antibodies that cross-react with DEN serotypes not covered and predispose the immunized person to dengue hemorrhagic fever.

Experiments with a dengue candidate vaccine developed from a Puerto Rican viral isolate also illustrate the cross reactions possible between vaccines to different flavivirus species, in this case between YF virus and DEN virus. Schlesinger et al. (1956) reported that only 61% of recipients of a DEN-2 vaccine who had not previously had yellow fever developed anti-DEN-2 viral antibody while 90% of those who had previously had yellow fever developed anti-DEN-2 viral antibody. Apparently, infection with YF virus provides some protection from DEN virus which the candidate vaccine boosts. The reverse, yellow fever immunization by dengue infection, has been advanced as the reason for the absence of yellow fever from Southeast Asia, where dengue is thought to have originated.

Tickborne encephalitis

The Austrian-produced inactivated virus vaccine for tickborne encephalitis (TBE) has been used in eastern European countries since the early 1980s. The 3 doses (2 primary and one booster after a year) protect 98-99% of recipients from TBE virus infection. A national immunization program in Austria in 1981 has reduced the number of cases steadily from 632 in 1982 to 201 in 1988 and to an eventual 50 (Brandt 1990). McNeil et al. (1985) found that the rate of exposure and illness among American soldiers serving in the area, and among travelers, is too small to justify the general use of this vaccine in non-indigenous individuals.

Other arboviral diseases

Vaccines that protect against infection by a number of other arboviruses have been developed, including WEE, SLE, Venezuelan equine encephalomyelitis (VEE) and eastern equine encephalomyelitis (EEE) viruses. Vaccines to protect horses from WEE, VEE and EEE viruses are highly effective and used routinely. However, because of cost, none of these is available for human use except for laboratory and field personnel whose work puts them at risk to infection. Many arboviral diseases still have no vaccines at all. Many of these vaccines have not been subjected to wide-scale phase III trials, so it is not known how effective or safe they would be if used in the larger human population. For instance, on the question of effectiveness, Strizki and Repik (1995) reported that the PE-6 inactivated virus vaccine used to protect research workers against EEE virus is less effective against the South American strain of EEE virus than the North American strain. This difference appears to be due to previously unrecognized variation in the envelope proteins of these viral strains.

Bacterial and Rickettsial Diseases

Surprisingly, no vaccines are routinely available for arthropodborne bacterial and rickettsial diseases. In part, this lack reflects the nonprofitability of developing vaccines and testing them for use in humans and domestic animals when the disease in question is not widespread. Also, bacterial diseases are treated easily with antibiotics. For instance, an attenuated E strain vaccine developed for epidemic typhus *(Rick-*

ettsia prowazekii) no longer is available in the USA since it results in mild illness in 10–15% of recipients and the infection can be treated with a short course of doxycycline (Strickland 1991).

Plague

A plague vaccine used in the USA is prepared from cultured plague bacilli that subsequently are inactivated in formaldehyde. After 3 doses, a small but significant proportion of recipients still have antibody levels too low to afford protection. Even those who are protected may still become ill when exposed to the bacteria, although the severity and length of the illness usually is reduced. Thus, whether vaccinated or not, US Centers for Disease Control and Prevention guidelines recommend that in cases of known exposure, prophylactic antibiotics be given anyway. This vaccine makes immunization of entire populations in endemic areas ineffective and even potentially dangerous. First, it is expensive. Second, it gives incomplete and temporary protection. Finally, vaccinated individuals, mistakenly thinking they are completely protected, might not seek antibiotic prophylaxis after suspected contact or even after the appearance of symptoms. Plague vaccine is reserved for those working in the laboratory or field directly with the pathogen or known reservoir animals. Every vaccine comes with its own set of characteristics that, along with the features of the disease and the epidemiological situation, profoundly affect how they should be used.

Lyme disease

The best candidate immunogenic components being considered for Lyme disease vaccines are the outer surface proteins (Osp) that *B. burgdorferi* expresses. Greatest attention has been focussed on OspA-based vaccines since Fikrig et al. (1990) showed OspA-immunized mice were protected from Lyme infection. Fikrig et al. (1992) also showed that the blood of OspA-immunized mice is lethal for the Lyme organisms in the tick. According to Shih et al. (1995), the OspA antibodies travel with the blood into the tick's gut, killing the bacteria before they are transmitted. Not only is transmission prevented but the normally lifelong infection also is cleared from the tick, thereby sparing future hosts the risk of illness, and the small but real potential for transovarial transmission of the spirochete is eliminated. Another target immunogen for a Lyme disease vaccine might be OspC since Schwan et al. (1995) showed that *B. burgdorferi* changes protein coats from OspA in the tick to OspC prior to transfer during feeding.

Various human Lyme disease vaccines currently are being tested but, at the time of writing, only one is available for general use. This is a recombinant OspA vaccine licensed under the name LYMErix^č to SmithKline Beecham by the FDA in December of 1998. According to company-supplied information summarized in Centers for Disease Control and Prevention (1999), this vaccine achieves from 59-80% (95% confidence limits) if given in 3 doses over 12 months. Vaccines for dogs also are available.

Orally administered Lyme disease vaccines also are under study. These would eliminate the need for injections and could have important epidemiological implications beyond those of injectable Lyme vaccines. In particular, it might be possible to immunize rodent reservoirs by distributing vaccine-treated food baits in appropriate habitats. Some studies (e.g., Dunne et al. 1995, Luke et al. 1997) have shown that recombinant OspA vaccines administered orally to mice prevent *B. burgdorferi* infection.

Protozoan Diseases

To date, there are no effective vaccines for diseases caused by eukaryotic parasites. Many factors account for this absence. An essential feature of virtually all eukaryotic life cycles (though not of prokaryotic ones) is regular, obligatory sexual reproduction, which introduces novel reassortments of genetic traits that may allow the parasite to sidestep the immunologic response. The sheer complexity of the eukaryotic genome is another factor. Unlike prokaryotes with a smaller genome, eukaryotic parasites can "afford" to carry several variants of gene encoding (e.g., surface proteins). This redundancy allows these parasites to change external protein coats regularly (see "Trypanosomiases"), meaning that the immune system is always at least a step behind and that simple single component vaccines will not work. Making vaccines sufficiently complex to provide protection against potentially hundreds of parasite variants is neither technically feasible nor affordable.

Malaria

Of all of the arthropodborne diseases, malaria has received the most attention aimed at vaccine development. The malaria vaccine has been the Holy Grail of tropical medicine for decades, but like the real one, this grail is elusive and some are convinced it never will be found. Those who search for malaria vaccines take heart from the fact that people living in malaria hyperendemic regions, if continually exposed, gradually develop immunity to the parasite by the time they become adolescents. Thus, adults in hyperendemic areas rarely die of malaria, although they may contract the disease repeatedly. Those who doubt that an effective malaria vaccine will ever be developed note that naturally-acquired immunities are partial, have to be reinforced by constant malaria challenge and that partial immunity provides little or no protection from new variants of the parasite introduced into an area or encountered while travelling.

Attempts to formulate malaria vaccines started under the impetus of quinine's unacceptability as a chemoprophylactic and the need to protect Allied soldiers in North Africa, Italy and the South Pacific during World War II. The initial efforts at developing a malaria vaccine involved injecting humans and various test animal hosts with parasite material from host sources (e.g., from infected red blood cells). These materials failed to bring about complete protection, or often, any protection at all from malaria challenge. Only when they were combined with adjuvants (in this case, Freund's adjuvant consisting of a mixture of mineral oil and a nonpathogenic species of *Mycobacterium*) could these vaccines protect birds or monkeys from infection with bird or monkey malarias. Unfortunately, the adjuvants caused liver degeneration in birds and serious skin and nervous system reactions in monkeys. With the end of World War II and the introduction of the miracle chemicals chloroquine and DDT, much of the incentive to develop a malaria vaccine dissipated. Little happened in malaria vaccine research until, in the mid-1960s, the wake-up call came in the form of increases in the malaria organism's resistance to chloroquine and the vector mosquito's resistance to DDT. In the ensuing 30+yr, a remarkable story unfolded around the development of a vaccine for malaria (Desowitz 1991). Researchers from all over the world have become involved in the search, vaccines have been promised several times only to fail to materialize, good and bad science has been done, public funds have been spent and misspent and the odd felony has been committed.

Unlike bacteria and viruses, the malaria parasite goes through several forms in the course of its complex life cycle in the vertebrate and mosquito host. Each of these forms is immunologically distinct and a possible target for a malaria vaccine. Three prospective vaccine types, **sporozoite vaccines**, **merozoite vaccines** and **gametocyte vaccines**, are discussed below (see also Table 15.7).

Anti-sporozoite vaccines

The sporozoite is the stage of the parasite that the mosquito passes on to the human host and, as the agent that infects the hepatocytes, is considered the logical vaccine target by many. However, one of the problems with sporozoites as a target is that they are free in circulation for only a few minutes before entering hepatocytes. Thus, an anti-sporozoite vaccine either must work quickly, or the vaccine must be able to recognize and kill infected hepatocytes through a T-cell response, or both. The potential for an anti-sporozoite vaccine was shown in the early and mid-1970s when human volunteers bitten by several hundred *Anopheles* mosquitoes infected with irradiated *P. falciparum* sporozoites were completely protected from falciparum malaria when bitten by normally-infected mosquitoes (Nardin et al. 1982). This protection also was found when human volunteers were vaccinated with irradiated sporozoites dissected from infective mosquitoes (Rieckmann et al. 1974). Though these experiments demonstrate the potential for an anti-sporozoite vaccine, neither technique is practical to prepare vaccines for large numbers of people. If malaria were a virus or bacterium, there would be the possibility of using culture methods to get enough sporozoites with which to make vaccines. Unfortunately, only the blood stages of P. *falciparum* have been cultured, so some other source of circumsporozoite antigen was needed. Molecular tools identified and sequenced several potentially immunogenic proteins in various *Plasmodium* species, and located the genes encoding them (Table 15.7). This discovery opened the way to producing synthetic vaccine candidates based on the native proteins and transferring the genes or parts of them into bacteria or yeast cells where their expression could provide ample amounts of the proteins or peptides for making recombinant vaccine candidates.

The most thoroughly studied of the sporozoite proteins is the circumsporozoite protein (CSP-1) which coats much of the outer surface of the sporozoite and binds the sporozoite to the hepatocyte. Parts of this molecule induce antibody responses of the immune system while others induce cellular responses. Vaccination with whole CSP-1 or CSP-1 components has provided some protection in small human trials, but overall its performance has been disappointing (Ballou et al. 1987). Yet, a sporozoite vaccine in the future still is considered a strong possibility. Perhaps the most effective vaccines will be based on more than one parasite protein (multi-antigen vaccine). Khusmith et al. (1991) achieved 100% protection of mice against *Plasmodium yoeli* by injecting them with a recombinant vaccine containing both CSP-1 and sporozoite surface protein-2 (SSP-2). On the other hand, perhaps different adjuvants, a common component of many vaccines, will improve effectiveness. Stoute et al. (1997) tested the protection conferred on humans to falciparum challenge by a recombinant CSP-1-hepatitis B vaccine conjugate administered with various adjuvants. Six of 7 subjects given the vaccine in an oil-water emulsion with the immune stimulants monophosphoryl lipid A and QS21 were fully protected.

Vaccines against blood stages

The merozoite stage of the malaria parasite, along with the erythrocytic stages of the parasite (collectively called the blood stages), also have been promoted as a target for vaccines, though less effort is devoted to blood stage vaccines than to sporozoite vaccines.

For reviews of the prospects and status for blood stage vaccines for malaria, see Pasloske and Howard (1994). There is reason to suspect that blood stage vaccines for malaria are realistic since monkeys can be protected from the symptoms of malaria with injections of attenuated merozoites of simian malarias. In these cases, even if total protection cannot always be achieved, merozoite or merozoite component vaccines often reduce the severity of malaria episodes in monkeys by reducing the peak parasitemias that are responsible for the pathological effects of the disease. Although the blood stages of *P. falciparum* have been cultured, advancement of blood stage vaccine development relies on the identification of blood stage proteins so they can be cloned in various vectors or synthesized. Laboratory reared antigen is not available on a cost-effective basis and would not be acceptable in human trials because it is impossible to eliminate the chance of contamination of the human blood components that go into the growth medium.

Several proteins of the blood stages of various *Plasmodium* species have been identified and tested for their immunogenicity and protection conferred on mice and monkeys (Table 15.7). Pasloske and Howard (1994) summarized some of these findings. One of the more promising candidates for blood stage vaccine research is merozoite surface protein-1 (MSP-1), a molecule that mediates the interaction between the merozoite and the red blood cell. MSP-1 is considered a good vaccine candidate. Although it varies from one *P. falciparum* strain to another, the central repetitive sequence is highly conserved, perhaps because variation would render the protein nonfunctional in some way. This conservation is an essential feature of the antigens upon which any vaccine is based. If highly conserved areas of the protein do not exist, vaccines made from that protein will protect against a limited subset of existing malaria strains. Another essential feature of the antigen is that the highly conserved areas must be strongly immunogenic since this is the basis of the vaccine's effect. At the other end of the spectrum, a molecule like erythrocyte membrane protein-1 (EMP-1), a parasite protein that studs the membrane of infected red blood cells, is a poor candidate for vaccine work. Su et al. (1995) showed that EMP-1 is a highly variable family of proteins coded for by up to 150 genes in the same *P. falciparum* clone. These genes cycle so that the EMP-1 protein changes regularly, staying ahead of the immune response. An EMP-1 based vaccine would have to contend with this ultra-variability.

The only blood stage vaccine to have been extensively tested on humans is a synthetic product designated SPf66, developed by Dr. Manuel Patarroyo, a Colombian physician and immunologist. SPf66 is composed of parts of several merozoite proteins including MSP-1 and portions of other less well understood merozoite proteins. Curiously, SPf66 is considered a blood stage vaccine, but the construct also has portions of CSP-1, a sporozoite antigen.

The history of the so-called "Patarroyo vaccine," summarized up to 1995 by Maurice (1995), illustrates the highs and the lows a candidate malaria vaccine can go through in the course of development and testing. The Patarroyo vaccine first came to world attention in the late 1980s when it suppressed the parasitemias associated with falciparum malaria in a small group of Colombian soldier volunteers (Patarroyo et al. 1988). This success encouraged Amador et al. (1992) to undertake a larger trial with a group of 399 soldiers, which achieved up to 82% protection (reduction in episodes of symptomatic falciparum malaria relative to the control group). Of course, the ability of young, healthy male soldiers to respond to a vaccine may not typify the response of the general population, so even larger studies were undertaken on groups of civilians in Venezuela, Ecuador and Colombia. As expected, rates of protection in these studies were lower though still impressive, ranging from 39–67% (Maurice 1995).

When the Patarroyo vaccine hit the world stage in the early 1990s, it received mixed reviews. Some of the objections were solid. For instance, methodological problems in the South American trials that had not been properly "blinded" were a concern as was the problem of duplicating the Spf66 vaccine in the USA. Other objections were based on skepticism and perhaps even jealousy that a then unknown sci-

Table 15.7. Some *Plasmodium* spp. and other antigens under investigation for use in development of malarial vaccines. (Table contined on next page).

Antigen	Gene/antigen expressed in:	Antigen function	Remarks
	Antigens of pre-erythrocytic stages		
Circumsporozoite protein-1 (CSP-1)	Surface protein of sporozoite of *P. falciparum*, *P. vivax* and others	Mediates attachment of sporozoite to hepatocyte	Induces cytotoxic T-cells and inhibits sporozoite invasion *in vitro*; partial to complete protection of mice, monkeys and humans with native, recombinant and synthetic vaccines
Glutamate-rich protein (GLURP)	*P. falciparum* protein in infected hepatocytes and erythrocytes	Unknown	Elicits lymphocyte proliferation *in vitro*
Liver stage-specific antigen-1 (LSA-1)	*P. falciparum* protein on infected hepatocytes	Unknown	Lymphocyte and cytotoxic T-cell responses *in vitro*; partial protection of *Aotus*; some protection from severe falciparum malaria by 17 amino acid repeat region
Liver stage-specific antigen-3 (LSA-3)	*P. falciparum* protein of sporozoites and infected hepatocytes	Unknown	Lymphocyte and T-cell response; partial inhibition of sporozoite invasion *in vitro*
	Antigens of erythrocytic stages		
Apical membrane antigen-1 (AMA-1, PF-83)	Apical region of merozoite and infected erythrocytes	May mediate erythrocyte invasion by merozoites	Humoral response *in vitro*; prevents symptomatic falciparum malaria in *Aotus*
Erythrocyte-binding antigen (EBA-175)	Merozoite surface protein in *P. falciparum*	Mediates merozoite attachment to erythrocytes	Protection of rabbits from blood stages
Erythrocyte membrane protein-1 (EMP-1)	Parasite protein on infected erythrocytes	Promotes agglutination of infected erythrocytes	Highly variable protein coded by a family of genes in *P. falciparum*; not a good vaccine candidate because of variability
Glycophorin-binding protein (GBP-130, PF-96, PF-101)	On surface of *P. falciparum* merozoites	Binds to glycophorin of erythrocyte membrane	Elicits humoral response and partially prevents erythrocyte invasion *in vitro*
Histidine-rich protein-2 (HRP-2)	*P. falciparum* merozoites	Strong chelator of bivalent metal ions	Elicits humoral response and provides partial protection from erythrocyte invasion *in vitro*

Table 15.7 (Continued). Some *Plasmodium* spp. and other antigens under investigation for use in development of malarial vaccines. (Table contined on next page).

Antigen	Gene/antigen expressed in:	Antigen function	Remarks
	Antigens of erythrocytic stages (continued)		
Merozoite surface protein-1 (MSP-1, MSA-1)	*P. falciparum* merozoite and hepatocyte surface protein; equivalent in *P. vivax* and others	Binds spectrin on erythrocyte membrane	Elicits toxic T-cell and lymphocyte response and protects from erythrocyte invasion *in vitro;* various native protein and recombinant vaccines prevent symptomatic malaria in hosts, including humans
Merozoite surface protein-2 (MSP-2, MSA-2, GP195)	*P. falciparum* merozoite surface protein	Unknown	Elicits lymphocyte response and partially protects erythrocytes *in vitro*; no protection in trials with monkeys
Merozoite surface protein-3 (MSP-3, MSA-3)	*P. falciparum* merozoite surface protein	Unknown	Elicits T-cell and lymphocyte responses; protects erythrocytes *in vitro*; prevents invasion of erythrocytes *in vitro*
Ring-infected erythrocyte surface antigen (RESA, PF155)	*P. falciparum* protein on ring stage-infected erythrocytes	Unknown	Elicits humoral and lymphocyte response *in vitro*; partial protection in *Aotus* monkeys from symptomatic falciparum malaria with purified antigens and recombinants
Sequestrin	*P. falciparum* protein on infected erythrocyte surface	Ligand for vascular epithelium receptors	
Serine repeat antigen (SERA, P140, P126)	*P. falciparum* merozoite antigen	Protease enzyme	Partial protection from erythrocyte invasion *in vitro*; partial protection of *Aotus* from symptomatic falciparum malaria
Rhoptry-associated protein-1 (RAP-1)	Protein on apical region of *P. falciparum* merozoite	Unknown	Partial protection from erythrocyte invasion *in vitro*; partial protection of *Saimiri* monkeys from falciparum malaria

Table 15.7 (Continued).

Antigen	Gene/antigen expressed in:	Antigen function	Remarks
		Sporogonic antigens	
Chitinase	*P. falciparum* ookinete	Aids in penetration of peritrophic membrane	Inclusion of chitin inhibitors in bloodmeal prevents *P. falciparum* oocyst formation in mosquitoes
PFS25	Surface protein of ookinetes and zygotes of *P. falciparum* and other species	Unknown	Recombinant version blocks transmission of *P. falciparum* to *Aotus*; inhibits development in mosquitoes
PFS40	*P. falciparum* gametocyte protein	Unkown	Purified antigen and recombinant prevents *P. falciparum* transmission to mice
PFS45-48	*P. falciparum* gametocyte surface protein	Unknown	
		Human antigens	
Falhesin I & II	Human erythrocyte protein	Erythrocyte anion transport and exchange	Partial and complete (resp.) protection from symptomatic falciparum malaria in *Aotus* and *Saimiri* monkeys by synthetic and recombinant proteins
		Mosquito antigens	
Midgut protease protein (MPP)	*Ae. aegypti* and *Anopheles* spp.	Blood protein digestion; induction of parasite chitinase	Inhibition of mosquito proteases may prevent chitinase induction and ookinete penetration of peritrophic membrane

entist working in the relative isolation of Colombia should be able to accomplish what others in the research centers of the world had been unable to do.

To address the substantive objections and to continue the requisite vaccine testing, Patarroyo worked hard to get human trials of his vaccine approved in Africa where, unlike South America, malaria transmission is continuous, affects mainly children, and results from different strains of *P. falciparum*. This effort eventually lead to trials in Tanzania involving almost 600 children aged 1–5 yr. The Tanzanian trial produced a 31% protection rate in vaccinated children (Alonso et al. 1994) and, despite its rather low efficacy (measles vaccine has an efficacy of between 80% and 95%), was heralded as a great success by many. The reason for this adulation is that even a 30% efficacious vaccine could prevent 100 million cases of malaria per year! However, others worried that country-wide vaccination programs by their nature cannot be as tightly controlled as vaccine trials and would not achieve similar levels of protection. There also was concern that developing countries where the malaria problem is most widespread would abandon other methods of malaria control (vector control, bednets) once they had spent their limited budgets on the vaccine.

Since the success of the Tanzanian trials, the Patarroyo vaccine has received 2 serious, potentially fatal, blows. Trials with a group of 600 Gambian children aged 6–11 months (D'Alessandro et al. 1995) showed no effect of the vaccine, while a study involving over 1,200 Karen refugee children in Thailand (Nosten et al. 1996) produced somewhat higher rates of malaria in treatment groups. In the high cost, high risk world of vaccine development, these results may kill what was once considered by many to be a sure bet as the first practical, affordable malaria vaccine.

Anti-gametocyte vaccines

The sexual stage of the malaria parasite also is a likely target for a vaccine. Vaccines that affect the sexual stage (gametocytes) sometimes are called "transmission blocking" vaccines since the gametocytes are the stage of the parasite that infects the mosquito. Transmission blocking vaccines may be easier to develop than vaccines to sporozoites or blood stages because the gametocytes, being adapted to the mosquito gut, do not have to cope with the complex mammalian immune system. Accordingly, the gametocytes should be easier targets than the other life stages.

Transmission blocking vaccines present a problem not encountered with other malaria vaccines: the individual receiving the immunization will not directly benefit from it since the gametocytes result from the erythrocytic cycle, which is responsible for the symptoms and pathology of malaria. For this reason, transmission blocking vaccines sometimes are referred to as altruistic vaccines. Whether a malaria sufferer is willing to be given a transmission blocking vaccine or not begs the ethical question of whether it is right to administer the vaccine without a curative dose of an antimalarial drug. The question then is, if the antimalarial drug clears the infection (which produces the gametocytes), what good is the vaccine? Similarly, what good is a transmission blocking vaccine if given in combination with a sporozoite or blood stage vaccine, either of which would prevent the erythrocytic cycle and the appearance of gametocytes? In fact, the transmission blocking vaccine will affect only the parasite if the drugs or other vaccines with which it is given fail. Clearly, a transmission blocking vaccine could be an important public health tool to limit the spread of drug or vaccine resistant malaria strains, but it will not help the recipient.

Several approaches to transmission blocking vaccines are under active investigation. Various surface proteins of the gametocytes have been identified and their genes sequenced (Table 15.7). Pfs48/45 is a glycoprotein of the plasma membrane of *P. falciparum* gametes that fits many of the criteria needed in a transmission blocking vaccine. Addition of Pfs48/45 mouse and rabbit antigens to bloodmeals prevents oocyst formation in mosquitoes (Drakeley et al. 1996) and many of its immunogenic portions (epitopes) were invariant in strains of *P. falciparum* from South America, Africa, Southeast Asia and Indonesia (Kocken et al. 1995). This protein, and others like it associated with the sexual stages, are prime candidates for malaria transmission blocking vaccines.

Vaccines against mosquito stages

Other transmission blocking vaccines could target enzymes used by the sexual stages of the parasite in the mosquito, or the mosquito itself. Fusion of male and female gametes in the peritrophic membrane-wrapped blood bolus in the mosquito's midgut forms the ookinetes of *Plasmodium* species. Penetration of the peritrophic membrane by the ookinete apparently is effected by chitinases it secretes (Huber et al. 1991). Shahabuddin et al. (1993) established the potential of an anti-chitinase vaccine in transmission blocking by demonstrating that addition of specific chitinase inhibitors to mosquito bloodmeals containing malaria gametocytes prevented oocyst development. These authors also showed that the mosquito protease activates the parasite chitinase. This suggests that antimosquito protease or midgut epithelium vaccines also may be transmission blockers.

The roles for malaria vaccines in the future may depend on who uses them. Vaccines with efficacies of 30% will not replace drug prophylaxis for travelers and other occasional, short-term visitors to endemic areas. However, they may play an important part in integrated malaria prevention programs for indigenous people. These programs also would use impregnated bednets and effective treatment of symptomatic malaria cases. Indeed, the added protection of a vaccine may be essential in some areas where bednet programs change stable malaria transmission patterns into unstable ones. Levels of acquired immunity in such areas will drop when the challenge that maintains immunity to this parasite no longer exists. Effective malaria vaccines probably also will have to protect against more than one life stage of the parasite. Failure of a sporozoite vaccine, for instance, will leave the victim susceptible to the ravaging effects of the erythrocytic cycle. Finally, to meet the need where it is greatest, future malaria vaccines will have to be safe and effective for all segments of the population including those hardest hit by the disease: children <5 yr old and pregnant women.

Leishmaniasis

The range of diseases and conditions caused by *Leishmania* spp. is broad and varies considerably within species as well as between species in the nature and severity of disease produced. See Leiby et al. (1994) for descriptions of clinical manifestations of the various species and strains of *Leishmania*. This variation also makes vaccine development against many of these parasites difficult.

The only effective immunization method for leishmanial disease comes not from modern biomedical research but from ancient mid-eastern cultural practices. Some middle eastern countries have long inoculated children against Oriental Sore, a potentially disfiguring disease, by scratching material from lesions of infected individuals into the skin of the buttock or upper leg. Typically, a single lesion spontaneously heals and produces long-lasting immunity to homologous strains in about 90% of treated individuals (Leiby et al. 1994). Leishmaniasis recidivans, a non-healing relapse of the Oriental Sore parasite, also occurs in a small percentage of vaccinated individuals years after the initial infection.

Tesh (1995) reviewed current control methods for zoonotic leishmaniasis, a rapidly expanding disease of children in many parts of the developing world for which domestic dogs form a link between sylvatic transmission cycles and the human population. Current recommendations in many endemic areas are that infected dogs be destroyed. This policy is fraught with problems of owner compliance and difficulties in determining which animals are infected. Tesh concludes that reservoir dog control is not feasible nor effective and urges that efforts be directed at developing a live attenuated vaccine to control infection in the domestic canine population of endemic countries.

Trypanosomiasis

The extreme variability of the parasite's surface coat, normally a prime target for vaccines, has hindered development of vaccines for African trypanosomiasis (Sileghem et al. 1994). These variable surface glycoproteins (VSGs) shift regularly during infections, effectively outrunning the immune response. Individual strains of trypanosomes may have a repertoire of thousands of VSGs, meaning that at any given time an individual may be infected by parasites bearing one of many possible variants of surface protein. It is neither technically nor practically feasible to make vaccines sufficiently polyvalent to protect against all of the possibilities.

Recent efforts to identify suitable targets for a trypanosome vaccine have been directed at proteins in the flagellar pocket of the parasite. This possible Achilles' heel of the trypanosome is not covered by surface glycoproteins and is the location of a number of relatively invariant proteins. Whether antibodies generated in response to immunization by these proteins, or elements of them, will destroy the parasite remains to be seen.

The prospects for a human vaccine against *Trypanosoma cruzi*, causative agent of Chagas disease, are just as poor as for other trypanosomes. However, as with zoonotic leishmaniasis, Chagas disease may best be tackled through its domestic canine reservoir. Domestic dogs are highly susceptible to *T. cruzi* and proximity of infected dogs increases the chances of infection in children (Gurtler et al. 1987). Basombrio et al. (1993) reported that a field trial in which 73 dogs vaccinated with an attenuated TCC strain of *T. cruzi* showed a 50% reduction in infection in experimental versus control dogs after one year. In addition, several dogs that were not protected from infection nonetheless were noninfective for *Triatoma infestans*. Parasite material to make these animal vaccines must be collected from the blood of infected reservoir animals or humans, so immunization of humans with the attenuated parasite is not acceptable. However, use of such vaccines in reservoir animals appears to have potential in controlling human cases although issues of cost still must be addressed.

Filariasis

The occurrence of naturally-acquired immunity to filarial nematodes sustains hope that vaccines for filarial diseases may be possible. Selkirk et al. (1992) referred to the natural immunity to *W. bancrofti* and *B. malayi* as **concomitant immunity**, meaning that infection with adult worms acquired early in life occurs concomitantly with gradually developing resistance to L3 (infective) larvae. Adult worms usually escape or tolerate host immune responses. Breakdown of this tolerance leads to die-off of adults and microfilariae, which in turn produces the serious tissue damage sometimes observed in filarial infections. Evidence for naturally-acquired resistance to *O. volvulus* L3-challenge is weaker.

The development of human vaccines for filarial nematodes is hampered by the absence of laboratory culture systems for these organisms and by the fact that the animal model systems in use incompletely mimic disease biology and human immunology. Nonetheless, several attempts to immunize animals against human and animal filarial infections have succeeded. In general, the immunogenic life stage is the L3 larva. Irradiated L3 larvae have immunized Mongolian jirds, BALB/c mice, cats and dogs against some or all of *B. malayi*, *Brugia pahangi*, *Dirofilaria immitis* and *O. volvulus* (see Selkirk et al. 1992 for greater detail). Lange et al. (1993) also reported immunizing mice against *O. volvulus* L3 larvae. Immunity to the infective stage of these parasites would effectively block transmission, but immunization against microfilariae also could help existing infections. Immunization against developing microfilariae has been studied less thoroughly because many of the model hosts do not permit adult establishment or reproduction. The association of severe pathology with mass-microfilarial death, especially in onchocerciasis, might limit usefulness of such vaccines.

Determining which molecules associated with the L3 larvae are the effective immunogens is the next step in this effort. Selkirk et al. (1992) pointed out that the relative success of living (irradiated) L3s over killed L3s suggests that secreted antigens may be more important as immunogens than surface proteins. Despite this fact, considerable effort also has been put into identifying surface proteins that if sufficiently immunogenic could be cloned for vaccine use. Various candidate glycoproteins have been identified, several occurring in all life stages of the parasite. Some of these surface glycoproteins occur with only minor changes in different species. For instance, a pair of proteins of 70 and 75 kDa molecular weight have almost identical makeups in *B. malayi*, *B. pahangi* and *W. bancrofti*. This similarity hints that filarial vaccines may work across species.

Vaccine development considers the desirability as well as the technical feasibility of the goal. This concern has particular significance for filarial diseases since the introduction of ivermectin, an effective filaricide for virtually all filarial species. Has ivermectin pulled the rug out from under this line of vaccine research? There is no easy answer. Chloroquine was going to rid the world of malaria 40 yr ago, so researchers largely stopped looking into malaria vaccines. Now there is more malaria than ever, there are few effective prophylactics, and effective vaccines are still years away.

For a detailed discussion of filarial immunology, see Williams et al. (1994). For a review of filarial vaccine research, see Selkirk et al. (1992).

Vaccines Against Arthropods

Arthropods that are biological vectors of disease all transmit pathogens by biting the host and taking a bloodmeal. In the process of feeding, these arthropods usually secrete saliva and other materials that elicit various immunologic responses (allergic, inflammatory) in the vertebrate host. These immunologic responses are being investigated as the bases for anti-arthropod vaccines. Kay and Kemp (1994) reviewed this topic. Many vaccines based on whole organism antigen or antigen from midgut, muscle, salivary glands or their secretions have been tested for their effects on vector insects. Results range from none to reduced fecundity and/or increased mortality. In some cases, the ability of the arthropod to support the development of pathogens is decreased. In general, anti-arthropod vaccines are more effective against ticks than insects and against insects such as lice that spend long periods in close association with the host.

The only commercial anti-arthropod vaccine currently in sight is to the cattle tick, *Boophilus microplus*. It is based on so-called "Bm86" antigen derived from fractions of partially engorged *B. microplus*. Tellam et al. (1992) reported that native Bm86 protein reduced the number of ticks engorging by 61% and reduced tick egg production by 91% while recombinant Bm86 achieved between 24–27% reductions in feeding and 77–89% reductions in tick fecundity. Kay and Kemp (1994) considered the prospects for other tick vaccines, especially for use in livestock, to be very good.

SURVEILLANCE

Surveillance programs are important adjuncts to prophylaxis because they help public health and agricultural authorities direct prevention and vector control efforts to where and when they are most effective. Chapter 13 covered the methods used for surveillance and will not be repeated here. A basic reference for surveillance for arboviruses is that of Moore et al. (1993). They defined surveillance for arboviruses as "...the organized monitoring of levels of virus activity, vector populations, infections in vertebrate hosts, human cases and other factors to detect or predict changes in the transmission dynamics of arboviruses." This description is equally apt and more inclusive if "arboviruses" is replaced with "arthropodborne pathogens." Some transmission cycles (e.g., many of the viral encephalitides) involve animal reservoirs and different vector species maintaining the pathogen within the reservoir than are transmitting it to humans or domestic animals. Other cycles (e.g., dengue and malaria) have fewer components. In either case, effective surveillance for a given disease requires a thorough understanding of that disease's transmission cycle.

QUARANTINE

The restriction of movement of animals and animal products from an area where an arthropodborne disease has been detected is an important method of disease prevention. These actions are carried out under the authority of the International Animal Health Code, issued periodically by the International Office of Epizootics (OIE). Vesicular stomatitis and Rift Valley fever are OIE List A diseases having serious potential for spread across national boundaries. Such diseases elicit strict restrictions in trade. Individual countries enforce other quarantines involving infected vertebrates or potential vectors.

[1]Drs. Donald A. Weidhaas and Dana A. Focks contributed material for this section.

VECTOR CONTROL[1]

Chapter 14 discussed the general principles of vector control. Some arthropodborne diseases rely heavily on vector control for prevention and control. This method is valuable especially where other preventive measures, such as vaccines, either are unavailable or ineffective.

Malaria

The development of residual treatments of DDT, applied to walls inside houses in the 1940s, led to programs of malaria control based almost entirely on indoor house spraying (Fig. 15.14). DDT persisted and could be applied just twice a year to prevent transmission. The basis for this approach to malaria control was that the continual toxic levels of DDT prevented transmission to humans by killing adult mosquitoes before they had lived long enough to bite a second time (the infective bite). This time is the earliest a female mosquito could be infected with the malaria parasite. Thus, a female must be at least 13 days old before it can transmit. Very few mosquitoes survive to that age. The indoor spray also was believed to have a "space repellent" effect that reduced the numbers of anophelines coming indoors and biting humans.

After resistance to chlorinated hydrocarbons occurred, other types of compounds (particularly organophosphorus, carbamate and synthetic pyrethroid materials) were developed as residual treatments. Because of the many problems associated with vector control (insecticide resistance, environmental and safety concerns, resistance to continued spraying of interiors of houses, loss of interest and funding), malaria control will require an integrated approach utilizing both vector and disease control methods in organized programs involving community participation. Prophylactic antimalarial drugs and personal protective devices such as repel-

Figure 15.14. Residual spraying of walls in a Guyanese village (courtesy of the Pan American Health Organization).

lents and bednets must supplement vector control programs.

Onchocerciasis

Elimination of onchocerciasis through control of the black fly vectors is both difficult and complex. Black flies breed and develop in streams and rivers and the winged, biting adults remain in the vicinity of their larval breeding sites. Reduction of vector populations must be maintained to keep the biting rate below the threshold of transmission and to allow the infection to die out of the human population.

At the start of the Onchocerciasis Control Program, the only means of managing the disease was through vector control. An elaborate system of continuous survey and reporting of the presence and density of black flies along the rivers of 7 countries allowed preparation of weekly maps of rivers that guided the fixed wing and helicopter aircraft personnel for the weekly treatments of larval breeding areas in the rivers. The program ceased transmission according to extensive medical surveys of individuals in villages. Expansion of the program to additional countries eventually led to resistance to temephos, the original larvicide of choice, in the black fly vectors and the development of alternative larvicides such as chlorphoxim, phoxim, pyraclofos, carbosulfan, permethrin and *Bacillus thuringiensis israelensis*. Since 1988, ivermectin has been incorporated into the African and Central and South American efforts to eliminate onchocerciasis.

Dengue

Aedes aegypti breed in artificial containers near human habitations, making it an ideal target for control through environmental management, i.e., the cleanup and destruction of artificial containers holding water that allows the mosquito eggs, larvae and pupae to develop. Such containers can include a large variety of discarded or useful items: bottles, cans, tires, water collecting devices and containers, vases and bird feeders. Although this approach appears logical and relatively simple, even community-based or urban sanitation programs may not completely prevent mosquito production and transmission. In Africa, *Ae. aegypti* breeds in natural water holding sites, as do *Ae. albopictus* and *Aedes polynesiensis* in other areas. Insecticides reduce or prevent disease. Safe larvicides are available for treating larval breeding water. Insecticides in ultra-low volume (ULV) applications against adults often are used, but are of doubtful value.

Filariasis

The most common vector of Bancroftian and Brugian filariasis is *Cx. quinquefasciatus* associated with polluted water in crowded urban areas. The most obvious control methods would be adequate and closed water, sanitation and sewage systems capable of preventing and eliminating open sewage ditches, latrines, privies, and collections of polluted water. Because they are not possible, control of the vector generally involves the use of larvicides. Control activities must be maintained for as long as the filarial worms persist in humans, and best results are obtained when vector control is combined with disease control.

Tickborne Diseases

Acaricides have been the basis for a variety of management strategies for ticks. Cattle can be sprayed or dipped in acaricidal solutions. Spraying vegetation or tick habitats with acaricides reduces tick populations. Non-chemical methods of pasture rotation, vegetation or animal management also are effective measures.

Louseborne Diseases

Louseborne typhus and relapsing fever generally are associated with famines, disasters, wars and unsanitary conditions. In such situations, the only means of disease control may be the use of insecticidal powders or shampoos to kill the high populations of lice that can develop quickly. In mass disasters, specially designed power dusters have been used to delouse many people in a short time. Where possible, personal hygiene, washing, sterilizing and ironing of personal clothing and bedding and impregnation of clothing with safe and approved insecticides are useful additions to the use of insecticidal powders and shampoos on the body.

Leishmaniasis

Phlebotomine sand flies are weak fliers and are highly susceptible to most insecticides. Residual spraying where the sand fly vectors rest and bite inside houses or on bednets can be highly effective. When sand flies rest outdoors, space sprays repeated often would not be economical. Because sand fly larvae usually congregate in animal burrows and other subterranean habitats, larval control is impractical.

Chagas Disease

Schofield (1994) summarized the problems associated with controlling Chagas disease and the triatomid bugs that transmit it. He pointed out that most of the approaches to control triatomids (insecticides, housing modifications, juvenile hormone mimics and other insect growth regulators, traps, chemosterilants, genetic manipulations and various biological control agents) have been tested. Insecticides and housing modifications are the methods of choice. Chlorinated hydrocarbons such as benzene hexachloride and dieldrin were used originally to spray houses for triatomid control, and later, organophosphorus or carbamate compounds were substituted. Pyrethroids (e.g., permethrin and deltamethrin) are now the materials of choice for house spraying. Changes such as housing improvements that minimize places for triatomids to hide and removal of domestic animals from houses effectively control disease where cultural tolerances and resources permit.

REFERENCES

Addiss, D.G., Eberhard, M.L., Lammie, P.J., McNeeley, M.B. and Spencer, H.C. 1993. Comparative efficacy of clearing-dose and single high-dose ivermectin and diethylcarbamazine against *Wuchereria bancrofti* microfilaremia. Am. J. Trop. Med. Hyg. **48:**178–185.

Alonso, P.L., Smith, T., Armstrong-Schellenberg, J.R.M., Masanja, H., Mwankusye, S., Urassa, H., Bastos de Azevedo, I., Chongela, J., Kobera, S., Menendez, C., Hurt, N., Thomas, M.C., Lyimo, E., Weiss, N.A., Hayes, R., Kitua, A.Y., Lopez, M.C., Kilama, W.L., Teuscher, T. and Tanner, M. 1994. Randomized trial of efficacy of Spf66 vaccine against *Plasmodium falciparum* malaria in children in southern Tanzania. The Lancet **344:**1175–1181.

Amador, R., Moreno, A., Valero, V., Murillo, L., Mora, A.L., Rojas, M., Rocha, C., Salcedo, M., Guzman, F., Espejo, F., Nuñez, F., Patarroyo, M.E. 1992. The first field trials of the chemically synthesized malaria vaccine SPf66: safety, immunogenicity and protectivity. Vaccine **10:**179–184.

Arredondo-Jimenez, J.L., Rodrigues, M.H., Loyola, E.G. and Bown, D.N. 1997. Behaviour of *Anopheles albimanus* in relation to pyrethroid-treated bed nets. Med. Vet. Entomol. **11:**87–94.

Bagla, P. 1994. India scrambles to distribute vaccine. Science **266:**23.

Ballou, W.R., Hoffman, S.L., Sherwood, J.A., Hollingdale, M.R., Neva, F.A., Hockmeyer, W.T., Gordon, D.M., Schnieider, I., Wirtz, R.A., Young, J.F., Wasserman, G.F., Reeve, P., Diggs, C.L. and Chulay, J.D. 1987. Safety and efficacy of a recombinant DNA *Plasmodium falciparum* sporozoite vaccine. Lancet **1:** 1277–1281.

Basch, P.F. 1994. Vaccines and world health. Science, policy and practice. Oxford University Press, New York. 274 pp.

Basombrio, M.A., Segura, M.A., Mora, M.C. and Gomez, L. 1993. Field trial of vaccination against American trypanosomiasis (Chagas' disease) in dogs. Am. J. Trop. Med. Hyg. **49:**143–151.

Beach, R.F., Ruebush, T.K., II, Sexton, J.D., Bright, P.L., Hightower, A.W., Breman, J.G., Mount, D.L. and Oloo, A.J. 1993. Effectiveness of permethrin-impregnated bed nets and curtains for malaria control in a holoendemic areas of western Kenya. Am. J. Trop. Med. Hyg. **49:**290–300.

Ben Saleh, B., Zakraoui, H., Zaatour, A., Ftaiti, A., Zaafouri, B., Garraoui, A., Olliaro, P.L., Dellagin, K. and Ben Ismail, R. 1995. A randomized placebo-controlled trial in Tunisia treating cutaneous leishmaniasis with parmomycin ointment. Am. J. Trop. Med. Hyg. **53:**162–166.

Boussinesq, M., Chippaux, J.P., Ernould, J.C., Quillevere, D. and Prod'hon, J. 1995. Effect of repeated treatments with ivermectin on the incidence of onchocerciasis in northern Cameroon. Am. J. Trop. Med. Hyg. **53:**63–67.

Brandt, W.E. 1990. Development of dengue and Japanese encephalitis vaccines. J. Infect. Dis. **162:**577–583.

Brasseur, P., Kouamouo, J., Moyou-Somo, R. and Druilhe, P. 1992. Multi-drug resistant falciparum malaria in Cameroon in 1997-88. II. Mefloquine resistance confirmed *in vitro* and its correlation with quinine resistance. Am. J. Trop. Med. Hyg. **46:**8–14.

Brewer, J.G., Grate, S.J., Peggins, J.O., Weina, P.J., Petras, J.M., Levine, B.S., Heiffer, M.H. and Schuster, B.G. 1994. Fatal neurotoxicity of arteether and artemether. Am. J. Trop. Med. Hyg. **51:**251–259.

Brinkmann, U. and Brinkmann, A. 1995. Economic aspects of the use of impregnated bed nets for malaria control. Bull. Wld. Hlth. Organiz. **73:**648-651.

Callahan, H.L. and Beverley, S.M. 1991. Heavy metal resistance: a new role of P-glycoproteins in *Leishmania*. J. Biol. Chem. **266:**18427–18430.

Canfield, C.J., Milhous, W.K., Ager, A.L., Rossan, R.N., Sweeney, T.R., Lewis, N.J. and Jacobus, D.P. 1993. PS-15: a potent, orally active antimalarial from a new class of folic acid antagonists. Am. J. Trop. Med. Hyg. **49:**121–126.

Centers for Disease Control and Prevention. 1999. Notice to Readers: availability of Lyme disease vaccine. Morb. Mort. Weekly Rep. **48:**35–42.

Chadee, D.D., Tilluckdharry, S.C., Rawlins, R., Doon, R. and Nathan, M.B. 1995. Mass chemotherapy with diethylcarbamazine for the control of Bancroftian filariasis: a twelve-year follow-up in northern Trinidad, including observations on *Mansonella ozzardi*. Am. J. Trop. Med. Hyg. **52:**174–176.

Choi, H.W., Breman, J.G., Teutsch, S.M., Liu, S., Hightower, A.W. and Sexton, J.D. 1995. The effectiveness of insecticide-impregnated bed nets in reducing cases of malaria infection: a meta-analysis of published results. Am. J. Trop. Med. Hyg. **5:**377–382.

Cilek, J.E. and Schreiber, E.T. 1994. Failure of the mosquito plant *Pelargonium citrosum* 'Van Leenii', to repel adult *Aedes albopictus* and *Culex quinquefasciatus* in Florida. J. Am. Mosq. Contr. Assoc. **10:**473–476.

Cohen, J. 1994. Bumps on the vaccine road. Science **265:**1371–1373.

Cowman, A.F., Galatis, D. and Thompson, J.K. 1994. Selection for mefloquine resistance in *Plasmodium falciparum* is linked to amplification of the pfmdrl gene and cross-resistance to halofantrine and quinine. Proc. Nat. Acad. Sci. **91:**1143–1147.

Crook, S.E. and Baptista, A. 1995. The effect of permethrin-impregnated wall-curtains on malaria transmission and mortality in the suburbs of Maputo, Mozambique. Trop. Geogr. Med. **47:**64–67.

Cupp, E.W., Mare, C.J., Cupp, M.S. and Ramberg, F.B. 1992. Biological transmission of vesicular stomatitis virus (New Jersey) by *Simulium vittatum* (Diptera: Simuliidae). J. Med. Entomol. **29:**137–140.

Curtis, C.F., Lines, J.D., Renz, L.B. and Renz, A. 1990. Natural and synthetic repellents. Pp. 75–92 *in* C.F. Curtis (ed.), Appropriate technology in vector control. CRC Press, Boca Raton, Florida.

D'Alessandro, U., Leach, A., Drakeley, C.J., Bennett, S., Olaleye, B.O., Fegan, G.W., Jawara, M., Langerock, P., George, M.O. and Targett, G.A. 1995. Efficacy trial of malaria vaccine SPf66 in Gambian infants. Lancet **346:**462–467.

Davies, D.M. 1972. The landing of blood-sucking female blackflies (Simuliidae: Diptera) on coloured materials. Proc. Entomol. Soc. Ont. **102:**124–155.

Davis, E.E. 1985. Insect repellents: concepts of their mode of action relative to potential sensory mechanisms in mosquitoes (Diptera: Culicidae). J. Med. Entomol. **22:**237–243.

De, D., Bhaduri, A.P. and Milhous, W.K. 1993. A novel pyrrolidoalkaneamine (WR268954) that modulates chloroquine resistance of *Plasmodium falciparum in vitro*. Am. J. Trop. Med. Hyg. **49:**113–120.

Desowitz, R.S. 1991. The malaria capers. Tales of parasites and people. W.W. Norton and Co., New York. 208 pp.

Domb, A.J., Marlinsky, A., Maniar, M. and Teomim, L. 1995. Insect repellent formulation of n,n-diethyl-m-toluamide (DEET) in a liposphere system: efficacy and skin uptake. J. Am. Mosq. Contr. Assoc. **11:**29–34.

Drakeley, C.J., Duraisingh, M.T., Povoa, M., Conway, D.J., Targett, G.A.T. and Baker, D.A. 1996. Geographical distribution of a variant epitope of Pfs48/45, a *Plasmodium falciparum* transmission-blocking vaccine candidate. Molec. Biochem. Parasitol. **81:**253–257.

Duke, B.O.L., Soula, G., Zea-Flores, G., Bratthauer, G.L. and Doumbo, O. 1991a. Migration and death of skin-dwelling *Onchocerca volvulus* microfilariae after treatment with ivermectin. Trop. Med. Parasitol. **42:**25–30.

Duke, B.O.L., Zea-Flores, J.C., Cupp, E.W. and Munoz, B. 1991b. Comparison of the effects of a single dose and of four six-monthly doses of ivermectin on adult *Onchocerca volvulus*. Am. J. Trop. Med. Hyg. **45:**132–137.

Duke, B.O.L., Zea-Flores, J.C., Castro, J., Cupp, E.W. and Munoz, B. 1992. Effects of three-month doses of ivermectin on adult *Onchocerca volvulus*. Am. J. Trop. Med. Hyg. **46:**189–194.

Dunne, M., Al-Ramadi, B.K., Barthold, S.W., Flavell, R.A. and Fikrig, E. 1995. Oral vaccination with an attenuated *Salmonella typhimurium* expressing *Borrelia burgdorferi* Osp A prevents murine Lyme borreliosis. Infect. Immun. **63:**1611–1614.

Fehrer, S.L. and Halliwell, R.E. 1987. Effectiveness of Avon's Skin-So-Soft® as a flea repellent on dogs. J. Am. Anim. Hosp. Assoc. **23:**217–220.

Fikrig, E., Barthold, S.W., Cantor, F.S. and Flavell, R.A. 1990. Protection of mice against the Lyme disease agent by immunizing with recombinant OspA. Science **250:**553–556.

Fikrig, E., Telford, S.R., III, Barthold, S.W., Kantor, F.S., Spielman, A. and Flavell, R.A. 1992. Elimination of *Borrelia burgdorferi* from vector ticks feeding on OspA-immunized mice. Proc. Nat. Acad. Sci. **89:**5418–5421.

Foil, L.D. and Hogsette, J.E. 1994. Biology and control of tabanids, stable flies and horn flies. Sci. Technol. Rev., Off. Intern. Epizoot. **13:**1125–1158.

Folz, S.D., Ash, K.A., Conder, G.A. and Rector, D.L. 1986. Amitraz: a tick and flea repellent and tick detachment drug. J. Vet. Pharm. Therap. **9:**150–156.

Fontanet, A.L., Johnston, D.B., Walker, A.M., Rooney, W., Thimasarn, K., Sturchler, D., Macdonald, M., Hours, M. and Wirth, D.F. 1993. High prevalence of mefloquine-resistant falciparum malaria in eastern Thailand. Bull. Wld. Hlth. Organiz. **71**:377–383.

Franco, M.A., Barbosa, A.C., Rath, S. and Dorea, J.G. 1995. Antimony oxidation states in antileishmanial drugs. Am. J. Trop. Med. Hyg. **52**:435–437.

Gahlinger, P.M., Reeves, W.C. and Milby, M.M. 1986. Air conditioning and television as protective factors in arboviral surveillance risk. Am. J. Trop. Med. Hyg. **35**:601–610.

Geier, M., Sass, H. and Boeckh, J. 1996. A search for components in human body odour that attract females of *Aedes aegypti*. Pp. 132–148 *in* G.R. Bock and G. Cardew (eds.), Olfaction in mosquito-host interactions. John Wiley and Sons, Chichester, West Sussex, England.

Gordeuk, V.R., Thuma, P.E., Brittenham, G.M., Biemba, G., Zulu, S., Simwanza, G., Kalense, P., M'Hango, A., Parry, D., Poltera, A.A. and Aikawa, M. 1993. Iron chelation as a chemotherapeutic strategy for falciparum malaria. Am. J. Trop. Med. Hyg. **48**:193–197.

Grothaus, R.H., Haskins, J.R., Schreck, C.E. and Gouck, H.K. 1976. Insect repellent jacket: status, value and potential. Mosq. News **36**:11–18.

Guiguemende, T.r., Aouba, A., Ouedraogo, J.B. and Lamizana, L. 1994. Ten-year surveillance of drug-resistant malaria in Burkina Faso (1982-1992). Am. J. Trop. Med. Hyg. **50**:699–704.

Gupta, R.K. and Rutledge, L.C. 1994. Role of repellents in vector control and disease prevention. Am. J. Trop. Med. Hyg. **50 (6, Suppl.)**:82–85.

Gupta, R.K. and Rutledge, L.C. 1991. Controlled release repellent formulations on human volunteers under three climatic regimens. J. Am. Mosq. Contr. Assoc. **7**:490–493.

Gurtler, R.E., Peterson, R.M., Lauricella, M.A. and Wisnivesky-Colli, C. 1987. Infectivity to vector *Triatoma infestans* of dogs infected with *Trypanosoma cruzi* in north-western Argentina. Ann. Trop. Med. Parasitol. **86**:111–119.

Harlan, H.J., Schreck, C.E. and Kline, D.L. 1983. Insect repellent jacket tests against biting midges. Am. J. Trop. Med. Hyg. **32**:185–188.

Hien, T.T. and White, N.J. 1993. Qinghaosu. Lancet **341**:603–608.

Hossain, M.I. and Curtis, C.F. 1989. Laboratory evaluation of deet and permethrin wide mesh netting against the mosquitoes. Entomol. Exp. Appl. **52**:93–102.

Huber, M., Cabib, E. and Miller, L.H. 1991. Malaria parasite chitinase and penetration of the mosquito peritrophic membrane. Proc. Nat. Acad. Sci. **88**:2807–2810.

Jong, E.C. and McMullen, R. 1995. The travel and tropical medicine manual. W.B. Saunders Co., Toronto. 532 pp.

Kay, B.H. and Kemp, D.H. 1994. Vaccines against arthropods. Special Symposium on Vector Control. Am. J. Trop. Med. Hyg. **50(6, Suppl.)**:87–96.

Kazura, J., Greenberg, J., Perry, R., Weil, G., Day, K. and Alpers, M. 1993. Comparison of single-dose diethylcarbamazine and ivermectin for treatment on Bancroftian filariasis in Papua New Guinea. Am. J. Trop. Med. Hyg. **49**:804–811.

Khusmith, S., Charoenvit, Y., Kumar, S., Sedegah, M., Beaudoin, R.L. and Hoffman, S.L. 1991. Protection against malaria by vaccination with sporozoite surface protein 2 plus CS protein. Science **252**:715–718.

Kocken, C.H.M., Milek, R.L.B., Lensen, T.H.W., Kaslow, D.C., Schoenmakers, J.G.G. and Konings, R.N.H. 1995. Minimal variation in the transmission-blocking vaccine candidate Pfs48/45 of the human malaria parasite *Plasmodium falciparum*. Molec. Biochem. Parasitol. **68**:115–118.

Kroeger, A., Gerhardus, A., Kruger, G., Mancheno, M. and Pesse, K. 1997. The contribution of repellent soap to malaria control. Am. J. Trop. Med. Hyg. **56**:580–584.

Kuthiala, A., Gupta, R. and Davis, E.E. 1992. The effect of repellents on the responsiveness of the antennal chemoreceptors for oviposition of the mosquito *Aedes aegypti*. J. Med. Entomol. **29**:639–643.

Kyle, D.E., Milhousa, W.K. and Rossan, R.N. 1993. Reversal of *Plasmodium falciparum* resistance to chloroquine in Panamanian *Aotus* monkeys. Am. J. Trop. Med. Hyg. **49**:783–788.

Lange, A.M., Yutanawiboonchai, W., Lok, J.B., Trpis, M. and Abraham, D. 1993. Induction of protective immunity against larval *Onchocerca volvulus* in a mouse model. Am. J. Trop. Med. Hyg. **49**:783–788.

Lange, R.W., Frankenfield, D.L., Moriarty-Sheehan, M., Contoreggi, C.S. and Frame, J.D. 1994. No evidence for chloroquine-associated retinopathy among missionaries on long-term malaria chemoprophylaxis. Am. J. Trop. Med. Hyg. **51**:389–392.

Leiby, D.A., Kanesa-thasan, N., Scott, P. and Nacy, C.A. 1994. Leishmaniasis. Pp. 87–118 *in* F. Keirszenbaum (ed.), Parasitic infections and the immune system. Academic Press, San Diego.

Lillie, T.H., Schreck, C.E. and Rahe, A.J. 1988. Effectiveness of personal protection against mosquitoes in Alaska. J. Med. Entomol. **25**:475–478.

Lindsay, L.R., Surgeoner, G.A., Heal, J.D. and Gallivan, G.J. 1996. Evaluation of the efficacy of 3% citronella candles and 5% citronella incense for protection against field populations of *Aedes* mosquitoes. J. Am. Mosq. Contr. Assoc. **12**:293–294.

Luke, C.J., Huebner, R.C., Kasmiersky, V. and Barbour, A.G. 1997. Oral delivery of purified lipoprotein OspA protects mice from systemic infection with *Borrelia burgdorferi*. Vaccine **15:**739–746.

Mani, T.R., Reuben, R. and Akiyama, J. 1991. Field efficacy of Mosbar mosquito repellent soap against vectors of Bancroftian filariasis and Japanese encephalitis in southern India. J. Am. Mosq. Contr. Assoc. **7:**565–568.

Marshall, E. 1995. Lyme disease: NIH gears up to test hotly disputed theory. Science **270:**228–229.

Maurice, J. 1995. Malaria vaccine raises a dilemma. Science **267:**320–323.

McIver, S.B. 1981. A model for the mechanism of action of the repellent DEET in *Aedes aegypti* (Diptera: Culicidae). J. Med. Entomol. **18:**357–361.

McKelvey, J.J., Jr. 1973. Man against tsetse: struggle for Africa. Cornell University Press, Ithaca, New York. 306 pp.

McNeil, J.G., Lednar, W.M., Stansfield, S.K., Prier, R.E. and Miller, R.N. 1985. Central European tick-borne encephalitis: assessment of risk for persons in the armed forces and vacationers. J. Infect. Dis. **152:**650–651.

Meyer, J.A. and Hunter, J.S., III. 1991. Residual activity of microencapsulated permethrin against stable flies on lactating dairy cows. Med. Vet. Entomol. **5:**359–362.

Mogi, M. and Sota, T. 1991. Towards integrated control of mosquitoes and mosquito-borne diseases in ricelands. Pp. 47–75 *in* K.F. Harris (ed.), Advances in disease vector research, Vol. 8. Springer-Verlag, New York.

Monath, T.P. (ed.) 1988, 1989. The arboviruses: epidemiology and ecology. 5 vol. CRC Press, Boca Raton, Florida. 1,319 pp.

Monath, T.P. and Nasidi, A. 1993. Should yellow fever vaccine be included in the extended program of immunization in Africa? A cost-effectiveness analysis for Nigeria. Am. J. Trop. Med. Hyg. **48:**274–299.

Moore, D.V. and Lanier, J.E. 1961. Observations on two *Plasmodium falciparum* infections with an abnormal response to chloroquine. Am. J. Trop. Med. Hyg. **10:**5–9.

Moore, C.G., McLean, R.G., Mitchell, C.J., Nasci, R.S., Tsai, T.F., Calisher, C.H., Marfin, A.A., Moore, P.S. and Gubler, D.J. 1993. Guidelines for arbovirus surveillance programs in the United States. Centers for Disease Control and Prevention, Division of Vector-borne Infectious Diseases, Fort Collins, Colorado. 81 pp.

Murai, K., Okayama, A., Horinouchi, H., Oshikawa, T., Tachibana, N. and Tsubouchi, H. 1995. Eradication of *Rickettsia tsutsugamushi* from patients' blood by chemotherapy, as assessed by the polymerase chain reaction. Am. J. Trop. Med. Hyg. **52:**325–327.

Nardin, E.H., Nussenzweig, V., Nussenzweig, R.S., Collins, W.E., Tranakchit Harinasuta, P., Tapchaisri, P. and Chomcharn, Y. 1982. Circumsporozoite proteins of human malaria parasites, *Plasmodium falciparum* and *Plasmodium vivax*. J. Exp. Med. **156:**20–30.

Nguyen, D.S., Dao, B.H., Ngugen, P.D., Nguyen, V.H., Le, N.B., Mai, V.S. and Meshnick, S.R. 1993. Treatment of malaria in Vietnam with oral artemisinin. Am. J. Trop. Med. Hyg. **48:**398–402.

Nosten, F., Luxemburger, C., Kyle, D.E., Ballou, W.R., Wittes, J., Wah, E., Chongsuphajaisiddhi, T., Gordon, D.M., White, N.J., Sadoff, J.C. and Heppner, D.G. 1996. Randomised double-blind placebo-controlled trial of SPf66 malaria vaccine in children in northwestern Thailand. Lancet **348:**701–707.

Osimitz, T.G. and Grothaus, R.H. 1995. The present safety assessment of DEET. J. Am. Mosq. Contr. Assoc. **11:**274–278.

Ottesen, E.A. and Campbell, W.C. 1994. Ivermectin in human medicine. J. Antimicrob. Chemother. **34:**195–203.

Parashar, B.D., Gupta, G.P. and Rao, K.M. 1989. Control of haematophagous flies on equines with permethrin-impregnated eartags. Med. Vet. Entomol. **3:**137–140.

Parashar, B.D., Gupta, G.P. and Rao, K.M. 1991. Control of the haematophagous fly *Hippobosca maculata*, a serious pest of equines, by deltamethrin. Med. Vet. Entomol. **5:**363–367.

Parkman, P.D. and Hardegree, M.C. 1994. Regulation and testing of vaccines. Pp. 889–901 *in* S.A. Plotkin and E.A. Mortimer (eds.), Vaccines, 2nd ed. W.B. Saunders, Toronto.

Pasloske, B.L. and Howard, R.J. 1994. The promise of asexual malaria vaccine development. Am. J. Trop. Med. Hyg. **50(4, Suppl.):**3–10.

Patarroyo, M.E., Amador, R., Clavijo, P., Moreno, A., Guzman, F., Romero, P., Tascon, R., Franco, A., Murillo, L.A., Ponton, G. and Trujillo, G. 1988. A synthetic vaccine protects humans against challenge with asexual blood stages of *Plasmodium falciparum* malaria. Nature **332:**158–161.

Paul, R.E.L., Packer, M.J., Walmsley, M., Lagog, M., Ranford-Cartwright, L.C., Paru, R. and Day, K.P. 1995. Mating patterns in malaria parasite populations of Papua New Guinea. Science **269:**1709–1711.

Poland, J.D., Cropp, C.B., Craven, R.B. and Monath, T.P. 1990. Evaluation of the potency and safety of inactivated Japanese encephalitis vaccine in U.S. inhabitants. J. Infect. Dis. **161:**878–882.

Quinones, M.L., Lines, J.D., Thomson, M.C., Jawara, M., Morris, J. and Greenwood, B.M. 1997. *Anopheles gambiae* gonotrophic cycle duration, biting and exiting behaviour unaffected by permethrin-impregnated bednets in The Gambia. Med. Vet. Entomol. **11:**71–78.

Remme, J. and Zongo, J.B. 1989. Demographic aspects of the epidemiology and control of onchocerciasis in West Africa. Pp. 367–386 *in* M.W. Service (ed.), Demography and vector-borne diseases. CRC Press, Boca Raton, Florida.

Renz, A. and Enyong, P. 1983. Trials of garments impregnated with "Deet" repellent as an individual repellent against *Simulium damnosum s.l.*, the vector of onchocerciasis in the savanna and forest regions of Cameroon. Z. Angew. Entomol. **95**:92–102.

Reuben, R. 1989. Obstacles to malaria control in India — the human factor. Pp. 143–154 *in* M.W. Service (ed.), Demography and vector-borne diseases. CRC Press, Boca Raton, Florida.

Richards, F.O., Jr., Klein, R.E., Zea Flores, R., Weller, S., Gatica, M., Zeisig, R. and Sexton, J. 1993. Permethrin-impregnated bed nets for malaria control in northern Guatemala: epidemiologic impact and community acceptance. Am. J. Trop. Med. Hyg. **49**:410–418.

Rieckmann, K.H., Carson, P.E., Beaudoin, R.L., Cassells, J.S. and Sell, K.W. 1974. Sporozoite induced immunity in man against an Ethiopian strain of *Plasmodium falciparum*. Trans. Royal Soc. Trop. Med. Hyg. **68**:258–259.

Rodriguez-Perez, M.A., Rodriguez, M.H., Margeli-Perez, H.M. and Rivas-Alcala, A.R. 1995. Effect of semiannual treatments of ivermectin on the prevalence and intensity of *Onchocerca volvulus* skin infection, ocular lesions, and infectivity of *Simulium ochraceum* populations in southern Mexico. Am. J. Trop. Med. Hyg. **52**:429–434.

Rutledge, L.C., Collister, D.M., Meixsell, V.E. and Eisenberg, G.H.G. 1983. Comparative sensitivity of representative mosquitoes (Diptera: Culicidae) to repellents. J. Med. Entomol. **20**:506–510.

Rutledge, L.C., Gupta, R.K., Wirtz, R.A. and Buescher, M.D. 1994. Evaluation of the laboratory mouse model for screening topical mosquito repellents. J. Am. Mosq. Contr. Assoc. **10**:565–571.

Schlesinger, R.W., Gordon, I., Frankel, J.W., Winter, J.W., Patterson, P.R. and Dorrance, W.R. 1956. Clinical and serologic response of man to immunization with attenuated dengue and yellow fever viruses. J. Immunol. **77**:352–364.

Schofield, G.J. 1994. Triatominae: biology and control. Eurocommunica Publ., West Sussex, UK. 80 pp.

Schreck, C.E. 1977. Techniques for evaluation of insect repellents: a critical review. Annu. Rev. Entomol. **22**:101–119.

Schreck, C.E. and Kline, D.L. 1989. Personal protection afforded by controlled-release topical repellents and permethrin-treated clothing against natural populations of *Aedes taeniorhynchus*. J. Am. Mosq. Contr. Assoc. **4**:77–80.

Schreck, C.E., Posey, K. and Smith, D. 1978. Durability of permethrin as a potential clothing treatment to protect against blood-feeding arthropods. J. Econ. Entomol. **71**:397–400.

Schreck, C.E., Snoddy, E.L. and Mount, G.A. 1980. Permethrin and repellents as clothing impregnants for protection from the lone star tick. J. Econ. Entomol. **73**:436–439.

Schwan, T.G., Piesman, J., Golde, W.T., Dolan, M.C. and Rosa, P.A. 1995. Induction of an outer surface protein on *Borrelia burgdorferi* during tick feeding. Proc. Nat. Acad. Sci. **92**:2909–2913.

Selkirk, M.E., Maizels, R.M. and Yazdanbakhsh, M. 1992. Immunity and the prospects for vaccination against filariasis. Immunobiology **184**:263–281.

Sexton, J.D. 1994. Impregnated bed nets for malaria control: biological success and social responsibility. Am. J. Trop. Med. Hyg. **50(6, Suppl.)**:72–81.

Shahabuddin, M., Toyoshima, T., Aikawa, M. and Kaslow, D.C. 1993. Transmission-blocking activity of a chitinase inhibitor and activation of malarial parasite chitinase by mosquito protease. Proc. Nat. Acad. Sci. **90**:4266–4270.

Sharp, T.W., Wallace, M.R., Hayes, C.G., Sanchez, J.L., DeFraites, R.F., Arthur, R.R., Thornton, S.A., Batchelor, R.A., Rozmajzl, R., Hanson, K., Wu, S.J., Iriye, C. and Burans, J.P. 1995. Dengue fever in U.S. troops during Operation Restore Hope, Somalia, 1992–1993. Am. J. Trop. Med. Hyg. **53**:89–94.

Shih, C.-M., Spielman, A. and Telford, S.R., III. 1995. Short report: mode of action of protective immunity to Lyme disease spirochetes. Am. J. Trop. Med. Hyg. **52**:72–74.

Sileghem, M., Flynn, J.N., Darji, A., De Baetselier, P. and Naessens, J. 1994. African trypanosomiasis. Pp. 1–51 *in* F. Kierszenbaum (ed.), Parasitic infections and the immune system. Academic Press, San Diego, California.

Slater, A.F.G. 1993. Chloroquine: mechanism of drug action and resistance in *Plasmodium falciparum*. Pharmacol. Theory **57**:203–235.

Snow, R.W., Lindsay, S.W., Hayes, R.J. and Greenwood, B.M. 1988. Permethrin-treated bed nets (mosquito nets) prevent malaria in Gambian children. Trans. Royal Soc. Trop. Med. Hyg. **82**:838–842.

Somo, R.M., Ngosso, A., Dinga, J.S., Enyong, P.A. and Fobi, G. 1993. A community-based trial of ivermectin for onchocerciasis control in the forest of southwestern Cameroon: clinical and parasitologic findings after three treatments. Am. J. Trop. Med. Hyg. **48**:9–13.

Stage, H.H. 1944. Saboteur mosquitoes. Nat. Geogr. Mag. **85**:165–179.

Stoute, J.A., Slaoui, M., Heppner, D.G., Momin, P., Kester, K.E., Desmons, P., Wellde, B.T., Carcon, N., Krzych, U., Marchand, M., Ballou, W.R. and Cohen, J. 1997. A preliminary evaluation of a recombinant circumsporozoite protein vaccine against *Plasmodium falciparum* malaria. New England J. Med. **336:**86–91.

Strickland, G.T. 1991. Hunter's tropical medicine, 7th ed. W.S. Saunders Co., Toronto. 1153 pp.

Strizki, J.M. and Repik, P.M. 1995. Differential reactivity of immune sera from human vaccines with field strains of eastern equine encephalitis virus. Am. J. Trop. Med. Hyg. **53:**565–570.

Su, X.Z., Heatwole, V.M., Wertheimer, S.P., Guinet, F., Herrfeldt, J.A., Peterson, D.S., Ravetch, J.A. and Wellems, T.A. 1995. The large diverse gene family Var encodes proteins involved in cytoadherence and antigenic variation of *Plasmodium falciparum*-infected erythrocytes. Cell **82:**89–100.

Tellam, R.L., Smith, D., Kemp, D.H. and Willadsen, P. 1992. Vaccination against ticks. Pp. 301–331 *in* W.K. Young (ed.), Animal parasite control using technology. CRC Press, Boca Raton, Florida.

Tesh, R.B. 1995. Control of zoonotic visceral leishmaniasis: is it time to change strategies? Am. J. Trop. Med. Hyg. **52:**287–292.

Thomson, M.C., Adiamah, J.H., Connor, S.J., Jawara, M., Bennett, S., D'Allesandro, U., Quinones, M., Langerock, P. and Greenwood, B.M. 1995. Entomological evaluation of The Gambia's national impregnated bednet programme. Ann. Trop. Med. Parasitol. **89:**229–241.

Trape, J.-F., Lefebvre-Zante, E., Legros, F., Druilhe, P., Rogier, C., Bouganali, H. and Salem, G. 1993. Malaria morbidity among children exposed to low seasonal transmission in Dakar, Senegal and its implications for malaria control in tropical Africa. Am. J. Trop. Med. Hyg. **48:**748–756.

Trpis, M., Childs, J.E., Fryauff, D.J., Greene, B.M., Williams, P.N., Munoz, B.E., Pacque, M.C. and Taylor, H.R. 1990. Effect of mass treatment of a human population on transmission of *Onchocerca volvulus* by *Simulium yahense* in Liberia, West Africa. Am. J. Trop. Med. Hyg. **42:**148–156.

Uilenberg, G. 1994. Ectoparasites of animals and control methods. Sci. Technol. Rev., Off. Intern. Epizoot. **13:**979–1416.

Williams, J.F., MacKenzie, C.D. and El Khalifa, M. 1994. African trypanosomiasis. Pp. 225–247 *in* F. Kierszenbaum (ed.), Parasitic infections and the immune system. Academic Press, San Diego, California.

Wyler, D.J. 1993. Malaria chemoprophylaxis for the traveler. New England J. Med. **329:**31–37.

Subject Index

B

D

L

M

Q

R

Scientific Names Index

CPSIA information can be obtained
at www.ICGtesting.com
Printed in the USA
LVHW101549271220
675126LV00011B/462

9 781402 014130